CALCULUS

Second Edition

Differentiation Formulas

1. $\left(f(x) \pm g(x)\right)' = f'(x) \pm g'(x)$

2. $\left(kf(x)\right)' = kf'(x)$

3. $\left(f(x)g(x)\right)' = f(x)g'(x) + g(x)f'(x)$

4. $\left(\dfrac{f(x)}{g(x)}\right)' = \dfrac{g(x)f'(x) - f(x)g'(x)}{\left(g(x)\right)^2}$

5. $\left(f\left(g(x)\right)\right)' = f'\left(g(x)\right) \cdot g'(x)$

6. $\dfrac{d}{dx}(x^n) = nx^{n-1}$

7. $\dfrac{d}{dx}(e^x) = e^x$

8. $\dfrac{d}{dx}(a^x) = a^x \ln a \quad (a > 0)$

9. $\dfrac{d}{dx}(\ln x) = \dfrac{1}{x}$

10. $\dfrac{d}{dx}(\sin x) = \cos x$

11. $\dfrac{d}{dx}(\cos x) = -\sin x$

12. $\dfrac{d}{dx}(\tan x) = \dfrac{1}{\cos^2 x}$

A Short Table of Indefinite Integrals

I. Basic Functions

1. $\displaystyle\int x^n \, dx = \dfrac{1}{n+1}x^{n+1} + C, \quad n \neq -1$

2. $\displaystyle\int \dfrac{1}{x} \, dx = \ln|x| + C$

3. $\displaystyle\int a^x \, dx = \dfrac{1}{\ln a}a^x + C, \quad a > 0$

4. $\displaystyle\int \ln x \, dx = x \ln x - x + C$

5. $\displaystyle\int \sin x \, dx = -\cos x + C$

6. $\displaystyle\int \cos x \, dx = \sin x + C$

7. $\displaystyle\int \tan x \, dx = -\ln|\cos x| + C$

II. Products of e^x, $\cos x$, and $\sin x$

8. $\displaystyle\int e^{ax} \sin(bx) \, dx = \dfrac{1}{a^2 + b^2}e^{ax}[a\sin(bx) - b\cos(bx)] + C$

9. $\displaystyle\int e^{ax} \cos(bx) \, dx = \dfrac{1}{a^2 + b^2}e^{ax}[a\cos(bx) + b\sin(bx)] + C$

10. $\displaystyle\int \sin(ax)\sin(bx) \, dx = \dfrac{1}{b^2 - a^2}[a\cos(ax)\sin(bx) - b\sin(ax)\cos(bx)] + C, \quad a \neq b$

11. $\displaystyle\int \cos(ax)\cos(bx) \, dx = \dfrac{1}{b^2 - a^2}[b\cos(ax)\sin(bx) - a\sin(ax)\cos(bx)] + C, \quad a \neq b$

12. $\displaystyle\int \sin(ax)\cos(bx) \, dx = \dfrac{1}{b^2 - a^2}[b\sin(ax)\sin(bx) + a\cos(ax)\cos(bx)] + C, \quad a \neq b$

III. Product of Polynomial $p(x)$ with $\ln x$, e^x, $\cos x$, $\sin x$

13. $\displaystyle\int x^n \ln x \, dx = \dfrac{1}{n+1}x^{n+1} \ln x - \dfrac{1}{(n+1)^2}x^{n+1} + C, \quad n \neq -1$

14. $\displaystyle\int p(x)e^{ax} \, dx = \dfrac{1}{a}p(x)e^{ax} - \dfrac{1}{a}\int p'(x)e^{ax} \, dx$

$$= \dfrac{1}{a}p(x)e^{ax} - \dfrac{1}{a^2}p'(x)e^{ax} + \dfrac{1}{a^3}p''(x)e^{ax} - \cdots$$
$$(+ - + - \ldots)$$

(signs alternate)

15. $\displaystyle \int p(x)\sin ax\,dx = -\frac{1}{a}p(x)\cos ax + \frac{1}{a}\int p'(x)\cos ax\,dx$

$\displaystyle \qquad = -\frac{1}{a}p(x)\cos ax + \frac{1}{a^2}p'(x)\sin ax + \frac{1}{a^3}p''(x)\cos ax - \cdots$

$\qquad (-++--++\ldots)$

(signs alternate in pairs after first term)

16. $\displaystyle \int p(x)\cos ax\,dx = \frac{1}{a}p(x)\sin ax - \frac{1}{a}\int p'(x)\sin ax\,dx$

$\displaystyle \qquad = \frac{1}{a}p(x)\sin ax + \frac{1}{a^2}p'(x)\cos ax - \frac{1}{a^3}p''(x)\sin ax - \cdots$

$\qquad (++--++--\ldots)$ (signs alternate in pairs)

IV. Integer Powers of $\sin x$ and $\cos x$

17. $\displaystyle \int \sin^n x\,dx = -\frac{1}{n}\sin^{n-1}x\cos x + \frac{n-1}{n}\int \sin^{n-2}x\,dx, \quad n \text{ positive}$

18. $\displaystyle \int \cos^n x\,dx = \frac{1}{n}\cos^{n-1}x\sin x + \frac{n-1}{n}\int \cos^{n-2}x\,dx, \quad n \text{ positive}$

19. $\displaystyle \int \frac{1}{\sin^m x}\,dx = \frac{-1}{m-1}\frac{\cos x}{\sin^{m-1}x} + \frac{m-2}{m-1}\int \frac{1}{\sin^{m-2}x}\,dx, \quad m \neq 1, m \text{ positive}$

20. $\displaystyle \int \frac{1}{\sin x}\,dx = \frac{1}{2}\ln\left|\frac{(\cos x)-1}{(\cos x)+1}\right| + C$

21. $\displaystyle \int \frac{1}{\cos^m x}\,dx = \frac{1}{m-1}\frac{\sin x}{\cos^{m-1}x} + \frac{m-2}{m-1}\int \frac{1}{\cos^{m-2}x}\,dx, \quad m \neq 1, m \text{ positive}$

22. $\displaystyle \int \frac{1}{\cos x}\,dx = \frac{1}{2}\ln\left|\frac{(\sin x)+1}{(\sin x)-1}\right| + C$

23. $\displaystyle \int \sin^m x\cos^n x\,dx$: If m is odd, let $w = \cos x$. If n is odd, let $w = \sin x$. If both m and n are even and non-negative, convert all to $\sin x$ or all to $\cos x$ (using $\sin^2 x + \cos^2 x = 1$), and use IV-17 or IV-18. If m and n are even and one of them is negative, convert to whichever function is in the denominator and use IV-19 or IV-21. If both m and n are even and negative, the substitution $w = \cos x$ converts the integral into a rational function which can be integrated by the method of partial fractions.

V. Quadratic in the Denominator

24. $\displaystyle \int \frac{1}{x^2+a^2}\,dx = \frac{1}{a}\arctan\frac{x}{a} + C, \quad a \neq 0$

25. $\displaystyle \int \frac{bx+c}{x^2+a^2}\,dx = \frac{b}{2}\ln|x^2+a^2| + \frac{c}{a}\arctan\frac{x}{a} + C, \quad a \neq 0$

26. $\displaystyle \int \frac{1}{(x-a)(x-b)}\,dx = \frac{1}{a-b}(\ln|x-a| - \ln|x-b|) + C, \quad a \neq b$

27. $\displaystyle \int \frac{cx+d}{(x-a)(x-b)}\,dx = \frac{1}{a-b}\left[(ac+d)\ln|x-a| - (bc+d)\ln|x-b|\right] + C, \quad a \neq b$

VI. Integrands Involving $\sqrt{a^2+x^2}$, $\sqrt{a^2-x^2}$, $\sqrt{x^2-a^2}$, $\quad a > 0$

28. $\displaystyle \int \frac{1}{\sqrt{a^2-x^2}}\,dx = \arcsin\frac{x}{a} + C$

29. $\displaystyle \int \frac{1}{\sqrt{x^2\pm a^2}}\,dx = \ln\left|x + \sqrt{x^2\pm a^2}\right| + C$

30. $\displaystyle \int \sqrt{a^2\pm x^2}\,dx = \frac{1}{2}\left(x\sqrt{a^2\pm x^2} + a^2\int \frac{1}{\sqrt{a^2\pm x^2}}\,dx\right) + C$

31. $\displaystyle \int \sqrt{x^2-a^2}\,dx = \frac{1}{2}\left(x\sqrt{x^2-a^2} - a^2\int \frac{1}{\sqrt{x^2-a^2}}\,dx\right) + C$

Lines

Slope of line through (x_1, y_1) and (x_2, y_2):

$$m = \frac{y_2 - y_1}{x_2 - x_1}$$

Point-slope equation of line through (x_1, y_1) with slope m:

$$y - y_1 = m(x - x_1)$$

Slope-intercept equation of line with slope m and y-intercept b:

$$y = b + mx$$

Rules of Exponents

$$a^x a^t = x^{x+t}$$
$$\frac{a^x}{a^t} = a^{x-t}$$
$$(a^x)^t = a^{xt}$$

Definition of Natural Log

$y = \ln x$ means $e^y = x$
ex: $\ln 1 = 0$ since $e^0 = 1$

Identities

$$\ln e^x = x$$
$$e^{\ln x} = x$$

Rules of Natural Logarithms

$$\ln(AB) = \ln A + \ln B$$
$$\ln\left(\frac{A}{B}\right) = \ln A - \ln B$$
$$\ln A^p = p \ln A$$

Distance And Midpoint Formulas

Distance D between (x_1, y_1) and (x_2, y_2):

$$D = \sqrt{(x_2 - x_1)^2 + (y_2 - y_1)^2}$$

Midpoint of (x_1, y_1) and (x_2, y_2):

$$\left(\frac{x_1 + x_2}{2}, \frac{y_1 + y_2}{2}\right)$$

Quadratic Formula

If $ax^2 + bx + c = 0$, then

$$x = \frac{-b \pm \sqrt{b^2 - 4ac}}{2a}$$

Factoring Special Polynomials

$$x^2 - y^2 = (x + y)(x - y)$$
$$x^3 + y^3 = (x + y)(x^2 - xy + y^2)$$
$$x^3 - y^3 = (x - y)(x^2 + xy + y^2)$$

Circles

Center (h, k) and radius r:

$$(x - h)^2 + (y - k)^2 = r^2$$

Ellipse

$$\frac{x^2}{a^2} + \frac{y^2}{b^2} = 1$$

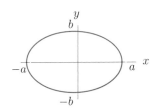

Hyperbola

$$\frac{x^2}{a^2} - \frac{y^2}{b^2} = 1$$

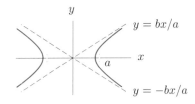

Geometric Formulas

Conversion between Radians and Degrees: $\quad \pi$ radians $= 180°$

Triangle	**Circle**	**Sector of Circle**
$A = \frac{1}{2}bh$	$A = \pi r^2$	$A = \frac{1}{2}r^2\theta \quad (\theta \text{ in radians})$
$\quad = \frac{1}{2}ab\sin\theta$	$C = 2\pi r$	$s = r\theta \quad (\theta \text{ in radians})$

  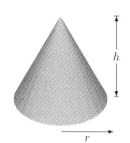

Sphere	**Cylinder**	**Cone**
$V = \frac{4}{3}\pi r^3 \quad A = 4\pi r^2$	$V = \pi r^2 h$	$V = \frac{1}{3}\pi r^2 h$

Trigonometric Functions

$$\sin\theta = \frac{y}{r}$$

$$\cos\theta = \frac{x}{r}$$

$$\tan\theta = \frac{y}{x}$$

$$\tan\theta = \frac{\sin\theta}{\cos\theta}$$

$$\cos^2\theta + \sin^2\theta = 1$$

The Binomial Theorem

$$(x+y)^n = x^n + nx^{n-1}y + \frac{n(n-1)}{1\cdot 2}x^{n-2}y^2 + \frac{n(n-1)(n-2)}{1\cdot 2\cdot 3}x^{n-3}y^3 + \cdots + nxy^{n-1} + y^n$$

$$(x-y)^n = x^n - nx^{n-1}y + \frac{n(n-1)}{1\cdot 2}x^{n-2}y^2 - \frac{n(n-1)(n-2)}{1\cdot 2\cdot 3}x^{n-3}y^3 + \cdots \pm nxy^{n-1} \mp y^n$$

CALCULUS

Second Edition

Produced by the Consortium based at Harvard and funded by a National Science Foundation Grant.

Deborah Hughes-Hallett
Harvard University

William G. McCallum
University of Arizona

Andrew M. Gleason
Harvard University

Douglas Quinney
University of Keele

Daniel E. Flath
University of South Alabama

Brad G. Osgood
Stanford University

Patti Frazer Lock
St. Lawrence University

Andrew Pasquale
Chelmsford High School

Sheldon P. Gordon
Suffolk County Community College

Jeff Tecosky-Feldman
Haverford College

David O. Lomen
University of Arizona

Joe B. Thrash
University of Southern Mississippi

David Lovelock
University of Arizona

Karen R. Thrash
University of Southern Mississippi

Thomas W. Tucker
Colgate University

with the assistance of
Otto K. Bretscher
Harvard University

WILEY

John Wiley & Sons, Inc.

New York Chichester Weinheim Brisbane Singapore Toronto

Cover Photo: Dennis O'Clair/Tony Stone Images/New York, Inc.

Problems from *Calculus: The Analysis of Functions,* by Peter D. Taylor
(Toronto: Wall & Emerson, Inc., 1992). Reprinted with permission of the publisher.

EXECUTIVE EDITOR Ruth Baruth
DEVELOPMENTAL EDITOR Nancy Perry
MARKETING MANAGER Leslie Hines
SENIOR PRODUCTION MANAGER Lucille Buonocore
SENIOR PRODUCTION EDITOR Monique Calello
SENIOR DESIGNER Laura Nicholls Boucher
ILLUSTRATION DIRECTOR Ishaya Monokoff
MANUFACTURING MANAGER Monique Calello

This book was set in Times Roman by the Consortium based at Harvard using LATEX,
Mathematica, and the package AsTeX, which was written by Alex Kasman for
this project.

This book was printed and bound by Courier Westford, Inc. This cover was printed by
Phoenix Color Corporation.

This book is printed on acid-free paper. ∞

The paper in this book was manufactured by a mill whose forest management programs
include sustained yield harvesting of its timberlands. Sustained yield harvesting principles
ensure that the number of trees cut each year does not exceed the amount of new growth.

ISBN 0-471-16442-9 (paperback)

Printed in the United States of America.

10 9 8 7 6

This project was supported, in part,
by the
National Science Foundation
Opinions expressed are those of the authors
and not necessarily those of the Foundation

PREFACE

Calculus is one of the greatest achievements of the human intellect. Inspired by problems in astronomy, Newton and Leibniz developed the ideas of calculus 300 years ago. Since then, each century has demonstrated the power of calculus to illuminate questions in mathematics, the physical sciences, engineering, and the social and biological sciences.

Calculus has been so successful because of its extraordinary power to reduce complicated problems to simple rules and procedures. Therein lies the danger in teaching calculus: it is possible to teach the subject as nothing but the rules and procedures—thereby losing sight of both the mathematics and of its practical value. This edition of *Calculus* continues our effort to refocus the teaching of calculus on concepts as well as procedures.

A Focused Vision: Conceptual Understanding

Our goal is to provide students with a clear understanding of the ideas of calculus as a solid foundation for subsequent courses in mathematics and other disciplines. When we designed this curriculum we started with a clean slate. We included some new topics, such as differential equations, and omitted some traditional topics whose inclusion we could not justify after discussions with mathematicians, engineers, physicists, chemists, biologists, and economists. We focused on a small number of key concepts, emphasizing depth of understanding rather than breadth of coverage.

The First Edition: A New Curriculum

The first edition of this book was the work of faculty at a consortium of eleven institutions, generously supported by the National Science Foundation. It represents the first consensus between such a diverse group of research mathematicians and instructors to have shaped a mainstream calculus text.

The Second Edition: Expanded Choices

The second edition has the same vision as the first edition and provides instructors with additional choices through the *Focus On* sections. Instructors can select a focus for their course which reflects their interests and the needs of their students. In particular:

- All *Focus On* sections are optional.
- Chapter 3 (the definite integral) may be covered immediately before Chapter 6 (the antiderivative).
- In Chapters 5 and 8, instructors may select the sections relevant to their students.
- Chapters 9 and 10 may be covered in either order.

Guiding Principles

Since students usually learn most when they are active, we feel that the exercises in a text are of central importance. In addition, we have found that multiple representations encourage students to reflect on the meaning of the material. Consequently, we have been guided by the following principles.

- Our problems are varied and some are challenging. Most cannot be done by following a template in the text.
- The Rule of Four: Where appropriate, topics should be presented geometrically, numerically, analytically, and verbally.

The Development of Mathematical Thinking

The first stage in the development of mathematical thinking is the acquisition of a clear intuitive picture of the central ideas. In the next stage, the student learns to reason with the intuitive ideas and explain the reasoning clearly in plain English. After this foundation has been laid, there is a choice of direction. Some students, for example mathematics majors, may benefit from a more theoretical approach, while others, for example science and engineering majors, may benefit from a further exploration of modeling. New supplementary sections, *Focus on Theory* and *Focus on Modeling*, provide material to support either choice.

Focus on Theory

Calculus as a logical structure of theorem and proof is a masterpiece of mathematics. Its beauty, if not glossed over in a broad but superficial treatment, attracts mathematically inclined students to their vocation. In the *Focus on Theory* sections we have chosen a few topics to cover in depth. We show how axioms, definitions, and theorems are formulated, and how proofs are constructed. Since we believe that to understand and appreciate this material students must work with it themselves, we have included challenging exercises that guide students to construct definitions and proofs on their own.

Focus on Modeling

Calculus is a powerful tool for analyzing the real world. Students gain an understanding of the power of calculus by focusing on its use in an extended problem. The *Focus on Modeling* sections take the time to explore selected applications of calculus in depth.

The Development of Mathematical Skills

To use calculus effectively, students need skill in both symbolic manipulation and the use of technology. The exact proportions of each may vary widely, depending on the preparation of the student and the wishes of the instructor. The book is adaptable to many different combinations.

Focus on Practice

These new sections increase students' skills in the mechanics of differentiation and integration.

Technology

The book does not require any specific software or technology. It has been used with graphing calculators, graphing software, and computer algebra systems. Any technology with the ability to graph functions and perform numerical integration will suffice. Students are expected to use their own judgement to determine where technology is useful.

Content

This content represents our vision of how calculus can be taught. It is flexible enough to accommodate individual course needs and requirements. Topics can easily be added or deleted, or the order changed.

Chapter 1: A Library of Functions

This chapter introduces all the elementary functions to be used in the book. Although the functions are probably familiar, the graphical, numerical, verbal, and modeling approach to them is likely to be new. We introduce exponential functions at the earliest possible stage, since they are fundamental to the understanding of real-world processes.

We encourage you to cover this chapter thoroughly, as the time spent on it will pay off when you get to the calculus.

Focus on Theory

These sections discuss the theoretical underpinnings of calculus, illustrate the process of formulating and proving a theorem using the example of the binomial theorem, and show how axioms are developed using the example of the completeness axiom.

Chapter 2: Key Concept---The Derivative

The purpose of this chapter is to give the student a practical understanding of the definition of the derivative and its interpretation as an instantaneous rate of change. The power rule is introduced; other rules are introduced in Chapter 4.

Focus on Theory

These sections present the definition of limit and continuity, and analyze the connection between differentiability and local linearity. Exercises lead students to a proof of the Intermediate Value Theorem.

Chapter 3: Key Concept---The Definite Integral

The purpose of this chapter is to give the student a practical understanding of the definite integral as a limit of Riemann sums and to bring out the connection between the derivative and the definite integral in the Fundamental Theorem of Calculus. Some instructors using the first edition of the book have delayed covering Chapter 3 until after Chapter 5 without any difficulty.

Focus on Theory

This section explores the formal definition of the definite integral using upper and lower sums.

Chapter 4: Short-Cuts to Differentiation

The derivatives of all the functions in Chapter 1 are introduced, as well as the rules for differentiating products, quotients, and composite functions.

Focus on Practice

This section provides a battery of drill problems on differentiation for students to practice their skills.

Chapter 5: Using the Derivative

The aim of this chapter is to enable the student to use the derivative in solving problems. It is not meant to be comprehensive and you do not need to cover all the sections.

Focus on Theory

This section pulls together many of the strands of the earlier theory sections, showing how the completeness axiom and the definition of limit and continuity can be used to prove the Extreme Value Theorem, the Increasing Function Theorem, and the Mean Value Theorem. Some of the proofs are given in the text, others are developed through the exercises.

Chapter 6: Constructing Antiderivatives

This chapter focuses on going backward from a derivative to the original function, first graphically and numerically, then analytically. It introduces the Second Fundamental Theorem of Calculus and the concept of a differential equation.

Focus on Modeling

This section discusses why the equations of motion are based on acceleration rather than velocity.

Chapter 7: Integration

This chapter includes several techniques of integration; others are included in the table of integrals. There is a discussion of numerical methods and of improper integrals.

Focus on Practice

This section presents a battery of integration problems for skill building.

Chapter 8: Using the Definite Integral

This chapter emphasizes the idea of subdividing a quantity to produce Riemann sums which, in the limit, yield a definite integral. It aims to show how the integral is used in geometry, physics, and economics. You do not need to cover all the sections.

Focus on Modeling

These sections study applications of integration to probability distributions.

Chapter 9: Approximations and Series

This chapter introduces Taylor Series and Fourier series via the idea of approximating functions with simpler functions. Geometric series, the ratio test, the harmonic series, and alternating series are discussed.

Focus on Theory

These sections present convergence tests in more detail, sketch a proof of the ratio test, and study bounds on the error in the Taylor approximation.

Chapter 10: Differential Equations

This chapter introduces differential equations. The emphasis is on qualitative solutions, modeling, and interpretation.

Focus on Modeling

These sections show applications of the logistic equation and systems of differential equations to population models, the spread of disease, predator-prey interactions, and competitive exclusion.

Appendices

The appendices contain material on roots and accuracy, compound interest, polar coordinates, complex numbers, Newton's method, and parametric equations.

Our Experiences

In the process of developing the ideas incorporated in this book, we were conscious of the need to test the materials thoroughly in a wide variety of institutions serving many different types of students. Before the first edition was produced, consortium members and colleagues at over one hundred schools around the country class-tested preliminary versions of the book at large and small public universities, liberal arts colleges, two-year institutions, and high schools. The first edition was used by a very large and diverse group of schools in semester and quarter systems, in large lectures and small classes, in computer labs, small groups, and traditional settings, and with a number of different technologies. In preparing the second edition, we solicited comments from a large number of mathematicians who had used the text. We continued to engage our colleagues in client disciplines to determine the mathematical needs of their students. This included careful reviewing by a group of engineering faculty from highly regarded engineering programs around the country. We were offered many valuable suggestions, which we have tried to incorporate into this second edition of the text, while maintaining our original commitment to a focused treatment of a limited number of topics.

Changes from the First Edition

We have streamlined some topics and added new sections on theory and on skill-building; we have moved some material into separate sections on modeling. The new arrangement divides the curriculum into a basic core and a set of supplementary sections from which the instructor can make choices to support a wide variety of courses.

General Changes

- There are more easy and medium level problems in each section, and a battery of drill problems in the *Focus on Practice* sections at the end of Chapters 4 (differentiation) and 7 (integration).

- There are short answers in the back of the book to those odd-numbered problems that have them.
- Each chapter concludes with a review of the main points, and there is a list of useful formulas on the endpapers.
- One or more projects are included at the end of each chapter; a wider selection is in the appendix.

Summary of New Material

- New *Focus on Theory* sections discuss the concepts of axiom, definition, and mathematical proof, and give a more theoretical treatment of selected topics, such as limits, differentiability, and the definite integral. The first edition's emphasis on concepts and on geometric and numerical reasoning serves as a natural introduction to this thread. These sections are optional and may be selected by instructors who prefer a more theoretical course.
- Chapter 4 contains a new section on local linearity and limits, which includes l'Hopital's rule.
- Chapter 5 contains a new section on the hyperbolic functions.
- Chapter 7 contains an expanded treatment of partial fractions.
- Chapter 8 contains a new subsection on center of mass.
- Chapter 9 contains new material on using the ratio test to find the radius of convergence of a power series, as well as new material on the harmonic and alternating series.
- There are two new appendices, one on parametric equations, and one containing additional projects.

Chapter-by-chapter description of changes

- Chapter 1. Notes on Compound Interest has been moved to an appendix. The introduction to continuity, formerly in an appendix, has been moved to this chapter. There are three new *Focus on Theory* sections: Underpinnings of Calculus, The Binomial Theorem, and Completeness of the Real Numbers.
- Chapter 2. The material on derivatives of constant and linear functions, and the statement of the power rule have been moved to Section 2.3. Approximations and Local Linearity, Notes on the Limit, and Notes on Differentiability have been replaced by two *Focus on Theory* sections: Limits and Continuity, and Differentiability and Linear Approximation.
- Chapter 3. The discussion in Section 3.2 of the relation between the value of n and the convergence of Riemann sums has been shortened and the material on visualizing the integral as area has been moved to this section. The applications of the definite integral formerly in Section 6.1 have been moved to Section 3.3. Section 3.4 has been expanded to include both the Fundamental Theorem and the properties of definite integrals formerly in Section 6.2. Further Notes on the Limit has been replaced by a *Focus on Theory* section on the Definite Integral.
- Chapter 4. Notes on the Tangent Line Approximation has been incorporated into a new section on Local Linearity and Limits, which contains a treatment of l'Hopital's rule. More problems on the applications of rates have been included. A new *Focus on Practice* section provides drill problems in differentiation.
- Chapter 5. The material on local extrema has been shortened and put into Section 5.1. The material on families of functions has been simplified and is in Section 5.2. Section 5.3 is on global extrema and includes the first section on optimization (formerly Section 5.5). The section on Newton's Method has been moved to an appendix, and a new section on Hyperbolic Functions has been added. There is a new *Focus on Theory* section on Theorems about Continuous and Differentiable Functions.
- Chapter 6. This chapter has been refocused on antiderivatives; the material about definite integrals has been moved to Chapter 3. The chapter starts with a section on finding antiderivatives graphically and numerically. The material on finding antiderivatives analytically (formerly Section 7.1) has been moved to Section 6.2. The material on constructing antiderivatives using the definite integral has been moved here to form a new section on The Second Fundamental Theorem of Calculus. The section on equations of motion has become a *Focus on Modeling* section.
- Chapter 7 now starts with substitutions.
- Chapter 8. Applications to geometry have been put first, and new material on center of mass has been added to the material on density in Section 8.2. The sections on distributions and probability have been

moved to *Focus on Modeling* sections.

- Chapter 9. This chapter now covers Approximations and Series. There is new material on the ratio test for determining the radius of convergence of a power series, as well as on harmonic and alternating series. A new *Focus on Theory* section deals with the convergence of series. The Error in Taylor Approximations is now a *Focus on Theory* section.

- Chapter 10. This chapter now covers Differential Equations. The sections on Models of Population Growth, Systems of Differential Equations, and Analyzing the Phase Plane have become *Focus on Modeling* sections.

- Appendices. The material on continuity has been moved to Chapter 1. The sections on Compound Interest and Newton's Method have been moved to appendices. A new section on Parametric Equations and a new section containing projects have been added.

Supplementary Materials

The following supplementary materials are available for the Second Edition.
- **Instructor's Manual with Sample Exam Questions** containing teaching tips, calculator programs, some overhead transparency masters and test questions arranged according to section.

- **Instructor's Solution Manual** with complete solutions to all problems.

- **Student's Solution Manual** with complete solutions to half the odd-numbered problems.

- **Instructor's Resource CD-ROM** which contains the Instructor's Manual, Instructor's Solutions Manual, additional projects, as well as other valuable resources.

Acknowledgements

First and foremost, we want to express our appreciation to the National Science Foundation for their faith in our ability to produce a revitalized calculus curriculum and, in particular, to Louise Raphael, John Kenelly, John Bradley, and James Lightbourne. We also want to thank the members of our Advisory Board, Benita Albert, Lida Barrett, Simon Bernau, Robert Davis, M. Lavinia DeConge-Watson, John Dossey, Ron Douglas, Eli Fromm, William Haver, Seymour Parter and Stephen Rodi for their ongoing guidance and advice.

In addition, a host of other people around the country and abroad deserve our thanks for all that they did to help our project succeed. They include: Leonid Andreev, Huriye Arikan, Ruth Baruth, Martin Batey, Maria Betkowski, Laura Boucher, Melkana Brakalova, John Bravman, Stephen Boyd, Jackie Boyd-DeMarzio, Otto Bretscher, Mark Bridger, Morton Brown, Lucille Buonocore, Monique Calello, Ray Cannon, Marilyn Carlson, Phil Cheifetz, Bob Condon, Jurie Conradie, Sterling G. Crossley, Ehud de Shalit, Bob Decker, Persi Diaconis, Tom Dick, Srdjan Divac, Wade Ellis, Paul Feehan, Hermann Flaschka, Leonid Fridman, Lynn Garner, Tom Gearhart, James F. Gibbons, Allan Gleason, Florence Gordon, Danny Goroff, Robin Gottlieb, Biddy Greene, Dudley Herschbach, JoEllen Hillyer, Leslie Hines, Luke Hunsberger, Richard Iltis, Rob Indik, Adrian Iovita, Pete Janzow, Mary Johenk, Jerry Johnson, Mille Johnson, Matthias Kawski, David Kazhdan, Mike Klucznik, Donna Krawczyk, Mark Kunka, Carl Leinbach, David Levermore, Don Lewis, Jim Lewis, John Lucas, Reginald Luke, Tom McMahon, Dan Madden, Brad Mann, Barry Mazur, Eric Mazur, Georgia Mederer, Ansie Meiring, Dave Meredith, Andrew Metrick, Ishaya Monokoff, Lang Moore, David Mumford, Don Myers, Bridget Neale, Alan Newell, Steve Olson, John Orr, Arnie Ostebee, Jose Padro, Mike Pavloff, Nancy Perry, Tony Phillips, John Prados, Nancy Prinz, Ted Pyne, Amy Radunskaya, Wayne Raskind, Gabriella Ratay, Janet Ray, George Rublein, Anneliese Schauerte, Wilfried Schmid, Walter Seaman, Marilyn Semrau, Alida Setford, Pat Shure, Esther Silberstein, Lynne Small, David Smith, Sharon Smith, Don Snow, Bob Speiser, Howard Stone, Steve Strogatz, "Suds" Sudholz, Cliff Taubes, Peter Taylor, Tom Timchek, Alan Tucker, Jerry Uhl, Bill Vélez, Charles Walter, Mary Jean Winter, Chris Yetman, Debbie Yoklic, Lee Zia, Paul Zorn, and all the people in the Harvard mathematics department who shared their computers and their space with us.

Reports from the following reviewers were most helpful in shaping the second edition. Mathematics faculty and students: Frank Beatrous, Dennis De Turck, G. Martin Forrest, John Hagood, Kenneth Hannsgen, Kerry Johnson, Lisa A. Mantini, Joe A. Marlin, Grace Orzech, Howard Penn, Anthony L. Peressini, David R. Pitts, Lynne Small, Michell Anne Stillman, Brian Tankersley.

Engineering faculty: Howie Choset, Michael R. Foster, Terry L. Kohutek, Dorian Liepmann, Howard Littman, Vijay Modi, J. W. Rogers, Jr., John Sanchez.

Most of all, to the remarkable team that worked day and night to get the text into the computer (and out again), to get the solutions written and the pictures labeled: we greatly appreciate your ingenuity, energy and dedication. Thanks to: Ebo Bentil, Duff Campbell, Kenny Ching, John Cho, Dave Chua, Eric Connally, Jie Cui, Mike Esposito, Greg Fung, David Grenda, David Harris, Alex Kasman, Misha Kazhdan, Alex Mallozzi, Elliot Marks, Jean Morris, Kyle Niedzwiecki, Ed Park, Dave Richards, Natalka Roshak, Ann Ryu, Adrian Vajiac, Mihaela Vajiac, Alice Wang, Eric Wepsic, Gang Zhang, and the team from Arizona, Josh Cowley, Noah Syroid, and Xianbao Xu.

Deborah Hughes-Hallett   David O. Lomen   Douglas Quinney

Andrew M. Gleason   David Lovelock   Jeff Tecosky-Feldman

Daniel E. Flath   William G. McCallum   Joe B. Thrash

Patti Frazer Lock   Brad G. Osgood   Karen R. Thrash

Sheldon P. Gordon   Andrew Pasquale   Thomas W. Tucker

To Students: How to Learn from this Book

- This book may be different from other math textbooks that you have used, so it may be helpful to know about some of the differences in advance. At every stage, this book emphasizes the *meaning* (in practical, graphical or numerical terms) of the symbols you are using. There is much less emphasis on "plug-and-chug" and using formulas, and much more emphasis on the interpretation of these formulas than you may expect. You will often be asked to explain your ideas in words or to explain an answer using graphs.

- The book contains the main ideas of calculus in plain English. Success in using this book will depend on reading, questioning, and thinking hard about the ideas presented. It will be helpful to read the text in detail, not just the worked examples.

- There are few examples in the text that are exactly like the homework problems, so homework problems can't be done by searching for similar–looking "worked out" examples. Success with the homework will come by grappling with the ideas of calculus.

- Many of the problems in the book are open-ended. This means that there is more than one correct approach and more than one correct solution. Sometimes, solving a problem relies on common sense ideas that are not stated in the problem explicitly but which you know from everyday life.

- This book assumes that you have access to a calculator or computer that can graph functions, find (approximate) roots of equations, and compute integrals numerically. There are many situations where you may not be able to find an exact solution to a problem, but can use a calculator or computer to get a reasonable approximation. An answer obtained this way is usually just as useful as an exact one. However, the problem does not always state that a calculator is required, so use your own judgement.

    If you mistrust technology, listen to this student, who started out the same way:

    > Using computers is strange, but surprisingly beneficial, and in my opinion is what leads to success in this class. I have difficulty visualizing graphs in my head, and this has always led to my downfall in calculus. With the assistance of the computers, that stress was no longer a factor, and I was able to concentrate on the concepts behind the shapes of the graphs, and since these became gradually more clear, I got increasingly better at picturing what the graphs should look like. It's the old story of not being able to get a job without previous experience, but not being able to get experience without a job. Relying on the computer to help me avoid graphing, I was tricked into focusing on what the graphs meant instead of how to make them look right, and what graphs symbolize is the fundamental basis of this class. By being able to see what I was trying to describe and learn from, I could understand a lot more about the concepts, because I could change the conditions and see the results. For the first time, I was able to see how everything works together . . . .

    That was a student at the University of Arizona who took calculus in Fall 1990, the first time we used the text. She was terrified of calculus, got a C on her first test, but finished with an A for the course.

- This book attempts to give equal weight to three methods for describing functions: graphical (a picture), numerical (a table of values) and algebraic (a formula). Sometimes it's easier to translate a problem given in one form into another. For example, you might replace the graph of a parabola with its equation, or plot a table of values to see its behavior. It is important to be flexible about your approach: if one way of looking at a problem doesn't work, try another.

- Students using this book have found discussing these problems in small groups helpful. There are a great many problems which are not cut-and-dried; it can help to attack them with the other perspectives your colleagues can provide. If group work is not feasible, see if your instructor can organize a discussion session in which additional problems can be worked on.

- You are probably wondering what you'll get from the book. The answer is, if you put in a solid effort, you will get a real understanding of one of the most important accomplishments of the millennium – calculus – as well as a real sense of how mathematics is used in the age of technology.

Table of Contents

6 CONSTRUCTING ANTIDERIVATIVES   **293**

7 INTEGRATION   **321**

8 USING THE DEFINITE INTEGRAL   **375**

9 APPROXIMATIONS AND SERIES   425

10 DIFFERENTIAL EQUATIONS   489

APPENDIX   **579**

ANSWERS TO ODD NUMBERED PROBLEMS   **623**

INDEX   **639**

CALCULUS

Second Edition

CHAPTER ONE

A LIBRARY OF FUNCTIONS

Functions are truly fundamental to mathematics. In everyday language we say, "The price of a ticket is a function of where you sit," or "The fuel needed to launch a rocket is a function of its payload." In each case, the word *function* expresses the idea that knowledge of one fact tells us another. In mathematics, the most important functions are those in which knowledge of one number tells us another number. If we know the length of the side of a square, its area is determined. If the circumference of a circle is known, its radius is determined.

Calculus starts with the study of functions. This chapter will lay the foundation for calculus by surveying the behavior of the most common functions, including powers, exponentials, logarithms, and the trigonometric functions. We will also explore ways of handling the graphs, tables, and formulas that represent these functions.

1.1   WHAT'S A FUNCTION?

In mathematics, a *function* is used to represent the dependence of one quantity upon another.

Let's look at an example. In the summer of 1990, the temperatures in Arizona reached an all-time high (so high, in fact, that some airlines decided it might be unsafe to land their planes there). The daily high temperatures in Phoenix for June 19–29 are given in Table 1.1.

TABLE 1.1   *Daily High Temperature in Phoenix, Arizona, June 19–29, 1990*

Date (t, June 1990)	19	20	21	22	23	24	25	26	27	28	29
Temperature (H, °F)	109	113	114	113	113	113	120	122	118	118	108

Although you may not have thought of something so unpredictable as temperature as being a function, the temperature *is* a function of date, because each day gives rise to one and only one high temperature. There is no formula for temperature (otherwise we would not need the weather bureau), but nevertheless the temperature does satisfy the definition of a function: Each input date, t, has a unique output temperature, H, associated with it.

We define a function as follows:

> A **function** is a rule that takes certain numbers as inputs and assigns to each a definite output number. The set of all input numbers is called the **domain** of the function and the set of resulting output numbers is called the **range** of the function.

The input is called the *independent variable* and the output is called the *dependent variable*. In the temperature example, the domain is the set of dates $t = \{19, 20, 21, 22, 23, 24, 25, 26, 27, 28, 29\}$ and the range is the set of temperatures $H = \{109, 113, 114, 120, 122, 118, 108\}$. We call the function f and write $H = f(t)$. Notice that a function may have identical outputs for different inputs (June 22, 23, and 24, for example).

Some quantities, such as date, are *discrete*, meaning they take only certain isolated values (dates must be integers). Other quantities, such as length, are *continuous* as they can be any number. For a continuous variable, domains and ranges are often written using interval notation:

$$a \leq t \leq b \text{ is written } [a, b]$$
$$a < t < b \text{ is written } (a, b).$$

Representation of Functions: Tables, Graphs, Formulas, and Words

Functions can be represented by tables, graphs, formulas, and descriptions in words. For example, the function giving the daily high temperatures in Phoenix, Arizona as a function of time can be represented by each of the graphs in Figure 1.1, as well as by Table 1.1.

Figure 1.1: Phoenix temperatures, June 1990

Healthy   Sick

Figure 1.2: EKG readings on two patients

Other functions arise naturally as graphs. Figure 1.2 contains electrocardiogram (EKG) pictures showing the heartbeat patterns of two patients, one normal and one not. Although it is possible to construct a formula to approximate an EKG function, this is seldom done. The pattern of repetitions is what a doctor needs to know, and these are much more easily seen from a graph than from a formula. However, each EKG represents a function which gives electrical activity as a function of time.

As another example of a function, consider the snow tree cricket. Surprisingly enough, all such crickets chirp at essentially the same rate if they are at the same temperature. That means that the chirp rate is a function of temperature. In other words, if we know the temperature, we can determine the chirp rate. Even more surprisingly, the chirp rate, C, in chirps per minute, increases steadily with the temperature, T, in degrees Fahrenheit, and can be computed by the formula

$$C = 4T - 160$$

to a fair degree of accuracy. We write $C = f(T)$ to express the fact that we think of C as a function of T. The graph of this function is in Figure 1.3.

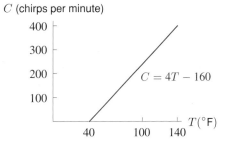

Figure 1.3: Cricket chirp rate versus temperature

Examples of Domain and Range

If the domain of a function is not specified, we usually take it to be the largest possible set of real numbers. For example, we usually think of the domain of the function $f(x) = x^2$ as all real numbers. However, the domain of the function $g(x) = 1/x$ is all real numbers except zero, since we cannot divide by zero.

Example 1 Find the domain and range of the function $y = f(x) = 1/(x-3)^2$.

Solution Since the formula makes sense for all values of x except 3, the domain is all x-values with $x \neq 3$. The range of f is all positive y-values.

Sometimes we specify the domain to be something smaller than the largest possible set of real numbers. This is called restricting the domain. For example, if the function $f(x) = x^2$ is used to represent the area of a square of side x, we consider only nonnegative values of x and restrict the domain to nonnegative numbers.

Example 2   Consider the function $C = f(T)$ giving chirp rate as a function of temperature. We restrict this function to temperatures for which the predicted chirp rate is positive, and up to the highest temperature ever recorded at a weather station, namely, $136°F$. What is the domain of this function f?

Solution   If we consider the equation

$$C = 4T - 160$$

simply as a mathematical relationship between two variables C and T, any T value is possible. However, if we think of it as a relationship between cricket chirps and temperature, then C cannot be less than 0. Since $C = 0$ leads to $0 = 4T - 160$, and so $T = 40°F$, we see that T cannot be less than $40°F$. (See Figure 1.3.) In addition, we are told that the function is not defined for temperatures above $136°$. Thus, for the function $C = f(T)$ we have

$$\text{Domain} = \text{All } T \text{ values between } 40°F \text{ and } 136°F$$
$$= \text{All } T \text{ values with } 40 \leq T \leq 136$$
$$= [40, 136].$$

Therefore, we say that the function $C = f(T)$ is represented by the formula

$$C = f(T) = 4T - 160 \quad \text{on the domain} \quad 40 \leq T \leq 136.$$

Example 3   Find the range of the function f, given the domain from Example 2. In other words, find all possible values of the chirp rate, C, in the equation $C = f(T)$.

Solution   Again, if we consider $C = 4T - 160$ simply as a mathematical relationship, its range is all real C values. However, when thinking of the meaning of $C = f(T)$ for crickets, we see that the function will predict cricket chirps per minute between 0 (when $T = 40°F$) and 384 (when $T = 136°F$). Hence,

$$\text{Range} = \text{All } C \text{ values from 0 to 384}$$
$$= \text{All } C \text{ values with } 0 \leq C \leq 384$$
$$= [0, 384].$$

So far we have used the temperature to predict the chirp rate and have thought of the temperature as the *independent variable* and the chirp rate as the *dependent variable*. However, we could do this backwards, and calculate the temperature from the chirp rate. From this point of view, the temperature is dependent on the chirp rate. Thus, which variable is dependent and which is independent may depend on your viewpoint.

Thinking of temperature as a function of chirp rate would enable us (in theory, at least) to use the chirp rate to measure temperature. The way we actually do measure temperature is based on another function: the relationship between the height of the liquid in a thermometer and temperature. Although the height of the mercury is certainly a function of temperature, we actually always use this relation the other way around and determine the temperature as a function of the height of the mercury.

Proportionality

A common functional relationship occurs when one quantity is *proportional* to another. For example, if apples are 60 cents a pound, we say the price you pay, p cents, is proportional to the weight you buy, w pounds, because

$$p = f(w) = 60w.$$

As another example, the area, A, of a circle is proportional to the square of the radius, r:

$$A = f(r) = \pi r^2.$$

> We say y is (directly) **proportional** to x if there is a constant k such that
>
> $$y = kx.$$
> This k is called the constant of proportionality.

We also say that one quantity is *inversely proportional* to another if one is proportional to the reciprocal of the other. For example, the speed, v, at which you make a 50-mile trip is inversely proportional to the time, t, taken, because v is proportional to $1/t$:

$$v = 50 \left(\frac{1}{t} \right) = \frac{50}{t}.$$

Notice that if y is directly proportional to x, then the magnitude of one variable increases (decreases) when the magnitude of the other increases (decreases). If, however, y is inversely proportional to x, then the magnitude of one variable increases when the value of the other decreases.

Problems for Section 1.1

1. Which of the graphs in Figure 1.4 best matches each of the following three stories?[1] Write a story for the remaining graph.

    (a) I had just left home when I realized I had forgotten my books, and so I went back to pick them up.
    (b) Things went fine until I had a flat tire.
    (c) I started out calmly but sped up when I realized I was going to be late.

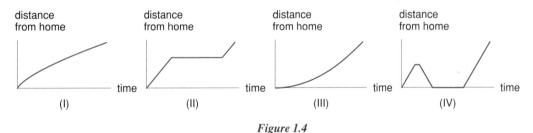

Figure 1.4

2. Match each story to the most appropriate graph. Write a story for the remaining graph.

    (a) During my journey, I had a flat tire. Then after fixing the flat tire, I had to speed up to avoid being late.
    (b) My car broke down and I left it parked at the road side.
    (c) As soon as I'd dropped off the package, I turned around and drove home.

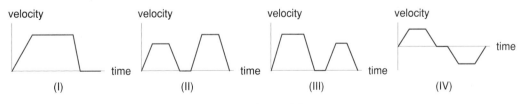

Figure 1.5

3. The population of a city, P (in millions), is a function of t, the number of years since 1950. We have $P = f(t)$. Explain the meaning of the statement $P(35) = 12$ in terms of the population of this city.

---

[1]Adapted from Jan Terwel, "Real Math in Cooperative Groups in Secondary Education." *Cooperative Learning in Mathematics*, ed. Neal Davidson (Reading: Addison Wesley, 1990), 234.

In Problems 4–7, write a formula representing the function described.

4. Kinetic energy, K, is proportional to the square of velocity, v.

5. The gravitational force, F, between two bodies is inversely proportional to the square of the distance d between them.

6. The average velocity, v, for a trip over a fixed distance is inversely proportional to the time of travel.

7. The volume of a sphere is proportional to the cube of its radius, r.

8. It warmed up throughout the morning, and then suddenly got much cooler around noon, when a storm came through. After the storm, it warmed up before cooling off at sunset. Sketch a possible graph of this day's temperature as a function of time.

9. A doctor rides his bicycle from his home to the clinic, a trip of about ten miles. He goes along at a constant speed until he reaches the big hill which starts at mile 4. He slows down as he goes up the hill, which peaks at mile 5. He is barely moving when he reaches the top and stops to rest for a bit. He then goes down the hill (fairly quickly thanks to the steep grade), reaching the bottom at mile 7. The last three miles to his office are flat. Sketch a graph which shows the doctor's *speed* as a function of the *distance* from home.

10. When a nerve impulse reaches the end of a neuron, neurotransmitter is released, initiating a nerve impulse in the next neuron. The neurotransmitter is then enzymatically degraded. Sketch a graph showing the concentration of neurotransmitter in the space between the neurons as a function of time.

11. A flight from Dulles Airport in Washington, DC to LaGuardia Airport in New York City has to circle LaGuardia several times before being allowed to land. Plot a graph of the distance of the plane from Washington against time, from the moment of takeoff until landing.

12. Plot the distance of the plane in Problem 11 from LaGuardia as a function of time, from the moment of take-off until landing.

13. An object is put outside on a cold day and its temperature, H, in °C is a function of the time, t, in minutes since it was put outside. A graph of the function $H = f(t)$ is given in Figure 1.6.

    (a) What does the statement $f(30) = 10$ mean in terms of temperature? Include units for 30 and for 10 in your answer.

    (b) Explain what the vertical intercept, a, and the horizontal intercept, b, represent in terms of temperature of the object and time outside.

Problems 14–15 are about supply and demand curves. Economists are interested in how the quantity, q, of an item which is manufactured and sold depends on the price, p, per unit. They think of quantity as a function of price. However, for historical reasons[2] the economists put price (the independent variable) on the vertical axis and quantity (the dependent variable) on the horizontal axis. Since manufacturers and customers react differently to changes in price, there are two functions relating p and q. The *supply curve* represents how the quantity of an item that manufacturers are willing to supply depends on the price for which the item can be sold. The *demand curve* represents how the quantity of an item demanded by consumers depends on its price.

Figure 1.6

Figure 1.7: Supply and demand curves

14. One of the graphs in Figure 1.7 is a supply curve, and the other is a demand curve. Which is which? Why?

15. The price p_0 in Figure 1.7(a) represents the price below which the manufacturers are unwilling to produce any of the item. What do the price p_1 and the quantity q_1 in Figure 1.7(b) represent in practical economic terms?

---

[2]Originally, the economists thought of price as the dependent variable and put it on the vertical axis. Unfortunately, when their point of view changed, the axes did not.

For Problems 16–18, give the approximate domain and range of each function. Assume the entire graph is shown.

16.

17.

18.

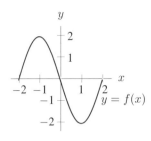

Find the domain and range of the functions in Problems 19–21.

19. $y = x^2 + 2$

20. $y = \dfrac{1}{x - 2}$

21. $y = \dfrac{1}{x^2 + 2}$

22. If $f(t) = \sqrt{t^2 - 16}$, find all values of t for which $f(t)$ is a real number. Solve $f(t) = 3$.

23. If $g(x) = (4 - x^2)/(x^2 + x)$, find the domain of $g(x)$. Solve $g(x) = 0$.

24. When Galileo was formulating the laws of motion, he considered the motion of a body starting from rest and falling under gravity. He originally thought that the velocity of such a falling body was proportional to the distance it had fallen. What do the experimental data in Table 1.2 tell you about Galileo's hypothesis? What alternative hypothesis is suggested by the two sets of data in Table 1.2 and Table 1.3?

TABLE 1.2

Distance (ft)	0	1	2	3	4
Velocity (ft/sec)	0	8	11.3	13.9	16

TABLE 1.3

Time (sec)	0	1	2	3	4
Velocity (ft/sec)	0	32	64	96	128

1.2 LINEAR FUNCTIONS

Probably the most commonly used functions are the *linear functions*. These are functions that have a constant rate of increase or decrease. A function is linear if its slope, or rate of change, is the same everywhere. For a function that is not linear, the rate of change will vary from point to point.

The Olympic Pole Vault

During the early years of the Olympics, the height of the winning pole vault increased approximately 8 inches every four years. Table 1.4 shows that the height started at 130 inches in 1900, and increased by the equivalent of 2 inches every year. So the height was a linear function of time from 1900 to 1912. If y is the winning height in inches and t is the number of years since 1900, we can write

$$y = f(t) = 130 + 2t.$$

Since $y = f(t)$ increases with t, we say that f is *an increasing function*. The coefficient 2 tells us the rate, in inches per year, at which the height increases. This rate is the *slope* of the line $f(t) = 130 + 2t$.

TABLE 1.4 *Olympic pole vault records (approximate)*

Year	1900	1904	1908	1912
Height (inches)	130	138	146	154

You can visualize the slope in Figure 1.8 as the ratio

$$\text{Slope} = \frac{\text{Rise}}{\text{Run}} = \frac{138 - 130}{1904 - 1900} = \frac{8}{4} = 2\text{inches/year.}$$

Calculating the slope (rise/run) using any other two points on the line gives the same value.

What about the constant 130? This represents the initial height in 1900, when $t = 0$. Geometrically, 130 is the *intercept* on the vertical axis.

You may wonder whether the linear trend continues beyond 1912. Not surprisingly, it doesn't exactly. The formula $y = 130 + 2t$ predicts that the height in the 1996 Olympics would be 322 inches or 26 feet 10 inches, which is considerably higher than the actual value of 19 feet 5.25 inches. There is clearly a danger in *extrapolating* too far from the given data. You should also observe that the data in Table 1.4 is *discrete*, because it is given only at specific points (every four years). However, we have treated the variable t as though it were *continuous*, because the function $y = 130 + 2t$ makes sense for all values of t. The graph in Figure 1.8 is of the continuous function because it is a solid line, rather than four separate points representing the years in which the Olympics were held.

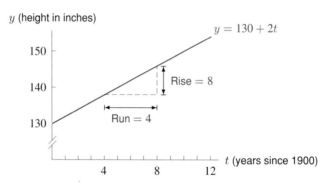

Figure 1.8: Olympic pole vault records

Linear Functions in General

> A **linear function** has the form
>
> $$y = f(x) = b + mx.$$
>
> Its graph is a line such that
> - m is the **slope**, or rate of change of y with respect to x.
> - b is the **vertical intercept**, or value of y when x is zero.

Notice that if the slope, m, is zero, we have $y = b$, a horizontal line.

> To recognize that a table of x and y values comes from a linear function, $y = b + mx$, look for differences in y-values that are constant for equal differences in x.

Difference Quotients and Delta Notation

We use the symbol Δ (the Greek letter capital delta) to mean "change in," so Δx means change in x and Δy means change in y.

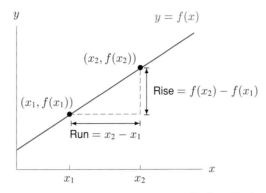

Figure 1.9: Difference quotient $= \dfrac{f(x_2) - f(x_1)}{x_2 - x_1}$

The slope of a linear function $y = f(x)$ can be calculated from values of the function at two points, given by x_1 and x_2, using the formula

$$m = \frac{\text{Rise}}{\text{Run}} = \frac{\Delta y}{\Delta x} = \frac{f(x_2) - f(x_1)}{x_2 - x_1}.$$

The quantity $(f(x_2) - f(x_1))/(x_2 - x_1)$ is called a *difference quotient* because it is the quotient of two differences. (See Figure 1.9). In Chapter 2, you will see that difference quotients play an important role in calculus.

The Success of Search and Rescue Teams

Consider the problem of the "search and rescue" teams working to find lost hikers in remote areas. To search for an individual, members of the search team separate and walk parallel to one another through the area to be searched. Experience has shown that the team's chance of finding a lost individual is related to the distance, d, by which team members are separated. For a particular type of terrain, the percent found[3] for various separations is recorded in Table 1.5.

TABLE 1.5 *Success rate versus separation of searchers*

Separation distance, d (ft)	Approximate percent found, P
20	90
40	80
60	70
80	60
100	50

Figure 1.10: Success rate versus separation of searchers

As we would expect, the data in the table indicates that as the separation distance decreases, a larger percentage of the lost hikers is found. Since $P = f(d)$ decreases as d increases, we say that P is a *decreasing function* of d. In addition, the data shows that for each 20-foot increase in distance, the percentage found drops by 10. This indicates that the graph of P against d is a line. (See Figure 1.10.) Notice that the slope is $-40/80 = -1/2$. The negative sign shows that P decreases as d increases. The magnitude of the slope is the rate at which P is decreasing as d increases.

[3]From *An Experimental Analysis of Grid Sweep Searching*, by J. Wartes (Explorer Search and Rescue, Western Region, 1974).

What about the vertical intercept? If $d = 0$, the searchers are walking shoulder to shoulder and we'd expect everyone to be found, so $P = 100$. This is exactly what happens if the line is continued to the vertical axis (a decrease of 20 in d causes an increase of 10 in P). Therefore, the equation of the line is

$$P = f(d) = 100 - \frac{1}{2}d.$$

What about the horizontal intercept? When $P = 0$, or $0 = 100 - \frac{1}{2}d$, then $d = 200$. The value $d = 200$ represents the separation distance at which, according to the model, no one is found. This is unreasonable, because even when the searchers are far apart, the search will sometimes be successful. This suggests that at some point, the linear relationship ceases to hold. As in the pole vault example, extrapolating too far beyond the given data may not give accurate answers.

Increasing versus Decreasing Functions

The terms increasing and decreasing can be applied to other functions, not just linear ones. See Figure 1.11. In general,

A function f is **increasing** if the values of $f(x)$ increase as x increases.
A function f is **decreasing** if the values of $f(x)$ decrease as x increases.

The graph of an *increasing* function *climbs* as we move from left to right.
The graph of a *decreasing* function *descends* as we move from left to right.

Increasing   Decreasing

Figure 1.11: Increasing and decreasing functions

Families of Linear Functions

Formulas such as $f(x) = b + mx$, in which the constants m and b can take on various values, are said to define a *family of functions*. All the functions in a family share certain properties—in this case, all the graphs are straight lines. Each of the functions in this section belongs to the linear family $f(x) = b + mx$. The constants m and b are called *parameters*; their meaning is shown in Figures 1.12 and 1.13. Notice the greater the magnitude of m is, the steeper the line is.

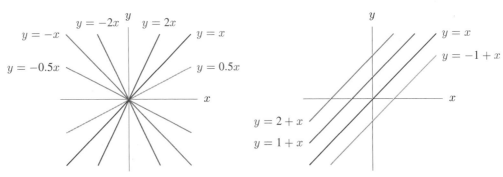

Figure 1.12: The family $y = mx$
(with $b = 0$)

Figure 1.13: The family $y = b + x$
(with $m = 1$)

Problems for Section 1.2

For Problems 1–3, determine the slope and the y-intercept of the line whose equation is given.

1. $7y + 12x - 2 = 0$
2. $-4y + 2x + 8 = 0$
3. $12x = 6y + 4$

For Problems 4–6, find the equation of the line that passes through the given points.

4. $(0, 0)$ and $(1, 1)$
5. $(0, 2)$ and $(2, 3)$
6. $(-2, 1)$ and $(2, 3)$

For Problems 7–9, use the facts that parallel lines have equal slopes and that the slopes of perpendicular lines are negative reciprocals of one another.

7. Find the equation of the line through the point $(2, 1)$ which is perpendicular to the line $y = 5x - 3$.
8. Find the equations of the lines through the point $(1, 5)$ that are parallel to and perpendicular to the line with equation $y + 4x = 7$.
9. Find the equations of the lines through the point (a, b) that are parallel and perpendicular to the line $y = mx + c$, assuming $m \neq 0$.
10. Match the graphs in Figure 1.14 with the equations below. (Note that the x and y scales may be unequal.)
    (a) $y = x - 5$ (c) $5 = y$ (e) $y = x + 6$
    (b) $-3x + 4 = y$ (d) $y = -4x - 5$ (f) $y = x/2$

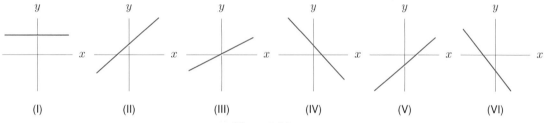

(I)   (II)   (III)   (IV)   (V)   (VI)

Figure 1.14

11. Match the graphs in Figure 1.15 with the equations below. (Note that the x and y scales may be unequal.)
    (a) $y = -2.72x$ (c) $y = 27.9 - 0.1x$ (e) $y = -5.7 - 200x$
    (b) $y = 0.01 + 0.001x$ (d) $y = 0.1x - 27.9$ (f) $y = x/3.14$

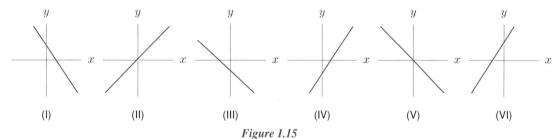

(I)   (II)   (III)   (IV)   (V)   (VI)

Figure 1.15

12. An equation of a line is $3x + 4y = -12$. Find the length of the portion of the line that lies between the x and y intercepts of the equation.
13. The table below gives values for a linear function. Find a formula for W in terms of R.

R	6	9	12	15	18
W	20	25	30	35	40

14. Suppose you are driving at a constant speed from Chicago to Detroit, a distance of 275 miles. About 120 miles from Chicago you pass through Kalamazoo, Michigan. Sketch a graph of your distance from Kalamazoo as a function of time.

15. Residents of the town of Maple Grove who are connected to the municipal water supply are billed a fixed amount yearly plus a charge for each cubic foot of water used. A household using 1000 cubic feet was billed $90, while one using 1600 cubic feet was billed $105.

(a) What is the charge per cubic foot?
(b) Write an equation for the total cost of a resident's water as a function of cubic feet of water used.
(c) How many cubic feet of water used would lead to a bill of $130?

16. You are knitting a scarf of constant width. Let $s(y) = b + ay$ be a function that gives the length $s(y)$, in feet, of the scarf after y skeins of yarn have been used. Here a and b are constants.

(a) What is the value of b? Why?
(b) Is a positive, negative, or zero?
(c) Your friend is also knitting a scarf using a pattern that alternates between stitches and empty spaces. Your own scarf has no such empty spaces, only solid stitches. Your friend's length formula is $p(y) = cy$, where c is a constant. Which constant is larger, a or c?

17. The graph of Fahrenheit temperature, °F, as a function of Celsius temperature, °C, is a line. You know that 212°F and 100°C both represent the temperature at which water boils. Similarly, 32°F and 0°C both represent water's freezing point.

(a) What is the slope of the graph?
(b) What is the equation of the line?
(c) Use the equation to find what Fahrenheit temperature corresponds to 20°C.
(d) What temperature is the same number of degrees in both Celsius and Fahrenheit?

18. Let p represent the selling price in dollars of a certain type of toy and q the number of toys (in thousands) which sell at that price. Market research indicates the following relationship between p and q:

p	1	2	3	4
q	950	900	850	800

(a) Find a formula for q as a linear function of p.
(b) Find a formula for p as a linear function of q.
(c) According to this model, if toys are given away for free, how many will be taken?

19. When a solid is subjected to small stresses (forces of tension or compression), the resulting strain (the fractional change in length) is proportional to the stress. The elasticity of the material is represented by the constant E, called Young's modulus, where

$$E = \frac{\text{Stress}}{\text{Strain}}.$$

Figure 1.16 shows the stress-strain curve for bone. Estimate Young's modulus for bone.

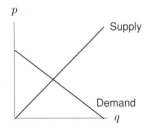

stress (10^7 newton/meter^2)

Figure 1.16   *Figure 1.17*

20. Linear supply and demand curves[4] are shown in Figure 1.17, with price on the vertical axis. The equilibrium point is defined to be where the supply and demand curves cross.

(a) Label the equilibrium price p^* and the equilibrium quantity q^* on the axes.
(b) Explain the effect on equilibrium price and quantity if the slope of the supply curve increases. Illustrate your answer graphically.

[4]For a discussion of supply and demand curves, see page 6.

(c) Explain the effect on equilibrium price and quantity if the slope of the demand curve becomes more negative, keeping the p-intercept constant. Illustrate your answer graphically.

21. Since the opening up of the West, the US population has moved westward. To observe this, we look at the "population center" of the US, which is the point at which the country would balance if it were a flat plate with no weight, and every person had equal weight. In 1790 the population center was east of Baltimore, Maryland. It has been moving westward ever since, and in 1990 it crossed the Mississippi river to Steelville, Missouri (southwest of St. Louis). During the second half of this century, the population center has moved about 50 miles west every 10 years.

   (a) Let us measure position westward from Steelville along the line running through Baltimore. Express the approximate position of the population center as a function of time, measured in years from 1990.
   (b) The distance from Baltimore to Steelville is a bit over 700 miles. Could the population center have been moving at roughly the same rate for the last two centuries?
   (c) Could the function in part (a) continue to apply for the next three centuries? Why or why not? [Hint: You may want to look at a map. Note that distances are in air miles and are not driving distances.]

22. For small changes in temperature, the formula for the expansion of a metal rod under a change in temperature is:
$$l - l_0 = al_0(t - t_0),$$
where l is the length of the object at temperature t, and l_0 is the initial length at temperature t_0, and a is a constant which depends on the type of metal.

   (a) Express l as a linear function of t. Find the slope and vertical intercept. [Hint: Treat the other quantities as constants.]
   (b) Suppose you had a rod which was initially 100 cm long at $60°$F and made of a metal with a equal to 10^{-5}. Write an equation giving the length of this rod at temperature t.
   (c) What does the sign of the slope of the graph tell you about the expansion of a metal under a change in temperature?

23. When a cold yam is put into a hot oven to bake, the temperature of the yam rises. The rate, R (in degrees per minute), at which the temperature of the yam rises is governed by Newton's Law of Heating, which says that the rate is proportional to the temperature difference between the yam and the oven. If the oven is at $350°$F and the temperature of the yam is $H°$F,

   (a) Write a formula giving R as a function of H.
   (b) Sketch the graph of R against H.

24. A body of mass m is falling downward with velocity v. Newton's Second Law of Motion, $F = ma$, says that the net downward force, F, on the body is proportional to its downward acceleration, a. The net force, F, consists of the force due to gravity, F_g, which acts downward, minus the air resistance, F_r, which acts upward. The force due to gravity is mg, where g is a constant. Assume the air resistance is proportional to the velocity of the body.

   (a) Write an expression for the net force, F, as a function of the velocity, v.
   (b) Write a formula giving a as a function of v.
   (c) Sketch a against v.

1.3 EXPONENTIAL FUNCTIONS

Population Growth

Consider the data for the population of Mexico in the early 1980s given in Table 1.6. To see how the population is growing, we might look at the increase in population from one year to the next, as shown in the third column. If the population had been growing linearly, all the numbers in the third column would be the same. But populations usually grow faster as they get bigger, because there are more people to have babies. So we shouldn't be surprised to see the numbers in the third column increasing.

TABLE 1.6   *Population of Mexico (estimated), 1980–1986*

Year	Population (millions)	Change in population (millions)
1980	67.38	
		1.75
1981	69.13	
		1.80
1982	70.93	
		1.84
1983	72.77	
		1.89
1984	74.66	
		1.94
1985	76.60	
		1.99
1986	78.59	

Suppose we divide each year's population by the previous year's population. We get, approximately,

$$\frac{\text{Population in 1981}}{\text{Population in 1980}} = \frac{69.13 \text{ million}}{67.38 \text{ million}} = 1.026$$

$$\frac{\text{Population in 1982}}{\text{Population in 1981}} = \frac{70.93 \text{ million}}{69.13 \text{ million}} = 1.026.$$

The fact that both calculations give 1.026 shows the population grew by about 2.6% between 1980 and 1981 *and* between 1981 and 1982. If we do similar calculations for other years, we find that the population grew by a factor of about 1.026, or 2.6%, every year. Whenever we have a constant growth factor (here 1.026), we have *exponential growth*. If t is the number of years since 1980,

When $t = 0$,   population $= 67.38 = 67.38(1.026)^0$.

When $t = 1$,   population $= 69.13 = 67.38(1.026)^1$.

When $t = 2$,   population $= 70.93 = 69.13(1.026) = 67.38(1.026)^2$.

When $t = 3$,   population $= 72.77 = 70.93(1.026) = 67.38(1.026)^3$.

So in general, the population t years after 1980 is given by

$$P = 67.38(1.026)^t.$$

This is an *exponential function* with base 1.026. It is called exponential because the independent variable, t, is in the exponent. The base represents the factor by which the population grows each year. If we assume that the same formula will hold for the next 50 years, the population graph has the shape shown in Figure 1.18. Since the population is growing, the function is increasing. Notice also that the population grows faster and faster as time goes on. This behavior is typical of an exponential function. Compare this with the behavior of a linear function, which climbs at the same rate everywhere and so has a straight-line graph. Because the graph of an exponential function is bending upward, we say it is *concave up*. Even exponential functions which climb slowly at first, such as this one, climb extremely quickly eventually. That is why exponential population growth is considered by some to be such a threat to the world.

Even if it represents reliable data, the smooth graph in Figure 1.18 is actually only an approximation to the true graph of the population of Mexico. Since we can't have fractions of people, the graph should really be jagged, jumping up or down by one each time someone is born or dies. However, with a population in the millions, the jumps are so small as to be invisible at the scale we are using. Therefore, the smooth graph is an extremely good approximation.

Example 1   Predict the population of Mexico in the year
(a)   2007 (when $t = 27$).   (b)   2034 (when $t = 54$).   (c)   2061 (when $t = 81$).

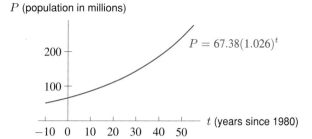

Figure 1.18: Population of Mexico (estimated): Exponential growth

Solution Extrapolating far into the future can be risky because it assumes that the population continues to grow exponentially with the same constant growth factor. (There could, for example, be a medical breakthrough that would increase the growth factor, or an epidemic that would decrease it.) Writing "≈" to represent approximately equal, the model we are using predicts

(a) $P = 67.38(1.026)^{27} \approx 67.38(2) = 134.76$ million.

(b) $P = 67.38(1.026)^{54} \approx 67.38(4) = 269.52$ million.

(c) $P = 67.38(1.026)^{81} \approx 67.38(8) = 539.04$ million.

If we look at the answers to Example 1, we see that after 27 years the population has doubled; after another 27 years (at $t = 54$), it has doubled again. In yet another 27 years (when $t = 81$), the population has doubled yet again. As a result, we say that the *doubling time* of the population of Mexico is 27 years. Every exponentially growing population has a fixed doubling time.

Concavity

We have used the term concave up to describe the graph in Figure 1.18. In Figure 1.22 on page 18, we will see a graph which is concave down [5]. In words:

> The graph of any function is **concave up** if it bends upward as we move left to right; it is **concave down** if it bends downward. (See Figure 1.19.) A line is neither concave up nor concave down.

Concave up   Concave down

Figure 1.19: Concavity of a graph

Musical Pitch

The pitch of a musical note is determined by the frequency of the vibration which causes it. Middle C on the piano, for example, corresponds to a vibration of 263 hertz (cycles per second). A note one octave above middle C vibrates at 526 hertz, and a note two octaves above middle C vibrates at 1052 hertz. (See Table 1.7.)

TABLE 1.7 *Pitch of notes above middle C*

Number, n, of octaves above middle C	Number of hertz, $V = f(n)$
0	263
1	526
2	1052
3	2104
4	4208

TABLE 1.8 *Pitch of notes below middle C*

n	$V = 263 \cdot 2^n$
-3	$263 \cdot 2^{-3} = 263(1/2^3) = 32.875$
-2	$263 \cdot 2^{-2} = 263(1/2^2) = 65.75$
-1	$263 \cdot 2^{-1} = 263(1/2) = 131.5$
0	$263 \cdot 2^0 = 263$

[5]In Chapter 5 we consider concavity in more depth.

Notice that the ratios of successive values of V are

$$\frac{526}{263} = 2 \quad \text{and} \quad \frac{1052}{526} = 2 \quad \text{and} \quad \frac{2104}{1052} = 2,$$

and so on. In other words, each value of V is twice the value before, so

$$f(1) = 526 = 263 \cdot 2 = 263 \cdot 2^1$$
$$f(2) = 1052 = 526 \cdot 2 = 263 \cdot 2^2$$
$$f(3) = 2104 = 1052 \cdot 2 = 263 \cdot 2^3.$$

In general

$$V = f(n) = 263 \cdot 2^n,$$

where n is the number of octaves above middle C. The base 2 represents the fact that as we go up an octave, the frequency of vibrations doubles. Indeed, our ears hear a note as one octave higher than another precisely because it vibrates twice as fast. For the negative values of n in Table 1.8, this function represents the octaves below middle C. The notes on a piano are represented by values of n between -3 and 4, and the human ear finds values of n between -4 and 7 audible.

Although $V = f(n) = 263 \cdot 2^n$ makes sense in musical terms only for certain values of n, values of the function $f(x) = 263 \cdot 2^x$ can be calculated for all real x. The graph of $f(x)$ has the typical exponential shape, as can be seen in Figure 1.20. It is increasing and concave up, climbing faster and faster as x increases.

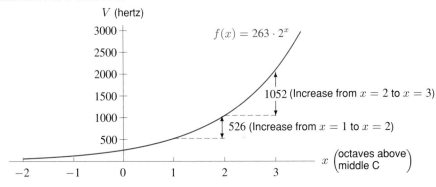

Figure 1.20: Pitch as a function of number of octaves above middle C

Removal of Pollutants from Jet Fuel

Now we look at an example in which a quantity is decreasing instead of increasing. Before kerosene can be used as jet fuel, federal regulations require that the pollutants in it be removed by passing the kerosene through clay. We will suppose the clay is in a pipe and that each foot of the pipe removes 20% of the pollutants that enter it. Therefore each foot leaves 80% of the pollution. If P_0 is the initial quantity of pollutant and $P = f(n)$ is the quantity left after n feet of pipe, then

$$f(0) = P_0$$
$$f(1) = (0.8)P_0$$
$$f(2) = (0.8)(0.8)P_0 = (0.8)^2 P_0$$
$$f(3) = (0.8)(0.8)^2 P_0 = (0.8)^3 P_0.$$

So, after n feet,

$$P = f(n) = P_0(0.8)^n.$$

In this example, n must be nonnegative. However, the *exponential decay function*

$$P = f(x) = P_0(0.8)^x$$

makes sense for any real x. We'll plot it with $P_0 = 1$ in Figure 1.21; some values of the function are shown in Table 1.9.

TABLE 1.9 *Values of decay function*

x	$P = (0.8)^x$
-2	1.56
-1	1.25
0	1
1	0.8
2	0.64
3	0.51
4	0.41

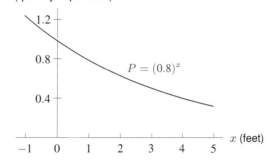

Figure 1.21: Pollutant removal: Exponential decay

Notice the way the function in Figure 1.21 is decreasing. Each additional foot of clay causes a smaller quantity of pollutant to be removed than the previous foot. This is because as the kerosene gets cleaner, there's less pollutant to remove, and so each foot of clay takes out less pollutant than the previous one. Compare this to the exponential growth in Figures 1.18 and 1.20 on pages 15 and 16, where each step upward is larger than the one before. Notice, however, that all three graphs are concave up.

The General Exponential Function

We say P is an **exponential function** of t with base a if

$$P = P_0 a^t,$$

where P_0 is the initial quantity (when $t = 0$) and a is the factor by which P changes when t increases by 1. If $a > 1$, we have exponential growth; if $0 < a < 1$, we have exponential decay.

The largest possible domain for the exponential function is all real numbers, provided $a > 0$. The reason we do not want $a \leq 0$ is that, for example, we cannot define $a^{1/2}$ if $a < 0$. Also, we do not usually have $a = 1$, since $P = P_0 a^t = P_0 1^t = P_0$ is then a constant function.

To recognize that a table of t and P values comes from an exponential function $P = P_0 a^t$, look for ratios of P values that are constant for equally spaced t values.

Radioactive Decay

Radioactive substances, such as uranium, decay by a certain percentage of their mass in a given unit of time. The most common way to express this rate of decay is to give the time period it takes for half the mass to decay. This period of time is called the *half-life* of the substance.

One of the most well-known radioactive substances is carbon-14, which is used to date organic objects. When an object, such as a piece of wood or bone, was part of a living organism, it accumulated small amounts of radioactive carbon-14. Once the organism dies, it no longer picks up carbon-14 through interaction with its environment (for example, through respiration). By measuring the proportion of carbon-14 in the object and comparing that to the proportion in living material, we can estimate how much of the original carbon-14 has decayed.

The half-life of carbon-14 is about 5730 years. We can write an exponential function for the amount of carbon-14 left after a period of t years. If $T = t/5730$ is the number of half-lives which have elapsed and C_0 is the original amount of carbon-14, then the amount, C, of carbon-14 left is given by

$$C = C_0 \left(\frac{1}{2}\right)^T = C_0 \left(\frac{1}{2}\right)^{(t/5730)}.$$

In general, suppose a substance has a half-life of h years (or minutes or seconds). Then, if Q_0 was the original quantity of the substance, the quantity, Q, of the substance left after t units of time is given by

$$Q = Q_0 \left(\frac{1}{2}\right)^{(t/h)}.$$

In summary, we use the following definitions:

The **doubling time** of an exponentially increasing population is the time required for the population to double.

The **half-life** of an exponentially decaying quantity is the time required for the quantity to be reduced by a factor of one half.

Drug Buildup

Suppose that we want to model the amount of a certain drug in the body. Imagine that initially there is none, but that the quantity slowly starts to increase via a continuous intravenous injection. As the quantity of the drug in the body increases, so does the rate at which the body excretes the drug, so that eventually the quantity levels off at a saturation level, S. The graph of quantity against time will look something like Figure 1.22.

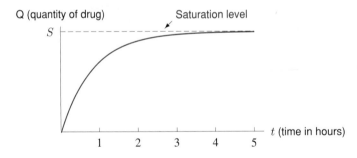

Figure 1.22: Buildup of drug in body

Suppose we want to construct a mathematical model of this situation; that is, suppose we want to find a formula giving the quantity, Q, in terms of time, t. Making a mathematical model often involves looking at a graph and deciding what kind of function has that shape. The graph in Figure 1.22 looks like an exponential decay function, but upside down. What actually decays is the difference between the saturation level, S, and the quantity, Q, in the blood. Suppose the difference between the saturation level and the quantity in the body is given by the formula

$$\text{Difference} = (\text{Initial difference}) \cdot (0.3)^t$$

with t in hours. Since the difference is $S - Q$, and the initial value of this difference is $S - 0 = S$, we have

$$S - Q = S \cdot (0.3)^t.$$

Solving for Q as a function of t gives

$$Q = S - S \cdot (0.3)^t$$
$$Q = f(t) = S \cdot \left(1 - (0.3)^t\right).$$

The graph of this function looks like exponential decay, but upside-down. Notice that the quantity, Q, starts at zero and increases toward S. Since the rate at which the quantity of the drug increases slows as the quantity approaches S, this graph is bending downward; hence it is increasing and *concave down*. It is an example of a function of the form

$$\boxed{Q = S(1 - a^t) \quad \text{with} \quad 0 < a < 1.}$$

Asymptotes

We say that the line representing the saturation level is a *horizontal asymptote*, because the graph gets arbitrarily close to it as time increases. As t gets larger, $(0.3)^t$ gets smaller, so Q gets closer to S. Using "$\to$" to mean "tends to," we can write $(0.3)^t \to 0$ as $t \to \infty$. Then

$$Q = S(1 - (0.3)^t) \to S(1 - 0) = S \quad \text{as} \quad t \to \infty,$$

so the graph of $Q = S(1 - (0.3)^t)$ has a horizontal asymptote at $Q = S$.

If the graph of $y = f(x)$ approaches a horizontal line $y = L$ as $x \to \infty$ or $x \to -\infty$, then the line $y = L$ is called a **horizontal asymptote**.[6] This occurs when

$$f(x) \to L \quad \text{as} \quad x \to \infty \qquad \text{or} \qquad f(x) \to L \quad \text{as} \quad x \to -\infty.$$

If the graph of $y = f(x)$ approaches the vertical line $x = K$ as $x \to K$ from one side or the other, that is, if

$$y \to \infty \quad \text{or} \quad y \to -\infty \quad \text{when} \quad x \to K,$$

then the line $x = K$ is called a **vertical asymptote**.

As an example, the graph of $f(x) = 2 + 1/(x - 3)$ has a vertical asymptote at $x = 3$ (See Figure 1.23.)

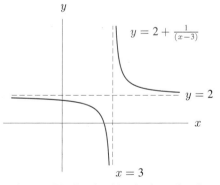

Figure 1.23: Graph with a horizontal and a vertical asymptote

---

[6]We are assuming that $f(x)$ gets arbitrarily close to L as $x \to \infty$.

The Family of Exponential Functions

The formula $P = P_0 a^t$ gives a family of exponential functions with parameters P_0 (the initial quantity) and a (the base, or growth/decay factor). The base tells us whether the function is increasing $(a > 1)$ or decreasing $(0 < a < 1)$. Since a is the factor by which P changes when t is increased by 1, large values of a mean fast growth; values of a near 0 mean fast decay. (See Figures 1.24 and 1.25.) All members of the family $P = P_0 a^t$ are concave up.

Alternative Formula for the Exponential Function

Exponential growth is often described in terms of growth rates in percent. For example, the population of Mexico is growing at 2.6% per year; in other words, the growth factor is $a = 1 + 0.026 = 1.026$. Similarly, each foot of clay removes 20% of the pollution from jet fuel, so the decay factor is $a = 1 - 0.20 = 0.8$. In general, the following formulas apply.

If r is the *growth* rate, then $a = 1 + r$, and

$$P = P_0 a^t = P_0 (1 + r)^t.$$

If r is the *decay* rate, then $a = 1 - r$, and

$$P = P_0 a^t = P_0 (1 - r)^t.$$

Note, for example, that $r = 0.05$ when the growth rate is 5%.

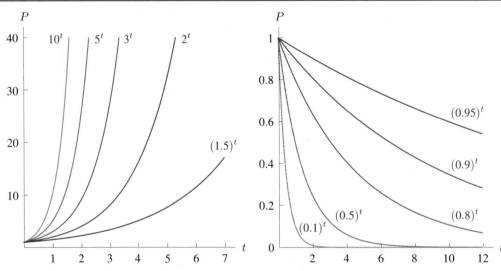

Figure 1.24: Exponential growth: $P = a^t$, for $a > 1$

Figure 1.25: Exponential decay: $P = a^t$, for $0 < a < 1$

Example 2   Suppose that $Q = f(t)$ is an exponential function of t. If $f(20) = 88.2$ and $f(23) = 91.4$:

(a) Find the base.   (b) Find the growth rate.   (c) Evaluate $f(25)$.

Solution   (a) Let

$$Q = Q_0 a^t.$$

Substituting $t = 20, Q = 88.2$ and $t = 23, Q = 91.4$ gives two equations for Q_0 and a:

$$88.2 = Q_0 a^{20} \quad \text{and} \quad 91.4 = Q_0 a^{23}.$$

Dividing the two equations enables us to eliminate Q_0:

$$\frac{91.4}{88.2} = \frac{Q_0 a^{23}}{Q_0 a^{20}} = a^3.$$

Solving for the base, a, gives

$$a = \left(\frac{91.4}{88.2}\right)^{1/3} \approx 1.012.$$

(b)   Since $a \approx 1.012$, the growth rate is $r \approx 0.012 = 1.2\%$.
(c)   We want to evaluate $f(25) = Q_0 a^{25} = Q_0(1.012)^{25}$. First we find Q_0 from the equation

$$88.2 = Q_0(1.012)^{20}.$$

Solving gives $Q_0 \approx 69.48$. Thus,

$$f(25) = 69.48(1.012)^{25} = 93.62.$$

Review: Definition and Properties of Exponents

Below we list the definitions and properties that are used to manipulate exponents.

Definition of Zero, Negative, and Fractional Exponents

$$a^0 = 1, \quad a^{-1} = \frac{1}{a}, \quad \text{and, in general, } a^{-x} = \frac{1}{a^x}$$

$$a^{1/2} = \sqrt{a}, \quad a^{1/3} = \sqrt[3]{a}, \quad \text{and, in general, } a^{1/n} = \sqrt[n]{a}.$$

Also, $a^{m/n} = \sqrt[n]{a^m} = (\sqrt[n]{a})^m$.

Properties of Exponents

1. $a^x \cdot a^t = a^{x+t}$ For example, $2^4 \cdot 2^3 = (2 \cdot 2 \cdot 2 \cdot 2) \cdot (2 \cdot 2 \cdot 2) = 2^7$.

2. $\dfrac{a^x}{a^t} = a^{x-t}$ For example, $\dfrac{2^4}{2^3} = \dfrac{2 \cdot 2 \cdot 2 \cdot 2}{2 \cdot 2 \cdot 2} = 2^1$.

3. $(a^x)^t = a^{xt}$ For example, $(2^3)^2 = 2^3 \cdot 2^3 = 2^6$.

Problems for Section 1.3

In Problems 1–4, decide whether each graph is concave up, concave down, or neither.

1.   2.   3.   4.

In Problems 5–6, explain whether the given function represents exponential growth or decay.

5. $p = 100(7)^t$

6. $p = 75(0.25)^t$

For Problems 7–9, say whether the data set appears to show exponential growth, exponential decay, or neither. If it does show exponential behavior, give a possible formula for the function.

7. **TABLE 1.10**

t	$g(t)$
0	1
1	2
2	4
3	8
4	16

8. **TABLE 1.11**

t	$f(t)$
2	9
3	18
4	81
5	243
6	486

9. **TABLE 1.12**

t	$h(t)$
3	2096
4	1048
5	524
6	262
7	131

10. If a yam is put in the oven, the temperature of the yam increases quickly at first, and then increases more and more slowly. Draw a graph of the temperature of the yam against time.

11. Each year the world's annual consumption of electricity rises. In addition, each year the increase in annual consumption also rises. Sketch a possible graph of the annual world consumption of electricity as a function of time.

12. A drug is injected into a patient's bloodstream over a five-minute interval. During this time, the quantity in the blood increases linearly. After five minutes the injection is discontinued, and the quantity then decays exponentially. Sketch a graph of the quantity versus time.

13. The number of cancer cells in a tumor grows slowly at first but then grows with increasing rapidity. Draw a possible graph of the number of cancer cells against time.

Decide if each of the functions in Problems 14–17 has a horizontal asymptote. If there is an asymptote, give its equation.

14. $y = 5(0.8)^x$ 15. $P = 5(1 - (0.8)^t)$ 16. $Q = 2^t + 2^{-t}$ 17. $Z = 3 + (0.1)^t$

In Problems 18–19, assume that $f(t) = Q_0 a^t = Q_0(1 + r)^t$. Given two values of f:

(a) Find the base, a.

(b) Find the percentage growth rate, r.

18. $f(5) = 75.94$ and $f(7) = 170.86$

19. $f(0.02) = 25.02$ and $f(0.05) = 25.06$

Give a possible formula for each function graphed in Problems 20–23.

20.

21.

22.

23.
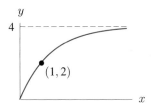

24. Each of the functions in Table 1.13 is increasing, but each increases in a different way. Which of the graphs in Figure 1.26 below best fits each function?

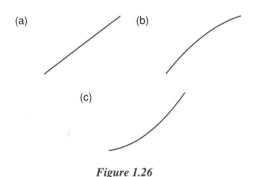

(a)   (b)

(c)

Figure 1.26

TABLE 1.13

t	$g(t)$	$h(t)$	$k(t)$
1	23	10	2.2
2	24	20	2.5
3	26	29	2.8
4	29	37	3.1
5	33	44	3.4
6	38	50	3.7

25. Match up the functions $h(s)$, $f(s)$, and $g(s)$, whose values are in Table 1.14, with the formulas

$$y = a(1.1)^s, \quad y = b(1.05)^s, \quad y = c(1.03)^s,$$

assuming a, b, and c are constants. Note that the function values have been rounded to two decimal places.

TABLE 1.14

s	$h(s)$	s	$f(s)$	s	$g(s)$
2	1.06	1	2.20	3	3.47
3	1.09	2	2.42	4	3.65
4	1.13	3	2.66	5	3.83
5	1.16	4	2.93	6	4.02
6	1.19	5	3.22	7	4.22

[handwritten: — divide → (1.05)]

26. A certain region has a population of 10,000,000 and an annual growth rate of 2%. Estimate the doubling time by guessing and checking.

27. (a) The half-life of radium-226 is 1620 years. Write a formula for the quantity, Q, of radium left after t years, if the initial quantity is Q_0.
    (b) What percentage of an original amount of radium is left after 500 years?

28. In the early 1960s, radioactive strontium-90 was released during atmospheric testing of nuclear weapons and got into the bones of people alive at the time. If the half-life of strontium-90 is 29 years, what fraction of the strontium-90 absorbed in 1960 remained in people's bones in 1990?

29. When the Olympic Games were held outside Mexico City in 1968, there was much discussion about the effect the high altitude (7340 feet) would have on the athletes. Assuming air pressure decays exponentially by 0.4% every 100 feet, by what percentage is air pressure reduced by moving from sea level to Mexico City?

30. (a) During 1996, the US population grew by 0.9% to 266.5 million. Assuming the population continues to grow at the same rate, what will it be at the start of the year 2000?
    (b) What was the annual percentage growth rate of the US population over the period from the start of 1990, when the population was 248.7 million, until the end of 1996?

31. During 1988, Nicaragua's inflation rate averaged 1.3% a day. This means that, on average, prices went up by 1.3% from one day to the next.

    (a) By what percentage did Nicaraguan prices increase in June of 1988?
    (b) What was Nicaragua's annual inflation rate during 1988?

32. Assume that the median price, P, of a home rose from \$50,000 in 1970 to \$100,000 in 1990. Let t be the number of years since 1970.

(a) Assume the increase in housing prices has been linear. Give an equation for the line representing price, P, in terms of t. Use this equation to complete column (a) of Table 1.15. Work with the price in units of \$1000.

(b) If instead the housing prices have been rising exponentially, determine an equation of the form $P = P_0 a^t$ which would represent the change in housing prices from 1970–1990, and complete column (b) of Table 1.15.

(c) On the same set of axes, sketch the functions represented in column (a) and column (b) of Table 1.15.

(d) Which model for the price growth do you think is more realistic?

TABLE 1.15

t	(a) Linear growth price in \$1000 units	(b) Exponential growth price in \$1000 units
0	50	50
10		
20	100	100
30		
40		

1.4 POWER FUNCTIONS

A *power function* is a function in which the dependent variable is proportional to a power of the independent variable. For example, the area, A, of a square of side s is given by

$$A = f(s) = s^2.$$

The volume, V, of a sphere of radius r is given by

$$V = g(r) = \frac{4}{3}\pi r^3.$$

Both of these are *power functions*. Another example is the function which describes how the gravitational attraction of the earth varies with distance. If F is the force of gravitational attraction on a unit mass at a distance r from the earth, Newton's Inverse Square Law of Gravitation says that

$$F = \frac{k}{r^2} \qquad \text{or} \qquad F = kr^{-2}$$

where k is a positive constant.

> A **power function** has the form
>
> $$f(x) = kx^p$$
>
> where k and p are any constants.

In this section we compare various power functions with one another and with the exponential functions.

Positive Integral Powers: $y = x$, $y = x^2$, $y = x^3$, ...

First, we look at functions of the form $f(x) = x^n$, with n a positive integer. Figures 1.27 and 1.28 show that the graphs of these functions fall into two groups: the odd powers and the even powers. All the odd powers (x, x^3, x^5, and so on) are increasing everywhere and their graphs are symmetric about the origin. All odd powers above $n = 1$ have a "seat" at the origin. The even powers, on the other hand, are first decreasing and then increasing, making them ⋃-shaped with symmetry about the y-axis. The even powers are concave up everywhere, whereas those with odd powers greater than 1 are concave down for negative x and concave up for positive x. For large x, the higher the power of x, the faster the function climbs. (See Figure 1.29.) For large values of x (in fact, for all $x > 1$), $y = x^5$ is above $y = x^4$, which is above $y = x^3$, and so on. Not only are the higher powers larger, but they are *much* larger. This is because if $x = 100$, for example, 100^5 is one hundred times as big as 100^4 which is one hundred times as big as 100^3. As x gets larger (written as $x \to \infty$), any positive power of x completely swamps all lower powers of x. We say that, as $x \to \infty$, higher powers of x *dominate* lower powers.

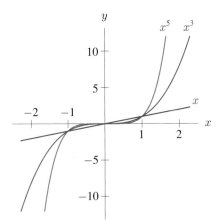

Figure 1.27: Odd powers of x: "Seat" shaped for $n > 1$

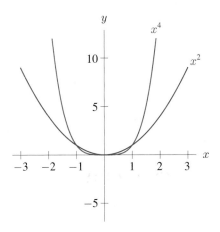

Figure 1.28: Even powers of x: ⋃-shaped

As $x \to 0$, the story is entirely different. See Figure 1.30, which is a close-up view near the origin. For x between 0 and 1, x^3 is bigger than x^4, which is bigger than x^5. (Try $x = 0.1$ to confirm this.) For values of x near zero, smaller powers dominate.

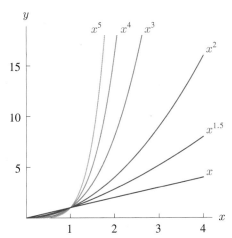

Figure 1.29: For large x: Large powers of x dominate

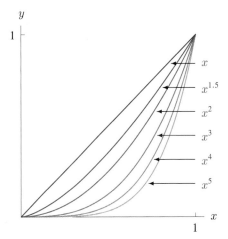

Figure 1.30: For $0 \le x \le 1$: Small powers of x dominate

Zero and Negative Integral Powers: $y = x^0, y = x^{-1}, y = x^{-2}, \ldots$

The function $y = x^0 = 1$ has a graph that is a horizontal line. For negative powers, rewriting

$$y = x^{-1} = \frac{1}{x} \quad \text{and} \quad y = x^{-2} = \frac{1}{x^2}$$

makes it clear that as $x > 0$ increases, the denominators increase and the functions decrease. The graphs of $y = x^{-1}$ and $y = x^{-2}$ have both the x- and y-axes as asymptotes. (See Figure 1.31.)

Figure 1.31: Comparison of zero and negative powers of x

Example 1   Plot a graph of pressure against volume for a fixed quantity of gas at a constant temperature. Use the fact that pressure is inversely proportional to volume.

Solution   Think of a fixed quantity of air—for example, inside a cylinder in a car engine. If the volume, V, of the air is decreased (by moving the pistons), the pressure, P, of the air is increased. Conversely, if the volume is increased, the pressure will decrease. For an ideal gas, Boyle's law gives the exact relationship between pressure and volume, provided the temperature is constant. It says P is inversely proportional to V, so

$$P = \frac{k}{V} = kV^{-1}, \quad \text{where } k \text{ is a positive constant.}$$

Both axes are asymptotes to the graph in Figure 1.32, because as the volume tends to infinity, the pressure tends to zero, and vice versa. The shape is known as a *hyperbola*. The power function $P = k/V = kV^{-1}$ differs from exponential decay in that it is undefined for $V = 0$, so this graph does not cross the vertical axis. In addition, it approaches the horizontal axis more slowly than the exponential function.

Figure 1.32: Graph of pressure, P, against volume, V, for Boyle's law

Example 2   Quantum mechanics predicts that the force between two gas molecules has two components: an attractive force which is proportional to r^{-7} (where r is the distance between the molecules) and a repulsive force which is proportional to r^{-13}. How does the net force vary with r?

Solution   A repulsive force is usually considered to be positive, whereas an attractive force is negative. So we have

$$F = -ar^{-7} + br^{-13} = -\frac{a}{r^7} + \frac{b}{r^{13}}, \quad \text{where } a, b \text{ are positive constants.}$$

If r is very small, $1/r^{13}$ is much larger than $1/r^7$, so

$$F \approx \frac{b}{r^{13}}$$

and the net force is repulsive. (This occurs when the molecules are so close together that the protons in the nuclei repel one another.) For very big r, $1/r^7$ is much larger than $1/r^{13}$, and so

$$F \approx -\frac{a}{r^7},$$

making the net force attractive. As $r \to \infty$, all the forces die away to zero.

Positive Fractional Powers: $y = x^{1/2}$, $y = x^{1/3}$, $y = x^{3/2}$, . . .

The function giving the side of a square, s, in terms of its area, A, involves a root, or fractional power:

$$s = \sqrt{A} = A^{1/2}.$$

Similarly, the equation relating the average number of species on an island to its area, A, is approximately

$$N = k\sqrt[3]{A} = kA^{1/3},$$

where k is a constant depending on the region of the world in which the island is found.[7]

We now look at functions of the form $y = x^{m/n} = \sqrt[n]{x^m}$. Since some fractional powers such as $x^{1/2}$ involve roots and are defined only for positive x and 0, we restrict the domain of these functions to $x \geq 0$.

Figure 1.33 shows that for large x (in fact, all $x > 1$), the graph of $y = x^{1/2}$ is below the graph of $y = x$, and $y = x^{1/3}$ is below $y = x^{1/2}$. This is reasonable since, for $x > 1$, squaring x makes it bigger and cubing it makes it bigger still; thus taking the square root of x makes it smaller and taking its cube root makes it smaller still. Between $x = 0$ and $x = 1$, the situation is reversed and $y = x^{1/3}$ is on top. (Why?) Not surprisingly, $y = x^{3/2}$ is between $y = x$ and $y = x^2$ for all x.

The other important feature to notice about the graphs of $y = x^{1/2}$ and $y = x^{1/3}$ is that they bend in a direction opposite to that of the graphs of $y = x^2$ and x^3. For example, the graph of $y = x^2$ is concave up. On the other hand, the graphs of $y = x^{1/2}$ and $y = x^{1/3}$ are concave down. Despite this, all these functions do become infinitely large as x increases. Notice also that all these functions pass through the point $(1, 1)$.

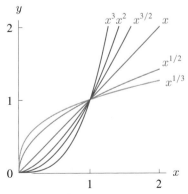

Figure 1.33: Comparison of some fractional powers of x

---

[7]*Scientific American*, September 1989, p. 112.

What Effect Do Coefficients Have?

We know that $x^2 < x^3$ for all $x > 1$. But which is larger, $50x^2$ or x^3? For moderate values of x, like $x = 10$, we have $50x^2 > x^3$. But eventually, $50x^2 < x^3$. In fact, $50x^2 < x^3$ for all $x > 50$. The graphs of $y = x^2$ and $y = x^3$ cross at $x = 1$, whereas the graphs of $y = 50x^2$ and $y = x^3$ cross at $x = 50$. (See Figure 1.34.) Thus, the effect of the factor of 50 is to change the point at which the graphs cross. However, x^3 ends up on top in both cases; provided the coefficients are positive, as $x \to \infty$, the higher power is always larger eventually.

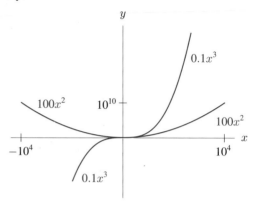

Figure 1.34: Graph of $y = x^3$ lies above graph of $y = 50x^2$ for large positive x

Figure 1.35: For large positive x, $y = 0.1x^3$ dominates $y = 100x^2$

Example 3   Which of $y = 100x^2$ and $y = 0.1x^3$ is larger as $x \to \infty$?

Solution   Since $x \to \infty$, we are looking at large positive values of x, where the higher power will eventually be larger, or dominate. This suggests that $y = 0.1x^3$ will be larger. In fact, $0.1x^3 > 100x^2$ when $x > 1000$. (See Figure 1.35.)

Exponentials and Power Functions: Which Dominate?

In everyday language, the word exponential is often used to imply very fast growth. But do exponential functions always grow faster than power functions? To determine what happens "in the long run," we often want to know which functions dominate as $x \to \infty$.

Let's consider $y = 2^x$ and $y = x^3$. The close-up view in Figure 1.36(a) shows that between $x = 2$ and $x = 4$, the graph of $y = 2^x$ lies below the graph of $y = x^3$. But the far-away view in Figure 1.36(b) shows that the exponential function $y = 2^x$ eventually overtakes $y = x^3$. And Figure 1.36(c), which gives a very far–away view, shows that, for large x, x^3 is insignificant compared to 2^x. Indeed, 2^x is growing so much faster than x^3 that the graph of 2^x appears almost vertical in comparison to the more leisurely climb of x^3.

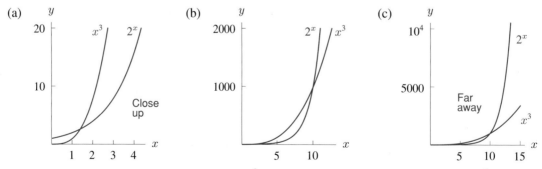

Figure 1.36: Comparison of $y = 2^x$ and $y = x^3$: Notice that $y = 2^x$ eventually dominates $y = x^3$

In fact, *every* exponential growth function eventually dominates *every* power function. Although an exponential function may be below a power function for some values of x, if we look at large enough x-values, a^x (with $a > 1$) will eventually dominate x^n, no matter what n is. Two more examples are presented in Figure 1.37 and Table 1.16.

Figure 1.37: Exponential function eventually dominates power function

TABLE 1.16 *Comparison of* x^{100} *and* 1.01^x

x	x^{100}	1.01^x
10^4	10^{400}	$1.6 \cdot 10^{43}$
10^5	10^{500}	$1.4 \cdot 10^{432}$
10^6	10^{600}	$2.4 \cdot 10^{4321}$

Can you guess what happens in the case of negative powers and negative exponents? For example, consider $y = 2^{-x}$ and $y = x^{-2}$. Since $y = 2^{-x} = 1/2^x$ and $y = x^{-2} = 1/x^2$, knowing that 2^x is eventually larger than x^2 tells you that $y = 2^{-x}$ is eventually smaller than $y = x^{-2}$. Hence $y = 2^{-x}$ is eventually below $y = x^{-2}$. (See Figure 1.38.) This behavior is also typical: Every exponential decay function will eventually approach 0 faster than every power function with a negative exponent.

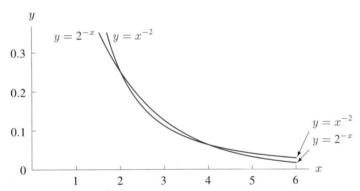

Figure 1.38: Comparison of $y = 2^{-x}$ and $y = x^{-2}$: Exponential dies away faster in the long run

Problems for Section 1.4

For Problems 1–8, simplify the given expression.

1. $8^{4/3}$
2. $27^{1/3}$
3. $10,000^{5/4}$
4. $16^{1/4}$
5. $27^{-1/3}$
6. $16^{-0.25}$
7. $8^{-2/3}$
8. $4^{-3/2}$

For Problems 9–11, determine which function dominates as $x \to \infty$.

9. $10 \cdot 2^x$ or $72,000x^{12}$
10. $0.1x^2$ or $10^{10}x$
11. $0.25\sqrt{x}$ or $25,000x^{-3}$

For Problems 12–14, determine what happens to the value of the function as $x \to \infty$ and as $x \to -\infty$.

12. $y = x^6$
13. $y = 0.25x^3 + 3$
14. $y = 2 \cdot 10^{4x}$

15. On a graphing calculator or computer, plot graphs of the following functions, first for $-5 \leq x \leq 5$, $-100 \leq y \leq 100$, and then for $-1.2 \leq x \leq 1.2$, $-2 \leq y \leq 2$.
    (a) $y = x, y = x^3, y = x^6, y = x^9$

(b) $y = x$, $y = x^4$, $y = x^7$, $y = x^{10}$

Observe the general shape of these functions: Do the odd powers have the same general shape? What about the even powers? Which function is largest in magnitude for big x? For x near 0? Is this what you expected?

16. (a) Use a graphing calculator or a computer to plot the graphs of $y = x^3$, $y = x^4$, and $y = x^5$ on the interval $-0.1 \leq x \leq 0.1$. Determine an appropriate range for y so that all powers will be distinguishable in the viewing rectangle.

    (b) Plot the same graphs for $-100 \leq x \leq 100$, and determine an appropriate range for y.

17. By hand, sketch global graphs of $f(x) = x^3$ and $g(x) = 20x^2$ on the same axes. Which function has larger values as $x \to \infty$?

18. By hand, sketch graphs of $f(x) = x^5$, $g(x) = -x^3$, and $h(x) = 5x^2$ on the same axes. Which has the largest positive values as $x \to \infty$? As $x \to -\infty$?

19. In physiology, the DuBois formula relates a person's surface area, s, in m^2, to weight w, in kg, and height h, in cm, by
$$s = 0.01 w^{0.25} h^{0.75}.$$

    (a) What is the surface area of a person who weighs 65 kg and is 160 cm tall?
    (b) What is the weight of a person whose height is 180 cm and who has a surface area of 1.5 m^2?
    (c) For people of fixed weight 70 kg, solve for h as a function of s. Simplify your answer.

20. Suppose a force between two atoms a distance r apart is given by
$$F = \frac{A}{r^3} - \frac{B}{r^2} \quad \text{where } A, B \text{ are positive constants.}$$

    (a) For small positive r, is F positive or negative?
    (b) For large positive r, is F positive or negative?
    (c) For what, if any, value(s) of r is F zero?
    (d) What are the vertical and horizontal asymptotes of F?

21. According to the April 1991 issue of *Car and Driver*, an Alfa Romeo going at 70 mph requires 177 feet to stop. Assuming that the stopping distance is proportional to the square of velocity, find the stopping distances required by an Alfa Romeo going at 35 mph and at 140 mph (its top speed).

22. Poiseuille's law gives the rate of flow, R, of a gas through a cylindrical pipe in terms of the radius of the pipe, r, for a fixed drop in pressure. Assume a constant drop in pressure throughout the remainder of this problem.

    (a) Determine a formula for Poiseuille's Law, given that the rate of flow is proportional to the fourth power of the radius.
    (b) If $R = 400$ cm^3/sec in a pipe of radius 3 cm for a certain gas, determine an explicit formula for the rate of flow of that gas through a pipe of radius r cm.
    (c) What is the rate of flow of the same gas through a pipe with a 5 cm radius?

23. Values of three functions are given in Table 1.17, rounded to two decimal places. One function is of the form $y = ab^t$, one is of the form $y = ct^2$, and one is of the form $y = kt^3$. Which function is which?

TABLE 1.17

t	$f(t)$	t	$g(t)$	t	$h(t)$
2.0	4.40	1.0	3.00	0.0	2.04
2.2	5.32	1.2	5.18	1.0	3.06
2.4	6.34	1.4	8.23	2.0	4.59
2.6	7.44	1.6	12.29	3.0	6.89
2.8	8.62	1.8	17.50	4.0	10.33
3.0	9.90	2.0	24.00	5.0	15.49

TABLE 1.18

x	$f(x)$	x	$g(x)$	x	$k(x)$
8.4	5.93	5.0	3.12	0.6	3.24
9.0	7.29	5.5	3.74	1.0	9.01
9.6	8.85	6.0	4.49	1.4	17.66
10.2	10.61	6.5	5.39	1.8	29.19
10.8	12.60	7.0	6.47	2.2	43.61
11.4	14.82	7.5	7.76	2.6	60.91

24. Values of three functions are contained in Table 1.18. (The numbers have been rounded to two decimal places.) Two are power functions and one is an exponential. One of the power functions is a quadratic and one a cubic. Which one is exponential? Which one is quadratic? Which one is cubic?

25. Use a graphing calculator or a computer to graph $y = x^4$ and $y = 3^x$. Determine approximate domains and ranges that give each of the graphs in Figure 1.39.

Figure 1.39

1.5 NOTES ON INVERSE FUNCTIONS

From Distance to Time and Back

On August 18, 1989, the runner Arturo Barrios of Mexico set a world record for 10,000-meters. His times, in seconds, at 2000-meter intervals are recorded in Table 1.19, where $f(d)$ is the number of seconds Barrios took to complete the first d meters of the race. For example, Barrios ran the first 4000 meters in 650.1 seconds, so $f(4000) = 650.1$. The function f is useful to athletes planning to compete with Barrios.

Let us now change our point of view and ask for distances rather than times. If we ask how far Barrios ran during the first 650.1 seconds of his race, the answer is clearly 4000 meters. Going backwards in this way from numbers of seconds to numbers of meters gives a function called the *inverse function* of f, denoted by f^{-1}. Thus, $f^{-1}(t)$ is the number of meters that Barrios ran during the first t seconds of his race. To find values of f^{-1}, we can either read Table 1.19 backwards, or use Table 1.20 which contains values of f^{-1}.

The two functions f and f^{-1} convey the same information, but they express it differently. For example, the fact that Barrios ran the first 6000 meters in 975.5 seconds can be written with either f or f^{-1}:

$$f(6000) = 975.5 \quad \text{or} \quad f^{-1}(975.5) = 6000.$$

The independent variable for f is the dependent variable for f^{-1}, and vice versa. The domains and ranges of f and f^{-1} are also interchanged. The domain of f is all distances d such that $0 \leq d \leq 10000$, which is the range of f^{-1}. The range of f is all times t, such that $0 \leq t \leq 1628.23$, which is the domain of f^{-1}.

TABLE 1.19 *Barrios's running time*

d (meters)	$f(d)$ (seconds)
0	0.00
2000	325.90
4000	650.10
6000	975.50
8000	1307.00
10000	1628.23

TABLE 1.20 *Distance run by Barrios*

t (seconds)	$f^{-1}(t)$ (meters)
0.00	0
325.90	2000
650.10	4000
975.50	6000
1307.00	8000
1628.23	10000

Which Functions Have Inverses?

If a function has an inverse, we say it is *invertible*. Let's first look at a function which is not invertible.

Consider the flight of the Mercury spacecraft *Freedom 7*, which carried Alan Shepard, Jr. into space in May 1961. Shepard was the first American to journey into space. After launch, his spacecraft rose to an altitude of 116 miles, and then came down into the sea. The function $f(t)$ giving the altitude in miles t minutes after lift-off does not have an inverse. To see why not, try to decide on a value for $f^{-1}(100)$. Clearly, $f^{-1}(100)$ should be the time when the altitude of the spacecraft was 100 miles.

However, there are two such times, one when the spacecraft was ascending and one when it was descending. (See Figure 1.40.)

The reason the altitude function does not have an inverse is that the altitude first increases and then decreases. The reason the Barrios time function did have an inverse is that it increased everywhere. Each running time, t, corresponds to some unique distance, d.

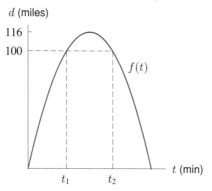

Figure 1.40: Two times, t_1 and t_2, at which altitude of spacecraft is 100 miles

Figure 1.41: An increasing function has an inverse

A function does not have to be increasing everywhere to have an inverse. Figure 1.41 suggests when an inverse will exist. The original function, f, takes us from an x-value to a y-value, as shown in Figure 1.41. Since having an inverse means there is a function going from a y-value to an x-value, the crucial question is whether we can get back. In other words, does each y-value correspond to a unique x-value? If so, there's an inverse; if not, there isn't. This principle may be stated geometrically, as follows:

A function has an inverse if (and only if) its graph intersects any horizontal line at most once.

For example, the function $f(x) = x^2$ does not have an inverse because many horizontal lines intersect the parabola twice.

Definition of an Inverse Function

If the function f is invertible, its inverse is defined as follows:

$$f^{-1}(t) = d \quad \text{means} \quad f(d) = t.$$

The notation f^{-1} for an inverse function is perhaps misleading, as it is easy to confuse with a reciprocal, which the inverse function is not. However, there's no changing such well-established notation!

Example 1   The temperature at which water boils decreases as altitude increases, a fact important to cooks. Let $f(h)$ be the boiling point of water (in $^\circ$C) at an altitude of h meters above sea level during standard atmospheric conditions. What is the meaning in practical terms of $f^{-1}(90)$ and of $f^{-1}(90) = 3000$? Evaluate $f^{-1}(100)$.

Solution   The function f goes from altitude to temperature, so f^{-1} goes from temperature to altitude. The expression $f^{-1}(90)$ represents the altitude in meters at which the boiling point of water is 90°C. The equation $f^{-1}(90) = 3000$ means that the boiling point of water is 90°C at an altitude of 3000 meters. The equation $f(3000) = 90$ has the same meaning. Since the boiling point of water is 100°C at sea level (where the altitude is 0 meters), we must have $f^{-1}(100) = 0$.

Formulas for Inverse Functions

If a function is defined by a formula, it is sometimes possible to find a formula for the inverse function as well. In Section 1.1, we looked at the snow tree cricket, whose chirp rate, C, in chirps per minute, is approximated by the formula

$$C = f(T) = 4T - 160,$$

where T is the temperature in degrees Fahrenheit. So far we have used this formula to predict the chirp rate from the temperature. But it is also possible to use this formula backwards to calculate the temperature from the chirp rate.

Example 2   Find the formula for the function giving temperature in terms of the number of cricket chirps per minute; that is, find the inverse function f^{-1} such that

$$T = f^{-1}(C).$$

Solution   Since C is an increasing function, f is invertible. We know $C = 4T - 160$. We solve for T, giving

$$T = \frac{C}{4} + 40,$$

so

$$f^{-1}(C) = \frac{C}{4} + 40.$$

Graphs of Inverse Functions

The function $f(x) = x^3$ is increasing everywhere and so has an inverse. To find the inverse, we solve

$$y = x^3$$

for x, giving

$$x = y^{1/3}.$$

The inverse function is

$$f^{-1}(y) = y^{1/3}$$

or, if we want to call the variable x,

$$f^{-1}(x) = x^{1/3}.$$

The graphs of $y = x^3$ and $y = x^{1/3}$ are shown in Figure 1.42. Notice that these graphs are the reflections of one another about the line $y = x$. For example, $(8, 2)$ is on the graph of $y = x^{1/3}$ because $2 = 8^{1/3}$, and $(2, 8)$ is on the graph of $y = x^3$ because $8 = 2^3$. The points $(8, 2)$ and $(2, 8)$ are reflections of one another about the line $y = x$. In general, if the x- and y-axes have the same scales:

> The graph of f^{-1} is the reflection of the graph of f about the line $y = x$.

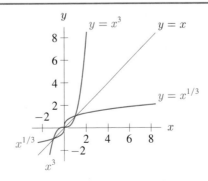

Figure 1.42: Graphs of inverse functions, $y = x^3$ and $y = x^{1/3}$, are reflections about the line $y = x$

Problems for Section 1.5

1. Let p be the price of an item and q be the number of items sold at that price. Assume $q = f(p)$. Explain what the following quantities mean in terms of prices and quantities sold.

    (a) $f(25)$

    (b) $f^{-1}(30)$

2. Suppose $C = f(A)$, where C is the cost in dollars, of building a store of area A square feet. Explain in terms of cost and square feet what the following quantities represent.

    (a) $f(10{,}000)$

    (b) $f^{-1}(20{,}000)$

3. Let $f(x)$ equal the temperature (°F) when the column of mercury in a particular thermometer is x inches long. What is the meaning of $f^{-1}(75)$ in practical terms?

For Problems 4–7, decide whether the function f is invertible.

4. $f(t)$ equals the number of customers in Macy's department store at t minutes past noon on December 18, 1993.

5. $f(n)$ is the number of students in your calculus class whose birthday is on the nth day of the year.

6. $f(x)$ is the volume in liters of x kilograms of water at 4°C.

7. $f(w)$ is the cost in cents of mailing a letter that weighs w grams.

For Problems 8–11, determine whether the functions $y = f(x)$ are invertible.

8.

9.

10.

11.

12. Graph each function below, and decide if it has an inverse. If so, approximate the inverse at $x = 20$.

    (a) $f(x) = x^2 + 3^x$

    (b) $g(x) = x^3 + 3^x$

13. The cost of producing q articles is given by the function $C = f(q) = 100 + 2q$.

    (a) Find a formula for the inverse function.
    (b) Explain in practical terms what the inverse function tells you.

14. A kilogram weighs about 2.2 pounds.

    (a) Write a formula for the function, f, which gives an object's mass in kilograms, k, as a function of its weight in pounds, p.
    (b) Find a formula for the inverse function of f. What does this inverse function tell you, in practical terms?

15. A function $y = t(x)$ is always increasing, has a domain of all real numbers, and has a range of $0 < t(x) < 4$. Sketch a possible graph of $t^{-1}(x)$.

16. Determine if the following statement is true or false, and explain your answer: If f is an increasing function, then f^{-1} is an increasing function.

17. Consider the function $y = f(x)$ whose graph is given in Figure 1.43.

    (a) Estimate $f^{-1}(2)$.
    (b) Sketch a graph of f^{-1} on the same axes.

Figure 1.43

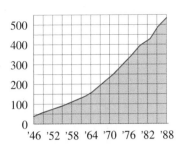

Figure 1.44

18. Figure 1.44 is the graph[8] of the function f, where $f(t)$ is the number (in millions) of motor vehicles registered in the world in the year t. (In 1988, one-third of the registered vehicles in the world were in the United States.)

 (a)   Is f invertible? Explain.
 (b)   What is the meaning of $f^{-1}(400)$ in practical terms? Evaluate $f^{-1}(400)$.
 (c)   Sketch the graph of f^{-1}.

1.6 LOGARITHMS

In Section 1.3, we set up a function approximating the population of Mexico (in millions) as

$$P = f(t) = 67.38(1.026)^t,$$

where t is the number of years since 1980. Writing the function this way shows that we are thinking of the population as a function of time, that the population was 67.38 million in 1980, and that it grows by 2.6% every year.

Now suppose that instead of calculating the population, we want to find when the population is expected to reach 200 million. This means we want to find the value of t for which

$$200 = f(t) = 67.38(1.026)^t.$$

Since the exponential function is always increasing and is eventually more than 200, there's exactly one value of t making $P = 200$. How should we find it? A reasonable way to start is trial and error. Taking $t = 40$ and $t = 50$ we get

$$P = f(40) = 67.38(1.026)^{40} = 188.115\ldots \quad \text{(so } t = 40 \text{ is too small),}$$
$$P = f(50) = 67.38(1.026)^{50} = 243.163\ldots \quad \text{(so } t = 50 \text{ is too large).}$$

Some more experimenting leads to

$$P = f(42) = 67.38(1.026)^{42} \approx 198.0$$
$$P = f(43) = 67.38(1.026)^{43} \approx 203.2$$

so t is between 42 and 43. In other words, the population is projected to reach 200 million sometime during the year 2022.

Although it is always possible to approximate t by trial and error like this, it would clearly be better to have a formula that gives t in terms of P. The *logarithm* function will enable us to do this.

---

[8]From D. Blerics and P. Walzer, "Energy for Motor Vehicles," *Scientific American*, September 1990.

Definition and Properties of Logs to Base 10

We define the *logarithm* function, $\log_{10} x$, to be the inverse of the exponential function, 10^x. We call 10 the *base*. So

> The **logarithm** to base 10 of x, written **log**$_{10} x$, is the power of 10 we need to get x. In other words,
>
> $$\log_{10} x = c \quad \text{means} \quad 10^c = x,$$
>
> We often write $\log x$ in place of $\log_{10} x$.

For example, $\log 1000 = \log 10^3 = 3$ since 3 is the power of 10 needed to get 1000. Similarly, $\log(0.1) = -1$ because $0.1 = 1/10 = 10^{-1}$. However, if we try to find $\log(-7)$ on a calculator, we get an error message because 10 to any power is never negative (or 0). In general,

> $\log x$ is not defined if x is negative or 0.

In working with logarithms, we need the following properties:

> ### Properties of Logarithms
> 1. $\log(AB) = \log A + \log B$
> 2. $\log\left(\frac{A}{B}\right) = \log A - \log B$
> 3. $\log(A^p) = p \log A$
> 4. $\log(10^x) = x$
> 5. $10^{\log x} = x$
>
> In addition, $\log 1 = 0$ because $10^0 = 1$.

The fact that $\log x$ is the power of 10 giving x justifies rule 5.

Solving Equations Using Logarithms

Logs are frequently useful when we have to solve for unknown exponents, as in the next examples.

Example 1   Find t such that $2^t = 7$.

Solution   First, notice that we expect t to be between 2 and 3 (because $2^2 = 4$ and $2^3 = 8$). To calculate t, take logs to base 10:

$$\log(2^t) = \log 7.$$

Then use the third property of logs, which says $\log(2^t) = t \log 2$, and get:

$$t \log 2 = \log 7.$$

Using a calculator to find the logs gives

$$t = \frac{\log 7}{\log 2} \approx 2.81.$$

Example 2   We return to the question of when the population of Mexico reaches 200 million. To get an answer, solve $200 = 67.38(1.026)^t$ for t, using logs.

Solution   Dividing both sides of the equation by 67.38, we get

$$\frac{200}{67.38} = (1.026)^t.$$

Now take logs of both sides:

$$\log\left(\frac{200}{67.38}\right) = \log(1.026^t).$$

Using the fact that $\log(A^t) = t \log A$, we get

$$\log\left(\frac{200}{67.38}\right) = t \log(1.026).$$

Solving this equation using a calculator to find the logs, we get

$$t = \frac{\log(200/63.78)}{\log(1.026)} \approx 42.4 \text{ years}$$

which is between $t = 42$ and $t = 43$, as we found at the beginning of this section. This value of t corresponds to the year 2022.

Example 3   Give a formula for the inverse of the following function (that is, solve for t in terms of P):

$$P = f(t) = 67.38(1.026)^t.$$

Solution   We want a formula expressing t as a function of P. Take logs:

$$\log P = \log(67.38(1.026)^t).$$

Since $\log(AB) = \log A + \log B$, we have

$$\log P = \log 67.38 + \log((1.026)^t).$$

Now use $\log(A^t) = t \log A$:

$$\log P = \log 67.38 + t \log 1.026.$$

Solve for t in two steps, using a calculator at the final stage:

$$t \log 1.026 = \log P - \log 67.38$$

$$t = \frac{\log P}{\log 1.026} - \frac{\log 67.38}{\log 1.026} \approx 89.7 \log P - 164.0.$$

Thus,

$$f^{-1}(P) = 89.7 \log P - 164.0.$$

Note that

$$f^{-1}(200) = 89.7(\log 200) - 164.0 \approx (89.7)(2.301) - 164.0 \approx 42.4,$$

which agrees with the result of Example 2.

Example 4   Find the half-life of the decaying exponential $P = P_0(0.8)^x$ that we used to model the removal of pollutants in jet fuel on page 16. What does your answer mean in practical terms?

Solution   We seek the value of x such that

$$P = \frac{1}{2}P_0,$$

so we must solve the equation

$$P_0(0.8)^x = \frac{1}{2}P_0.$$

Dividing both sides by P_0 leaves

$$(0.8)^x = \frac{1}{2}.$$

Taking logs of both sides gives

$$\log(0.8)^x = \log\left(\frac{1}{2}\right)$$

$$x(\log 0.8) = \log\left(\frac{1}{2}\right)$$

$$x = \frac{\log(1/2)}{\log 0.8}$$

$$x \approx 3.1 \text{ feet.}$$

In practical terms, this tells us that forcing kerosene through 3.1 feet of clay pipe removes half of its impurities.

The Graph of $\log x$

Example 5   Graph $f(x) = \log x$, and compare it to the graph of $g(x) = 10^x$.

Solution   The values of $\log x$ in Table 1.21 were found using a calculator. Because no power of 10 gives 0, $\log 0$ is undefined. See Figure 1.45. The domain of $y = \log x$ is positive x-values; the range is all y values. In contrast, $y = 10^x$ has domain all x-values and range all positive y-values. The graph of $y = \log x$ has a vertical asymptote at $x = 0$, whereas $y = 10^x$ has a horizontal asymptote at $x = 0$.

One of the big differences between $g(x) = 10^x$ and $f(x) = \log x$ is that the exponential function grows extremely quickly whereas the log function grows extremely slowly. However, the log function does go to infinity, albeit slowly, as x increases.

TABLE 1.21   *Values for $\log x$ and 10^x*

x	$\log x$	x	10^x
0	undefined	0	1
1	0	1	10
2	0.3	2	100
3	0.5	3	10^3
4	0.6	4	10^4
$\vdots$	$\vdots$	$\vdots$	$\vdots$
10	1	10	10^{10}

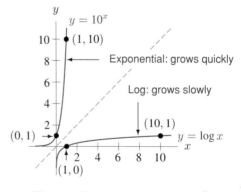

Figure 1.45: Graphs of $\log x$ and 10^x

Since $f(x) = \log x$ and $g(x) = 10^x$ are inverse functions, the graphs of the two functions are reflections of one another about the line $y = x$, provided the scales along the x- and y-axes are equal. (See Figure 1.45.)

Comparison between Logarithms and Power Functions: $A \log x$ and x^p

Although there is a superficial resemblance between the graphs of $y = x^{1/3}$ and $y = \log x$, shown in Figure 1.46, there are important differences. The graph of $x^{1/3}$ includes the origin $(0,0)$, whereas the graph of $\log x$ never reaches the vertical axis $x = 0$, which is an asymptote. Another difference is the rate of growth of the two functions for large x. The fact that $\log(1{,}000{,}000) = 6$ and $\log(10{,}000{,}000) = 7$ shows how slowly the logarithm climbs when x is large. Indeed, the logarithm climbs more slowly than any positive power of x.

Figure 1.47 compares $x^{1/3}$ and $100 \log x$; Table 1.22 compares $x^{0.001}$ and $1000 \log x$. In both cases, the logarithmic function is eventually less than the power function. In fact, x^p dominates $A \log x$ for large x no matter what the values of $p > 0$ and $A > 0$. This follows from the fact that 10^x dominates Bx^q for large x for any $q > 0$ and $B > 0$.

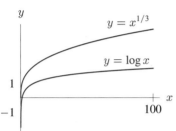

Figure 1.46: Comparison of $y = \log x$ and $y = x^{1/3}$

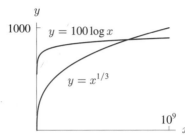

Figure 1.47: Comparison of $y = x^{1/3}$ and $y = 100 \log x$

TABLE 1.22 *Comparison of $x^{0.001}$ and $1000 \log x$*

x	$x^{0.001}$	$1000 \log x$
10^{5000}	10^5	$5 \cdot 10^6$
10^{6000}	10^6	$6 \cdot 10^6$
10^{7000}	10^7	$7 \cdot 10^6$

Problems for Section 1.6

1. Construct a table of values to compare the values of $f(x) = \log x$ and $g(x) = \sqrt{x}$ for $x = 1, 2, \ldots, 10$. Round to two decimal places. Use these values to graph both functions.

For Problems 2–8, solve for x using logs. (You can check your answer using a graphing calculator or computer.)

2. $3^x = 11$ 3. $17^x = 2$ 4. $10 = 4^x$ 5. $20 = 50(1.04)^x$

6. $25 = 2(5)^x$ 7. $2 \cdot 5^x = 11 \cdot 7^x$ 8. $4 \cdot 3^x = 7 \cdot 5^x$

In Problems 9–12, solve for t.

9. $a = b^t$ 10. $P = P_0 a^t$ 11. $Q = Q_0 a^{nt}$ 12. $P_0 a^t = Q_0 b^t$

For Problems 13–17, simplify the expression as much as possible.

13. $\log A^3 - \log B^{2/3} + \log A^{1/3} + \log B^{5/3}$ 14. $\dfrac{\log(ABC)}{\log\left(\frac{1}{ABC}\right)}$

15. $\dfrac{\log A^2 - 2 \log B}{\log A^2 + \log\left(\frac{1}{AB}\right)}$ 16. $10^{\log(AB)}$

17. $100^{(\log A - \log B)}$

Plot a graph of the functions in Problems 18–19 on a calculator or computer. Explain what you see.

18. $y = \log(10^x)$ 19. $y = 10^{\log x}$

For Problems 20–21, determine which function has larger values as $x \to \infty$.

20. $x^{1/3}$ or $75{,}000 \log x$ 21. $\log x$ or $10^{-0.01x}$

22.   Find the equation of the line l in Figure 1.48.

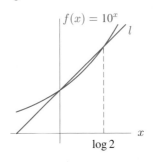

Figure 1.48

23.   What is the doubling time of prices which are increasing by 5% a year?

24.   The population of a region is growing exponentially. If there were 40,000,000 people in 1980 ($t = 0$) and 56,000,000 in 1990, find an expression for the population at any time t, in years. What would you predict for the year 2000? What is the doubling time?

25.   In 1980, there were about 170 million vehicles (cars and trucks) and about 227 million people in the United States. If the number of vehicles has been growing at 4% a year, while the population has been growing at 1% a year, in what year will there be, on average, one vehicle per person?

26.   Owing to an innovative rural public health program, infant mortality in Senegal, West Africa, is being reduced at a rate of 10% per year. How long will it take for infant mortality to be reduced by 50%?

27.   The half-life of a certain radioactive substance is 12 days. If there are 10.32 grams initially,

   (a)   Write an equation to determine the amount, A, of the substance as a function of time.
   (b)   When will the substance be reduced to 1 gram?

28.   (a)   Find the doubling time, D, for annual growth rates, i%, of 2%, 3%, 4%, and 5%.
   (b)   Since D decreases as i increases, we might guess that D is inversely proportional to i, that is, $D = k/i$. Use your answers to part (a) to confirm that, approximately, $D = 70/i$. This is the "Rule of 70" used by bankers. To estimate the doubling time of an investment, the banker divides 70 by the yearly interest rate.

29.   Find the inverse function of $p(t) = (1.04)^t$.

1.7 THE NUMBER e AND NATURAL LOGARITHMS

The calculations in the previous section were made with *common logarithms*, which means logarithms with base 10. However, in science, the most frequently used base is the famous number $e = 2.71828\dots$. In fact, this base is used so often that the logarithm with base e is called the *natural logarithm* and is denoted by "ln." You will find a ln button on most scientific calculators as well as an e^x button; that should be some indication of how important the base e is. At first glance, this is all somewhat mysterious. What can possibly be natural about using logarithms to the base $2.71828\dots$? The full answer to that question must wait until Chapter 4, where we show that many calculus formulas come out neater when e is used as the base rather than any other base.

Definition and Properties of the Natural Logarithm

The natural logarithm of x, written $\ln x$, is defined to be the logarithm to base e of x. This means $\ln x$ is the inverse function of e^x. So

> The **natural logarithm** of x, written $\ln x$, is the power of e needed to get x. In other words,
>
> $$\ln x = c \quad \text{means} \quad e^c = x.$$

The rules for manipulating natural logs are the same as those for logs to base 10, and $\ln x$ is not defined when x is negative or 0.

Properties of Natural Logarithms

1. $\ln (AB) = \ln A + \ln B$

2. $\ln \left(\dfrac{A}{B} \right) = \ln A - \ln B$

3. $\ln (A^p) = p \ln A$

4. $\ln e^x = x$

5. $e^{\ln x} = x$

In addition, $\ln 1 = 0$ because $e^0 = 1$.

Using the $\boxed{\text{LN}}$ button on a calculator to plot a graph of $f(x) = \ln x$ for $0 < x \le 10$, we get Figure 1.49. Observe that the graph of $y = \ln x$, like the graph of $y = \log x$, climbs very slowly as x increases. Also, note the x-axis intercept is $x = 1$, since $\ln 1 = 0$.

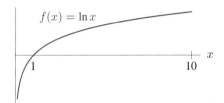

Figure 1.49*:* Graph of the natural logarithm

Using Exponential Functions with Base e

We can show that any exponential function can be written using e.

Any **exponential growth** function can be written in the form

$$P = P_0 e^{kt}$$

and any **exponential decay** function can be written as

$$Q = Q_0 e^{-kt},$$

where P_0 and Q_0 are the initial quantities and k is a positive constant.

We say that P and Q are growing or decaying at a *continuous rate* of k. (For example, $k = 0.02$ corresponds to a continuous rate of 2%.)

The reason that k is called the continuous rate is explored in detail in Chapter 10.

Example 1   Sketch the graphs of $P = e^{0.5t}$ and $Q = e^{-0.2t}$.

Solution

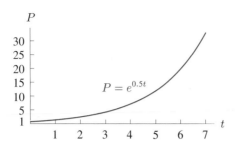

Figure 1.50: An exponential growth function   **Figure 1.51:** An exponential decay function

The graph of $P = e^{0.5t}$ is in Figure 1.50. Notice that the graph is the same shape as the previous exponential growth curves: increasing and concave up. Indeed, the families a^t, $a > 1$, and e^{kt}, $k > 0$, are one and the same. The graph of $Q = e^{-0.2t}$ is in Figure 1.51; it has the same shape as other exponential decay functions. The families a^t, $0 < a < 1$, and e^{kt}, $k < 0$, are one and the same.

Example 2   The release of chlorofluorocarbons used in air conditioners and, to a lesser extent, in household sprays (hair spray, shaving cream, etc.) destroys the ozone in the upper atmosphere. At the present time, the amount of ozone, Q, is decaying exponentially at a continuous rate of 0.25% per year. What is the half-life of ozone? In other words, at this rate, how long will it take for half the ozone to disappear?

Solution   If Q_0 is the initial quantity of ozone, then

$$Q = Q_0 e^{-0.0025t}.$$

We want to find T, the value of t making $Q = Q_0/2$, so

$$Q_0 e^{-0.0025T} = \frac{Q_0}{2}.$$

Dividing by Q_0 and then taking natural logs yields

$$\ln\left(e^{-0.0025T}\right) = -0.0025T = \ln\left(\frac{1}{2}\right) \approx -0.6931,$$

so

$$T \approx 277 \text{ years.}$$

In Example 2 the decay rate was given. However, in many situations where we expect to find exponential growth or decay, the rate is not given. To find it, we must know the quantity at two different times and then solve for the growth or decay rate, as in the next example.

Example 3   The population of Kenya was 19.5 million in 1984 and 21.2 million in 1986. Assuming it increases exponentially, find a formula for the population of Kenya as a function of time.

Solution   If we measure the population, P, in millions and time, t, in years since 1984, we can say
$$P = P_0 e^{kt} = 19.5 e^{kt},$$

where $P_0 = 19.5$ is the initial value of P. We find k by using the fact that $P = 21.2$ when $t = 2$, so
$$21.2 = 19.5 e^{k \cdot 2}.$$

To find k, we divide both sides by 19.5, giving

$$\frac{21.2}{19.5} = 1.087 = e^{2k}.$$

Now take natural logs of both sides:

$$\ln(1.087) = \ln(e^{2k}).$$

Using a calculator and the fact that $\ln(e^{2k}) = 2k$, this becomes

$$0.0834 = 2k.$$

So

$$k \approx 0.042,$$

and therefore

$$P = 19.5e^{0.042t}.$$

Since $k = 0.042 = 4.2\%$, we say the population of Kenya was growing at a continuous rate of 4.2% per year.

In Example 3 we chose to use e for the base of the exponential function representing Kenya's population, making clear that the continuous growth rate was 4.2%. If, however, we had wanted to emphasize the annual growth rate, we could have expressed the exponential function in the form

$$P = P_0 a^t.$$

We now show in Example 4 that the annual growth rate is about 4.3%, which is just slightly more than the continuous growth rate of 4.2% found in Example 3.

Example 4   Express the population of Kenya in the form

$$P = P_0 a^t.$$

Solution   Since the population grew from 19.5 to 21.2 million in 2 years, we know that

$$21.2 = 19.5a^2,$$

so

$$a = \left(\frac{21.2}{19.5}\right)^{1/2} \approx 1.043.$$

This gives

$$P = P_0(1.043)^t.$$

Relationship between a^t and e^{kt}

In general, an exponential function of the form $P_0 e^{kt}$ can always be written in the form $P_0 a^t$. The reason is that

$$P_0 e^{kt} = P_0(e^k)^t,$$

which suggests taking

$$a = e^k, \quad \text{so} \quad k = \ln a.$$

The two different formulas, $P = P_0 e^{kt}$ and $P = P_0 a^t$, have the same graph and represent the same function.

> Exponential growth or decay can always be written in two ways,
>
> $$P = P_0 a^t \quad \text{or} \quad P = P_0 e^{kt}.$$
>
> Here a is the growth factor per unit time, $a - 1$ is the *growth rate* per unit time, and k is the *continuous growth rate*. Then
>
> $$a = e^k, \quad \text{so} \quad k = \ln a.$$
>
> Note $k > 0$ gives exponential growth; $k < 0$ gives decay.

If a comes from a percentage growth rate r, that is, $a = 1 + r$, then the continuous growth rate $k = \ln(1 + r)$ will be slightly less than, but very close to, r, provided r is small.

Time Constants and Half Lives

Engineers measure the speed at which an exponentially decaying quantity decreases using a *time constant*. This is the time taken for the quantity to decrease to $1/e$ times its initial value.

Example 5   Find the time constant of a circuit in which the charge, Q, is given by

$$Q = Q_0 e^{-t/RC}.$$

Here Q_0 is the initial charge, t is time, and R and C are constants which depend on the circuit.

Solution   If T is the time constant, then $Q = Q_0/e$ at time T, so

$$\frac{Q_0}{e} = Q_0 e^{-T/RC}.$$

Canceling Q_0 gives

$$e^{-T/RC} = \frac{1}{e} = e^{-1}.$$

Taking natural logs gives

$$-\frac{T}{RC} = -1,$$

so

$$T = RC.$$

Since $1/2 > 1/e$, there is greater decay during a time constant than a half-life, so the time constant is longer than the half-life. See Figure 1.52.

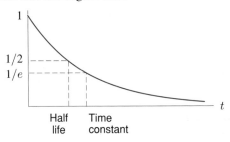

Figure 1.52: Comparing the half-life and the time constant

Problems for Section 1.7

1. Construct a table of values to compare $f(x) = \log x$ and $g(x) = \ln x$ for $x = 1, 2, \ldots, 10$. Round to two decimal places. Use these values to graph both functions.

Solve for x in Problems 2–7.

2. $2^x = e^{x+1}$

3. $2e^{3x} = 4e^{5x}$

4. $7^{x+2} = e^{17x}$

5. $10^{x+3} = 5e^{7-x}$

6. $2x - 1 = e^{\ln x^2}$

7. $9^x = 2e^{x^2}$

For Problems 8–10, solve for t. Assume a and b are positive constants and k is nonzero.

8. $a = be^t$

9. $P = P_0 e^{kt}$

10. $ae^{kt} = e^{bt}$, where $k \neq b$

Simplify the expressions in Problems 11–17 completely.

11. $e^{\ln(1/2)}$

12. $5e^{\ln(A^2)}$

13. $\ln(e^{2AB})$

14. $\ln e^{\sqrt{A}}$

15. $\ln\left(\dfrac{1}{e}\right) + \ln AB$ 16. $2\ln\left(e^A\right) + 3\ln B^e$ 17. $2\ln(AB) - \ln\left(\dfrac{B}{A}\right)$

Convert the functions in Problems 18–23 into the form $P = P_0 a^t$. Which represent exponential growth and which represent exponential decay?

18. $P = 15e^{0.25t}$ 19. $P = 2e^{-0.5t}$ 20. $P = 10e^{0.917t}$

21. $P = 79e^{-2.5t}$ 22. $P = P_0 e^{0.2t}$ 23. $P = 7e^{-\pi t}$

Convert the functions in Problems 24–27 into the form $P = P_0 e^{kt}$.

24. $P = 15(1.5)^t$ 25. $P = 10(1.7)^t$ 26. $P = 174(0.9)^t$ 27. $P = 4(0.55)^t$

28.   (a) Evaluate the quantity $(\ln x)/(\log x)$ for several values of x (say $x = 0.5, 5, 10, 100$). What do you notice?

    (b) Plot a graph of $y = (\ln x)/(\log x)$ on a computer or calculator. What do you see? How does it relate to your results in part (a)?

    (c) Take the natural logs of both sides of the equation $x = 10^{\log x}$ to show that $\dfrac{\ln x}{\log x} = \ln 10$.

29. Find the inverse function of $f(t) = 50e^{0.1t}$.

30. Find the inverse function of $f(t) = 1 + \ln t$.

31. Define $g(x) = \dfrac{2}{e^x + 5}$.

    (a) Is g increasing or decreasing?

    (b) Explain why g is invertible and find a formula for g^{-1}.

32. Define $f(x) = \dfrac{1}{1 + e^{-x}}$.

    (a) Is f increasing or decreasing?

    (b) Explain why f is invertible, and find a formula for $f^{-1}(x)$.

    (c) What is the domain of f^{-1}?

    (d) Sketch the graphs of f and f^{-1} on the same axes, and explain their relationship.

33.   (a) A population grows according to the equation $P = P_0 e^{kt}$ (with P_0, k constants). Find the population as a function of time, t, if it grows at a continuous rate of 2% a year and starts at 1 million.

    (b) Plot a graph of the population you found in part (a) against time.

34. The air in a factory is being filtered so that the quantity of a pollutant, P (measured in mg/liter), is decreasing according to the equation $P = P_0 e^{-kt}$, where t represents time in hours . If 10% of the pollution is removed in the first five hours:

    (a) What percentage of the pollution is left after 10 hours?

    (b) How long will it take before the pollution is reduced by 50%?

    (c) Plot a graph of pollution against time. Show the results of your calculations on the graph.

    (d) Explain why the quantity of pollutant might decrease in this way.

35. Air pressure, P, decreases exponentially with the height above the surface of the earth, h:

$$P = P_0 e^{-0.00012h}$$

where P_0 is the air pressure at sea level and h is in meters.

    (a) If you go to the top of Mount McKinley, height 6198 meters (about 20,320 feet), what is the air pressure, as a percent of the pressure at sea level?

    (b) The maximum cruising altitude of an ordinary commercial jet is around 12,000 meters (about 40,000 feet). At that height, what is the air pressure, as a percent of the sea level value?

36. Under certain circumstances, the velocity, V, of a falling raindrop is given by $V = V_0(1 - e^{-t})$, where t is time and V_0 is a positive constant.

    (a) Sketch a rough graph of V against t, for $t \geq 0$.

    (b) What does V_0 represent?

37. The population of the United States was 226.5 million in 1980 and 248.7 million in 1990. In what year is the population expected to go over 300 million?

38. The half-life of radioactive strontium-90 is 29 years. In 1960, radioactive strontium-90 was released into the atmosphere during testing of nuclear weapons, and was absorbed into people's bones. How many years does it take until only 10% of the original amount absorbed remains?

39. If the size of a bacteria colony doubles in 5 hours, how long will it take for the number of bacteria to triple?

40. Predict the earth's population in the year 2000. An almanac lists the population in 1980 as 4.478 billion, and in 1994 as 5.642 billion. What is the doubling time of the world's population?

41. Suppose that a certain radioactive substance has a half-life of 5 years. An object starts with 20 kg of the radioactive material.

    (a) How much of the radioactive material is left after 10 years?
    (b) The object can be moved safely when the quantity of the radioactive material is 0.1 kg or less. How much time must pass before the object can be moved?

42. One of the main contaminants of a nuclear accident, such as that at Chernobyl, is strontium-90, which decays exponentially at a continuous rate of approximately 2.47% per year. Preliminary estimates after the Chernobyl disaster suggested that it would be about 100 years before the region would again be safe for human habitation. What percent of the original strontium-90 would still remain by this time?

43. The population, P, in millions, of Nicaragua was 3.6 million in 1990 and growing at an annual rate of 3.4%. Let t be time in years since 1990.

    (a) Express P as a function in the form $P = P_0 a^t$.
    (b) Express P as an exponential function using the base e.
    (c) Compare the annual and continuous growth rates.

44. Geological dating of rocks is done using potassium-40 rather than carbon-14 because potassium has a longer half-life. The potassium decays to argon, which remains trapped in the rocks and can be measured; thus the original quantity of potassium can be calculated. The half-life of potassium-40 is $1.28 \cdot 10^9$ years. Find a formula giving the quantity, P, of potassium-40 remaining as a function of time in years, assuming the initial quantity is P_0:

    (a) Using base 1/2.
    (b) Using base e.

45. A picture supposedly painted by Vermeer (1632–1675) contains 99.5% of its carbon-14 (half-life 5730 years). From this information decide whether the picture is a fake. Explain your reasoning.

46. Assume that the amount of electric charge on a capacitor is an exponentially decaying function of time:

$$Q = Q_0 e^{-t/k}.$$

    (a) What is the time constant?
    (b) Show that after a period of 3 time constants, over 95% of the charge has decayed.
    (c) Show that after 5 time constants, less than 1% of the charge remains.

1.8 NEW FUNCTIONS FROM OLD

Shifts and Stretches

The graph of a constant multiple of a given function is easy to visualize: each y-value is stretched or shrunk by that multiple. For example, consider the function $f(x)$ and its multiples $y = 3f(x)$ and $y = -2f(x)$. Their graphs are shown in Figure 1.53. The factor 3 in the function $y = 3f(x)$ stretches each $f(x)$ value by multiplying it by 3; the factor -2 in the function $y = -2f(x)$ stretches $f(x)$ by multiplying by 2 and reflects it about the x-axis. You can think of the multiples of a given function as a family of functions.

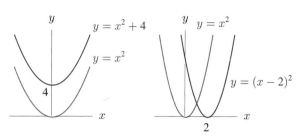

Figure 1.53: Multiples of the function $f(x)$

Figure 1.54: Graphs of $y = x^2$ with $y = x^2 + 4$ and $y = (x-2)^2$

It is also easy to create families of functions by shifting graphs. For example, $y - 4 = x^2$ is the same as $y = x^2 + 4$, which is the graph of $y = x^2$ shifted up by 4. Similarly, $y = (x-2)^2$ is the graph of $y = x^2$ shifted right by 2. (See Figure 1.54.)

- Multiplying a function by a constant, c, stretches the graph vertically (if $c > 1$) or shrinks the graph vertically (if $0 < c < 1$). A negative sign (if $c < 0$) reflects the graph about the x-axis, in addition to shrinking or stretching.
- Replacing y by $(y - k)$ moves a graph up by k (down if k is negative).
- Replacing x by $(x - h)$ moves a graph to the right by h (to the left if h is negative).

We can sketch the graph of the sum of two functions by imagining the y-values of the two graphs stacked on top of each other.

Example 1 Graph the function $y = 2x^2 + 1/x$ for $x > 0$.

Solution We graph the functions $y = 2x^2$ and $y = 1/x$ separately first. Now imagine the corresponding y-values stacked on top of one another, and you will get the graph of $y = 2x^2 + 1/x$. (See Figure 1.55.)

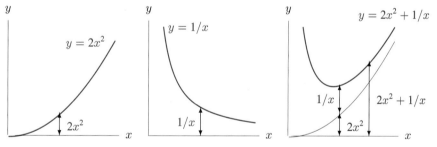

Figure 1.55: Summing two functions from their graphs

Notice that, for x near 0, the graph of $y = 2x^2 + 1/x$ looks much like the graph of $y = 1/x$ (because $1/x$ is much larger than $2x^2$ there). For large x, the graph of $y = 2x^2 + 1/x$ looks like the graph of $y = 2x^2$.

Composite Functions

If oil is spilled from a tanker, the area of the oil slick will grow with time. Suppose that the oil slick is always a perfect circle. (In practice, this does not happen because of winds and tides and the location of the coastline.) The area of the oil slick is a function of its radius. This function is

$$A = f(r) = \pi r^2.$$

The radius is also a function of time, because the radius increases as more oil spills. Thus, the area, being a function of the radius, is also a function of time. If, for example, the radius is given by

$$r = g(t) = 1 + t,$$

then the area is given as a function of time by substitution:

$$A = \pi r^2 = \pi(1 + t)^2.$$

We say that A is a *composite function* or a "function of a function," which is written

$$A = \underbrace{f(g(t))}_{} = \pi(g(t))^2 = \pi(1 + t)^2.$$

Composite function;
f is outside function,
g is inside function

Let's look at this example and think about how we would calculate A using the formula $\pi(1 + t)^2$. For any given t, the first step is to find $1 + t$, and the second step is to square and multiply by π. So the first step corresponds to the inside function $g(t) = 1 + t$, and the second step corresponds to the outside function $f(r) = \pi r^2$.

Example 2   If $f(x) = x^2$ and $g(x) = x + 1$, find each of the following:
(a) $f(g(2))$ (b) $g(f(2))$ (c) $f(g(x))$ (d) $g(f(x))$

Solution   (a)   Since $g(2) = 3$, we have $f(g(2)) = f(3) = 9$.
(b)   Since $f(2) = 4$, we have $g(f(2)) = g(4) = 5$. Notice that $f(g(2)) \neq g(f(2))$.
(c) $f(g(x)) = f(x + 1) = (x + 1)^2$.
(d) $g(f(x)) = g(x^2) = x^2 + 1$. Again, notice that $f(g(x)) \neq g(f(x))$.

Example 3   Express each of the following functions as a composition:

(a) $h(t) = (1 + t^3)^{27}$ (b) $k(x) = \log(x^2)$ (c) $l(x) = (\log x)^2$ (d) $n(y) = e^{-y^2}$

Solution   In each case think about how you would calculate a value of the function. The first stage of the calculation will give you the inside function, and the second stage will give you the outside one.
(a)   For $(1 + t^3)^{27}$, the first stage is cubing and adding 1, so the inside function is $g(t) = 1 + t^3$. The second stage is taking the 27th power, so the outside function is $f(y) = y^{27}$. Then

$$f(g(t)) = f(1 + t^3) = (1 + t^3)^{27}.$$

In fact, there are lots of different answers: $g(t) = t^3$ and $f(y) = (1+y)^{27}$ is another possibility.
(b)   Evaluating this function involves squaring x first and then taking the log. So if $g(x) = x^2$ is the inside function and $f(y) = \log y$ is the outside function, then

$$f(g(x)) = \log(x^2).$$

(c)   This function involves taking the log first and then squaring. Using the same definitions of f and g as in part (b), namely $f(x) = \log x$ and $g(t) = t^2$, the composition is

$$g(f(x)) = (\log x)^2.$$

By evaluating the functions in parts (b) and (c) for $x = 2$, say, giving $\log(2^2) = 0.602$, and $\log^2(2) = 0.091$, you can see that the order in which two functions are composed can make a difference.
(d)   To calculate e^{-y^2} you square y, take its negative, and then take e to that power. So if we set $g(y) = -y^2$ and $f(z) = e^z$, then we can write

$$f(g(y)) = e^{-y^2}.$$

Alternatively, you could take $g(y) = y^2$ and $f(z) = e^{-z}$.

Odd and Even Functions: Symmetry

There is a certain symmetry apparent in the graphs of $f(x) = x^2$ and $g(x) = x^3$ in Figure 1.56. For each point (x, x^2) on the graph of f, the point $(-x, x^2)$ is also on the graph; for each point (x, x^3) on the graph of g, the point $(-x, -x^3)$ is also on the graph. The graph of $f(x) = x^2$ is symmetric about the y-axis, whereas the graph of $g(x) = x^3$ is symmetric about the origin. The graph of any polynomial involving only even powers has symmetry about the y-axis, while polynomials with only odd powers are symmetric about the origin. Consequently, any functions with these symmetry properties are called *even* and *odd*.

Figure 1.56: Symmetry of even and odd functions

For any function f,
  f is an **even** function if $f(-x) = f(x)$ for all x.
  f is an **odd** function if $f(-x) = -f(x)$ for all x.

For example, $g(x) = e^{x^2}$ is even and $h(x) = x^{1/3}$ is odd. However, you should be aware that many functions do not have any symmetry and are neither even nor odd.

Problems for Section 1.8

1. The graph of $y = f(x)$ is shown in Figure 1.57. Sketch graphs of each of the following. Label any intercepts or asymptotes that can be determined.

   (a) $y = f(x) + 3$ (b) $y = 2f(x)$ (c) $y = f(x + 4)$ (d) $y = 4 - f(x)$

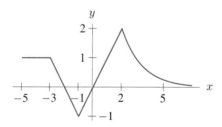

Figure 1.57

2. (a) Write an equation for a graph obtained by vertically stretching the graph of $y = x^2$ by a factor of 2, followed by a vertical upward shift of 1 unit. Sketch the graph.

   (b) What is the equation if the order of the transformations (stretching and shifting) in part (a) is interchanged?

   (c) Are the two graphs the same? Explain the effect of reversing the order of transformations.

For Problems 3–11, let $f(x) = 2x + 1$, $g(x) = \ln(x + 3)$, and $h(x) = e^{4x+7}$. Find the following:

3. $f(g(x))$ 4. $g(f(x))$ 5. $f(h(x))$

6. $h(f(x))$ 7. $g(h(x))$ 8. $h(g(x))$

9. $g(h(x) - 3)$ 10. $f(g(h(x)))$ 11. $h(g(f(x)))$

For Problems 12–15, does the graph suggest the function is even, odd or neither?

12. 13.   14.   15.

Are the functions in Problems 16–17 even, odd, or neither?

16. $f(x) = x^6 + x^3 + 1$ 17. $f(x) = x^3 + x^2 + x$

For Problems 18–21, determine functions f and g such that $h(x) = f(g(x))$. [Note: There is more than one correct answer. Do not choose $f(x) = x$ or $g(x) = x$.]

18. $h(x) = (x + 1)^3$ 19. $h(x) = x^3 + 1$ 20. $h(x) = \ln(x^3)$ 21. $h(x) = (\ln x)^3$

For Problems 22–25, let $m(z) = z^2$. Find and simplify the following quantities:

22. $m(z + 1) - m(z)$ 23. $m(z + h) - m(z)$

24. $m(z) - m(z - h)$ 25. $m(z + h) - m(z - h)$

26.   Find a calculator window in which the graph of $f(x) = 9x^2 + 5$ looks exactly like the graph of $g(x) = x^2$ in the window $-3 < x < 3$, $-1 < y < 9$.

27.   Find a calculator window in which the graph of $f(x) = 10x^2 + 1000x$ looks exactly like the graph of $g(x) = x^2$ in the window $-10 < x < 10$, $-10 < y < 100$. [Hint: Use completing the square.]

28.   Graph $f(x) = 3^x$ in the calculator window $0 < x < 3$, $0 < y < 27$. Then graph $g(x) = 2(3^{x/2})$ in the window $0 < x < 6$, $0 < y < 54$. Explain why the two graphs look exactly the same.

29.   (a)   You are given the following table of values for a function f. For what values of x can we evaluate $g(x) = f(x/2)$ exactly?

x	0	1	2	3	4	5	6
$f(x)$	0	2	6	12	20	30	42

(b)   Construct a table of values for $g(x) = f(x/2)$.
(c)   How will the graph of g differ from the graph of f?

For Problems 30–35, suppose that f and g are given by the graphs in Figure 1.58:

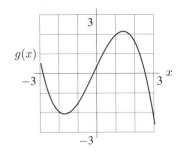

Figure 1.58

30.   Estimate $f(g(1))$.   31.   Estimate $g(f(2))$.
32.   Estimate $f(f(1))$.   33.   Sketch a graph of $f(g(x))$.
34.   Sketch a graph of $g(f(x))$.   35.   Sketch a graph of $f(f(x))$.

36. Let $h(x)$ and $j(x)$ be the functions graphed in Figures 1.59 and 1.60.

  (a) Sketch graphs of $h(j(x))$ and $j(h(x))$. Label all intercepts.
  (b) If possible, sketch a graph of a function $k(x)$ such that $j(k(x)) = k(j(x)) = x$. If this is not possible, explain why.

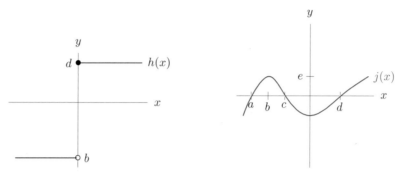

Figure 1.59   Figure 1.60

37. The Heaviside step function, H, is graphed in Figure 1.61. Sketch graphs of the following functions.

  (a) $2H(x)$
  (b) $H(x) + 1$
  (c) $H(x + 1)$
  (d) $-H(x)$
  (e) $H(-x)$

Figure 1.61

For Problems 38–40, use the graph of $y = f(x)$ in Figure 1.62 to graph the indicated functions:

38. $y = 2f(x)$ 39. $y = 2 - f(x)$ 40. $y = 1/f(x)$

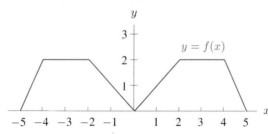

Figure 1.62

41. Complete Table 1.23 to show values for functions f, g, and h, given the following conditions:

  (a) f is symmetric about the y-axis.
  (b) g is symmetric about the origin.
  (c) h is the composition of g with f [that is, $h(x) = g(f(x))$].

TABLE 1.23

x	$f(x)$	$g(x)$	$h(x)$
-3	0	0	
-2	2	2	
-1	2	2	
0	0	0	
1			
2			
3			

1.9 THE TRIGONOMETRIC FUNCTIONS

Trigonometry originated as part of the study of triangles. Indeed, the name *tri-gon-o-metry* means the measurement of three-cornered figures. Thus, the first definitions of the trigonometric functions were in terms of triangles. However, the trigonometric functions can also be defined using the unit circle, a definition that makes them periodic, or repeating. Many naturally occurring processes are also periodic. The water level in a tidal basin, the blood pressure in a heart, an alternating current, and the position of the air molecules transmitting a musical note all fluctuate regularly. Such phenomena can be represented by trigonometric functions.

We will use only the three trigonometric functions found on a calculator: the sine, the cosine, and the tangent.

Radians

There are two commonly used ways to represent the input of the trigonometric functions: radians and degrees. The formulas of calculus, as you will see, are neater in radians than in degrees.

> An angle of 1 **radian** is defined to be the angle at the center of a unit circle which cuts off an arc of length 1, measured counterclockwise. (See Figure 1.63(a).) A unit circle has radius 1.

An angle of 2 radians cuts off an arc of length 2 on a unit circle. A negative angle, such as $-1/2$ radians, cuts off an arc of length $1/2$, but measured clockwise. (See Figure 1.63(b).)

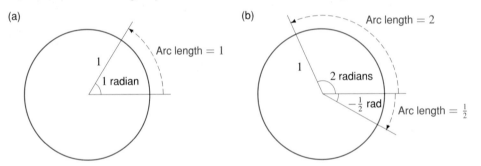

Figure 1.63: Radians defined using unit circle

It is useful to think of angles as rotations, since then we can make sense of angles of over $360°$. Since one full rotation of $360°$ cuts off an arc of length 2π, the circumference of the unit circle, it follows that

$$360° = 2\pi \text{ radians,} \quad \text{so} \quad 180° = \pi \text{ radians.}$$

In other words, 1 radian = $180°/\pi$, so one radian is about $60°$. The word radians is often dropped, so if an angle or rotation is referred to without units, it is understood to be in radians.

Arc Length

Radians are also useful for computing the length of an arc of a circle of radius other than 1. If the circle has radius r and the arc cuts off an angle θ, as in Figure 1.64, then we have the following relation:

> Arc length $= s = r\theta$.

Figure 1.64: Arc length of a sector of a circle

The Sine and Cosine

The two basic trigonometric functions—the sine and cosine—are defined using a unit circle. In Figure 1.65, an angle of t radians is measured counterclockwise around the circle from the point $(1, 0)$. If P has coordinates (x, y), we define

$$\cos t = x \quad \text{and} \quad \sin t = y.$$

Note that we will assume that the angles are *always* in radians unless specified otherwise.

Since the equation of the unit circle is $x^2 + y^2 = 1$, we have the following fundamental identity

$$\cos^2 t + \sin^2 t = 1.$$

As t increases and P moves around the circle, the values of $\sin t$ and $\cos t$ oscillate between 1 and -1, and eventually repeat as P moves through points where it has been before. If t is negative, the angle is measured clockwise around the circle.

Amplitude, Period, and Phase

The graphs of sine and cosine are shown in Figure 1.66. Notice that sine is an odd function, and cosine is even. The maximum and minimum values of sine and cosine are $+1$ and -1, because those are the maximum and minimum values of y and x on the unit circle. After the point P has moved around the complete circle once, the values of $\cos t$ and $\sin t$ start to repeat; we say the functions are *periodic*.

> For any periodic function of time:
> - The **amplitude** is half the distance between maximum and minimum values (if it exists).
> - The **period** is the time needed for the function to execute one complete cycle.

The amplitude of $\cos t$ and $\sin t$ is 1, and the period is 2π. Why 2π? Because that's the value of t when the point P has gone exactly once around the circle. (Remember that $360° = 2\pi$ radians.)

In looking at Figure 1.66 you can see that the sine and cosine graphs are exactly the same shape, only shifted horizontally. Since the cosine is the sine shifted $\pi/2$ to the left,

$$\cos t = \sin(t + \pi/2).$$

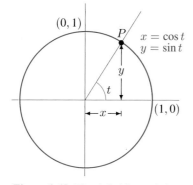

Figure 1.65: The definitions of $\sin t$ and $\cos t$

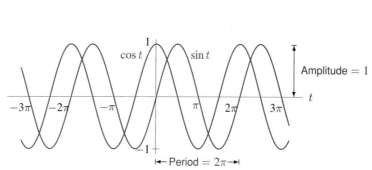

Figure 1.66: Graphs of $\cos t$ and $\sin t$

This says that the cosine of any number is the same as the sine of the number that is $\pi/2$ further to the right—which is exactly what the graphs show. Equivalently, the sine is the cosine shifted $\pi/2$ to the right:

$$\sin t = \cos(t - \pi/2).$$

We say that the *phase difference* between $\sin t$ and $\cos t$ is $\pi/2$.

Functions whose graphs are the shape of a sine or cosine curve are called *sinusoidal* functions.

> To describe arbitrary amplitudes and periods of sinusoidal functions, we use functions of the form
>
> $$f(t) = A\sin(Bt) \qquad \text{and} \qquad g(t) = A\cos(Bt),$$
>
> where $|A|$ is the amplitude and $2\pi/|B|$ is the period.
> To represent arbitrary phase differences, we shift a graph of the correct amplitude and period horizontally by replacing t with $t - h$ or $t + h$.

Example 1 Find and show on a graph the amplitude and period of the functions

(a) $y = 5\sin(2t)$ (b) $y = -5\sin\left(\dfrac{t}{2}\right)$ (c) $y = 1 + 2\sin t$

Solution (a) From Figure 1.67, you can see that the amplitude of $y = 5\sin(2t)$ is 5 because the factor of 5 stretches the oscillations up to 5 and down to -5. The period of $y = \sin(2t)$ is π, because when t changes from 0 to π, the quantity $2t$ changes from 0 to 2π, and so the sine function will have gone through one complete oscillation.

(b) Figure 1.68 shows that the amplitude of $y = -5\sin\left(t/2\right)$ is again 5, because the negative sign reflects the oscillations in the t-axis, but does not change how far up or down they go. The period of $y = -5\sin\left(t/2\right)$ is 4π because when t changes from 0 to 4π, the quantity $t/2$ changes from 0 to 2π, and so the sine function goes through one complete oscillation.

(c) The 1 shifts the graph $y = 2\sin t$ up by 1. Since $y = 2\sin t$ has an amplitude of 2 and a period of 2π, the graph of $y = 1 + 2\sin t$ goes up to 3 and down to -1, and has a period of 2π. (See Figure 1.69.) Thus, $y = 1 + 2\sin t$ also has amplitude 2.

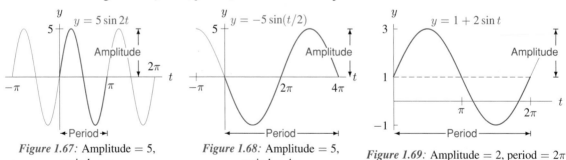

Figure 1.67: Amplitude = 5, period = π

Figure 1.68: Amplitude = 5, period = 4π

Figure 1.69: Amplitude = 2, period = 2π

Example 2 Find formulas for the functions describing the oscillations in Figures 1.70–1.72.

Figure 1.70

Figure 1.71

Figure 1.72

Solution (a) The function in Figure 1.70 looks like a sine function of amplitude 3, so $g(t) = 3\sin(Bt)$. Since the function executes one full oscillation between $t = 0$ and $t = 12\pi$, when t changes by 12π, the quantity Bt changes by 2π. This means $B \cdot 12\pi = 2\pi$, so $B = 1/6$. Therefore, $g(t) = 3\sin(t/6)$ has the graph shown.

(b) The function in Figure 1.71 looks like an upside down cosine function with amplitude 2, so $f(t) = -2\cos(Bt)$. The function completes one oscillation between $t = 0$ and $t = 4$. Thus, when t changes by 4, the quantity Bt changes by 2π, so $B \cdot 4 = 2\pi$, or $B = \pi/2$. Therefore, $f(t) = -2\cos(\pi t/2)$ has the graph shown.

(c) The function in Figure 1.72 looks like the function $g(t)$ in Figure 1.70, but shifted a distance of π to the right. Since $g(t) = 3\sin(t/6)$, we replace t by $(t-\pi)$ to obtain $h(t) = 3\sin[(t-\pi)/6]$.

There are many possible equations for these graphs; you may be able to find others.

Example 3 On February 10, 1990, high tide in Boston was at midnight. The water level at high tide was 9.9 feet; later, at low tide, it was 0.1 feet. Assuming the next high tide is at exactly 12 noon and that the height of the water is given by a sine or cosine curve, find a formula for the water level in Boston as a function of time.

Solution Let y be the water level in feet, and let t be the time measured in hours from midnight. Then the oscillations are to have amplitude 4.9 feet ($= (9.9 - 0.1)/2$) and period 12, so $12B = 2\pi$ and $B = \pi/6$. Since the water is highest at midnight, when $t = 0$, the oscillations are best represented by a cosine, because the cosine is at its maximum at the beginning of the cycle. (See Figure 1.73.) We can say

$$\text{Height above average} = 4.9\cos\left(\frac{\pi}{6}t\right).$$

Since the average level of the water was 5 feet ($= (9.9 + 0.1)/2$), we want a cosine curve shifted up by 5. We get this by adding 5:

$$y = 5 + 4.9\cos\left(\frac{\pi}{6}t\right).$$

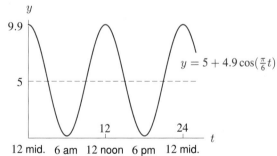

Figure 1.73: Function approximating the tide in Boston on February 10, 1990

Example 4 Of course, there's something wrong with the assumption in Example 3 that the next high tide will be at noon. If so, the high tide would always be at noon or midnight, instead of progressing slowly through the day, as in fact it does. The interval between successive high tides actually averages about 12 hours 24 minutes. Using this, give a more accurate formula for the height of the water as a function of time.

Solution The period is 12 hours 24 minutes $= 12.4$ hours, so $B = 2\pi/12.4$, giving

$$y = 5 + 4.9\cos\left(\frac{2\pi}{12.4}t\right) = 5 + 4.9\cos(0.507t).$$

Example 5 Use the information from Example 3 to write a formula for the water level in Boston on a day when the high tide is at 2 pm.

Solution When the high tide is at midnight

$$y = 5 + 4.9 \cos(0.507t).$$

Since 2 pm is 14 hours after midnight, we replace t by $(t - 14)$. Therefore, on a day when the high tide is at 2 pm,

$$y = 5 + 4.9 \cos(0.507(t - 14)).$$

Example 6 Explain why the following function represents an oscillation which is dying out, called a *damped oscillation*:

$$f(x) = (0.9)^x \sin x.$$

Solution Think of $f(x)$ as a sine function with an amplitude of $(0.9)^x$, which decays as x increases. See Figure 1.74. This is the kind of motion we'd expect from a weight on a spring, where the oscillations are dying out because of friction.

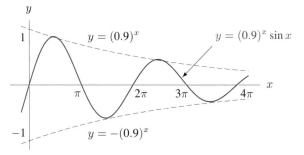

Figure 1.74: A damped oscillation

The Tangent Function

If t is any number with $\cos t \neq 0$, we define the tangent function by

$$\tan t = \frac{\sin t}{\cos t}.$$

Figure 1.65 on page 53 shows the geometrical meaning of the tangent function: $\tan t$ is the slope of the line through the origin $(0,0)$ and the point $P = (\cos t, \sin t)$ on the unit circle.

The tangent function is undefined wherever $\cos t = 0$, namely, at $t = \pm\pi/2, \pm 3\pi/2, \ldots$, and it has a vertical asymptote at each of these points. Now think about where $\tan t$ will be positive and where it will be negative. It will be positive where $\sin t$ and $\cos t$ have the same sign. The graph of the tangent is shown in Figure 1.75.

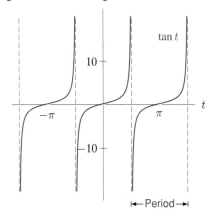

Figure 1.75: The tangent function

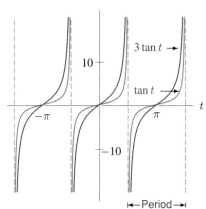

Figure 1.76: Multiple of tangent

Notice that the tangent function has period π, because it repeats every π units. Does it make sense to talk about the amplitude of the tangent function? Not if we're thinking of the amplitude as a measure of the size of the oscillation, because the tangent becomes infinitely large near each vertical asymptote. We can still multiply the tangent by a constant, but that constant no longer represents an amplitude. (See Figure 1.76.)

The Inverse Trigonometric Functions

On occasion, you may need to find a number with a given sine. For example, you might want to find x such that

$$\sin x = 0$$

or such that

$$\sin x = 0.3.$$

The first of these equations can be solved by inspection; the solutions are $x = 0, \pm\pi, \pm 2\pi, \ldots$. Solving the second equation requires a calculator, and it also has infinitely many solutions. They are

$$x \approx 0.305, 2.84, 0.305 \pm 2\pi, 2.84 \pm 2\pi, \ldots$$

For each equation, we pick out the only solution between $-\pi/2$ and $\pi/2$ as the preferred solution. For example, the preferred solution to $\sin x = 0$ is $x = 0$, and the preferred solution to $\sin x = 0.3$ is $x = 0.305$. We define the inverse sine, written "arcsin" or "$\sin^{-1}$," as the function which gives the preferred solution.

> For $-1 \le y \le 1$,
> $$\arcsin y = x$$
> means
> $$\sin x = y \quad \text{with} \quad -\frac{\pi}{2} \le x \le \frac{\pi}{2}.$$

Thus the arcsine is the inverse function to the piece of the sine function having domain $[-\pi/2, \pi/2]$. (See Table 1.24 and Figure 1.77.) On a calculator, the arcsine function[9] is usually denoted by $\boxed{\sin^{-1}}$.

TABLE 1.24   *Values of* $\sin x$ *and* $\sin^{-1} x$

x	$\sin x$	x	$\sin^{-1} x$
$-\frac{\pi}{2}$	-1.000	-1.000	$-\frac{\pi}{2}$
-1.0	-0.841	-0.841	-1.0
-0.5	-0.479	-0.479	-0.5
0.0	0.000	0.000	0.0
0.5	0.479	0.479	0.5
1.0	0.841	0.841	1.0
$\frac{\pi}{2}$	1.000	1.000	$\frac{\pi}{2}$

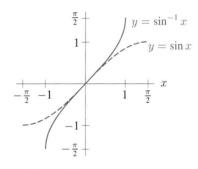

Figure 1.77: The arcsine function

The inverse tangent, written "arctan" or "$\tan^{-1}$," is the inverse function for the piece of the tangent function having the domain $-\pi/2 < x < \pi/2$. On a calculator, the inverse tangent is usually denoted by $\boxed{\tan^{-1}}$. The graph of the arctangent is shown in Figure 1.79.

---

[9]Note that $\sin^{-1} x = \arcsin x$ is not the same as $(\sin x)^{-1} = 1/\sin x$.

Figure 1.78: The tangent function   **Figure 1.79:** The arctangent function

For any y,
$$\arctan y = x$$
means
$$\tan x = y \quad \text{with} \quad -\frac{\pi}{2} < x < \frac{\pi}{2}.$$

The inverse cosine function, written "arccos" or "$\cos^{-1}$," is discussed in Problem 34.

Problems for Section 1.9

For Problems 1–9, draw the angle using a ray through the origin, and determine whether the sine, cosine, and tangent of that angle are positive, negative, zero, or undefined.

1. $\frac{3\pi}{2}$ 2. 2π 3. $\frac{\pi}{4}$ 4. 3π 5. $\frac{\pi}{6}$

6. $\frac{4\pi}{3}$ 7. $\frac{-4\pi}{3}$ 8. 4 9. -1

Given that $\sin(\pi/12) = 0.259$ and $\cos(\pi/5) = 0.809$, compute (without using the trigonometric functions on your calculator) the quantities in Problems 10–13. You may want to draw a picture showing the angles involved, and then check your answers on a calculator.

10. $\cos\left(-\frac{\pi}{5}\right)$ 11. $\sin\frac{11\pi}{12}$ 12. $\sin\frac{\pi}{5}$ 13. $\cos\frac{\pi}{12}$

14. Use the solution to Example 3 on page 55 to estimate the water level in Boston Harbor at 3:00 am, 4:00 am, and 5:00 pm on February 10, 1990.

15. A compact disk spins at a rate of 200 to 500 revolutions per minute. What are the equivalent rates measured in radians per second?

16. When a car's engine makes less than about 200 revolutions per minute, it stalls. What is the period of the rotation of the engine when it is about to stall?

17. What is the difference between $\sin x^2$, $\sin^2 x$, and $\sin(\sin x)$? Express each of the three as a composition. (Note: $\sin^2 x = (\sin x)^2$.)

18. Consider the function $y = 5 + \cos(3x)$.

   (a) What is its amplitude?   (b) What is its period?   (c) Sketch its graph.

19. Match the functions below with the graphs in Figure 1.80.

   (a) $y = 2\cos(t - \pi/2)$ (b) $y = 2\cos t$ (c) $y = 2\cos(t + \pi/2)$

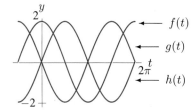

Figure 1.80

For Problems 20–28, find a possible formula for each graph.

20.

21.

22.

23.

24.

25.

26.

27.

28.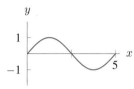

29. The voltage, V, of an electrical outlet in a home is given as a function of time, t (in seconds), by $V = V_0 \cos(120\pi t)$.

   (a) What is the period of the oscillation?
   (b) What does V_0 represent?
   (c) Sketch the graph of V against t. Label the axes.

30. A population of animals oscillates sinusoidally between a low of 700 on January 1 and a high of 900 on July 1.

   (a) Graph the population against time.
   (b) Find a formula for the population as a function of time, t, measured in months since the start of the year.

31. The visitors' guide to St. Petersburg, Florida, contains the chart shown in Figure 1.81 to advertise their good weather. Fit a trigonometric function approximately to the data. The independent variable should be time in months. In order to do this, you will need to estimate the amplitude and period of the data, and when the maximum occurs. (There are many possible answers to this problem, depending on how you read the graph.)

Temp(°F)	Jan	Feb	Mar	Apr	May	June	July	Aug	Sept	Oct	Nov	Dec
100°												
90°						●	●	●	●			
80°			●	●	●					●	●	
70°	●	●										●
60°												
50°												

Figure 1.81: "St. Petersburg...where we're famous for our wonderful weather and year-round sunshine." (Reprinted with permission)

32. The Bay of Fundy in Canada is reputed to have the largest tides in the world. The difference between low and high water levels is 15 meters (nearly 50 feet). Suppose at a particular point in the Bay of Fundy, the depth of the water, y meters, is given as a function of time, t, in hours since midnight on January 1, 1997, by

$$y = D + A \cos \left(B(t - C) \right).$$

   (a) What is the physical meaning of D?
   (b) What is the value of A?
   (c) What is the value of B? Assume the time between successive high tides is 12.4 hours.
   (d) What is the physical meaning of C?

33. (a) Using a calculator set in radians, make a table of values, to two decimal places, of $f(x) = \arcsin x$, for $x = -1, -0.8, -0.6, \ldots, 0, \ldots, 0.8, 1$. (Arcsine is denoted by $\boxed{\sin^{-1}}$ on most calculators).
   (b) Sketch $f(x) = \arcsin x$. Mark on the graph the domain and range of f.

34. This problem introduces the arccosine function, or inverse cosine, denoted by $\boxed{\cos^{-1}}$ on most calculators.

   (a) Using a calculator set in radians, make a table of values, to two decimal places, of $g(x) = \arccos x$, for $x = -1, -0.8, -0.6, \ldots, 0, \ldots, 0.8, 1$.
   (b) Sketch the graph of $g(x) = \arccos x$.
   (c) Based on your graph, what are the domain and range of the arccosine?
   (d) Why is the domain of the arccosine the same as the domain of the arcsine?
   (e) Why is the range of the arccosine *not* the same as the range of the arcsine? To answer this, look at how the domain of the original sine function was restricted to construct the arcsine. Why can't the domain of the cosine be restricted in exactly the same way to construct the arccosine? How should the domain of the cosine be restricted?

35. Using a calculator or computer, estimate all points of intersection of the graphs of $f(x) = x + \sin x$ and $g(x) = x^3$. How do you know you have found all of them?

36. (a) Use a graphing calculator or computer to estimate the period of $2 \sin \theta + 3 \cos(2\theta)$.
   (b) Explain your answer, given that the period of $\sin \theta$ is 2π and the period of $\cos(2\theta)$ is π.

37. A baseball hit at an angle of θ to the horizontal with initial velocity v_0 has horizontal range, R, given by

$$R = \frac{v_0^2}{g} \sin(2\theta).$$

   Here g is a constant, the acceleration due to gravity. Sketch R as a function of θ for $0 \leq \theta \leq \pi/2$. What angle gives the maximum range? What is the maximum range?

38. Consider a mass oscillating on the end of a spring. The distance, y, of the mass from its equilibrium position is given by

$$y = y_0 \cos(2\pi \omega t).$$

   Here y is in centimeters, t is time in seconds, and y_0 and ω are positive constants.

   (a) What is the meaning of y_0 in terms of the oscillation?
   (b) How many oscillations are completed in 1 second?

1.10 POLYNOMIALS AND RATIONAL FUNCTIONS

Polynomials

Some of the best-known functions for which there are formulas are the polynomials:

$$y = p(x) = a_n x^n + a_{n-1} x^{n-1} + \cdots + a_1 x + a_0.$$

Here n is a positive integer called the *degree* of the polynomial, and $a_n, a_{n-1}, \ldots, a_1, a_0$ are constants, with leading coefficient $a_n \neq 0$. An example of a polynomial of degree $n = 3$ is

$$y = p(x) = 2x^3 - x^2 - 5x - 7.$$

In this case $a_3 = 2, a_2 = -1, a_1 = -5$, and $a_0 = -7$. The shape of the graph of a polynomial

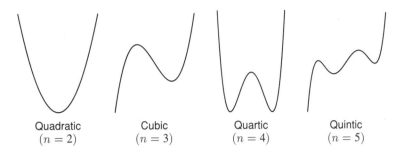

Quadratic	Cubic	Quartic	Quintic
$(n = 2)$	$(n = 3)$	$(n = 4)$	$(n = 5)$

Figure 1.82: Graphs of typical polynomials of degree n

depends on its degree; typical graphs are shown in Figure 1.82. These graphs correspond to a positive coefficient for x^n; a negative coefficient turns the graph upside down. Notice that the graph of the quadratic "turns around" once, the cubic "turns around" twice, and the quartic (fourth degree) "turns around" three times. An nth degree polynomial "turns around" at most $n - 1$ times (where n is a positive integer), but there may be fewer turns.

Example 1   Find possible formulas for the polynomials whose graphs are in Figure 1.83.

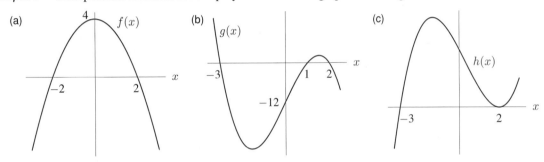

Figure 1.83: Graphs of polynomials

Solution   (a)   This graph appears to be a parabola, turned upside down, and moved up by 4, so
$$f(x) = -x^2 + 4.$$

The negative sign turns the parabola upside down and the $+4$ moves it up by 4. Notice that this formula does give the correct x-intercepts since $0 = -x^2 + 4$ has solutions $x = \pm 2$. These values of x are called *zeros* of f.

We can also solve this problem by looking at the x-intercepts first, which tell us that $f(x)$ has factors of $(x + 2)$ and $(x - 2)$. So
$$f(x) = k(x + 2)(x - 2).$$

To find k, use the fact that the graph has a y-intercept of 4, so $f(0) = 4$, giving
$$4 = k(0 + 2)(0 - 2),$$

or $k = -1$. Therefore, $f(x) = -(x + 2)(x - 2)$, which multiplies out to $-x^2 + 4$. Note that
$$f(x) = 4 - \frac{x^4}{4}$$

also fits the requirements, but its "shoulders" are sharper. There are many possible answers to these questions.

(b)   This looks like a cubic with factors $(x + 3)$, $(x - 1)$, and $(x - 2)$, one for each intercept:
$$g(x) = k(x + 3)(x - 1)(x - 2).$$

Since the y-intercept is -12,
$$-12 = k(0 + 3)(0 - 1)(0 - 2).$$

So $k = -2$, and
$$g(x) = -2(x + 3)(x - 1)(x - 2).$$

(c) This also looks like a cubic with zeros at $x = 2$ and $x = -3$. Notice that at $x = 2$ the graph of $h(x)$ touches the x-axis but doesn't cross it, whereas at $x = -3$ the graph crosses the x-axis. We say that $x = 2$ is a *double zero*, but that $x = -3$ is a single zero.

To find a formula for $h(x)$, first imagine the graph of $h(x)$ to be slightly lower down, so that the graph has one x-intercept near $x = -3$ and two near $x = 2$, say at $x = 1.9$ and $x = 2.1$. Then a formula would be

$$h(x) \approx k(x + 3)(x - 1.9)(x - 2.1).$$

Now move the graph back to its original position. The zeros at $x = 1.9$ and $x = 2.1$ move toward $x = 2$, giving

$$h(x) = k(x + 3)(x - 2)(x - 2) = k(x + 3)(x - 2)^2.$$

The double zero leads to a repeated factor, $(x - 2)^2$. Notice that when $x > 2$, the factor $(x - 2)^2$ is positive, and when $x < 2$, the factor $(x - 2)^2$ is still positive. This reflects the fact that $h(x)$ doesn't change sign near $x = 2$. Compare this with the behavior near the single zero at $x = -3$, where h does change sign.

We cannot find k, as no coordinates are given for points off of the x-axis. Inserting any positive value of k will vertically stretch the graph but not change the zeros, and so any positive k will work.

Example 2   Using a calculator or computer, sketch graphs of $y = x^4$ and $y = x^4 - 15x^2 - 15x$ for $-4 \leq x \leq 4$ and for $-20 \leq x \leq 20$. Set the y range to $-100 \leq y \leq 100$ for the first domain, and to $-100 \leq y \leq 200{,}000$ for the second. What do you observe?

Solution   From the graphs in Figure 1.84 we see that close up (for $-4 \leq x \leq 4$) the graphs look different; from far away, however, they are almost indistinguishable. The reason is that the leading terms (those with the highest power of x) are the same, namely x^4, and for large values of x, the leading term dominates the other terms.

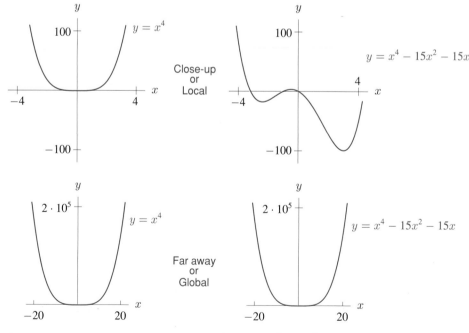

Figure 1.84: Local and global views of $y = x^4$ and $y = x^4 - 15x^2 - 15x$

When $x = \pm 20$, the differences in the values of the two functions, although large, are tiny compared with the vertical scale (-100 to $200{,}000$). See Table 1.25. So these differences can't be seen on the graph.

TABLE 1.25 *Numerical values of $y = x^4$ and $y = x^4 - 15x^2 - 15x$*

x	$y = x^4$	$y = x^4 - 15x^2 - 15x$	Difference
-20	160,000	154,300	5700
-15	50,625	47,475	3150
15	50,625	47,025	3600
20	160,000	153,700	6300

Rational Functions

Rational functions are those of the form

$$f(x) = \frac{p(x)}{q(x)},$$

where p and q are polynomials. Their graphs often have vertical asymptotes where the denominator is zero. If the denominator is nowhere zero, there are no vertical asymptotes. Rational functions may also have horizontal asymptotes; these occur if $f(x)$ approaches a finite number as $x \to \infty$ or $x \to -\infty$. We call the behavior of a function as $x \to \pm\infty$ its *end behavior*.

Example 3   Graph and describe the behavior of $y = \dfrac{1}{x^2 + 4}$.

Solution   This graph has no vertical asymptotes as the denominator is never zero. The graph is symmetric about the y-axis, and the x-axis is a horizontal asymptote since

$$y \to 0 \quad \text{as} \quad x \to \pm\infty.$$

See Figure 1.85.

Figure 1.85: Graph of $y = \frac{1}{x^2+4}$

Example 4   Plot and describe the graph of $y = \dfrac{x^2 - 4}{x^2 - 1}$, including end behavior.

Solution   Factoring gives

$$y = \frac{x^2 - 4}{x^2 - 1} = \frac{(x + 2)(x - 2)}{(x + 1)(x - 1)}$$

so $x = \pm 1$ are vertical asymptotes. If $y = 0$, then $(x + 2)(x - 2) = 0$ or $x = \pm 2$; these are the x-intercepts. Note that the vertical asymptotes arise from zeros of the denominator, while zeros of the numerator give rise to x-intercepts. Substituting $x = 0$ gives $y = 4$; this is the y-intercept. Notice that, in this example, positive and negative x's give the same y value, so the graph is symmetric about the y-axis. This is because $(-x)^2 = x^2$.

TABLE 1.26 *Values of* $y = \frac{x^2-4}{x^2-1}$

x	$y = \frac{x^2-4}{x^2-1}$
± 10	0.969697
± 100	0.999700
± 1000	0.999997

Figure 1.86: Graph of the function $y = \frac{x^2-4}{x^2-1}$

To see what happens as $x \to \pm\infty$, look at the y-values in Table 1.26. Clearly y is getting closer to 1 as x gets large positively or negatively. This can also be seen by realizing that as $x \to \pm\infty$, only the highest powers of x really matter. For large x, the 4 and the 1 are insignificant compared to x^2, so

$$y = \frac{x^2-4}{x^2-1} \approx \frac{x^2}{x^2} = 1 \quad \text{for large } x.$$

So $y \to 1$ as $x \to \pm\infty$, and therefore the horizontal asymptote is $y = 1$. Since, for $x > 1$, the denominator is positive and the numerator is less than the denominator, the graph lies *below* its asymptote. (Why doesn't the graph lie below $y = 1$ when $-1 < x < 1$?) See Figure 1.86.

Problems for Section 1.10

1. Determine the end behavior of each function below, as $x \to +\infty$ and as $x \to -\infty$.
    (a) $f(x) = x^7$
    (b) $f(x) = 5 + 24x + 78x^3 - 12x^4$
    (c) $f(x) = x^{-4}$
    (d) $f(x) = \left(6x^3 - 5x^2 + 12\right) / \left(x^3 - 8\right)$

2. Assume that each of the graphs in Figure 1.87 is of a polynomial. For each graph:
    (a) What is the minimum possible degree of the polynomial?
    (b) Is the *leading coefficient* of the polynomial positive or negative? (You may assume that the graphs are in windows large enough to show the global behavior.)

(I)   (II)   (III)   (IV)   (V)

Figure 1.87

For Problems 3–6, sketch graphs of the polynomials by hand. Check your work using a calculator or computer.

3. $f(x) = (x-3)(x-4)(x-5)$
4. $f(x) = (x+3)(x+4)(x+1)(x+2)$
5. $f(x) = 5(x^2-4)(x^2-25)$
6. $f(x) = -5(x^2-4)(25-x^2)$

7. Graph each of the following polynomials on a graphing calculator or computer and use the graphs to determine which ones are odd functions, even functions, or neither.

    (a) $a(x) = x^2$
    (b) $b(x) = x^3$
    (c) $c(x) = x^4$
    (d) $d(x) = -10x^5$
    (e) $e(x) = x^3 + 3x^2$
    (f) $f(x) = x^4 - x^2$
    (g) $g(x) = x^5 - 2x^3$
    (h) $h(x) = 2x^4 + 5$
    (i) $i(x) = 7x + 5$

8. Based on your answers to Problem 7, how can you tell whether a polynomial is an odd or even function by just looking at its formula?

For each of the rational functions in Problems 9–11, find asymptotes, the behavior of the function as $x \to \pm\infty$ and the behavior of the function near any vertical asymptotes. Then use this information to sketch a graph by hand. Check your work using a computer or graphing calculator.

9. $y = \dfrac{1 - x^2}{x - 2}$

10. $y = \dfrac{1 - 4x}{2x + 2}$

11. $y = \dfrac{x^2 + 2x + 1}{x^2 - 4}$

12. Which of the functions I–III meet each of the following descriptions? There may be more than one function for each description, or none at all.

    (a) Horizontal asymptote of $y = 1$.
    (b) The x-axis is a horizontal asymptote.
    (c) Symmetric about the y-axis.
    (d) An odd function.
    (e) Vertical asymptotes at $x = \pm 1$.

    I. $y = \dfrac{x - 1}{x^2 + 1}$
    II. $y = \dfrac{x^2 - 1}{x^2 + 1}$
    III. $y = \dfrac{x^2 + 1}{x^2 - 1}$

13. The height of an object above the ground at time t is given by

$$s = v_0 t - \frac{g}{2} t^2,$$

where v_0 represents the initial velocity and g is a constant called the acceleration due to gravity.

    (a) At what height is the object initially?
    (b) How long is the object in the air before it hits the ground?
    (c) When will the object reach its maximum height?
    (d) What is that maximum height?

14. A pomegranate is thrown from ground level straight up into the air at time $t = 0$ with velocity 64 feet per second. Its height at time t is $f(t) = -16t^2 + 64t$. Find the time it hits the ground and the instant that it reaches its highest point. What is the maximum height?

15. If $f(x) = ax^2 + bx + c$, what do you know about the values of a, b, and c if:

    (a) $(1, 1)$ is on the graph of $f(x)$?
    (b) $(1, 1)$ is the vertex of the graph of $f(x)$? (You may want to use the fact that the equation for the axis of symmetry of the parabola $y = ax^2 + bx + c$ is $x = -b/2a$.)
    (c) The y intercept of the graph is $(0, 6)$?
    (d) Find a quadratic function that satisfies all three conditions.

Determine cubic polynomials that represent each of the graphs in Problems 16–17.

16.

17.

For Problems 18–21:
(a) Find a possible formula for the graph.
(b) For each graph, read off approximate intervals over which the function is increasing and over which it is decreasing.

18.

19.

20.

21.

22. The graph of a rational function $y = f(x)$ is given in Figure 1.88. If $f(x) = g(x)/h(x)$ with $g(x)$ and $h(x)$ both quadratic functions, give possible formulas for $g(x)$ and $h(x)$. (There are many possible answers.)

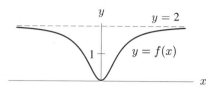

Figure 1.88

23. Match the following functions with the graphs in Figure 1.89. Assume $0 < b < a$.

(a) $y = \dfrac{a}{x} - x$

(b) $y = \dfrac{(x-a)(x+a)}{x}$

(c) $y = \dfrac{(x-a)(x^2+a)}{x^2}$

(d) $y = \dfrac{(x-a)(x+a)}{(x-b)(x+b)}$

(I)   (II)   (III)   (IV)

Figure 1.89

24. The rate, R, at which a population in a confined space increases is proportional to the product of the current population, P, and the difference between the *carrying capacity*, L, and the current population. (The carrying capacity is the maximum population the environment can sustain.)

    (a) Write R as a function of P.
    (b) Sketch R as a function of P.

25. Newton's Second Law of Motion, $F = ma$, tells us that the net force, F, on a train of mass m is proportional to its acceleration, a. Suppose that the only forces are those of the engine, which exerts a constant force, F_E, in the direction of motion, and the wind resistance, which exerts a force proportional to the square of the train's velocity, v, but in the opposite direction.

    (a) Write a formula giving a as a function of v.
    (b) Sketch a graph of a against v.

26. Consider a box-shaped package with square ends. The US postal service will accept such a package if the sum of its length plus girth is less than 108 inches. (The girth is the perimeter of the square cross-section perpendicular to the length.) Find a formula for the volume, V, of a package which just meets the post office criteria in terms of s, the length of the side of the square end. Sketch a graph of V against s.

27. A cylindrical can of fixed volume V has closed ends and radius r for $r > 0$.

    (a) Find the surface area, S, as a function of r.
    (b) What happens to the value of S as $r \to \infty$?
    (c) Sketch a graph of S against r, assuming $V = 10$ cm^3.

28. After running 3 miles at a speed of x mph, a man walked the next 6 miles at a speed that was 2 mph slower. Express the total time spent on the trip as a function of x. What horizontal and vertical asymptotes does the graph of this function have?

29. Consider the point P at the intersection of the circle $x^2 + y^2 = 2a^2$, and the parabola $y = x^2/a$ in Figure 1.90. If a is increased, the point P will trace out a curve. Find the equation of this curve.

Figure 1.90

30. Suppose c is the velocity of light. An object of mass m moving with a velocity v, which is small compared to c, has energy given approximately by

$$E \approx \frac{1}{2}mv^2.$$

If v is comparable in size to c, then the energy must be computed by the exact formula

$$E = mc^2 \left(\frac{1}{\sqrt{1 - v^2/c^2}} - 1 \right).$$

(a) Plot a graph of both formulas for E against v for $0 \le v \le 5 \cdot 10^8$ and $0 \le E \le 5 \cdot 10^{17}$. Take $m = 1$ kg and $c = 3 \cdot 10^8$ m/sec. Explain how you can predict from the exact formula the position of the vertical asymptote.

(b) What do the graphs tell you about the approximation? For what values of v does the first formula give a good approximation to E?

1.11 INTRODUCTION TO CONTINUITY

This section introduces the concept of *continuity* from graphical and numerical points of view. The section on continuity and limits on page 127 investigates the same idea in more depth.

Continuity of a Function on an Interval

Roughly speaking, a function is said to be *continuous* on an interval if its graph has no breaks, jumps, or holes in that interval. Continuity is important because, as we shall see, functions with this property have many other desirable properties.

For example, to locate the zeros of a function, we often look for intervals where the function changes sign. In the case of the function $f(x) = 3x^3 - x^2 + 2x - 1$, for instance, we expect[10] to find a zero between 0 and 1 because $f(0) = -1$ and $f(1) = 3$. (See Figure 1.91.) To be sure that $f(x)$ has a zero there, we need to know that the graph of the function has no breaks or jumps in it. Otherwise the graph could jump across the x-axis, changing sign but not creating a zero. For example, $f(x) = 1/x$ has opposite signs at $x = -1$ and $x = 1$, but no zeros because of the break at $x = 0$. (See Figure 1.92.) In order to be certain that a function has a zero in an interval on which it changes sign, we need to know that the function is defined and continuous.

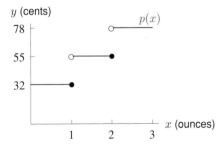

Figure 1.91: The graph of
$f(x) = 3x^3 - x^2 + 2x - 1$

Figure 1.92: No zero although
$f(-1)$ and $f(1)$ have opposite signs

Figure 1.93: Cost of mailing a letter

[10]This is due to the Intermediate Value Theorem, which is discussed on page 84.

What Does Continuity Mean Graphically?

A continuous function has a graph which can be drawn without lifting the pencil from the paper.

Example: The function $f(x) = 3x^2 - x^2 + 2x + 1$ is continuous on any interval. (See Figure 1.91.)

Example: The function $f(x) = 1/x$ is not defined at $x = 0$. It is continuous on any interval not containing the origin. (See Figure 1.92.)

Example: Suppose $p(x)$ is the price of mailing a first-class letter weighing x ounces. It costs 32¢ for one ounce or less, 55¢ between the first and second ounces, and so on. So the graph (in Figure 1.93) is a series of steps. This function is not continuous on any interval containing a positive integer because the graph jumps at these points.

What Does Continuity Mean Numerically?

A function is continuous if nearby values of the independent variable give nearby values of the function. In practical work, continuity is important because it means that small errors in the independent variable lead to small errors in the value of the function.

Example: Suppose that $f(x) = x^2$ and that we want to compute $f(\pi)$. Knowing f is continuous tells us that taking $x = 3.14$ should give a good approximation to $f(\pi)$, and that we can get a better approximation to $f(\pi)$ by using more decimals of π.

Example: If $p(x)$ is the cost of mailing a letter weighing x ounces, then $p(0.99) = p(1) = 32$¢, whereas $p(1.01) = 55$¢, because as soon as we get over 1 ounce, the price jumps up to 55¢. So a small difference in the weight of a letter can lead to a significant difference in its mailing cost. Hence p is not continuous.

Which Functions are Continuous?

Requiring a function to be continuous on an interval is not asking very much, as any function whose graph is an unbroken curve over the interval is continuous. For example, exponential functions, polynomials, and sine and cosine are continuous on every interval. Rational functions are continuous on any interval in which their denominators are not zero. Functions created by adding, multiplying, or composing continuous functions are also continuous.

Example 1 What do the values in Table 1.27 tell you about the zeros of $f(x) = \cos x - 2x^2$?

TABLE 1.27

x	$f(x)$
0	1.00
0.2	0.90
0.4	0.60
0.6	0.11
0.8	−0.58
1.0	−1.46

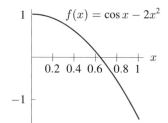

Figure 1.94: Zeros occur where the graph of a continuous function crosses the horizontal axis

Solution Since $f(x)$ is the difference of the two continuous functions, it is continuous. We conclude that $f(x)$ has at least one zero in the interval $0.6 < x < 0.8$, since $f(x)$ changes from positive to negative on that interval. The graph of $f(x)$ in Figure 1.94 suggests that there is only one zero in the interval $0 \leq x \leq 1$, but we cannot be sure of this from the graph or the table of values.

Problems for Section 1.11

Are the functions in Problems 1–8 continuous on the given intervals?

1. $\dfrac{1}{x-2}$ on $[0,3]$

2. $\dfrac{1}{x-2}$ on $[-1,1]$

3. $\dfrac{x}{x^2+2}$ on $[-2,2]$

4. $\dfrac{1}{\sqrt{2x-5}}$ on $[3,4]$

5. $\dfrac{1}{\sin x}$ on $\left[-\frac{\pi}{2},\frac{\pi}{2}\right]$

6. $\dfrac{1}{\cos x}$ on $[0,\pi]$

7. $\dfrac{e^{\sin\theta}}{\cos\theta}$ on $\left[-\frac{\pi}{4},\frac{\pi}{4}\right]$

8. $\dfrac{e^x}{e^x-1}$ on $[-1,1]$

9. Discuss the continuity of the function g graphed in Figure 1.95 and defined as follows:

$$g(\theta)=\begin{cases}\dfrac{\sin\theta}{\theta} & \text{for }\theta\neq 0 \\[2mm] 1/2 & \text{for }\theta=0.\end{cases}$$

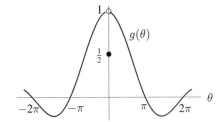

Figure 1.95

10. Sketch the graphs of three different functions that are continuous on $0\le x\le 1$ and that have the values given in the table. The first function is to have exactly one zero in $[0,1]$, the second is to have at least two zeros in the interval $[0.6,0.8]$, and the third is to have at least two zeros in the interval $[0,0.6]$.

x	0	0.2	0.4	0.6	0.8	1.0
$f(x)$	1.00	0.90	0.60	0.11	−0.58	−1.46

11. Use a computer or calculator to sketch the functions $y(x)=\sin x$ and $z_k(x)=ke^{-x}$ for $k=1,2,4,6,8,10$. In each case find the smallest positive solution of the equation $y(x)=z_k(x)$. Now define a new function f by

$$f(k)=\{\text{Smallest positive solution of }y(x)=z_k(x)\}.$$

Explain why the function $f(k)$ is not continuous on the interval $0\le k\le 10$.

CHAPTER SUMMARY

- **Function terminology**

  Domain/range, increasing/decreasing, concavity, zeros (roots), even/odd, end behavior, asymptotes.

- **Linear Functions**

  Slope, y-intercept. Grow by equal amounts in equal times.

- **Exponential Functions**

  Exponential growth and decay, growth rate, continuous growth rate, doubling time, half life. Grow by equal percentages in equal times.

- **Power Functions**

  Fractional powers, negative powers.

- **Logarithmic Functions**

  Log base 10, natural logarithm.

- **Trigonometric Functions**

  Sine, cosine, tangent, amplitude, period, arcsine, arctangent.

- **Polynomials and Rational Functions**

- **New Functions from Old**

  Inverse functions, composition of functions, shifting, stretching, shrinking, flipping.

- **Working with Functions**

  Find a formula for a linear, exponential, power, logarithmic, or trigonometric function, given graph, table of values, or verbal description. Fit functions to data. Find roots, vertical and horizontal asymptotes.

- **Comparisons Between Functions**

  Exponential functions dominate power and linear functions, higher powers dominate lower powers, power and linear functions dominate logarithmic functions.

- **Continuity**

REVIEW PROBLEMS FOR CHAPTER ONE

1. The entire graph of $y = f(x)$ is shown in Figure 1.96.

   (a) What is the domain of $f(x)$?
   (b) What is the range of $f(x)$?
   (c) List all roots of $f(x)$.
   (d) List all intervals on which $f(x)$ is decreasing.
   (e) Is $f(x)$ concave up or concave down at $x = 6$?
   (f) What is $f(4)$?
   (g) Is this function invertible? Explain.

Figure 1.96

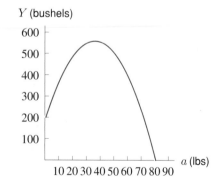

Figure 1.97

2. A graph of the yield, Y, of an apple orchard (in bushels) against the amount, a, of fertilizer (in pounds) used on the orchard is shown in Figure 1.97.

   (a) Describe the effect of the amount of fertilizer on the yield of the orchard.
   (b) What is the vertical intercept? Explain what it means in terms of apples and fertilizer.
   (c) What is the horizontal intercept? Explain what it means in terms of apples and fertilizer.
   (d) What is the range of this function for $0 \le a \le 80$?
   (e) Is the function increasing or decreasing at $a = 60$?
   (f) Is the function concave up or concave down at $a = 40$?

3. When a new product is advertised, more and more people try it. However, the rate at which new people try it slows as time goes on.

   (a) Sketch a graph of the total number of people who have tried such a product against time.
   (b) What do you know about the concavity of the graph?

4. A car starts out slowly and then goes faster and faster until a tire blows out. Sketch a possible graph of the distance the car has traveled as a function of time.

5. Sketch reasonable graphs for the following. Pay particular attention to the concavity of the graphs, and explain your reasoning.

   (a) The total revenue generated by a car rental business, plotted against the amount spent on advertising.
   (b) The temperature of a cup of hot coffee standing in a room, plotted as a function of time.

6. In each pair below, which function will eventually be larger as x goes to infinity?

   (a) $2x^5$ or $200x^4$
   (b) $10x^3$ or e^x
   (c) x^{-2} or x^{-5}
   (d) $x^{1/2}$ or $\ln x$

Find a possible equation involving an exponential for the graph given in Problems 7–8.

7.

8.

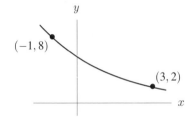

9. By shifting the graph of $y = x^3$, find a cubic polynomial with the graph in Figure 1.98.

10. A population is known to be growing exponentially. Estimate the doubling time of the population shown by the graph in Figure 1.99, and verify graphically that the doubling time is independent of where you start on the graph. Show algebraically that if $P = P_0 a^t$ doubles between time t and time $t + d$, then d is the same number no matter what t is.

Figure 1.98

population

80,000
60,000
40,000
20,000

1 2 3 4 5 6 7 8 9   time (years)

Figure 1.99

TABLE 1.28

x	$f(x)$	$g(x)$	$h(x)$
-2	12	16	37
-1	17	24	34
0	20	36	31
1	21	54	28
2	18	81	25

11. Table 1.28 contains values for three different functions.

   (a) Which (if any) of these functions are linear functions? For those functions which are linear, find the formula.
   (b) Which (if any) of these functions are exponential functions? For those functions which are exponential, find the formula.

12. Draw the graph of a function $f(x)$ satisfying the following conditions:
    - As $x \to \infty$, $f(x) \to 0$.
    - As $x \to -\infty$, $f(x) \to +\infty$.
    - The roots of $f(x)$ are $-3, 1$, and 2.

13. An airplane uses a fixed amount of fuel for takeoff, a (different) fixed amount for landing, and a third fixed amount per mile when it is in the air. How does the total quantity of fuel required depend on the length of the trip? Write a formula for the function involved. Explain the meaning of the constants in your formula.

14. For tax purposes, you may have to report the value of your assets, such as cars or refrigerators. The value you report depreciates, or drops, with time. The idea is that a car you originally paid $10,000 for may be worth only $5000 a few years later. The simplest way to calculate the value of your asset is using "straight-line depreciation," which assumes that the value is a linear function of time. If a $950 refrigerator depreciates completely in seven years, find a formula for its value as a function of time.

15. For $g(x) = x^2 + 2x + 3$, find and simplify:
    (a) $g(2+h)$ (b) $g(2)$ (c) $g(2+h) - g(2)$

16. If $f(x) = x^2 + 1$, find and simplify:
    (a) $f(t+1)$ (b) $f(t^2+1)$ (c) $f(2)$ (d) $2f(t)$ (e) $[f(t)]^2 + 1$

17. For $f(n) = 3n^2 - 2$ and $g(n) = n + 1$, find and simplify:
    (a) $f(n) + g(n)$ (d) $f(g(n))$
    (b) $f(n)g(n)$ (e) $g(f(n))$
    (c) The domain of $f(n)/g(n)$.

State whether the functions in Problems 18–21 are continuous on the interval $[-1, 1]$.

18. $f(x) = |x|$ 19. $g(x) = \dfrac{|x|}{x}$ 20. $h(\theta) = \theta \sin \theta$ 21. $f(t) = \dfrac{\sin t}{t^2}$

Convert the functions in Problems 22–23 into the form $P = P_0 a^t$.

22. $P = 2.91 e^{0.55t}$ 23. $P = (5 \cdot 10^{-3}) e^{-1.9 \cdot 10^{-2} t}$

24. Different kinds of the same element (called different *isotopes*) can have very different half-lives. The decay of plutonium-240 is described by the formula

$$Q = Q_0 e^{-0.00011t},$$

    whereas the decay of plutonium-242 is described by

$$Q = Q_0 e^{-0.0000018t}.$$

    Find the half-lives of plutonium-240 and plutonium-242.

25. (a) Use the data from Table 1.29 to determine a formula of the form

$$Q = Q_0 e^{rt}$$

    which would give the number of rabbits, Q, at time t (in months).
    (b) What is the approximate doubling time for this population of rabbits?
    (c) Use your equation to predict when the rabbit population will reach 1000.

TABLE 1.29

t	0	1	2	3	4	5
Q	25	43	75	130	226	391

26. A culture of bacteria originally numbers 500. After 2 hours there are 1500 bacteria in the culture. Assuming exponential growth, find how many are present after 6 hours.

27. A culture of 100 bacteria doubles after 2 hours. How long will it take for the number of bacteria to reach 3,200?

28. One hundred kilograms of a certain radioactive substance decay to 40 kg after 10 years. Find how much remains after 20 years.

29. Find the half-life of a radioactive substance that is reduced by 30% in 20 hours.

30. A radioactive substance has a half-life of 8 years. If 200 gm are present initially, how much remains at the end of 12 years? How long will it be until 90% of the original amount has decayed?

31. Suppose prices are increasing by 0.1% a day.

    (a) By what percent do prices increase a year?
    (b) Looking at your answer to part (a), guess the approximate doubling time of prices increasing at this rate. Check your guess.

32. In the early 1920s, Germany had tremendously high inflation, called hyperinflation. Photographs of the time show people going to the store with wheelbarrows full of money. If a loaf of bread cost 1/4 RM in 1919 and 2,400,000 RM in 1922, what was the average yearly inflation rate between 1919 and 1922?

33. Each planet moves around the sun in an elliptical orbit. The orbital period, T, of a planet is the time it takes the planet to go once around the sun. The semimajor axis of each planet's orbit is the average of the largest and the smallest distances between the planet and the sun. Johannes Kepler (1571-1630) discovered that the period of a planet is proportional to the $\frac{3}{2}$ power of its semimajor axis. What is the orbiting period (in days) of Mercury, the closest planet to the sun, with a semimajor axis of 58 million km? What is the period (in years) of Pluto, the farthest planet, with a semimajor axis of 6000 million km? The semimajor axis of the earth is 150 million km. [Hint: What is the earth's period?]

34. (a) Consider the functions graphed in Figure 1.100(a). Find the coordinates of C.
    (b) Consider the functions in Figure 1.100(b). Find the coordinates of C in terms of b.

Figure 1.100

Find possible formulas for the graphs in Problems 35–37.

35.

36.

37.

Find possible formulas for the graphs in Problems 38–43.

38.

39.

40.

41.

42.

43.

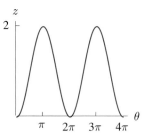

44. The depth of water in a tank oscillates sinusoidally once every 6 hours. If the smallest depth is 5.5 feet and the largest depth is 8.5 feet, find a formula for the depth in terms of time, measured in hours. (There are many possible answers.)

45. Given the graph of $y = h(x)$ in Figure 1.101:

   (a) Sketch a graph of:

      (i) $y = h^{-1}(x)$

      (ii) $y = \dfrac{1}{h(x)}$

   (b) What becomes of the asymptote when you sketch the inverse function?

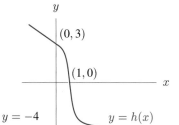

Figure 1.101

46. Each of the functions described by the data in Table 1.30 is increasing over its domain, but each increases in a different way. Which of the graphs in Figure 1.102 best fits each function?

TABLE 1.30

x	$f(x)$	x	$g(x)$	x	$h(x)$
1	1	3.0	1	10	1
2	2	3.2	2	20	2
4	3	3.4	3	28	3
7	4	3.6	4	34	4
11	5	3.8	5	39	5
16	6	4.0	6	43	6
22	7	4.2	7	46.5	7
29	8	4.4	8	49	8
37	9	4.6	9	51	9
47	10	4.8	10	52	10

(a)   (b)   (c)

Figure 1.102

47.

Figure 1.103 is a graph of the function $f(t)$. Here $f(t)$ is the depth in meters below the Atlantic Ocean floor where t million-year-old rock can be found.[11]

(a) Evaluate $f(15)$, and say what it means in practical terms.

(b) Is f invertible? Explain.

(c) Evaluate $f^{-1}(120)$, and say what it means in practical terms.

(d) Sketch a graph of f^{-1}.

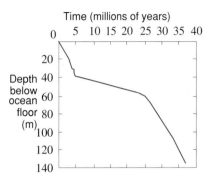

Figure 1.103

48. A fish population is reproducing at an annual rate equal to 5% of the current population, P. Meanwhile, fish are being caught by fishermen at a constant rate, Y (measured in fish per year).

(a) Write a formula for the rate, R, at which the fish population is increasing as a function of P.
(b) Sketch a graph of R against P.

49. Glucose is fed by intravenous injection at a constant rate, k, into a patient's bloodstream. Once there, the glucose is removed at a rate proportional to the amount of glucose present. If R is the net rate at which the quantity, G, of glucose in the blood is increasing:

(a) Write a formula giving R as a function of G.
(b) Sketch a graph of R against G.

50. A *catalyst* in a chemical reaction is a substance which speeds up the reaction but which does not itself change. If the product of a reaction is itself a catalyst, the reaction is said to be *autocatalytic*. Suppose the rate, r, of a particular autocatalytic reaction is proportional to the quantity of the original material remaining times the quantity of product, p, produced. If the initial quantity of the original material is A and the amount remaining is $A - p$:

(a) Express r as a function of p.
(b) What is the value of p when the reaction is proceeding fastest?

51. What approximate domains and ranges make the graphs of $y = x^2$ and $y = 0.01e^{0.01x}$ look like each of the graphs in Figure 1.104?

Figure 1.104

---

[11]Data of Dr. Murlene Clark based on core samples drilled by the research ship *Glomar Challenger*, taken from *Initial Reports of the Deep Sea Drilling Project*.

PROJECTS

1. **Matching Functions to Data**

    From the data in Table 1.31, determine a possible formula for each function.[12] Write an explanation of your reasoning.

    TABLE 1.31

x	$f(x)$	$g(x)$	$h(x)$	$F(x)$	$G(x)$	$H(x)$
−5	−10	20	25	0.958924	0.544021	2.958924
−4.5	−9	19	20.25	0.97753	−0.412118	2.97753
−4	−8	18	16	0.756802	−0.989358	2.756802
−3.5	−7	17	12.25	0.350783	−0.656987	2.350783
−3	−6	16	9	−0.14112	0.279415	1.85888
−2.5	−5	15	6.25	−0.598472	0.958924	1.401528
−2	−4	14	4	−0.909297	0.756802	1.090703
−1.5	−3	13	2.25	−0.997495	−0.14112	1.002505
−1	−2	12	1	−0.841471	−0.909297	1.158529
−0.5	−1	11	0.25	−0.479426	−0.841471	1.520574
0	0	10	0	0	0	2
0.5	1	9	0.25	0.479426	0.841471	2.479426
1	2	8	1	0.841471	0.909297	2.841471
1.5	3	7	2.25	0.997495	0.14112	2.997495
2	4	6	4	0.909297	−0.756802	2.909297
2.5	5	5	6.25	0.598472	−0.958924	2.598472
3	6	4	9	0.14112	−0.279415	2.14112
3.5	7	3	12.25	−0.350783	0.656987	1.649217
4	8	2	16	−0.756802	0.989358	1.243198
4.5	9	1	20.25	−0.97753	0.412118	1.02247
5	10	0	25	−0.958924	−0.544021	1.041076

2. **Compound Interest**

    The newspaper article below is from *The New York Times*, May 27, 1990. Fill in the three blanks. (For the first blank, assume that daily compounding is essentially the same as continuous compounding. For the last blank, assume the interest has been compounded yearly, and give your answer in dollars. Exclude the occurrence of leap years.)

213 Years After Loan, Uncle Sam Is Dunned
By LISA BELKIN
Special to The New York Times

SAN ANTONIO, May 26 — More than 200 years ago, a wealthy Pennsylvania merchant named Jacob DeHaven lent $450,000 to the Continental Congress to rescue the troops at Valley Forge. That loan was apparently never repaid.

So Mr. DeHaven's descendants are taking the United States Government to court to collect what they believe they are owed.

The total: ___ in today's dollars if the interest is compounded daily at 6 percent, the going rate at the time. If compounded yearly, the bill is only ___.

Family Is Flexible

The descendants say that they are willing to be flexible about the amount of a settlement and that they might even accept a heartfelt thank you or perhaps a DeHaven statue. But they also note that interest is accumulating at ___ a second.

---
[12]Based on a problem by Lee Zia

FOCUS ON THEORY

UNDERPINNINGS OF CALCULUS

Mathematics in its applied form has existed from time immemorial. Commercial arithmetic and the geometry of land-surveying and building construction were well-developed by 1500 B.C. Gradually people realized that simple mathematical facts may be interrelated in non-obvious ways, and that the interrelationships themselves were worthy of study. Thales (640-546 B.C.) is said to have *proved* that the sum of the angles of a triangle is two right angles. This is the oldest indication we have of the idea of proof in geometry. In the next section we look at an example of a proof.

Over the next few centuries, people began to think of geometry as being not about points, lines, and circles drawn with chalk on a slate, but about abstract entities: points tinier than the smallest speck, lines that are perfectly straight, and circles that are perfectly round. In other words, what one draws on a slate or carves into stone is merely an imperfect model of the abstract reality. Plato (427-347 B.C.) extended this view. He believed that the entire world of experience was an imperfect shadow of the true reality. However, one cannot reason about abstract things without taking some of their properties for granted. In mathematics, these assumptions are called *axioms* or *postulates*.

About 300 B.C. Euclid wrote a textbook, *The Elements*, covering a good deal of geometry, some number theory, and some more advanced topics concerning irrational numbers. It was and is the most successful textbook ever written. (It is still in print.) Euclid begins his treatment of geometry by stating several axioms concerning lines and circles. For example:

If A and B are two points, there is a circle having center A that passes through B.

This is surely a reasonable property to ascribe to the abstract points and circles that we imagine. He goes on to prove many facts about figures in the plane and in space, including the celebrated Pythagorean theorem: The area of the square drawn on the hypotenuse of a right triangle is equal to the sum of the areas of the squares drawn on the two legs.

For many years *The Elements* was revered as the pinnacle of logical reasoning. It is quite rightly regarded as a masterpiece, but its reasoning is no longer thought to be airtight. In fact, there is an error (by modern standards) in the proof of the very first proposition. The argument goes as follows.

Starting from two points A and B, consider the circle with center A passing through B and the circle with center B passing through A. (See Figure 1.105.) Let these circles cut one another at C...

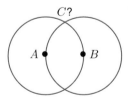

Figure 1.105

But why must there be a point at which these two circles intersect? It is clear from the figure that they do, but the figure is drawn in the real world, not the abstract world of pure geometry. Perhaps when the circles we have drawn are replaced by their abstract representations, it might turn out that there isn't any point where C ought to be. Here and in several other places Euclid seems to have relied on a figure. Perhaps he did not know how to describe clearly those properties of drawings that he believed carried over to abstract geometry and guaranteed the existence of C, and so he left it to his readers to decide whether they believed that C would exist in the ideal realm. Although these deficiencies were noted in classical times, *The Elements* retained its status as the ultimate example of mathematical rigor until well into the nineteenth century. Finally, after centuries of study by many mathematicians, Hilbert (1863-1943) gave what is regarded today as the definitive treatment of Euclidean geometry. It is important to realize that none of the theorems stated by Euclid have been found to be wrong in substance; the difficulties lie entirely in Euclid's incomplete statement of the axioms on which he was relying.

Calculus belongs to a different branch of mathematics than geometry. Instead of lines, circles, and angles, calculus studies the behavior of numerical functions: specifically, functions that represent a rate of change. Calculus also provides a language for expressing laws of nature that govern everything from the behavior of the atomic nucleus to the life cycles of stars.

Some anticipations of calculus can be seen in Euclid and other classical writers, but most of the ideas appear first in the seventeenth century. Newton (1642-1727) and Leibniz (1646-1716) are generally credited with shaping the subject into a coherent theory. Newton's most famous work, *Philosophiae Naturalis Principia Mathematica* (in three volumes) appeared in 1686-1687. Its best known result is that the Laws of Planetary Motion, which had been announced by Kepler (1571-1630) on purely empirical evidence, can be deduced from simpler universal laws, such as the Law of Gravity. In addition, Newton's theory explained other astronomical phenomena, such as the irregularities in the motion of the moon, and terrestrial phenomena, such as the tides. The real significance of *Principia* lies in its demonstration that very complicated physical systems can be successfully modeled by pure mathematics. Although *Principia* uses geometrical arguments, not calculus, the ideas in it were, by Newton's own statement, generated with the aid of calculus.

After its start in the seventeenth century, calculus went for over a century without a proper axiomatic foundation. Newton wrote that it could be rigorously founded on the idea of *limits*, but he never presented his ideas in detail. A limit is, roughly speaking, the value approached by a function near a given point. During the eighteenth century many mathematicians based their work on limits, but their definition of limit was not clear. In 1784, Lagrange (1736-1813) at the Berlin Academy proposed a prize for a successful axiomatic foundation for calculus. He and others were interested in being as certain of the internal consistency of calculus as they were about algebra and geometry. No one was able to successfully respond to the challenge. It remained for Cauchy (1789-1857) to show, around 1820, that limits can be defined rigorously by means of inequalities. The modern definition of the limit, given on page 128, is essentially due to Cauchy.[13]

This rigorous definition of the limit was the advance that was needed in order to begin the axiomatic foundations of calculus, where every result is carefully proved from axioms, or from theorems that have already been proved. Courses in *analysis* follow this chain of logical reasoning.

In this book, we concentrate on developing the solid intuitive understanding on which the rigorous approach depends. We emphasize plausibility arguments, not proofs, but give some glimpses of the theoretical underpinnings of calculus in the "Focus on Theory" sections. We hope that these brief excursions into a more theoretical world encourage you to investigate further.

THE BINOMIAL THEOREM

In everyday life we are often content to believe things simply by observing that they seem to be true. In mathematics, however, we decide what is true by means of logical arguments. Mathematicians attempt to eliminate all possible sources of disagreement by carefully stating *axioms* (assumptions) and *definitions*, formulating precise *theorems* (statements to be proved), and using strict rules of logic. In this section we illustrate how theorems are formulated and proved by studying the example

---

[13]Grabiner, Judith V. "Who Gave You the Epsilon? Cauchy and the Origins of Rigorous Calculus." *American Mathematical Monthly* 90 (1983) pp. 185-194.

of the binomial theorem. In the next section we see how, and why, an axiom is introduced, using as an example the completeness of the real numbers.

Recall the algebraic formulas for squaring and cubing $x + y$:

$$(x + y)^2 = x^2 + 2xy + y^2$$
$$(x + y)^3 = x^3 + 3x^2y + 3xy^2 + y^3.$$

We want to find a general formula for $(x + y)^n$ for any positive integer n. We do this in three stages:

- Find a pattern.
- Formulate a statement, called a *conjecture*, describing the pattern.
- Prove the conjecture.

Once we have proved the statement, it becomes a *theorem*.

Finding a pattern

First we look at some more examples. Multiplying out $(x + y)^n$ for $n = 4, 5, 6$ gives:

$$(x + y)^4 = x^4 + 4x^3y + 6x^2y^2 + 4xy^3 + y^4$$
$$(x + y)^5 = x^5 + 5x^4y + 10x^3y^2 + 10x^2y^3 + 5xy^4 + y^5$$
$$(x + y)^6 = x^6 + 6x^5y + 15x^4y^2 + 20x^3y^3 + 15x^2y^4 + 6xy^5 + y^6.$$

Notice that the exponents of x and y in each term on the right always add up to n. The reason for this is that in the expansion of

$$(x + y)^n = \underbrace{(x + y)(x + y) \cdots (x + y)}_{n \text{ times}},$$

each term comes from choosing x's from some of the factors and y's from the others. The total number of x's and y's chosen equals the total number of $(x + y)$'s, which is n. For example, in the expansion of $(x + y)^3$, choosing x from one of the factors and y from the other two yields a term xy^2. There are three different ways of doing this (depending on which factor x is chosen from), so there are three terms of this form, giving $3xy^2$.

We arrange the coefficients in the expansion of $(x + y)^n$ in a triangle called Pascal's triangle, after the French mathematician Blaise Pascal.

$$
\begin{array}{ccccccccccccc}
 & & & & & 1 & & 1 & & & & & \\
 & & & & 1 & & 2 & & 1 & & & & \\
 & & & 1 & & 3 & & 3 & & 1 & & & \\
 & & 1 & & 4 & & 6 & & 4 & & 1 & & \\
 & 1 & & 5 & & 10 & & 10 & & 5 & & 1 & \\
1 & & 6 & & 15 & & 20 & & 15 & & 6 & & 1
\end{array}
$$

The second row in this triangle gives the coefficients in the expansion of $(x+y)^2 = x^2 + 2xy + y^2$, namely 1, 2, and 1. The next row gives the coefficients for $(x + y)^3$, and so on. The top row gives the coefficients for the expansion $(x + y)^1 = x + y$.

There appears to be a pattern to the triangle: The outside entries are all 1s; each inside entry is equal to the sum of the entries immediately to its left and right in the row above. For example, each 10 in the fourth row has a 4 and a 6 immediately above it, and $10 = 4 + 6$.

Formulating the theorem

We want to prove that, for any n, the coefficients in the expansion of $(x + y)^n$ satisfy the pattern we have observed for $n = 1, \ldots, 6$. The general case is made easier by writing C_k^n for the coefficient of $x^{n-k}y^k$ in the expansion of $(x+y)^n$, so

$$(x+y)^n = C_0^n x^n + C_1^n x^{n-1}y + C_2^n x^{n-2}y^2 + \cdots + C_{n-1}^n xy^{n-1} + C_n^n y^n.$$

Thus, for example, $C_3^5 = 10$ because the x^2y^3 term in the expansion of $(x+y)^5$ is $10x^2y^3$.

Now

$$C_0^n \quad C_1^n \quad C_2^n \ldots \quad C_{n-1}^n \quad C_n^n$$

is the n-th row in Pascal's triangle. There are two rules that described the pattern we have observed: first, the outside entries are all 1s; second, each inside entry is the sum of the two above it. Thus, we must show, first, that $C_0^n = 1$ and $C_n^n = 1$ for all n, and, second, that

$$C_k^n = C_{k-1}^{n-1} + C_k^{n-1}, \quad 0 < k < n.$$

Notice, if $0 < k < n$, then C_k^n is an inside entry in the triangle, and C_{k-1}^{n-1} and C_k^{n-1} are the entries immediately above it. Now we can state the theorem we want to prove as follows:

The Binomial Theorem

If n is a positive integer and we write

$$(x+y)^n = C_0^n x^n + C_1^n x^{n-1}y + C_2^n x^{n-2}y^2 + \cdots + C_{n-1}^n xy^{n-1} + C_n^n y^n,$$

then

$$C_0^n = C_n^n = 1, \quad \text{for } n \geq 1,$$

and

$$C_k^n = C_{k-1}^{n-1} + C_k^{n-1}, \quad \text{for} \quad n \geq 2 \quad \text{and} \quad 0 < k < n.$$

Proof   In the expansion of

$$(x+y)^n = \underbrace{(x+y)(x+y)\cdots(x+y)}_{n \text{ times}},$$

there is only one way of getting the term x^n, and that is by choosing an x from each factor. So, the coefficient of x^n is 1. By the same argument, the coefficient of y^n is also 1, so

$$C_0^n = C_n^n = 1.$$

To prove $C_k^n = C_{k-1}^{n-1} + C_k^{n-1}$, we write

$$(x+y)^n = C_0^n x^n + C_1^n x^{n-1}y + C_2^n x^{n-2}y^2 + \cdots + C_{n-1}^n xy^{n-1} + C_n^n y^n,$$

and we write

$$(x+y)^{n-1} = C_0^{n-1}x^{n-1} + C_1^{n-1}x^{n-2}y + C_2^{n-1}x^{n-3}y^2 + \cdots + C_{n-2}^{n-1}xy^{n-2} + C_{n-1}^{n-1}y^{n-1}.$$

Now, we will use the fact that

$$(x+y)^n = (x+y)(x+y)^{n-1}.$$

Substituting in the expressions for $(x+y)^{n-1}$ and $(x+y)^n$ gives

$$C_0^n x^n + C_1^n x^{n-1}y + \quad \cdots \quad + C_{n-1}^n xy^{n-1} + C_n^n y^n$$
$$= (x+y)\left(C_0^{n-1}x^{n-1} + C_1^{n-1}x^{n-2}y + \quad \cdots \quad + C_{n-2}^{n-1}xy^{n-2} + C_{n-1}^{n-1}y^{n-1}\right)$$
$$= x\left(C_0^{n-1}x^{n-1} + C_1^{n-1}x^{n-2}y + \quad \cdots \quad + C_{n-2}^{n-1}xy^{n-2} + C_{n-1}^{n-1}y^{n-1}\right)$$
$$\qquad + y\left(C_0^{n-1}x^{n-1} + C_1^{n-1}x^{n-2}y + \quad \cdots \quad + C_{n-2}^{n-1}xy^{n-2} + C_{n-1}^{n-1}y^{n-1}\right)$$
$$= C_0^{n-1}x^n + \left(C_1^{n-1} + C_0^{n-1}\right)x^{n-1}y + \cdots + \left(C_{n-1}^{n-1} + C_{n-2}^{n-1}\right)xy^{n-1} + C_{n-1}^{n-1}y^n.$$

The inside terms in this expression have the form $(C_k^{n-1} + C_{k-1}^{n-1})x^{n-k}y^k$, for $k = 1, \ldots, n-1$. The corresponding term in the expansion of $(x+y)^n$ is $C_k^n x^{n-k}y^k$. Since the expressions are equal, the coefficients of like terms must be equal, so

$$C_k^n = C_k^{n-1} + C_{k-1}^{n-1} = C_{k-1}^{n-1} + C_k^{n-1},$$

which is what we wanted to show.

A Formula for the Binomial Coefficients

The numbers C_k^n are called *binomial coefficients*. They are usually computed using the following formula, rather than by writing out Pascal's triangle. (Note that $k! = k(k-1)\cdots 3 \cdot 2 \cdot 1$.)

$$\boxed{C_k^n = \frac{n!}{k!(n-k)!} = \frac{n(n-1)\cdots(n-k+1)}{k!}}$$

This formula holds for $k = 0$ and $k = n$ if we adopt the convention that $0! = 1$:

$$C_0^n = \frac{n!}{0!(n-0)!} = \frac{n!}{n!} = 1 \qquad \text{and} \qquad C_n^n = \frac{n!}{n!(n-n)!} = \frac{n!}{n!} = 1.$$

To prove the formula in general, we use an important technique called *mathematical induction*. There are two steps:
- Prove the formula in the case $n = 1$.
- Prove that if it holds for a specific positive integer n then it holds for $n + 1$.

The second step, called the induction step, enables us to deduce that the formula is true for all n, as follows. Since we know by the first step that it is true for $n = 1$, by the induction step it is true for $n = 2$. But then, by the induction step again, it is true for $n = 3$, and so on.

Proof We have already proved that the formula holds for $n = 1$ since the only binomial coefficients in that case are C_0^1 and C_1^1.

Now we prove the induction step. Suppose that our formula is true for n. That is, suppose that

$$C_k^n = \frac{n!}{k!(n-k)!}, \qquad 0 \le k \le n.$$

We want to deduce the formula for $n + 1$. That is, we want to show that

$$C_k^{n+1} = \frac{(n+1)!}{k!(n+1-k)!}, \qquad 0 \le k \le n+1.$$

We already know that this is true if $k = 0$ or $k = n + 1$. If $0 < k < n + 1$, then, using the Binomial Theorem,

$$C_k^{n+1} = C_k^n + C_{k-1}^n = \frac{n!}{k!(n-k)!} + \frac{n!}{(k-1)!(n-k+1)!}$$

$$= \frac{n!}{k(k-1)!(n-k)!} + \frac{n!}{(k-1)!(n-k+1)(n-k)!}$$

$$= \frac{n!}{(k-1)!(n-k)!}\left(\frac{1}{k} + \frac{1}{n-k+1}\right)$$

$$= \frac{n!}{(k-1)!(n-k)!}\left(\frac{n-k+1+k}{k(n-k+1)}\right)$$

$$= \frac{n!}{(k-1)!(n-k)!}\frac{(n+1)}{k(n-k+1)} = \frac{(n+1)!}{k!(n-k+1)!},$$

which is what we wanted to prove.

Problems on the Binomial Theorem

1. The formula for the binomial coefficients gives C_k^n as a ratio of integers. Is C_k^n necessarily an integer? Could C_k^n be a fraction? Justify your answer.

2. Look at the entries in the first few rows of Pascal's triangle on page 79. You should see a pattern of symmetry.

   (a) Describe the pattern in words.
   (b) Formulate a conjecture about the binomial coefficients C_k^n that describes the pattern mathematically.
   (c) Prove your conjecture.

3. Add the entries across the rows in Pascal's triangle for the first six rows. You should notice a pattern to the sequence of numbers that you obtain.

   (a) Formulate a general conjecture that describes the pattern.
   (b) Prove your conjecture.

COMPLETENESS OF THE REAL NUMBERS

If two people argue about something long enough, they may eventually reveal the hidden assumptions that are the root of the argument. In the same way, mathematicians arrive at axioms by a process which is somewhat like arguing with themselves. In attempting to understand something, they question every seemingly obvious statement, hoping to eventually arrive at fundamental axioms.

We apply this method to the process of finding a root of a polynomial by zooming in on its graph. This will lead us to a subtle property of the real numbers, called *completeness*. Many proofs involving limits depend on this property.

Case Study: Finding the Roots of a Polynomial

Consider the polynomial $f(x) = 3x^3 - x^2 + 2x - 1$ on the interval $[0, 1]$. Since $f(0) = -1$ and $f(1) = 3$, we expect that the graph of f crosses the x-axis at some point $x = r$ between $x = 0$ and $x = 1$. Since the coordinates of this point are $(r, 0)$, we have $f(r) = 0$. Suppose we estimate this root by graphing the polynomial on a calculator or computer and zooming in. We start by knowing that

$$0 \le r \le 1.$$

By zooming in, we find $f(0.4) < 0$ and $f(0.5) > 0$, so r is trapped in a smaller interval

$$0.4 \le r \le 0.5.$$

Successive zooming in shows that

$$0.45 \le r \le 0.46$$

$$0.459 \le r \le 0.460$$

$$0.4598 \le r \le 0.4599.$$

At each stage, we divide the interval into tenths and pick any one for which f is negative at the left end point and positive at the right. (If f is 0 at any of the endpoints, we have found r and can stop.) Continuing this way we obtain a sequence of intervals, each one one-tenth the length of the previous one and each one containing r. (See Figure 1.106.) Although a calculator will only give us a finite number of digits, we could in principle continue forever, generating an infinite sequence of intervals.

Figure 1.106: Zooming in on a zero of $f(x) = 3x^3 - x^2 + 2x - 1$

It seems that this process leads us to a number r such that $f(r) = 0$. However, there are two questions that can be raised:

- How do we know[14] that this process of zooming in really does close in on a number, r?
- How do we know that $f(r) = 0$?

The Completeness Axiom

Consider the first question above (the answer to the second question is worked out in Problem 26 on page 135). The left hand endpoints of the nested intervals form an ever increasing sequence of decimals; clearly the number r that we are looking for is the smallest number that is greater than all these decimals. The completeness axiom says that, given any nonempty set of numbers, if there is any number which is greater than or equal to all of the numbers in the set, then there is a smallest such number.[15] A number which is greater than or equal to all the numbers in a set is called an *upper bound* for the set. We have the following:

> ### The Completeness Axiom
>
> Any nonempty set of real numbers which has an upper bound has a *least* upper bound.

Example 1   For each of the following sets, say whether it has an upper bound. If so, give the least upper bound.

    (a)   The set of x such that $-2 < x < 3$.
    (b)   The set of x such that $-2 \leq x \leq 3$.
    (c)   The set of all integers.
    (d)   The sequence $0.9, \ 0.99, \ 0.999, \ldots$.

Solution   (a)   The numbers 3, 4, and π are all upper bounds; 3 is the least upper bound.
    (b)   Same as part (a); an upper bound for a set can be in the set, since it only has to be greater than *or equal to* each number in the set.
    (c)   There is no upper bound for this set; no matter how large a number we choose for the upper bound, there will always be some integer bigger than it.
    (d)   All the numbers are less than 1, and 1 is the smallest number with this property. Thus, this sequence has a least upper bound of 1.

In the previous example we could see directly what the least upper bounds were. In other situations it may not be obvious. The completeness axiom guarantees the existence of the least upper bound but offers no help in finding it.

---

[14]This is related to the question raised on page 77 about the existence of an intersection point C between two circles.

[15]It is possible to give a definition of the real numbers in which the completeness axiom becomes a theorem.

The Nested Interval Theorem

Now we see how the completeness axiom ensures that the process of zooming in closes in on a specific number.

> ## Nested Interval Theorem
>
> Given an infinite sequence of closed intervals, $[a_n, b_n]$, each one contained within the previous one, then there is at least one number in all the intervals.

Proof   Because each interval $[a_n, b_n]$ is contained in the previous one, each a_n must be at least as large as the previous one, so

$$a_1 \leq a_2 \leq a_3 \leq \cdots \leq a_n \leq \cdots.$$

Similarly, each b_n is no larger than its predecessor, so

$$\cdots \leq b_n \leq \cdots \leq b_3 \leq b_2 \leq b_1.$$

All the a_n are bounded above by b_1, and in fact by every b_n. Therefore, by the completeness axiom, we know that the a_n have a least upper bound; call it r. Since r is an upper bound, $r \geq a_n$ for all n. Since r is the least upper bound, r must be less than or equal to each of the upper bounds b_n. Thus, r is in all the intervals.

Note that in our statement of the Nested Interval Property, we did not assume that the lengths of the intervals approached zero, so in general there may be more than one number r in all the intervals. However, when the lengths do approach zero, as in the case of finding a root by zooming in, there is a *unique* number r (see Problem 2).

The Intermediate Value Theorem

When we considered the polynomial $3x^3 - x^2 + 2x - 1$, we assumed it must have a root between $x = 0$ and $x = 1$ because it was negative at $x = 0$ and positive at $x = 1$. More generally, our intuitive notion of continuity tells us that, as we follow the graph of a continuous function f from some point $(a, f(a))$ to another point $(b, f(b))$, then f must take on all intermediate values between $f(a)$ and $f(b)$. (See Figure 1.107.) This is:

> ## Intermediate Value Theorem
>
> Suppose f is continuous on a closed interval $[a, b]$. If k is any number between $f(a)$ and $f(b)$, then there is at least one number c in $[a, b]$ such that $f(c) = k$.

Problems 26 and 27 on page 135 suggest a way of proving the Intermediate Value Theorem using the Nested Interval Theorem.

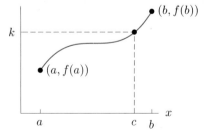

Figure 1.107: The Intermediate Value Theorem

Problems on the Completeness of Real Numbers

1. (a) Using the definitions in this section as a guide, define the following terms:
   (i) A lower bound of a set of numbers
   (ii) The greatest lower bound of a set of numbers
   (b) State the completeness axiom in terms of lower bounds.

2. Let r be a number contained in each of a sequence of nested intervals $[a_n, b_n]$. Suppose that the width of the intervals, $|b_n - a_n|$, goes to 0 as $n \to \infty$. Prove that r is unique. [Hint: Suppose there were two such numbers, and argue to a contradiction.]

3. In this problem we will use the completeness axiom to show that an infinite decimal expansion actually defines a real number, and that the first n digits of the expansion give the number accurate to n decimal places. Let x_n be the number defined by the first n digits of the expansion; we call x_n the n-th *truncation* of the expansion.

   (a) For any n, show that $x_n + (1/10^n)$ is an upper bound for the set of all the truncations.
   (b) Deduce that there is a real number, c, such that $x_n \leq c \leq x_n + (1/10)^n$ for all n. Thus x_n represents c accurate to n places, so it's reasonable to say that c is the number represented by the infinite decimal expansion.

CHAPTER TWO

KEY CONCEPT: THE DERIVATIVE

We begin this chapter by investigating the problem of speed: How can we measure the speed of a moving object at a given instant in time? Or, more fundamentally, what do we mean by the term *speed*? We'll come up with a definition of speed that has wide-ranging implications — not just for the speed problem, but for measuring the rate of change of any quantity. Our journey will lead us to the key concept of *derivative*, which forms the basis for our study of calculus.

The derivative can be interpreted geometrically as the slope of a curve, and physically as a rate of change. Because derivatives can be used to represent everything from fluctuations in interest rates to the rates at which fish populations vary and gas molecules move, they have applications throughout the sciences.

2.1 HOW DO WE MEASURE SPEED?

The speed of an object at an instant in time is surprisingly difficult to define precisely. Consider the statement "At the instant it crossed the finish line, the horse was traveling at 42 mph." How can such a claim be substantiated? A photograph taken at that instant will show the horse motionless—it is no help at all. There is some paradox in trying to study the horse's motion at a particular instant in time, since by focusing on a single instant we stop the motion!

Problems of motion were of central concern to Zeno and other philosophers as early as the fifth century B.C. The modern approach, made famous by Newton's calculus, is to stop looking for a simple notion of speed at an instant, and instead to look at speed over small intervals containing the instant. This method sidesteps the philosophical problems mentioned earlier but introduces new ones of its own.

We illustrate the ideas discussed above by an idealized example, called a thought experiment. It is idealized in the sense that we assume that we can make measurements of distance and time as accurately as we wish.

A Thought Experiment: Average and Instantaneous Velocity

We look at the speed of a small object (say, a grapefruit) that is thrown straight upward into the air at $t = 0$ seconds. The grapefruit leaves the thrower's hand at high speed, slows down until it reaches its maximum height, and then gradually speeds up in the downward direction and finally, "Splat!" (See Figure 2.1.)

But suppose that we would like to be more precise in our determination of the speed, say, at $t = 1$ second. We assume that we can measure the height of the grapefruit above the ground at any time t; we think of the height, y, as a function of time. (See Table 2.1.) "Splat!" comes sometime between 6 and 7 seconds. The numbers show the behavior noted above: During the first second the grapefruit travels $90 - 6 = 84$ feet, and during the second second it travels only $142 - 90 = 52$ feet. Hence the grapefruit traveled faster over the first interval, $0 \leq t \leq 1$, than the second interval, $1 \leq t \leq 2$.

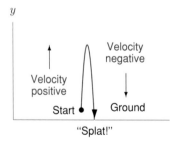

Figure 2.1: The grapefruit's path is straight up and down

TABLE 2.1   *Height of the grapefruit above the ground*

t (sec)	0	1	2	3	4	5	6
y (feet)	6	90	142	162	150	106	30

Velocity versus Speed

From now on, we will make a distinction between velocity and speed. Suppose an object moves along a line. We pick one direction to be positive and say that the *velocity* is positive if it is in the same direction, and negative if it is in the opposite direction. For the grapefruit, upward is positive and downward is negative. (See Figure 2.1.) *Speed* is the magnitude of the velocity and so is always positive or zero.

> If $s(t)$ is the position of an object at time t, then the **average velocity** of the object over the interval $a \leq t \leq b$ is
>
> $$\text{Average velocity} = \frac{\text{Change in position}}{\text{Change in time}} = \frac{s(b) - s(a)}{b - a}.$$
>
> In words, the **average velocity** of an object over an interval is the net change in position during the interval divided by the change in time.

Example 1   Compute the average velocity of the grapefruit over the interval $4 \leq t \leq 5$. What is the significance of the sign of your answer?

Solution   During this interval, the grapefruit moves $(106 - 150) = -44$ feet. Therefore the average velocity is -44 ft/sec. The negative sign means the height is decreasing and the grapefruit is moving downward.

Example 2   Compute the average velocity of the grapefruit over the interval $1 \leq t \leq 3$.

Solution   Average velocity $= (162 - 90)/(3 - 1) = 72/2 = 36$ ft/sec.

The average velocity is a useful concept since it gives a rough idea of the behavior of the grapefruit: if two grapefruits are hurled into the air, and one has an average velocity of 10 ft/sec over the interval $0 \leq t \leq 1$ while the second has an average velocity of 100 ft/sec over the same interval, clearly the second one is moving faster.

But average velocity over an interval doesn't solve the problem of measuring the velocity of the grapefruit at *exactly* $t = 1$ second. To get closer to an answer to that question, we have to look at what happens near $t = 1$ in more detail. The data[1] in Figure 2.2 shows the average velocity over small intervals on either side of $t = 1$.

Notice that the average velocity before $t = 1$ is slightly more than the average velocity after $t = 1$. We expect to define the velocity *at* $t = 1$ to be something between these two average velocities. As the size of the interval shrinks, the values of the velocity before $t = 1$ and the velocity after $t = 1$ get closer together. In the smallest interval in Figure 2.2, both velocities are 68.0 ft/sec (to one decimal place), so we will define the velocity at $t = 1$ to be 68.0 ft/sec (to one decimal place).

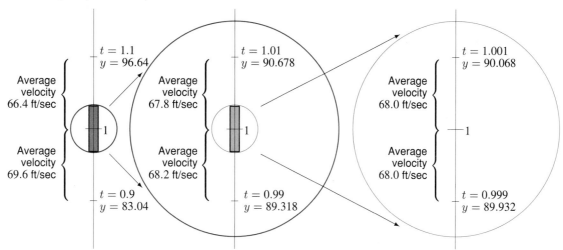

Figure 2.2: Average velocities over intervals on either side of $t = 1$: Showing successively smaller intervals

[1]The data is in fact calculated from the formula $y = 6 + 100t - 16t^2$.

Of course, if we calculate to more decimal places, the average velocities before and after $t = 1$ would no longer agree. To calculate the velocity at $t = 1$ to more decimal places of accuracy, we would take smaller and smaller intervals on either side of $t = 1$ until the average velocities agree to the number of decimal places we wanted. In this way, we could estimate the velocity at $t = 1$ to any accuracy.

Defining Instantaneous Velocity Using Limit Notation

When we take smaller and smaller intervals, it turns out that the average velocities are always just above or just below 68 ft/sec. It seems natural, then, to define velocity at the instant $t = 1$ to be 68 ft/sec. This is called the *instantaneous velocity* at this point. Its definition depends on our being convinced that smaller and smaller intervals will provide average speeds that come arbitrarily close to 68. This process is called *taking the limit*.

Notice how we have replaced the original difficulty of computing velocity at a point by a search for an argument to convince ourselves that the average velocities do approach a number as the time intervals shrink in size. In a sense, we have traded one hard question for another, since we don't yet have any idea how to be certain what number the average velocities are approaching. In the thought experiment, the number seems to be exactly 68, but what if it were 68.000001? How can we be sure that we have taken small enough intervals? Showing that the limit is exactly 68 requires more precise knowledge of the limiting process.

We now define instantaneous velocity at an arbitrary point $t = a$. We use the same method as for $t = 1$: we look at smaller and smaller intervals of size h around $t = a$. Then, over the interval $a \leq t \leq a + h$,

$$\text{Average velocity} = \frac{s(a + h) - s(a)}{h}.$$

The same formula holds when $h < 0$. The instantaneous velocity is the number that the average velocities approach as the intervals decrease in size, that is, as h becomes smaller. So we define

$$\text{Instantaneous velocity} = \text{Limit, as } h \text{ approaches } 0, \text{ of } \frac{s(a + h) - s(a)}{h}.$$

This is written more compactly using limit notation, as follows:

Let $s(t)$ give the position at time t. Then the **instantaneous velocity** at $t = a$ is

$$\begin{array}{c} \text{Instantaneous velocity} \\ \text{at } t = a \end{array} = \lim_{h \to 0} \frac{s(a + h) - s(a)}{h}.$$

In words, the **instantaneous velocity** of an object at time $t = a$ is given by the limit of the average velocity over an interval, as the interval shrinks around a.

This expression forms the foundation of the rest of calculus. Be sure that you are not confused by the notation and recognize it for what it is: the number that the average velocities approach as the intervals shrink. To estimate that number, the limit, we look at intervals of smaller and smaller, but never zero, length.

Visualizing Velocity: Slope of Curve

Now we will see how to visualize velocity using a graph of height. Let's go back to the grapefruit. Suppose that Figure 2.3 shows the height of the grapefruit plotted against time. (Note that this is not a picture of the grapefruit's path, which is straight up and down.)

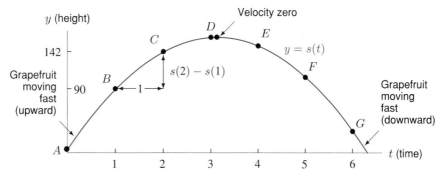

Figure 2.3: The height, y, of the grapefruit at time t

How can we visualize the average velocity on this graph? Suppose $y = s(t)$. Let's consider the interval $1 \leq t \leq 2$ and the expression

$$\text{Average velocity} = \frac{\text{Change in position}}{\text{Change in time}} = \frac{s(2) - s(1)}{2 - 1} = \frac{142 - 90}{1} = 52 \text{ ft/sec.}$$

Now $s(2) - s(1)$ is the change in position over the interval, and it is marked vertically in Figure 2.3. The 1 in the denominator is the time elapsed and is marked horizontally in Figure 2.3. Therefore,

$$\text{Average velocity} = \frac{\text{Change in position}}{\text{Change in time}} = \text{Slope of line joining } B \text{ and } C.$$

(See Figure 2.3.) A similar argument shows the following:

The **average velocity** over any time interval $a \leq t \leq b$ is the slope of the line joining the points on the graph of $s(t)$ corresponding to $t = a$ and $t = b$.

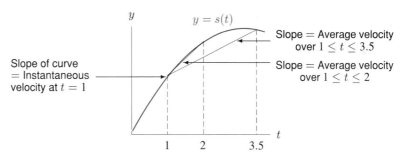

Figure 2.4: Average velocities over small intervals

The next question is how to visualize the instantaneous velocity. Let's think about how we found the instantaneous velocity. We took average velocities over smaller and smaller intervals beginning at the point $t = 1$. Two such velocities are represented by the slopes of the lines in Figure 2.4. As the length of the interval shrinks, the slope of the line gets closer to the slope of the curve at $t = 1$.

The cornerstone of the idea is the fact that, on a very small scale, most functions look almost like straight lines. Imagine taking the graph of a function near a point and "zooming in" to get a close-up view. (See Figure 2.5.) The more we zoom in, the more the curve will appear to be a straight line. In other words, if we repeatedly zoom in on a section of the curve centered at a point of interest, the section of curve will eventually look like a straight line. We call the slope of this line the *slope of the curve* at the point. Therefore, the slope of the magnified line is the instantaneous velocity. Thus, we can say:

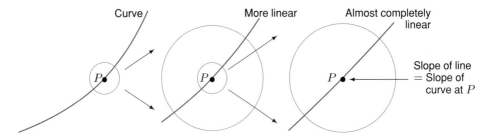

Figure 2.5: Estimating the slope of the curve at the point by "zooming in"

> The **instantaneous velocity** is the slope of the curve at a point.

Look back at the graph of the grapefruit's height as a function of time in Figure 2.3. If we think of the velocity at any point as the slope of the curve there, we can see how the grapefruit's velocity varies during its journey. At points A and B the curve has a large positive slope, indicating that the grapefruit is traveling up rapidly. Point D is almost at the top: the grapefruit is slowing down. At the peak, the slope of the curve is zero: the fruit has slowed to zero velocity for an instant in preparation for its return to earth. At point E the curve has a small negative slope, indicating a slow velocity of descent. Finally, the slope of the curve at point G is large and negative, indicating a large downward velocity that is responsible for the "Splat."

The Idea of a Limit

In the process of defining the instantaneous velocity, we obverved average velocities as time intervals shrank around a point and introduced the notation for a limit. Now we look a bit more at the idea of a limit, which is considered in more detail in the Focus on Theory section on page 127. We talk about the *limit* of the function at the point c.

> We write $\lim_{x \to c} f(x)$ to represent the number L approached by $f(x)$ as x approaches c.

Example 3   Investigate $\lim_{x \to 2} x^2$.

Solution   Notice that we can make x^2 as close to 4 as we like by taking x sufficiently close to 2. (Look at the values of 1.9^2, 1.99^2, 1.999^2, and 2.1^2, 2.01^2, 2.001^2; they seem to be approaching 4 in Table 2.2.) We write

$$\lim_{x \to 2} x^2 = 4,$$

which is read "the limit, as x approaches 2, of x^2 is 4." Notice that the limit doesn't ask what happens *at* $x = 2$, so it is not sufficient to substitute 2 to find the answer. The limit describes behavior of a function *near* a point, not *at* the point.

TABLE 2.2   *Values of x^2*

x	1.9	1.99	1.999	2.001	2.01	2.1
x^2	3.61	3.96	3.996	4.004	4.04	4.41

Example 4   Use a graph to estimate $\lim\limits_{\theta \to 0} \dfrac{\sin \theta}{\theta}$. (Be sure to use radians.)

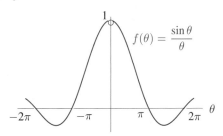

Figure 2.6: Find the limit as $\theta \to 0$

Solution   Figure 2.6 shows that as θ approaches 0 from either side, the value of $\dfrac{\sin \theta}{\theta}$ appears to approach 1, suggesting that $\lim\limits_{\theta \to 0} \dfrac{\sin \theta}{\theta} = 1$. Zooming in on the graph near $\theta = 0$ provides further support for this conclusion. Notice that $\dfrac{\sin \theta}{\theta}$ is undefined at $\theta = 0$.

Example 5   Estimate $\lim\limits_{h \to 0} \dfrac{(3+h)^2 - 9}{h}$ numerically.

Solution   The limit is the value approached by this expression as h approaches 0. The values in Table 2.3 seem to be converging to 6 as $h \to 0$. So it is a reasonable guess that

$$\lim_{h \to 0} \frac{(3+h)^2 - 9}{h} = 6.$$

However, we cannot be sure that the limit is *exactly* 6 by looking at the table. To calculate the limit exactly requires algebra.

TABLE 2.3   *Values of* $\left((3+h)^2 - 9\right)/h$

h	-0.1	-0.01	-0.001	0.001	0.01	0.1
$\left((3+h)^2 - 9\right)/h$	5.9	5.99	5.999	6.001	6.01	6.1

Example 6   Use algebra to find $\lim\limits_{h \to 0} \dfrac{(3+h)^2 - 9}{h}$.

Solution   Expanding the numerator gives

$$\frac{(3+h)^2 - 9}{h} = \frac{9 + 6h + h^2 - 9}{h} = \frac{6h + h^2}{h}.$$

Since taking the limit as $h \to 0$ means looking at values of h near, but not equal, to 0, we can cancel h, giving

$$\lim_{h \to 0} \frac{(3+h)^2 - 9}{h} = \lim_{h \to 0} (6 + h).$$

As h approaches 0, the values of $(6 + h)$ approach 6, so

$$\lim_{h \to 0} \frac{(3+h)^2 - 9}{h} = \lim_{h \to 0} (6 + h) = 6.$$

Problems for Section 2.1

1. A car is driven at a constant speed. Sketch a graph of the distance the car has traveled as a function of time.

2. A car is driven at an increasing speed. Sketch a graph of the distance the car has traveled as a function of time.

3. A car starts at a high speed, and its speed then decreases slowly. Sketch a graph of the distance the car has traveled as a function of time.

4. For the function shown in Figure 2.7, at what labeled points is the slope of the graph positive? Negative? At which labeled point does the graph have the greatest (i.e., most positive) slope? The least slope (i.e., negative and with the largest magnitude)?

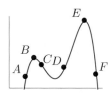

Figure 2.7

5. Match the points labeled on the curve in Figure 2.8 with the given slopes.

Slope	Point
-3	
-1	
0	
$1/2$	
1	
2	

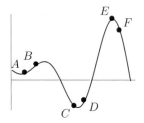

Figure 2.8

6. For the graph $y = f(x)$ shown in Figure 2.9, arrange the following numbers in ascending (i.e., smallest to largest) order:

    -   The slope of the graph at A.
    -   The slope of the graph at B.
    -   The slope of the graph at C.
    -   The slope of the line AB.
    -   The number 0.
    -   The number 1.

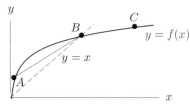

Figure 2.9

7. The graph of $f(t)$ in Figure 2.10 gives the position of a particle at time t. List the following quantities in order, smallest to largest.
    -   A, the average velocity between $t = 1$ and $t = 3$,
    -   B, the average velocity between $t = 5$ and $t = 6$,
    -   C, the instantaneous velocity at $t = 1$,
    -   D, the instantaneous velocity at $t = 3$,
    -   E, the instantaneous velocity at $t = 5$,
    -   F, the instantaneous velocity at $t = 6$.

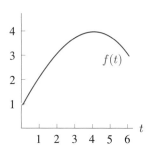

Figure 2.10

8.   Suppose a particle is moving at varying velocity along a straight line and that $s = f(t)$ represents the distance of the particle from a point as a function of time, t. Sketch a possible graph for f if the average velocity of the particle between $t = 2$ and $t = 6$ is the same as the instantaneous velocity at $t = 5$.

Estimate the limits in Problems 9–12 by substituting smaller and smaller values of h. Give your answers to one decimal place.

9. $\lim\limits_{h \to 0} \dfrac{(3 + h)^3 - 27}{h}$ 10. $\lim\limits_{h \to 0} \dfrac{\cos h - 1}{h}$ 11. $\lim\limits_{h \to 0} \dfrac{7^h - 1}{h}$ 12. $\lim\limits_{h \to 0} \dfrac{e^{1+h} - e}{h}$

For each of the functions in Problems 13–22, do the following four things:

(a)   Make a table of values of $f(x)$ for $x = 0.1, 0.01, 0.001, 0.0001, -0.1, -0.01, -0.001,$ and -0.0001.

(b)   Make a conjecture about the value of $\lim\limits_{x \to 0} f(x)$.

(c)   Graph the function to see if it is consistent with your answers to parts (a) and (b).

(d)   Find an interval for x near 0 such that the difference between your conjectured limit and the value of the function is less than 0.01. (In other words, find a window of height 0.02 such that the graph exits the sides of the window and not the top or bottom of the window.)

13. $f(x) = 3x + 1$

14. $f(x) = x^2 - 1$

15. $f(x) = \sin 2x$

16. $f(x) = \sin 3x$

17. $f(x) = \dfrac{\sin 2x}{x}$

18. $f(x) = \dfrac{\sin 3x}{x}$

19. $f(x) = \dfrac{e^x - 1}{x}$

20. $f(x) = \dfrac{e^{2x} - 1}{x}$

21. $f(x) = \dfrac{\cos 2x - 1 + 2x^2}{x^3}$

22. $f(x) = \dfrac{\cos 3x - 1 + 4.5x^2}{x^3}$

2.2   THE DERIVATIVE AT A POINT

Average Rate of Change

Now we will apply the analysis of Section 2.1 to any function $y = f(x)$, not just to height as a function of time. In the case of height, we looked at the change in height divided by the change in time, which tells us

$$\begin{array}{c} \text{Average rate of change of height} \\ \text{with respect to time} \end{array} = \frac{s(a + h) - s(a)}{h}.$$

This ratio is called the *difference quotient*. For any function f, we say:

$$\begin{array}{c} \text{Average rate of change} \\ \text{of } f \text{ over the interval} \\ \text{from } a \text{ to } a + h \end{array} = \frac{f(a + h) - f(a)}{h}.$$

The numerator, $f(a + h) - f(a)$, measures the change in the value of f over the interval from a to $a + h$. Therefore, the difference quotient is the change in f divided by the change in x. See Figure 2.11.

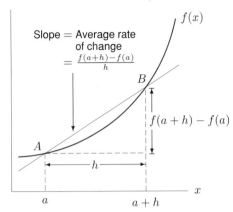

Figure 2.11: Visualizing the average rate of change of f

Although the interval is no longer necessarily a time interval, we still talk about the *average rate of change* of f over the interval. If we want to emphasize the independent variable, we talk about the average rate of change of f *with respect to x*.

Average Rate of Change versus Absolute Change

The average rate of change of a function over an interval is not the same as the absolute change. Absolute change is just the difference in the values of f at the ends of the interval, that is,

$$f(a + h) - f(a).$$

The average rate of change, however, is the absolute change divided by the size of the interval,

$$\frac{f(a + h) - f(a)}{h}.$$

The average rate of change tells how quickly (or slowly) the function changes from one end of the interval to the other, relative to the size of the interval. It is often more useful to know the rate of change than the absolute change. For example, if someone offers you a $150 return on a $100 investment, you will want to know how long it is going to take to make that money. Just knowing the absolute change in your money, $50, is not enough, but knowing the rate of change (i.e., $50 divided by the time it takes to make it) will help you decide whether or not to make the investment.

Blowing Up a Balloon

Consider the function which gives the radius of a sphere in terms of its volume. For example, think of blowing air into a balloon. You've probably noticed that a balloon seems to blow up faster at the start and then slows down as you blow more air into it. What you're seeing is variation in the rate of change of the radius with respect to volume.

Example 1   The volume, V, of a sphere is given by $V = 4\pi r^3/3$. Solving for r in terms of V gives

$$r = f(V) = \left(\frac{3V}{4\pi}\right)^{1/3}.$$

Calculate the average rate of change of r with respect to V over the intervals $0.5 \leq V \leq 1$ and $1 \leq V \leq 1.5$.

Solution   Using the formula for the average rate of change gives

$$\begin{array}{l} \text{Average rate of change} \\ \text{of radius for } 0.5 \leq V \leq 1 \end{array} = \frac{f(1) - f(0.5)}{0.5} = 2\left(\left(\frac{3}{4\pi}\right)^{1/3} - \left(\frac{1.5}{4\pi}\right)^{1/3}\right) \approx 0.26$$

$$\begin{array}{l} \text{Average rate of change} \\ \text{of radius for } 1 \leq V \leq 1.5 \end{array} = \frac{f(1.5) - f(1)}{0.5} = 2\left(\left(\frac{4.5}{4\pi}\right)^{1/3} - \left(\frac{3}{4\pi}\right)^{1/3}\right) \approx 0.18.$$

So we see that the rate decreases as the volume increases.

Instantaneous Rate of Change: The Derivative

We can also define the *instantaneous rate of change* of a function at a point in the same way that we defined instantaneous velocity: we look at the average rate of change over smaller and smaller intervals. This instantaneous rate of change is so important that it is given its own name, the *derivative of f at a*, denoted by $f'(a)$. We define it as follows:

> The **derivative of f at a**, written $f'(a)$, is defined as
>
> $$\begin{array}{l} \text{Rate of change} \\ \text{of } f \text{ at } a \end{array} = f'(a) = \lim_{h \to 0} \frac{f(a+h) - f(a)}{h}.$$
>
> If the limit exists, then f is said to be **differentiable at a**.

To emphasize that $f'(a)$ is the rate of change of $f(x)$ as the variable x changes, we call $f'(a)$ the derivative of f *with respect to x* at $x = a$. When the function $y = s(t)$ represents the position of an object, the derivative $s'(t)$ is the velocity.

Example 2   By choosing small values for h, estimate the instantaneous rate of change of the radius of a sphere with respect to change in volume at $V = 1$.

Solution   With $h = 0.01$ and $h = -0.01$, we have the difference quotients

$$\frac{f(1.01) - f(1)}{0.01} \approx 0.2061 \quad \text{and} \quad \frac{f(0.99) - f(1)}{-0.01} \approx 0.2075.$$

With $h = 0.001$ and $h = -0.001$,

$$\frac{f(1.001) - f(1)}{0.001} \approx 0.2067 \quad \text{and} \quad \frac{f(0.999) - f(1)}{-0.001} \approx 0.2069.$$

The values of these difference quotients suggest that the limit is between 0.2067 and 0.2069. We conclude that the value is about 0.207; taking smaller h values confirms this. So we say

$$f'(1) = \begin{array}{l} \text{Instantaneous rate of change} \\ \text{of radius with respect} \\ \text{to volume at } V = 1 \end{array} \approx 0.207.$$

In this example we found an approximation to the instantaneous rate of change, or derivative, by substituting in smaller and smaller values of h. Now we see how to visualize the derivative.

Visualizing the Derivative: Slope of Curve and Slope of Tangent

As with velocity, we can visualize the derivative $f'(a)$ as the slope of the graph of f at $x = a$. In addition, there is another way to think of $f'(a)$. Consider the difference quotient $(f(a+h) - f(a))/h$. The numerator, $f(a+h) - f(a)$, is the vertical distance marked in Figure 2.12 and h is the horizontal distance, so

$$\text{Average rate of change of } f = \frac{f(a+h) - f(a)}{h} = \text{Slope of line } AB.$$

As h becomes smaller, the line AB approaches the tangent line to the curve at A. (See Figure 2.13.) We say

$$\begin{array}{c}\text{Instantaneous} \\ \text{rate of change of } f \\ \text{at } a\end{array} = \lim_{h \to 0} \frac{f(a+h) - f(a)}{h} = \text{Slope of tangent at } A.$$

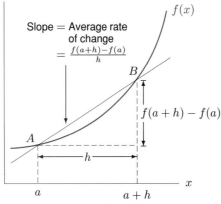

Figure 2.12: Visualizing the average rate of change of f

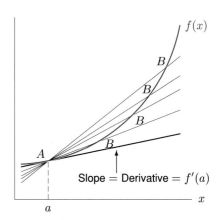

Figure 2.13: Visualizing the instantaneous rate of change of f

The derivative at point A can be interpreted as:

- The slope of the curve at A.
- The slope of the tangent line to the curve at A.

The slope interpretation is often useful in gaining rough information about the derivative, as the following examples show.

Example 3 Is the derivative of $\sin x$ at $x = \pi$ positive or negative?

Solution Looking at a graph of $\sin x$ in Figure 2.14 (remember, x is in radians), we see that a tangent line drawn at $x = \pi$ has negative slope. So the derivative at this point is negative.

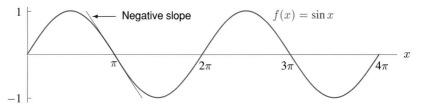

Figure 2.14: Tangent line to $\sin x$ at $x = \pi$

Recall that if we zoom in on the graph of a function $y = f(x)$ at the point where $x = a$, we usually find that the graph looks more and more like a straight line with slope $f'(a)$.

Example 4   By zooming in on the point $(0,0)$ on the graph of the sine function, estimate the value of the derivative of $\sin x$ at $x = 0$, with x in radians.

Solution   Figure 2.15 shows successive graphs of $\sin x$, with smaller and smaller scales. On the interval $-0.1 \leq x \leq 0.1$, the graph looks like a straight line of slope 1. Thus, the derivative of $\sin x$ at $x = 0$ is about 1.

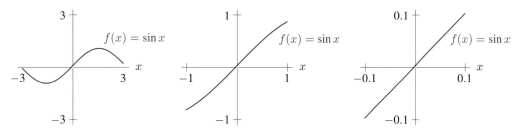

Figure 2.15: Zooming in on the graph of $\sin x$ near $x = 0$

Later we will show that the derivative of $\sin x$ at $x = 0$ is exactly 1. (See page 216 in Section 4.5.) From now on we will assume that this is so.

Example 5   Use the tangent line at $x = 0$ to estimate values of $\sin x$ near $x = 0$.

Solution   In the previous example we see that near $x = 0$, the graph of $y = \sin x$ looks like the graph of the straight line $y = x$; we can use this line to estimate values of $\sin x$ when x is close to 0. For example, the point on the straight line $y = x$ with x coordinate 0.32 is $(0.32, 0.32)$. Since the line is close to the graph of $y = \sin x$, we estimate that $\sin 0.32 \approx 0.32$. (See Figure 2.16.) Checking on a calculator, we find that $\sin 0.32 \approx 0.3146$, so our estimate is quite close. Notice that the graph suggests that the real value of $\sin 0.32$ is slightly less than 0.32.

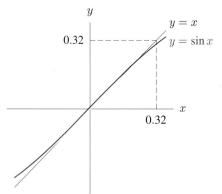

Figure 2.16: Approximating $y = \sin x$ by $y = x$

Why Do We Use Radians and Not Degrees?

After Example 4 we stated that the derivative of $\sin x$ at $x = 0$ is 1, when x is in radians. This is the reason we choose to use radians. If we had done Example 4 in degrees, the derivative of $\sin x$ would have turned out to be a much messier number. (See Problem 25, page 104.)

Estimating the Derivative of an Exponential Function

Example 6   Estimate the value of the derivative of $f(x) = 2^x$ at $x = 0$ graphically and numerically.

Solution   Graphically: Figure 2.17 indicates that the graph is concave up. Assuming this, the slope at A is between the slope of BA and the slope of AC. Since

$$\text{Slope of line } BA = \frac{(2^0 - 2^{-1})}{(0 - (-1))} = \frac{1}{2} \quad \text{and} \quad \text{Slope of line } AC = \frac{(2^1 - 2^0)}{(1 - 0)} = 1,$$

we know that the derivative is between $1/2$ and 1.

Numerically: To estimate the derivative at $x = 0$, we need to look at values of the difference quotient

$$\frac{f(0 + h) - f(0)}{h} = \frac{2^h - 2^0}{h} = \frac{2^h - 1}{h}$$

for small h. Table 2.4 shows some values of 2^h together with values of the difference quotients. (See Problem 33 on page 104 for what happens for very small values of h.)

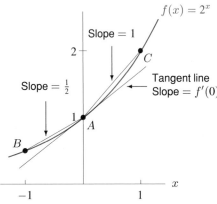

Figure 2.17: Graph of $y = 2^x$ showing the derivative at $x = 0$

TABLE 2.4   *Numerical values for difference quotient of 2^x at $x = 0$*

h	2^h	Difference quotient: $\frac{2^h - 1}{h}$
-0.0003	0.999792078	0.693075
-0.0002	0.999861380	0.693099
-0.0001	0.999930688	0.693123
0	1	
0.0001	1.00006932	0.693171
0.0002	1.00013864	0.693195
0.0003	1.00020797	0.693219

Our concavity assumption tells us that difference quotients calculated with negative h's are smaller than the derivative, and those calculated with positive h's are larger. From Table 2.4 we see that the derivative is between 0.693123 and 0.693171. To three decimal places, $f'(0) = 0.693$.

Example 7   Find an approximate equation for the tangent line to $f(x) = 2^x$ at $x = 0$.

Solution   From the previous example, we know the slope of the tangent line is about 0.693. Since we also know the line has y-intercept 1, its equation is

$$y = 0.693x + 1.$$

Computing the Derivative of x^2

In previous examples we got an approximation to the derivative by using smaller and smaller values of h. We now see how to find the derivative of x^2 exactly.

Example 8   Find the derivative of the function $f(x) = x^2$ at the point $x = 1$.

Solution   We need to look at

$$f'(1) = \lim_{h \to 0} \frac{f(1+h) - f(1)}{h}.$$

This is the same as

$$\lim_{h \to 0} \frac{(1+h)^2 - 1^2}{h} = \lim_{h \to 0} \frac{(1 + 2h + h^2) - 1}{h} = \lim_{h \to 0} \frac{2h + h^2}{h}.$$

Since the limit only examines values of h close to, but not equal to zero, we can divide by h in the expression $(2h + h^2)/h$. We get

$$\lim_{h \to 0} \frac{h(2 + h)}{h} = \lim_{h \to 0} (2 + h).$$

This limit is 2, so $f'(1) = 2$. At $x = 1$ the rate of change of x^2 is 2.

Since the derivative is the rate of change, $f'(1) = 2$ means that for small changes in x near $x = 1$, the change in $f(x) = x^2$ is about twice as big as the change in x. As an example, if x changes from 1 to 1.1, a net change of 0.1, then $f(x)$ changes by about 0.2. Figure 2.18 shows this geometrically.

Table 2.5 shows the derivative of $f(x) = x^2$ numerically. Notice that near $x = 1$, every time the value of x increases by 0.001, the value of x^2 increases by approximately 0.002. Near $x = 1$ the function is approximately linear with slope $0.002/0.001 = 2$.

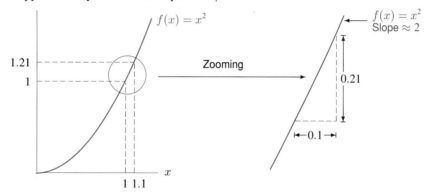

Figure 2.18: Graph of $f(x) = x^2$ near $x = 1$ has slope ≈ 2

TABLE 2.5   *Values of $f(x) = x^2$ near $x = 1$*

x	x^2	Difference in successive x^2 values
0.998	0.996004	
		0.001997
0.999	0.998001	
		0.001999
1.000	1.000000	
		0.002001
1.001	1.002001	
		0.002003
1.002	1.004004	
↑		↑
x increments of 0.001		All approximately 0.002

Problems for Section 2.2

1. Sketch a rough graph of $f(x) = \sin x$, and use the graph to decide whether the derivative of $f(x)$ at $x = 3\pi$ is positive or negative. Give reasons for your decision.

2. (a) Make a table of values rounded to two decimal places for the function $f(x) = e^x$ for $x = 1$, 1.5, 2, 2.5, and 3. Then use the table to answer parts (b) and (c).
    (b) Find the average rate of change of $f(x)$ between $x = 1$ and $x = 3$.
    (c) Use average rates of change to approximate the instantaneous rate of change of $f(x)$ at $x = 2$.

3. Label points A, B, C, D, E, and F on the graph of $y = f(x)$ in Figure 2.19.

    (a) Point A is a point on the curve where the derivative is negative.
    (b) Point B is a point on the curve where the value of the function is negative.
    (c) Point C is a point on the curve where the derivative is largest.
    (d) Point D is a point on the curve where the derivative is zero.
    (e) Points E and F are different points on the curve where the derivative is about the same.

Figure 2.19

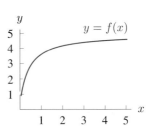

Figure 2.20

4. The graph of $y = f(x)$ is shown in Figure 2.20. Which is larger in each of the following pairs?

    (a) Average rate of change: Between $x = 1$ and $x = 3$? Or between $x = 3$ and $x = 5$?
    (b) $f(2)$ or $f(5)$?
    (c) $f'(1)$ or $f'(4)$?

5. On a copy of Figure 2.21, mark lengths that represent the quantities in parts (a) – (d). (Pick any convenient x, and assume $h > 0$.)
    (a) $f(x)$ (b) $f(x + h)$ (c) $f(x + h) - f(x)$ (d) h

    (e) Using your answers to parts (a)–(d), show how the quantity $\dfrac{f(x + h) - f(x)}{h}$ can be represented as the slope of a line on the graph.

Figure 2.21

Figure 2.22

6. On a copy of Figure 2.22, mark lengths that represent the quantities in parts (a) – (d). (Pick any convenient x, and assume $h > 0$.)
    (a) $f(x)$ (b) $f(x + h)$ (c) $f(x + h) - f(x)$ (d) h

    (e) Using your answers to parts (a)–(d), show how the quantity $\dfrac{f(x + h) - f(x)}{h}$ can be represented as the slope of a line on the graph.

7. Show how you can represent the following on a copy of Figure 2.23.

    (a) $f(4)$ (b) $f(4) - f(2)$ (c) $\dfrac{f(5) - f(2)}{5 - 2}$ (d) $f'(3)$

8. Consider the function $y = f(x)$ shown in Figure 2.23. For each of the following pairs of numbers, decide which is larger. Explain your answer.

    (a) $f(3)$ or $f(4)$?   (b) $f(3) - f(2)$ or $f(2) - f(1)$?

    (c) $\dfrac{f(2) - f(1)}{2 - 1}$ or $\dfrac{f(3) - f(1)}{3 - 1}$?   (d) $f'(1)$ or $f'(4)$?

9. With the function f given by Figure 2.23, arrange the following quantities in ascending order:

    $$0, \quad 1, \quad f'(2), \quad f'(3), \quad f(3) - f(2)$$

10. Suppose $y = f(x)$ graphed in Figure 2.23 represents the cost of manufacturing x kilograms of a chemical. Then $f(x)/x$ represents the average cost of producing 1 kilogram when x kilograms are made. This problem asks you to visualize these averages graphically.

    (a) Show how to represent $f(4)/4$ as the slope of a line.
    (b) Which is larger, $f(3)/3$ or $f(4)/4$?

Figure 2.23

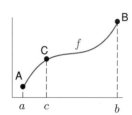

Figure 2.24

11. Consider the function shown in Figure 2.24.

    (a) Write an expression involving f for the slope of the line joining A and B.
    (b) Draw the tangent line at C. Compare its slope to the slope of the line in part (a).
    (c) Are there any other points on the curve at which the slope of the tangent line is the same as the slope of the tangent line at C? If so, mark them on the graph. If not, why not?

12. Table 2.6 shows values of $f(x) = x^3$ near $x = 2$ (to three decimal places). Use it to estimate $f'(2)$.

    TABLE 2.6

x	1.998	1.999	2.000	2.001	2.002
x^3	7.976	7.988	8.000	8.012	8.024

Find the derivatives in Problem 13–18 algebraically.

13. $g(t) = 3t^2 + 5t$ at $t = -1$ 14. $f(x) = 5x^2$ at $x = 10$ 15. $f(x) = x^3$ at $x = -2$

16. $f(x) = x^3 + 5$ at $x = 1$ 17. $g(x) = 1/x$ at $x = 2$ 18. $g(z) = z^{-2}$, find $g'(2)$

For Problems 19–22, find the equation of the line tangent to the function at the given point.

19. $f(x) = x^3$ at $x = -2$ 20. $f(x) = 5x^2$ at $x = 10$

21. $f(x) = x$ at $x = 20$ 22. $f(x) = 1/x^2$ at $(1, 1)$

23. Find the value of the derivative of $f(x) = x^2 + 1$ at $x = 3$ algebraically. Find the equation of the tangent line to f at $x = 3$.

24. Find the equation of the tangent line to $f(x) = x^2 + x$ at $x = 3$.
    Sketch a graph of the function and this tangent line.

25. (a) Estimate $f'(0)$ if $f(x) = \sin x$, with x in degrees.
    (b) In Example 4 on page 99, we found that the derivative of $\sin x$ at $x = 0$ was 1. Why do we get a different result here? (This problem shows why radians are almost always used in calculus.)

26. Estimate the derivative of $f(x) = x^x$ at $x = 2$.

27. For $y = f(x) = 3x^{3/2} - x$, use your calculator to construct a graph of $y = f(x)$, for $0 \le x \le 2$. From your graph, estimate $f'(0)$ and $f'(1)$.

28. Let $f(x) = \ln(\cos x)$. Use your calculator to approximate the instantaneous rate of change of f at the point $x = 1$. Do the same thing for $x = \pi/4$. (Note: Be sure that your calculator is set in radians.)

29. There is a function called the error function, $y = \text{erf}(x)$. Suppose your calculator has a button for $\text{erf}(x)$ that gives the following values:

    $$\text{erf}(0) = 0 \quad \text{erf}(1) = 0.84270079 \quad \text{erf}(0.1) = 0.11246292 \quad \text{erf}(0.01) = 0.01128342.$$

    (a) Use all this information to determine your best estimate for $\text{erf}'(0)$. (Give only those digits of which you feel reasonably certain.)
    (b) Suppose you find that $\text{erf}(0.001) = 0.00112838$. How does this extra information change your answer to part (a)?

30. (a) Use your calculator to approximate the derivative of the hyperbolic sine function (written $\sinh x$) at the points 0, 0.3, 0.7, and 1.
    (b) Can you find a relation between the values of this derivative and the values of the hyperbolic cosine (written $\cosh x$)?

31. The population, P, of China, in billions, can be approximated by the function

    $$P = 1.15(1.014)^t,$$

    where t is the number of years since the start of 1993. According to this model, how fast is the population growing at the start of 1993 and at the start of 1995? Give your answers in millions of people per year.

32. (a) Sketch graphs of the functions $f(x) = \frac{1}{2}x^2$ and $g(x) = f(x) + 3$ on the same set of axes. What can you say about the slopes of the tangent lines to the two graphs at the point $x = 0$? $x = 2$? Any point $x = x_0$?
    (b) Explain why adding a constant value, C, to any function does not change the value of the slope of its graph at any point. [Hint: Let $g(x) = f(x) + C$, and calculate the difference quotients for f and g.]

33. Suppose Table 2.4 on page 100 is continued with smaller values of h. A particular calculator gives the results in Table 2.7. (Your calculator may give slightly different results.) Comment on the values of the difference quotient in Table 2.7. In particular, why is the last value of $(2^h - 1)/h$ zero? What do you expect the calculated value of $(2^h - 1)/h$ to be when $h = 10^{-20}$?

TABLE 2.7 *Questionable values of difference quotients of 2^x near $x = 0$*

h	Difference quotient: $(2^h - 1)/h$
10^{-4}	0.6931712
10^{-6}	0.693147
10^{-8}	0.6931
10^{-10}	0.69
10^{-12}	0

2.3 THE DERIVATIVE FUNCTION

In the last section we looked at the derivative of a function at a fixed point. Now we consider what happens at a variety of points. The derivative generally takes on different values at different points and is itself a function.

First, remember that the derivative of a function at a point tells us the rate at which the value of the function is changing at that point. Geometrically, if we "zoom in" on a point in the graph until it looks like a straight line, the slope of that line is the derivative at that point. Equivalently, we can think of the derivative as the slope of the tangent line at the point, because as we "zoom in," the curve and the tangent line become indistinguishable.

Example 1 Estimate the derivative of the function $f(x)$ graphed in Figure 2.25 at $x = -2, -1, 0, 1, 2, 3, 4, 5$.

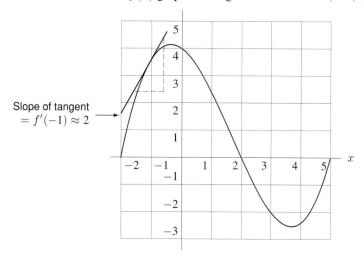

Slope of tangent
$= f'(-1) \approx 2$

Figure 2.25: Estimating the derivative graphically as the slope of the tangent line

Solution From the graph we estimate the derivative at any point by placing a straightedge so that it forms the tangent line at that point, and then using the grid squares to estimate the slope of the straightedge. For example, the tangent at $x = -1$ is drawn in Figure 2.25, and has a slope of about 2, so $f'(-1) \approx 2$. Notice that the slope at $x = -2$ is positive and fairly large; the slope at $x = -1$ is positive but smaller. At $x = 0$, the slope is negative, by $x = 1$ it has become more negative, and so on. Some estimates of the derivative are listed in Table 2.8. You should check these values. Are they reasonable? Is the derivative positive where you expect? Negative?

TABLE 2.8 *Estimated values of derivative of function in Figure 2.25*

x	-2	-1	0	1	2	3	4	5
$f'(x)$	6	2	-1	-2	-2	-1	1	4

The important point to notice is that for every x-value, there's a corresponding value of the derivative. Therefore, the derivative is itself a function of x.

> For any function f, we define the **derivative function**, f', by
> $$f'(x) = \text{Rate of change of } f \text{ at } x = \lim_{h \to 0} \frac{f(x+h) - f(x)}{h}.$$

For every x-value for which this limit exists, we say f is *differentiable at* that x-value. If the limit exists for all x in the domain of f, we say f is *differentiable everywhere*. The functions we shall deal with will be differentiable at every point in their domain, except perhaps for a few isolated points.

The Derivative Function: Graphically

Example 2   Sketch the graph of the derivative of the function shown in Figure 2.25.

Solution   We plot the values of this derivative given in Table 2.8. We obtain Figure 2.26, which shows a graph of the derivative (the black curve), along with the original function (color).

You should check for yourself that this graph of f' makes sense. The values of f' are positive where f is increasing ($x < -0.3$ or $x > 3.8$) and negative where f is decreasing. Notice that at the points where f has large positive slope, such as $x = -2$, the graph of the derivative is far above the x-axis, as it should be, since the value of the derivative is large there. On the other hand, at points where the slope is gentler, such as $x = -1$, the graph of f' is closer to the x-axis, since the derivative is smaller.

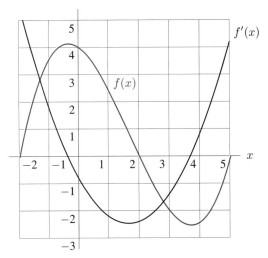

Figure 2.26: Function (colored) and derivative (black) from Example 1

What Does the Derivative Tell Us Graphically?

When f' is positive, the tangent line to f is sloping up; when f' is negative, the tangent line to f is sloping down. If $f' = 0$ everywhere, then the tangent line to f is horizontal everywhere, and so f is constant. We see that the sign of f' tells us whether f is increasing or decreasing.

> If $f' > 0$ on an interval, then f is *increasing* over that interval.
> If $f' < 0$ on an interval, then f is *decreasing* over that interval.

Moreover, the magnitude of the derivative gives us the magnitude of the rate of change; so if f' is large (positive or negative), then the graph of f is steep (up or down), whereas if f' is small the graph of f slopes gently. With this in mind, we can deduce a lot about the behavior of a function from the behavior of its derivative.

The Derivative Function: Numerically

If we are given a table of function values instead of a graph of the function, we can estimate values of the derivative.

Example 3   Table 2.9 gives values of $c(t)$, the concentration (μg/cm^3) of a drug in the bloodstream at time t (min). Construct a table of estimated values for $c'(t)$, the rate of change of $c(t)$ with respect to time.

TABLE 2.9   *Concentration as a function of time*

t (min)	0	0.1	0.2	0.3	0.4	0.5	0.6	0.7	0.8	0.9	1.0
$c(t)$ (μg/cm^3)	0.84	0.89	0.94	0.98	1.00	1.00	0.97	0.90	0.79	0.63	0.41

Solution   We want to estimate values of c' using the values in the table. To do this, we have to assume that the data points are close enough together that the concentration doesn't change wildly between them. From the table, we can see that the concentration is increasing between $t = 0$ and $t = 0.4$, so we expect a positive derivative there. However, the increase is quite slow, so we expect the derivative to be small. The concentration doesn't change between 0.4 and 0.5, so we expect the derivative to be roughly 0 there. From $t = 0.5$ to $t = 1.0$, the concentration starts to decrease, and the rate of decrease gets larger and larger, so we expect the derivative to be negative and of greater and greater magnitude.

Using the data in the table, we can estimate the derivative using the difference quotient:

$$c'(t) \approx \frac{c(t + h) - c(t)}{h}.$$

Since the data points are 0.1 apart, we use $h = 0.1$. We estimate that,

$$c'(0) \approx \frac{c(0.1) - c(0)}{0.1} = \frac{0.89 - 0.84}{0.1} = 0.5 \ \mu\text{g/cm}^3\text{/min}$$

$$c'(0.1) \approx \frac{c(0.2) - c(0.1)}{0.1} = \frac{0.94 - 0.89}{0.1} = 0.5 \ \mu\text{g/cm}^3\text{/min}$$

$$c'(0.2) \approx \frac{c(0.3) - c(0.2)}{0.1} = \frac{0.98 - 0.94}{0.1} = 0.4 \ \mu\text{g/cm}^3\text{/min}$$

$$c'(0.3) \approx \frac{c(0.4) - c(0.3)}{0.1} = \frac{1.00 - 0.98}{0.1} = 0.2 \ \mu\text{g/cm}^3\text{/min}$$

$$c'(0.4) \approx \frac{c(0.5) - c(0.4)}{0.1} = \frac{1.00 - 1.00}{0.1} = 0.0 \ \mu\text{g/cm}^3\text{/min}$$

and so on. These values are tabulated in Table 2.10. Notice that the derivative has small positive values up until $t = 0.4$, where it is roughly 0, and then it gets more and more negative, as we expected. The slopes are shown on the graph of $c(t)$ in Figure 2.27.

TABLE 2.10
Estimated derivative of concentration

t	$c'(t)$
0	0.5
0.1	0.5
0.2	0.4
0.3	0.2
0.4	0.0
0.5	-0.3
0.6	-0.7
0.7	-1.1
0.8	-1.6
0.9	-2.2

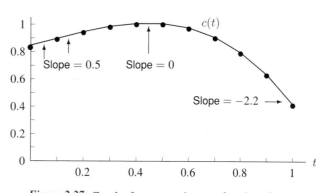

Figure 2.27: Graph of concentration as a function of time

Improving Numerical Estimates for the Derivative

In the previous example, the estimate for the derivative at 0.2 used the interval to the right; we found the average rate of change between $t = 0.2$ and $t = 0.3$. However, we could equally well have gone to the left and used the rate of change between $t = 0.1$ and $t = 0.2$ to approximate the derivative at 0.2. For a more accurate result, we could average these slopes and say

$$c'(0.2) \approx \frac{1}{2} \left(\begin{array}{c} \text{Slope to left} \\ \text{of } 0.2 \end{array} + \begin{array}{c} \text{Slope to right} \\ \text{of } 0.2 \end{array} \right) = \frac{0.5 + 0.4}{2} = 0.45.$$

In general, averaging the slopes leads to a more accurate answer.

Derivative Function: From a Formula

If we are given a formula for f, can we come up with a formula for f'? We often can, as shown in the next example. Indeed, much of the power of calculus depends on our ability to find formulas for the derivatives of all the functions we described earlier. This is done systematically in Chapter 4.

Derivative of a Constant Function

The graph of a constant function $f(x) = k$ is a horizontal line, with a slope of 0 everywhere. Therefore, its derivative is 0 everywhere. (See Figure 2.28.)

$$\boxed{\text{If } f(x) = k, \text{ then } f'(x) = 0.}$$

Figure 2.28: A constant function

Derivative of a Linear Function

We already know that the slope of a straight line is constant. This tells us that the derivative of a linear function is constant.

$$\boxed{\text{If } f(x) = b + mx, \text{ then } f'(x) = \text{Slope} = m.}$$

Derivative of a Power Function

Example 4   Find a formula for the derivative of $f(x) = x^2$.

Solution   Before computing the formula for $f'(x)$ algebraically, let's try to guess the formula by looking for a pattern in the values of $f'(x)$. Table 2.11 contains values of $f(x) = x^2$ (rounded to three decimals), which we can use to estimate the values of $f'(1)$, $f'(2)$, and $f'(3)$.

TABLE 2.11   *Values of $f(x) = x^2$ near $x = 1$, $x = 2$, $x = 3$ (rounded to three decimals)*

x	x^2		x	x^2		x	x^2
0.999	0.998		1.999	3.996		2.999	8.994
1.000	1.000		2.000	4.000		3.000	9.000
1.001	1.002		2.001	4.004		3.001	9.006
1.002	1.004		2.002	4.008		3.002	9.012

Near $x = 1$, the value of x^2 increases by about 0.002 each time x increases by 0.001, so

$$f'(1) \approx \frac{0.002}{0.001} = 2.$$

Similarly, near $x = 2$ and $x = 3$, the value of x^2 increases by about 0.004 and 0.006, respectively, when x increases by 0.001. So

$$f'(2) \approx \frac{0.004}{0.001} = 4 \qquad \text{and} \qquad f'(3) \approx \frac{0.006}{0.001} = 6.$$

Knowing the value of f' at specific points can never tell us the formula for f', but it certainly can be suggestive: Knowing $f'(1) \approx 2$, $f'(2) \approx 4$, $f'(3) \approx 6$ suggests that $f'(x) = 2x$.

The derivative is calculated by forming the difference quotient and taking the limit as h goes to zero. The difference quotient is

$$\frac{f(x+h) - f(x)}{h} = \frac{(x+h)^2 - x^2}{h} = \frac{x^2 + 2xh + h^2 - x^2}{h} = \frac{2xh + h^2}{h}.$$

Since h never actually reaches zero, we can divide it out in the last expression to get $2x + h$. The limit of this as h goes to zero is $2x$, so

$$f'(x) = \lim_{h \to 0}(2x + h) = 2x.$$

Example 5   Calculate $f'(x)$ if $f(x) = x^3$.

Solution   We look at the difference quotient

$$\frac{f(x+h) - f(x)}{h} = \frac{(x+h)^3 - x^3}{h}.$$

Multiplying out gives $(x + h)^3 = x^3 + 3x^2h + 3xh^2 + h^3$, so

$$f'(x) = \lim_{h \to 0} \frac{x^3 + 3x^2h + 3xh^2 + h^3 - x^3}{h} = \lim_{h \to 0} \frac{3x^2h + 3xh^2 + h^3}{h}.$$

Since in taking the limit as $h \to 0$, we consider values of h near, but not equal to, zero, we can divide by h giving

$$f'(x) = \lim_{h \to 0} \frac{3x^2h + 3xh^2 + h^3}{h} = \lim_{h \to 0}(3x^2 + 3xh + h^2).$$

As $h \to 0$, the value of $(3xh + h^2) \to 0$ so

$$f'(x) = \lim_{h \to 0}(3x^2 + 3xh + h^2) = 3x^2.$$

The previous two examples show how to compute the derivatives of power functions of the form $f(x) = x^n$, when n is 2 or 3. We can use the Binomial Theorem (from the Focus on Theory section on page 78) to show that, for n a positive integer,

$$\boxed{\text{If } f(x) = x^n \text{ then } f'(x) = nx^{n-1}.}$$

This result is in fact valid for any real value of n.

Problems for Section 2.3

For Problems 1–9, sketch a graph of the derivative function of each of the given functions.

1.

2.

3.

4.

5.

6.

7.

8.

9.

10. (a) Sketch a smooth curve whose slope is both everywhere positive and increasing gradually.
    (b) Sketch a smooth curve whose slope is both everywhere positive and decreasing gradually.
    (c) Sketch a smooth curve whose slope is both everywhere negative and increasing gradually (i.e., becoming less and less negative).
    (d) Sketch a smooth curve whose slope is both everywhere negative and decreasing gradually (i.e., becoming more and more negative).

11. Given the numerical values shown, find approximate values for the derivative of $f(x)$ at each of the x-values given. Where is the rate of change of $f(x)$ positive? Where is it negative? Where does the rate of change of $f(x)$ seem to be greatest?

x	0	1	2	3	4	5	6	7	8
$f(x)$	18	13	10	9	9	11	15	21	30

12. The values of x and the corresponding values of $g(x)$ are given in the table. For what value of x is $g'(x)$ closest to 3?

x	2.7	3.2	3.7	4.2	4.7	5.2	5.7	6.2
$g(x)$	3.4	4.4	5.0	5.4	6.0	7.4	9.0	11.0

Find a formula for the derivatives of the functions in Problems 13–16 algebraically.

13. $g(x) = 2x^2 - 3$ 14. $k(x) = 1/x$ 15. $l(x) = 1/x^2$ 16. $m(x) = 1/(x+1)$

17. Draw the graph of a continuous function $y = f(x)$ that satisfies the following three conditions.
   - $f'(x) > 0$ for $x < -2$,
   - $f'(x) < 0$ for $-2 < x < 2$,
   - $f'(x) = 0$ for $x > 2$.

18. Draw the graph of a continuous function $y = f(x)$ that satisfies the following three conditions.
   - $f'(x) > 0$ for $-\frac{\pi}{2} < x < \frac{\pi}{2}$,
   - $f'(x) < 0$ for $-\pi < x < -\frac{\pi}{2}$ and $\frac{\pi}{2} < x < \pi$,
   - $f'(x) = 0$ at $x = -\frac{\pi}{2}$ and $x = \frac{\pi}{2}$.

For Problems 19–22, sketch the graph of $f(x)$, and use this graph to sketch the graph of $f'(x)$.

19. $f(x) = x^2$ 20. $f(x) = x(x-1)$ 21. $f(x) = \cos x$ 22. $f(x) = \log x$

For Problems 23–28, sketch the graph of $y = f'(x)$ for the function given.

23.

24.

25.

26.

27.

28.

29. In the graph of f in Figure 2.29, at which of the labeled x-values is

   (a) $f(x)$ greatest?   (b) $f(x)$ least?
   (c) $f'(x)$ greatest?   (d) $f'(x)$ least?

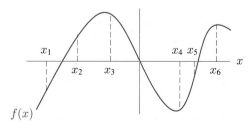

Figure 2.29

30. Consider a vehicle moving along a straight road. Suppose $f(t)$ gives the vehicle's distance from its starting point at time t. Which of the graphs in Figure 2.30 could be $f'(t)$ for the following scenarios? (Assume the scales on the vertical axes are all the same.)

    (a) A bus on a popular route, with no traffic
    (b) A car with no traffic and all green lights
    (c) A car in heavy traffic conditions

Figure 2.30

31. A child inflates a balloon, admires it for a while and then lets the air out at a constant rate. If $V(t)$ gives the volume of the balloon at time t, then Figure 2.31 shows $V'(t)$ as a function of t. At what time does the child:

    (a) Begin to inflate the balloon?
    (b) Finish inflating the balloon?
    (c) Begin to let the air out?
    (d) What would the graph of $V'(t)$ look like if the child had alternated between pinching and releasing the open end of the balloon, instead of letting the air out at a constant rate?

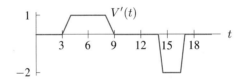

Figure 2.31

32. The population of a herd of deer is modeled by

    $$P(t) = 4000 + 500 \sin\left(2\pi t - \frac{\pi}{2}\right)$$

    where t is measured in years from January 1.

    (a) How does this population vary with time? Sketch a graph of $P(t)$ for one year.
    (b) Use the graph to decide when in the year the population is a maximum. What is that maximum? Is there a minimum? If so, when?
    (c) Use the graph to decide when the population is growing fastest. When is it decreasing fastest?
    (d) Estimate roughly how fast the population is changing on the first of July.

33. Looking at the graph, explain why if $f(x)$ is an even function, then $f'(x)$ is odd.

34. Looking at the graph, explain why if $g(x)$ is an odd function, then $g'(x)$ is even.

2.4   INTERPRETATIONS OF THE DERIVATIVE

We have already seen how the derivative can be interpreted as a slope and as a rate of change. In this section, we will see examples of other interpretations. The purpose of these examples is not to make a catalog of interpretations but to illustrate the process of obtaining them.

An Alternative Notation for the Derivative

So far we have used the notation f' to stand for the derivative of the function f. An alternative notation for derivatives was introduced by the German mathematician Wilhelm Gottfried Leibniz (1646–1716) when calculus was first being developed. If the variable y depends on the variable x, that is, if

$$y = f(x),$$

then he wrote dy/dx for the derivative, so

$$\frac{dy}{dx} = f'(x).$$

Leibniz's notation is quite suggestive, especially if we think of the letter d in dy/dx as standing for "small difference in . . . ." The notation dy/dx reminds us that the derivative is a limit of ratios of the form

$$\frac{\text{Difference in } y\text{-values}}{\text{Difference in } x\text{-values}}.$$

The notation dy/dx is useful for determining the units for the derivative: The units for dy/dx are the units for y divided by the units for x. The separate entities dy and dx officially have no independent meaning: they are all part of one notation. In fact, a good formal way to view the notation dy/dx is to think of d/dx as a single symbol meaning "the derivative with respect to x of . . .". So dy/dx can be viewed as

$$\frac{d}{dx}(y), \quad \text{meaning "the derivative with respect to } x \text{ of } y\text{."}$$

On the other hand, many scientists and mathematicians really do think of dy and dx as separate entities representing "infinitesimally" small differences in y and x, even though it is difficult to say exactly how small "infinitesimal" is. Although not formally correct, it can be very helpful to think intuitively of dy/dx as a very small change in y divided by a very small change in x.

For example, recall that if $s = f(t)$ is the position of a moving object at time t, then $v = f'(t)$ is the velocity of the object at time t. Writing

$$v = \frac{ds}{dt}$$

reminds us that v is a velocity, since the notation suggests a distance, ds, over a time, dt, and we know that distance over time is velocity. Similarly, we recognize

$$\frac{dy}{dx} = f'(x)$$

as the slope of the graph of $y = f(x)$ by remembering that slope is vertical rise, dy, over horizontal run, dx.

The disadvantage of Leibniz's notation is that it is awkward to specify the x-value at which we are evaluating the derivative. To specify $f'(2)$, for example, we have to write

$$\left.\frac{dy}{dx}\right|_{x=2}.$$

Using Units to Interpret the Derivative

The following examples illustrate how useful units can be in suggesting interpretations of the derivative.

For example, suppose $s = f(t)$ gives the distance, in meters, of a body from a fixed point as a function of time, t, in seconds. Then knowing that

$$\left.\frac{ds}{dt}\right|_{t=2} = f'(2) = 10 \text{ meters/sec}$$

tells us that when $t = 2$ sec, the body is moving at an instantaneous velocity of 10 meters/sec. This means that if the body continued to move at this speed for a whole second, it would cover 10 meters. In practice, however, the velocity of the body may be changing and so may not remain 10 meters/sec for long. Notice that the units of instantaneous velocity and of average velocity are the same. The units of the instantaneous and the average rate of change are *always* the same.

Example 1   The cost C (in dollars) of building a house A square feet in area is given by the function $C = f(A)$. What is the practical interpretation of the function $f'(A)$?

Solution   In the alternative notation,

$$f'(A) = \frac{dC}{dA}.$$

This is a cost divided by an area, so it is measured in dollars per square foot. You can think of dC as the extra cost of building an extra dA square feet of house. Then you can think of dC/dA as the additional cost per square foot. So if you are planning to build a house roughly A square feet in area, $f'(A)$ is the cost per square foot of the *extra* area involved in building a slightly larger house, and is called the *marginal cost*. The marginal cost is probably smaller than the average cost per square foot for the entire house, since once you are already set up to build a large house, the cost of adding a few square feet is likely to be small.

Example 2   The cost of extracting T tons of ore from a copper mine is $C = f(T)$ dollars. What does it mean to say that $f'(2000) = 100$?

Solution   In the alternative notation,

$$f'(2000) = \left.\frac{dC}{dT}\right|_{T=2000} = 100.$$

Since C is measured in dollars and T is measured in tons, dC/dT must be measured in dollars per ton. So the statement $f'(2000) = 100$ says that when 2000 tons of ore have already been extracted from the mine, the cost of extracting the next ton is approximately \$100.

Example 3   If $q = f(p)$ gives the number of pounds of sugar produced when the price per pound is p dollars, then what are the units and the meaning of

$$f'(3) = 50?$$

Solution   Since $f'(3)$ is the limit as $h \to 0$ of

$$\frac{f(3+h) - f(3)}{h}$$

and $f(3 + h) - f(3)$ is in pounds, whereas h is in dollars, the units of the difference quotient are pounds/dollar. Since $f'(3)$ is the limit of the difference quotient, its units are also pounds/dollar. The statement

$$f'(3) = 50 \text{ pounds/dollar}$$

tells us that the rate of change of q with respect to p is 50 when $p = 3$. Rephrasing, this means that when the price is \$3, the quantity produced is increasing at 50 pounds/dollar. This is an instantaneous rate of change, meaning that, if the rate were to remain 50 pounds/dollar, and if the price were to increase by a whole dollar, the quantity produced would increase by approximately 50 pounds. In fact, the rate probably doesn't remain constant and so the quantity produced would probably not be exactly 50 pounds. Notice that the units of the derivative and of the average rate of change are again the same.

Example 4   You are told that water is flowing through a pipe at a constant rate of 10 cubic feet per second. Interpret this rate as the derivative of some function.

Solution   You might think at first that the statement has something to do with the velocity of the water, but in fact a flow rate of 10 cubic feet per second could be achieved either with very slowly moving water through a large pipe, or with very rapidly moving water through a narrow pipe. If we look at the units—cubic feet per second—we realize that we are being given the rate of change of a quantity measured in cubic feet. But a cubic foot is a measure of volume, so we are being told the rate of change of a volume. One way to visualize this is to imagine all the water that is flowing through the pipe ending up in a tank somewhere. Let $V(t)$ be the volume of water in the tank at time t. Then we are being told that the rate of change of $V(t)$ is 10, or

$$V'(t) = \frac{dV}{dt} = 10.$$

Example 5   Suppose $P = f(t)$ is the population of Mexico in millions, where t is the number of years since 1980. Explain the meaning of the statements:

(a) $f'(6) = 2$ (b) $f^{-1}(95.5) = 16$ (c) $(f^{-1})'(95.5) = 0.46$

Solution   (a)   The units of P are millions of people, the units of t are years, so the units of $f'(t)$ are millions of people per year. Therefore the statement $f'(6) = 2$ tells us that at $t = 6$ (that is, in 1986), the population of Mexico was increasing at 2 million people per year.

(b)   The statement $f^{-1}(95.5) = 16$ tells us that the year when the population was 95.5 million was $t = 16$ (that is, in 1996).

(c)   The units of the derivative $(f^{-1})'(P)$ are years per million of population. The statement $(f^{-1})'(95.5) = 0.46$ tells us that when the population was 95.5 million, it took about 0.46 years for the population to increase by 1 million.

Problems for Section 2.4

1.   Let $f(x)$ be the elevation in feet of the Mississippi river x miles from its source. What are the units of $f'(x)$? What can you say about the sign of $f'(x)$?

2.   The temperature, T, in degrees Fahrenheit, of a cold yam placed in a hot oven is given by $T = f(t)$, where t is the time in minutes since the yam was put in the oven.

(a)   What is the sign of $f'(t)$? Why?
(b)   What are the units of $f'(20)$? What is the practical meaning of the statement $f'(20) = 2$?

3.   Suppose $P(t)$ is the monthly payment, in dollars, on a mortgage which will take t years to pay off. What are the units of $P'(t)$? What is the practical meaning of $P'(t)$? What is its sign?

4.   Suppose $C(r)$ is the total cost of paying off a car loan borrowed at an annual interest rate of $r\%$. What are the units of $C'(r)$? What is the practical meaning of $C'(r)$? What is its sign?

5.   After investing \$1000 at an annual interest rate of 7% compounded continuously for t years, your balance is \$$B$, where $B = f(t)$. What are the units of dB/dt? What is the financial interpretation of dB/dt?

6.   Investing \$1000 at an annual interest rate of $r\%$, compounded continuously, for 10 years gives you a balance of \$$B$, where $B = g(r)$. What is a financial interpretation of the statements

(a) $g(5) \approx 1649$?
(b) $g'(5) \approx 165$? What are the units of $g'(5)$?

7.   An ice cream company knows that the cost, C (in dollars), to produce g quarts of cookie dough ice cream is a function of g, so $C = f(g)$.

(a)   If $f(200) = 70$, what are the units of the 200? What are the units of the 70? Explain clearly what this equation is telling you.
(b)   If $f'(200) = 3$, what are the units of the 200? What are the units of the 3? Explain clearly what this equation is telling you.

8. An economist is interested in how the price of a certain item affects its sales. Suppose that at a price of $\$p$, a quantity, q, of the item is sold. If $q = f(p)$, explain the meaning of each of the following statements:

   (a) $f(150) = 2000$ \qquad\qquad\qquad (b) $f'(150) = -25$

9. Let $p(h)$ be the pressure in dynes per cm^2 on a diver at a depth of h meters below the surface of the ocean. What do each of the following quantities mean to the diver? Give units for the quantities.

   (a) $p(100)$ \qquad\qquad\qquad\qquad (b) h such that $p(h) = 1.2 \cdot 10^6$

   (c) $p(h) + 20$ \qquad\qquad\qquad\qquad (d) $p(h + 20)$

   (e) $p'(100)$ \qquad\qquad\qquad\qquad (f) h such that $p'(h) = 20$

10. Let $f(t)$ be the number of centimeters of rainfall that has fallen since midnight, where t is the time in hours. Interpret the following in practical terms, giving units.

    (a) $f(10) = 3.1$ \qquad (b) $f^{-1}(10) = 16$ \qquad (c) $f'(8) = 0.4$ \qquad (d) $(f^{-1})'(5) = 2$

11. Let W be the amount of water, in gallons, in a bathtub at time t, in minutes.

    (a) What are the meaning and units of dW/dt?
    (b) Suppose the bathtub is full of water at time t_0, so that $W(t_0) > 0$. Subsequently, at time $t_p > t_0$, the plug is pulled. Is dW/dt positive, negative, or zero:
      (i) For $t_0 < t < t_p$?
      (ii) After the plug is pulled?
      (iii) When all the water has drained from the tub?

12. Suppose w is the weight of a stack of papers in an open container, and t represents time. A lighted match is thrown into the stack at time $t = 0$.

    (a) What is the sign of dw/dt for t in the period of time while the stack of paper is in the process of burning?
    (b) What behavior of dw/dt indicates that the fire is no longer burning?
    (c) If the fire starts small but increases in vigor over a certain time interval, then does dw/dt increase or decrease over that interval? What about $|dw/dt|$?

13. If $g(v)$ is the fuel efficiency, in miles per gallon, of a car going at v miles per hour, what are the units of $g'(55)$? What is the practical meaning of the statement $g'(55) = -0.54$?

14. (a) If you jump out of an airplane without a parachute, you fall faster and faster until wind resistance causes you to approach a steady velocity, called the *terminal* velocity. Sketch a graph of your velocity against time.
    (b) Explain the concavity of your graph.
    (c) Assuming wind resistance to be negligible at $t = 0$, what natural phenomenon is represented by the slope of the graph at $t = 0$?

15. A company's revenue from car sales, C (measured in thousands of dollars), is a function of advertising expenditure, a, also measured in thousands of dollars. Suppose $C = f(a)$.

    (a) What does the company hope is true about the sign of f'?
    (b) What does the statement $f'(100) = 2$ mean in practical terms? How about $f'(100) = 0.5$?
    (c) Suppose the company plans to spend about $\$100,000$ on advertising. If $f'(100) = 2$, should the company spend more or less than $\$100,000$ on advertising? What if $f'(100) = 0.5$?

16. Let $P(x) =$ the number of people in the US of height $\leq x$ inches. What is the meaning of $P'(66)$? What are its units? Estimate $P'(66)$ (using common sense). Is $P'(x)$ ever negative? [Hint: You may want to approximate $P'(66)$ by a difference quotient, using $h = 1$. Also, you may assume the US population is about 250 million, and that 66 inches $= 5$ feet 6 inches.]

2.5 THE SECOND DERIVATIVE

Since the derivative is itself a function, we can consider its derivative. For a function f, the derivative of its derivative is called the *second derivative*, and written f'' (read "f double-prime"). If $y = f(x)$, the second derivative can also be written as $\dfrac{d^2y}{dx^2}$, which means $\dfrac{d}{dx}\left(\dfrac{dy}{dx}\right)$, the derivative of $\dfrac{dy}{dx}$.

What Do Derivatives Tell Us?

Recall that the derivative of a function tells you whether a function is increasing or decreasing:

- If $f' > 0$ on an interval, then f is *increasing* over that interval.

- If $f' < 0$ on an interval, then f is *decreasing* over that interval.

Since f'' is the derivative of f',

- If $f'' > 0$ on an interval, then f' is *increasing* over that interval.

- If $f'' < 0$ on an interval, then f' is *decreasing* over that interval.

What does it mean for f' to be increasing or decreasing? An example in which f' is increasing is shown in Figure 2.32, where the curve is bending upward, or is *concave up*. In the example shown in Figure 2.33, in which f' is decreasing, the graph is bending downward, or is *concave down*. These figures suggest

---

If $f'' > 0$ on an interval, then f' is increasing, so the graph of f is concave up there.

If $f'' < 0$ on an interval, then f' is decreasing, so the graph of f is concave down there.

Figure 2.32: Meaning of f'': The slope increases from left to right, so f'' is positive and f is concave up

Figure 2.33: Meaning of f'': The slope decreases from left to right, so f'' is negative and f is concave down

---

Example 1   For the functions whose graphs are given in Figure 2.34, decide where their second derivatives are positive and where they are negative.

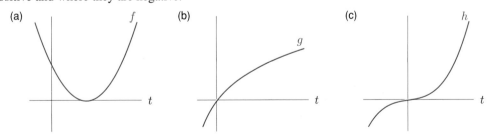

Figure 2.34: What signs do the second derivatives have?

Solution   From the graphs it appears that:

(a) $f'' > 0$ everywhere, because the graph of f is concave up everywhere.

(b) $g'' < 0$ everywhere, because the graph is concave down everywhere.

(c) $h'' < 0$ for $t < 0$, because the graph of h is concave down there; $h'' > 0$ for $t > 0$, because the graph of h is concave up there.

Interpretation of the Second Derivative as a Rate of Change

If we think of the derivative as a rate of change, then the second derivative is a rate of change of a rate of change. If the second derivative is positive, the rate of change of f is increasing; if the second derivative is negative, the rate of change of f is decreasing.

The second derivative can be a matter of practical concern. In 1985 a newspaper headline reported the Secretary of Defense as saying that Congress had cut the defense budget. As his opponents pointed out, however, Congress had merely cut the rate at which the defense budget was increasing.[2] In other words, the derivative of the defense budget was still positive (the budget was increasing), but the second derivative was negative (the budget's rate of increase had slowed).

Example 2   A population, P, growing in a confined environment often follows a *logistic* growth curve, like that shown in Figure 2.35. Use d^2P/dt^2 to describe how the rate at which the population is increasing changes over time. What are practical interpretations of t_0 and L?

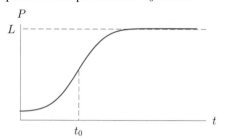

Figure 2.35: Logistic growth curve

Solution   Initially, the population is increasing, and at an increasing rate. So, initially dP/dt is increasing and so $d^2P/dt^2 > 0$. At t_0, the rate at which the population is increasing is a maximum. In other words, at time t_0 the population is growing fastest. Beyond t_0, the rate at which the population is growing is decreasing, and so $d^2P/dt^2 < 0$. At t_0, the curve changes from concave up to concave down, and $d^2P/dt^2 = 0$ there.

The quantity L represents the limiting value of the population as $t \to \infty$. Biologists call L the *carrying capacity* of the environment.

Example 3   Tests on the C5 Chevy Corvette sports car gave the results[3] in Table 2.12.

(a)   Estimate dv/dt for the time intervals shown.

(b)   What can you say about the sign of d^2v/dt^2 over the period shown?

TABLE 2.12   *Velocity of C5 Chevy Corvette*

Time, t (sec)	0	3	6	9	12
Velocity, v (meters/sec)	0	20	33	43	51

Solution   (a)   For each time interval we can calculate the average rate of change of velocity. For example, from $t = 0$ to $t = 3$ we have

$$\frac{dv}{dt} \approx \text{Average rate of change of } v = \frac{20 - 0}{3 - 0} = 6.67 \, \frac{\text{m/sec}}{\text{sec}}.$$

Estimated values of dv/dt are in Table 2.13.

(b)   Since the values of dv/dt are decreasing, $d^2v/dt^2 < 0$. The graph of v against t in Figure 2.36 confirms this; it is concave down. The fact that $dv/dt > 0$ tells us that the car is speeding up;

---

[2]In the *Boston Globe*, March 13, 1985, Representative William Gray (D–Pa.) was reported as saying: "It's confusing to the American people to imply that Congress threatens national security with reductions when you're really talking about a reduction in the increase."

[3]Adapted from the report in *Car and Driver*, February 1997.

the fact that $d^2v/dt^2 < 0$ tells us that the rate of increase decreased over this time period.

TABLE 2.13 *Estimates for dv/dt (meters/sec/sec)*

Time interval (sec)	$0 - 3$	$3 - 6$	$6 - 9$	$9 - 12$
Average rate of change (dv/dt)	6.67	4.33	3.33	2.67

Figure 2.36: Velocity of C5 Chevy Corvette

Velocity and Acceleration

The car in Example 3 is speeding up; we say that it is accelerating. We define *acceleration* as the rate of change of velocity with respect to time. If $v(t)$ is the velocity of an object at time t, we have

$$\begin{array}{c}\text{Average acceleration} \\ \text{from } t \text{ to } t + h\end{array} = \frac{v(t + h) - v(t)}{h},$$

and

$$\text{Instantaneous acceleration} = v'(t) = \lim_{h \to 0} \frac{v(t + h) - v(t)}{h}.$$

If the term velocity or acceleration is used alone, it is assumed to be instantaneous. Since velocity is the derivative of position, acceleration is the second derivative of position. Summarizing:

> If $y = s(t)$ is the position of an object at time t, then
>
> - Velocity: $v(t) = \dfrac{dy}{dt} = s'(t)$.
>
> - Acceleration: $a(t) = \dfrac{d^2y}{dt^2} = s''(t) = v'(t)$.

Example 4   A particle is moving along a straight line. If its distance, s, to the right of a fixed point is given by Figure 2.37, estimate:

(a)   When the particle is moving to the right and when it is moving to the left.

(b)   When the particle has positive acceleration and when it has negative acceleration.

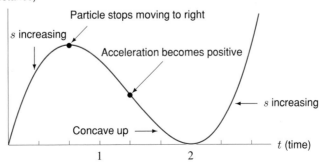

Figure 2.37: Distance of particle to right of a fixed point

Solution (a) The particle is moving to the right whenever s is increasing. From the graph, this appears to be for $0 < t < \frac{2}{3}$ and for $t > 2$. For $\frac{2}{3} < t < 2$, the value of s is decreasing, so the particle is moving to the left.

(b) The particle has positive acceleration whenever the curve is concave up, which appears to be for $t > \frac{4}{3}$. The particle has negative acceleration when the curve is concave down, for $t < \frac{4}{3}$.

Problems for Section 2.5

For Problems 1–6, give the signs of the first and second derivatives for each of the following functions.

1.

2.

3.

4.

5.

6.

7. An accelerating sports car goes from 0 mph to 60 mph in five seconds. Its velocity is given in Table 2.14, converted from miles per hour to feet per second, so that all time measurements are in seconds. (Note: 1 mph is 22/15 ft/sec.) Find the average acceleration of the car over each of the first two seconds.

TABLE 2.14

Time, t (sec)	0	1	2	3	4	5
Velocity, $v(t)$ (ft/sec)	0	30	52	68	80	88

8. Let $P(t)$ represent the price of a share of stock of a corporation at time t. What does each of the following statements tell us about the signs of the first and second derivatives of $P(t)$?

(a) "The price of the stock is rising faster and faster."
(b) "The price of the stock is close to bottoming out."

9. "Winning the war on poverty" has been described cynically as slowing the rate at which people are slipping below the poverty line. Assuming that this is happening:

(a) Sketch a graph of the total number of people in poverty against time.
(b) If N is the number of people below the poverty line at time t, what are the signs of dN/dt and d^2N/dt^2?

10. In economics, *total utility* refers to the total satisfaction from consuming some commodity. According to the economist Samuelson[4]:

> As you consume more of the same good, the total (psychological) utility increases. However, ... with successive new units of the good, your total utility will grow at a slower and slower rate because of a fundamental tendency for your psychological ability to appreciate more of the good to become less keen.

(a) Sketch the total utility as a function of the number of units consumed.
(b) In terms of derivatives, what is Samuelson telling us?

[4]From Paul A. Samuelson, *Economics*, 11th edition (New York: McGraw-Hill, 1981).

11. In April 1991, the *Economist* carried an article[5] which said:

> Suddenly, everywhere, it is not the rate of change of things that matters, it is the rate of change of rates of change. Nobody cares much about inflation; only whether it is going up or down. Or rather, whether it is going up fast or down fast. "Inflation drops by disappointing two points," cries the billboard. Which roughly translated means that prices are still rising, but less fast than they were, though not quite as much less fast as everybody had hoped.

In the last sentence, there are three statements about prices. Rewrite these as statements about derivatives.

12. IBM-Peru uses second derivatives to assess the relative success of various advertising campaigns. They assume that all campaigns produce some increase in sales. If a graph of sales against time shows a positive second derivative during a new advertising campaign, what does this suggest to IBM management? Why? What does a negative second derivative during a new campaign suggest?

13. An industry is being charged by the Environmental Protection Agency (EPA) with dumping unacceptable levels of toxic pollutants in a lake. Over a period of several months, an engineering firm makes daily measurements of the rate at which pollutants are being discharged into the lake.

Suppose the engineers produce a graph similar to either Figure 2.38(a) or Figure 2.38(b). For each case, give an idea of what argument the EPA might make in court against the industry and of the industry's defense.

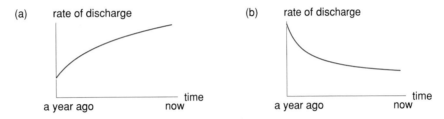

Figure 2.38: Discharge of pollutants

14. Given the following data:

x	0	0.2	0.4	0.6	0.8	1.0
$f(x)$	3.7	3.5	3.5	3.9	4.0	3.9

(a) Estimate $f'(0.6)$ and $f'(0.5)$.   (b) Estimate $f''(0.6)$.

(c) Where do you think the maximum and minimum values of f occur in the interval $0 \le x \le 1$?

15. The graph of a function $f(x)$ is shown in Figure 2.39. On a copy of the table indicate whether f, f', f'' at each marked point is positive, negative, or zero.

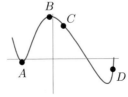

Figure 2.39

Point	f	f'	f''
A			
B			
C			
D			

---

[5]From "The Tyranny of Differential Calculus: $d^2P/dt^2 > 0 =$ misery." *The Economist* (London: April 6, 1991).

16. The graph of f' (not f) is given in Figure 2.40. At which of the marked values of x is
    (a) $f(x)$ greatest?   (b) $f(x)$ least?   (c) $f'(x)$ greatest?   (d) $f'(x)$ least?
    (e) $f''(x)$ greatest?   (f) $f''(x)$ least?

Figure 2.40: Note that this is a graph of f'

Figure 2.41

17. Which of the points labeled by letters in the graph of f in Figure 2.41 have

    (a) f' and f'' nonzero and of the same sign?
    (b) At least two of f, f', and f'' equal to zero?

18. The distance of a car from its initial position t minutes after setting out is given by $s(t) = 5t^2 + 3$ kilometers. What are the car's velocity and acceleration at time t? Give units.

CHAPTER SUMMARY

- **Rate of change**
    Average, instantaneous
- **Definition of Derivative**
    Difference quotient, limit
- **Estimating and computing derivatives**
    Estimate derivatives from a graph, table of values, or formula. Use definition to find derivatives of simple functions algebraically. Know derivatives of constant, linear, and power functions.
- **Interpretation of Derivatives**
    Rate of change, instantaneous velocity, slope, using units
- **Second Derivative**
    Concavity, acceleration
- **Working with Derivatives**
    Understand relation between sign of f' and whether f is increasing or decreasing. Sketch graph of f' from graph of f.

REVIEW PROBLEMS FOR CHAPTER TWO

1. A bicyclist pedals at a fairly constant rate, with evenly spaced intervals of coasting. Sketch a graph of the distance she has traveled as a function of time.

2. As you drive away from home, your speed is fast, then slow, then fast again. Draw a graph of your distance from home as a function of time.

3. (a) Make a table of values, rounded to two decimal places, for $f(x) = \log x$ (that is, log base 10) with $x = 1, 1.5, 2, 2.5, 3$. Then use this table to answer parts (b) and (c).
   (b) Find the average rate of change of $f(x)$ between $x = 1$ and $x = 3$.
   (c) Use average rates of change to approximate the instantaneous rate of change of $f(x)$ at $x = 2$.

4. For the function $f(x) = \log x$, estimate $f'(1)$. From the graph of $f(x)$, would you expect your estimate to be greater than or less than $f'(1)$?

5. For $f(x) = \ln x$, construct tables, rounded to four decimals, near $x = 1$, $x = 2$, $x = 5$, and $x = 10$. Use the tables to estimate $f'(1)$, $f'(2)$, $f'(5)$, and $f'(10)$. Then guess a general formula for $f'(x)$.

6. Given the values for a Bessel function $J_0(x)$ in Table 2.15, what is your best estimate for the derivative of $J_0(x)$ at $x = 0.5$?

TABLE 2.15

x	0	0.1	0.2	0.3	0.4	0.5
$J_0(x)$	1.0	0.9975	0.9900	0.9776	0.9604	0.9385

x	0.6	0.7	0.8	0.9	1.0	
$J_0(x)$	0.9120	0.8812	0.8463	0.8075	0.7652	

Sketch the graphs of the derivatives of the functions shown in Problems 7–12. Be sure your sketches are consistent with the important features of the graphs of the original functions.

7.

8.

9.

10.

11.

12.
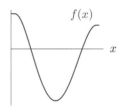

Find a formula for the derivatives of the functions in Problems 13–14 algebraically.

13. $f(x) = 5x^2 + x$

14. $n(x) = \dfrac{1}{x} + 1.$

15. (a) On the same set of axes, graph $f(x) = \sin x$ and $g(x) = \sin 2x$ from $x = 0$ to $x = 2\pi$.
     (b) On a second set of axes, sketch the graphs of $f'(x)$ and $g'(x)$ and compare them. (Be careful when comparing the slopes of $f(x)$ and $g(x)$ at each point.)

16. Do the values for the function $y = k(x)$ in Table 2.16 suggest that the graph of $k(x)$ is concave up or concave down for $1 \le x \le 3.3$? Write a sentence in support of your conclusion.

TABLE 2.16

x	1.0	1.2	1.5	1.9	2.5	3.3
$k(x)$	4.0	3.8	3.6	3.4	3.2	3.0

17. A yam has just been taken out of the oven and is cooling off before being eaten. The temperature, T, of the yam (measured in degrees Fahrenheit) is a function of how long it has been out of the oven, t (measured in minutes). Thus, we have $T = f(t)$.

    (a) Is $f'(t)$ positive or negative? Why?

    (b) What are the units for $f'(t)$?

18. An economist is interested in how the price of a certain commodity affects its sales. Suppose that at a price of \$$p$, a quantity q of the commodity is sold. If $q = f(p)$, explain in economic terms the meaning of the statements $f(10) = 240,000$ and $f'(10) = -29,000$.

19. Students were asked to evaluate $f'(4)$ from the following table which shows values of the function f:

x	1	2	3	4	5	6
$f(x)$	4.2	4.1	4.2	4.5	5.0	5.7

    • Student A estimated the derivative as $f'(4) \approx \dfrac{f(5) - f(4)}{5 - 4} = 0.5$.

    • Student B estimated the derivative as $f'(4) \approx \dfrac{f(4) - f(3)}{4 - 3} = 0.3$.

    • Student C suggested that they should split the difference and estimate the average of these two results, that is, $f'(4) \approx \frac{1}{2}(0.5 + 0.3) = 0.4$.

    (a) Sketch the graph of f, and indicate how the three estimates are represented on the graph.

    (b) Explain which answer is likely to be best.

    (c) Use Student C's method to find an algebraic formula which approximates $f'(x)$ using increments of size h.

20. Each of the graphs in Figure 2.42 shows the position of a particle moving along the x-axis as a function of time, $0 \le t \le 5$. The vertical scales of the graphs are the same. During this time interval, which particle has

    (a) Constant velocity?   (d) Zero average velocity?

    (b) The greatest initial velocity?   (e) Zero acceleration?

    (c) The greatest average velocity?   (f) Positive acceleration throughout?

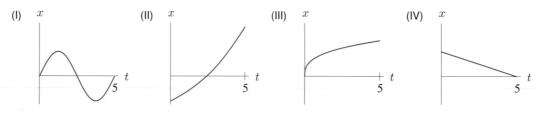

Figure 2.42

21. A person with a certain liver disease first exhibits larger and larger concentrations of certain enzymes (called SGOT and SGPT) in the blood. As the disease progresses, the concentration of these enzymes drops, first to the predisease level and eventually to zero (when almost all of the liver cells have died). Monitoring the levels of these enzymes allows doctors to track the progress of a patient with this disease. If $C = f(t)$ is the concentration of the enzymes in the blood as a function of time,

    (a) Sketch a possible graph of $C = f(t)$.

    (b) Mark on the graph the intervals where $f' > 0$ and where $f' < 0$.

    (c) What does $f'(t)$ represent, in practical terms?

22. The population of a herd of deer is modeled by

$$P(t) = 4000 + 400 \sin\left(\frac{\pi}{6}t\right) + 180 \sin\left(\frac{\pi}{3}t\right)$$

where t is measured in months from the first of April.

    (a) Use a calculator or computer to sketch a graph showing how this population varies with time.

Use the graph to answer the following questions.

(b) When is the herd largest? How many deer are in it at that time?
(c) When is the herd smallest? How many deer are in it then?
(d) When is the herd growing the fastest? When is it shrinking the fastest?
(e) How fast is the herd growing on April 1?

23. Let $g(x) = \sqrt{x}$ and $f(x) = kx^2$, where k is a constant.

(a) Find the slope of the tangent line to the graph of g at the point $(4, 2)$.
(b) Find the equation of this tangent line.
(c) If the graph of f contains the point $(4, 2)$, find k.
(d) Where does the graph of f intersect the tangent line?

24. A circle with center at the origin and radius of length $\sqrt{19}$ has equation $x^2 + y^2 = 19$. Graph the circle.

(a) Just from looking at the graph, what can you say about the slope of the line tangent to the circle at the point $(0, \sqrt{19})$? What about the slope of the tangent at $(\sqrt{19}, 0)$?
(b) Estimate the slope of the tangent to the circle at the point $(2, -\sqrt{15})$ by graphing the tangent carefully at that point.
(c) Use the result of part (b) and the symmetry of the circle to find slopes of the tangents drawn to the circle at $(-2, \sqrt{15})$, $(-2, -\sqrt{15})$, and $(2, \sqrt{15})$.

25. A continuous function defined for all x has the following properties:
- f is increasing.   • f is concave down.   • $f(5) = 2$.   • $f'(5) = \frac{1}{2}$.

(a) Sketch a possible graph for f.   (b) How many zeros does f have?
(c) What can you say about the location of the zeros?   (d) What is $\lim\limits_{x \to -\infty} f(x)$?
(e) Is it possible that $f'(1) = 1$?   (f) Is it possible that $f'(1) = \frac{1}{4}$?

26. The number of hours, H, of daylight in Madrid is a function of t, the number of days since the start of the year. Figure 2.43 shows a one-month portion of the graph of H.

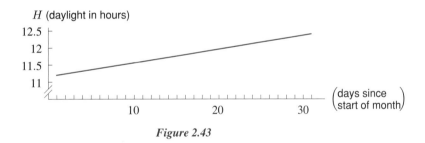

Figure 2.43

(a) Comment on the shape of the graph. Why does it look like a straight line?
(b) What month does this graph show? How do you know?
(c) What is the approximate slope of this line? What does the slope represent in practical terms?

27. Suppose you put a yam in a hot oven, maintained at a constant temperature of $200°$C. As the yam picks up heat from the oven, its temperature rises.[6]

(a) Draw a possible graph of the temperature T of the yam against time t (minutes) since it is put into the oven. Explain any interesting features of the graph, and in particular explain its concavity.
(b) Suppose that, at $t = 30$, the temperature T of the yam is $120°$ and increasing at the (instantaneous) rate of $2°$/min. Using this information, plus what you know about the shape of the T graph, estimate the temperature at time $t = 40$.
(c) Suppose in addition you are told that at $t = 60$, the temperature of the yam is $165°$. Can you improve your estimate of the temperature at $t = 40$?
(d) Assuming all the data given so far, estimate the time at which the temperature of the yam is $150°$.

---

[6]From Peter D. Taylor, *Calculus: The Analysis of Functions* (Toronto: Wall & Emerson, Inc., 1992)

PROJECTS

1. **Hours of Daylight as a Function of Latitude**

   Let $S(x)$ be the number of sunlight hours on a cloudless June 21, as a function of latitude, x, measured in degrees.

   (a) What is $S(0)$?

   (b) Let x_0 be the latitude of the Arctic Circle ($x_0 \approx 66°30'$). In the northern hemisphere, $S(x)$ is given, for some constants a and b, by the formula:

   $$S(x) = \begin{cases} a + b \arcsin\left(\dfrac{\tan x}{\tan x_0}\right) & \text{for } 0 \leq x < x_0 \\ 24 & \text{for } x_0 \leq x \leq 90. \end{cases}$$

   Find a and b so that $S(x)$ is continuous.

   (c) Calculate $S(x)$ for Tucson, Arizona ($x = 32°13'$) and Walla Walla, Washington ($46°4'$).

   (d) Graph $S(x)$, for $0 \leq x \leq 90$.

   (e) Does $S(x)$ appear to be differentiable?

2. **US Population**

   Census figures for the US population (in millions) are listed in Table 2.17. Let f be the function such that $P = f(t)$ is the population (in millions) in year t.

TABLE 2.17 *US population (in millions), 1790–1990*

Year	Population	Year	Population	Year	Population	Year	Population
1790	3.9	1850	23.1	1910	92.0	1970	205.0
1800	5.3	1860	31.4	1920	105.7	1980	226.5
1810	7.2	1870	38.6	1930	122.8	1990	248.7
1820	9.6	1880	50.2	1940	131.7		
1830	12.9	1890	62.9	1950	150.7		
1840	17.1	1900	76.0	1960	179.0		

   (a) (i)   Estimate the rate of change of the population for the years 1900, 1945, and 1990.

   (ii) When, approximately, was the rate of change of the population greatest?

   (iii) Estimate the US population in 1956.

   (iv) Based on the data from the table, what would you predict for the census in the year 2000?

   (b) Assume that f is increasing (as the values in the table suggest). Then f is invertible.

   (i)   What is the meaning of $f^{-1}(100)$?

   (ii) What does the derivative of $f^{-1}(P)$ at $P = 100$ represent? What are its units?

   (iii) Estimate $f^{-1}(100)$.

   (iv) Estimate the derivative of $f^{-1}(P)$ at $P = 100$.

   (c) (i)   Usually we think the US population $P = f(t)$ as a smooth function of time. To what extent is this justified? What happens if we zoom in at a point of the graph? What about events such as the Louisiana Purchase? Or the moment of your birth?

   (ii) What do we in fact mean by the rate of change of the population at a particular time t?

   (iii) Give another example of a real-world function which is not smooth but is usually treated as such.

FOCUS ON THEORY

LIMITS AND CONTINUITY

In this section we use the example of limits and continuity to illustrate how formal definitions are developed from intuitive ideas.

Definition of Limit

By the beginning of the 19th century, calculus had proved its worth, and there was no doubt about the correctness of its answers. However, it was not until the work of the French mathematician Augustin Cauchy (1789–1857) that a formal definition of the limit was given, similar to the following:

> Suppose a function f, is defined on an interval around c, except perhaps not at the point $x = c$. We define the **limit** of the function $f(x)$ as x approaches c, written $\lim_{x \to c} f(x)$, to be a number L (if one exists) such that $f(x)$ is as close to L as we please whenever x is sufficiently close to c (but $x \neq c$). If L exists, we write
>
> $$\lim_{x \to c} f(x) = L.$$

Shortly, we will see how "as close as we please" and "sufficiently close" can be given a precise meaning using inequalities. First, we look at $\lim_{\theta \to 0}(\sin\theta/\theta)$ more closely (see Example 4 on page 93).

Example 1   By graphing $y = (\sin\theta)/\theta$ in an appropriate window, find how close θ should be to 0 in order to make $(\sin\theta)/\theta$ within 0.01 of 1.

Solution   Since we want $(\sin\theta)/\theta$ to be within 0.01 of 1, we set the y-range on the graphing window to go from 0.99 to 1.01. Our first attempt with $-0.5 \leq \theta \leq 0.5$ yields the graph in Figure 2.44. Since we want the y-values to stay within the range $0.99 < y < 1.01$, we do not want the graph to leave the window through the top or bottom. By trial and error, we find that changing the θ-range to $-0.2 \leq \theta \leq 0.2$ gives the graph in Figure 2.45. Thus, the graph suggests that $(\sin\theta)/\theta$ will be within 0.01 of 1 whenever θ is within 0.2 of 0. Proving this requires an analytical argument, not just graphs from a calculator.

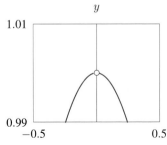

Figure 2.44: $(\sin\theta)/\theta$ with $-0.5 \leq \theta \leq 0.5$

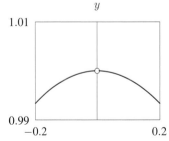

Figure 2.45: $(\sin\theta)/\theta$ with $-0.2 \leq \theta \leq 0.2$

When we say "$f(x)$ is as close to L as we please," we mean that we can specify a maximum distance between $f(x)$ and L. We express the distance using absolute values:

$$|f(x) - L| = \text{Distance between } f(x) \text{ and } L.$$

Using ϵ (the Greek letter epsilon) to stand for the distance we have specified, we write

$$|f(x) - L| < \epsilon$$

to indicate that the maximum distance between $f(x)$ and L is less than ϵ. In Example 1 we used $\epsilon = 0.01$. In a similar manner we interpret "x is sufficiently close to c" as specifying a maximum distance between x and c:

$$|x - c| < \delta,$$

where δ (the Greek letter delta) tells us how close x should be to c. In Example 1 we found $\delta = 0.2$.

If $\lim\limits_{x \to c} f(x) = L$, we know that no matter how narrow the band determined by ϵ in Figure 2.46, there's always a δ which makes the graph stay within the band for $c - \delta < x < c + \delta$.

Thus we restate the definition of a limit, using symbols:

Definition of Limit

We define $\lim\limits_{x \to c} f(x)$ to be the number L (if one exists) such that for any $\epsilon > 0$ (as small as we want), there is a $\delta > 0$ (sufficiently small) such that if $|x - c| < \delta$ and $x \neq c$, then $|f(x) - L| < \epsilon$.

Realize that the point of this definition is that for any ϵ we are given, we need to be able to determine a corresponding δ. One way to do this is to give an explicit expression for δ in terms of ϵ.

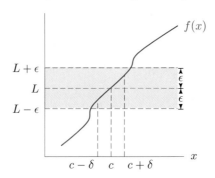

Figure 2.46: What the definition of the limit means practically

---

Example 2   Use algebra to find a maximum distance between x and 2 which ensures that x^2 is within 0.1 of 4. Use a similar argument to show that $\lim\limits_{x \to 2} x^2 = 4$.

Solution   We write $x = 2 + h$. We want to find the values of h making x^2 within 0.1 of 4. We know

$$x^2 = (2 + h)^2 = 4 + 4h + h^2,$$

so x^2 differs from 4 by $4h + h^2$. Since we want x^2 to be within 0.1 of 4, we need

$$|x^2 - 4| = |4h + h^2| = |h| \cdot |4 + h| < 0.1.$$

Assuming $0 < |h| < 1$, we know $|4 + h| < 5$, so we need

$$|x^2 - 4| < 5|h| < 0.1.$$

Thus, if we choose h such that $0 < |h| < 0.1/5 = 0.02$, then x^2 is less than 0.1 from 4.

An analogous argument using any small ϵ instead of 0.1 shows that if we take $\delta = \epsilon/5$, then

$$|x^2 - 4| < \epsilon \quad \text{for all} \quad |x - 2| < \epsilon/5.$$

Thus, we have used the definition to show that

$$\lim_{x \to 2} x^2 = 4.$$

It is important to understand that the ϵ, δ definition by itself does not make it easier to calculate limits. The advantage of the ϵ, δ definition is that it makes it possible to put calculus on a rigorous foundation. From this foundation, we can prove the following properties. See Problems 13–16.

Theorem: Properties of Limits

Assuming all the limits on the right hand side exist:

1. If b is a constant, then $\lim_{x \to c} \left(b f(x) \right) = b \left(\lim_{x \to c} f(x) \right)$.

2. $\lim_{x \to c} \left(f(x) + g(x) \right) = \lim_{x \to c} f(x) + \lim_{x \to c} g(x)$.

3. $\lim_{x \to c} \left(f(x) g(x) \right) = \left(\lim_{x \to c} f(x) \right) \left(\lim_{x \to c} g(x) \right)$.

4. $\lim_{x \to c} \dfrac{f(x)}{g(x)} = \dfrac{\lim_{x \to c} f(x)}{\lim_{x \to c} g(x)}$, provided $\lim_{x \to c} g(x) \neq 0$.

5. For any constant k, $\lim_{x \to c} k = k$.

6. $\lim_{x \to c} x = c$.

These properties underlie many limit calculations, though we seldom acknowledge them explicitly.

Example 3   Explain how the limit properties are used in the following calculation:

$$\lim_{x \to 3} \frac{x^2 + 5x}{x + 9} = \frac{3^2 + (5)(3)}{3 + 9} = 2.$$

Solution   We calculate this limit in stages, using the limit properties to justify each step:

$$\lim_{x \to 3} \frac{x^2 + 5x}{x + 9} = \frac{\lim_{x \to 3} (x^2 + 5x)}{\lim_{x \to 3} (x + 9)} \qquad \text{Property 4 (since } \lim_{x \to 3}(x + 9) \neq 0)$$

$$= \frac{\lim_{x \to 3} (x^2) + \lim_{x \to 3} (5x)}{\lim_{x \to 3} x + \lim_{x \to 3} 9} \qquad \text{Property 2}$$

$$= \frac{\left(\lim_{x \to 3} x \right)^2 + 5 \left(\lim_{x \to 3} x \right)}{\lim_{x \to 3} x + \lim_{x \to 3} 9} \qquad \text{Properties 1 and 3}$$

$$= \frac{3^2 + (5)(3)}{3 + 9} = 2. \qquad \text{Properties 5 and 6}$$

One- and Two-Sided Limits

When we write

$$\lim_{x \to 2} f(x),$$

we mean the number that $f(x)$ approaches as x approaches 2 *from both sides*. We examine values of $f(x)$ as x approaches 2 through values greater than 2 (such as 2.1, 2.01, 2.003) and values less than 2 (such as 1.9, 1.99, 1.994). If we want x to approach 2 only through values greater than 2, we write

$$\lim_{x \to 2^+} f(x)$$

for the number that $f(x)$ approaches (assuming such a number exists). Similarly,

$$\lim_{x \to 2^-} f(x)$$

denotes the number (if it exists) obtained by letting x approach 2 through values less than 2. We call $\lim_{x \to 2^+} f(x)$ a *right-hand limit* and $\lim_{x \to 2^-} f(x)$ a *left-hand limit*.

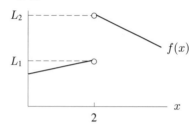

Figure 2.47: Left- and right-hand limits
at $x = 2$

For the function graphed in Figure 2.47, we have

$$\lim_{x \to 2^-} f(x) = L_1 \qquad \lim_{x \to 2^+} f(x) = L_2.$$

If the left- and right-hand limits were equal, that is, if $L_1 = L_2$, then it could easily be proved that $\lim_{x \to 2} f(x)$ exists and $\lim_{x \to 2} f(x) = L_1 = L_2$. Since $L_1 \neq L_2$ in Figure 2.47, we see that $\lim_{x \to 2} f(x)$ does not exist in this case.

When Limits Do Not Exist

Whenever there is not a number L such that $\lim_{x \to c} f(x) = L$, we say $\lim_{x \to c} f(x)$ does not exist. In addition to cases in which the left- and right-hand limits are not equal, there are some other cases in which limits fail to exist. Here are three examples.

Example 4   Explain why $\lim_{x \to 2} \dfrac{|x - 2|}{x - 2}$ doesn't exist.

Solution   Figure 2.48 shows the problem: The right-hand limit and the left-hand limit are different. For $x > 2$, we have $|x - 2| = x - 2$, so as x approaches 2 from the right,

$$\lim_{x \to 2^+} \frac{|x - 2|}{x - 2} = \lim_{x \to 2^+} \frac{x - 2}{x - 2} = \lim_{x \to 2^+} 1 = 1.$$

Similarly, if $x < 2$, then $|x - 2| = 2 - x$ so

$$\lim_{x \to 2^-} \frac{|x - 2|}{x - 2} = \lim_{x \to 2^-} \frac{2 - x}{x - 2} = \lim_{x \to 2^-} (-1) = -1.$$

So if $\lim_{x \to 2} \dfrac{|x - 2|}{x - 2} = L$ then L would have to be both 1 and -1. Since L cannot have two different values, the limit does not exist.

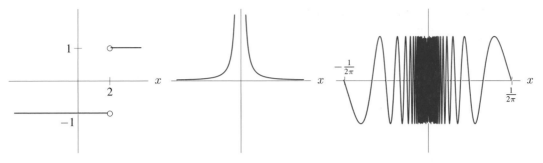

Figure 2.48: Graph of $\frac{|x-2|}{x-2}$ *Figure 2.49:* Graph of $\frac{1}{x^2}$ *Figure 2.50:* Graph of $\sin\left(\frac{1}{x}\right)$

Example 5   Explain why $\lim\limits_{x \to 0} \dfrac{1}{x^2}$ doesn't exist.

Solution   As x approaches zero, $1/x^2$ becomes arbitrarily large, so it can't stay close to any finite number L. See Figure 2.49. Therefore we say $1/x^2$ has no limit as $x \to 0$.

Example 6   Explain why $\lim\limits_{x \to 0} \sin\left(\dfrac{1}{x}\right)$ doesn't exist.

Solution   We know that the sine function has values between -1 and 1. The graph in Figure 2.50 oscillates more and more rapidly as $x \to 0$. There are x-values as close to 0 as we like where $\sin(1/x) = 0$. There are also x-values as close to 0 as we like where $\sin(1/x) = 1$. So if the limit existed, it would have to be both 0 and 1. Thus, the limit does not exist.

Limits at Infinity

Sometimes we want to know what happens to $f(x)$ as x gets large, that is, the end behavior of f.

> If $f(x)$ gets as close to a number L as we please when x gets sufficiently large, then we write
>
> $$\lim_{x \to \infty} f(x) = L.$$
>
> Similarly, if $f(x)$ approaches L as x gets more and more negative, then we write
>
> $$\lim_{x \to -\infty} f(x) = L.$$

The symbol ∞ does not represent a number. Writing $x \to \infty$ means that we consider arbitrarily large values of x. If the limit of $f(x)$ as $x \to \infty$ or $x \to -\infty$ is L, we say that the graph of f has a *horizontal asymptote* $y = L$.

Example 7   Investigate $\lim\limits_{x \to \infty} \dfrac{1}{x}$ and $\lim\limits_{x \to -\infty} \dfrac{1}{x}$.

Solution   A graph of $f(x) = \frac{1}{x}$ in a large window shows $1/x$ approaching zero as x increases in either the positive or the negative direction (See Figure 2.51). This is as we would expect, since dividing 1 by larger and larger numbers yields answers which are smaller and smaller. This suggests that

$$\lim_{x \to \infty} \frac{1}{x} = \lim_{x \to -\infty} \frac{1}{x} = 0,$$

and $f(x) = 1/x$ has $y = 0$ as a horizontal asymptote as $x \to \pm\infty$.

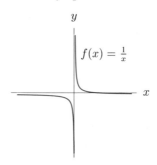

Figure 2.51: The end behavior of $f(x) = \frac{1}{x}$

Definition of Continuity

We can now define continuity. Recall that the idea of continuity rules out breaks, jumps, or holes by demanding that the behavior of a function *near* a point be consistent with its behavior *at* the point:

> The function f is **continuous** at $x = c$ if f is defined at $x = c$ and
>
> $$\lim_{x \to c} f(x) = f(c).$$
>
> In other words, $f(x)$ is as close as we please to $f(c)$ provided x is close enough to c. The function is **continuous on an interval** $[a, b]$ if it is continuous at every point in the interval.[7]

Constant functions and $f(x) = x$ are continuous. (See Problem 16.) Using the continuity of sums and products, we can show that any polynomial is continuous. Proving that $\sin x$, $\cos x$, and e^x are continuous is more difficult. The following theorem, based on the properties of limits on page 129, makes it easier to decide whether a given function is continuous.

> ### Theorem: Continuity of Sums, Products, and Quotients of Functions
>
> Suppose that f and g are continuous on an interval and that b is a constant. Then, on that same interval,
>
>   1. $bf(x)$ is continuous
>   2. $f(x) + g(x)$ is continuous
>   3. $f(x)g(x)$ is continuous
>   4. $f(x)/g(x)$ is continuous, provided $g(x) \neq 0$ on the interval.

We prove the first of these properties.

---

[7]If c is an endpoint of the interval, we define continuity at $x = c$ using one-sided limits at c.

Proof   To prove that $bf(x)$ is continuous, pick any point c in the interval. We must show that $\lim_{x \to c} bf(x) = bf(c)$. Since $f(x)$ is continuous, we already know that $\lim_{x \to c} f(x) = f(c)$. So, by the first property of limits,

$$\lim_{x \to c} (bf(x)) = b \left(\lim_{x \to c} f(x) \right) = bf(c).$$

Since c was chosen arbitrarily, we have shown that $bf(x)$ is continuous at every point in the interval.

Theorem: Continuity of Composite Functions

If f and g are continuous, and if the composite function $f(g(x))$ is defined on an interval, then $f(g(x))$ is continuous on that interval.

Assuming the continuity of $\sin x$ and e^x, this result shows us, for example, that $\sin(e^x)$ and $e^{\sin x}$ are both continuous. A proof of the continuity of composite functions is outlined in Problem 17.

Problems on Limits and Continuity

1. Consider the function $(\sin \theta)/\theta$. Estimate how close θ should be to 0 to make $(\sin \theta)/\theta$ stay within 0.001 of 1.

2. The function $g(\theta) = (\sin \theta)/\theta$ is not defined at $\theta = 0$. Is it possible to define $g(0)$ in such a way that g is continuous at $\theta = 0$? Explain your answer.

Use a graph to estimate each of the limits in Problems 3–6.

3. $\displaystyle \lim_{\theta \to 0} \frac{\sin (2\theta)}{\theta}$ (use radians)

4. $\displaystyle \lim_{\theta \to 0} \frac{\cos \theta - 1}{\theta}$ (use radians)

5. $\displaystyle \lim_{\theta \to 0} \frac{\sin \theta}{\theta}$ (use degrees)

6. $\displaystyle \lim_{\theta \to 0} \frac{\theta}{\tan(3\theta)}$ (use radians)

7. Consider the limit

$$\lim_{x \to 0^+} x^x.$$

   Estimate this limit either by evaluating x^x for smaller and smaller positive values of x (say $x = 0.1, 0.01, 0.001, \ldots$) or by zooming in on the graph of $y = x^x$ near $x = 0$.

8. (a) Give an example of a function such that $\lim_{x \to 2} f(x) = \infty$.

   (b) Give an example of a function such that $\lim_{x \to 2} f(x) = -\infty$.

9. Consider the function $f(x) = \sin(1/x)$.

   (a) Find a sequence of x-values that approach 0 such that $\sin(1/x) = 0$.
   [Hint: Use the fact that $\sin(\pi) = \sin(2\pi) = \sin(3\pi) = \ldots = \sin(n\pi) = 0$.]

   (b) Find a sequence of x-values that approach 0 such that $\sin(1/x) = 1$.
   [Hint: Use the fact that $\sin(n\pi/2) = 1$ if $n = 1, 5, 9, \ldots$.]

   (c) Find a sequence of x-values that approach 0 such that $\sin(1/x) = -1$.

   (d) Explain why your answers to any two of parts (a)–(c) show that $\lim_{x \to 0} \sin(1/x)$ does not exist.

10. Write the definition of the following statement both in words and in symbols:

$$\lim_{h \to a} g(h) = K.$$

11. For each of the following functions do the following:

 (i) Make a table of values of $f(x)$ for $x = a + 0.1$, $a + 0.01$, $a + 0.001$, $a + 0.0001$, $a - 0.1$, $a - 0.01$, $a - 0.001$, and $a - 0.0001$.

 (ii) Make a conjecture about the value of $\lim_{x \to a} f(x)$.

 (iii) Graph the function to see if it is consistent with your answers to parts (i) and (ii).

 (iv) Find an interval for x containing a such that the difference between your conjectured limit and the value of the function is less than 0.01 on that interval. (In other words, find a window of height 0.02 such that the graph exits the sides of the window and not the top or bottom of the window.)

 (a) $f(x) = \dfrac{x^2 - 4}{x - 2}$, $a = 2$

 (b) $f(x) = \dfrac{x^2 - 9}{x - 3}$, $a = 3$

 (c) $f(x) = \dfrac{\sin x - 1}{x - \pi/2}$, $a = \dfrac{\pi}{2}$

 (d) $f(x) = \dfrac{\sin 5x - 1}{x - \pi/2}$, $a = \dfrac{\pi}{2}$

 (e) $f(x) = \dfrac{e^{2x-2} - 1}{x - 1}$, $a = 1$

 (f) $f(x) = \dfrac{e^{0.5x-1} - 1}{x - 2}$, $a = 2$

12. Assuming that limits as $x \to \infty$ have the properties listed for limits as $x \to c$ on page 129, use algebraic manipulations to evaluate $\lim_{x \to \infty}$ for the following functions:

 (a) $f(x) = \dfrac{x + 3}{2 - x}$

 (b) $f(x) = \dfrac{x^2 + 2x - 1}{3 + 3x^2}$

 (c) $f(x) = \dfrac{x^2 + 4}{x + 3}$

 (d) $f(x) = \dfrac{2x^3 - 16x^2}{4x^2 + 3x^3}$

 (e) $f(x) = \dfrac{x^4 + 3x}{x^4 + 2x^5}$

 (f) $f(x) = \dfrac{3e^x + 2}{2e^x + 3}$

 (g) $f(x) = \dfrac{2e^{-x} + 3}{3e^{-x} + 2}$

13. This problem suggests a proof of the first property of limits on page 129: $\lim_{x \to c} bf(x) = b \lim_{x \to c} f(x)$.

 (a) First, prove the property in the case $b = 0$.
 (b) Now suppose that $b \neq 0$. Let $\epsilon > 0$. Show that if $|f(x) - L| < \epsilon/|b|$, then $|bf(x) - bL| < \epsilon$.
 (c) Finally, prove that if $\lim_{x \to c} f(x) = L$ then $\lim_{x \to c} bf(x) = bL$. [Hint: Choose δ so that if $|x - c| < \delta$, then $|f(x) - L| < \epsilon/|b|$.]

14. Prove the second property of limits: $\lim_{x \to c} \big(f(x) + g(x)\big) = \lim_{x \to c} f(x) + \lim_{x \to c} g(x)$. Assume that the limits on the right exist.

15. This problem suggests a proof of the third property of limits:

$$\lim_{x \to c} \big(f(x)g(x)\big) = \left(\lim_{x \to c} f(x)\right)\left(\lim_{x \to c} g(x)\right)$$

(assuming the limits on the right exist). Let $L_1 = \lim_{x \to c} f(x)$ and $L_2 = \lim_{x \to c} g(x)$.

 (a) First, show that if $\lim_{x \to c} f(x) = \lim_{x \to c} g(x) = 0$, then $\lim_{x \to c} \big(f(x)g(x)\big) = 0$.
 (b) Show algebraically that $f(x)g(x) = \big(f(x) - L_1\big)\big(g(x) - L_2\big) + L_1 g(x) + L_2 f(x) - L_1 L_2$.
 (c) Use the second limit property (see Problem 14) to explain why
   $\lim_{x \to c} \big(f(x) - L_1\big) = \lim_{x \to c} \big(g(x) - L_2\big) = 0$.
 (d) Use parts (a) and (c) to explain why $\lim_{x \to c} \big(f(x) - L_1\big)\big(g(x) - L_2\big) = 0$.
 (e) Finally, use parts (b) and (d) and the first and second limit properties to show that
   $\lim_{x \to c} \big(f(x)g(x)\big) = \left(\lim_{x \to c} f(x)\right)\left(\lim_{x \to c} g(x)\right)$.

16. Show that the following functions are both continuous everywhere.
 (a) $f(x) = k$ (a constant)   (b) $g(x) = x$

17. This problem suggests a proof of the theorem on continuity of composite functions on page 133: If f and g are continuous and the composite function $f\left(g(x)\right)$ is defined on an interval, then $f\left(g(x)\right)$ is continuous on that interval.

    Let c be a point inside the interval where $f\left(g(x)\right)$ is defined. We must show that $\lim_{x \to c} f\left(g(x)\right) = f\left(g(c)\right)$. Let $d = g(c)$. Then the continuity of f at d means that $\lim_{y \to d} f(y) = f(d)$. Thus, for a given $\epsilon > 0$, we can choose $\delta > 0$ so that $|y - d| < \delta$ implies $|f(y) - f(d)| < \epsilon$.

    Now, take $y = g(x)$ and show that the continuity of g means that we can find a $\delta_1 > 0$ such that, if $|x - c| < \delta_1$, then $|g(x) - d| < \delta$. Explain how this establishes the continuity of $f\left(g(x)\right)$ at $x = c$.

For each value of ϵ in Problems 18–19, find a positive value of δ such that the graph of the function leaves the window $a - \delta < x < a + \delta, b - \epsilon < y < b + \epsilon$ by the sides and not through the top or bottom.

18. $f(x) = -2x + 3; a = 0; b = 3; \epsilon = 0.2, 0.1, 0.02, 0.01, 0.002, 0.001$.

19. $g(x) = -x^3 + 2; a = 0; b = 2; \epsilon = 0.1, 0.01, 0.001$.

20. Show that $\lim_{x \to 0} (-2x + 3) = 3$. You may use the result of Problem 18.

21. Show that $\lim_{x \to 0} (-x^3 + 2) = 2$. [Hint: Try $\delta = \epsilon^{1/3}$.]

In Problems 22–24, modify the definition of limit on page 128 to give a definition of each of the following.

22. A right-hand limit   23. A left-hand limit   24. $\lim_{x \to \infty} f(x) = L$

25. Consider the function $f(x) = \begin{cases} x \sin\left(\dfrac{1}{x}\right) & x \neq 0 \\ 0 & x = 0. \end{cases}$

    Show that f is continuous everywhere, but that it is neither always increasing nor always decreasing on the interval $[0, \epsilon]$ for any $\epsilon > 0$, no matter how small.

26. On page 82 we showed how to find a sequence of intervals $[a_n, b_n]$ that close in on a root r of $f(x) = 3x^3 - x^2 + 2x - 1$. In this problem we use the continuity of f to prove that r is in fact a root, that is, that $f(r) = 0$. You may assume that all the intervals have been chosen so that $f(a_n) < 0$ and $f(b_n) > 0$.

    (a) Suppose $f(r) = L > 0$. Use the definition of continuity with any ϵ such that $\epsilon < L$ to choose a δ such that

    $$|f(x) - L| < \epsilon \quad \text{for all} \quad |x - r| < \delta.$$

    Find an a_n in the interval $[r - \delta, r + \delta]$ and arrive at a contradiction involving $f(a_n)$.
    (b) Suppose $f(r) = L < 0$. Make a similar argument to arrive at a contradiction involving a b_n.
    (c) Conclude that $f(r) = 0$.

27. Adapt the zooming argument on page 82 and the argument in Problem 26 to prove the Intermediate Value Theorem: If f is continuous on $[a, b]$ and k is between $f(a)$ and $f(b)$, there is a point c in $[a, b]$ with $f(c) = k$. [Hint: Consider $g(x) = f(x) - k$ and look for a zero of g.]

DIFFERENTIABILITY AND LINEAR APPROXIMATION

In this section we analyze the tangent line approximation and its error. This leads us to a different way of looking at differentiability. Recall that:

> A function f is said to be **differentiable at** $x = a$ if $f'(a)$ exists.

Most functions we deal with have a derivative at every point in their domain; they are said to be *differentiable everywhere*.

How Can We Recognize Whether a Function Is Differentiable?

If a function has a derivative at a point, its graph must have a tangent line there; the slope of the tangent line is the derivative. When we zoom in on the graph of the function, we see a nonvertical straight line.

Occasionally we meet a function which fails to have a derivative at a few points. For example, a discontinuous function whose graph has a break at some point cannot have a derivative at that point. Some of the ways in which a function can fail to be differentiable at a point are if:

- The function is not continuous at the point.

- The graph has a sharp corner at that point.

- The graph has a vertical tangent line.

Figure 2.52 shows a function which appears to be differentiable at all points except $x = a$ and $x = b$. There is no tangent at A because the graph has a corner there. As x approaches a from the left, the slope of the line joining P to A converges to some positive number. As x approaches a from the right, the slope of the line joining P to A converges to some negative number. Thus the slopes approach different numbers as we approach $x = a$ from different sides. Therefore the function is not differentiable at $x = a$. At B, there is no sharp corner, but as x approaches b, the slope of the line joining B to Q does not converge; it just keeps growing larger and larger. This reflects the fact that the graph has a vertical tangent at B. Since the slope of a vertical line is not defined, the function is not differentiable at $x = b$.

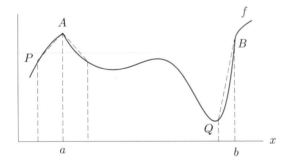

Figure 2.52: A function which is not differentiable at A or B

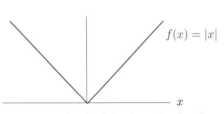

Figure 2.53: Graph of absolute value function, showing point of non-differentiability at $x = 0$

Examples of Nondifferentiable Functions

The best-known function with a corner is the *absolute value* function defined as follows:

$$f(x) = |x| = \begin{cases} x & \text{if } x \geq 0, \\ -x & \text{if } x < 0. \end{cases}$$

The graph of this function is in Figure 2.53. Near $x = 0$, even close-up views of the graph of $f(x)$ look the same, so this is a corner which can't be straightened out by zooming in.

Example 8   Try to compute the derivative of the function $f(x) = |x|$ at $x = 0$. Is f differentiable there?

Solution   To find the slope at $x = 0$, we want to look at

$$\lim_{h \to 0} \frac{f(h) - f(0)}{h} = \lim_{h \to 0} \frac{|h| - 0}{h} = \lim_{h \to 0} \frac{|h|}{h}.$$

As h approaches 0 from the right, h is always positive, so $|h| = h$, and the ratio is always 1. As h approaches 0 from the left, h is negative, so $|h| = -h$, and the ratio is -1. Since the limits are different from each side, the limit of the difference quotient does not exist. Thus, the absolute value function is not differentiable at $x = 0$. The limits of 1 and -1 correspond to the fact that the slope of the right-hand part of the graph is 1, and the slope of the left-hand part is -1.

Example 9   Investigate the differentiability of $f(x) = x^{1/3}$ at $x = 0$.

Solution   This function is smooth at $x = 0$ (no sharp corners) but appears to have a vertical tangent there. (See Figure 2.54.) Looking at the difference quotient at $x = 0$, we see

$$\lim_{h \to 0} \frac{(0 + h)^{1/3} - 0^{1/3}}{h} = \lim_{h \to 0} \frac{h^{1/3}}{h} = \lim_{h \to 0} \frac{1}{h^{2/3}}.$$

As $h \to 0$ the denominator becomes small, so the fraction grows without bound. Hence, the function fails to have a derivative at $x = 0$.

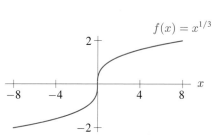

Figure 2.54: Continuous function not differentiable at $x = 0$: Vertical tangent

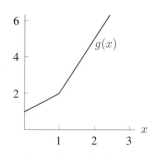

Figure 2.55: Continuous function not differentiable at $x = 1$

Example 10   Consider the function given by the formulas

$$g(x) = \begin{cases} x + 1 & \text{if} \quad x \leq 1 \\ 3x - 1 & \text{if} \quad x > 1. \end{cases}$$

This kind of function is called *piecewise linear* because each part of it is linear. Draw the graph of g. Is g continuous? Is g differentiable at $x = 1$?

Solution   The graph in Figure 2.55 has no breaks in it, so the function is continuous. However, the graph has a corner at $x = 1$ which no amount of magnification will remove. To the left of $x = 1$, the slope is 1; to the right of $x = 1$, the slope is 3. Thus, the difference quotient at $x = 1$ will fail to have a limit, and so the function g is not differentiable at $x = 1$.

A great deal of interest has been sparked in the last few years in the study of curves which do not possess derivatives *anywhere*. These curves, known as *fractals*, arise in the modeling of natural processes that are random and chaotic, such as the path of a water molecule in a glass of water. As the molecule bounces off of its neighbors haphazardly, it traces out a path with many jagged, nondifferentiable corners. Although the path may be smooth between collisions, it can be modeled effectively by a curve that is not differentiable anywhere. The coastlines of Maine and Washington are also examples. They never straighten out, no matter how close we look.

Differentiability and Linear Approximation

When we zoom in on the graph of a differentiable function, it looks like a straight line. In fact, the graph is not exactly a straight line when we zoom in; however, its deviation from straightness is so small that it can't be detected by the naked eye. Let's examine what this means. The straight line that we think we see when we zoom in on the graph of $f(x)$ at $x = a$ has slope equal to the derivative, $f'(a)$, so the equation is

$$y = f(a) + f'(a)(x - a).$$

The fact that the graph looks like a line means that y is a good approximation to $f(x)$. (See Figure 2.56.) This suggests the following definition:

The Tangent Line Approximation

Suppose f is differentiable at a. Then, for values of x near a, the tangent line approximation to $f(x)$ is

$$f(x) \approx f(a) + f'(a)(x - a).$$

The expression $f(a) + f'(a)(x - a)$ is called the *local linearization* of f near $x = a$. We are thinking of a as fixed, so that $f(a)$ and $f'(a)$ are constant.
The **error**, $E(x)$, in the approximation is defined by

$$E(x) = f(x) - f(a) - f'(a)(x - a).$$

It can be shown that the tangent line approximation is the best linear approximation to f near a. See Problem 15.

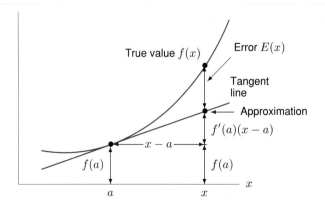

Figure 2.56: The tangent line approximation and its error

Example 11   What is the tangent line approximation for $f(x) = \sin x$ near $x = 0$? Assume that $f'(0) = 1$.

Solution The tangent line approximation of f near $x = 0$ is

$$f(x) \approx f(0) + f'(0)(x - 0).$$

If $f(x) = \sin x$, then $f(0) = \sin 0 = 0$. Using the given fact that $f'(0) = 1$, the approximation is

$$\sin x \approx x.$$

This means that, near $x = 0$, the function $f(x) = \sin x$ is well approximated by the function $y = x$. If we zoom in on the graphs of the functions $\sin x$ and x near the origin, we won't be able to tell them apart. (See Figure 2.57.)

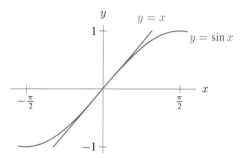

Figure 2.57: Tangent line approximation to $y = \sin x$

Estimating the Error in the Approximation

Let us look at the error, $E(x)$, which is the difference between $f(x)$ and the local linearization. (Look back at Figure 2.56.) The fact that the graph of f looks like a line as we zoom in means that not only is $E(x)$ small for x near a, but also that $E(x)$ is small relative to $(x - a)$. To demonstrate this, we prove the following theorem about the ratio $E(x)/(x - a)$.

> ### Theorem: Differentiability and Local Linearity
>
> Suppose f is differentiable at $x = a$ and $E(x)$ is the error in the tangent line approximation, that is:
>
> $$E(x) = f(x) - f(a) - f'(a)(x - a).$$
>
> Then
>
> $$\lim_{x \to a} \frac{E(x)}{x - a} = 0.$$

Proof Using the definition of $E(x)$, we have

$$\frac{E(x)}{x - a} = \frac{f(x) - f(a) - f'(a)(x - a)}{x - a} = \frac{f(x) - f(a)}{x - a} - f'(a).$$

Taking the limit as $x \to a$ and using the definition of the derivative, we see that

$$\lim_{x \to a} \frac{E(x)}{x - a} = \lim_{x \to a} \left(\frac{f(x) - f(a)}{x - a} - f'(a) \right) = f'(a) - f'(a) = 0.$$

Why Differentiability Makes A Graph Look Straight

We can use the error $E(x)$ to understand why differentiability makes a graph look straight when we zoom in.

Example 12   Consider the graph of $f(x) = \sin x$ near $x = 0$, and its linear approximation computed in Example 11. Show that there is an interval around 0 with the property that the distance from $f(x) = \sin x$ to the linear approximation is less than $0.1|x|$ for all x in the interval.

Solution   The linear approximation of $f(x) = \sin x$ near 0 is $y = x$, so we write

$$\sin x = x + E(x).$$

Since $\sin x$ is differentiable at $x = 0$, the theorem tells us that

$$\lim_{x \to 0} \frac{E(x)}{x} = 0.$$

If we take $\epsilon = 1/10$, then the definition of limit guarantees that there is a $\delta > 0$ such that

$$\left| \frac{E(x)}{x} \right| < 0.1 \quad \text{for all} \quad |x| < \delta.$$

In other words, for x in the interval $(-\delta, \delta)$, we have $|x| < \delta$, so

$$|E(x)| < 0.1|x|.$$

(See Figure 2.58.)

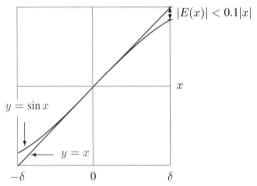

Figure 2.58: Graph of $y = \sin x$ and its linear approximation $y = x$, showing a window in which the magnitude of the error, $|E(x)|$, is less than $0.1|x|$ for all x in the window

We can generalize from this example to explain why differentiability makes the graph of f look straight when viewed over a small graphing window. Suppose f is differentiable at $x = a$. Then we know $\lim\limits_{x \to a} \left| \dfrac{E(x)}{x - a} \right| = 0$. So, for any $\epsilon > 0$, we can find a δ small enough so that

$$\left| \frac{E(x)}{x - a} \right| < \epsilon, \qquad \text{for} \quad a - \delta < x < a + \delta.$$

So, for any x in the interval $(a - \delta, a + \delta)$, we have

$$|E(x)| < \epsilon|x - a|.$$

Thus, the error, $E(x)$, is less than ϵ times $|x - a|$, the distance between x and a. So, as we zoom in on the graph by choosing smaller ϵ, the deviation, $|E(x)|$, of f from its tangent line shrinks, even relative to the scale on the x-axis. So, zooming makes a differentiable function look straight.

Differentiability and Continuity

The fact that a function which is differentiable at a point has a tangent line suggests that the function is continuous there, as the next theorem shows.

> ### Theorem: A Differentiable Function Is Continuous
>
> If $f(x)$ is differentiable at a point $x = a$, then $f(x)$ is continuous at $x = a$.

Proof   We assume $f(x)$ is differentiable at $x = a$. Then we know

$$f'(a) = \lim_{x \to a} \frac{f(x) - f(a)}{x - a}.$$

So we must have

$$\lim_{x \to a}(f(x) - f(a)) = \lim_{x \to a}\left((x - a)\frac{f(x) - f(a)}{x - a}\right) = \left(\lim_{x \to a}(x - a)\right) \cdot \left(\lim_{x \to a}\frac{f(x) - f(a)}{x - a}\right)$$
$$= 0 \cdot f'(a) = 0.$$

Then,

$$\lim_{x \to a} f(x) = f(a),$$

which means $f(x)$ is continuous at $x = a$.

Problems on Differentiability and Linear Approximation

1. For each of the graphs in Figure 2.59, list the x-values for which the function appears to be
   (i) Not continuous and (ii) Not differentiable.

Figure 2.59

2. Look at the graph of $f(x) = (x^2 + 0.0001)^{1/2}$ shown in Figure 2.60. The graph of f appears to have a sharp corner at $x = 0$. Do you think f has a derivative at $x = 0$?

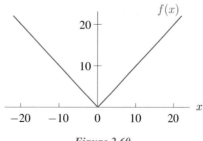

Figure 2.60

Decide if the functions in Problems 3–5 are differentiable at $x = 0$. Try zooming in on a graphing calculator, or calculating the derivative $f'(0)$ from the definition.

3. $f(x) = (x + |x|)^2 + 1$

4. $f(x) = \begin{cases} x \sin(1/x) + x & \text{for } x \neq 0 \\ 0 & \text{for } x = 0 \end{cases}$

5. $f(x) = \begin{cases} x^2 \sin(1/x) & \text{for } x \neq 0 \\ 0 & \text{for } x = 0 \end{cases}$

6.   An electric charge, Q, in a circuit is given as a function of time, t, by

$$Q = \begin{cases} C & \text{for } t \leq 0 \\ Ce^{-t/RC} & \text{for } t > 0, \end{cases}$$

where C and R are positive constants. The electric current, I, is the rate of change of charge, so

$$I = \frac{dQ}{dt}.$$

(a)   Is the charge, Q, a continuous function of time?

(b)   Do you think the current, I, is defined for all times, t? [Hint: To graph this function, take, for example, $C = 1$ and $R = 1$.]

7.   A magnetic field, B, is given as a function of the distance, r, from the center of a wire as follows:

$$B = \begin{cases} \dfrac{r}{r_0} B_0 & \text{for } r \leq r_0 \\[2mm] \dfrac{r_0}{r} B_0 & \text{for } r > r_0. \end{cases}$$

(a)   Sketch a graph of B against r. What is the meaning of the constant B_0?

(b)   Is B continuous at $r = r_0$? Give reasons.

(c)   Is B differentiable at $r = r_0$? Give reasons.

8.   A cable is made of an insulating material in the shape of a long, thin cylinder of radius r_0. It has electric charge distributed evenly throughout it. The electric field, E, at a distance r from the center of the cable is given by

$$E = \begin{cases} kr & \text{for } r \leq r_0 \\[1mm] k\dfrac{r_0^2}{r} & \text{for } r > r_0. \end{cases}$$

(a)   Is E continuous at r_0?

(b)   Is E differentiable at r_0?

(c)   Sketch a graph of E as a function of r.

9.   Graph the function defined by

$$g(r) = \begin{cases} 1 + \cos(\pi r/2) & \text{for } -2 \leq r \leq 2 \\ 0 & \text{for } r < -2 \text{ or } r > 2. \end{cases}$$

(a)   Is g continuous at $r = 2$? Explain your answer.

(b)   Do you think g is differentiable at $r = 2$? Explain your answer.

10.   The potential, ϕ, of a charge distribution at a point on the y-axis is given by

$$\phi = \begin{cases} 2\pi\sigma \left(\sqrt{y^2 + a^2} - y \right) & \text{for } y \geq 0 \\[2mm] 2\pi\sigma \left(\sqrt{y^2 + a^2} + y \right) & \text{for } y < 0 \end{cases}$$

where σ and a are positive constants. [Hint: To graph this function, take, for example, $2\pi\sigma = 1$ and $a = 1$.]

(a)   Is ϕ continuous at $y = 0$?

(b)   Do you think ϕ is differentiable at $y = 0$?

11. Consider the function $f(x) = \sqrt{x}$. Assume $f'(4) = 1/4$.

   (a) Find and sketch $f(x)$ and the tangent line approximation to $f(x)$ near $x = 4$.
   (b) Compare the true value of $f(4.1)$ with the value obtained by using the tangent line approximation.
   (c) Compare the true and approximate values of $f(16)$.
   (d) Using a graph, explain why the tangent line approximation is a good one when $x = 4.1$ but not when $x = 16$.

12. Local linearization will give values too small for the function x^2 and too large for the function $\sqrt{x}$. Draw pictures to explain why.

13. Find the local linearization of $f(x) = x^2$ near $x = 1$.

14. Consider the graph of $f(x) = x^2$ near $x = 1$. Find an interval around $x = 1$ with the property that in any smaller interval, the graph of $f(x) = x^2$ never diverges from its local linearization by more than $0.1|x - 1|$ for all x in the interval.

15. Consider a function f and a point a. Suppose there is a number L such that the linear function g

$$g(x) = f(a) + L(x - a)$$

is a good approximation to f. By good approximation, we mean that

$$\lim_{x \to a} \frac{E_L(x)}{x - a} = 0,$$

where $E_L(x)$ is the approximation error defined by

$$f(x) = g(x) + E_L(x) = f(a) + L(x - a) + E_L(x).$$

Show that f is differentiable at $x = a$ and that $f'(a) = L$. Thus the tangent line approximation is the only good linear approximation.

CHAPTER THREE

KEY CONCEPT: THE DEFINITE INTEGRAL

We started Chapter 2 by calculating the velocity from the distance traveled. This led us to the notion of the derivative, or rate of change, of a function. Now we consider the reverse problem: given the velocity, how can we calculate the distance traveled? This will lead us to our second key concept, the *definite integral*, which computes the total change in a function from its rate of change. We will then discover that the definite integral can be used to calculate not only distance but also other quantities, such as the area under a curve and the average value of a function.

We end this chapter with the Fundamental Theorem of Calculus, which tells us that we can use a definite integral to get information about a function from its derivative. Calculating derivatives and calculating definite integrals are, in a sense, reverse processes.

3.1 HOW DO WE MEASURE DISTANCE TRAVELED?

If the velocity of a moving object is a constant, we can find the distance it travels using the formula

$$\text{Distance} = \text{Velocity} \times \text{Time}.$$

In this section we see how to estimate the distance when the velocity is not a constant.

A Thought Experiment: How Far Did the Car Go?

Velocity Data Every Two Seconds

Suppose a car is moving with increasing velocity. Suppose we measure the car's velocity every two seconds, and obtain the data in Table 3.1:

TABLE 3.1 *Velocity of car every two seconds*

Time (sec)	0	2	4	6	8	10
Velocity (ft/sec)	20	30	38	44	48	50

How far has the car traveled? Since we don't know how fast the car is moving at every moment, we can't calculate the distance exactly, but we can make an estimate. The velocity is increasing, so the car is going at least 20 ft/sec for the first two seconds. Since Distance = Velocity × Time, the car goes at least $(20)(2) = 40$ feet during the first two seconds. Likewise, it goes at least $(30)(2) = 60$ feet during the next two seconds, and so on. During the ten-second period it goes at least

$$(20)(2) + (30)(2) + (38)(2) + (44)(2) + (48)(2) = 360 \text{ feet}.$$

Thus, 360 feet is an underestimate of the total distance traveled during the ten seconds.

To get an overestimate, we can reason this way: During the first two seconds, the car's velocity is at most 30 ft/sec, so it moved at most $(30)(2) = 60$ feet. In the next two seconds it moved at most $(38)(2) = 76$ feet, and so on. Therefore, over the ten-second period it moved at most

$$(30)(2) + (38)(2) + (44)(2) + (48)(2) + (50)(2) = 420 \text{ feet}.$$

Therefore,

$$360 \text{ feet} \leq \text{Total distance traveled} \leq 420 \text{ feet}.$$

There is a difference of 60 feet between the upper and lower estimates.

Velocity Data Every One Second

What if we want a more accurate estimate? Then we should make more frequent velocity measurements, say every second. The data is in Table 3.2.

As before, we get a lower estimate for each second by using the velocity at the beginning of that second. During the first second the velocity is at least 20 ft/sec, so the car travels at least $(20)(1) = 20$ feet. During the next second the car moves at least 26 feet, and so on. So now we can say

$$
\begin{aligned}
\text{New lower estimate} ={}& (20)(1) + (26)(1) + (30)(1) + (35)(1) + (38)(1) \\
& + (42)(1) + (44)(1) + (46)(1) + (48)(1) + (49)(1) \\
={}& 378 \text{ feet}.
\end{aligned}
$$

TABLE 3.2 *Velocity of car every second*

Time (sec)	0	1	2	3	4	5	6	7	8	9	10
Velocity (ft/sec)	20	26	30	35	38	42	44	46	48	49	50

Notice that this is greater than the old lower estimate of 360 feet.

We get a new upper estimate by considering the velocity at the end of each second. During the first second the velocity is at most 26 ft/sec, and so the car moves at most $(26)(1) = 26$ feet; in the next second it moves at most 30 feet, and so on.

$$\begin{aligned}
\text{New upper estimate} &= (26)(1) + (30)(1) + (35)(1) + (38)(1) + (42)(1) \\
&\quad + (44)(1) + (46)(1) + (48)(1) + (49)(1) + (50)(1) \\
&= 408 \text{ feet.}
\end{aligned}$$

This is less than the old upper estimate of 420 feet. Now we know that

$$378 \text{ feet} \leq \text{Total distance traveled} \leq 408 \text{ feet.}$$

Notice that the difference between the new upper and lower estimates is now 30 feet, half of what it was before. By halving the interval of measurement, we have halved the difference between the upper and lower estimates.

Visualizing Distance on the Velocity Graph: Two Second Data

We can represent both upper and lower estimates on a graph of the velocity. On such a graph we can also see how changing the time interval between velocity measurements changes the accuracy of our estimates.

The velocity can be graphed by plotting the two second data in Table 3.1 and drawing a smooth curve through the data points. (See Figure 3.1.) The area of the first dark rectangle is $(20)(2) = 40$, the lower estimate of the distance moved during the first two seconds. The area of the second dark rectangle is $(30)(2) = 60$, the lower estimate for the distance moved in the next two seconds. The total area of the dark rectangles represents the lower estimate for the total distance moved during the ten seconds.

If the dark and light rectangles are considered together, the first area is $(30)(2) = 60$, the upper estimate for the distance moved in the first two seconds. The second area is $(38)(2) = 76$, the upper estimate for the next two seconds. Continuing this calculation suggests that the upper estimate for the total distance is represented by the sum of the areas of the dark and light rectangles. Therefore, the area of the light rectangles alone represents the difference between the two estimates.

To calculate the difference between the two estimates, look at Figure 3.1 and imagine the light rectangles all pushed to the right and stacked on top of each other. This gives a rectangle of width 2 and height 30. Notice that the height, 30, is just the difference between the initial and final values of the velocity: $30 = 50 - 20$. The width, 2, is the time interval between velocity measurements.

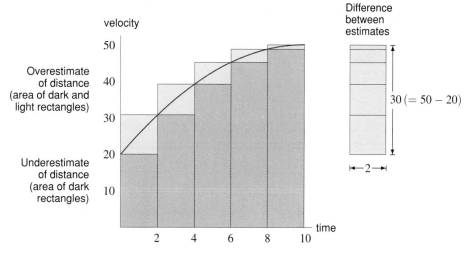

Figure 3.1: Velocity measured every 2 seconds

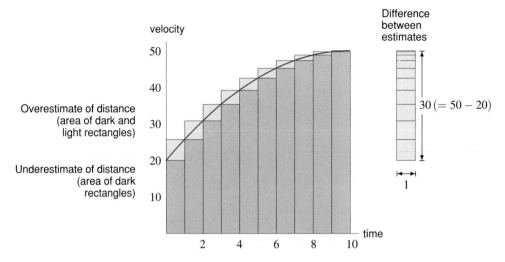

Figure 3.2: Velocity measured every second

Visualizing Distance on the Velocity Graph: One Second Data

The data for the velocities measured every second is graphed in Figure 3.2. The area of the dark rectangles again represents the lower estimate, and the dark and light rectangles together represent the upper estimate. As before, the difference between the two estimates is represented by the area of the light rectangles. This difference can be calculated by stacking the light rectangles vertically, giving a rectangle of the same height as before but of half the width. Its area is therefore half what it was before. Again, the height of this stack is $50 - 20 = 30$, but its width is the time interval between measurements, which is 1.

Example 1 What would be the difference between the upper and lower estimates if the velocity were given every tenth of a second? Every hundredth of a second? Every thousandth of a second?

Solution Every tenth of a second: Difference between estimates $= (50 - 20)(1/10) = 3$ feet.
Every hundredth of a second: Difference between estimates $= (50 - 20)(1/100) = 0.3$ feet.
Every thousandth of a second: Difference between estimates $= (50 - 20)(1/1000) = 0.03$ feet.

Example 2 How frequently must the velocity be recorded in order to estimate the total distance traveled to within 0.1 feet?

Solution The difference between the velocity at the beginning and end of the observation period is $50 - 20 = 30$. If the time between the measurements is h, then the difference between the upper and lower estimates is $(30)h$. We want

$$(30)h < 0.1,$$

or

$$h < \frac{0.1}{30} \approx 0.0033.$$

So if the measurements are made less than 0.0033 seconds apart, the distance estimate will be accurate to within 0.1 feet.

Making the Estimates for Distance Precise

We now obtain an exact expression for the total distance traveled. We express the exact total distance traveled as a limit of upper or lower estimates.

We want to know the distance traveled by a moving object over the time interval $a \leq t \leq b$. Let the velocity at time t be given by the function $v = f(t)$. We take measurements of $f(t)$ at equally spaced times $t_0, t_1, t_2, \ldots, t_n$, with time $t_0 = a$ and time $t_n = b$. The time interval between any two consecutive measurements is

$$\Delta t = \frac{b-a}{n},$$

where Δt means the change, or increment, in t.

During the first time interval, the velocity can be approximated by $f(t_0)$, so the distance traveled is approximately

$$f(t_0)\Delta t.$$

During the second time interval, the velocity is about $f(t_1)$, so the distance traveled is about

$$f(t_1)\Delta t.$$

Continuing in this way and adding up all the estimates, we get an estimate for the total distance. In the last interval, the velocity is approximately $f(t_{n-1})$, so the last term is $f(t_{n-1})\Delta t$:

$$\text{Total distance traveled} \atop \text{between } t = a \text{ and } t = b \qquad \approx f(t_0)\Delta t + f(t_1)\Delta t + f(t_2)\Delta t + \cdots + f(t_{n-1})\Delta t.$$

This is called a *left-hand sum* because we used the value of velocity from the left end of each time interval. It is represented by the sum of the areas of the rectangles in Figure 3.3.

We can also calculate a *right-hand sum* by using the value of the velocity at the right end of each time interval. In that case the estimate for the first interval is $f(t_1)\Delta t$, for the second interval it is $f(t_2)\Delta t$, and so on. The estimate for the last interval is now $f(t_n)\Delta t$, so

$$\text{Total distance traveled} \atop \text{between } t = a \text{ and } t = b \qquad \approx f(t_1)\Delta t + f(t_2)\Delta t + f(t_3)\Delta t + \cdots + f(t_n)\Delta t.$$

The right-hand sum is represented by the area of the rectangles in Figure 3.4.

If f is an increasing function, the left-hand sum is an underestimate of the total distance traveled. The reason is that a left-hand sum uses the velocity at the start of each interval to compute the distance

Figure 3.3: Left-hand sums

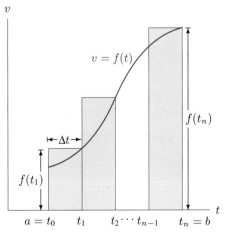

Figure 3.4: Right-hand sums

traveled, whereas in fact the velocity continues to increase after that measurement. Similarly, if f is increasing, the right-hand sum is an overestimate. If f is decreasing, as in Figure 3.5, then the roles of the two sums are reversed.

For either increasing or decreasing functions, the exact value of the distance traveled lies somewhere between the two estimates. Thus the accuracy of our estimate depends on how close these two sums are. For a function which is increasing throughout or decreasing throughout the interval $[a, b]$:

$$\left| \begin{array}{c} \text{Difference between} \\ \text{upper and lower estimates} \end{array} \right| = \left| \begin{array}{c} \text{Difference between} \\ f(a) \text{ and } f(b) \end{array} \right| \times \Delta t = |f(b) - f(a)| \cdot \Delta t.$$

(Absolute values are used to make the difference nonnegative.) In Figure 3.5, the area of the light rectangles is the difference between estimates. By making the time interval, Δt, between measurements small enough, we can make this difference between lower and upper estimates as small as we like.

In the car example, as n increased, the overestimates of the distance traveled decreased and the underestimates increased, trapping the exact distance between them. This suggests that to find exactly the total distance traveled between $t = a$ and $t = b$, we take the limit of the sums, as n goes to infinity.

$$\begin{array}{ll} \text{Total distance traveled} \\ \text{between } t = a \text{ and } t = b \end{array} = \lim_{n \to \infty} (\text{Left-hand sum})$$

$$= \lim_{n \to \infty} \left[f(t_0)\Delta t + f(t_1)\Delta t + \cdots + f(t_{n-1})\Delta t \right]$$

$$= \text{Area under curve } f(t) \text{ from } t = a \text{ to } t = b$$

and

$$\begin{array}{ll} \text{Total distance traveled} \\ \text{between } t = a \text{ and } t = b \end{array} = \lim_{n \to \infty} (\text{Right-hand sum})$$

$$= \lim_{n \to \infty} \left[f(t_1)\Delta t + f(t_2)\Delta t + \cdots + f(t_n)\Delta t \right]$$

$$= \text{Area under curve } f(t) \text{ from } t = a \text{ to } t = b.$$

Provided f is continuous, the limits of the left and the right sums are both equal to the total distance traveled. This method of calculating the distance by taking the limit of a sum works even if the velocity is not increasing throughout, or decreasing throughout, the interval.

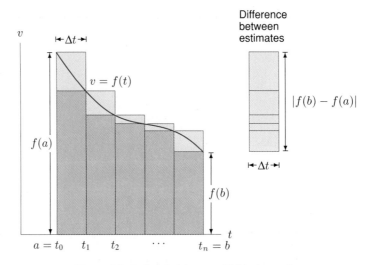

Figure 3.5: Left and right sums if f is decreasing

Problems for Section 3.1

1. A car comes to a stop six seconds after the driver applies the brakes. While the brakes are on, the following velocities are recorded:

Time since brakes applied (sec)	0	2	4	6
Velocity (ft/sec)	88	45	16	0

   (a) Give lower and upper estimates for the distance the car traveled after the brakes were applied.
   (b) On a sketch of velocity against time, show the lower and upper estimates of part (a).

2. A student is speeding down Route 11 in his fancy red Porche when his radar system warns him of an obstacle 400 feet ahead. He immediately applies the brakes, starts to slow down, and spots a skunk in the road directly ahead of him.

   Suppose that the "black box" in the Porche records the car's speed every two seconds, producing the following table. Assume that the speed decreases throughout the 10 seconds it takes to stop, although not necessarily at a uniform rate.

Time since brakes applied (sec)	0	2	4	6	8	10
Speed (ft/sec)	100	80	50	25	10	0

   (a) Using the information in this table, what is your best estimate of the total distance that the student's car traveled before coming to rest?
   (b) Which statement below can you justify from the information given in the story and data table? (Choose one and justify it.)
      (i) The car stopped before getting to the skunk.
      (ii) The "black box" data is inconclusive. The skunk may or may not have been hit.
      (iii) The unfortunate skunk was hit by the car.

3. Roger decides to run a marathon. Roger's friend Jeff rides behind him on a bicycle and clocks his pace every 15 minutes. Roger starts out strong, but after an hour and a half he is so exhausted that he has to stop. The data Jeff collected are summarized below:

Time spent running (min)	0	15	30	45	60	75	90
Speed (mph)	12	11	10	10	8	7	0

   (a) Assuming that Roger's speed is never increasing, give upper and lower estimates for the distance Roger ran during the first half hour.
   (b) Give upper and lower estimates for the distance Roger ran in total during the entire hour and a half.
   (c) How often would Jeff have needed to measure Roger's pace in order to find lower and upper estimates within 0.1 mile of the actual distance that he ran?

4. Coal gas is produced at a gasworks. Pollutants in the gas are removed by scrubbers, which become less and less efficient as time goes on. The following measurements, made at the start of each month, show the rate at which pollutants are escaping in the gas:

Time (months)	0	1	2	3	4	5	6
Rate pollutants are escaping (tons/month)	5	7	8	10	13	16	20

   (a) Make an overestimate and an underestimate of the total quantity of pollutants that escaped during the first month.

(b)   Make an overestimate and an underestimate of the total quantity of pollutants that escaped during the six months.

(c)   How often would measurements have to be made in order to find overestimates and underestimates which differ by less than 1 ton from the exact quantity of pollutants that escaped during the first six months?

5.   Suppose time, t, is given in seconds and your velocity, v, in meters/second, is given by

$$v(t) = 1 + t^2 \quad \text{for} \quad 0 \le t \le 6.$$

Use $\Delta t = 2$ to estimate the distance traveled during this time. Find the left- and right-hand sums, and then average the two.

6.   For $0 \le t \le 1$, a bug is crawling at a velocity, v, determined by the formula

$$v = \frac{1}{1+t},$$

where t is in hours and v is in meters/hour. Use $\Delta t = 0.2$ to estimate the distance that the bug crawls during this hour. Find an overestimate and an underestimate. Then average the two to get a new estimate.

7.   In Figure 3.6, use the grid to estimate the area of the region bounded by the curve, the horizontal axis and the vertical lines $x = 3$ and $x = -3$. Get an upper and a lower estimate that are within 4 square units of one another. Explain your answer.

Figure 3.6

Figure 3.7

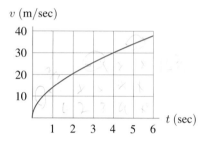

Figure 3.8

8.   (a)   In Figure 3.7 estimate the shaded area with an error of at most 0.1.
      (b)   How can you approximate this shaded area to any desired degree of accuracy?

9.   Figure 3.8 shows the graph of the velocity, v, of an object (in m/sec). Estimate the total distance the object traveled between $t = 0$ and $t = 6$.

10.   You jump out of an airplane. Before your parachute opens you fall faster and faster, but your acceleration decreases as you fall because of air resistance. The table below gives your acceleration, a (in m/sec^2), after t seconds.

t	0	1	2	3	4	5
a	9.81	8.03	6.53	5.38	4.41	3.61

(a)   Give upper and lower estimates of your speed at $t = 5$.

(b)   Get a new estimate by taking the average of your upper and lower estimates. What does the concavity of the graph of acceleration tell you about your new estimate?

11.   When an aircraft attempts to climb as rapidly as possible, its climb rate decreases with altitude. (This occurs because the air is less dense at higher altitudes.) Table 3.3 shows performance data for a certain single-engine aircraft.

TABLE 3.3

Altitude (1000 ft)	0	1	2	3	4	5	6	7	8	9	10
Climb rate (ft/min)	925	875	830	780	730	685	635	585	535	490	440

(a)   Calculate upper and lower estimates for the time required for this aircraft to climb from sea level to 10,000 ft.

(b)   If climb rate data were available in increments of 500 ft, what would be the difference between a lower and upper estimate of climb time based on 20 subdivisions?

3.2 THE DEFINITE INTEGRAL

In Section 3.1 we saw how distance traveled can be approximated by sums and expressed exactly as the limit of a sum. In this section we show how these sums can be defined for any function f, whether or not it represents a velocity. Suppose we have a function $f(t)$ which is continuous for $a \leq t \leq b$. We divide the interval from a to b into n equal subdivisions, and we call the width of an individual subdivision Δt, so

$$\Delta t = \frac{b - a}{n}.$$

We let $t_0, t_1, t_2, \ldots, t_n$ be endpoints of the subdivisions, as in Figures 3.9 and 3.10. As before, we construct two sums:

$$\text{Left-hand sum} = f(t_0)\Delta t + f(t_1)\Delta t + \cdots + f(t_{n-1})\Delta t$$

and

$$\text{Right-hand sum} = f(t_1)\Delta t + f(t_2)\Delta t + \cdots + f(t_n)\Delta t.$$

These sums represent the areas in Figures 3.9 and 3.10, provided $f(t) \geq 0$. In Figure 3.9, the first rectangle has width Δt and height $f(t_0)$, since the top of its left edge just touches the curve, and hence it has area $f(t_0)\Delta t$. The second rectangle has width Δt and height $f(t_1)$, and hence has area $f(t_1)\Delta t$, and so on. The sum of all these areas is the left-hand sum. The right-hand sum, shown in Figure 3.10, is constructed in the same way, except that each rectangle touches the curve on its right edge instead of its left.

Figure 3.9: Left-hand sum

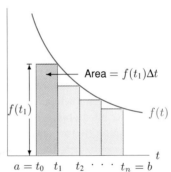

Figure 3.10: Right-hand sum

Writing Left and Right Sums Using Sigma Notation

Both the left-hand and right-hand sums can be written more compactly using *sigma*, or summation, notation. The symbol $\sum$ is a capital sigma, or Greek letter "S." We write

$$\text{Right-hand sum} = \sum_{i=1}^{n} f(t_i)\Delta t = f(t_1)\Delta t + f(t_2)\Delta t + \cdots + f(t_n)\Delta t.$$

The $\sum$ tells us to add terms of the form $f(t_i)\Delta t$. The "$i = 1$" at the base of the sigma sign tells us to start at $i = 1$, and the "n" at the top tells us to stop at $i = n$.

In the left-hand sum we start at $i = 0$ and stop at $i = n - 1$, so we write

$$\text{Left-hand sum} = \sum_{i=0}^{n-1} f(t_i)\Delta t = f(t_0)\Delta t + f(t_1)\Delta t + \cdots + f(t_{n-1})\Delta t.$$

Taking the Limit to Obtain the Definite Integral

In the previous section we took the limit of these sums as n went to infinity, and we do the same here. If f is continuous for $a \leq t \leq b$, the limits of the left- and right-hand sums exist and are equal. The *definite integral* is the limit of these sums. A formal definition of the definite integral is given in the Focus on Theory section on page 181.

Suppose f is continuous for $a \leq t \leq b$. The **definite integral** of f from a to b, written

$$\int_a^b f(t)\, dt,$$

is the limit of the left-hand or right-hand sums with n subdivisions of $[a, b]$ as n gets arbitrarily large. In other words,

$$\int_a^b f(t)\, dt = \lim_{n \to \infty} (\text{Left-hand sum}) = \lim_{n \to \infty} \left(\sum_{i=0}^{n-1} f(t_i) \Delta t \right)$$

and

$$\int_a^b f(t)\, dt = \lim_{n \to \infty} (\text{Right-hand sum}) = \lim_{n \to \infty} \left(\sum_{i=1}^{n} f(t_i) \Delta t \right).$$

Each of these sums is called a *Riemann sum*, f is called the *integrand*, and a and b are called the *limits of integration*.

The "$\int$" notation comes from an old-fashioned "S," which stands for "sum" in the same way that $\sum$ does. The "dt" in the integral comes from the factor Δt. Notice that the limits on the $\sum$ symbol are 0 and $n - 1$ for the left-hand sum, and 1 and n for the right-hand sum, whereas the limits on the $\int$ sign are a and b.

Computing a Definite Integral

In practice, we often approximate definite integrals numerically using a calculator or computer. They use programs which compute sums for larger and larger values of n, and eventually give a value for the integral. Different calculators and computers may give slightly different estimates, owing to round-off error and the fact that they may use different approximation methods. Some (but not all) definite integrals can be computed exactly. However, any definite integral can be approximated numerically.

In the next example, we see how numerical approximation works. For each value of n, we calculate an over- and an under-estimate for the integral. As we increase the value of n the over- and under-estimates get closer together, trapping the value of the integral between them. By increasing the value of n sufficiently, we can calculate the integral to any desired accuracy.

Example 1   Calculate the left-hand and right-hand sums with $n = 2$ and $n = 10$ for $\int_1^2 \dfrac{1}{t}\, dt$. How do the values of these sums compare with the exact value of the integral?

Solution   Here $a = 1$ and $b = 2$, so for $n = 2$, $\Delta t = (2 - 1)/2 = 0.5$. Therefore, $t_0 = 1$, $t_1 = 1.5$ and $t_2 = 2$. (See Figure 3.11.) We have

$$\begin{aligned}
\text{Left-hand sum} &= f(1)\Delta t + f(1.5)\Delta t \\
&= 1(0.5) + \frac{1}{1.5}(0.5) \\
&\approx 0.8333, \\
\text{Right-hand sum} &= f(1.5)\Delta t + f(2)\Delta t \\
&= \frac{1}{1.5}(0.5) + \frac{1}{2}(0.5) \\
&\approx 0.5833.
\end{aligned}$$

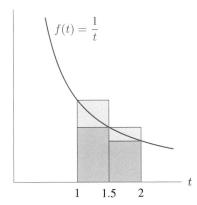

Figure 3.11: Approximating $\int_1^2 \frac{1}{t}\, dt$
with $n = 2$

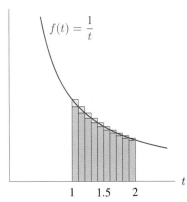

Figure 3.12: Approximating $\int_1^2 \frac{1}{t}\, dt$
with $n = 10$

From Figure 3.11 we see that the left-hand sum is bigger than the area under the curve and the right-hand sum is smaller. So the area under the curve $f(t) = 1/t$ from $t = 1$ to $t = 2$ is between 0.5833 and 0.8333 :

$$0.5833 < \int_1^2 \frac{1}{t}\, dt < 0.8333.$$

When $n = 10$, we have $\Delta t = (2 - 1)/10 = 0.1$ (see Figure 3.12), so

$$\text{Left-hand sum} = f(1)\Delta t + f(1.1)\Delta t + \cdots + f(1.9)\Delta t$$
$$= \left(1 + \frac{1}{1.1} + \cdots + \frac{1}{1.9}\right) 0.1$$
$$\approx 0.7188,$$
$$\text{Right-hand sum} = f(1.1)\Delta t + f(1.2)\Delta t + \cdots + f(2)\Delta t$$
$$= \left(\frac{1}{1.1} + \frac{1}{1.2} + \cdots + \frac{1}{2}\right) 0.1$$
$$\approx 0.6688.$$

From Figure 3.12 you can see that the left-hand sum is larger than the area under the curve, and the right-hand sum smaller, so

$$0.6688 < \int_1^2 \frac{1}{t}\, dt < 0.7188.$$

Notice that the left- and right-hand sums trap the exact value of the integral between them. As the subdivisions become finer, the left- and right-hand sums get closer together.

Example 2 Use left and right sums with $n = 250$ for $\int_1^2 \frac{1}{t}\, dt$ to estimate the value of the integral.

Solution Using a program on a calculator or computer, we see that

$$0.6921 < \int_1^2 \frac{1}{t}\, dt < 0.6941.$$

So we can say that

$$\int_1^2 \frac{1}{t}\, dt \approx 0.69.$$

to two decimal places. The exact value is known to be $\int_1^2 \frac{1}{t}\, dt = \ln 2 = 0.693147\ldots\,.$

The Definite Integral as an Area

If $f(x)$ is positive we can interpret each term $f(x_0)\Delta x$, $f(x_1)\Delta x$, ... in a left- or right-hand Riemann sum as the area of a rectangle. See Figure 3.13. As the width Δx of the rectangles approaches zero, the rectangles fit the curve of the graph more exactly, and the sum of their areas gets closer and closer to the area under the curve shaded in Figure 3.14. This suggests that:

When $f(x)$ is positive and $a < b$:

$$\text{Area under graph of } f \atop \text{between } a \text{ and } b = \int_a^b f(x)\,dx.$$

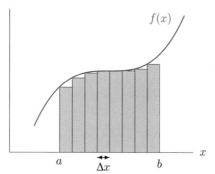

Figure 3.13: Area of rectangles approximating the area under the curve

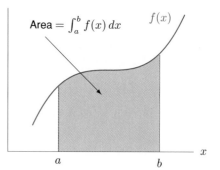

Figure 3.14: The definite integral $\int_a^b f(x)\,dx$

Example 3   Consider the integral $\displaystyle\int_{-1}^1 \sqrt{1 - x^2}\,dx$.

(a)   Interpret the integral as an area, and find its exact value.

(b)   Estimate the integral using a calculator or computer. Compare your answer to the exact value.

Solution   (a)   The integral is the area under the graph of $y = \sqrt{1 - x^2}$ between -1 and 1. Rewriting this equation as $x^2 + y^2 = 1$, we see that the graph is a semicircle of radius 1 and area $\pi/2$ (see Figure 3.15).

(b)   A calculator estimates the integral as $1.5707966\ldots$. For comparison, $\pi/2 = 1.5707963\ldots$.

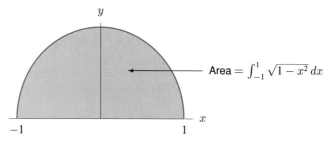

Figure 3.15: Area interpretation of $\int_{-1}^1 \sqrt{1 - x^2}\,dx$

When $f(x)$ is not positive

We have assumed in drawing Figure 3.14 that the graph of $f(x)$ lies above the x-axis. If the graph lies below the x-axis, then each value of $f(x)$ is negative, so each $f(x)\Delta x$ is negative, and the area gets counted negatively. In that case, the definite integral is the negative of the area.

Example 4   How does the definite integral $\int_{-1}^{1} (x^2 - 1)\, dx$ relate to the area between the parabola $y = x^2 - 1$ and the x-axis?

Solution   Using a calculator, we find $\int_{-1}^{1} (x^2 - 1)\, dx \approx -1.33$. The parabola lies below the axis between $x = -1$ and $x = 1$. (See Figure 3.16.) So the area between the parabola and the x-axis is approximately 1.33.

When $f(x)$ is positive for some x values and negative for others, and $a < b$:

$\int_{a}^{b} f(x)\, dx$ is the sum of the areas above the x-axis, counted positively, and the areas below the x-axis, counted negatively.

Example 5   Interpret the definite integral $\int_{0}^{\sqrt{2\pi}} \sin(x^2)\, dx$ in terms of areas.

Solution   The integral is the area above the x-axis, A_1, minus the area below the x-axis, A_2. See Figure 3.17. Approximating the integral with a calculator gives

$$\int_{0}^{\sqrt{2\pi}} \sin(x^2)\, dx \approx 0.43.$$

The graph of $y = \sin(x^2)$ crosses the x-axis where $x^2 = \pi$, that is, at $x = \sqrt{\pi}$. The next crossing is at $x = \sqrt{2\pi}$. Breaking the integral into two parts and calculating each one separately gives

$$\int_{0}^{\sqrt{\pi}} \sin(x^2)\, dx \approx 0.89 \quad \text{and} \quad \int_{\sqrt{\pi}}^{\sqrt{2\pi}} \sin(x^2)\, dx \approx -0.46.$$

So $A_1 \approx 0.89$ and $A_2 \approx 0.46$. Then, as we would expect,

$$\int_{0}^{\sqrt{2\pi}} \sin(x^2)\, dx = A_1 - A_2 \approx 0.89 - 0.46 = 0.43.$$

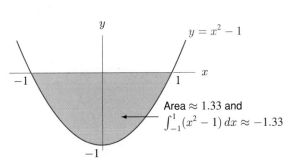

Figure 3.16: Integral $\int_{-1}^{1} (x^2 - 1)\, dx$ is negative of shaded area

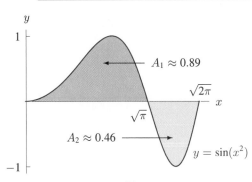

Figure 3.17: Integral $\int_{0}^{\sqrt{2\pi}} \sin(x^2)\, dx = A_1 - A_2$

Problems for Section 3.2

1. On a copy of Figure 3.18, draw rectangles representing each of the following Riemann sums for the function f on the interval $0 \le t \le 8$. Calculate the value of each sum.

    (a)   Left-hand sum with $\Delta t = 4$
    (b)   Right-hand sum with $\Delta t = 4$
    (c)   Left-hand sum with $\Delta t = 2$
    (d)   Right-hand sum with $\Delta t = 2$

Figure 3.18

2. Write out the terms of the right-hand sum with $n = 5$ that could be used to approximate $\displaystyle\int_3^7 \frac{1}{1+x}\,dx$. Do not evaluate the terms or the sum.

For Problems 3–8, construct a table of left- and right-hand sums with 2, 10, 50, and 250 subdivisions. Observe the limit to which your sums are tending as the number of subdivisions gets larger, and hence estimate the value of the definite integral.

3. $\displaystyle\int_0^1 x^3\,dx$

4. $\displaystyle\int_0^{\pi/2} \cos x\,dx$

5. $\displaystyle\int_2^3 \sin(t^2)\,dt$

6. $\displaystyle\int_0^1 e^{t^2}\,dt$

7. $\displaystyle\int_{0.2}^3 \sin\left(1/x\right)\,dx$

8. $\displaystyle\int_1^2 x^x\,dx$

9. Estimate $\displaystyle\int_1^2 x^2\,dx$ using left- and right-hand sums with four subdivisions. How far from the true value of the integral could your estimate be?

10. (a)   Use a calculator or computer to find $\displaystyle\int_0^6 (x^2 + 1)\,dx$. Represent this value as the area under a curve.

    (b)   Estimate $\displaystyle\int_0^6 (x^2 + 1)\,dx$ using a left-hand sum with $n = 3$. Represent this sum graphically on a sketch of $f(x) = x^2 + 1$. Is this sum an overestimate or underestimate of the true value found in part (a)?

    (c)   Estimate $\displaystyle\int_0^6 (x^2 + 1)\,dx$ using a right-hand sum with $n = 3$. Represent this sum on your sketch. Is this sum an overestimate or underestimate?

11. A table of values for $f(t)$ is given. Estimate $\displaystyle\int_0^{100} f(t)\,dt$.

t	0	20	40	60	80	100
$f(t)$	1.2	2.8	4.0	4.7	5.1	5.2

Estimate the area of the regions in Problems 12–17.

12. Under the curve $y = \cos t$ for $0 \le t \le \pi/2$.

13. Under the curve $y = 7 - x^2$ and above the x-axis.

14. Under the curve $y = \cos \sqrt{x}$ for $0 \le x \le 2$.

15. Under the curve $y = e^x$ and above the line $y = 1$ for $0 \le x \le 2$.

16. Between $y = x^2$ and $y = x^3$ for $0 \le x \le 1$.

17. Between $y = x^{1/2}$ and $y = x^{1/3}$ for $0 \le x \le 1$.

18. In Problems 16 and 17 we calculated the areas between $y = x^2$ and $y = x^3$ and between $y = x^{1/2}$ and $y = x^{1/3}$ on $0 \le x \le 1$. Explain why you would expect these two areas to be equal.

19. (a) Sketch a graph of $f(x) = x(x + 2)(x - 1)$.
    (b) Find the total area between the graph and the x-axis between $x = -2$ and $x = 1$.
    (c) Find $\displaystyle\int_{-2}^{1} f(x)\, dx$ and interpret it in terms of areas.

20. Compute the definite integral $\displaystyle\int_{0}^{4} \cos \sqrt{x}\, dx$ and interpret the result in terms of areas.

21. Without computing the integral, decide if

$$\int_{0}^{2\pi} e^{-x} \sin x \, dx$$

is positive or negative, and explain your decision. [Hint: Sketch $e^{-x} \sin x$.]

22. Suppose that we use $n = 500$ subintervals to approximate $\displaystyle\int_{-1}^{1} (2x^3 + 4)\, dx$. Without computing the Riemann sums, find the difference between the right- and left-hand Riemann sums.

23. The graph of a function $f(t)$ is given in Figure 3.19. Which of the following four numbers could be an estimate of $\displaystyle\int_{0}^{1} f(t)\,dt$ accurate to two decimal places? Explain how you chose your answer.

    (a) -98.35 (b) 71.84 (c) 100.12 (d) 93.47

Figure 3.19

Figure 3.20

24. The graph of $y = f(x)$ is given in Figure 3.20.

    (a) What is $\displaystyle\int_{-3}^{0} f(x)\, dx$?

    (b) If the area of the shaded region is A, estimate $\displaystyle\int_{-3}^{4} f(x)\, dx$.

25. Consider the definite integral $\int_0^1 x^4 dx$.

    (a) Write an expression for a right-hand Riemann sum approximation for this integral using n subdivisions. Express each x_i, $i = 1,2,\ldots,n$, in terms of i.
    (b) Use a computer algebra system to obtain a formula for the sum you wrote in part (a) in terms of n.
    (c) Take the limit of this expression for the sum as $n \to \infty$, thereby finding the exact value of this integral.

26. Repeat Problem 25, using the definite integral $\int_0^1 x^5 dx$.

27. Three terms of a left-hand sum used to approximate a definite integral $\int_a^b f(x)\, dx$ are as follows.

$$\left(2 + 0 \cdot \frac{4}{3}\right)^2 \cdot \frac{4}{3} + \left(2 + 1 \cdot \frac{4}{3}\right)^2 \cdot \frac{4}{3} + \left(2 + 2 \cdot \frac{4}{3}\right)^2 \cdot \frac{4}{3}.$$

    Find possible values for a and b and a possible formula for $f(x)$.

28. Consider the integral $\int_1^2 \frac{1}{t} dt$. In Example 1, by dividing the interval $1 \le t \le 2$ into 10 equal parts, we showed that

$$0.1 \left[\frac{1}{1.1} + \frac{1}{1.2} + \ldots + \frac{1}{2.0}\right] \le \int_1^2 \frac{1}{t} dt \le 0.1 \left[\frac{1}{1} + \frac{1}{1.1} + \ldots + \frac{1}{1.9}\right].$$

    (a) Now divide the interval $1 \le t \le 2$ into n equal parts to show that

$$\sum_{r=1}^{n} \frac{1}{n+r} < \int_1^2 \frac{1}{t} dt < \sum_{r=0}^{n-1} \frac{1}{n+r}.$$

    (b) Show that the difference between the upper and lower sums in part (a) is $1/2n$.
    (c) The exact value of $\int_1^2 (1/t)\, dt$ is $\ln(2)$. How large should n be to approximate $\ln(2)$ with an error of at most $5 \cdot 10^{-6}$, using one of the sums in part (a)?

3.3 INTERPRETATIONS OF THE DEFINITE INTEGRAL

The Notation and Units for the Definite Integral

Just as the Leibniz notation dy/dx for the derivative reminds us that the derivative is the limit of a ratio of differences, the notation for the definite integral helps us recall the meaning of the integral. The symbol

$$\int_a^b f(x)\, dx$$

reminds us that an integral is a limit of sums (the integral sign is an old-fashioned S) of terms of the form "$f(x)$ times a small difference of x." Officially, dx is not a separate entity, but a part of the whole integral symbol. Just as one thinks of d/dx as a single symbol meaning "the derivative with respect to x of$\ldots$," one can think of $\int_a^b \ldots dx$ as a single symbol meaning "the integral of $\ldots$ with respect to x."

However, many scientists and mathematicians informally think of dx as an "infinitesimally" small bit of x which in this context is multiplied by a function value $f(x)$. This viewpoint is often the key to interpreting the meaning of a definite integral. For example, if $f(t)$ is the velocity of a moving particle at time t, then $f(t)dt$ may by thought of informally as velocity $\times$ time, giving the distance traveled by the particle during a small bit of time dt. The integral $\int_a^b f(t)\, dt$ may then be thought of as the sum of all these small distances, giving us the net change in position of the particle between $t = a$ and $t = b$.

The notation for the integral also helps us determine what units should be used for the value of the integral. Since the terms being added up are products of the form "$f(x)$ times a difference in x," the unit of measurement for $\int_a^b f(x)\,dx$ is the product of the units for $f(x)$ and the units for x. For example, if $f(t)$ is velocity measured in meters/second and t is time measured in seconds, then

$$\int_a^b f(t)\,dt$$

has units of (meters/sec)×(sec) = meters. This is what we expect, since the value of this integral represents change in position.

As another example, graph $y = f(x)$ with the same units of measurement of length along the x- and y-axes, say cm. Then $f(x)$ and x are measured in the same units, so

$$\int_a^b f(x)\,dx$$

is measured in square units of cm × cm = cm^2. Again, this is what we would expect since in this context the integral represents an area.

The Definite Integral as an Average

We know how to find the average of n numbers: Add them and divide by n. But how do we find the average value of a continuously varying function? Let us consider an example. Suppose $f(t)$ is the temperature at time t, measured in hours since midnight, and that we want to calculate the average temperature over a 24-hour period. One way to start would be to average the temperatures at n equally spaced times, $t_1, t_2, \ldots, t_n$, during the day.

$$\text{Average temperature} \approx \frac{f(t_1) + f(t_2) + \cdots + f(t_n)}{n}.$$

The larger we make n, the better the approximation. We can rewrite this expression as a Riemann sum over the interval $0 \leq t \leq 24$ if we use the fact that $\Delta t = 24/n$, so $n = 24/\Delta t$:

$$\begin{aligned}
\text{Average temperature} &\approx \frac{f(t_1) + f(t_2) + \cdots + f(t_n)}{24/\Delta t} \\
&= \frac{f(t_1)\Delta t + f(t_2)\Delta t + \cdots + f(t_n)\Delta t}{24} \\
&= \frac{1}{24}\sum_{i=1}^{n} f(t_i)\Delta t.
\end{aligned}$$

As $n \to \infty$, the Riemann sum tends towards an integral, and $1/24$ of the sum also approximates the average temperature better. It makes sense, then, to write

$$\begin{aligned}
\text{Average temperature} &= \lim_{n \to \infty} \frac{1}{24}\sum_{i=1}^{n} f(t_i)\Delta t \\
&= \frac{1}{24}\int_0^{24} f(t)\,dt.
\end{aligned}$$

We have found a way of expressing the average temperature over an interval in terms of an integral. Generalizing for any function f, if $a < b$, we define

$$\boxed{\begin{array}{l} \text{Average value of } f \\ \text{from } a \text{ to } b \end{array} = \frac{1}{b-a}\int_a^b f(x)\,dx.}$$

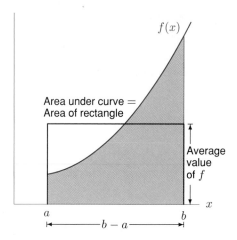

Figure 3.21: Area and average value

How to Visualize the Average on a Graph

The definition of average value tells us that

$$\text{(Average value of } f) \cdot (b - a) = \int_a^b f(x)\, dx.$$

Let's interpret the integral as the area under the graph of f. Then the average value of f is the height of a rectangle whose base is $(b - a)$ and whose area is the same as the area under the graph of f. (See Figure 3.21.)

Example 1 Suppose that $C(t)$ represents the daily cost of heating your house, measured in dollars per day, where t is time measured in days and $t = 0$ corresponds to January 1, 1997. Interpret $\int_0^{90} C(t)\, dt$ and $\dfrac{1}{90 - 0} \int_0^{90} C(t)\, dt$.

Solution The units for the integral $\int_0^{90} C(t)\, dt$ are (dollars/day) × (days) = dollars. The integral represents the total cost in dollars to heat your house for the first 90 days of 1997, namely the months of January, February, and March. The second expression is measured in (1/days)(dollars) or dollars per day, the same units as $C(t)$. It represents the average cost per day to heat your house during the first 90 days of 1997.

Example 2 On page 14, we saw that the population of Mexico could be modeled by the function

$$P = f(t) = 67.38(1.026)^t,$$

where P is in millions of people and t is in years since 1980. Use this function to predict the average population of Mexico between the years 2000 and 2020.

Solution We want the average value of $f(t)$ between $t = 20$ and $t = 40$. This is given by

$$\text{Average population} = \frac{1}{40 - 20} \int_{20}^{40} f(t)\, dt \approx \frac{1}{20}(2942.66) = 147.1.$$

We used a calculator to evaluate the integral. We see that the average population of Mexico between 2000 and 2020 is predicted to be about 147 million people.

Applications of the Definite Integral

The total distance traveled by a moving object in a given time interval may be represented by a definite integral of the velocity. The following examples show how representing a quantity as a definite integral, and thereby as an area, can be helpful even if we don't evaluate the integral.

Example 3   Two cars start from rest at a traffic light and accelerate for several minutes. Figure 3.22 shows their velocities as a function of time.   (a) Which car is ahead after one minute?   (b) Which car is ahead after two minutes?

Figure 3.22: Velocities of two cars. Which is ahead when?

Solution   (a)   For the first minute car 1 goes faster than car 2, and therefore car 1 must be ahead at the end of one minute.

(b)   At the end of two minutes the situation is less clear, since car 1 was going faster for the first minute and car 2 for the second. However, if $v = f(t)$ is the velocity of a car after t minutes, then we know that

$$\text{Distance traveled in two minutes} = \int_0^2 f(t)\,dt,$$

since the integral of velocity is distance traveled. This definite integral may also be interpreted as the area under the graph of f between 0 and 2. Since the area representing the distance traveled by car 2 is clearly larger than the area for car 1 (see Figure 3.22), we know that car 2 has traveled farther than car 1.

Example 4   A car starts at noon and travels with the velocity shown in Figure 3.23. A truck starts at 1 pm from the same place and travels at a constant velocity of 50 mph.

(a)   How far away is the car when the truck starts?

(b)   During the period when the car is ahead of the truck, when is the distance between them greatest, and what is that greatest distance?

(c)   When does the truck overtake the car, and how far have both traveled then?

Figure 3.23: Velocity of car for Example 4

Solution   To find distances from the velocity graph, we use the fact that if t is the time measured from noon, and v is the velocity, then

$$\begin{array}{ccc} \text{Distance traveled} \\ \text{by car up to time } T \end{array} = \int_0^T v\, dt = \begin{array}{c} \text{Area under velocity} \\ \text{graph between 0 and } T. \end{array}$$

The truck's motion can be represented on the same graph by the horizontal line $v = 50$, starting at $t = 1$. The distance traveled by the truck is then the rectangular area under this line, and the distance between the two vehicles is the difference between these areas. Note each small rectangle on the graph corresponds to moving at 10 mph for a half hour (i.e., to a distance of 5 miles).

(a)   The distance traveled by the car when the truck starts is represented by the shaded area in Figure 3.24, which totals about seven rectangles or about 35 miles.

Figure 3.24: Shaded area represents distance traveled by car from noon to 1pm

(b)   The car starts ahead of the truck, and the distance between them increases as long as the velocity of the car is greater than the velocity of the truck. Later, when the truck's velocity exceeds the car's, the truck starts to gain on the car. In other words, the distance between the car and the truck will increase as long as $v_{\text{car}} > v_{\text{truck}}$, and it will decrease when $v_{\text{car}} < v_{\text{truck}}$. Therefore, the maximum distance occurs when $v_{\text{car}} = v_{\text{truck}}$, that is, when $t \approx 4.3$ hours (at about 4:20 pm). (See Figure 3.25.) The distance traveled by the car is the area under the v_{car} graph between $t = 0$ and $t = 4.3$; the distance traveled by the truck is the area under the v_{truck} line between $t = 1$ (when it started) and $t = 4.3$. So the distance between the car and truck is represented by the shaded area in Figure 3.25, which is approximately

$$35 \text{ miles} + 50 \text{ miles} = 85 \text{ miles}.$$

(c)   The truck overtakes the car when both have traveled the same distance. This occurs when the area under the curve (v_{car}) up to that time equals the area under the line (v_{truck}) up to that time. Since the areas under v_{car} and v_{truck} overlap (see Figure 3.26), they are equal when the lightly shaded area equals the heavily shaded area (which we know is about 85 miles). This happens when $t \approx 8.3$ hours, or about 8:20 pm.

Figure 3.25: Shaded area = distance by which car is ahead at about 4:20 pm

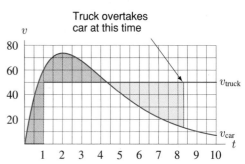

Figure 3.26: Truck overtakes car when dark and light shaded areas are equal

Problems for Section 3.3

1. For the two cars in Example 3, page 163, estimate:

   (a) The distances moved by car 1 and car 2 during the first minute.
   (b) The time at which the two cars have gone the same distance.

2. Consider the car and the truck in Example 4, page 163.

   (a) How fast is the distance between the car and the truck increasing or decreasing at 3 pm?
   (b) What is the practical significance (in terms of the distance between the car and the truck) of the fact that the car's velocity is maximized at about 2 pm?

3. Consider the car and the truck in Example 4, page 163, but suppose the truck starts at noon. (Everything else remains the same.)

   (a) Sketch a new graph showing the velocities of both car and truck against time.
   (b) How many times do the two graphs intersect? What does each intersection mean in terms of the distance between the two?

4. If $f(t)$ is measured in meters/second^2 and t is measured in seconds, what are the units of $\int_a^b f(t)\,dt$?

5. If $f(t)$ is measured in dollars per year and t is measured in years, what are the units of $\int_a^b f(t)\,dt$?

6. If $f(x)$ is measured in pounds and x is measured in feet, what are the units of $\int_a^b f(x)\,dx$?

7. Oil is leaking out of a ruptured tanker at a rate of $r = f(t)$ gallons per minute, where t is in minutes. Write a definite integral expressing the total quantity of oil which leaks out of the tanker in the first hour.

In Problems 8–10, find the average value of the function over the given interval.

8. $g(t) = 1 + t$ over $[0, 2]$ 9. $f(x) = 4x + 7$ over $[1, 3]$ 10. $g(t) = e^t$ over $[0, 10]$

11. If the graph of f is in Figure 3.27:

    (a) What is $\int_1^6 f(x)\,dx$?   (b) What is the average value of f on $[1, 6]$?

12. A two-day environmental clean up operation started at 9 am on the first day. The number of workers fluctuated as shown in Figure 3.28. If the workers were paid $10 per hour, how much was the total personnel cost of the operation?

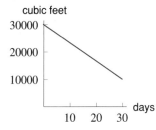

Figure 3.27   *Figure 3.28*   *Figure 3.29*

13. Suppose in Problem 12 that the workers were paid $10 per hour for work during the time period 9 am to 5 pm and were paid $15 per hour for work during the rest of the day. What would the total personnel costs of the operation have been under these conditions?

14. A warehouse charges it customers $5 per day for every 10 cubic feet of space used for storage. Figure 3.29 records the storage used by one company over a month. How much will the company have to pay?

15. If $f(x) = 2$, show that the average value of $f(x)$ over the interval $[a, b]$ is 2.

16. (a) Without computing any integrals, explain why the average value of $f(x) = \sin x$ on $[0, \pi]$ must be between 0.5 and 1.
     (b) Compute this average. Give 2 decimal places in your answer.

17. (a) What is the average value of $f(x) = \sqrt{1 - x^2}$ over the interval $0 \leq x \leq 1$?
     (b) How can you tell whether this average value is more or less than 0.5 without doing any calculations?

18. How do the units for the average value relate to the units for $f(x)$ and the units for x?

19. A bar of metal is cooling from $1000°C$ to room temperature, $20°C$. The temperature, H, of the bar t minutes after it starts cooling is given, in $°C$, by

$$H = 20 + 980e^{-0.1t}.$$

     (a) Find the temperature of the bar at the end of one hour.
     (b) Find the average value of the temperature over the first hour.
     (c) Is your answer to part (b) greater or smaller than the average of the temperatures at the beginning and the end of the hour? Explain this in terms of the concavity of the graph of H.

20. The value, V, of a Tiffany lamp, worth \$225 in 1965, increases at 15% per year. Its value in dollars t years after 1965 is given by
$$V = 225(1.15)^t.$$
     Find the average value of the lamp over the period 1965–2000.

21. The number of hours, H, of daylight in Madrid as a function of date is approximated by the formula

$$H = 12 + 2.4 \sin[0.0172(t - 80)],$$

     where t is the number of days since the start of the year. Find the average number of hours of daylight in Madrid:

     (a) in January ≈ 9.88 (b) in June   (c) over a whole year

     (d) Comment on the relative magnitudes of your answers to parts (a), (b), and (c). Why are they reasonable?

22. A bicyclist is pedaling along a straight road with velocity, v, given in Figure 3.30. Suppose the cyclist starts 5 miles from a lake, and that positive velocities take her away from the lake and negative velocities towards the lake. When is the cyclist farthest from the lake, and how far away is she then?

23. A force F parallel to the x-axis is given by the graph in Figure 3.31. Estimate the work, W, done by the force, where $W = \int_0^{16} F(x)\,dx$.

Figure 3.30

Figure 3.31

Figure 3.32

24. For the even function f in Figure 3.32, write an expression involving one or more definite integrals that denotes:
     (a) The average value of f for $0 \leq x \leq 5$.   (b) The average value of $|f|$ for $0 \leq x \leq 5$.

25. For the even function f in Figure 3.32, consider the average value of f over the following intervals:

     (I) $0 \leq x \leq 1$ (II) $0 \leq x \leq 2$ (III) $0 \leq x \leq 5$ (IV) $-2 \leq x \leq 2$

     (a) For which interval is the average value of f least?
     (b) For which interval is the average value of f greatest?
     (c) For which pair of intervals are the average values equal?

3.4 THEOREMS ABOUT DEFINITE INTEGRALS

The Definite Integral of a Rate Gives Total Change

We have seen how the definite integral of a velocity function can be interpreted as total distance traveled. If $v(t)$ is velocity and $s(t)$ is position, then $v(t) = s'(t)$ and we know that

$$\text{Total change in position} = s(b) - s(a) = \int_a^b s'(t)\, dt.$$

In this section we generalize this result to explain why the integral of the rate of change of any quantity gives the total change in that quantity.

Suppose $F'(t)$ is the rate of change of some quantity $F(t)$ with respect to time, and that we are interested in the total change in $F(t)$ between $t = a$ and $t = b$. We divide the interval $a \leq t \leq b$ into n subintervals, each of length Δt. For each small interval, we estimate the change in $F(t)$, written ΔF, and then add all these up. In each subinterval we assume the rate of change of $F(t)$ is approximately constant, so that we can say

$$\Delta F \approx \text{Rate of change of } F \times \text{Time elapsed.}$$

For the first subinterval, from t_0 to t_1, the rate of change of $F(t)$ is approximately $F'(t_0)$, so

$$\Delta F \approx F'(t_0)\,\Delta t.$$

Similarly, for the second interval

$$\Delta F \approx F'(t_1)\,\Delta t.$$

Summing over all the subintervals, we get

$$\text{Total change in } F = \sum_{i=0}^{n-1} \Delta F \approx \sum_{i=0}^{n-1} F'(t_i)\,\Delta t.$$

We have approximated the change in $F(t)$ as a left-hand sum.

However, the total change in $F(t)$ between the times $t = a$ and $t = b$ is simply $F(b) - F(a)$. Taking the limit as n goes to infinity suggests the following result:[1]

$$F(b) - F(a) = \begin{array}{c} \text{Total change in } F(t) \\ \text{between } t = a \text{ and } t = b \end{array} = \lim_{n \to \infty} \sum_{i=0}^{n-1} F(t_i)\Delta t = \int_a^b F'(t)\, dt.$$

This result is one of the most important in calculus because it makes the connection between the derivative and the definite integral. It is called the Fundamental Theorem of Calculus and is often stated as follows:

The Fundamental Theorem of Calculus

If f is continuous on the interval $[a, b]$ and $f(t) = F'(t)$, then

$$\int_a^b f(t)\, dt = F(b) - F(a).$$

In words:

   The definite integral of a rate of change gives total change.

---

[1] We could equally well have used a right-hand sum, since the definite integral is their common limit.

What Is Involved in Proving the Fundamental Theorem?

The argument we have given makes the Fundamental Theorem plausible. However, it is not a mathematical proof. Each of the approximations

$$\Delta F \approx F'(t_i)\,\Delta t, \quad i = 1, \ldots, n,$$

involves a small error. To prove that the total change in F is well-approximated by the Riemann sum $\sum_{i=0}^{n-1} F'(t_i)\,\Delta t$, we must show that the sum of all the errors is small (in fact, that it can be made as small as we like by choosing n large enough). We will not give the details of this argument. However, the key idea brings out an important fact about local linearization: the error in the local linearization $\Delta F \approx F'(t_i)\,\Delta t$ is not merely small, it is small relative to the size of Δt. This follows from the definition of the derivative: if we choose Δt small enough, we can ensure that the error in the approximation

$$F'(t_i) \approx \frac{\Delta F}{\Delta t}$$

is as small as we like. In other words,

$$F'(t_i) = \frac{\Delta F}{\Delta t} + \epsilon,$$

where ϵ can be made as small as we like. Multiplying through by Δt, we get an approximation

$$F'(t_i)\Delta t = \Delta F + \epsilon \Delta t.$$

Thus, in the approximation $F'(t_i)\Delta t \approx \Delta F$, the error is $\epsilon \Delta t$, which is small even relative to Δt. When we add these small errors, we still get a small total error. For a proof of the Fundamental Theorem from another point of view, see Problem 14 on page 291.

Using the Fundamental Theorem

The Fundamental Theorem provides a precise way of computing certain definite integrals.

Example 1 Compute $\displaystyle\int_1^3 2x \, dx$ by two different methods.

Solution Using left- and right-hand sums, we can approximate this integral as accurately as we want. With $n = 100$, for example, the left-sum is 7.96 and the right sum is 8.04. Using $n = 500$ we learn

$$7.992 < \int_1^3 2x \, dx < 8.008.$$

The Fundamental Theorem, on the other hand, allows us to compute the integral exactly. We take $f(x) = 2x$. By Example 4, on page 108, we know that if $F(x) = x^2$, then $F'(x) = 2x$. So we use $f(x) = 2x$ and $F(x) = x^2$ and obtain

$$\int_1^3 2x \, dx = F(3) - F(1) = 3^2 - 1^2 = 8.$$

The Fundamental Theorem can also be used when the rate, $F'(t)$, is known and we want to find the total change $F(b) - F(a)$. If we also know $F(a)$, the theorem enables us to reconstruct the function F from knowledge about its derivative $F' = f$.

Example 2 Let $F(t)$ represent a bacteria population which is 5 million at time $t = 0$. Suppose that after t hours, the population is growing at an instantaneous rate of 2^t million bacteria per hour. Estimate the total increase in the bacteria population during the first hour, and the population at $t = 1$.

Solution Since the rate at which the population is growing is $F'(t) = 2^t$, we have

$$\text{Change in population} = F(1) - F(0) = \int_0^1 2^t \, dt.$$

Using a calculator gives

$$\text{Change in population} = \int_0^1 2^t \, dt \approx 1.44 \text{ million bacteria.}$$

Since $F(0) = 5$, the population at $t = 1$ is given by

$$\text{Population} = F(1) = F(0) + \int_0^1 2^t \, dt \approx 5 + 1.44 = 6.44 \text{ million.}$$

Properties of the Definite Integral

For the definite integral $\int_a^b f(x) \, dx$, we have so far only considered the case $a < b$. We now allow $a \geq b$. We still set $x_0 = a$, $x_n = b$, and $\Delta x = (b - a)/n$. As before, we have $\int_a^b f(x)dx = \lim_{n \to \infty} \sum_{i=1}^n f(x_i)\Delta x$.

Theorem: Properties of Limits of Integration

If a, b, and c are any numbers and f is a continuous function, then

1. $\displaystyle\int_b^a f(x) \, dx = -\int_a^b f(x) \, dx.$

2. $\displaystyle\int_a^c f(x) \, dx + \int_c^b f(x) \, dx = \int_a^b f(x) \, dx.$

In words:
1.   The integral from b to a is the negative of the integral from a to b.
2.   The integral from a to c plus the integral from c to b is the integral from a to b.

By interpreting the integrals as areas, we can justify these results for $f \geq 0$. In fact, they are true for all functions for which the integrals make sense.

Why is $\displaystyle\int_b^a f(x) \, dx = -\int_a^b f(x) \, dx$?

By definition, both integrals are approximated by sums of the form $\sum f(x_i)\Delta x$. The x_is are the same in each case: the only difference in the sums for $\int_b^a f(x) \, dx$ and $\int_a^b f(x) \, dx$ is that in the first $\Delta x = (a - b)/n = -(b - a)/n$ and in the second $\Delta x = (b - a)/n$. Since everything else about the sums is the same, we must have $\int_b^a f(x) \, dx = -\int_a^b f(x) \, dx$.

Why is $\displaystyle\int_a^c f(x) \, dx + \int_c^b f(x) \, dx = \int_a^b f(x) \, dx$?

Suppose $a < c < b$. Figure 3.33 suggests that $\int_a^c f(x) \, dx + \int_c^b f(x) \, dx = \int_a^b f(x) \, dx$ since the area under f from a to c plus the area under f from c to b together make up the whole area under f from a to b.

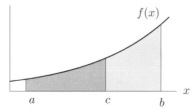

Figure 3.33: Additivity of the definite integral $(a < c < b)$

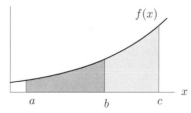

Figure 3.34: Additivity of the definite integral $(a < b < c)$

Actually, this property holds for all numbers a, b, and c, not just those satisfying $a < c < b$. (See Figure 3.34.) For example, the area under f from 3 to 6 is equal to the area from 3 to 8 *minus* the area from 6 to 8, so

$$\int_3^6 f(x)\, dx = \int_3^8 f(x)\, dx - \int_6^8 f(x)\, dx = \int_3^8 f(x)\, dx + \int_8^6 f(x)\, dx.$$

Example 3   Suppose you know that $\int_0^{1.25} \cos(x^2)\, dx = 0.98$ and $\int_0^1 \cos(x^2)\, dx = 0.90$. (See Figure 3.35.) What are the values of the following integrals?

(a) $\displaystyle \int_1^{1.25} \cos(x^2)\, dx$ (b) $\displaystyle \int_{-1}^1 \cos(x^2)\, dx$ (c) $\displaystyle \int_{1.25}^{-1} \cos(x^2)\, dx$

Figure 3.35: Graph of $f(x) = \cos(x^2)$

Solution   (a)   Since $\int_0^{1.25} \cos(x^2)\, dx = \int_0^1 \cos(x^2)\, dx + \int_1^{1.25} \cos(x^2)\, dx$ by the additivity property, we get
$0.98 = 0.90 + \int_1^{1.25} \cos(x^2)\, dx$, so $\int_1^{1.25} \cos(x^2)\, dx = 0.08$.

(b) $\int_{-1}^1 \cos(x^2)\, dx = \int_{-1}^0 \cos(x^2)\, dx + \int_0^1 \cos(x^2)\, dx$.
By the symmetry of $\cos(x^2)$ about the y-axis, $\int_{-1}^0 \cos(x^2)\, dx = \int_0^1 \cos(x^2)\, dx = 0.90$.
So $\int_{-1}^1 \cos(x^2)\, dx = 0.90 + 0.90 = 1.80$.

(c) $\int_{1.25}^{-1} \cos(x^2)\, dx = -\int_{-1}^{1.25} \cos(x^2)\, dx = -\left(\int_{-1}^0 \cos(x^2)\, dx + \int_0^{1.25} \cos(x^2)\, dx\right) =$
$-(0.90 + 0.98) = -1.88$.

Theorem: Properties of Sums and Constant Multiples of the Integrand

Let f and g be continuous functions and let c be a constant.

1. $\displaystyle \int_a^b (f(x) \pm g(x))\, dx = \int_a^b f(x)\, dx \pm \int_a^b g(x)\, dx.$

2. $\displaystyle \int_a^b cf(x)\, dx = c \int_a^b f(x)\, dx.$

In words:
1.   The integral of the sum (or difference) of two functions is the sum (or difference) of their integrals.
2.   The integral of a constant times a function is that constant times the integral of the function.

Why Do these Properties Hold?

Both can be visualized by thinking of the definition of the definite integral as the limit of a sum of areas of rectangles.

For property 1, let's suppose that f and g are positive on the interval $[a, b]$ so that the area under $f(x) + g(x)$ is approximated by the sum of the areas of rectangles like the one shaded in Figure 3.36. The area of this rectangle is

$$[f(x_i) + g(x_i)]\Delta x = f(x_i)\Delta x + g(x_i)\Delta x.$$

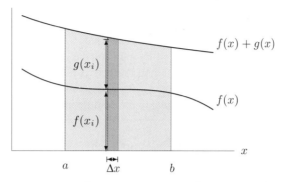

Figure 3.36: Area $= \int_a^b [f(x) + g(x)]\, dx = \int_a^b f(x)\, dx + \int_a^b g(x)\, dx$

Since $f(x_i)\Delta x$ is the area of a rectangle under the graph of f, and $g(x_i)\Delta x$ is the area of a rectangle under the graph of g, the area under $f(x) + g(x)$ is the sum of the areas under $f(x)$ and $g(x)$.

For property 2, notice that multiplying a function by c stretches or flattens the graph in the vertical direction by a factor of c. Thus, it stretches or flattens the height of each approximating rectangle by c, and hence multiplies the area by c.

Example 4   Evaluate the definite integral $\displaystyle\int_0^2 (1 + 3x)\, dx$ exactly.

Solution   We can break this integral up as follows:

$$\int_0^2 (1 + 3x)\, dx = \int_0^2 1\, dx \;+\; \int_0^2 3x\, dx = \int_0^2 1\, dx \;+\; 3\int_0^2 x\, dx.$$

This expresses our original integral in terms of two simpler integrals. From the area interpretation of the integral, we see that

$$\int_0^2 1\, dx = 2,$$

since it represents the area under the horizontal line $y = 1$ between $x = 0$ and $x = 2$ (see Figure 3.37). Similarly,

$$\int_0^2 x\, dx = \frac{1}{2} \cdot 2 \cdot 2 = 2$$

because it is the area of the triangle in Figure 3.38. Therefore,

$$\int_0^2 (1 + 3x)\, dx = \int_0^2 1\, dx + 3\int_0^2 x\, dx = 2 + 3(2) = 8.$$

Figure 3.37: Area representing $\int_0^2 1\, dx$

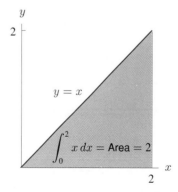

Figure 3.38: Area representing $\int_0^2 x\, dx$

Comparing Integrals

Suppose we have constants m and M such that $m \leq f(x) \leq M$ for $a \leq x \leq b$. We say f is *bounded above* by M and *bounded below* by m. Then the graph of f lies between the horizontal lines $y = m$ and $y = M$. So the definite integral lies between $m(b - a)$ and $M(b - a)$. See Figure 3.39.

Suppose $f(x) \leq g(x)$ for $a \leq x \leq b$, as in Figure 3.40. By a similar argument, the definite integral of f is less than or equal to the definite integral of g. This leads us to the following results:

Theorem: Results About Comparison of Definite Integrals

Let f and g be continuous functions.

1. If $m \leq f(x) \leq M$ for $a \leq x \leq b$, then $\quad m(b - a) \leq \int_a^b f(x)\, dx \leq M(b - a)$.

2. If $f(x) \leq g(x)$ for $a \leq x \leq b$, then $\quad \int_a^b f(x)\, dx \leq \int_a^b g(x)\, dx$.

Figure 3.39: The area under the graph of f lies between the areas of the rectangles

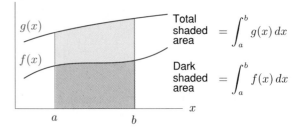

Figure 3.40: If $f(x) \leq g(x)$ then $\int_a^b f(x)\, dx \leq \int_a^b g(x)\, dx$

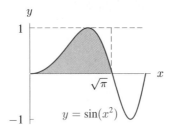

Figure 3.41: Graph showing that $\int_0^{\sqrt{\pi}} \sin(x^2)\, dx < \sqrt{\pi}$

Example 5 Explain why $\displaystyle\int_0^{\sqrt{\pi}} \sin(x^2)\, dx \leq \sqrt{\pi}$.

Solution Since $\sin(x^2) \leq 1$ for all x (see Figure 3.41), part 2 of the theorem gives

$$\int_0^{\sqrt{\pi}} \sin(x^2)\, dx \leq \int_0^{\sqrt{\pi}} 1\, dx = \sqrt{\pi}.$$

Example 6 Show that $\displaystyle 2 \leq \int_0^2 \sqrt{1 + x^3}\, dx \leq 6$.

Solution Notice that $f(x) = \sqrt{1 + x^3}$ is increasing for $0 \leq x \leq 2$, since x^3 gets bigger as x increases. This means that $f(0) \leq f(x) \leq f(2)$. For this function, $f(0) = 1$ and $f(2) = 3$. Thus we can imagine the area under $f(x)$ as lying between the area under the line $y = 1$ and the area under the line $y = 3$ on the interval $0 \leq x \leq 2$. That is,

$$1(2 - 0) \ \leq \ \int_0^2 \sqrt{1 + x^3}\, dx \ \leq \ 3(2 - 0).$$

Problems for Section 3.4

1. The graph of a derivative $f'(x)$ is shown in Figure 3.42. Fill in the table of values for $f(x)$ given that $f(0) = 2$.

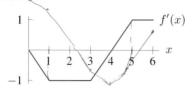

x	0	1	2	3	4	5	6
$f(x)$	2	1.5	0.5	-0.5	-1	-0.5	0.5

Figure 3.42: Graph of f', not f

2. Water is leaking out of a tank at a rate of $R(t)$ gallons/hour, where t is measured in hours.

    (a) Write a definite integral that expresses the total amount of water that leaks out in the first two hours.

    (b) Figure 3.43 is a graph of $R(t)$. On a sketch, shade in the region whose area represents the total amount of water that leaks out in the first two hours.

    (c) Give an upper and lower estimate of the total amount of water that leaks out in the first two hours.

3. Figure 3.44 shows the graph of f. If $F' = f$ and $F(0) = 0$, find $F(b)$ for $b = 1, 2, 3, 4, 5, 6$.

    1 1.5 1 0 -1 -1.5

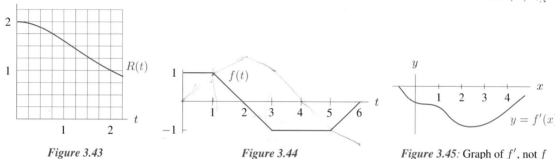

Figure 3.43   **Figure 3.44**   **Figure 3.45:** Graph of f', not f

Problems 4–5 concern the graph of f' in Figure 3.45.

4. Which is greater, $f(0)$ or $f(1)$?

5. List the following in increasing order: $\dfrac{f(4) - f(2)}{2}$, $f(3) - f(2)$, $f(4) - f(3)$.

6. A news broadcast in early 1993 said the average American's annual income is changing at a rate of $r(t) = 40(1.002)^t$ dollars per month, where t is in months from January 1, 1993. How much did the average American's income change during 1993?

7. A cup of coffee at $90°$C is put into a $20°$C room when $t = 0$. If the coffee's temperature is changing at a rate given in $°$C per minute by

$$r(t) = -7e^{-0.1t}, \quad t \text{ in minutes,}$$

    estimate, to one decimal place, the coffee's temperature when $t = 10$.

8. The rate at which the world's oil is being consumed is continuously increasing. Suppose the rate (in billions of barrels per year) is given by the function $r = f(t)$, where t is measured in years and $t = 0$ is the start of 1990.

    (a) Write a definite integral which represents the total quantity of oil used between the start of 1990 and the start of 1995.

    (b) Suppose $r = 32e^{0.05t}$. Using a left-hand sum with five subdivisions, find an approximate value for the total quantity of oil used between the start of 1990 and the start of 1995.

    (c) Interpret each of the five terms in the sum from part (b) in terms of oil consumption.

9. The graph of a function $y = f(x)$ is given in Figure 3.46. Suppose $f(x)$ is the rate (in thousands of algae per hour) at which a population of algae is growing, where x is in hours.

    (a) Estimate the average value of the rate of growth of this population over the interval $x = -1$ to $x = 3$. Explain how you arrived at your answer.

    (b) Estimate the total change in the algae population over the interval $x = -3$ to $x = 3$.

Figure 3.46

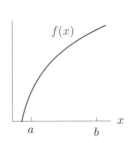

Figure 3.47

For Problems 10–13, mark the following quantities on a copy of the graph of f in Figure 3.47.

10. A length representing $f(b) - f(a)$.

11. A slope representing $\dfrac{f(b) - f(a)}{b - a}$.

12. An area representing $F(b) - F(a)$, where $F' = f$.

13. A length roughly approximating

$$\frac{F(b) - F(a)}{b - a}, \text{ where } F' = f.$$

Suppose $\int_a^b f(x)\, dx = 8$, $\int_a^b (f(x))^2\, dx = 12$, $\int_a^b g(t)\, dt = 2$, and $\int_a^b (g(t))^2\, dt = 3$. Find the integrals in Problems 14–19.

14. $\int_a^b (f(x) + g(x))\, dx$

15. $\int_a^b \left((f(x))^2 - (g(x))^2\right) dx$

16. $\int_a^b (f(x))^2\, dx - (\int_a^b f(x)\, dx)^2$

17. $\int_a^b cf(z)\, dz$

18. $\int_a^b \left(c_1 g(x) + (c_2 f(x))^2\right) dx$

19. $\int_{a+5}^{b+5} f(x - 5)\, dx$

20. The function for the *standard normal distribution*, which is often used in statistics, has the formula

$$\frac{1}{\sqrt{2\pi}} e^{-x^2/2}$$

and the graph in Figure 3.48. Statistics books usually contain tables such as the one below, showing only the area under the curve from 0 to b, for different values of b.

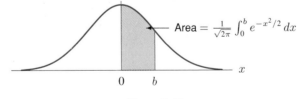

Area $= \dfrac{1}{\sqrt{2\pi}} \int_0^b e^{-x^2/2}\, dx$

Figure 3.48

b	$\dfrac{1}{\sqrt{2\pi}} \int_0^b e^{-x^2/2}\, dx$
1	0.3413
2	0.4772
3	0.4987
4	0.5000

Use the information given in the table and the symmetry of the standard normal curve about the y-axis to find:

(a) $\dfrac{1}{\sqrt{2\pi}} \displaystyle\int_1^3 e^{-x^2/2}\, dx$.

(b) $\dfrac{1}{\sqrt{2\pi}} \displaystyle\int_{-2}^3 e^{-x^2/2}\, dx$.

21. (a) Is $\int_{-1}^{1} e^{x^2}\, dx$ positive, negative, or zero? Explain.

   (b) Explain why $0 < \int_{0}^{1} e^{x^2}\, dx < 3$.

22. Without calculating the integral, explain why the following statements are false.

   (a) $\int_{-2}^{-1} e^{x^2}\, dx = -3$

   (b) $\int_{-1}^{1} \left| \dfrac{\cos(x+2)}{1 + \tan^2 x} \right|\, dx = 0$

23. Without doing any computations, find the values of

   (a) $\int_{-2}^{2} \sin x\, dx,$

   (b) $\int_{-\pi}^{\pi} x^{113}\, dx.$

24. Use the property $\int_{b}^{a} f(x)\, dx = -\int_{a}^{b} f(x)\, dx$ to show that $\int_{a}^{a} f(x)\, dx = 0.$

25. The average value of $y = v(x)$ equals 4 for $1 \le x \le 6$, and equals 5 for $6 \le x \le 8$. What is the average value of $v(x)$ for $1 \le x \le 8$?

CHAPTER SUMMARY

- **Definite integral as limit of right-hand or left-hand sums**
- **Interpretations of the definite integral**
  Total change from rate of change, change in position given velocity, area, $(b - a)\cdot$ Average value
- **Properties of the definite integral**
  Properties involving integrand, properties involving limits, comparison between integrals.
- **Fundamental Theorem of Calculus**
- **Working with the definite integral**
  Estimate definite integral from graph, table of values, or formula.

REVIEW PROBLEMS FOR CHAPTER THREE

1. A car going 80 ft/sec (about 55 mph) brakes to a stop in 8 seconds. Its velocity is recorded every 2 seconds and is given in the following table.

   (a) Give your best estimate of the distance traveled by the car during the 8 seconds.

   (b) To estimate the distance traveled accurate to within 20 feet, how often should you record the velocity?

t (seconds)	0	2	4	6	8
$v(t)$ (ft/sec)	80	52	28	10	0

2. Fill in Table 3.4 with the appropriate left- and right-hand sums with n subdivisions for the integral $\int_{-1}^{2} 5e^{-x^2}\, dx$. Use the table to estimate the value of the definite integral.

TABLE 3.4

n	Left	Right
5		
25		
500		

3. A graph of $y = f(x)$ is given in Figure 3.49. Estimate $\int_0^{20} f(x)\,dx$.

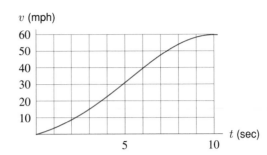

Figure 3.49   **Figure 3.50**

4. A car accelerates smoothly from 0 to 60 mph in 10 seconds. Suppose the car's velocity as a function of time is given in Figure 3.50. Estimate how far the car travels during the 10-second period.

5. Your velocity is given by

$$v(t) = \sin(t^2) \quad \text{for } 0 \le t \le 1.1.$$

Represent the distance traveled during this time by an integral, and use a calculator or computer to find the distance traveled.

Find the area of the regions in Problems 6–9.

6. Between $y = x^2 - 9$ and the x-axis.

7. Under one arch of the curve $y = \sin x$.

8. Between the parabola $y = 4 - x^2$ and the x-axis.

9. Between the line $y = 1$ and one arch of the curve $y = \sin \theta$.

10. If $\int_2^5 (2f(x) + 3)\,dx = 17$, find $\int_2^5 f(x)\,dx$.

11. The graph of a continuous function f is given in Figure 3.51. Rank the following integrals in ascending numerical order. Explain your reasons.

(i) $\int_0^2 f(x)\,dx$ (ii) $\int_0^1 f(x)\,dx$ (iii) $\int_0^2 (f(x))^{1/2}\,dx$ (iv) $\int_0^2 (f(x))^2\,dx$.

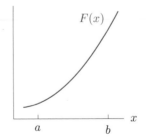

Figure 3.51   **Figure 3.52**

For Problems 12–14, assuming $F' = f$, mark the following quantities on a copy of Figure 3.52.

12. A slope representing $f(a)$.

13. A length representing $\int_a^b f(x)\,dx$.

14. A slope representing $\dfrac{1}{b - a}\int_a^b f(x)\,dx$.

15. (a) Sketch a graph of $f(x) = \sin(x^2)$ and mark on it the points $x = \sqrt{\pi}, \sqrt{2\pi}, \sqrt{3\pi}, \sqrt{4\pi}$.

(b) Use your graph to decide which of the four numbers

$$\int_0^{\sqrt{n\pi}} \sin(x^2)\, dx \quad n = 1, 2, 3, 4$$

is largest. Which is smallest? How many of the numbers are positive?

16. The Environmental Protection Agency was recently asked to investigate a spill of radioactive iodine. Measurements showed the ambient radiation levels at the site to be four times the maximum acceptable limit, so the EPA ordered an evacuation of the surrounding neighborhood.

It is known that the level of radiation from an iodine source decreases according to the formula

$$R(t) = R_0 e^{-0.004t}$$

where R is the radiation level (in millirems/hour) at time t, R_0 is the initial radiation level (at $t = 0$), and t is the time measured in hours.

(a) How long will it take for the site to reach an acceptable level of radiation?

(b) How much total radiation (in millirems) will have been emitted by that time, assuming the maximum acceptable limit is 0.6 millirems/hour?

17. Assume the population, P, of Mexico (in millions), is given by

$$P = 67.38(1.026)^t ,$$

where t is the number of years since 1980.

(a) What was the average population of Mexico between 1980 and 1990?

(b) What is the average of the population of Mexico in 1980 and the population in 1990?

(c) Explain, in terms of the concavity of the graph of P (see Figure 1.18 on page 15), why your answer to part (b) is larger or smaller than your answer to part (a).

18. The graphs in Figure 3.53 represent the velocity, v, of a particle moving along the x-axis for time $0 \le t \le 5$. The vertical scales of all graphs are the same. Identify the graph showing which particle

(a) has a constant acceleration.

(b) ends up farthest to the left of where it started.

(c) ends up the farthest from its starting point.

(d) experiences the greatest initial acceleration.

(e) has the greatest average velocity.

(f) has the greatest average acceleration.

(I)   (II)   (III)   (IV)   (V)

Figure 3.53

19. A new sales agent finds that as she gains experience, she increases the number of large appliances she sells each month. In the first month she sells only seven, but each month she sells two more than the month before, so that the number she sells in month t is $2t + 5$.

(a) Find the average number of large appliances she sells per month over the first year arithmetically, by calculating the number of appliances sold each month and then taking the average over 12 months.

(b) Now find the average by integration as though the sales function applied for all values of t (instead of just for integers).

(c) How well do the two results compare?

(d) If the answer you found in part (a) is viewed as the true answer, and the integral answer as an approximation, why would anyone want to use the integral answer instead of the true answer?

(e) Draw a picture showing both answers as the area of some region. Mark on your picture a region representing the error in the integral answer.

20. Water is run into a large tank through a hose at a constant rate. After 5 minutes a hole is opened in the bottom of the tank, and water starts to flow out. Initially the flow rate through the hole is twice as great as the rate through the hose, but as the water level in the tank goes down, the flow rate through the hole decreases; after another 10 minutes the water level in the tank appears to be constant. Plot graphs of the flow rates through the hose and through the hole against time on the same pair of axes. Show how the volume of water in the tank at any time can be interpreted as an area (or the difference between two areas) on the graph. In particular, interpret the steady-state volume of water in the tank.[2]

21. The Glen Canyon Dam at the top of the Grand Canyon prevents natural flooding. In 1996, scientists decided an artificial flood was necessary to restore the environmental balance. Water was released through the dam at a controlled rate[3] shown in Figure 3.54. The figure also shows the rate of flow of the last natural flood in 1957.

(a) At what rate was water passing through the dam in 1996 before the artificial flood?
(b) At what rate was water passing down the river in the pre-flood season in 1957?
(c) Estimate the maximum rates of discharge for the 1996 and 1957 floods.
(d) Approximately how long did the 1996 flood last? How long did the 1957 flood last?
(e) Estimate how much additional water passed down the river in 1996 as a result of the artificial flood.
(f) Estimate how much additional water passed down the river in 1957 as a result of the flood.

Figure 3.54

22. The Montgolfier brothers (Joseph and Etienne) were eighteenth-century pioneers in the field of hot-air ballooning. Had they had the appropriate instruments, they might have left us a record of one of their early experiments, like that shown in Figure 3.55. The graph shows their vertical velocity, v, with upward as positive.

(a) Over what intervals was the acceleration positive? Negative?
(b) What was the greatest altitude achieved, and at what time?
(c) At what time was the upward acceleration greatest?
(d) At what time was the deceleration greatest?
(e) What might have happened during this flight to explain the answer to part (d)?
(f) This particular flight ended on top of a hill. How do you know that it did, and what was the height of the hill above the starting point?

---

[2]From *Calculus: The Analysis of Functions*, by Peter D. Taylor (Toronto: Wall & Emerson, Inc., 1992)

[3]Adapted from M. Collier, R. Webb, E. Andrews "Experimental Flooding in Grand Canyon" in *Scientific American* (January 1997).

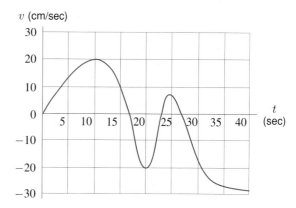

Figure 3.55   *Figure 3.56*

23. A mouse moves back and forth in a tunnel, attracted to bits of cheddar cheese alternately introduced to and removed from the ends (right and left) of the tunnel. The graph of the mouse's velocity, v, is given in Figure 3.56, with positive velocity corresponding to motion toward the right end. Assuming that the mouse starts $(t = 0)$ at the center of the tunnel, use the graph to estimate the time(s) at which

  (a) The mouse changes direction.
  (b) The mouse is moving most rapidly to the right; to the left.
  (c) The mouse is farthest to the right of center; farthest to the left.
  (d) The mouse's speed (i.e., the magnitude of its velocity) is decreasing.
  (e) The mouse is at the center of the tunnel.

24. For the even function f graphed in Figure 3.57:

  (a) Suppose you know $\int_0^2 f(x)\,dx$. What is $\int_{-2}^2 f(x)\,dx$?
  (b) Suppose you know $\int_0^5 f(x)\,dx$ and $\int_2^5 f(x)\,dx$. What is $\int_0^2 f(x)\,dx$?
  (c) Suppose you know $\int_{-2}^5 f(x)\,dx$ and $\int_{-2}^2 f(x)\,dx$. What is $\int_0^5 f(x)\,dx$?

25. For the even function f graphed in Figure 3.57:

  (a) Suppose you know $\int_{-2}^2 f(x)\,dx$ and $\int_0^5 f(x)\,dx$. What is $\int_2^5 f(x)\,dx$?
  (b) Suppose you know $\int_{-2}^5 f(x)\,dx$ and $\int_{-2}^0 f(x)\,dx$. What is $\int_2^5 f(x)\,dx$?
  (c) Suppose you know $\int_2^5 f(x)\,dx$ and $\int_{-2}^5 f(x)\,dx$. What is $\int_0^2 f(x)\,dx$?

26. The graph of some function f is given in Figure 3.58. List, from *least* to *greatest*,

  (a) $f'(1)$.
  (b) The average value of $f(x)$, $0 \le x \le a$.
  (c) The average value of the rate of change of $f(x)$, for $0 \le x \le a$.
  (d) $\displaystyle\int_0^a f(x)\,dx$.

Figure 3.57   *Figure 3.58*   *Figure 3.59*

27. A force, F, acts on a particle of mass m, moving along a line. The velocity v of the particle is shown in Figure 3.59. Using $F = ma$, where a is the acceleration of the particle, decide if the work done by the force on the particle in each of the intervals $[0, t_1]$, $[t_1, t_2]$, $[t_2, t_3]$, $[t_3, t_4]$, is positive, negative, or zero. (Work done = Force $\cdot$ Distance, when force is constant.)

PROJECTS

1. **Carbon Dioxide in Pond Water**

    Biological activity in a pond is reflected in the rate at which carbon dioxide, CO_2, is added to or withdrawn from the water. Plants take CO_2 out of the water during the day for photosynthesis and put CO_2 into the water at night. Animals put CO_2 into the water all the time as they breathe. Biologists are interested in how the net rate at which CO_2 enters a pond varies during the day. Figure 3.60 shows this rate as a function of time of day.[4] The rate is measured in millimoles (mmol) of CO_2 per liter of water per hour; time is measured in hours past dawn. At dawn, there were 2.600 mmol of CO_2 per liter of water.

    (a) What can be concluded from the fact that the rate is negative during the day and positive at night?

    (b) Some scientists have suggested that plants respire (breathe) at a constant rate at night, and that they photosynthesize at a constant rate during the day. Does Figure 3.60 support this view?

    (c) When was the CO_2 content of the water at its lowest? How low did it go?

    (d) How much CO_2 was released into the water during the 12 hours of darkness? Compare this quantity with the amount of CO_2 withdrawn from the water during the 12 hours of daylight. How can you tell by looking at the graph whether the CO_2 in the pond is in equilibrium?

    (e) Estimate the CO_2 content of the water at three hour intervals throughout the day. Use your estimates to plot a graph of CO_2 content throughout the day.

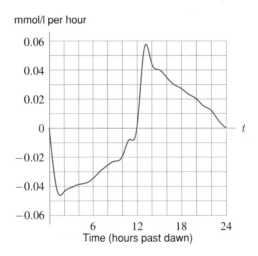

Figure 3.60: Rate at which CO_2 is entering the pond

2. **Average Velocity**

    In Section 2.1, page 89, we defined the average velocity of a particle over the interval $a \leq t \leq b$ as the change in position during this interval divided by $(b - a)$. If the velocity of the particle is $v = f(t)$, the average velocity is also given by $\frac{1}{b-a} \int_a^b f(t)dt$. Explain why these two ways of calculating the average velocity always lead to the same answer.

---

[4]Data from R. J. Beyers, *The Pattern of Photosynthesis and Respiration in Laboratory Microsystems* (Mem. 1st. Ital. Idrobiol., 1965).

FOCUS ON THEORY

THE DEFINITE INTEGRAL

Recall that if f is continuous on $[a, b]$ the definite integral is given by a limit of left or right sums:

$$\int_a^b f(x)\,dx = \lim_{n \to \infty} \sum_{i=0}^{n-1} f(x_i)\Delta x = \lim_{x \to \infty} \sum_{i=1}^{n} f(x_i)\Delta x.$$

This provides a method for approximating definite integrals numerically.[5] In this section we give a formal definition of the definite integral that makes use of more general sums.

A Special Case: Monotonic Functions

A function which is either increasing throughout an interval or decreasing throughout that interval is said to be *monotonic* on the interval. In Section 3.1 we saw that if f is monotonic, the left and right sums trap the exact value of the integral between them. Let us continue Example 1 on page 154, which looks at the value of

$$\int_1^2 \frac{1}{t}\,dt.$$

The left- and right-hand sums for $n = 2, 10, 50$, and 250 are listed in Table 3.5.

Because the function $f(t) = 1/t$ is decreasing, the left-hand sums converge to the integral from above, and the right-hand sums converge from below. From the last row of the table we can deduce that

$$0.6921 < \int_1^2 \frac{1}{t}\,dt < 0.6941,$$

so $\int_1^2 \frac{1}{t}\,dt \approx 0.69$ to two decimal places.

The Difference Between the Upper and Lower Estimates

To be sure that the left- and right-hand sums trap a unique number between them, we need to know that the difference between them approaches zero. On page 150 we saw that for a monotonic function f on the interval $[a, b]$:

$$\left| \begin{array}{c} \text{Difference between} \\ \text{upper and lower estimates} \end{array} \right| = |f(b) - f(a)| \cdot \Delta t,$$

where $\Delta t = (b - a)/n$. We can make this difference as small as we like by choosing Δt small enough.

TABLE 3.5 *Left- and right-hand sums for $\int_1^2 \frac{1}{t}\,dt$*

n	Left-hand sum	Right-hand sum
2	0.8333	0.5833
10	0.7188	0.6688
50	0.6982	0.6882
250	0.6941	0.6921

---

[5]In practice, we often approximate integrals using more sophisticated numerical methods.

TABLE 3.6   *Left- and right-hand sums for $\int_0^{2.5} \sin(t^2)\, dt$*

n	Left-hand sum	Right-hand sum
2	1.2500	1.2085
10	0.4614	0.4531
50	0.4324	0.4307
250	0.4307	0.4304
1000	0.4306	0.4305

When f Is Not Monotonic

If f is not monotonic, the definite integral is not always bracketed between the left- and right-hand sums. For example, Table 3.6 gives sums for the integral $\int_0^{2.5} \sin(t^2)\, dt$. Although $\sin(t^2)$ is certainly not monotonic on $[0, 2.5]$, by the time we get to $n = 250$, it is pretty clear that $\int_0^{2.5} \sin(t^2)\, dt \approx 0.43$ to two decimal places. Notice, however, that 0.43 does not lie between 1.2500 and 1.2085, the left- and right-hand sums for $n = 2$, or even between 0.4614 and 0.4531, the two sums for $n = 10$. If the integrand is not monotonic, the left- and right-hand sums may both be larger (or smaller) than the integral. (See Problems 2 and 3 for more examples of this behavior.)

Defining The Definite Integral by Upper and Lower Sums

When f is not monotonic, it is difficult to get upper and lower bounds for $\int_a^b f(x)\, dx$ from left and right sums. So instead we take the following approach for any function, f. If f is continuous, this new approach agrees with the previous approach. As before, we consider a subdivision of $[a, b]$ into n intervals; however, now we allow the subintervals to have different lengths. We let Δx_i be the length of the i-th interval, and make the following definition:

Suppose that f is bounded above and below on $[a, b]$. A **lower sum** for f on the interval $[a, b]$ is a sum

$$\sum_{i=1}^n m_i \Delta x_i,$$

where m_i is the greatest lower bound for f on the i-th interval. An **upper sum** is

$$\sum_{i=1}^n M_i \Delta x_i,$$

where M_i is the least upper bound for f on the i-th interval.

See Figure 3.61. Now instead of taking a limit as $n \to \infty$, we consider the least upper and greatest lower bounds of these sums. We make the following definition.

Definition of the Definite Integral

Suppose that f is bounded above and below on $[a, b]$. Let L be the least upper bound for all the lower sums for f on $[a, b]$, and let U be the greatest lower bound for all the upper sums. If $L = U$, then we say that f is *integrable* and we define $\int_a^b f(x)\, dx$ to be equal to the common value of L and U.

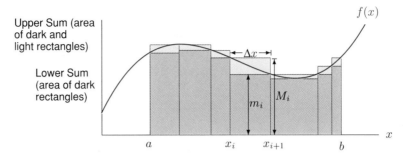

Figure 3.61: Lower and upper sums approximating $\int_a^b f(x)\,dx$

Using the Definition in a Proof

As an example, we will prove the theorem stated on page 172:

Theorem: The Mean Value Inequality for Integrals

If $m \leq f(x) \leq M$ for all x in $[a, b]$, and if f is integrable on $[a, b]$, then

$$m(b - a) \leq \int_a^b f(x)\,dx \leq M(b - a).$$

Geometrically, if f is positive, this theorem says that the area under the graph of f is less than the area of the rectangle of height M, and greater than the area of the rectangle of height m. See Figure 3.39 on page 172.

Proof   The simplest subdivision of $[a, b]$ is the one that consists of one subinterval, namely, $[a, b]$ itself. Although it does not give a very good approximation for the definite integral, it still counts as a subdivision. The least upper bound for f on $[a, b]$ is less than or equal to M, and the length of the only subinterval in the subdivision is $b - a$. So the upper sum for this subdivision is less than or equal to $M(b - a)$. Since every upper sum is an upper estimate for $\int_a^b f(x)\,dx$, we have

$$\int_a^b f(x)\,dx \leq \text{Upper sum} \leq M(b - a).$$

The argument for the other inequality is similar, using lower sums. (See Problem 19.)

Problem 20 gives another example of a proof using the definition of the definite integral.

Continuous Functions Are Integrable

In this section we will prove the following:

Theorem: Continuous Functions are Integrable

If f is continuous on $[a, b]$, then $\int_a^b f(x)\,dx$ exists.

The Key Question

It can be shown (for example, using the Extreme Value Theorem in Chapter 5 on page 286) that a continuous function on a finite interval is bounded above and below. Since any lower sum is less than or equal to any upper sum (see Problems 13–17), it follows that $L \leq U$. (See Problem 18.) To show that f is integrable, we need only show that $L = U$, so the question is the following:

> Can we find a subdivision where the lower sum is as close as we like to the upper sum?

If we could, then $L < U$ would not be a possibility, since then $U - L$ would be a positive number, and we would be able to find lower and upper sums less than $U - L$ apart. In that case, either the lower sum would be bigger than L or the upper sum less than U. This can't happen, since L is an upper bound for the lower sums, and U is a lower bound for the upper sums.

We have already seen that if f is monotonic we can find upper and lower sums that are arbitrarily close; just take the left and right sums. This proves that monotonic functions are integrable. If f is not monotonic, we proceed differently.

Squeezing the Integral Between Lower and Upper Sums

For a subdivision of $[a, b]$ we have

$$\text{Difference between upper and lower sums} = \sum_{i=1}^{n} (M_i - m_i)\Delta x_i,$$

where M_i is the least upper bound for f on the i-th subinterval and m_i is the greatest lower bound. The number $M_i - m_i$ represents the amount by which f varies on the i-th subinterval; we call $M_i - m_i$ the variation[6] of f on this subinterval. Suppose that we could choose the subdivision so that the variation on each subinterval was less than some small positive number ϵ. Then

$$\text{Difference between upper and lower sums} < \sum_{i=1}^{n} \epsilon \Delta x_i = \epsilon \sum_{i=1}^{n} \Delta x_i = \epsilon(b - a).$$

By choosing ϵ small enough we would be able to make the difference as small as we liked. Thus the next question is:

> Can we find a subdivision where the maximum variation of f on each of the subintervals is as small as we like?

Making the Variation Small on Each Subinterval

Let ϵ be a positive number, as small as we like. We want to prove that there is a subdivision of $[a, b]$ such that the variation of f on each subinterval is less than ϵ. We will give an indirect proof: we assume that there is no such subdivision, and show that this leads to impossible consequences.

Divide the interval $[a, b]$ into two halves. If each half has a subdivision where the maximum variation on subintervals is less than ϵ, we can put the two subdivisions together to form a subdivision of $[a, b]$ with the same property.

So if $[a, b]$ fails to have such a subdivision, then so does one of the halves. Choose a half that does not have such a subdivision and divide it in half again. Once more, one of the halves must fail to have a subdivision where the variation of f on each subinterval is less than ϵ. Continuing in this way we find a nested sequence of intervals, each of which fails to have such a subdivision.

---

[6]This is not the same as the total variation used in more advanced texts.

By the Nested Interval Theorem on page 84, these intervals all contain some number c. Since f is continuous at c, we can find an interval around c on which the variation is less than ϵ. One of our nested intervals must be contained in this interval, since they get arbitrarily small. So the variation of f on one of the nested intervals is less than ϵ. This is impossible, given the way we chose each nested interval. So our supposition that $[a, b]$ fails to have the required subdivision is false; there must be a subdivision of $[a, b]$ such that the variation of f on each subinterval is less than ϵ.

Summary

We have shown by contradiction that for every positive number ϵ, no matter how small, there is a subdivision of $[a, b]$ such that the variation of f on each subinterval is less than ϵ. This means we can make the upper and lower sums as close as we like; hence $L = U$ and $\int_a^b f(x)\,dx = U = L$. That is, the continuous function f is integrable.

More General Riemann Sums

Left- and right-hand sums are special cases of Riemann sums. For a general Riemann sum, as with upper and lower sums, we allow subdivisions to have different lengths. Also, instead of evaluating f only at the left or right endpoint of each subdivision, we allow it to be evaluated anywhere in the subdivision. Thus, a general Riemann sum has the form

$$\sum_{i=1}^{n} \text{(Value of } f \text{ at some point in } i\text{-th subdivision)} \cdot \text{(Length of } i\text{-th subdivision)}.$$

(See Figure 3.62.) As before, we let $x_0, x_1, \ldots, x_n$ be the endpoints of the subdivisions, so the length of the i-th subdivision is $\Delta x_i = x_i - x_{i-1}$. For each i we choose a point c_i in the i-th subinterval at which to evaluate f, leading to the following definition:

A general Riemann sum for f on the interval $[a, b]$ is a sum of the form

$$\sum_{i=1}^{n} f(c_i)\Delta x_i,$$

where $a = x_0 < x_1 < \cdots < x_n = b$, and, for $i = 1, \ldots, n$, $\Delta x_i = x_i - x_{i-1}$, and $x_{i-1} \leq c_i \leq x_i$.

We define the *error* in an approximation to be the magnitude of the difference between the approximate and the true values. (Notice that error doesn't mean mistake here.) Since the true value

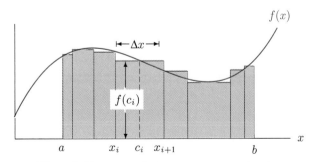

Figure 3.62: A general Riemann sum approximating
$\int_a^b f(x)\,dx$

of the integral is between any upper estimate and any lower estimate, the error in approximating a definite integral by a Riemann sum must be less than the difference between the upper and lower sums using the same subdivision. If $\int_a^b f(x)\,dx$ exists, there is a subdivision for which the upper and lower sums are as close as we like. So we can approximate the integral arbitrarily closely by Riemann sums.

Problems on the Definite Integral

1. Write a few sentences in support of or in opposition to the following statement:
   "If a left-hand sum underestimates a definite integral by a certain amount, then the corresponding right-hand sum will overestimate the integral by the same amount."

2. Using the graph of $2 + \cos x$, for $0 \le x \le 4\pi$, list the following quantities in increasing order: the value of the integral $\int_0^{4\pi}(2 + \cos x)\,dx$, the left-hand sum with $n = 2$ subdivisions, and the right-hand sum with $n = 2$ subdivisions.

3. Sketch the graph of a function f (you do not need to give a formula for f) on an interval $[a, b]$ with the property that with $n = 2$ subdivisions,

$$\int_a^b f(x)\,dx < \text{Left-hand sum} < \text{Right-hand sum}.$$

For Problems 4–12, find a subdivision using subintervals of equal length for which the lower and upper sums differ by less than 0.1. Give these sums and an estimate for the integral which is within 0.05 of the true value. Explain your reasoning. [Except for Problem 12, each function is monotonic over the given interval.]

4. $\displaystyle\int_0^5 x^2\,dx$

5. $\displaystyle\int_1^2 2^x\,dx$

6. $\displaystyle\int_1^4 \frac{1}{\sqrt{1+x^2}}\,dx$

7. $\displaystyle\int_1^{1.5} \sin x\,dx$

8. $\displaystyle\int_0^{\pi/4} \frac{d\theta}{\cos\theta}$

9. $\displaystyle\int_{-2}^{-1} \cos^3 y\,dy$

10. $\displaystyle\int_1^5 (\ln x)^2\,dx$

11. $\displaystyle\int_{1.1}^{1.7} e^t \ln t\,dt$

12. $\displaystyle\int_{-3}^3 e^{-t^2}\,dt$

In Problems 13–17 you will show that every lower sum for a given bounded function f on an interval $[a, b]$ is less than every upper sum, using the idea of a *refinement* of a subdivision. Given a subdivision of the interval $[a, b]$, we can subdivide one or more of its subintervals to obtain a new subdivision. We say that the new subdivision is a *refinement* of the old one. Notice that if one subdivision's set of endpoints contains another's, then the first is a refinement of the second.

13. Show that the lower sum for f on $[a, b]$ using a given subdivision is less than or equal to the upper sum using the same subdivision.

14. In this problem we will show that refining a subdivision results in a lower sum which is larger than the original. Let f be a function defined and bounded from below on $[a, b]$, and choose a subdivision of $[a, b]$, with endpoints $a = x_0 < x_1 < \cdots < x_{n-1} < x_n = b$.

   (a) Suppose $x_{i-1} \le y \le x_i$. Let m_i be the greatest lower bound for f on $[x_{i-1}, x_i]$. Show that m_i is less than or equal to the greatest lower bound for f on $[x_{i-1}, y]$ and the greatest lower bound for f on $[y, x_i]$.

   (b) Show that the lower sum for f using the subdivision $a = x_0 < x_1 < \cdots < x_{n-1} < x_n = b$ is less than or equal to the lower sum using the same subdivision with y included.

   (c) Show that the lower sum for f using the subdivision $a = x_0 < x_1 < \cdots < x_{n-1} < x_n = b$ is less than or equal to the lower sum using any refinement of the subdivision.

15. In this problem we will show that refining a subdivision results in an upper sum which is smaller than the original. Let f be a function defined and bounded from above on $[a, b]$, and choose a subdivision of $[a, b]$, with endpoints $a = x_0 < x_1 < \cdots < x_{n-1} < x_n = b$.

    (a) Suppose $x_{i-1} \le y \le x_i$. Let M_i be the least upper bound for f on $[x_{i-1}, x_i]$. Show that M_i is greater than or equal to the least upper bound for f on $[x_{i-1}, y]$ and the least upper bound for f on $[y, x_i]$.

    (b) Show that the upper sum for f using the subdivision $a = x_0 < x_1 < \cdots < x_{n-1} < x_n = b$ is greater than or equal to the upper sum using the same subdivision with y included.

    (c) Show that the upper sum for f using the subdivision $a = x_0 < x_1 < \cdots < x_{n-1} < x_n = b$ is greater than or equal to the upper sum using any refinement of the subdivision.

16. Given two subdivisions of $[a, b]$, show that there is a third one which is a refinement of both.

17. Show that any lower sum for f on $[a, b]$ is less than or equal to any upper sum. [Hint: The lower sum uses one subdivision of $[a, b]$; the upper sum uses another. Use Problem 16 to choose a common refinement of the two subdivisions and then use Problems 13–15.]

18. Let f be a function defined and bounded on $[a, b]$, let L be the least upper bound for all the lower sums for f on $[a, b]$, and let U be the greatest lower bound for all the upper sums.

    (a) Show that if L were strictly greater than U, then there would be a lower sum that was strictly greater than an upper sum. [Hint: Let $\epsilon = L - U$, and find a lower sum within $\epsilon/3$ of L and an upper sum within $\epsilon/3$ of U.]

    (b) Deduce that $L \le U$.

19. On page 183 we proved one half of the Mean Value Inequality for Integrals. Prove the other half: that is, if f is continuous on $[a, b]$ and $f(x) \ge m$ for x in $[a, b]$, then $m(b - a) \le \int_a^b f(x)\, dx$.

20. In this problem we will prove that if f is continuous on $[a, b]$ and if c is in $[a, b]$, then

$$\int_a^b f(x)\, dx = \int_a^c f(x)\, dx + \int_c^b f(x)\, dx.$$

    (a) Show that if ℓ_1 is a lower sum for f on $[a, c]$, and if ℓ_2 is a lower sum for f on $[c, b]$, then $\ell_1 + \ell_2$ is a lower sum for f on $[a, b]$.

    (b) Show that if ℓ is a lower sum for f on $[a, b]$, then there is a lower sum ℓ_1 for f on $[a, c]$ and a lower sum ℓ_2 for f on $[c, b]$ such that $\ell \le \ell_1 + \ell_2$.

    (c) Let L be the least upper bound of all the lower sums on $[a, b]$, let L_1 be the least upper bound of all the lower sums on $[a, c]$, and let L_2 be the least upper bound of all the lower sums on $[c, a]$. Use parts (a) and (b) to show that $L = L_1 + L_2$.

Since f is continuous on $[a, b]$, $L = \int_a^b f(x)\, dx$, $L_1 = \int_a^c f(x)\, dx$, and $L_2 = \int_c^b f(x)\, dx$. Thus you have proved the required statement.

CHAPTER FOUR

SHORT-CUTS TO DIFFERENTIATION

In Chapter 2, we defined the derivative function

$$f'(x) = \lim_{h \to 0} \frac{f(x+h) - f(x)}{h}$$

and saw how the derivative represents a slope and a rate of change. We learned how to approximate the derivative of a function given graphically (by estimating the slope of the tangent at each point) and numerically (by finding the average rate of change of the function between data values). We calculated the derivatives of x^2 and x^3 exactly using the definition.

In this chapter we make a systematic study of the derivatives of functions given by formulas. These functions include powers, polynomials, exponential, logarithmic, and trigonometric functions. The chapter also contains general rules, such as the product, quotient, and chain rules, which allow us to differentiate combinations of functions.

Useful Notation: We write $\dfrac{d}{dx}(x^3)$, for example, to mean the derivative of x^3 with respect to x. Similarly, $\dfrac{d}{d\theta}\left(\sin(\theta^2)\right)$ denotes the derivative of $\sin(\theta^2)$ with θ as the variable.

4.1 POWERS AND POLYNOMIALS

Derivative of a Constant Times a Function

Figure 4.1 shows the graph of $y = f(x)$ and of three multiples: $y = 3f(x)$, $y = \frac{1}{2}f(x)$, and $y = -2f(x)$. What is the relationship between the derivatives of these functions? In other words, for a particular x-value, how are the slopes of these graphs related?

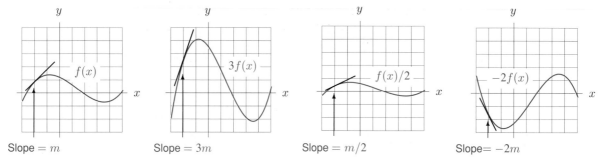

Figure 4.1: A function and its multiples: Derivative of multiple is multiple of derivative

Multiplying by a constant stretches or shrinks the graph (and reflects it in the x-axis if the constant is negative). This changes the slope of the curve at each point. If the graph has been stretched, the "rises" have all been increased by the same factor, whereas the "runs" remain the same. Thus, the slopes are all steeper by the same factor. If the graph has been shrunk, the slopes are all smaller by the same factor. If the graph has been reflected in the x-axis, the slopes will all have their signs reversed. In other words, if a function is multiplied by a constant, c, so is its derivative:

Derivative of a Constant Multiple

If f is differentiable and c is a constant, then

$$\frac{d}{dx}\left[cf(x)\right] = cf'(x).$$

This result can be obtained algebraically, too:

$$\frac{d}{dx}\left[cf(x)\right] = \lim_{h \to 0} \frac{cf(x+h) - cf(x)}{h} = \lim_{h \to 0} c\,\frac{f(x+h) - f(x)}{h}$$
$$= c \lim_{h \to 0} \frac{f(x+h) - f(x)}{h} = cf'(x).$$

You may wonder why c can be taken across the limit sign. The reason is that c is a constant. If a function's value gets close to a certain number, then c times that function gets close to c times that number. The function in this case is $\left[f(x+h) - f(x)\right]/h$.

Derivatives of Sums and Differences

Suppose we have two functions, $f(x)$ and $g(x)$, with the values listed in Table 4.1. Values of the sum $f(x) + g(x)$ are given in the same table.

TABLE 4.1 *Sum of Functions*

x	$f(x)$	$g(x)$	$f(x) + g(x)$
0	100	0	100
1	110	0.2	110.2
2	130	0.4	130.4
3	160	0.6	160.6
4	200	0.8	200.8

We see that adding the increments of $f(x)$ and the increments of $g(x)$ gives the increments of $f(x) + g(x)$. For example, as x increases from 0 to 1, $f(x)$ increases by 10 and $g(x)$ increases by 0.2, while $f(x) + g(x)$ increases by $110.2 - 100 = 10.2$. Similarly, as x increases from 3 to 4, $f(x)$ increases by 40 and $g(x)$ by 0.2, while $f(x) + g(x)$ increases by $200.8 - 160.6 = 40.2$.

From this example, we see that the rate at which $f(x) + g(x)$ is increasing is the sum of the rates at which $f(x)$ and $g(x)$ are increasing. Similar reasoning applies to the difference, $f(x) - g(x)$. In terms of derivatives:

Derivative of Sum and Difference

If f and g are differentiable, then

$$\frac{d}{dx}\left[f(x) + g(x)\right] = f'(x) + g'(x) \quad \text{and} \quad \frac{d}{dx}\left[f(x) - g(x)\right] = f'(x) - g'(x).$$

We can justify the sum rule using the definition of the derivative:

$$\frac{d}{dx}\left[f(x) + g(x)\right] = \lim_{h \to 0}\frac{\left[f(x+h) + g(x+h)\right] - \left[f(x) + g(x)\right]}{h}$$

$$= \lim_{h \to 0}\left[\underbrace{\frac{f(x+h) - f(x)}{h}}_{\text{Limit of this is } f'(x)} + \underbrace{\frac{g(x+h) - g(x)}{h}}_{\text{Limit of this is } g'(x)}\right]$$

$$= f'(x) + g'(x).$$

Powers of x

In Chapter 2 we showed that

$$f'(x) = \frac{d}{dx}(x^2) = 2x \quad \text{and} \quad g'(x) = \frac{d}{dx}(x^3) = 3x^2.$$

The graphs of $f(x) = x^2$ and $g(x) = x^3$ and their derivatives are shown in Figures 4.2 and 4.3. Notice $f'(x) = 2x$ has the behavior we expect. It is negative for $x < 0$ (when f is decreasing), zero for $x = 0$, and positive for $x > 0$ (when f is increasing). Similarly, $g'(x) = 3x^2$ is zero when $x = 0$, but positive everywhere else, as g is increasing everywhere else.

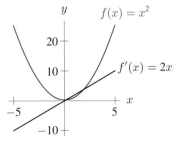

Figure 4.2: Graphs of $f(x) = x^2$ and its derivative $f'(x) = 2x$

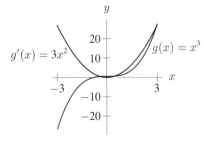

Figure 4.3: Graphs of $g(x) = x^3$ and its derivative $g'(x) = 3x^2$

These examples are special cases of the power rule which we justify for any positive integer n on page 193:

The Power Rule

For any constant real number n,

$$\frac{d}{dx}(x^n) = nx^{n-1}$$

Problem 52 asks you to show that this rule holds for negative integral powers. In Section 4.6 we indicate how to justify it for powers of the form $1/n$.

Example 1   Use the power rule to differentiate (a) $\dfrac{1}{x^3}$, (b) $x^{1/2}$, (c) $\dfrac{1}{\sqrt[3]{x}}$.

Solution   (a)   For $n = -3$: $\dfrac{d}{dx}\left(\dfrac{1}{x^3}\right) = \dfrac{d}{dx}(x^{-3}) = -3x^{-3-1} = -3x^{-4} = -\dfrac{3}{x^4}.$

(b)   For $n = 1/2$: $\dfrac{d}{dx}\left(x^{1/2}\right) = \dfrac{1}{2}x^{(1/2)-1} = \dfrac{1}{2}x^{-1/2} = \dfrac{1}{2\sqrt{x}}.$

(c)   For $n = -1/3$: $\dfrac{d}{dx}\left(\dfrac{1}{\sqrt[3]{x}}\right) = \dfrac{d}{dx}\left(x^{-1/3}\right) = -\dfrac{1}{3}x^{(-1/3)-1} = -\dfrac{1}{3}x^{-4/3} = -\dfrac{1}{3x^{4/3}}.$

Example 2   Use the definition of the derivative to justify the power rule for $n = -2$: Show $\dfrac{d}{dx}(x^{-2}) = -2x^{-3}.$

Solution   Provided $x \neq 0$, we have

$$\frac{d}{dx}\left(x^{-2}\right) = \frac{d}{dx}\left(\frac{1}{x^2}\right) = \lim_{h \to 0}\left(\frac{\frac{1}{(x+h)^2} - \frac{1}{x^2}}{h}\right) = \lim_{h \to 0}\frac{1}{h}\left[\frac{x^2 - (x+h)^2}{(x+h)^2x^2}\right] \quad \text{(Combining fractions over a common denominator)}$$

$$= \lim_{h \to 0}\frac{1}{h}\left[\frac{x^2 - (x^2 + 2xh + h^2)}{(x+h)^2x^2}\right] \quad \text{(Multiplying out)}$$

$$= \lim_{h \to 0}\frac{-2xh - h^2}{h(x+h)^2x^2} \quad \text{(Simplifying numerator)}$$

$$= \lim_{h \to 0}\frac{-2x - h}{(x+h)^2x^2} \quad \text{(Dividing numerator and denominator by h)}$$

$$= \frac{-2x}{x^4} \quad \text{(Letting $h \to 0$)}$$

$$= -2x^{-3}.$$

The graphs of x^{-2} and its derivative, $-2x^{-3}$, are shown in Figure 4.4. Does the graph of the derivative have the features you expect?

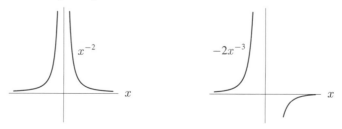

Figure 4.4: Graphs of x^{-2} and its derivative, $-2x^{-3}$

Justification of $\dfrac{d}{dx}(x^n) = nx^{n-1}$, for n a Positive Integer

To calculate the derivatives of x^2 and x^3, we had to expand $(x+h)^2$ and $(x+h)^3$. To calculate the derivative of x^n, we must expand $(x+h)^n$. Let's look back at the previous expansions:

$$(x+h)^2 = x^2 + 2xh + h^2, \qquad (x+h)^3 = x^3 + 3x^2h + 3xh^2 + h^3,$$

Using the Binomial Theorem on page 78 gives

$$(x+h)^n = x^n + nx^{n-1}h + \underbrace{\cdots\cdots + h^n}.$$

$$\text{Terms involving } h^2 \text{ and higher powers of } h$$

Now to find the derivative,

$$\frac{d}{dx}(x^n) = \lim_{h \to 0} \frac{(x+h)^n - x^n}{h}$$

$$= \lim_{h \to 0} \frac{(x^n + nx^{n-1}h + \cdots + h^n) - x^n}{h}$$

$$\text{Terms involving } h^2 \text{ and higher powers of } h$$

$$= \lim_{h \to 0} \frac{nx^{n-1}h + \overbrace{\cdots + h^n}}{h}.$$

When we factor out h from terms involving h^2 and higher powers of h, each term will still have an h in it. Factoring and dividing, we get:

$$\text{Terms involving } h \text{ and higher powers of } h$$

$$\frac{d}{dx}(x^n) = \lim_{h \to 0} \frac{h(nx^{n-1} + \cdots + h^{n-1})}{h} = \lim_{h \to 0}(nx^{n-1} + \overbrace{\cdots + h^{n-1}}).$$

But as $h \to 0$, all terms involving an h will go to 0, so

$$\frac{d}{dx}(x^n) = \lim_{h \to 0}(nx^{n-1} + \underbrace{\cdots + h^{n-1}}) = nx^{n-1}.$$

$$\text{These terms go to 0}$$

Derivatives of Polynomials

Now that we know how to differentiate powers, constant multiples, and sums, we can differentiate any polynomial.

Example 3 Find the derivatives of (a) $5x^2 + 3x + 2,$ (b) $\sqrt{3}x^7 - \dfrac{x^5}{5} + \pi.$

Solution (a)

$$\frac{d}{dx}(5x^2 + 3x + 2) = 5\frac{d}{dx}(x^2) + 3\frac{d}{dx}(x) + \frac{d}{dx}(2)$$

$$= 5 \cdot 2x + 3 \cdot 1 + 0 \qquad \text{(Since the derivative of a constant, } \tfrac{d}{dx}(2), \text{ is zero.)}$$

$$= 10x + 3.$$

(b)

$$\frac{d}{dx}\left(\sqrt{3}x^7 - \frac{x^5}{5} + \pi\right) = \sqrt{3}\frac{d}{dx}(x^7) - \frac{1}{5}\frac{d}{dx}(x^5) + \frac{d}{dx}(\pi)$$

$$= \sqrt{3} \cdot 7x^6 - \frac{1}{5} \cdot 5x^4 + 0 \qquad \text{(Since } \pi \text{ is a constant, } d\pi/dx = 0.)$$

$$= 7\sqrt{3}x^6 - x^4.$$

We can also use the rules we have seen so far to differentiate expressions which are not polynomials.

Example 4   Differentiate (a) $5\sqrt{x} - \dfrac{10}{x^2} + \dfrac{1}{2\sqrt{x}}$ (b) $0.1x^3 + 2x^{\sqrt{2}}$

Solution   (a) $\dfrac{d}{dx}\left(5\sqrt{x} - \dfrac{10}{x^2} + \dfrac{1}{2\sqrt{x}}\right) = \dfrac{d}{dx}\left(5x^{1/2} - 10x^{-2} + \dfrac{1}{2}x^{-1/2}\right)$

$$= 5 \cdot \dfrac{1}{2}x^{-1/2} - 10(-2)x^{-3} + \dfrac{1}{2}\left(-\dfrac{1}{2}\right)x^{-3/2}$$

$$= \dfrac{5}{2\sqrt{x}} + \dfrac{20}{x^3} - \dfrac{1}{4x^{3/2}}.$$

(b) $\dfrac{d}{dx}(0.1x^3 + 2x^{\sqrt{2}}) = 0.1\dfrac{d}{dx}(x^3) + 2\dfrac{d}{dx}(x^{\sqrt{2}}) = 0.3x^2 + 2\sqrt{2}x^{\sqrt{2}-1}.$

Example 5   Find the second derivative and interpret its sign for
(a) $f(x) = x^2$,   (b) $g(x) = x^3$,   (c) $k(x) = x^{1/2}$.

Solution   (a) If $f(x) = x^2$, then $f'(x) = 2x$, so $f''(x) = \dfrac{d}{dx}(2x) = 2$. Since f'' is always positive, f is concave up, as expected for a parabola opening upwards. (See Figure 4.5.)

(b) If $g(x) = x^3$, then $g'(x) = 3x^2$, so $g''(x) = \dfrac{d}{dx}(3x^2) = 3\dfrac{d}{dx}(x^2) = 3 \cdot 2x = 6x$. This is positive for $x > 0$ and negative for $x < 0$, which means x^3 is concave up for $x > 0$ and concave down for $x < 0$. (See Figure 4.6.)

(c) If $k(x) = x^{1/2}$, then $k'(x) = \frac{1}{2}x^{(1/2)-1} = \frac{1}{2}x^{-1/2}$, so

$$k''(x) = \dfrac{d}{dx}\left(\dfrac{1}{2}x^{-1/2}\right) = \dfrac{1}{2} \cdot \left(-\dfrac{1}{2}\right)x^{-(1/2)-1} = -\dfrac{1}{4}x^{-3/2}.$$

Now k' and k'' are only defined on the domain of k, that is, $x \geq 0$. When $x > 0$, we see that $k''(x)$ is negative, so k is concave down. (See Figure 4.7.)

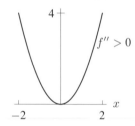

Figure 4.5: $f(x) = x^2$ and $f''(x) = 2$

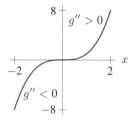

Figure 4.6: $g(x) = x^3$ and $g''(x) = 6x$

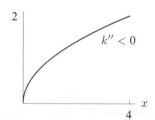

Figure 4.7: $k(x) = x^{1/2}$ and $k''(x) = -\frac{1}{4}x^{-3/2}$

Example 6   If the position of a body, in meters, is given as a function of time t, in seconds, by

$$s = -4.9t^2 + 5t + 6,$$

find the velocity and acceleration of the body at time t.

Solution   The velocity, v, is the derivative of the position:

$$v = \dfrac{ds}{dt} = \dfrac{d}{dt}(-4.9t^2 + 5t + 6) = -9.8t + 5,$$

and the acceleration, a, is the derivative of the velocity:

$$a = \dfrac{dv}{dt} = \dfrac{d}{dt}(-9.8t + 5) = -9.8.$$

Notice that v is in meters/second and a is in meters/second².

Example 7   Figure 4.8 shows the graph of a cubic polynomial. Both graphically and algebraically, describe the behavior of the derivative of this cubic.

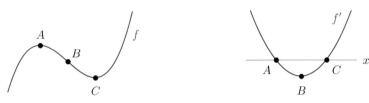

Figure 4.8: The cubic of Example 7   **Figure 4.9:** Derivative of the cubic of Example 7

Solution   Graphical approach: Suppose we move along the curve from left to right. To the left of A, the slope is positive; it starts very positive and decreases until the curve reaches A, where the slope is 0. Between A and C the slope is negative. Between A and B the slope is decreasing (getting more negative); it is most negative at B. Between B and C the slope is negative but increasing; at C the slope is zero. From C to the right, the slope is positive and increasing. The graph of the derivative function is shown in Figure 4.8.

Algebraic approach: f is a cubic that goes to $+\infty$ as $x \to +\infty$, so

$$f(x) = ax^3 + bx^2 + cx + d$$

with $a > 0$. Hence,

$$f'(x) = 3ax^2 + 2bx + c,$$

whose graph is a parabola opening upward, as in Figure 4.9.

Problems for Section 4.1

1. Let $f(x) = 7$. Using the definition of the derivative, show that $f'(x) = 0$ for all values of x.
2. Let $f(x) = -3x + 2$ and $g(x) = 2x + 1$.
    (a) If $k(x) = f(x) + g(x)$, find a formula for $k(x)$ and verify the sum rule by comparing $k'(x)$ with $f'(x) + g'(x)$.
    (b) If $j(x) = f(x) - g(x)$, find a formula for $j(x)$ and compare $j'(x)$ with $f'(x) - g'(x)$.
3. (a) If $f(x) = 5x - 3$, $g(x) = -2x + 1$, find the derivative of $f(g(x))$.
    (b) Use your answer to part (a) to make a conjecture about the derivative of the composition of two linear functions. Explain why what you say is true for any two linear functions.

For Problems 4–24, find the derivatives of the given functions.

4. $y = x^{11}$
5. $y = x^{12}$
6. $y = -x^{-11}$
7. $y = x^{3.2}$
8. $y = x^{-12}$
9. $y = x^{4/3}$
10. $y = x^{3/4}$
11. $y = x^{-3/4}$
12. $f(x) = \dfrac{1}{x^4}$
13. $f(x) = \sqrt[4]{x}$
14. $f(x) = x^e$
15. $y = 4x^{3/2} - 5x^{1/2}$
16. $y = 6x^3 + 4x^2 - 2x$
17. $y = -3x^4 - 4x^3 - 6x + 2$
18. $y = 3t^5 - 5\sqrt{t} + \dfrac{7}{t}$
19. $y = 3t^2 + \dfrac{12}{\sqrt{t}} - \dfrac{1}{t^2}$
20. $y = z^2 + \dfrac{1}{2z}$
21. $y = \dfrac{x^2 + 1}{x}$
22. $g(z) = \dfrac{z^7 + 5z^6 - z^3}{z^2}$
23. $f(t) = \dfrac{t^2 + t^3 - 1}{t^4}$
24. $y = \dfrac{\theta - 1}{\sqrt{\theta}}$

25. Which functions in Problems 4–15 have derivatives that do not exist at $x = 0$?

For Problems 26–34, determine if the derivative rules from this section apply. If they do, find the derivative. If they don't apply, indicate why.

26. $y = \sqrt{x}$

27. $y = (x + 3)^{1/2}$

28. $y = 3x^2 + 4$

29. $y = \dfrac{1}{3z^2} + \dfrac{1}{4}$

30. $y = \dfrac{1}{3x^2 + 4}$

31. $y = 3^x$

32. $y = \dfrac{1}{3\sqrt{x}} + \dfrac{1}{4}$

33. $g(x) = 72\sqrt[6]{x} + \dfrac{27}{x^{2/3}}$

34. $g(x) = x^\pi - x^{-\pi}$

35. If $f(t) = 2t^3 - 4t^2 + 3t - 1$, find $f'(t)$ and $f''(t)$.

36. If $f(x) = 4x^3 + 6x^2 - 23x + 7$, find the intervals on which $f'(x) \geq 1$.

37. On what intervals is the function $f(x) = x^4 - 4x^3$ both decreasing and concave up?

38. For what values of x is the graph of $y = x^5 - 5x$ both increasing and concave up?

39. If $f(x) = 13 - 8x + \sqrt{2}x^2$ and $f'(r) = 4$, find r.

40. (a) Find the *eighth* derivative of $f(x) = x^7 + 5x^5 - 4x^3 + 6x - 7$. Think ahead! (The nth derivative is the result of differentiating n times.)
    (b) Find the seventh derivative of $f(x)$.

41. Find the equation of the line tangent to the graph of f at $(1, 1)$, where f is given by $f(x) = 2x^3 - 2x^2 + 1$.

42. Show that for any power function $f(x) = x^n$, we have $f'(1) = n$.

43. Given a power function of the form $f(x) = ax^n$, with $f'(2) = 3$ and $f'(4) = 24$, find n and a.

44. Is there a value of n which makes $y = x^n$ a solution to the equation $13x\dfrac{dy}{dx} = y$? If so, what value?

45. Using a graph to help you, find the equations of all lines through the origin tangent to the parabola

$$y = x^2 - 2x + 4.$$

Sketch the lines on the graph.

46. A ball is dropped from the top of the Empire State building to the ground below. The height, y, of the ball above the ground (in feet) is given as a function of time, t, (in seconds) by

$$y = 1250 - 16t^2.$$

   (a) Find the velocity of the ball at time t. What is the sign of the velocity? Why is this to be expected?
   (b) Show that the acceleration of the ball is a constant. What are the value and sign of this constant?
   (c) When does the ball hit the ground, and how fast is it going at that time? Give your answer in feet per second and in miles per hour (1 ft/sec = 15/22 mph).

47. The gravitational attraction, F, between the earth and a satellite of mass m at a distance r from the center of the earth is given by

$$F = \frac{GMm}{r^2},$$

where M is the mass of the earth, and G is a constant. Find the rate of change of force with respect to distance.

48. The period, T, of a pendulum is given in terms of its length, l, by

$$T = 2\pi\sqrt{\frac{l}{g}},$$

where g is the acceleration due to gravity (a constant).

   (a) Find $\dfrac{dT}{dl}$.
   (b) What is the sign of $\dfrac{dT}{dl}$? What does this tell you about the period of pendulums?

49. We know that the graph of a tangent line lies close to the graph of the function near the point of tangency. However, usually the further we go from the point of tangency, the greater the distance between the graph of the function and the tangent line. Let's investigate this for $f(x) = 1/x$. Find the value of f at $x = 2$. Find the tangent line to the curve at $x = 1$, and use it to approximate the value of f at $x = 2$. Now find the tangent to the curve at $x = 100$ and use it to approximate the value of f at $x = 2$. Which tangent line lies closer to the curve at $x = 2$? Is there something wrong with our basic idea? Explain.

50. (a) Use the formula for the area of a circle of radius r, $A = \pi r^2$, to find $\dfrac{dA}{dr}$.

    (b) The result from part (a) should look familiar. What does $\dfrac{dA}{dr}$ represent geometrically? Draw a picture.

    (c) Use the difference quotient to explain the observation you made in part (b).

51. What is the formula for $V(r)$, the volume of a sphere of radius r? Find $\dfrac{dV}{dr}$. What is the geometrical meaning of $\dfrac{dV}{dr}$?

52. Using the definition of derivative, justify the formula $\dfrac{d}{dx}(x^n) = nx^{n-1}$.

    (a) For $n = -1$; for $n = -3$.   (b) For any negative integer n.

4.2 THE EXPONENTIAL FUNCTION

What would we expect the graph of the derivative of the exponential function $f(x) = a^x$ to look like? The graph of the exponential function is shown in Figure 4.10. The function increases slowly for $x < 0$ and more rapidly for $x > 0$, so the values of f' are small for $x < 0$ and larger for $x > 0$. Since the function is increasing for all values of x, the graph of the derivative must lie above the x-axis. In fact, it appears that the graph of f' must resemble the graph of f itself. We will see how this observation holds for $f(x) = 2^x$ and $g(x) = 3^x$.

Figure 4.10: $f(x) = a^x$, with $a > 1$

The Derivatives of 2^x and 3^x

In Chapter 2, we saw that the derivative of $f(x) = 2^x$ at $x = 0$ is given by

$$f'(0) = \lim_{h \to 0} \frac{2^h - 2^0}{h} = \lim_{h \to 0} \frac{2^h - 1}{h} \approx 0.6931.$$

The estimate for this limit is obtained by evaluating $(2^h - 1)/h$ for small values of h. Similarly, we can estimate the derivative at $x = 1$ and $x = 2$:

$$f'(1) = \lim_{h \to 0} \frac{2^{1+h} - 2^1}{h} \approx 1.3863,$$

$$f'(2) = \lim_{h \to 0} \frac{2^{2+h} - 2^2}{h} \approx 2.7726.$$

Can you see the relationship between these values of the derivative? If we notice that $1.3863 \approx 2(0.6931)$ and $2.7726 \approx 4(0.6931)$, we see

$$f'(0) \approx 0.6931 = 0.6931 \cdot 2^0,$$
$$f'(1) \approx 1.3863 \approx 0.6931 \cdot 2^1,$$
$$f'(2) \approx 2.7726 \approx 0.6931 \cdot 2^2.$$

It looks as though we have $f'(x) \approx 0.6931 \cdot 2^x$, which is in fact true. Let's look at the derivative function to see why this happens:

$$f'(x) = \lim_{h \to 0} \left(\frac{2^{x+h} - 2^x}{h} \right) = \lim_{h \to 0} \left(\frac{2^x 2^h - 2^x}{h} \right) = \lim_{h \to 0} 2^x \left(\frac{2^h - 1}{h} \right)$$

$$= 2^x \lim_{h \to 0} \left(\frac{2^h - 1}{h} \right) \qquad \text{(Since } x \text{ and } 2^x \text{ are fixed during this calculation)}$$

$$= f'(0) 2^x.$$

We have already estimated $f'(0) \approx 0.6931$, so we get

$$\boxed{\frac{d}{dx}(2^x) = f'(x) \approx (0.6931)2^x.}$$

The graphs of $f(x) = 2^x$ and $f'(x) \approx (0.6931)2^x$ are shown in Figure 4.11. Notice that they do indeed resemble one another.

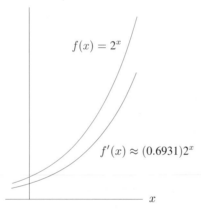

Figure 4.11: Graph of $f(x) = 2^x$ and its derivative

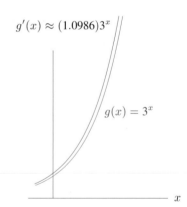

Figure 4.12: Graph of $g(x) = 3^x$ and its derivative

Example 1   Find the derivative of $g(x) = 3^x$ and plot g and g' on the same axes.

Solution   As before

$$g'(x) = \lim_{h \to 0} \frac{3^{x+h} - 3^x}{h} = \lim_{h \to 0} \frac{3^x 3^h - 3^x}{h} = 3^x \lim_{h \to 0} \left(\frac{3^h - 1}{h} \right).$$

Using a calculator gives

$$\lim_{h \to 0} \left(\frac{3^h - 1}{h} \right) \approx 1.0986,$$

so

$$g'(x) \approx (1.0986)3^x.$$

The graphs are shown in Figure 4.12.

Notice that for both $f(x) = 2^x$ and $g(x) = 3^x$, *the derivative is proportional to the original function*. For $f(x) = 2^x$, we have $f'(x) \approx (0.6931)2^x$, so the constant of proportionality is less than 1, and the graph of the derivative is below the graph of the original function. For $g(x) = 3^x$, we have $g'(x) \approx (1.0986)3^x$, so the constant is greater than 1 and the graph of the derivative is above that of the original function.

Note on Round-Off Error and Limits

If we try to evaluate $(2^h - 1)/h$ on a calculator by taking smaller and smaller values of h, the values of $(2^h - 1)/h$ will at first get closer to 0.6931. However, they will eventually move away from the correct value $0.6931\ldots$ because of the *round-off error* (i.e., errors introduced by the fact that the calculator can only hold a certain number of digits).

As we try smaller and smaller values of h, how do we know when to stop? Unfortunately, there is no fixed rule. A calculator can only suggest the value of a limit, but can never confirm that this value is correct. In this case, it looks like the limit is about 0.6931 because the values of $(2^h - 1)/h$ hover around 0.6931 for a while. To be sure this is correct, we would have to find the limit by theoretical means.

The Derivative of a^x and the Definition of e

The calculation of the derivative of $f(x) = a^x$, for $a > 0$, is similar to that of 2^x and 3^x:

$$f'(x) = \lim_{h \to 0} \frac{a^{x+h} - a^x}{h} = a^x \lim_{h \to 0} \frac{a^h - 1}{h}.$$

The quantity $\lim_{h \to 0} (a^h - 1)/h$ doesn't depend on x, and so is a constant for any particular a. Therefore the derivative is again proportional to the original function, with constant of proportionality

$$\lim_{h \to 0} \frac{a^h - 1}{h}.$$

We can't use a calculator to evaluate this limit without knowing the value of a. However, when $a = 2$, we know that the limit (0.6931) is less than 1, and the derivative is smaller than the original function. When $a = 3$, the limit (1.0986) is more than 1, and the derivative is greater than the original function. Is there an in-between case, when derivative and function are exactly equal? In other words:

Is there a value of a that makes $\dfrac{d}{dx}(a^x) = a^x$?

If so, we have found a function with the remarkable property that it is equal to its own derivative.

So let's look for such an a. This means we want to find a such that

$$\lim_{h \to 0} \frac{a^h - 1}{h} = 1, \qquad \text{or, for small } h, \qquad \frac{a^h - 1}{h} \approx 1.$$

Solving for a suggests that we can calculate a as follows:

$$a^h - 1 \approx h, \qquad \text{or} \qquad a^h \approx 1 + h, \qquad \text{so} \qquad a \approx (1 + h)^{1/h}.$$

Taking small values of h, as in Table 4.2, we can see $a \approx 2.718\ldots$, which looks like the number e introduced in Chapter 1.

TABLE 4.2

h	$(1 + h)^{1/h}$
0.001	2.7169239
0.0001	2.7181459
0.00001	2.7182682

In fact, it can be shown that

$$e = \lim_{h \to 0}(1 + h)^{1/h} = 2.718\ldots \quad \text{and} \quad \lim_{h \to 0}\frac{e^h - 1}{h} = 1.$$

This means that e^x is its own derivative:

$$\frac{d}{dx}(e^x) = e^x.$$

It turns out that the constants involved in the derivatives of 2^x and 3^x are natural logarithms. In fact, since $0.6931 \approx \ln 2$ and $1.0986 \approx \ln 3$, we (correctly) guess that

$$\frac{d}{dx}(2^x) = (\ln 2)2^x \quad \text{and} \quad \frac{d}{dx}(3^x) = (\ln 3)3^x.$$

In Section 4.6, we will show that, in general,

$$\frac{d}{dx}(a^x) = (\ln a)a^x.$$

Figure 4.13 shows the graph of the derivative of 2^x below the graph of the function, and the graph of the derivative of 3^x above the graph of the function. With $e \approx 2.718$, the function e^x and its derivative are identical.

Since $\ln a$ is a constant, the derivative of a^x is proportional to a^x. Many quantities have rates of change which are proportional to themselves; for example, the simplest model of population growth has this property. The fact that the constant of proportionality is 1 when $a = e$ makes e a particularly useful base for exponential functions.

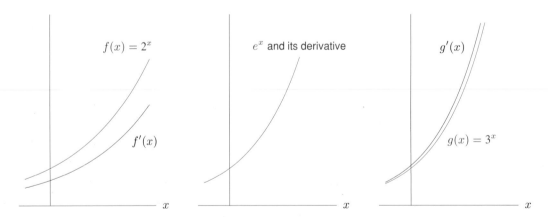

Figure 4.13: Graphs of the functions 2^x, e^x, and 3^x and their derivatives

Example 2   Differentiate $2 \cdot 3^x + 5e^x$.

Solution

$$\frac{d}{dx}(2 \cdot 3^x + 5e^x) = 2\frac{d}{dx}(3^x) + 5\frac{d}{dx}(e^x) = 2\ln 3 \cdot 3^x + 5e^x \approx (2.1972)3^x + 5e^x.$$

Problems for Section 4.2

Find the derivatives of the functions in Problems 1–21.

1. $f(x) = 2e^x + x^2$

2. $y = 5t^2 + 4e^t$

3. $y = 5^x + 2$

4. $f(x) = 2^x + 2 \cdot 3^x$

5. $y = 5x^2 + 2^x + 3$

6. $f(x) = 12e^x + 11^x$

7. $y = 4 \cdot 10^x - x^3$

8. $y = 3x - 2 \cdot 4^x$

9. $y = \dfrac{3^x}{3} + \dfrac{33}{\sqrt{x}}$

10. $f(x) = e^2 + x^e$

11. $f(x) = e^{1+x}$

12. $f(t) = e^{t+2}$

13. $y = e^{\theta - 1}$

14. $z = (\ln 4)e^x$

15. $z = (\ln 4)4^x$

16. $f(z) = (\ln 3)z^2 + (\ln 4)e^z$

17. $f(t) = (\ln 3)^t$

18. $f(x) = x^3 + 3^x$

19. $y = 5 \cdot 5^t + 6 \cdot 6^t$

20. $y = \pi^2 + \pi^x$

21. $f(x) = x^{\pi^2} + (\pi^2)^x$

Which of the functions in Problems 22–30 can be differentiated using the rules we have developed so far? Differentiate if you can; otherwise, indicate why the rules discussed so far do not apply.

22. $y = x^2 + 2^x$

23. $y = \sqrt{x} - \left(\frac{1}{2}\right)^x$

24. $y = x^2 \cdot 2^x$

25. $y = \dfrac{2^x}{x}$

26. $y = e^{x+5}$

27. $y = e^{5x}$

28. $y = 4^{(x^2)}$

29. $f(z) = (\sqrt{4})^z$

30. $f(\theta) = 4^{\sqrt{\theta}}$

31. Since January 1, 1960, the population of Slim Chance has been described by the formula

$$P = 35{,}000(0.98)^t,$$

where P is the population of the city t years after the start of 1960. At what rate was the population changing on January 1, 1983?

32. With a yearly inflation rate of 5%, prices are described by

$$P = P_0(1.05)^t,$$

where P_0 is the price in dollars when $t = 0$ and t is time in years. Suppose $P_0 = 1$. How fast (in cents/year) are prices rising when $t = 10$?

33. Certain pieces of antique furniture increased very rapidly in price in the 1970s and 1980s. For example, the value of a particular rocking chair is well approximated by

$$V = 75(1.35)^t,$$

where V is in dollars and t is the number of years since 1975. Find the rate, in dollars per year, at which the price is increasing.

34. The value of a certain automobile purchased in 1997 can be approximated by the function $V(t) = 25(0.85)^t$, where t is the time, in years, from the date of purchase, and V is the value, in thousands of dollars.

    (a) Evaluate and interpret $V(4)$.
    (b) Find an expression for $V'(t)$, including units.
    (c) Evaluate and interpret $V'(4)$.
    (d) Use $V(t)$, $V'(t)$, and any other considerations you think are relevant to write a paragraph in support of or in opposition to the following statement: "From a monetary point of view, it is best to keep this vehicle as long as possible."

35. (a) Find the slope of the graph of $f(x) = 1 - e^x$ at the point where it crosses the x-axis.
     (b) Find the equation of the tangent line to the curve at this point.
     (c) Find the equation of the line which is perpendicular to the tangent line at this point. (This line is known as the *normal* line.)

36. Find the value of c in Figure 4.14, where the line l tangent to the graph of $y = 2^x$ at $(0, 1)$ intersects the x-axis.

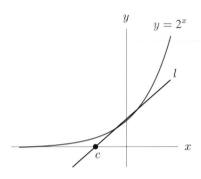

Figure 4.14

37. Find the quadratic polynomial $g(x) = ax^2 + bx + c$ which best fits the function $f(x) = e^x$ at $x = 0$, in the sense that

$$g(0) = f(0), \quad \text{and} \quad g'(0) = f'(0), \quad \text{and} \quad g''(0) = f''(0).$$

Using a computer or calculator, sketch graphs of f and g on the same axes. What do you notice?

38. Using the equation of the tangent line to the graph of e^x at $x = 0$, show that

$$e^x \geq 1 + x$$

for all values of x. A sketch may be helpful.

39. Find all solutions of the equation

$$2^x = 2x.$$

How do you know that you found all solutions?

4.3 THE PRODUCT AND QUOTIENT RULES

We now know how to find derivatives of powers and exponentials, and of sums and constant multiples of functions. This section will show how to find the derivatives of products and quotients.

Using Δ Notation

To express the difference quotients of general functions, some additional notation is helpful. We write Δf, read "delta f," for a small change in the value of f,

$$\Delta f = f(x + h) - f(x).$$

In this notation, the derivative is the limit of the ratio $\Delta f / h$:

$$f'(x) = \lim_{h \to 0} \frac{\Delta f}{h}.$$

The Product Rule

Suppose we know the derivatives of $f(x)$ and $g(x)$ and want to calculate the derivative of the product, $f(x)g(x)$. The derivative of the product is calculated by taking the limit, namely,

$$\frac{d[f(x)g(x)]}{dx} = \lim_{h \to 0} \frac{f(x+h)g(x+h) - f(x)g(x)}{h}.$$

To picture the quantity $f(x+h)g(x+h) - f(x)g(x)$, imagine the rectangle with sides $f(x+h)$ and $g(x+h)$ in Figure 4.15, where $\Delta f = f(x+h) - f(x)$ and $\Delta g = g(x+h) - g(x)$.

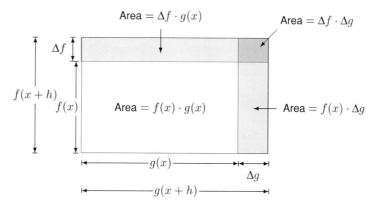

Figure 4.15: Illustration for the product rule (with Δf, Δg positive)

Then

$$f(x+h)g(x+h) - f(x)g(x) = (\text{Area of whole rectangle}) - (\text{Unshaded area})$$
$$= \text{Area of the three shaded rectangles}$$
$$= \Delta f \cdot g(x) + f(x) \cdot \Delta g + \Delta f \cdot \Delta g.$$

Now divide by h:

$$\frac{f(x+h)g(x+h) - f(x)g(x)}{h} = \frac{\Delta f}{h} \cdot g(x) + f(x) \cdot \frac{\Delta g}{h} + \frac{\Delta f \cdot \Delta g}{h}.$$

To evaluate the limit as $h \to 0$, we examine the three terms on the right separately. Notice that

$$\lim_{h \to 0} \frac{\Delta f}{h} \cdot g(x) = f'(x)g(x) \quad \text{and} \quad \lim_{h \to 0} f(x) \cdot \frac{\Delta g}{h} = f(x)g'(x).$$

In the third term we multiply the top and bottom by h to get $\frac{\Delta f}{h} \cdot \frac{\Delta g}{h} \cdot h$. Then,

$$\lim_{h \to 0} \frac{\Delta f \cdot \Delta g}{h} = \lim_{h \to 0} \frac{\Delta f}{h} \cdot \frac{\Delta g}{h} \cdot h = \lim_{h \to 0} \frac{\Delta f}{h} \cdot \lim_{h \to 0} \frac{\Delta g}{h} \cdot \lim_{h \to 0} h = f'(x) \cdot g'(x) \cdot 0 = 0.$$

Therefore, we conclude that

$$\lim_{h \to 0} \frac{f(x+h)g(x+h) - f(x)g(x)}{h} = \lim_{h \to 0} \left(\frac{\Delta f}{h} \cdot g(x) + f(x) \cdot \frac{\Delta g}{h} + \frac{\Delta f \cdot \Delta g}{h} \right)$$
$$= \lim_{h \to 0} \frac{\Delta f}{h} \cdot g(x) + \lim_{h \to 0} f(x) \cdot \frac{\Delta g}{h} + \lim_{h \to 0} \frac{\Delta f \cdot \Delta g}{h}$$
$$= f'(x)g(x) + f(x)g'(x).$$

Thus we have the following rule:

The Product Rule

If $u = f(x)$ and $v = g(x)$ are differentiable, then

$$(fg)' = f'g + fg'.$$

The product rule can also be written

$$\frac{d(uv)}{dx} = \frac{du}{dx} \cdot v + u \cdot \frac{dv}{dx}.$$

In words:

> The derivative of a product is the derivative of the first times the second, plus the first times the derivative of the second.

Another justification of the product rule is given in Problem 23 on page 233.

Example 1 Differentiate (a) $x^2 e^x$,   (b) $(3x^2 + 5x)e^x$,   (c) $\dfrac{e^x}{x^2}$.

Solution (a)

$$\frac{d(x^2 e^x)}{dx} = \left(\frac{d(x^2)}{dx}\right) e^x + x^2 \frac{d(e^x)}{dx} = 2xe^x + x^2 e^x = (2x + x^2)e^x.$$

(b)

$$\frac{d((3x^2 + 5x)e^x)}{dx} = \left(\frac{d(3x^2 + 5x)}{dx}\right) e^x + (3x^2 + 5x)\frac{d(e^x)}{dx}$$
$$= (6x + 5)e^x + (3x^2 + 5x)e^x = (3x^2 + 11x + 5)e^x.$$

(c) First we must write $\dfrac{e^x}{x^2}$ as the product $x^{-2}e^x$:

$$\frac{d}{dx}\left(\frac{e^x}{x^2}\right) = \frac{d(x^{-2}e^x)}{dx} = \left(\frac{d(x^{-2})}{dx}\right) e^x + x^{-2}\frac{d(e^x)}{dx}$$
$$= -2x^{-3}e^x + x^{-2}e^x = (-2x^{-3} + x^{-2})e^x.$$

The Quotient Rule

Suppose we want to differentiate a function of the form $Q(x) = f(x)/g(x)$. (Of course, we have to avoid points where $g(x) = 0$.) We want a formula for Q' in terms of f' and g'.

Assuming that $Q(x)$ is differentiable,[1] we can use the product rule on $f(x) = Q(x)g(x)$:

$$f'(x) = Q'(x)g(x) + Q(x)g'(x)$$
$$= Q'(x)g(x) + \frac{f(x)}{g(x)}g'(x).$$

Solving for $Q'(x)$ gives

$$Q'(x) = \frac{f'(x) - \dfrac{f(x)}{g(x)}g'(x)}{g(x)}.$$

---

[1] The method in Example 6 on page 211 can be used to explain why $Q(x)$ must be differentiable.

Multiplying the top and bottom by $g(x)$ to simplify gives

$$\left(\frac{f(x)}{g(x)}\right)' = \frac{f'(x)g(x) - f(x)g'(x)}{(g(x))^2}.$$

So we have the following rule:

The Quotient Rule

If $u = f(x)$ and $v = g(x)$ are differentiable, then

$$\left(\frac{f}{g}\right)' = \frac{f'g - fg'}{g^2},$$

or equivalently,

$$\frac{d}{dx}\left(\frac{u}{v}\right) = \frac{\dfrac{du}{dx}\cdot v - u\cdot \dfrac{dv}{dx}}{v^2}.$$

In words:

The derivative of a quotient is the derivative of the numerator times the denominator minus the numerator times the derivative of the denominator, all over the denominator squared.

Example 2   Differentiate (a) $\dfrac{5x^2}{x^3+1}$,   (b) $\dfrac{1}{1+e^x}$,   (c) $\dfrac{e^x}{x^2}$.

Solution   (a)

$$\frac{d}{dx}\left(\frac{5x^2}{x^3+1}\right) = \frac{\left(\dfrac{d}{dx}(5x^2)\right)(x^3+1) - 5x^2\dfrac{d}{dx}(x^3+1)}{(x^3+1)^2} = \frac{10x(x^3+1) - 5x^2(3x^2)}{(x^3+1)^2}$$

$$= \frac{-5x^4 + 10x}{(x^3+1)^2}.$$

(b)

$$\frac{d}{dx}\left(\frac{1}{1+e^x}\right) = \frac{\left(\dfrac{d}{dx}(1)\right)(1+e^x) - 1\dfrac{d}{dx}(1+e^x)}{(1+e^x)^2} = \frac{0(1+e^x) - 1(0+e^x)}{(1+e^x)^2}$$

$$= \frac{-e^x}{(1+e^x)^2}.$$

(c)   This is the same as part (c) of Example 1, but this time we will do it by the quotient rule.

$$\frac{d}{dx}\left(\frac{e^x}{x^2}\right) = \frac{\left(\dfrac{d(e^x)}{dx}\right)x^2 - e^x\left(\dfrac{d(x^2)}{dx}\right)}{(x^2)^2} = \frac{e^x x^2 - e^x(2x)}{x^4}$$

$$= e^x\left(\frac{x^2 - 2x}{x^4}\right) = e^x\left(\frac{x-2}{x^3}\right).$$

This is, in fact, the same answer as before, although it looks different. Can you show that it is the same?

Problems for Section 4.3

1. If $f(x) = x^2(x^3 + 5)$, find $f'(x)$ two ways: by using the product rule and by multiplying out before taking the derivative. Do you get the same result? Should you?

2. If $f(x) = 2^x \cdot 3^x$, find $f'(x)$ two ways: by using the product rule and by using the fact that $2^x \cdot 3^x = 6^x$. Do you get the same result?

For Problems 3–23, find the derivative. In some cases, it may be to your advantage to simplify first.

3. $f(x) = xe^x$

4. $y = x \cdot 2^x$

5. $y = \sqrt{x} \cdot 2^x$

6. $f(x) = (x^2 - \sqrt{x})3^x$

7. $z = (s^2 - \sqrt{s})(s^2 + \sqrt{s})$

8. $y = (t^2 + 3)e^t$

9. $w = (t^3 + 5t)(t^2 - 7t + 2)$

10. $y = (t^3 - 7t^2 + 1)e^t$

11. $f(x) = \dfrac{x}{e^x}$

12. $g(x) = \dfrac{25x^2}{e^x}$

13. $g(w) = \dfrac{w^{3.2}}{5^w}$

14. $h(t) = \dfrac{t+4}{t-4}$

15. $z = \dfrac{3t+1}{5t+2}$

16. $z = \dfrac{t^2 + 5t + 2}{t + 3}$

17. $f(x) = \dfrac{x^2 + 3}{x}$

18. $w = \dfrac{y^3 - 6y^2 + 7y}{y}$

19. $y = \dfrac{\sqrt{t}}{t^2 + 1}$

20. $f(z) = \dfrac{3z^2}{5z^2 + 7z}$

21. $w(x) = \dfrac{17e^x}{2^x}$

22. $h(p) = \dfrac{1 + p^2}{3 + 2p^2}$

23. $f(x) = \dfrac{1 + x}{2 + 3x + 4x^2}$

24. If $f(x) = (3x + 8)(2x - 5)$, find $f'(x)$ and $f''(x)$.

25. For what intervals is $f(x) = xe^{-x}$ concave down?

26. For what intervals is $g(x) = \dfrac{1}{x^2 + 1}$ concave down?

27. Differentiate $f(t) = e^{-t}$ by writing it as $f(t) = \dfrac{1}{e^t}$.

28. Differentiate $f(x) = e^{2x}$ by writing it as $f(x) = e^x \cdot e^x$.

29. Differentiate $f(x) = e^{3x}$ by writing it as $f(x) = e^x \cdot e^{2x}$ and using the result of Problem 28.

30. Based on your answers to Problems 28 and 29, guess the derivative of e^{4x}.

31. (a) Differentiate $y = \dfrac{e^x}{x}$, $y = \dfrac{e^x}{x^2}$, and $y = \dfrac{e^x}{x^3}$.

    (b) What do you anticipate the derivative of $y = \dfrac{e^x}{x^n}$ will be? Confirm your guess.

32. Using the product rule and the fact that $\dfrac{d(x)}{dx} = 1$, show that $\dfrac{d(x^2)}{dx} = 2x$ and $\dfrac{d(x^3)}{dx} = 3x^2$.

33. Use the product rule to show that $\dfrac{d}{dx}(x^{1/2}) = \dfrac{1}{2x^{1/2}}$. [Hint: Write $x = x^{1/2}x^{1/2}$.]

34. Suppose f and g are differentiable functions with the values shown in the following table. For each of the following functions h, find $h'(2)$.

    (a) $h(x) = f(x) + g(x)$

    (b) $h(x) = f(x)g(x)$

    (c) $h(x) = \dfrac{f(x)}{g(x)}$

x	$f(x)$	$g(x)$	$f'(x)$	$g'(x)$
2	3	4	5	-2

35. Given $\left\{\begin{array}{ll} H(3) = 1 & F(3) = 5 \\ H'(3) = 3 & F'(3) = 4 \end{array}\right\}$ find $\left\{\begin{array}{ll} \text{(a)} \ G'(3) & \text{if } G(z) = F(z) \cdot H(z) \\ \text{(b)} \ G'(3) & \text{if } G(w) = F(w)/H(w) \end{array}\right\}$.

36. Find a possible formula for a function $y = f(x)$ such that $f'(x) = 10x^9 e^x + x^{10} e^x$.

37. The quantity, q, of a certain skateboard sold depends on the selling price, p, in dollars, so we write $q = f(p)$. You are given that $f(140) = 15{,}000$ and $f'(140) = -100$.

    (a) What do $f(140) = 15{,}000$ and $f'(140) = -100$ tell you about the sales of skateboards?

    (b) The total revenue, R, earned by the sale of skateboards is given by $R = pq$. Find $\left.\dfrac{dR}{dp}\right|_{p=140}$.

    (c) What is the sign of $\left.\dfrac{dR}{dp}\right|_{p=140}$? If the skateboards are currently selling for \$140, what happens to revenue if the price is increased to \$141?

38. When an electric current passes through two resistors with resistance r_1 and r_2, connected in parallel, the combined resistance, R, can be calculated from the equation

$$\frac{1}{R} = \frac{1}{r_1} + \frac{1}{r_2}.$$

    Find the rate at which the combined resistance changes with respect to changes in r_1. Assume that r_2 is constant.

39. A museum has decided to sell one of its paintings and to invest the proceeds. If the picture is sold between the years 2000 and 2020 and the money from the sale is invested in a bank account earning 5% annual interest compounded once a year, then $B(t)$, the balance in the year 2020, depends on the year, t, in which the painting is sold and the sale price $P(t)$. If t is measured from the year 2000 so that $0 < t < 20$ then

$$B(t) = P(t)(1.05)^{20-t}.$$

    (a) Explain why $B(t)$ is given by this formula.
    (b) Show that the formula for $B(t)$ is equivalent to

$$B(t) = (1.05)^{20} \frac{P(t)}{(1.05)^t}.$$

    (c) Find $B'(10)$, given that $P(10) = 150{,}000$ and $P'(10) = 5000$.

40. Let $f(v)$ be the gas consumption (in liters/km) of a car going at velocity v (in km/hr). In other words, $f(v)$ tells you how many liters of gas the car uses to go one kilometer, if it is going at velocity v. You are told that

$$f(80) = 0.05 \text{ and } f'(80) = 0.0005.$$

    (a) Let $g(v)$ be the distance the same car goes on one liter of gas at velocity v. What is the relationship between $f(v)$ and $g(v)$? Find $g(80)$ and $g'(80)$.

    (b) Let $h(v)$ be the gas consumption in liters per hour. In other words, $h(v)$ tells you how many liters of gas the car uses in one hour if it is going at velocity v. What is the relationship between $h(v)$ and $f(v)$? Find $h(80)$ and $h'(80)$.

    (c) How would you explain the practical meaning of the values of these functions and their derivatives to a driver who knows no calculus?

41. The function

$$f(x) = e^x$$

    has the properties

$$f'(x) = f(x) \text{ and } f(0) = 1.$$

    Explain why $f(x)$ is the only function with both these properties.
    [Hint: Assume $g'(x) = g(x)$, and $g(0) = 1$, for some function $g(x)$. Define $h(x) = g(x)/e^x$, and compute $h'(x)$. Then use the fact that a function with a derivative of 0 must be a constant function.]

42. Find $f'(x)$ for the following functions with the product rule, rather than by multiplying out.

    (a) $f(x) = (x - 1)(x - 2)$.
    (b) $f(x) = (x - 1)(x - 2)(x - 3)$.
    (c) $f(x) = (x - 1)(x - 2)(x - 3)(x - 4)$.

43. Use the answer from Problem 42 to guess $f'(x)$ for the following function:

$$f(x) = (x - r_1)(x - r_2)(x - r_3) \cdots (x - r_n)$$

where $r_1, r_2, \ldots, r_n$ are any real numbers.

44. (a) Provide a three dimensional analogue for the geometrical demonstration of the formula for the derivative of a product, given in Figure 4.15 on page 203. In other words, find a formula for the derivative of $F(x) \cdot G(x) \cdot H(x)$ using Figure 4.16.

(b) Verify your results by writing $F(x) \cdot G(x) \cdot H(x)$ as $[F(x) \cdot G(x)] \cdot H(x)$ and using the product rule twice.

(c) Generalize your result to n functions: what is the derivative of

$$f_1(x) \cdot f_2(x) \cdot f_3(x) \cdots f_n(x)?$$

Figure 4.16: A graphical representation of the 3-dimensional product rule

4.4 THE CHAIN RULE

Composite functions such as $\sin(3t)$ or e^{-x^2} occur frequently in practice. In this section we will see how to differentiate such functions.

The Derivative of a Composition of Functions

Suppose $f(g(x))$ is a composite function, with f being the outside function and g being the inside. Let us write

$$z = g(x) \quad \text{and} \quad y = f(z), \quad \text{so} \quad y = f(g(x)).$$

Then a small change in x, called Δx, generates a small change in z, called Δz. In turn, Δz generates a small change in y called Δy. Provided Δx and Δz are not zero, we can say:

$$\frac{\Delta y}{\Delta x} = \frac{\Delta y}{\Delta z} \cdot \frac{\Delta z}{\Delta x}.$$

Since $\dfrac{dy}{dx} = \lim_{\Delta x \to 0} \dfrac{\Delta y}{\Delta x}$, this suggests that in the limit as Δx, Δy, and Δz get smaller and smaller, we have:

The Chain Rule

$$\frac{dy}{dx} = \frac{dy}{dz} \cdot \frac{dz}{dx}.$$

Since $\dfrac{dy}{dz} = f'(z)$ and $\dfrac{dz}{dx} = g'(x)$, we can also write

$$\frac{d}{dx} f(g(x)) = f'(z) \cdot g'(x).$$

Substituting $z = g(x)$, we can rewrite this as follows:

The Chain Rule

$$\frac{d}{dx}f(g(x)) = f'(g(x)) \cdot g'(x).$$

In words:

The derivative of a composite function is the product of the derivatives of the outside and inside functions. The derivative of the outside function must be evaluated at the inside function.

A justification of the chain rule is given in Problem 24 on page 233.

Example 1   Suppose that the length, L, in cm, of a steel bar depends on the air temperature, $H°$C, and that H depends on time, t, measured in hours. If the length increases by 2 cm for every degree increase in temperature, and the temperature is increasing at $3°$C per hour, how fast is the length of the bar increasing? What are the units for your answer?

Solution   We expect the rate at which the length is increasing to be in cm/hr. We are told that

$$\text{Rate length increasing with respect to temperature} = \frac{dL}{dH} = 2 \text{ cm/°C}$$

$$\text{Rate temperature increasing with respect to time} = \frac{dH}{dt} = 3°\text{C/hr.}$$

We want to calculate the rate at which the length is increasing with respect to time, or dL/dt. We think of L as a function of H, and H as a function of t. By the chain rule we know that

$$\frac{dL}{dt} = \frac{dL}{dH} \cdot \frac{dH}{dt} = \left(2\frac{\text{cm}}{°\text{C}}\right) \cdot \left(3\frac{°\text{C}}{\text{hr}}\right) = 6 \text{ cm/hr.}$$

Thus, the length is increasing at 6 cm/hr.

Example 1 above shows us how to interpret the chain rule in practical terms. The next examples show how it is used to compute derivatives of functions given by formulas.

Example 2   Find the derivatives of the following functions: (a) $(4x^2 + 1)^7$ (b) e^{3x}

Solution   (a)   Here $z = g(x) = 4x^2 + 1$ is the inside function; $f(z) = z^7$ is the outside function. Since $g'(x) = 8x$ and $f'(z) = 7z^6$, we have

$$\frac{d}{dx}\left[(4x^2 + 1)^7\right] = 7z^6 \cdot 8x = 7(4x^2 + 1)^6 \cdot 8x = 56x(4x^2 + 1)^6.$$

(b)   Let $z = g(x) = 3x$ and $f(z) = e^z$. Then $g'(x) = 3$ and $f'(z) = e^z$, so

$$\frac{d}{dx}\left(e^{3x}\right) = e^z \cdot 3 = 3e^{3x}.$$

Example 3   Differentiate

(a) $(x^2+1)^{100}$ (b) $\sqrt{3x^2 + 5x - 2}$ (c) $\dfrac{1}{x^2 + x^4}$ (d) $\sqrt{e^x + 1}$ (e) e^{x^2}

Solution   (a)   Here $z = g(x) = x^2 + 1$ is the inside function; $f(z) = z^{100}$ is the outside function. Now $g'(x) = 2x$ and $f'(z) = 100z^{99}$, so

$$\frac{d[(x^2 + 1)^{100}]}{dx} = 100z^{99} \cdot 2x = 100(x^2 + 1)^{99} \cdot 2x = 200x(x^2 + 1)^{99}.$$

(b)   Here $z = g(x) = 3x^2 + 5x - 2$ and $f(z) = \sqrt{z}$, so $g'(x) = 6x + 5$ and $f'(z) = \dfrac{1}{2\sqrt{z}}$. Hence

$$\frac{d(\sqrt{3x^2 + 5x - 2})}{dx} = \frac{1}{2\sqrt{z}} \cdot (6x + 5) = \frac{1}{2\sqrt{3x^2 + 5x - 2}} \cdot (6x + 5).$$

(c)   Let $z = g(x) = x^2 + x^4$ and $f(z) = 1/z$, so $g'(x) = 2x + 4x^3$ and $f'(z) = -z^{-2} = -\dfrac{1}{z^2}$. Then

$$\frac{d}{dx}\left(\frac{1}{x^2 + x^4}\right) = -\frac{1}{z^2}(2x + 4x^3) = -\frac{2x + 4x^3}{(x^2 + x^4)^2}.$$

We could have done this problem using the quotient rule. Try it and see that you get the same answer!

(d)   Let $z = g(x) = e^x + 1$ and $f(z) = \sqrt{z}$. Therefore $g'(x) = e^x$ and $f'(z) = \dfrac{1}{2\sqrt{z}}$. We get

$$\frac{d(\sqrt{e^x + 1})}{dx} = \frac{1}{2\sqrt{z}}e^x = \frac{e^x}{2\sqrt{e^x + 1}}.$$

(e)   In order to figure out which is the inside function and which is the outside, notice that to evaluate e^{x^2} we first evaluate x^2 and then take e to that power. This tells us that the inside function is $z = g(x) = x^2$ and the outside function is $f(z) = e^z$. Therefore, $g'(x) = 2x$, and $f'(z) = e^z$, giving

$$\frac{d(e^{x^2})}{dx} = e^z \cdot 2x = e^{x^2} \cdot 2x = 2xe^{x^2}.$$

Example 4   Find the derivative of e^{2x} by the chain rule and by the product rule.

Solution   Chain rule: Let the inside function be $z = g(x) = 2x$ and the outside function be $f(z) = e^z$. Then

$$\frac{d(e^{2x})}{dx} = f'(g(x)) \cdot g'(x) = e^{2x} \cdot 2 = 2e^{2x}.$$

Product rule: Write $e^{2x} = e^x \cdot e^x$. Then

$$\frac{d(e^{2x})}{dx} = \frac{d(e^x e^x)}{dx} = \left(\frac{d(e^x)}{dx}\right)e^x + e^x\left(\frac{d(e^x)}{dx}\right) = e^x \cdot e^x + e^x \cdot e^x = 2e^{2x}.$$

The chain rule can often be used to find rates, as in the following example.

Example 5   A circular oil spill is spreading. If the radius is increasing at 0.2 km/hr when the radius is 3 km, find the rate at which the area is increasing at that time. Give units in your answer.

Solution   If A is the area of the spill in km^2 and r is the radius in km, then

$$A = \pi r^2.$$

Differentiating using the chain rule gives

$$\frac{dA}{dt} = \frac{dA}{dr} \cdot \frac{dr}{dt} = 2\pi r \cdot \frac{dr}{dt}.$$

We know that $dr/dt = 0.2$ km/hr when $r = 3$, so

$$\frac{dA}{dt} = (2\pi \cdot 3 \text{ km})(0.2 \text{ km/hr}) = 1.2\pi \approx 3.77 \text{ km}^2/\text{hr}.$$

Notice that the units of dA/dt are (km)(km/hr)=km^2/hr, which is area per unit time, as expected.

Using the Product and Chain Rules to Differentiate a Quotient

If you prefer, you can differentiate a quotient by the product and chain rules, instead of by the quotient rule. The resulting formulas may look different, but they will be equivalent.

Example 6   Find $k'(x)$ if $k(x) = \dfrac{x}{x^2 + 1}$.

Solution   One way is to use the quotient rule:

$$k'(x) = \frac{1 \cdot (x^2 + 1) - x \cdot (2x)}{(x^2 + 1)^2}$$

$$= \frac{1 - x^2}{(x^2 + 1)^2}.$$

Alternatively, we can write the original function as a product,

$$k(x) = x\frac{1}{x^2 + 1} = x \cdot (x^2 + 1)^{-1},$$

and use the product rule:

$$k'(x) = 1 \cdot (x^2 + 1)^{-1} + x \cdot \frac{d}{dx}\left[(x^2 + 1)^{-1}\right].$$

Now use the chain rule to differentiate $(x^2 + 1)^{-1}$. Let $z = x^2 + 1$, and $f(z) = z^{-1}$, so

$$\frac{d}{dx}\left[(x^2 + 1)^{-1}\right] = -z^{-2} \cdot 2x = -(x^2 + 1)^{-2} \cdot 2x = \frac{-2x}{(x^2 + 1)^2}.$$

Therefore,

$$k'(x) = \frac{1}{x^2 + 1} + x \cdot \frac{-2x}{(x^2 + 1)^2} = \frac{1}{x^2 + 1} - \frac{2x^2}{(x^2 + 1)^2}.$$

If we put these two fractions over a common denominator, we will get the same answer as from the quotient rule.

Problems for Section 4.4

Find the derivatives of the functions in Problems 1–30.

1. $f(x) = (x + 1)^{99}$
2. $f(x) = \sqrt{1 - x^2}$
3. $w = (t^2 + 1)^{100}$
4. $w = (t^3 + 1)^{100}$
5. $w = (\sqrt{t} + 1)^{100}$
6. $f(t) = e^{3t}$
7. $f(x) = 2^{(x+2)}$
8. $g(x) = 3^{(2x+7)}$
9. $k(x) = (x^3 + e^x)^4$
10. $z(x) = \sqrt[3]{2x + 5}$
11. $y = \dfrac{\sqrt{z}}{2^z}$
12. $w = \sqrt{(x^2 \cdot 5^x)^3}$
13. $y = e^{3w/2}$
14. $y = e^{-4t}$
15. $y = \sqrt{s^3 + 1}$
16. $w = e^{\sqrt{s}}$
17. $y = te^{-t^2}$
18. $f(z) = \sqrt{z}\,e^{-z}$
19. $f(z) = \dfrac{\sqrt{z}}{e^z}$
20. $z = 2^{5t-3}$
21. $f(t) = te^{5-2t}$
22. $f(z) = \dfrac{1}{(e^z + 1)^2}$
23. $f(\theta) = \dfrac{1}{1 + e^{-\theta}}$
24. $f(x) = 6e^{5x} + e^{-x^2}$

25. $f(w) = (5w^2 + 3)e^{w^2}$ 26. $w = (t^2 + 3t)(1 - e^{-2t})$ 27. $f(y) = \sqrt{10^{(5-y)}}$

28. $f(x) = e^{-(x-1)^2}$ 29. $f(y) = e^{e^{(y^2)}}$ 30. $f(t) = 2 \cdot e^{-2e^{2t}}$

31. Find the equation of the line tangent to $y = f(x)$ at $x = 1$, where $f(x)$ is the function in Problem 24.

32. For what values of x is the graph of $y = e^{-x^2}$ concave down?

33. Suppose $f(x) = (2x + 1)^{10}(3x - 1)^7$. Find a formula for $f'(x)$. Then decide on a reasonable way to simplify your result, and find a formula for $f''(x)$.

34. Given
$$\begin{cases} F(2) = 1 & G(4) = 2 \\ F(4) = 3 & G(3) = 4 \\ F'(2) = 5 & G'(4) = 6 \\ F'(4) = 7 & G'(3) = 8 \end{cases}$$
find
$$\begin{cases} \text{(a)} & H(4) & \text{if } H(x) = F(G(x)) \\ \text{(b)} & H'(4) & \text{if } H(x) = F(G(x)) \\ \text{(c)} & H(4) & \text{if } H(x) = G(F(x)) \\ \text{(d)} & H'(4) & \text{if } H(x) = G(F(x)) \\ \text{(e)} & H'(4) & \text{if } H(x) = F(x)/G(x) \end{cases}$$

35. Suppose f and g are differentiable functions with the values given in the following table. For each of the following functions h, find $h'(2)$.

(a) $h(x) = f(g(x))$ (b) $h(x) = g(f(x))$ (c) $h(x) = f(f(x))$

x	$f(x)$	$g(x)$	$f'(x)$	$g'(x)$
2	5	5	e	$\sqrt{2}$
5	2	8	π	7

36. If the derivative of $y = k(x)$ equals 2 when $x = 1$, what is the derivative of
(a) $k(2x)$ when $x = \dfrac{1}{2}$?   (b) $k(x + 1)$ when $x = 0$?   (c) $k\left(\dfrac{1}{4}x\right)$ when $x = 4$?

37. Given $y = f(x)$ with $f(1) = 4$ and $f'(1) = 3$, find
(a) $g'(1)$ if $g(x) = \sqrt{f(x)}$.   (b) $h'(1)$ if $h(x) = f(\sqrt{x})$.

38. Is $x = \sqrt[3]{2t + 5}$ a solution to the equation $3x^2 \dfrac{dx}{dt} = 2$? Why or why not?

39. Find a possible formula for a function $m(x)$ such that $m'(x) = x^5 \cdot e^{(x^6)}$.

40. Suppose that the population of zebra mussels in a certain area of the St. Lawrence River is approximated by $P(t) = 10e^{0.6t}$, where t is measured in months since the zebra mussels first arrived in the area. Calculate the following quantities:
(a) $P(12)$.   (b) $P'(12)$.
Give units with your answers and explain what each quantity tells us in terms of zebra mussels.

41. One gram of radioactive carbon-14 decays according to the formula
$$Q = e^{-0.000121t}$$
where Q is the number of grams of carbon-14 remaining after t years.

(a) Find the rate at which carbon-14 is decaying (in grams/year).
(b) Sketch the rate you found in part (a) against time.

42. The temperature, H, in degrees Fahrenheit ($^\circ$F), of a can of soda that is put into a refrigerator to cool is given as a function of time, t, in hours, by
$$H = 40 + 30e^{-2t}.$$

(a) Find the rate at which the temperature of the soda is changing (in $^\circ$F/hour).
(b) What is the sign of $\dfrac{dH}{dt}$? Why does it have this sign?
(c) When, for $t \geq 0$, is the magnitude of $\dfrac{dH}{dt}$ largest? In terms of the can of soda, why is this?

43. If you invest P dollars in a bank account at an annual interest rate of $r\%$, then after t years you will have B dollars, where

$$B = P\left(1 + \frac{r}{100}\right)^t.$$

  (a) Find dB/dt, assuming P and r are constant. In terms of money, what does dB/dt represent?
  (b) Find dB/dr, assuming P and t are constant. In terms of money, what does dB/dr represent?

44. Suppose a pebble is dropped in still water, forming a circular ripple whose radius is expanding at a constant rate of 10 cm/sec. Find a formula giving the area enclosed by the ripple as a function of time. When the radius is 20 cm, how fast is the area enclosed by the ripple increasing?

45. The theory of relativity predicts that an object whose mass is m_0 when it is at rest will appear heavier when moving at speeds near the speed of light. When the object is moving at speed v, its mass m is given by

$$m = \frac{m_0}{\sqrt{1 - (v^2/c^2)}}, \qquad \text{where } c \text{ is the speed of light.}$$

  (a) Find $\dfrac{dm}{dv}$.   (b) In terms of physics, what does $\dfrac{dm}{dv}$ tell you?

46. The charge, Q, on a capacitor which starts discharging at time $t = 0$ is given by

$$Q = \begin{cases} Q_0 & \text{for } t \le 0 \\ Q_0 e^{-t/RC} & \text{for } t > 0, \end{cases}$$

  where R and C are positive constants depending on the circuit and Q_0 is the charge at $t = 0$, where $Q_0 \ne 0$. The current, I, flowing in the circuit is given by $I = dQ/dt$.

  (a) Find the current I for $t < 0$ and for $t > 0$.
  (b) Is it possible to define I at $t = 0$?
  (c) Is the function Q differentiable at $t = 0$?

47. An electric charge is decaying exponentially according to the formula

$$Q = Q_0 e^{-t/RC}.$$

  Show that the charge, Q, and the electric current, $I = dQ/dt$, have the same time constant. (The time constant is the time to decay to $1/e$ times the original value.)

48. A function f is said to have a *zero of multiplicity* m at $x = a$ if

$$f(x) = (x - a)^m h(x), \quad \text{with } h(a) \ne 0.$$

  Explain why a function having a zero of multiplicity m at $x = a$ satisfies $f^{(p)}(a) = 0$, for $p = 1, 2, \ldots m - 1$.

4.5 THE TRIGONOMETRIC FUNCTIONS

Derivatives of the Sine and Cosine

Since the sine and cosine functions are periodic, their derivatives must be periodic also. (Why?) Let's look at the graph of $f(x) = \sin x$ in Figure 4.17 and estimate the derivative function graphically.

Figure 4.17: The sine function

First we might ask where the derivative is zero. (At $x = \pm\pi/2, \pm3\pi/2, \pm5\pi/2$, etc.) Then ask where the derivative is positive and where it is negative. (Positive for $-\pi/2 < x < \pi/2$; negative for $\pi/2 < x < 3\pi/2$, etc.) Since the largest positive slopes are at $x = 0, 2\pi$, and so on, and the largest negative slopes are at $x = \pi, 3\pi$, and so on, we get something like the graph in Figure 4.18.

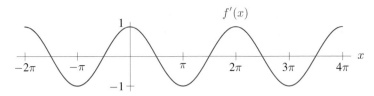

Figure 4.18: Derivative of $f(x) = \sin x$

The graph of the derivative in Figure 4.18 looks suspiciously like the graph of the cosine function. This might lead us to conjecture, quite correctly, that the derivative of the sine is the cosine.

Of course, we cannot be sure, just from the graphs, that the derivative of the sine really is the cosine. However, for now we'll assume that the derivative of the sine *is* the cosine and confirm the result at the end of the section.

One thing we can do now is to check that the derivative function in Figure 4.18 has amplitude 1 (as it ought to if it is the cosine). That means we have to convince ourselves that the derivative of $f(x) = \sin x$ is 1 when $x = 0$. The next example suggests that this is true when x is in radians.

Example 1   Using a calculator, estimate the derivative of $f(x) = \sin x$ at $x = 0$. Make sure your calculator is set in radians.

Solution   Since $f(x) = \sin x$,

$$f'(0) = \lim_{h \to 0} \frac{\sin(0 + h) - \sin 0}{h} = \lim_{h \to 0} \frac{\sin h}{h}.$$

Table 4.3 contains values of $(\sin h)/h$ which suggest that this limit is 1, so we estimate

$$f'(0) = \lim_{h \to 0} \frac{\sin h}{h} = 1.$$

TABLE 4.3

h (radians)	-0.1	-0.01	-0.001	-0.0001	0.0001	0.001	0.01	0.1
$(\sin h)/h$	0.99833	0.99998	1.0000	1.0000	1.0000	1.0000	0.99998	0.99833

Warning: It is important to notice that in the previous example h was in *radians*; any conclusions we have drawn about the derivative of $\sin x$ are valid *only* when x is in radians.

Example 2   Starting with the graph of the cosine function, sketch a graph of its derivative.

Solution   The graph of $g(x) = \cos x$ is in Figure 4.19(a). Its derivative is 0 at $x = 0, \pm\pi, \pm 2\pi$, and so on; it is positive for $-\pi < x < 0$, $\pi < x < 2\pi$, and so on; and it is negative for $0 < x < \pi$, $2\pi < x < 3\pi$, and so on. The derivative is in Figure 4.19(b).

(a)

(b)

Figure 4.19: $g(x) = \cos x$ and its derivative, $g'(x)$

As we did with the sine, we use the graphs to make a conjecture. The derivative of the cosine in Figure 4.19(b) looks exactly like the graph of sine, except reflected about the x-axis. But how can we be sure that the derivative is $-\sin x$?

Example 3 Use the relation $\dfrac{d}{dx}(\sin x) = \cos x$ to show that $\dfrac{d}{dx}(\cos x) = -\sin x$.

Solution Since the cosine function is the sine function shifted to the left by $\pi/2$, we would expect the derivative of the cosine to be the derivative of the sine, shifted to the left by $\pi/2$. Since

$$\cos x = \sin\left(x + \frac{\pi}{2}\right),$$

we can use the chain rule:

$$\frac{d}{dx}(\cos x) = \frac{d}{dx}\left(\sin\left(x + \frac{\pi}{2}\right)\right) = \cos\left(x + \frac{\pi}{2}\right).$$

But $\cos(x + \pi/2)$ is the cosine shifted to the left by $\pi/2$, which gives a sine curve reflected about the x-axis. So we have

$$\frac{d}{dx}(\cos x) = \cos\left(x + \frac{\pi}{2}\right) = -\sin x.$$

For x in radians, $\qquad \dfrac{d}{dx}(\sin x) = \cos x \quad$ and $\quad \dfrac{d}{dx}(\cos x) = -\sin x.$

Example 4 Differentiate (a) $2\sin(3\theta)$,   (b) $\cos^2 x$,   (c) $\cos(x^2)$,   (d) $e^{-\sin t}$.

Solution Use the chain rule:

(a) $\dfrac{d}{d\theta}(2\sin(3\theta)) = 2\dfrac{d}{d\theta}(\sin(3\theta)) = 2(\cos(3\theta))\dfrac{d}{d\theta}(3\theta) = 2(\cos(3\theta))3 = 6\cos(3\theta).$

(b) $\dfrac{d}{dx}(\cos^2 x) = \dfrac{d}{dx}\left((\cos x)^2\right) = 2(\cos x)\cdot\dfrac{d}{dx}(\cos x) = 2(\cos x)(-\sin x) = -2\cos x\sin x.$

(c) $\dfrac{d}{dx}\left(\cos(x^2)\right) = -\sin(x^2)\cdot\dfrac{d}{dx}(x^2) = -2x\sin(x^2).$

(d) $\dfrac{d}{dt}(e^{-\sin t}) = e^{-\sin t}\dfrac{d}{dt}(-\sin t) = -(\cos t)e^{-\sin t}.$

Derivative of the Tangent Function

Since $\tan x = \sin x/\cos x$, we differentiate $\tan x$ using the quotient rule. Writing $(\sin x)'$ for $d(\sin x)/dx$, we have:

$$\frac{d}{dx}(\tan x) = \frac{d}{dx}\left(\frac{\sin x}{\cos x}\right) = \frac{(\sin x)'(\cos x) - (\sin x)(\cos x)'}{\cos^2 x} = \frac{\cos^2 x + \sin^2 x}{\cos^2 x} = \frac{1}{\cos^2 x}.$$

For x in radians, $\qquad \dfrac{d}{dx}(\tan x) = \dfrac{1}{\cos^2 x}.$

The graphs of $f(x) = \tan x$ and $f'(x) = 1/\cos^2 x$ are in Figure 4.20. Is it reasonable that f' is always positive? Are the asymptotes of f' where we expect?

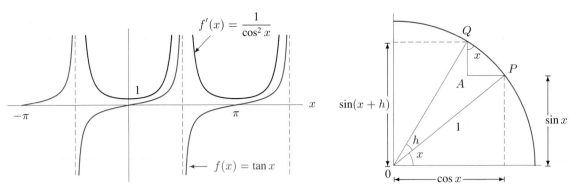

Figure 4.20: The function $\tan x$ and its derivative   **Figure 4.21:** Unit circle showing $\sin(x+h)$ and $\sin x$

Example 5   Differentiate (a) $2\tan(3t)$,   (b) $\tan(1-\theta)$,   (c) $\dfrac{1+\tan t}{1-\tan t}$.

Solution   (a)   Use the chain rule:

$$\frac{d}{dt}(2\tan(3t)) = 2\frac{1}{\cos^2(3t)}\frac{d}{dt}(3t) = \frac{6}{\cos^2(3t)}.$$

(b)   Use the chain rule:

$$\frac{d}{d\theta}(\tan(1-\theta)) = \frac{1}{\cos^2(1-\theta)}\cdot\frac{d}{d\theta}(1-\theta) = \frac{-1}{\cos^2(1-\theta)}.$$

(c)   Use the quotient rule:

$$\frac{d}{dt}\left(\frac{1+\tan t}{1-\tan t}\right) = \frac{\left(\dfrac{d(1+\tan t)}{dt}\right)(1-\tan t) - (1+\tan t)\dfrac{d(1-\tan t)}{dt}}{(1-\tan t)^2}$$

$$= \frac{\dfrac{1}{\cos^2 t}(1-\tan t) - (1+\tan t)\left(-\dfrac{1}{\cos^2 t}\right)}{(1-\tan t)^2}$$

$$= \frac{2}{\cos^2 t \cdot (1-\tan t)^2}.$$

Informal Justification of $\dfrac{d}{dx}(\sin x) = \cos x$

Consider the unit circle in Figure 4.21. To find the derivative of $\sin x$, we need to estimate

$$\frac{\sin(x+h) - \sin x}{h}.$$

In Figure 4.21, the quantity $\sin(x+h) - \sin x$ is represented by the length QA. The arc QP is of length h, so

$$\frac{\sin(x+h) - \sin x}{h} = \frac{QA}{\text{Arc } QP}.$$

Now, if h is small, QAP is approximately a right triangle because the arc QP is almost a straight line. Furthermore, using geometry, we can show that angle $AQP \approx x$. For small h, we have

$$\frac{\sin(x+h) - \sin x}{h} = \frac{QA}{\text{Arc } QP} \approx \cos x.$$

As $h \to 0$, the approximation gets better, so

$$\frac{d}{dx}(\sin x) = \lim_{h\to 0}\frac{\sin(x+h) - \sin x}{h} = \cos x.$$

Other derivations of this result are given in Problems 39 and 40 on page 218.

Problems for Section 4.5

1. Construct a table of values for $\cos x$, $x = 0, 0.1, 0.2, \ldots, 0.6$. Using the difference quotient, estimate the derivative at these points (use $h = 0.001$), and compare it with $(-\sin x)$.

Find the derivatives of the functions in Problems 2–29.

2. $r(\theta) = \sin \theta + \cos \theta$

3. $s(\theta) = \cos \theta \sin \theta$

4. $t(\theta) = \dfrac{\cos \theta}{\sin \theta}$

5. $z = \cos(4\theta)$

6. $f(x) = \sin(3x)$

7. $w = \sin(e^t)$

8. $f(x) = x^2 \cos x$

9. $f(x) = e^{\cos x}$

10. $f(y) = e^{\sin y}$

11. $f(x) = \sqrt{1 - \cos x}$

12. $f(x) = \cos(\sin x)$

13. $f(x) = \tan(\sin x)$

14. $k(x) = \sqrt{(\sin(2x))^3}$

15. $h(x) = 2^{\sin x}$

16. $w = 2^{2 \sin x + e^x}$

17. $z = \theta e^{\cos \theta}$

18. $f(x) = 2x \sin(3x)$

19. $f(x) = \sin(2x) \cdot \sin(3x)$

20. $y = e^{\theta} \sin(2\theta)$

21. $f(x) = e^{-2x} \cdot \sin x$

22. $z = \sqrt{\sin t}$

23. $y = \sin^5 \theta$

24. $g(z) = \tan(e^z)$

25. $z = \tan(e^{-3\theta})$

26. $w = e^{-\sin \theta}$

27. $h(t) = t \cos t + \tan t$

28. $f(\alpha) = \cos \alpha + 3 \sin \alpha$

29. $f(\theta) = \theta^2 \sin \theta + 2\theta \cos \theta - 2 \sin \theta$

30. Find the *fiftieth* derivative of $y = \cos x$.

31. Find a possible formula for the function $q(x)$ such that $q'(x) = \dfrac{e^x \cdot \sin x - e^x \cdot \cos x}{(\sin x)^2}$.

32. A boat at anchor is bobbing up and down in the sea. The vertical distance, y, in feet, between the sea floor and the boat is given as a function of time, t, in minutes, by

$$y = 15 + \sin(2\pi t).$$

(a) Find the vertical velocity, v, of the boat at time t.

(b) Make rough sketches of y and v against t.

33. On page 55 the depth, y, of water in Boston harbor was given by

$$y = 5 + 4.9 \cos\left(\frac{\pi}{6}t\right),$$

where t is the number of hours since midnight.

(a) Find $\dfrac{dy}{dt}$. What does $\dfrac{dy}{dt}$ represent, in terms of water level?

(b) For $0 \leq t \leq 24$, when is $\dfrac{dy}{dt}$ zero? (Figure 1.73 on page 55 may be helpful.) Explain what it means (in terms of water level) for $\dfrac{dy}{dt}$ to be zero.

34. The voltage, V, in volts, in an electrical outlet is given as a function of time, t, in seconds, by the function $V = 156 \cos(120\pi t)$.

(a) Give an expression for the rate of change of voltage with respect to time.

(b) Is the rate of change ever zero? Explain.

(c) What is the maximum value of the rate of change?

35. The function $y = A \sin\left(\left(\sqrt{\frac{k}{m}}\right)t\right)$ represents the oscillations of a mass m at the end of a spring. The constant k measures the stiffness of the spring.

(a) Find a time at which the mass is farthest from its equilibrium position. Find a time at which the mass is moving fastest. Find a time at which the mass is accelerating fastest.

(b) What is the period, T, of the oscillation?

(c) Find dT/dm. What does the sign of dT/dm tell you?

36. Find the equations of the tangent lines to the graph of $f(x) = \sin x$ at $x = 0$ and at $x = \pi/3$. Use each tangent line to approximate $\sin(\pi/6)$. Would you expect these results to be equally accurate, since they are taken equally far away from $x = \pi/6$ but on opposite sides? If the accuracy is different, can you account for the difference?

37. A lighthouse is 2 km from the long, straight coastline shown in Figure 4.22. Find the rate of change of the distance of the spot of light from the point O with respect to the angle θ.

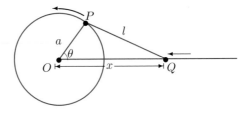

Figure 4.22   *Figure 4.23*

38. The metal bar of length l in Figure 4.23 has one end attached at the point P to a circle of radius a. Point Q at the other end can slide back and forth along the x-axis.

   (a) Find x as a function of θ.
   (b) Assume lengths are in centimeters and the angular speed $(d\theta/dt)$ is 2 radians/second counter-clockwise. Find the speed at which the point Q is moving when

   (i) $\theta = \pi/2$,   (ii) $\theta = \pi/4$.

39. We will use the following identities to calculate the derivatives of $\sin x$ and $\cos x$:

$$\sin(a + b) = \sin a \cos b + \sin b \cos a$$
$$\cos(a + b) = \cos a \cos b - \sin a \sin b.$$

   (a) Use the definition of the derivative to show that if $f(x) = \sin x$,

$$f'(x) = \sin x \lim_{h \to 0} \frac{\cos h - 1}{h} + \cos x \lim_{h \to 0} \frac{\sin h}{h}.$$

   (b) Estimate the limits in part (a) with your calculator to explain why $f'(x) = \cos x$.
   (c) If $g(x) = \cos x$, use the definition of the derivative to show that $g'(x) = -\sin x$.

40. In this problem you will calculate the derivative of $\tan \theta$ rigorously (and without using the derivatives of $\sin \theta$ or $\cos \theta$). You will then use your result for $\tan \theta$ to calculate the derivatives of $\sin \theta$ and $\cos \theta$. Figure 4.24 shows $\tan \theta$ and $\Delta(\tan \theta)$, which is the change in $\tan \theta$, namely $\tan(\theta + \Delta\theta) - \tan \theta$.

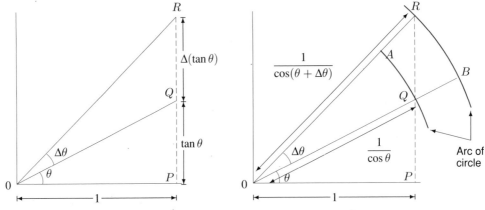

Figure 4.24: $\tan \theta$ and $\Delta(\tan \theta)$

(a) By paying particular attention to how the two figures relate and using the fact that

$$\text{Area of Sector OAQ} \le \text{Area of Triangle OQR} \le \text{Area of Sector OBR}$$

explain why

$$\frac{\Delta\theta}{2\pi} \cdot \pi \left(\frac{1}{\cos\theta}\right)^2 \le \frac{1}{2} \cdot 1 \cdot \Delta(\tan\theta) \le \frac{\Delta\theta}{2\pi} \cdot \pi \left(\frac{1}{\cos(\theta+\Delta\theta)}\right)^2 .$$

[Hint: A sector of a circle with angle α at the center has area $\alpha/(2\pi)$ times the area of the whole circle.]

(b) Use part (a) to show as $\Delta\theta \to 0$ that

$$\frac{\Delta\tan\theta}{\Delta\theta} \to \left(\frac{1}{\cos\theta}\right)^2 ,$$

and hence that $\dfrac{d(\tan\theta)}{d\theta} = \left(\dfrac{1}{\cos\theta}\right)^2 .$

(c) Derive the identity $(\tan\theta)^2 + 1 = \left(\dfrac{1}{\cos\theta}\right)^2$. Then differentiate both sides of this identity with respect to θ, using the chain rule and the result of part (b) to show that $\dfrac{d}{d\theta}(\cos\theta) = -\sin\theta$.

(d) Differentiate both sides of the identity $(\sin\theta)^2 + (\cos\theta)^2 = 1$ and use the result of part (c) to show that $\dfrac{d}{d\theta}(\sin\theta) = \cos\theta$.

4.6 APPLICATIONS OF THE CHAIN RULE

In this section we will use the chain rule to calculate the derivatives of fractional powers, logarithms, exponentials, and the inverse trigonometric functions.[2]

Finding the Derivative of an Inverse Function: Derivative of $x^{1/2}$

Earlier we calculated the derivative of x^n with n an integer, but we have been using the result for non-integral values of n as well. We now confirm that the power rule holds for $n = 1/2$ by calculating the derivative of $f(x) = x^{1/2}$ using the chain rule. Since

$$[f(x)]^2 = x,$$

the derivative of $[f(x)]^2$ and the derivative of x must be equal, so

$$\frac{d}{dx}\left[f(x)\right]^2 = \frac{d}{dx}(x).$$

We can use the chain rule with $f(x)$ as the inside function to obtain:

$$\frac{d}{dx}[f(x)]^2 = 2f(x) \cdot f'(x) = 1.$$

Solving for $f'(x)$ gives

$$f'(x) = \frac{1}{2f(x)} = \frac{1}{2x^{1/2}},$$

or

$$\frac{d}{dx}(x^{1/2}) = \frac{1}{2x^{1/2}} = \frac{1}{2}x^{-1/2}.$$

A similar calculation can be used to obtain the derivative of $x^{1/n}$ where n is a positive integer.

---

[2]It requires a separate justification, not given here, that these functions are differentiable.

Derivative of $\ln x$

We use the chain rule to differentiate an identity involving $\ln x$. Since $e^{\ln x} = x$, we have

$$\frac{d}{dx}(e^{\ln x}) = \frac{d}{dx}(x),$$

$$e^{\ln x} \cdot \frac{d}{dx}(\ln x) = 1. \qquad \text{(Since } e^x \text{ is outside function and } \ln x \text{ is inside function)}$$

Solving for $d(\ln x)/dx$ gives

$$\frac{d}{dx}(\ln x) = \frac{1}{e^{\ln x}} = \frac{1}{x},$$

so

$$\boxed{\frac{d}{dx}(\ln x) = \frac{1}{x}.}$$

Example 1   Differentiate   (a) $\ln(x^2 + 1)$,   (b) $t^2 \ln t$,   (c) $\sqrt{1 + \ln(1 - y)}$.

Solution   (a)   Using the chain rule:

$$\frac{d}{dx}\left(\ln(x^2 + 1)\right) = \frac{1}{x^2 + 1}\frac{d}{dx}(x^2 + 1) = \frac{2x}{x^2 + 1}.$$

(b)   Using the product rule:

$$\frac{d}{dt}(t^2 \ln t) = \frac{d}{dt}(t^2) \cdot \ln t + t^2 \frac{d}{dt}(\ln t) = 2t \ln t + t^2 \cdot \frac{1}{t} = 2t \ln t + t.$$

(c)   Using the chain rule:

$$\frac{d}{dy}\left(\sqrt{1 + \ln(1 - y)}\right) = \frac{d}{dy}\left(1 + \ln(1 - y)\right)^{1/2}$$

$$= \frac{1}{2}\left(1 + \ln(1 - y)\right)^{-1/2} \cdot \frac{d}{dy}\left(1 + \ln(1 - y)\right) \qquad \text{(Using the chain rule)}$$

$$= \frac{1}{2\sqrt{1 + \ln(1 - y)}} \cdot \frac{1}{1 - y} \cdot \frac{d}{dy}(1 - y) \qquad \text{(Using the chain rule again)}$$

$$= \frac{-1}{2(1 - y)\sqrt{1 + \ln(1 - y)}}.$$

Derivative of a^x

Earlier we showed that the derivative of a^x is proportional to a^x. Now we show that the constant of proportionality is $\ln a$. We use the identity

$$\ln(a^x) = x \ln a.$$

Differentiating, using $\frac{d}{dx}(\ln x) = \frac{1}{x}$ and the chain rule, and remembering that $\ln a$ is a constant, we obtain:

$$\frac{d}{dx}(\ln a^x) = \frac{1}{a^x} \cdot \frac{d}{dx}(a^x) = \ln a.$$

Solving gives the result we suggested in Section 4.2:

$$\boxed{\frac{d}{dx}(a^x) = (\ln a)a^x.}$$

Derivatives of Inverse Trigonometric Functions

In Section 1.9 we defined $\arcsin x$ as the angle between $-\pi/2$ and $\pi/2$ (inclusive) whose sine is x. Similarly, $\arctan x$ as the angle strictly between $-\pi/2$ and $\pi/2$ whose tangent is x. To find $\dfrac{d}{dx}(\arctan x)$ we use the identity $\tan(\arctan x) = x$. Differentiating using the chain rule gives

$$\frac{1}{\cos^2(\arctan x)} \cdot \frac{d}{dx}(\arctan x) = 1,$$

so

$$\frac{d}{dx}(\arctan x) = \cos^2(\arctan x).$$

Using the identity $1 + \tan^2\theta = \dfrac{1}{\cos^2\theta}$, and replacing θ by $\arctan x$, we have

$$\cos^2(\arctan x) = \frac{1}{1 + \tan^2(\arctan x)} = \frac{1}{1 + x^2}.$$

Thus we have

$$\boxed{\frac{d}{dx}(\arctan x) = \frac{1}{1 + x^2}.}$$

By a similar argument, we obtain the result:

$$\boxed{\frac{d}{dx}(\arcsin x) = \frac{1}{\sqrt{1 - x^2}}.}$$

Example 2   Differentiate (a) $\arctan(t^2)$,   (b) $\arcsin(\tan\theta)$.

Solution   Use the chain rule:

(a) $\dfrac{d}{dt}\left(\arctan(t^2)\right) = \dfrac{1}{1 + (t^2)^2} \cdot \dfrac{d}{dt}(t^2) = \dfrac{2t}{1 + t^4}.$

(b) $\dfrac{d}{dt}\left(\arcsin(\tan\theta)\right) = \dfrac{1}{\sqrt{1 - (\tan\theta)^2}} \cdot \dfrac{d}{d\theta}(\tan\theta) = \dfrac{1}{\sqrt{1 - \tan^2\theta}} \cdot \dfrac{1}{\cos^2\theta}.$

Example 3   An airplane, flying at 450 km/hr at a constant altitude of 5 km, is approaching a camera mounted on the ground. Let θ be the angle at which the camera is pointed. See Figure 4.25. When $\theta = \pi/3$, how fast does the camera have to rotate in order to keep the plane in view?

Solution   Suppose the plane is vertically above the point B. Let x be the distance between B and C. The fact that the plane is moving toward C at 450 km/hr means that x is decreasing and $dx/dt = -450$ km/hr. From Figure 4.25, we see that $\tan\theta = 5/x$.

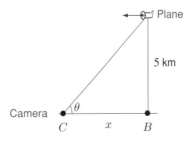

Figure 4.25: Plane approaching a camera at C

Differentiating $\tan\theta = 5/x$ with respect to t and using the chain rule gives

$$\frac{1}{\cos^2\theta}\frac{d\theta}{dt} = -5x^{-2}\frac{dx}{dt}.$$

We want to solve for $d\theta/dt$ when $\theta = \pi/3$. At that moment, $\cos\theta = 1/2$ and $\tan\theta = \sqrt{3}$; so $x = 5/\sqrt{3}$. Substituting gives

$$\frac{1}{(1/2)^2}\frac{d\theta}{dt} = -5\left(\frac{5}{\sqrt{3}}\right)^{-2}\cdot(-450)$$

$$\frac{d\theta}{dt} = 67.5 \text{ radians/hour.}$$

This answer tells us that the camera must turn at roughly 1 radian per minute if it is to remain pointed at the plane.

Problems for Section 4.6

For Problems 1–22, find the derivative of the given function. In some cases, it may be to your advantage to simplify before differentiating.

1. $f(t) = \ln(t^2 + 1)$
2. $f(x) = \ln(1 - x)$
3. $f(x) = \ln(e^{2x})$

4. $f(x) = e^{\ln(e^{2x^2+3})}$
5. $f(z) = \dfrac{1}{\ln z}$
6. $f(\theta) = \ln(\cos\theta)$

7. $f(x) = \ln(1 - e^{-x})$
8. $f(\alpha) = \ln(\sin\alpha)$
9. $f(x) = \ln(e^x + 1)$

10. $f(t) = \ln(\ln t) + \ln(\ln 2)$
11. $f(x) = \ln(e^{7x})$
12. $f(x) = e^{(\ln x)+1}$

13. $f(w) = \ln(\cos(w - 1))$
14. $f(t) = \ln(e^{\ln t})$
15. $f(y) = \arcsin(y^2)$

16. $g(t) = \arctan(3t - 4)$
17. $g(\alpha) = \sin(\arcsin\alpha)$
18. $g(t) = e^{\arctan(3t^2)}$

19. $g(t) = \cos(\ln t)$
20. $h(z) = z^{\ln 2}$
21. $h(w) = w\arcsin w$

22. $f(x) = \cos(\arcsin(x + 1))$

23. On what intervals is $\ln(x^2 + 1)$ concave up?

24. Using the chain rule, find $\dfrac{d}{dx}(\arcsin x)$.

25. Using the chain rule, find $\dfrac{d}{dx}(\log x)$. (Recall that $\log x = \log_{10} x$.)

26. To compare the acidity of different solutions, chemists use the pH (which is a single number, not the product of p and H). The pH is defined in terms of the concentration, x, of hydrogen ions in the solution as

$$\text{pH} = -\log x.$$

Find the rate of change of pH with respect to hydrogen ion concentration when the pH is 2. [Hint: Use the result of Problem 25.]

27. Hungary is one of the few countries in the world where the population is decreasing, currently at about 0.2% a year. Thus, if t is time in years since 1990, the population P, in millions, of Hungary can be approximated by

$$P = 10.8(0.998)^t.$$

(a) What does this model predict the population of Hungary will be in the year 2000?
(b) How fast (in people/year) does this model predict Hungary's population will be decreasing in the year 2000?

28. Imagine you are zooming in on the graph of each of the following functions near the origin:

$$y = x \qquad y = \sqrt{x} \qquad y = x^2 \qquad y = \sin x$$

$$y = x \sin x \qquad y = \tan x \qquad y = \sqrt{x/(x+1)} \qquad y = x^3$$

$$y = \ln(x+1) \qquad y = \tfrac{1}{2}\ln(x^2+1) \qquad y = 1 - \cos x \qquad y = \sqrt{2x - x^2}$$

   Which of them look the same? Group together those functions which become indistinguishable, and give the equations of the lines they look like.

29. (a) For $x > 0$, find the derivative of $f(x) = \arctan x + \arctan(1/x)$ and simplify.
    (b) What does your result tell you about f?

30. (a) Find the equation of the tangent line to $y = \ln x$ at $x = 1$.
    (b) Use it to calculate approximate values for $\ln(1.1)$ and $\ln(2)$.
    (c) Using a graph, explain whether the approximate values you have calculated are smaller or larger than the true values. Would the same result have held if you had used the tangent line to estimate $\ln(0.9)$ and $\ln(0.5)$? Why?

31. (a) Find the equation of the best quadratic approximation to $y = \ln x$ at $x = 1$. The best quadratic approximation has the same first and second derivatives as $y = \ln x$ at $x = 1$.
    (b) Use a computer or calculator to graph the approximation and $y = \ln x$ on the same set of axes. What do you notice?
    (c) Use your quadratic approximation to calculate approximate values for $\ln(1.1)$ and $\ln(2)$.

32. The gravitational force, F, on a rocket at a distance, r, from the center of the earth is given by

$$F = \frac{k}{r^2},$$

   where $k = 10^{13}$ newton $\cdot$ km^2. When the rocket is 10^4 km from the center of the earth, it is moving away at 0.2 km/sec. How fast is the gravitational force changing at that moment? Give units. (A newton is a unit of force.)

33. A train is traveling at 0.8 km/min along a long straight track, moving in the direction shown in Figure 4.26. A movie camera, 0.5 km away from the track, is focused on the train.

Figure 4.26

    (a) Express z, the distance between the camera and the train, as a function of x.
    (b) How fast is the distance from the camera to the train changing when the train is 1 km from the camera? Give units.
    (c) How fast is the camera rotating (in radians/min) at the moment when the train is 1 km from the camera?

34. The radius of a spherical balloon is increasing by 2 cm/sec. At what rate is air being blown into the balloon at the moment when the radius is 10 cm? Give units in your answer.

35. Coroners estimate time of death from body temperature using the simple rule of thumb that a body cools about 2°F during the first hour after death and about 1°F for each additional hour. (The temperature is measured using a small probe inserted into the liver which, being large and vascular, holds body heat the longest.)

   Assuming an air temperature of 68°F and a living body temperature of 98.6°F, the temperature $T(t)$ in °F at time t in hours is given by:

$$T(t) = 68 + 30.6e^{-kt}$$

   where $t = 0$ is the instant that death occurs.

    (a) For what value of k will the body cool by 2°F in the first hour?

(b)   Using the value of k found in part (a), after how many hours will the temperature of the body be decreasing at a rate of $1°F$ per hour?

(c)   Using the value of k found in part (a), show that, 24 hours after death, the coroner's rule of thumb gives approximately the same temperature as the formula.

36.   For the amusement of the guests, some hotels have elevators on the outside of the building. Suppose such a hotel is 300 feet high. You are standing by a window 100 feet above the ground and 150 feet away from the hotel, and the elevator descends at a constant speed of 30 ft/sec, starting at time $t = 0$, where t is time in seconds. Let θ be the angle between the line of your horizon and your line of sight to the elevator. (See Figure 4.27.)

(a)   Find a formula for $h(t)$, the elevator's height above the ground as it descends from the top of the hotel.

(b)   Using your answer to part (a), express θ as a function of time t and find the rate of change of θ with respect to t.

(c)   If the rate of change of θ is a measure of how fast the elevator appears to you to be moving, at what height will the elevator be when it appears to be moving fastest?

Figure 4.27: Descending elevator

4.7   IMPLICIT FUNCTIONS

In earlier chapters, most functions were written in the form $y = f(x)$; here y is said to be an *explicit* function of x. An equation such as

$$x^2 + y^2 = 4$$

is said to give y as an *implicit* function of x. Its graph is the circle in Figure 4.28. Since there are x-values which correspond to two y-values, y is not a function of x on the whole circle. Solving gives

$$y = \pm\sqrt{4 - x^2},$$

where $y = \sqrt{4 - x^2}$ represents the top half of the circle and $y = -\sqrt{4 - x^2}$ represents the bottom half. So y is a function of x on the top half, and y is a different function of x on the bottom half.

But let's consider the circle as a whole. The equation does represent a curve which has a tangent line at each point. The slope of this tangent can be found by differentiating the equation of the circle with respect to x:

$$\frac{d}{dx}(x^2) + \frac{d}{dx}(y^2) = \frac{d}{dx}(4).$$

If we think of y as a function of x and use the chain rule, we get

$$2x + 2y\frac{dy}{dx} = 0.$$

Solving gives

$$\frac{dy}{dx} = -\frac{x}{y}.$$

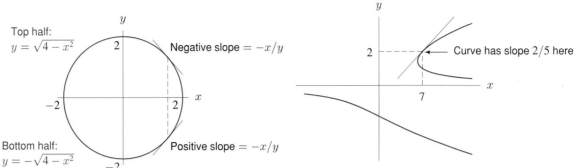

Figure 4.28: Graph of $x^2 + y^2 = 4$

Figure 4.29: Graph of $y^3 - xy = -6$ and its
tangent line at $(7, 2)$

The derivative here depends on both x and y (instead of just on x). This is because for each x-value (except for $x = \pm 2$) there are two y-values, and the curve has a different slope at each one. Figure 4.28 shows that for x and y both positive, we are on the top right quarter of the curve and the slope is negative (as the formula predicts). For x positive and y negative, we are on the bottom right quarter of the curve and the slope is positive (as the formula predicts). Differentiating the equation of the circle has given us the slope of the curve at all points except $(2, 0)$ and $(-2, 0)$, where the tangent is vertical. In general, this process of *implicit differentiation* leads to a derivative whenever the expression for the derivative does not have a zero in the denominator.

Example 1   Make a table of x and approximate y-values for the equation $y^3 - xy = -6$ near $x = 7, y = 2$. Your table should include the x-values $6.8, 6.9, 7.0, 7.1$, and 7.2.

Solution   We would like to solve for y in terms of x, but we cannot isolate y by factoring. There is a formula for solving cubics, somewhat like the quadratic formula, but it is too complicated to be useful here. Instead, first observe that $x = 7, y = 2$ does satisfy the equation. (Check this!) Now find dy/dx by implicit differentiation:

$$\frac{d}{dx}(y^3) - \frac{d}{dx}(xy) = \frac{d}{dx}(-6)$$

$$3y^2\frac{dy}{dx} - 1 \cdot y - x\frac{dy}{dx} = 0 \quad \text{(Differentiating with respect to x)}$$

$$3y^2\frac{dy}{dx} - x\frac{dy}{dx} = y$$

$$(3y^2 - x)\frac{dy}{dx} = y \quad \text{(Factoring out $\frac{dy}{dx}$)}$$

$$\frac{dy}{dx} = \frac{y}{3y^2 - x}.$$

When $x = 7$ and $y = 2$, we have

$$\frac{dy}{dx} = \frac{2}{12 - 7} = \frac{2}{5}.$$

(See Figure 4.29.) The equation of the tangent line at $(7, 2)$ is

$$y - 2 = \frac{2}{5}(x - 7)$$

or

$$y = 0.4x - 0.8.$$

Since the tangent lies very close to the curve near the point $(7, 2)$, we use the equation of the tangent line to calculate the following approximate y-values:

x	6.8	6.9	7.0	7.1	7.2
Approximate y	1.92	1.96	2.00	2.04	2.08

Notice that although the equation $y^3 - xy = -6$ leads to a curve which is difficult to deal with algebraically, it still looks like a straight line locally.

Example 2   Find all points where the tangent line to $y^3 - xy = -6$ is either horizontal or vertical.

Solution   From the previous example, $\dfrac{dy}{dx} = \dfrac{y}{3y^2 - x}$. The tangent is horizontal when the numerator of dy/dx equals 0, so $y = 0$. Since we also must satisfy $y^3 - xy = -6$, we get $0^3 - x \cdot 0 = -6$, which is impossible. We conclude that there are no points on the curve where the tangent line is horizontal.

The tangent is vertical when the denominator of dy/dx is 0, giving $3y^2 - x = 0$. Thus, $x = 3y^2$ at any point with a vertical tangent line. Again, we must also satisfy $y^3 - xy = -6$, so

$$y^3 - (3y^2)y = -6,$$
$$-2y^3 = -6,$$
$$y = \sqrt[3]{3} \approx 1.442.$$

We can then find x by substituting $y = \sqrt[3]{3}$ in $y^3 - xy = -6$. We get $3 - x(\sqrt[3]{3}) = -6$, so $x = 9/(\sqrt[3]{3}) \approx 6.240$. So the tangent line is vertical at $(6.240, 1.442)$.

Using implicit differentiation and the expression for dy/dx to locate the points where the tangent is vertical or horizontal, as in the previous example, is a first step in obtaining an overall picture of the curve $y^3 - xy = -6$. However, filling in the rest of the graph, even roughly, by using the sign of dy/dx to tell us where the curve is increasing or decreasing can be difficult.

Problems for Section 4.7

For Problems 1–12, find dy/dx.

1. $x^2 + y^2 = \sqrt{7}$
2. $x^2 + xy - y^3 = xy^2$
3. $\sqrt{x} = 5\sqrt{y}$
4. $\sqrt{x} + \sqrt{y} = 25$
5. $\ln x + \ln(y^2) = 3$
6. $e^{x^2} + \ln y = 0$
7. $\arctan(x^2 y) = xy^2$
8. $x \ln y + y^3 = \ln x$
9. $\sin(xy) = 2x + 5$
10. $x^{2/3} + y^{2/3} = a^{2/3}$ (a is a constant.)
11. $e^{\cos y} = x^3 \arctan y$
12. $\cos^2 y + \sin^2 y = y + 2$

For Problems 13–16, find the equations of the tangent lines to the following curves at the indicated points.

13. $xy^2 = 1$ at $(1, -1)$
14. $\ln(xy) = 2x$ at $(1, e^2)$
15. $y^2 = \dfrac{x^2}{xy - 4}$ at $(4, 2)$
16. $x^{2/3} + y^{2/3} = a^{2/3}$ at $(a, 0)$

17. Show that the power rule for derivatives applies to rational powers of the form $y = x^{m/n}$ by raising both sides to the nth power and using implicit differentiation.

18. (a) Find the equations of the tangent lines to the circle $x^2 + y^2 = 25$ at the points where $x = 4$.
    (b) Find the equations of the normal lines to this circle at the same points. (The normal line to a curve at a point is perpendicular to the tangent line at that point.)
    (c) At what point do the two normal lines intersect?

19. (a) Find the slope of the tangent line to the ellipse $\dfrac{x^2}{25} + \dfrac{y^2}{9} = 1$ at the point (x, y).
    (b) Are there any points where the slope is not defined?

20. Consider the equation $x^3 + y^3 - xy^2 = 5$.

    (a) Find dy/dx by implicit differentiation.
    (b) Give a table of approximate y-values near $x = 1, y = 2$ for $x = 0.96, 0.98, 1, 1.02, 1.04$.
    (c) Find the y-value for $x = 0.96$ by substituting $x = 0.96$ in the equation and solving for y using a computer or calculator. Compare with your answer in part (b).
    (d) Find all points where the tangent line is horizontal or vertical.

21. Find the equation of the tangent line to the curve $y = x^2$ at $x = 1$. Show that this line is also a tangent to a circle centered at $(8, 0)$ and find the equation of this circle.

22. Sketch the circles $y^2 + x^2 = 1$ and $y^2 + (x - 3)^2 = 4$. There is a line with positive slope that is tangent to both circles. Determine the points at which this tangent line touches each circle.

4.8 LINEAR APPROXIMATIONS AND LIMITS

The Tangent Line Approximation

Since the graph of a function and its tangent line have the same slope at the point of tangency (namely, the derivative at that point), the tangent line stays close to the graph of the function near that point. This means that we can approximate values of the function by values on the tangent line. (See Figure 4.30.) The slope of the tangent to the graph of $y = f(x)$ at $x = a$ is $f'(a)$, so the equation of the tangent line is

$$y = f(a) + f'(a)(x - a).$$

Now we approximate the values of f by the y values from the tangent line, giving the following result.

> **The Tangent Line Approximation**
>
> For values of x near a,
> $$f(x) \approx f(a) + f'(a)(x - a)$$
> We are thinking of a as fixed, so $f(a)$ and $f'(a)$ are constant.

The expression $f(a) + f'(a)(x - a)$ is a linear function which approximates $f(x)$ well near a; it is called the *local linearization* of f near $x = a$.

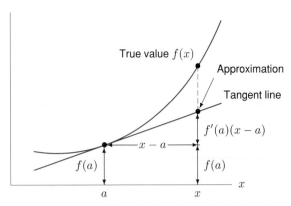

Figure 4.30: Local linearization: Approximation by the tangent line

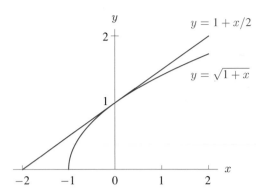

Figure 4.31: Tangent line approximation to $\sqrt{1+x}$ near $x = 0$

Example 1   Find the tangent line approximation for $\sqrt{1+x}$ near $x = 0$.

Solution   With $f(x) = \sqrt{1+x}$, the chain rule gives $f'(x) = 1/(2\sqrt{1+x})$, so $f(0) = 1$ and $f'(0) = 1/2$. Therefore the tangent line approximation of f near $x = 0$,

$$f(x) \approx f(0) + f'(0)(x - 0),$$

becomes

$$\sqrt{1+x} \approx 1 + \frac{x}{2}.$$

This means that, near $x = 0$, the function $\sqrt{1+x}$ can be approximated by its tangent line $y = 1 + x/2$. (See Figure 4.31.)

Example 2   What is the local linearization of e^{kx} near $x = 0$?

Solution   If $f(x) = e^{kx}$, then $f(0) = 1$ and, by the chain rule, $f'(x) = ke^{kx}$, so $f'(0) = ke^{k \cdot 0} = k$. Thus

$$f(x) \approx f(0) + f'(0)(x - 0)$$

becomes

$$e^{kx} \approx 1 + kx.$$

This is the tangent line approximation to e^{kx} near $x = 0$. In other words, if we zoom in on the functions $f(x) = e^{kx}$ and $y = 1 + kx$ near the origin, we won't be able to tell them apart.

Using Local Linearity to Find Limits

Suppose we want to calculate the exact value of the limit

$$\lim_{x \to 0} \frac{e^{2x} - 1}{x}.$$

Substituting $x = 0$ gives us $0/0$, which is undefined:

$$\frac{e^{2(0)} - 1}{0} = \frac{1 - 1}{0} = \frac{0}{0}.$$

Substituting values of x near 0 gives us an approximate value for the limit.

However, the limit can be calculated exactly using local linearity. Suppose we let $f(x)$ be the numerator, so $f(x) = e^{2x} - 1$, and $g(x)$ be the denominator, so $g(x) = x$. Then $f(0) = 0$ and $f'(x) = 2e^{2x}$, so $f'(0) = 2$. When we zoom in on the graph of $f(x) = e^{2x} - 1$ near the origin, we see its tangent line $y = 2x$ shown in Figure 4.32. We are interested in the ratio $(e^{2x} - 1)/x = f(x)/g(x)$, which is approximately the ratio of the y-values in Figure 4.32. This ratio of y-values is just the ratio of the slopes of the lines. So

$$\frac{f(x)}{g(x)} = \frac{e^{2x} - 1}{x} \approx \frac{2}{1} = \frac{f'(0)}{g'(0)}.$$

As $x \to 0$, this approximation gets better, and we have

$$\lim_{x \to 0} \frac{e^{2x} - 1}{x} = 2.$$

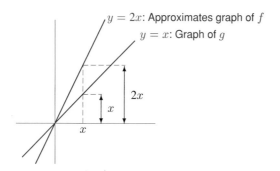

Figure 4.32: Ratio $(e^{2x} - 1)/x$ is approximated by ratio of slopes as we zoom in near the origin

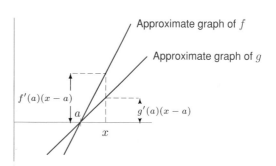

Figure 4.33: Ratio $f(x)/g(x)$ is approximated by ratio of slopes, $f'(a)/g'(a)$, as we zoom in at a

L'Hopital's Rule

If $f(a) = g(a) = 0$, we can use the same method to investigate limits of the form

$$\lim_{x \to a} \frac{f(x)}{g(x)}.$$

As in the previous case, we zoom in on the graphs of $f(x)$ and $g(x)$. Figure 4.33 shows that both graphs cross the x-axis at $x = a$. This suggests that the limit of $f(x)/g(x)$ as $x \to a$ is the ratio of slopes, giving the following result:

L'Hopital's rule: If f and g are differentiable, $f(a) = g(a) = 0$, and $g'(a) \neq 0$, then

$$\lim_{x \to a} \frac{f(x)}{g(x)} = \frac{f'(a)}{g'(a)}.$$

To justify this result, let us assume $g'(a) \neq 0$ and consider the quantity $f'(a)/g'(a)$. Using the definition of the derivative and the fact that $f(a) = g(a) = 0$, we have

$$\frac{f'(a)}{g'(a)} = \frac{\displaystyle\lim_{h \to 0} \frac{f(a+h) - f(a)}{h}}{\displaystyle\lim_{h \to 0} \frac{g(a+h) - g(a)}{h}}$$

$$= \frac{\displaystyle\lim_{h \to 0} (f(a+h)/h)}{\displaystyle\lim_{h \to 0} (g(a+h)/h)}$$

$$= \lim_{h \to 0} \frac{f(a+h)}{g(a+h)} = \lim_{x \to a} \frac{f(x)}{g(x)}.$$

Note that if $f'(a) \neq 0$ and $g'(a) = 0$, the limit of $f(x)/g(x)$ does not exist.

Example 3   Use l'Hopital's rule to confirm that $\lim\limits_{x \to 0} \dfrac{\sin x}{x} = 1$.

Solution   Let $f(x) = \sin x$ and $g(x) = x$. Then $f(0) = g(0) = 0$ and $f'(x) = \cos x$ and $g'(x) = 1$. Thus,

$$\lim_{x \to 0} \frac{\sin x}{x} = \frac{\cos 0}{1} = 1.$$

If we also have $f'(a) = g'(a) = 0$, then we can use the following result:

> More general form of **l'Hopital's rule:** If f and g are differentiable and $f(a) = g(a) = 0$, then
>
> $$\lim_{x \to a} \frac{f(x)}{g(x)} = \lim_{x \to a} \frac{f'(x)}{g'(x)},$$
>
> provided the limit on the right exists.

Example 4   Calculate $\lim\limits_{t \to 0} \dfrac{e^t - 1 - t}{t^2}$.

Solution   Let $f(t) = e^t - 1 - t$ and $g(t) = t^2$. Then $f(0) = e^0 - 1 - 0 = 0$ and $g(0) = 0$, and $f'(t) = e^t - 1$ and $g'(t) = 2t$. So

$$\lim_{t \to 0} \frac{e^t - 1 - t}{t^2} = \lim_{t \to 0} \frac{e^t - 1}{2t}.$$

Since $f'(0) = g'(0) = 0$, the ratio $f'(0)/g'(0)$ is not defined. So we use l'Hopital's rule again:

$$\lim_{t \to 0} \frac{e^t - 1 - t}{t^2} = \lim_{t \to 0} \frac{e^t - 1}{2t} = \lim_{t \to 0} \frac{e^t}{2} = \frac{1}{2}.$$

The following forms of l'Hopital's rule apply to limits involving infinity.

> L'Hopital's rule can also be used
> - When $\lim_{x \to a} f(x) = \pm\infty$ and $\lim_{x \to a} g(x) = \pm\infty$,
>
> or
>
> - When $a = \pm\infty$.
>
> It can be shown that under these circumstances:
>
> $$\lim_{x \to a} \frac{f(x)}{g(x)} = \lim_{x \to a} \frac{f'(x)}{g'(x)}$$
>
> (where a may be $\pm\infty$), provided the limit on the right-hand side exists.

Notice that we cannot evaluate $f'(x)/g'(x)$ directly when $a = \pm\infty$. The next example shows how this version of l'Hopital's rule is used.

Example 5   Calculate $\lim\limits_{x \to \infty} \dfrac{5x + e^{-x}}{7x}$.

Solution   Let $f(x) = 5x + e^{-x}$ and $g(x) = 7x$. Then $\lim\limits_{x \to \infty} f(x) = \lim\limits_{x \to \infty} g(x) = \infty$, and $f'(x) = 5 - e^{-x}$ and $g'(x) = 7$, so

$$\lim_{x \to \infty} \frac{5x + e^{-x}}{7x} = \lim_{x \to \infty} \frac{(5 - e^{-x})}{7} = \frac{5}{7}.$$

We can also use l'Hopital's rule to calculate some limits of the form $\lim\limits_{x \to \infty} f(x)g(x)$, providing we rewrite them appropriately.

Example 6 Calculate $\lim\limits_{x \to \infty} xe^{-x}$.

Solution Since $\lim\limits_{x \to \infty} x = \infty$ and $\lim\limits_{x \to \infty} e^{-x} = 0$, we see that

$$xe^{-x} \to 0 \cdot \infty \qquad \text{as } x \to \infty.$$

Since $0 \cdot \infty$ is undefined, we rewrite the function xe^{-x} as

$$xe^{-x} = \frac{x}{e^x}.$$

Now we use l'Hopital's rule, since

$$xe^{-x} = \frac{x}{e^x} \to \frac{\infty}{\infty} \qquad \text{as } x \to \infty.$$

Taking $f(x) = x$ and $g(x) = e^x$ gives $f'(x) = 1$ and $g'(x) = e^x$, so

$$\lim_{x \to \infty} xe^{-x} = \lim_{x \to \infty} \frac{x}{e^x} = \lim_{x \to \infty} \frac{1}{e^x} = 0.$$

Dominance: Powers, Polynomials, Exponentials, and Logarithms

In Chapter 1, we saw that some functions were much larger than others as $x \to \infty$. We say that g *dominates* f as $x \to \infty$ if $\lim\limits_{x \to \infty} \dfrac{f(x)}{g(x)} = 0$. L'Hopital's rule gives us an easy way of checking this.

Example 7 Show that $x^{1/2}$ dominates $\ln x$ as $x \to \infty$.

Solution We apply l'Hopital's rule to $(\ln x)/x^{1/2}$:

$$\lim_{x \to \infty} \frac{\ln x}{x^{1/2}} = \lim_{x \to \infty} \frac{1/x}{\frac{1}{2}x^{-1/2}}.$$

To evaluate this limit, we simplify and get

$$\lim_{x \to \infty} \frac{1/x}{\frac{1}{2}x^{-1/2}} = \lim_{x \to \infty} \frac{2x^{1/2}}{x} = \lim_{x \to \infty} \frac{2}{x^{1/2}} = 0.$$

Therefore we have shown that

$$\lim_{x \to \infty} \frac{\ln x}{x^{1/2}} = 0,$$

which tells us that $x^{1/2}$ dominates $\ln x$ as $x \to \infty$.

Example 8 Show that any exponential function of the form e^{kx} (with $k > 0$) dominates any power function of the form Ax^p (with A and p positive) as $x \to \infty$.

Solution We apply l'Hopital's rule repeatedly to Ax^p/e^{kx}:

$$\lim_{x \to \infty} \frac{Ax^p}{e^{kx}} = \lim_{x \to \infty} \frac{Apx^{p-1}}{ke^{kx}} = \lim_{x \to \infty} \frac{Ap(p-1)x^{p-2}}{k^2 e^{kx}} = \cdots$$

Keep applying l'Hopital's rule until the power of x is no longer positive. Then the limit of the numerator must be a finite number, while the limit of the denominator must be ∞. Therefore we have

$$\lim_{x \to \infty} \frac{Ax^p}{e^{kx}} = 0,$$

so e^{kx} dominates Ax^p.

Problems for Section 4.8

1. What is the tangent line approximation to $1/x$ near $x = 1$?

2. Show that $1 - x/2$ is the tangent line approximation to $1/\sqrt{1 + x}$ near $x = 0$.

3. Show that $e^{-x} \approx 1 - x$ near $x = 0$.

4. Interpret the linear approximation $e^{rt} \approx 1 + rt$ in terms of interest in a bank account if r is the continuous annual interest rate and t is the time in years.

5. (a) Show that $1 + kx$ is the local linearization of $(1 + x)^k$ near $x = 0$.
   (b) Someone claims that the square root of 1.1 is about 1.05. Without using a calculator, do you think that this estimate is about right?
   (c) Is the actual number above or below 1.05?

For Problems 6–9, find the sign of $\lim\limits_{x \to a} \dfrac{f(x)}{g(x)}$ from the figure.

6.   7.   8.   9.

10. Evaluate $\lim\limits_{x \to 0^+} x \ln x$. [Hint: Write $x \ln x = \dfrac{\ln x}{1/x}$.]

11. Based on your knowledge of the behavior of the numerator and denominator, first predict the value of the following limits, and then find each limit using l'Hopital's rule.

   (a) $\lim\limits_{x \to 0} \dfrac{\sin x}{x^2}$ (b) $\lim\limits_{x \to 0} \dfrac{\sin^2 x}{x}$ (c) $\lim\limits_{x \to 0} \dfrac{\sin x}{x^{1/3}}$ (d) $\lim\limits_{x \to 0} \dfrac{(\sin x)^{1/3}}{x}$

12. (a) What is the slope of $f(x) = \sin(3x)$ at $x = 0$?
   (b) What is the slope of $g(x) = 5x$ at $x = 0$?
   (c) Use the results of parts (a) and (b) to calculate $\lim\limits_{x \to 0} \dfrac{\sin(3x)}{5x}$.

In Problems 13–16, which function dominates as $x \to \infty$?

13. x^5 and $0.1x^7$ 14. $0.01x^3$ and $50x^2$

15. $\ln(x + 3)$ and $x^{0.2}$ 16. x^{10} and $e^{0.1x}$

17. The functions f and g and their tangent lines at $(4, 0)$ are shown in Figure 4.34. Find $\lim\limits_{x \to 4} \dfrac{f(x)}{g(x)}$.

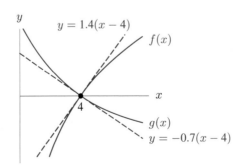

Figure 4.34

18. Explain why l'Hopital's rule cannot be used to calculate each of the following. Then evaluate the limit if it exists.

    (a) $\displaystyle\lim_{x\to 1}\frac{\sin(2x)}{x}$

    (b) $\displaystyle\lim_{x\to 0}\frac{\cos x}{x}$

    (c) $\displaystyle\lim_{x\to\infty}\frac{e^{-x}}{\sin x}$

19. Find the horizontal asymptote of $f(x)=\dfrac{2x^3+5x^2}{3x^3-1}$.

20. Multiply the local linearization of e^x near $x=0$ by itself to obtain an approximation for e^{2x}. Compare this with the actual local linearization of e^{2x}. Explain why these two approximations are consistent, and discuss which one is more accurate.

21. (a) Show that $1-x$ is the local linearization of $\dfrac{1}{1+x}$ near $x=0$.

    (b) From your answer to part (a), show that near $x=0$,

    $$\frac{1}{1+x^2}\approx 1-x^2.$$

    (c) Without differentiating, what do you think the derivative of $\dfrac{1}{1+x^2}$ is at $x=0$?

22. From the local linearizations of e^x and $\sin x$ near $x=0$, write down the local linearization of the function $e^x\sin x$. From this result, write down the derivative of $e^x\sin x$ at $x=0$. Using this technique, write down the derivative of $e^x\sin x/(1+x)$ at $x=0$.

23. Use local linearization to derive the product rule,

    $$[f(x)g(x)]'=f'(x)g(x)+f(x)g'(x).$$

    [Hint: Use the definition of the derivative and the local linearizations $f(x+h)\approx f(x)+f'(x)h$ and $g(x+h)\approx g(x)+g'(x)h$.]

24. Derive the chain rule using local linearization. [Hint: In other words, differentiate $f(g(x))$, using $g(x+h)\approx g(x)+g'(x)h$ and $f(z+k)\approx f(z)+f'(z)k$.]

CHAPTER SUMMARY

- **Derivatives of elementary functions**
    Powers, polynomials, rational functions, exponential functions, logarithms, trigonometric functions, inverse trigonometric functions

- **Derivatives of sums, differences, and constant multiples**

- **Product and quotient rules**

- **Chain rule**
    Differentiation of implicitly defined functions, inverse functions.

- **Working with the derivative**
    Tangent line approximation, local linearity, l'Hopital's rule

REVIEW PROBLEMS FOR CHAPTER FOUR

1. Find the slope of the curve $x^2+3y^2=7$ at the point $(2,-1)$.

2. Assume y is a differentiable function of x and that $y + \sin y + x^2 = 9$. Find dy/dx at the point $x = 3, y = 0$.

3. Suppose we are given the data in the table about the functions f and g and their derivatives. Find the following values:

   (a) $h(4)$ if $h(x) = f(g(x))$

   (b) $h'(4)$ if $h(x) = f(g(x))$

   (c) $h(4)$ if $h(x) = g(f(x))$

   (d) $h'(4)$ if $h(x) = g(f(x))$

   (e) $h'(4)$ if $h(x) = g(x)/f(x)$

   (f) $h'(4)$ if $h(x) = f(x)g(x)$

x	1	2	3	4
$f(x)$	3	2	1	4
$f'(x)$	1	4	2	3
$g(x)$	2	1	4	3
$g'(x)$	4	2	3	1

4. Given: $r(2) = 4$, $s(2) = 1$, $s(4) = 2$, $r'(2) = -1$, $s'(2) = 3$, $s'(4) = 3$. Compute the following derivatives, or state what additional information you would need to be able to compute the derivative.

   (a) $H'(2)$ if $H(x) = r(x) \cdot s(x)$

   (b) $H'(2)$ if $H(x) = \sqrt{r(x)}$

   (c) $H'(2)$ if $H(x) = r(s(x))$

   (d) $H'(2)$ if $H(x) = s(r(x))$

5. Imagine you are zooming in on the graphs of the following functions near the origin:

   $$y = \arcsin x \qquad y = \sin x - \tan x \qquad y = x - \sin x \qquad y = \arctan x$$

   $$y = \frac{\sin x}{1 + \sin x} \qquad y = \frac{x^2}{x^2 + 1} \qquad y = \frac{1 - \cos x}{\cos x} \qquad y = \frac{x}{x^2 + 1}$$

   $$y = \frac{\sin x}{x} - 1 \qquad y = -x \ln x \qquad y = e^x - 1 \qquad y = x^{10} + \sqrt[10]{x}$$

   $$y = \frac{x}{x + 1}$$

   Which of them look the same? Group together those functions which become indistinguishable, and give the equation of the line they look like. [Note: $(\sin x)/x - 1$ and $-x \ln x$ never quite make it to the origin.]

6. The graphs of $\sin x$ and $\cos x$ intersect once between 0 and $\pi/2$. What is the angle between the two curves at the point where they intersect? (You need to think about how the angle between two curves should be defined.)

7. The acceleration due to gravity, g, at a distance r from the center of the earth is given by

   $$g = \frac{GM}{r^2},$$

   where M is the mass of the earth and G is a constant.

   (a) Find $\dfrac{dg}{dr}$.

   (b) What is the practical interpretation (in terms of acceleration) of $\dfrac{dg}{dr}$? Why would you expect it to be negative?

   (c) You are told that $M = 6 \times 10^{24}$ and $G = 6.67 \times 10^{-20}$ where M is in kilograms and r in kilometers. What is the value of $\dfrac{dg}{dr}$ at the surface of the earth ($r = 6400$ km) ?

   (d) What does this tell you about whether or not it is reasonable to assume g is constant near the surface of the earth?

8. In 1990, the population of Mexico was about 84 million and growing at 2.6% annually, while the population of the US was about 250 million and growing at 0.7% annually. Which population was growing faster, if we measure growth rates in people/year? Explain your answer.

9. Suppose the distance, s, of a moving body from a fixed point is given as a function of time, t, by $s = 20e^{t/2}$.

   (a) Find the velocity, v, of the body as a function of t.

(b)   Find a relationship between v and s, then show that s satisfies the differential equation $s' = \frac{1}{2}s$.

10.   Air pressure at sea level is 30 inches of mercury. At an altitude of h feet above sea level, the air pressure, P, in inches of mercury, is given by

$$P = 30e^{-3.23 \times 10^{-5}h}$$

(a)   Sketch a rough graph of P against h.

(b)   Find the equation of the tangent line to the graph at $h = 0$.

(c)   A rough rule of thumb used by travelers is that air pressure drops about 1 inch for every 1000-foot increase in height above sea level. Write an approximate formula for the air pressure given by this rule of thumb.

(d)   What is the relation between your answers to parts (b) and (c)? Explain why the rule of thumb works.

(e)   Are the predictions made by the rule of thumb too large or too small? Why?

11.   Suppose the depth of the water, y, in meters, in the Bay of Fundy, Canada, is given as a function of time, t, in hours after midnight, by the function

$$y = 10 + 7.5\cos(0.507t).$$

How quickly is the tide rising or falling (in meters/hour) at each of the following times?

    (a)   6:00 am   (b)   9:00 am   (c)   Noon   (d)   6:00 pm

12.   A yam is put in a hot oven, maintained at a constant temperature $200°$C. Suppose that at time $t = 30$ minutes, the temperature T of the yam is $120°$ and is increasing at an (instantaneous) rate of $2°$/min. Newton's law of cooling (or, in our case, warming) implies that the temperature at time t will be given by a formula of the form

$$T(t) = 200 - ae^{-bt}.$$

Find a and b.

13.   An object is oscillating at the end of a spring. Its position, in centimeters, relative to a fixed point, is given as a function of time, t, in seconds, by

$$y = y_0 \cos(2\pi\omega t), \quad \text{with } \omega \text{ a constant.}$$

(a)   Find an expression for the velocity and acceleration of the object.

(b)   How do the amplitudes of the position, velocity, and acceleration functions compare? How do the periods of these functions compare?

(c)   Show that the function y satisfies the differential equation

$$\frac{d^2y}{dt^2} + 4\pi^2\omega^2 y = 0.$$

14.   Suppose the total number of people, N, who have contracted a disease by a time t days after its outbreak is given by

$$N = \frac{1{,}000{,}000}{1 + 5{,}000e^{-0.1t}}.$$

(a)   In the long run, how many people will have had the disease?

(b)   Is there any day on which more than a million people fall sick? Half a million? Quarter of a million? (Note: You do not have to try to find out on what days these things happen.)

15.   A spherical cell is growing at a constant rate of 400 μm^3/day (1 $\mu m= 10^{-6}$ m). At what rate is its radius increasing when the radius is 10 μm?

16.   When the growth of a spherical cell depends on the flow of nutrients through the surface, it is reasonable to assume that the growth rate, dV/dt, is proportional to the surface area, S. Assume that for a particular cell $dV/dt = \frac{1}{3} \cdot S$. At what rate is its radius r increasing?

17.   A radio navigation system used by aircraft gives a cockpit readout of the distance, s, in miles, between a fixed ground station and the aircraft. The system also gives a readout of the instantaneous rate of change, ds/dt, of this distance in miles/hour. An aircraft on a straight flight path at a constant altitude of 10,560 feet (2 miles) has passed directly over the ground station and is now flying away from it. What is the speed of this aircraft along its constant altitude flight path when the cockpit readouts are $s = 4.6$ miles and $ds/dt = 210$ miles/hour?

Problems 18–19 involve Boyle's Law which states that for a fixed quantity of gas at a constant temperature, the pressure, P, and the volume, V, are inversely related. Thus, for some constant k

$$PV = k.$$

18. Suppose a fixed quantity of gas is allowed to expand at constant temperature. Find the rate of change of pressure with respect to volume.

19. Suppose a certain quantity of gas occupies 10 cm^3 at a pressure of 2 atmospheres. Suppose the pressure is increased, while keeping the temperature constant.

    (a) Does the volume increase or decrease?
    (b) If the pressure is increasing at a rate of 0.05 atmospheres/minute at a time when the pressure is 2 atmospheres, find the rate at which the volume is changing at that moment. What are the units of your answer?

20. Find the nth derivative of each of the following functions:

    (a) $\ln x$ (b) xe^x (c) $e^x \cos x$

PROJECTS

1. **Rule of 70**

   The "Rule of 70" is a rule of thumb to estimate how long it takes money in a bank to double. Suppose the money is in an account earning $i\%$ annual interest, compounded yearly. The Rule of 70 says that the time it takes the amount of money to double is approximately $70/i$ years, assuming i is small. Find the local linearization of $\ln(1 + x)$, and use it to explain why this rule works.

2. **Newton's Method**

   Read Appendices A and E on finding roots using bisection and Newton's method.

   (a) What is the smallest positive zero of the function $f(x) = \sin x$? Apply Newton's method, with initial guess $x_0 = 3$, to see how fast it converges to $\pi = 3.1415926536\ldots$.

      (i) Compute the first two approximations, x_1 and x_2; compare x_2 with π.

      (ii) Newton's method works very well here. Explain why. To do this, you will have to outline the basic idea behind Newton's method.

      (iii) Estimate the location of the zero using bisection, starting with the interval $[3, 4]$. How does bisection compare to Newton's method in terms of accuracy?

   (b) Newton's method can be very sensitive to your initial estimate, x_0. For example, consider finding a zero of $f(x) = \sin x - \frac{2}{3}x$.

      (i) Use Newton's method with the following initial estimates to find a zero:

      $$x_0 = 0.904, \quad x_0 = 0.905, \quad x_0 = 0.906.$$

      (ii) What happens?

FOCUS ON PRACTICE

DIFFERENTIATION

Find derivatives for the functions in Problems 1–114. Assume a, b, c, and k are constants.

1. $f(t) = 3t^2 - 4t + 1$

2. $y = 17x + 24x^{1/2}$

3. $g(x) = -\frac{1}{2}(x^5 + 2x - 9)$

4. $g(t) = \dfrac{t^3 + k}{t}$

5. $f(x) = 5x^4 + \dfrac{1}{x^2}$

6. $z = \dfrac{t^2 + 3t + 1}{t + 1}$

7. $y = \dfrac{e^{2x}}{x^2 + 1}$

8. $f(x) = \dfrac{x^2 + 3x + 2}{x + 1}$

9. $y = \left(\dfrac{x^2 + 2}{3}\right)^2$

10. $g(\theta) = \sin^2(2\theta) - \pi\theta$

11. $g(x) = \sin(2 - 3x).$

12. $R(x) = 10 - 3\cos(\pi x)$

13. $f(z) = \dfrac{z^2 + 1}{3z}$

14. $q(r) = \dfrac{3r}{5r + 2}$

15. $h(z) = \sqrt{\dfrac{\sin(2z)}{\cos(2z)}}$

16. $y = x\ln x - x + 2$

17. $j(x) = \ln(e^{ax} + b)$

18. $y = 2x(\ln x + \ln 2) - 2x + e$

19. $g(\theta) = \sin(\tan\theta)$

20. $w(x) = \tan(x^2)$

21. $f(x) = \sin(\sin x + \cos x)$

22. $j(x) = \cos\left(\sin^{-1}x\right)$

23. $k(\alpha) = \sin^5\alpha\cos^3\alpha$

24. $f(w) = \cos^2 w + \cos(w^2)$

25. $g(t) = \dfrac{4}{3 + \sqrt{t}}$

26. $g(t) = \dfrac{t - 4}{t + 4}$

27. $y = \dfrac{1}{e^{3x} + x^2}$

28. $h(w) = (w^4 - 2w)^5$

29. $q(\theta) = \sqrt{4\theta^2 - \sin^2(2\theta)}$

30. $g(t) = (t\cos t + \tan^3(t^5))^4$

31. $h(w) = w^3\ln(10w)$

32. $f(x) = \ln(\sin x + \cos x)$

33. $g(x) = \arcsin(\sin\pi x)$

34. $r(t) = \arcsin(2t)$

35. $w(r) = \sqrt{r^4 + 1}$

36. $h(w) = -2w^{-3} + 3\sqrt{w}$

37. $h(x) = \sqrt{\dfrac{x^2 + 9}{x + 3}}$

38. $f(x) = \sqrt{\dfrac{1 - \sin x}{1 - \cos x}}$

39. $T(u) = \arctan\left(\dfrac{u}{1 + u}\right)$

40. $w = 2^{-4z}\sin(\pi z)$

41. $v(t) = t^2 e^{-ct}$

42. $f(x) = \pi^x + x^\pi$

43. $f(x) = \dfrac{x}{1 + \ln x}$

44. $G(x) = \dfrac{\sin^2 x + 1}{\cos^2 x + 1}$

45. $a(t) = \ln\left(\dfrac{1 - \cos t}{1 + \cos t}\right)^4$

46. $f(x) = e^{\ln(kx)}$

47. $R(\theta) = e^{\sin(3\theta)}$

48. $f(x) = e^\pi + \pi^x$

49. $y = \pi^{(x+2)}$

50. $g(x) = e^{\pi x}$

51. $g(\theta) = e^{\sin\theta}$

52. $f(\theta) = 2^{-\theta}$

53. $f(x) = e^{2x}\left(x^2 + 5^x\right)$

54. $h(x) = 2^{e^{3x}}$

55. $h(t) = \dfrac{4 - t}{4 + t}$

56. $r(y) = \dfrac{y}{\cos y + a}$

57. $h(z) = \left(\dfrac{b}{a + z^2}\right)^4$

58. $p(t) = e^{4t+2}$

59. $h(z) = (\ln 2)^z$

60. $j(x) = \dfrac{x^3}{a} + \dfrac{a}{b}x^2 - cx + k$

61. $f(x) = \cos(\arctan 3x)$

62. $f(x) = (3x^2 + \pi)(e^x - 4)$

63. $g(t) = e^{(1+3t)^2}$

64. $f(z) = \dfrac{z^2 + 1}{\sqrt{z}}$

65. $h(r) = \dfrac{r^2}{2r + 1}$

66. $g(x) = 2x - \dfrac{1}{\sqrt[3]{x}} + 3^x - e$

67. $f(t) = 2te^t - \dfrac{1}{\sqrt{t}}$

68. $w = \dfrac{5 - 3z}{5 + 3z}$

69. $g(w) = \dfrac{1}{2^w + e^w}$

70. $f(y) = \ln\left(\ln(2y^3)\right)$

71. $f(x) = \dfrac{x^3}{9}(3\ln x - 1)$

72. $g(x) = x^k + k^x$

73. $r(\theta) = \sin\left((3\theta - \pi)^2\right)$

74. $s(\theta) = \sin^2(3\theta - \pi)$

75. $h(t) = \ln\left(e^{-t} - t\right)$

76. $p(\theta) = \dfrac{\sin(5 - \theta)}{\theta^2}$

77. $w(\theta) = \dfrac{\theta}{\sin^2\theta}$

78. $g(x) = \dfrac{x^2 + \sqrt{x} + 1}{x^{3/2}}$

79. $s(x) = \arctan(2 - x)$

80. $r(\theta) = e^{\left(e^\theta + e^{-\theta}\right)}$

81. $m(n) = \sin(e^n)$

82. $k(\alpha) = e^{\tan(\sin\alpha)}$

83. $g(t) = t\cos(\sqrt{t}e^t)$

84. $f(r) = (\tan 2 + \tan r)^e$

85. $y = e^{-\pi} + \pi^{-e}$

86. $y = \left(x^2 + 5\right)^3 \left(3x^3 - 2\right)^2$

87. $h(x) = xe^{\tan x}$

88. $y = e^{2x}\sin^2(3x)$

89. $g(x) = \tan^{-1}(3x^2 + 1)$

90. $y = 2^{\sin x}\cos x$

91. $h(x) = \ln e^{ax}$

92. $k(x) = \ln e^{ax} + \ln b$

93. $f(\theta) = e^{k\theta} - 1$

94. $N(\theta) = \tan(\arctan(k\theta))$

95. $f(t) = e^{-4kt}\sin t$

96. $f(x) = a^{5x}$

97. $f(x) = \dfrac{a^2 - x^2}{a^2 + x^2}$

98. $w(r) = \dfrac{ar^2}{b + r^3}$

99. $f(s) = \dfrac{a^2 - s^2}{\sqrt{a^2 + s^2}}$

100. $h(t) = e^{kt}(\sin at + \cos bt)$

101. $H(t) = (at^2 + b)e^{-ct}$

102. $g(\theta) = \sqrt{a^2 - \sin^2\theta}$

103. $y = \arctan\left(\dfrac{2}{x}\right)$

104. $r(t) = \ln\left(\sin\left(\dfrac{t}{k}\right)\right)$

105. $g(u) = \dfrac{e^{au}}{a^2 + b^2}$

106. $g(w) = \dfrac{5}{(a^2 - w^2)^2}$

107. $y = \dfrac{e^x - e^{-x}}{e^x + e^{-x}}$

108. $y = \dfrac{e^{ax} - e^{-ax}}{e^{ax} + e^{-ax}}$

109. $f(x) = (2 - 4x - 3x^2)(6x^e - 3\pi)$

110. $f(t) = (\sin(2t) - \cos(3t))^4$

111. $s(y) = \sqrt[3]{(\cos^2 y + 3 + \sin^2 y)}$

112. $f(x) = (4 - x^2 + 2x^3)(6 - 4x + x^7)$

113. $h(x) = \left(\dfrac{1}{x} - \dfrac{1}{x^2}\right)(2x^3 + 4)$

114. $f(z) = \sqrt{5z} + 5\sqrt{z} + \dfrac{5}{\sqrt{z}} - \sqrt{\dfrac{5}{z}} + \sqrt{5}$

115. If $g(2) = 3$ and $g'(2) = -4$, find $f'(2)$ for the following:

    (a) $f(x) = x^2 - 4g(x)$

    (b) $f(x) = \dfrac{x}{g(x)}$

    (c) $f(x) = x^2 g(x)$

    (d) $f(x) = (g(x))^2$

    (e) $f(x) = x\sin(g(x))$

    (f) $f(x) = x^2\ln(g(x))$

116. For parts (a)–(f) of Problem 115, determine the equation of the line tangent to f at $x = 2$.

For Problems 117–122, assume that y is a differentiable function of x and find dy/dx.

117. $xy - x - 3y - 4 = 0$

118. $6x^2 + 4y^2 = 36$

119. $ax^2 - by^2 = c^2$

120. $x^2 y - 2y + 5 = 0$

121. $x^3 + y^3 - 4x^2 y = 0$

122. $\sin(ay) + \cos(bx) = xy$

CHAPTER FIVE

USING THE DERIVATIVE

In Chapter 2, we introduced the derivative and some of its interpretations. In Chapter 4, we saw how to differentiate all of the standard functions, including powers, exponentials, logarithms, and trigonometric functions. Now we will use first and second derivatives to analyze the behavior of families of functions and to solve optimization problems.

5.1   USING FIRST AND SECOND DERIVATIVES

What Derivatives Tell Us About a Function and its Graph

As we saw in Chapter 2, the connection between derivatives of a function and the function itself is given by the following:

- If $f' > 0$ on an interval, then f is increasing on that interval.
- If $f' < 0$ on an interval, then f is decreasing on that interval.
- If $f'' > 0$ on an interval, then the graph of f is concave up on that interval.
- If $f'' < 0$ on an interval, then the graph of f is concave down on that interval.

We can do more with these principles now than we could in Chapter 2 because we now have formulas for the derivatives of the elementary functions.

When we graph a function on a computer or calculator, we often see only part of the picture. Information given by the first and second derivatives can help identify regions with interesting behavior.

Example 1   Use a computer or calculator to sketch a useful graph of the function $f(x) = x^3 - 9x^2 - 48x + 52$.

Solution   Since f is a cubic polynomial, we expect a graph that is roughly S-shaped. Graphing this function with $-10 \leq x \leq 10$, $-10 \leq y \leq 10$, gives the two nearly vertical lines in Figure 5.1. We know that there is more going on than this, but how do we know where to look?

Figure 5.1: Unhelpful graph of
$f(x) = x^3 - 9x^2 - 48x + 52$

We use the derivative to determine where the function is increasing and where it is decreasing. The derivative of f is
$$f'(x) = 3x^2 - 18x - 48.$$
To find where $f' > 0$ or $f' < 0$, we first find where $f' = 0$, that is, where $3x^2 - 18x - 48 = 0$. Factoring, we get $3(x-8)(x+2) = 0$, so $x = -2$ or $x = 8$. Since $f' = 0$ *only* at $x = -2$ and $x = 8$, and since f' is continuous, f' cannot change sign on any of the three intervals $x < -2$, or $-2 < x < 8$, or $8 < x$. How can we tell the sign of f' on each of these intervals? The easiest way is to pick a point and substitute into f'. For example, since $f'(-3) = 33 > 0$, we know f' is positive for $x < -2$, so f is increasing for $x < -2$. Similarly, since $f'(0) = -48$ and $f'(10) = 72$, we know that f decreases between $x = -2$ and $x = 8$ and increases for $x > 8$. Summarizing:

We find that $f(-2) = 104$ and $f(8) = -396$. Hence on the interval $-2 < x < 8$ the function decreases from a high of 104 to a low of -396. (Now we see why not much showed up in our first calculator graph.) One more point on the graph is easy to get: the y intercept, $f(0) = 52$. With just these three points we can get a much more helpful graph. By setting the plotting window to

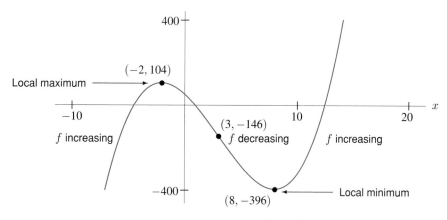

Figure 5.2: Useful graph of $f(x) = x^3 - 9x^2 - 48x + 52$

$-10 \leq x \leq 20$ and $-400 \leq y \leq 400$, we get Figure 5.2, which gives much more insight into the behavior of $f(x)$ than the graph in Figure 5.1.

In Figure 5.2, we see that part of the graph is concave up and part is concave down. We can use the second derivative to analyze concavity. We have

$$f''(x) = 6x - 18.$$

So, $f''(x) < 0$ when $x < 3$ and $f''(x) > 0$ when $x > 3$, and so the graph of f is concave down for $x < 3$ and concave up for $x > 3$. At $x = 3$, we have $f''(x) = 0$. Summarizing:

f concave down $\cap$ $x = 3$ f concave up $\cup$

$f'' < 0$ $f'' = 0$ $f'' > 0$

Local Maxima and Minima

We are often interested in points such as those marked local maximum and local minimum in Figure 5.2. We have the following definition:

> Suppose p is a point in the domain of f:
> - f has a **local minimum** at p if $f(p)$ is less than or equal to the values of f for points near p.
> - f has a **local maximum** at p if $f(p)$ is greater than or equal to the values of f for points near p.

We use the adjective "local" because we are describing only what happens near p.

How Do We Detect a Local Maximum or Minimum?

In the preceding example, the points $x = -2$ and $x = 8$, where $f'(x) = 0$, played a key role in leading us to local maxima and minima. We give a name to such points:

> For any function f, a point p in the domain of f where $f'(p) = 0$ or $f'(p)$ is undefined is called a **critical point** of the function. In addition, the point $\big(p, f(p)\big)$ on the graph of f is also called a **critical point**. A **critical value** of f is the value, $f(p)$, of the function at a critical point, p.

Notice that "critical point of f" can refer either to points in the domain of f or to points on the graph of f. You will know which meaning is intended from the context.

Geometrically, at a critical point where $f'(p) = 0$, the line tangent to the graph of f at p is horizontal. At a critical point where $f'(p)$ is undefined, there is no horizontal tangent to the graph—there's either a vertical tangent or no tangent at all. (For example, $x = 0$ is a critical point for the absolute value function $f(x) = |x|$.) However, most of the functions we will work with will be differentiable everywhere, and therefore most of our critical points will be of the $f'(p) = 0$ variety.

The critical points divide the domain of f into intervals on which the sign of the derivative remains the same, either positive or negative. Therefore, if f is defined on the interval between two successive critical points, its graph cannot change direction on that interval; it is either increasing or decreasing. We have the following result, which is equivalent to the theorem proved on page 287.

> **Theorem:** If a continuous function f has a local maximum or minimum at p, and if p is not an endpoint of the domain, then p is a critical point.

Warning! The sign of f' doesn't *have* to change at a critical point. Consider $f(x) = x^3$, which has a critical point at $x = 0$. (See Figure 5.3.) The derivative, $f'(x) = 3x^2$, is positive on both sides of $x = 0$, so f increases on both sides of $x = 0$, and there is neither a local maximum nor a local minimum at $x = 0$. In other words, not every critical point is a local maximum or local minimum.

Figure 5.3: Critical point which is not a local maximum or minimum

Testing For Local Maxima and Minima

If f' has different signs on either side of a critical point p, with $f'(p) = 0$, then the graph changes direction at p and looks like one of those in Figure 5.4. So we have the following criterion:

> ### The First-Derivative Test for Local Maxima and Minima
>
> Suppose p is a critical point of a continuous function f.
> - If f' changes from negative to positive at p, then f has a local minimum at p.
> - If f' changes from positive to negative at p, then f has a local maximum at p.

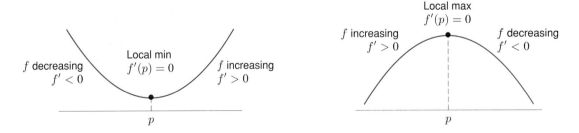

Figure 5.4: Changes in direction at a critical point, p: Local maxima and minima

---

Example 2   Use a graph of the function $f(x) = \dfrac{1}{x(x-1)}$ to observe its local maxima and minima. Confirm your observation analytically.

Solution   The graph in Figure 5.5 suggests that this function has no local minima but that there is a local maximum at about $x = \frac{1}{2}$. Confirming this analytically means using the formula for the derivative to show that what we expect is true. Since $f(x) = (x^2 - x)^{-1}$, we have

$$f'(x) = -1(x^2 - x)^{-2}(2x - 1) = -\frac{2x - 1}{(x^2 - x)^2}.$$

So $f'(x) = 0$ where $2x - 1 = 0$. Thus, the only critical point in the domain of f is $x = \frac{1}{2}$.

Furthermore, $f'(x) > 0$ where $0 < x < 1/2$, and $f'(x) < 0$ where $1/2 < x < 1$. Thus, f increases for $0 < x < 1/2$ and decreases for $1/2 < x < 1$. According to the first derivative test, the critical point $x = 1/2$ is a local maximum.

For $-\infty < x < 0$ or $1 < x < \infty$, there are no critical points and no local maxima or minima. Although $1/(x(x-1)) \to 0$ both as $x \to \infty$ and as $x \to -\infty$, we don't say 0 is a local minimum because $1/\left(x(x-1)\right)$ never actually *equals* 0.

Notice that although $f' > 0$ everywhere that it is defined for $x < \frac{1}{2}$, the function f is not increasing throughout this interval. The problem is that f and f' are not defined at $x = 0$, so we cannot conclude that f is increasing when $x < 1/2$.

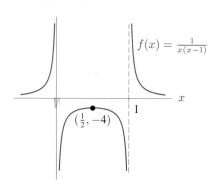

Figure 5.5: Find local maxima and minima

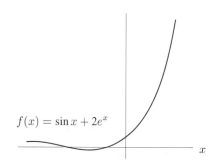

Figure 5.6: Explain the absence of local maxima and minima for $x \geq 0$

Example 3   The graph of $f(x) = \sin x + 2e^x$ is in Figure 5.6. Using the derivative, explain why there are no local maxima or minima for $x \geq 0$.

Solution   Local maxima and minima can occur only at critical points. Now $f'(x) = \cos x + 2e^x$, which is defined for all x. We know $\cos x$ is always between -1 and 1, and $2e^x \geq 2$ for $x \geq 0$, so $f'(x)$ cannot be 0 for any $x \geq 0$. Therefore there are no local maxima or minima there.

The Second-Derivative Test for Local Maxima and Minima

Knowing the concavity of a function can also be useful in testing if a critical point is a local maximum or a local minimum. Suppose p is a critical point of f, with $f'(p) = 0$, so that the graph of f has a horizontal tangent line at p. If the graph is concave up at p, then f has a local minimum at p. Likewise, if the graph is concave down, f has a local maximum. (See Figure 5.7.) This suggests:

Figure 5.7: Local maxima and minima and concavity

The Second-Derivative Test for Local Maxima and Minima

- If $f'(p) = 0$ and $f''(p) > 0$ then f has a local minimum at p.
- If $f'(p) = 0$ and $f''(p) < 0$ then f has a local maximum at p.
- If $f'(p) = 0$ and $f''(p) = 0$ then the test tells us nothing.

Example 4   Classify as local maxima or local minima the critical points of $f(x) = x^3 - 9x^2 - 48x + 52$.

Solution   As we saw in Example 1 on page 240,

$$f'(x) = 3x^2 - 18x - 48$$

and the critical points of f are $x = -2$ and $x = 8$. We have

$$f''(x) = 6x - 18.$$

Thus $f''(8) = 30 > 0$, so f has a local minimum at $x = 8$. Since $f''(-2) = -30 < 0$, f has a local maximum at $x = -2$.

Warning! The second-derivative test does not tell us anything if both $f'(p) = 0$ and $f''(p) = 0$. For example, if $f(x) = x^3$ and $g(x) = x^4$, both $f'(0) = g'(0) = 0$ and $f''(0) = g''(0) = 0$. The point $x = 0$ is a minimum for g but is neither a maximum nor a minimum for f. However, the first-derivative test is still useful. For example, g' changes sign from negative to positive at $x = 0$, so we know g has a local minimum there.

Concavity and Inflection Points

We have studied points where the slope changes sign, which led us to critical points. Now we look at points where the concavity changes.

A point at which the graph of a function changes concavity is called an **inflection point** of f.

The words "inflection point of f" can refer either to a point in the domain of f or to a point on the graph of f. The context of the problem will tell you which is meant.

How Do We Locate an Inflection Point?

Since the concavity changes at an inflection point, the sign of f'' changes there. It is positive on one side of the inflection point, and negative on the other; so at the inflection point, f'' is zero or undefined. (See Figure 5.8.)

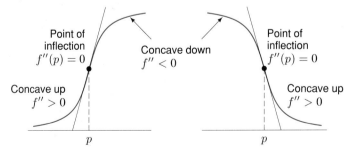

Figure 5.8: Change in concavity at p: Points of inflection are local
maxima or minima of f'

Example 5   Find the critical and inflection points for $g(x) = xe^{-x}$ and sketch the graph of g for $x \geq 0$.

Solution   Taking derivatives and simplifying, we have

$$g'(x) = (1 - x)e^{-x} \quad \text{and} \quad g''(x) = (x - 2)e^{-x}.$$

So $x = 1$ is a critical point, and $g' > 0$ for $x < 1$ and $g' < 0$ for $x > 1$. Hence g increases to a local maximum at $x = 1$ and then decreases. Also, $g(x) \to 0$ as $x \to \infty$. There is an inflection point at $x = 2$ since $g'' < 0$ for $x < 2$ and $g'' > 0$ for $x > 2$. The graph is sketched in Figure 5.9.

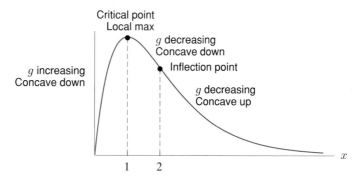

Figure 5.9: Graph of $g(x) = xe^{-x}$

Warning! Not every point x where $f''(x) = 0$ (or f'' is undefined) is an inflection point (just as not every point where $f' = 0$ is a local maximum or minimum). For instance $f(x) = x^4$ has $f''(x) = 12x^2$ so $f''(0) = 0$, but $f'' > 0$ when $x > 0$ and when $x < 0$, so there is *no* change in concavity at $x = 0$. See Figure 5.10.

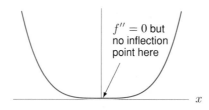

Figure 5.10: Graph of $f(x) = x^4$

Inflection Points and Local Maxima and Minima of the Derivative

Inflection points can also be interpreted in terms of first derivatives. If p is an inflection point, then $f''(p) = 0$ (or $f''(p)$ is undefined) and hence p is a critical point of the derivative function f'. If f' is continuous, this critical point is a local maximum or minimum of f', since f'' changes sign at p.

> A function f with a continuous derivative has an inflection point at p if either of the following conditions hold:
> - f' has a local minimum or a local maximum at p.
> - f'' changes sign at p.

Example 6   Graph $f(x) = x + \sin x$, and determine where f is increasing most rapidly, and least rapidly.

Solution   The graph of f is in Figure 5.11 and the graph of $f'(x) = 1 + \cos x$ is in Figure 5.12.
Where is f increasing most rapidly? At the points $x = \ldots, -2\pi, 0, 2\pi, 4\pi, 6\pi, \ldots$, because these points are local maxima for f', and f' has the same value at each of them. Likewise f is increasing least rapidly at the points $x = \ldots, -3\pi, -\pi, \pi, 3\pi, 5\pi, \ldots$, since these points are local minima for f'. Notice that the points where f is increasing most rapidly and the points where it is increasing least rapidly are inflection points of f.

Figure 5.11: Graph of $f(x) = x + \sin x$

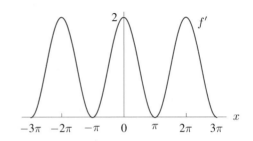

Figure 5.12: Graph of $f'(x) = 1 + \cos x$

Example 7   Suppose that water is being poured into the vase in Figure 5.13 at a constant rate, measured in volume per unit time. Graph $y = f(t)$, the depth of the water against time, t. Explain the concavity and indicate the inflection points.

Solution   At first the water level, y, rises quite slowly because the base of the vase is wide, and so it takes a lot of water to make the depth increase. However, as the vase narrows, the rate at which the water is rising increases. This means that initially y is increasing at an increasing rate and the graph is concave up. The rate of increase in the water level is at a maximum when the water reaches the middle of the vase, where the diameter is smallest; this is an inflection point. After that, the rate at which y increases starts to decrease again, and so the graph is concave down. (See Figure 5.14.)

Figure 5.13: A vase

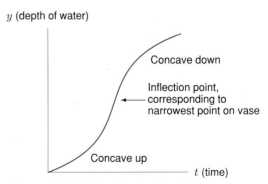

Figure 5.14: Graph of depth of water in the vase, y, against time, t

Problems for Section 5.1

1. Indicate all critical points on the graph of f in Figure 5.15 and determine which correspond to local maxima of f, which to local minima, and which to neither.

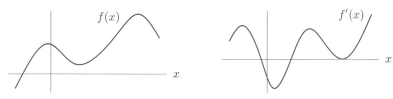

Figure 5.15   **Figure 5.16**

2. Indicate on the graph of the derivative function f' in Figure 5.16 the x-values that are critical points of the function f itself. At which critical points does f have local maxima, local minima, or neither?

3. Suppose f has a continuous derivative whose values are given in the following table.

   (a) Estimate the x-coordinates of critical points of f for $0 \le x \le 10$.
   (b) For each critical point, indicate if it is a local maximum of f, local minimum, or neither.

x	0	1	2	3	4	5	6	7	8	9	10
$f'(x)$	5	2	1	-2	-5	-3	-1	2	3	1	-1

4. Sketch graphs of two continuous functions f and g, each of which has exactly five critical points, the points A–E in Figure 5.17, and which satisfy the following conditions:

   (a) $\displaystyle \lim_{x \to -\infty} f(x) = \infty$ and
        $\displaystyle \lim_{x \to \infty} f(x) = \infty$

   (b) $\displaystyle \lim_{x \to -\infty} g(x) = -\infty$ and
        $\displaystyle \lim_{x \to \infty} g(x) = 0$

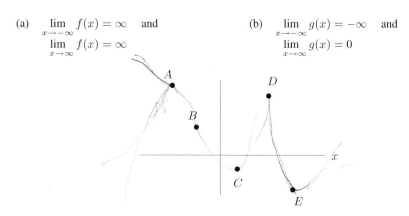

Figure 5.17

5. Indicate on the graph of the derivative f' in Figure 5.18 the x-values that are inflection points of the function f itself.

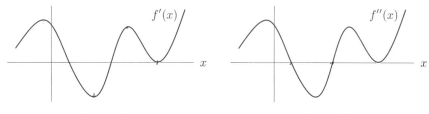

Figure 5.18   **Figure 5.19**

6. Indicate on the graph of the second derivative f'' in Figure 5.19 the x-values that are inflection points of the function f itself.

7. Find the inflection points of $f(x) = x^4 + x^3 - 3x^2 + 2$.

Using a calculator or computer, sketch the graph of each of the functions in Problems 8–15. Describe briefly in words the interesting features of the graph including the location of the critical points and where the function is increasing/decreasing. Then use the derivative and algebra to explain the shape of the graph.

8. $f(x) = x^3 - 6x + 1$ 9. $f(x) = x^3 + 6x + 1$ 10. $f(x) = 3x^5 - 5x^3$

11. $f(x) = x + 2\sin x$ 12. $f(x) = e^x - 10x$ 13. $f(x) = e^x + \sin x$

14. $f(x) = xe^{-x^2}$ 15. $f(x) = x\ln x, \quad x > 0$

16. Use the graphs you drew in Problems 8–15 to describe in words the concavity of each graph, including approximate x-coordinates for all points of inflection. Then use f'' and algebra to explain what you see.

17. You might think the graph of $f(x) = x^2 + \cos x$ should look like a parabola with some waves on it. Sketch the actual graph of $f(x)$ using a calculator or computer. Explain what you see using $f''(x)$.

18. Find values of a and b so that the function $f(x) = x^2 + ax + b$ has a local minimum at the point $(6, -5)$.

19. Find the value of a so that the function $f(x) = xe^{ax}$ has a critical point at $x = 3$.

20. Choose the constants a and b in the function

$$f(x) = axe^{bx}$$

such that $f(\frac{1}{3}) = 1$ and the function has a local maximum at $x = \frac{1}{3}$.

21. Assume the function f is differentiable everywhere and has just one critical point, at $x = 3$. In parts (a) through (d), you are given additional conditions. In each case decide whether $x = 3$ is a local maximum, a local minimum, or neither. Explain your reasoning. Also sketch possible graphs for all four cases.
     (a) $f'(1) = 3$ and $f'(5) = -1$ (c) $f(1) = 1, f(2) = 2, f(4) = 4, f(5) = 5$
     (b) $\lim_{x\to\infty} f(x) = \infty$ and   (d) $f'(2) = -1, f(3) = 1, \lim_{x\to\infty} f(x) = 3$
          $\lim_{x\to-\infty} f(x) = \infty$

22. Assume that the polynomial f has exactly two local maxima and one local minimum, and that these are the only critical points of f.

     (a) Sketch a possible graph of f.
     (b) What is the largest number of zeros f could have?
     (c) What is the least number of zeros f could have?
     (d) What is the least number of inflection points f could have?
     (e) What is the smallest degree f could have?
     (f) Find a possible formula for $f(x)$.

23. (a) Water is flowing at a constant rate into a cylindrical container standing vertically. Sketch a graph showing the depth of water against time.
     (b) Water is flowing at a constant rate into a cone-shaped container standing on its point. Sketch a graph showing the depth of the water against time.

24. If water is flowing at a constant rate (i.e., constant volume per unit time) into the Grecian urn in Figure 5.20, sketch a graph of the depth of the water against time. Mark on the graph the time at which the water reaches the widest point of the urn.

Figure 5.20

Figure 5.21

25. If water is flowing at a constant rate (i.e., constant volume per unit time) into the vase in Figure 5.21, sketch a graph of the depth of the water against time. Mark on the graph the time at which the water reaches the corner of the vase.

26. The rabbit population on a small Pacific island is approximated by

$$P(t) = \frac{2000}{1 + e^{(5.3 - 0.4t)}}$$

with t measured in years since 1774, when Captain James Cook left 10 rabbits on the island. Using a calculator or computer:

 (a) Graph P. Does the population level off?
 (b) Estimate when the rabbit population grew most rapidly. How large was the population at that time?
 (c) What natural causes could lead to the shape of the graph of P?

For Problems 27–32, sketch a possible graph of $y = f(x)$, using the given information about the derivatives $y' = f'(x)$ and $y'' = f''(x)$. Assume that the function is defined and continuous for all real x.

27.

28.

29.

30.

31.

32.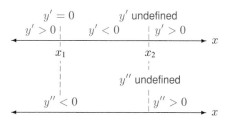

33. Let f be a function with $f(x) > 0$ for all x. Set $g = 1/f$.

 (a) If f is increasing in an interval around x_0, what about g?
 (b) If f has a local maximum at x_1, what about g?
 (c) If f is concave down at x_2, what about g?

34. What happens to concavity when functions are added?

 (a) If $f(x)$ and $g(x)$ are concave up for all x, is $f(x) + g(x)$ concave up for all x?
 (b) If $f(x)$ is concave up for all x and $g(x)$ is concave down for all x, what can you say about the concavity of $f(x) + g(x)$? For example, what happens if $f(x)$ and $g(x)$ are both polynomials of degree 2?
 (c) If $f(x)$ is concave up for all x and $g(x)$ is concave down for all x, is it possible for $f(x) + g(x)$ to change concavity infinitely often?

5.2 FAMILIES OF CURVES: A QUALITATIVE STUDY

We saw in Chapter 1 that knowledge of one function can provide knowledge of the graphs of many others. The shape of the graph of $y = x^2$ also tells us, indirectly, about the graphs of $y = x^2 + 2$, $y = (x + 2)^2$, $y = 2x^2$, and countless other functions. We say that all functions of the form $y = a(x + b)^2 + c$ form a *family of functions*; their graphs are like that of $y = x^2$, except for shifts and stretches determined by the values of $a, b,$ and c. The constants a, b, c are called *parameters*. Different values of the parameters give different members of the family.

One reason for studying families of functions is their use in mathematical modeling. Confronted with the problem of modeling some phenomenon, a crucial first step involves recognizing families of functions which might fit the available data.

Motion Under Gravity: $y = -4.9t^2 + v_0t + y_0$

The position of an object moving vertically under the influence of gravity can be described by a function in the two-parameter family

$$y = -4.9t^2 + v_0t + y_0$$

where t is time in seconds and y is the distance in meters above the ground. Why do we need the parameters v_0 and y_0 to describe all such motions? Notice that at time $t = 0$ we have $y = y_0$. Thus the parameter y_0 gives the height above ground of the object at time $t = 0$. Since $dy/dt = -9.8t + v_0$, the parameter v_0 gives the velocity of the object at time $t = 0$.

Curves of the Form $y = A \sin(Bx)$

This family is used to model a wave. We saw in Section 1.9 that $|A|$ is the amplitude of the wave and that $2\pi/|B|$ is its period. Figures 5.22 and 5.23 illustrate these facts.

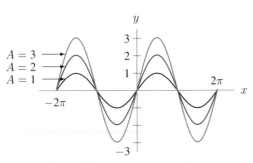

Figure 5.22: The family $y = A \sin x$
(with $B = 1$)

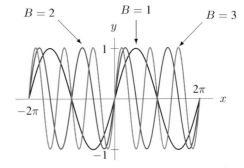

Figure 5.23: The family $y = \sin(Bx)$
(with $A = 1$)

Curves of the Form $y = e^{-(x-a)^2}$

The role of the parameter a in this family is to shift the graph of $y = e^{-x^2}$ to the right or left. Notice that the value of y is always positive. Since $y \to 0$ as $x \to \pm\infty$, the x-axis is a horizontal asymptote. Thus $y = e^{-(x-a)^2}$ is the family of horizontal shifts of the bell-shaped curve shown in Figure 5.24.

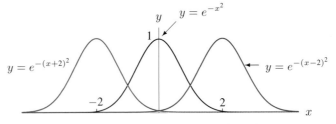

Figure 5.24: The family $y = e^{-(x-a)^2}$

Curves of the Form $y = e^{-(x-a)^2/b}$

This family is related to the *normal density* function, used in probability and statistics.[1] We assume that $b > 0$. The graph of a typical member of the family is a bell-shaped curve as in Figure 5.24. We know that the parameter a shifts the graph horizontally. We now consider the role of the parameter b by studying the family with $a = 0$:

$$y = e^{-x^2/b}.$$

To investigate the critical points and points of inflection, we calculate

$$\frac{dy}{dx} = -\frac{2x}{b}e^{-x^2/b}$$

and, using the product rule, we get

$$\frac{d^2y}{dx^2} = -\frac{2}{b}e^{-x^2/b} - \frac{2x}{b}\left(-\frac{2x}{b}e^{-x^2/b}\right) = \frac{2}{b}\left(\frac{2x^2}{b} - 1\right)e^{-x^2/b}.$$

There is a critical point where $dy/dx = 0$, that is, where

$$\frac{dy}{dx} = -\frac{2x}{b}e^{-x^2/b} = 0.$$

Since $e^{-x^2/b}$ is never zero, the only critical point is $x = 0$. At that point, $y = 1$ and $d^2y/dx^2 < 0$. Hence, by the second derivative test, there is a local maximum at $x = 0$.

Inflection points occur where the second derivative changes sign; thus, we start by finding values of x for which $d^2y/dx^2 = 0$. Since $e^{-x^2/b}$ is never zero, $d^2y/dx^2 = 0$ when

$$\frac{2x^2}{b} - 1 = 0.$$

Solving for x gives

$$x = \pm\sqrt{\frac{b}{2}}.$$

Looking at the expression for d^2y/dx^2, we see that d^2y/dx^2 is negative for $x = 0$, and positive as $x \to \pm\infty$. Therefore the concavity changes at $x = -\sqrt{b/2}$ and at $x = \sqrt{b/2}$, so we have inflection points there. Returning to the two-parameter family $y = e^{-(x-a)^2/b}$, we conclude that there is a maximum at $x = a$, obtained by horizontally shifting the maximum at $x = 0$ of $y = e^{-x^2/b}$ by a units. There are inflection points at $x = a \pm \sqrt{b/2}$ obtained by shifting the inflection points $x = \pm\sqrt{b/2}$ of $y = e^{-x^2/b}$ by a units. (See Figure 5.25.) At these points $y = e^{-1/2} \approx 0.6$.

With this information we can see the effect of the parameters. The parameter a determines the location of the center of the bell and the parameter b determines how narrow or wide the bell is. (See Figure 5.26.) If b is small, then the inflection points are close to a and the bell is sharply peaked near a; if b is large, the inflection points are further away from a and the bell is more spread out.

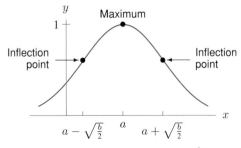

Figure 5.25: Graph of $y = e^{-(x-a)^2/b}$: bell-shaped curve with peak at $x = a$

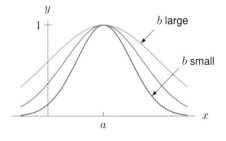

Figure 5.26: Graph of $y = e^{-(x-a)^2/b}$ for fixed a and various b

[1]Probabilists divide our function by a constant, $\sqrt{\pi b}$, to get the normal density.

Curves of the Form $y = a\left(1 - e^{-bx}\right)$

This is a two-parameter family. We will only consider $a, b > 0$. The graph of one member, with $a = 2$ and $b = 1$, is in Figure 5.27. Such a graph represents a quantity which is increasing but leveling off. For example, a body dropped in a thick fluid speeds up initially, but its velocity levels off as it approaches a terminal velocity. Similarly, if a pollutant pouring into a lake builds up toward a saturation level, its concentration may be described in this way. The graph might also represent the temperature of an object in an oven.

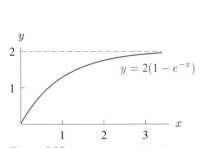

Figure 5.27: One member of the family $y = a(1 - e^{-bx})$, with $a = 2$

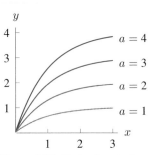

Figure 5.28: Fixing $b = 1$ gives $y = a(1 - e^{-x})$, graphed for various a

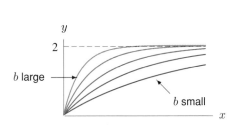

Figure 5.29: Fixing $a = 2$ gives $y = 2(1 - e^{-bx})$, graphed for various b

We examine the effect on the graph of varying a. Fix b at some positive number, say $b = 1$. Substitute different values for a and look at the graphs in Figure 5.28. We see that as x gets larger, y approaches a from below. The reason is $e^{-bx} \to 0$ as $x \to \infty$. Physically, the value of a represents the terminal velocity of a falling body or the saturation level of the pollutant in the lake.

Now examine the effect of varying b on the graph. Fix a at some positive number, say $a = 2$. Substitute different values for b and look at the graphs in Figure 5.29. The parameter b determines how sharply the curve rises and how quickly it gets close to the line $y = a$.

Let's confirm this last observation analytically. For $y = a(1 - e^{-bx})$, we have $dy/dx = abe^{-bx}$, so the slope of the tangent to the curve at $x = 0$ is ab. For larger b, the curve rises more rapidly at $x = 0$. How long does it take the curve to climb halfway up from $y = 0$ to $y = a$? When $y = a/2$, we have

$$a(1 - e^{-bx}) = \frac{a}{2}, \qquad \text{which leads to} \qquad x = \frac{\ln 2}{b}.$$

If b is large then $(\ln 2)/b$ is small, so in a short distance the curve is already half way up to a. If b is small, then $(\ln 2)/b$ is large and we have to go a long way out to get up to $a/2$. See Figure 5.30.

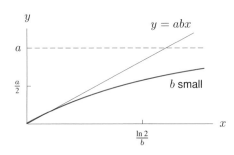

Figure 5.30: Tangent at $x = 0$ to $y = a(1 - e^{-bx})$, with fixed a, and large and small b

Problems for Section 5.2

1. Consider the function $p(x) = x^3 - ax$, where a is constant.

    (a) If $a < 0$, show that $p(x)$ is always increasing.
    (b) If $a > 0$, show that $p(x)$ has a local maximum and a local minimum.
    (c) Sketch and label typical graphs for the cases when $a < 0$ and when $a > 0$.

2. Consider the function $p(x) = x^3 - ax$, where a is constant and $a > 0$.

    (a) Find the local maxima and minima of p.
    (b) What effect does increasing the value of a have on the positions of the maxima and minima?
    (c) On the same axes, sketch and label the graphs of p for three positive values of a.

3. What effect does increasing the value of a have on the graph of $f(x) = x^2 + 2ax$? Consider roots, maxima and minima, and both positive and negative values of a.

4. The number, N, of people who have heard a rumor spread by mass media is modeled by the following function of time, t:
    $$N(t) = a(1 - e^{-kt}).$$
    Suppose that there are 200,000 people in the population who hear the rumor eventually. If 10% of them heard it the first day, find a and k, assuming t is measured in days.

5. Suppose the temperature, T, of a yam put into a hot oven maintained at 200°C is given as a function of time, t, by
    $$T = a(1 - e^{-kt}) + b$$
    where T is in degrees Celsius and t is in minutes.

    (a) If the yam starts at 20°C, find a and b.
    (b) If the temperature of the yam is initially increasing at 2°C per minute, find k.

6. Consider the family of functions $y = f(x) = x - k\sqrt{x}$, with k a positive constant and $x \geq 0$. Show that the graph of $f(x)$ has a local minimum at a point whose x-coordinate is $1/4$ of the way between its x-intercepts.

7. (a) Find all critical points of $f(x) = x^4 + ax^2 + b$.
    (b) Under what conditions on a and b will this function have exactly one critical point? What is the one critical point, and is it a local maximum, a local minimum, or neither?
    (c) Under what conditions on a and b will this function have exactly three critical points? What are they and which are local maxima and which are local minima?
    (d) Is it ever possible for this function to have two critical points? No critical points? More than three critical points? Give an explanation in each case.

8. Sketch several members of the family $y = e^{-ax} \sin bx$ for $b = 1$, and describe the graphical significance of the parameter a.

9. Sketch several members of the family $e^{-ax} \sin bx$ for $a = 1$, and describe the graphical significance of the parameter b.

10. Sketch graphs of $y = xe^{-bx}$ for $b = 1, 2, 3, 4$. Describe the graphical significance of b.

11. Find the coordinates of the critical point of $y = xe^{-bx}$ and use it to confirm your answer to Problem 10.

12. If $a > 0, b > 0$, confirm analytically that $f(x) = a(1 - e^{-bx})$ is everywhere increasing and everywhere concave down.

13. Consider the family of curves
    $$y = Ae^{-Bx^2}, \qquad \text{for positive } A, B.$$
    Analyze the effect of varying A and B on the shape of the curve. Illustrate your answer with sketches.

14. Consider the *surge function*
    $$y = axe^{-bx}, \qquad \text{for positive } a, b.$$

    (a) Find the local maxima, local minima, and points of inflection.
    (b) How does varying a and b affect the shape of the graph?
    (c) On one set of axes, graph this function for several values of a and b.

15. Consider the family of functions $f(x) = x^2 + \cos(kx)$, $k > 0$.

    (a) Using a calculator or computer, sketch the graph of f for $k = 0.5, 1, 3, 5$. Try to find the smallest number k at which you see points of inflection in the graph of f.
    (b) Explain why the graph of f has no points of inflection if $k \leq \sqrt{2}$, and infinitely many points of inflection if $k > \sqrt{2}$.
    (c) Explain why f has only a finite number of critical points, no matter what the value of k.

16. Consider the family of functions of the form $f(x) = e^x - kx$, for $k > 0$.

    (a) Using a calculator or computer, sketch the graph of f for $k = 1/4, 1/2, 1, 2, 4$. Describe what happens as k changes.
    (b) Show that f has a local minimum at $x = \ln k$.
    (c) Find the value of k for which the local minimum is the largest.

17. Consider the family

    $$y = \frac{A}{x + B}.$$

    (a) If $B = 0$, what is the effect of varying A on the graph?
    (b) If $A = 1$, what is the effect of varying B?
    (c) On one set of axes, graph the function for several values of A and B.

18. The potential energy, U, of a particle moving along the x-axis is given by

    $$U = b\left(\frac{a^2}{x^2} - \frac{a}{x}\right)$$

    where a and b are positive constants. Consider the graph of U against x for $x > 0$.

    (a) Find the intercepts and asymptotes.
    (b) Compute the local maxima and minima.
    (c) Sketch the graph.

19. The force, F, on a particle with potential energy U is given by

    $$F = -\frac{dU}{dx}.$$

    Using the expression for U in Problem 18, graph F and U on the same axes, labeling intercepts and local maxima and minima.

20. If the force between two atoms in a molecule is given as a function of the distance, r, between the molecules by

    $$f(r) = -\frac{A}{r^2} + \frac{B}{r^3} \qquad r > 0,$$

    where A, B are positive constants:

    (a) What are the zeros and asymptotes of f?
    (b) Find the coordinates of the critical points and inflection points of f.
    (c) Sketch a graph of f.
    (d) Illustrating your answers with a sketch, describe the effect on the graph of f of:
        (i) Increasing B, holding A fixed
        (ii) Increasing A, holding B fixed

5.3   OPTIMIZATION

It is often important to find the largest or smallest value of some quantity. For example, automobile engineers want to construct a car that uses the least amount of fuel, scientists want to calculate which wavelength carries the maximum radiation at a given temperature, and urban planners want to design traffic patterns to minimize delays. The techniques for solving such problems belong to the field of mathematics called *optimization*. The next three sections show how the derivative provides an efficient way of solving many optimization problems.

Global Maxima and Minima

The single greatest (or least) value of a function f over a specified domain is called the *global maximum* (or *minimum*) of f. Recall that the local maxima and minima tell us where a function is locally largest or smallest. Now we are interested in where the function is absolutely largest or smallest in a given domain. We define

> - f has a **global minimum** at p if $f(p)$ is less than or equal to all values of f.
> - f has a **global maximum** at p if $f(p)$ is greater than or equal to all values of f.

How Do We Find Global Maxima and Minima?

If f is a continuous function defined on a closed interval $a \leq x \leq b$ (that is, an interval containing its endpoints), Figure 5.31 illustrates that the global maximum or minimum of f occurs either at a local maximum or a local minimum, or at one of the endpoints of the interval, $x = a$ or $x = b$.

Figure 5.31: Global maximum and minimum on a closed interval $a \leq x \leq b$

> **To find the global maximum and minimum of a continuous function on a closed interval:** Compare values of the function at all the critical points in the interval and at the endpoints.

What if the function is defined on an open interval $a < x < b$ (that is, an interval not including its endpoints), or on all real numbers? Then there may or may not be a global maximum or minimum. For example, there is no global maximum in Figure 5.32 because the function has no actual largest value. The global minimum in Figure 5.32 coincides with the local minimum and is marked. There is a global minimum but no global maximum in Figure 5.33.

> **To find the global maximum and minimum of a continuous function on an open interval or on all real numbers:** Find the value of the function at all the critical points and sketch a graph. Look at the function values when x approaches the endpoints of the interval, or approaches $\pm\infty$, as appropriate.

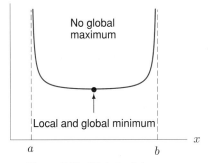

Figure 5.32: Global minimum on $a < x < b$

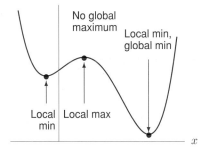

Figure 5.33: Global minimum when the domain is all real numbers

Example 1 Find the global maxima and minima of $f(x) = x^3 - 9x^2 - 48x + 52$ on the following intervals:

(a) $-5 \le x \le 12$ (b) $-5 \le x \le 14$ (c) $-5 \le x < \infty$.

Solution (a) We have previously obtained the critical points $x = -2$ and $x = 8$ using

$$f'(x) = 3x^2 - 18x - 48 = 3(x + 2)(x - 8).$$

We evaluate f at the critical points and the endpoints of the interval:

$$f(-5) = (-5)^3 - 9(-5)^2 - 48(-5) + 52 = -58$$
$$f(-2) = 104$$
$$f(8) = -396$$
$$f(12) = -92.$$

Comparing these function values, we see that the global maximum on $[-5, 12]$ is 104 and occurs at $x = -2$, and the global minimum on $[-5, 12]$ is -396 and occurs at $x = 8$.

(b) For the interval $[-5, 14]$, we compare

$$f(-5) = -58, \quad f(-2) = 104, \quad f(8) = -396, \quad f(14) = 360.$$

The global maximum is now 360 and occurs at $x = 14$, and the global minimum is still -396 and occurs at $x = 8$. Notice that since the function is increasing for $x > 8$, changing the right-hand end of the interval from $x = 12$ to $x = 14$ alters the global maximum but not the global minimum. See Figure 5.34.

(c) Figure 5.34 shows that for $-5 \le x < \infty$ there is no global maximum, because we can make $f(x)$ as large as we please by choosing x large enough. The global minimum remains -396 at $x = 8$.

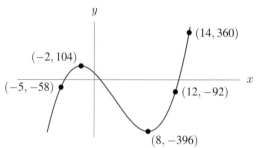

Figure 5.34: Graph of $f(x) = x^3 - 9x^2 - 48x + 52$

Example 2 When an arrow is shot into the air, its range, R, is defined as the horizontal distance from the archer to the point where the arrow hits the ground. If the ground is horizontal and we neglect air resistance, it can be shown that

$$R = \frac{v_0{}^2 \sin(2\theta)}{g},$$

where v_0 is the initial velocity of the arrow, g is the (constant) acceleration due to gravity, and θ is the angle above horizontal, so $0 \le \theta \le \pi/2$. (See Figure 5.35.) What initial angle maximizes R?

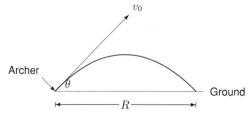

Figure 5.35: Arrow's path

Solution   We can find the maximum of this function without using calculus. The maximum value of R occurs when $\sin(2\theta) = 1$, so $\theta = \arcsin(1)/2 = \pi/4$, giving $R = v_0^2/g$.

Let's see how we can do the same problem with calculus. We want to find the global maximum of R for $0 \le \theta \le \pi/2$. First we look for critical points:

$$\frac{dR}{d\theta} = 2\frac{v_0^2 \cos(2\theta)}{g}.$$

Setting $dR/d\theta$ equal to 0, we get

$$0 = \cos(2\theta), \quad \text{or} \quad 2\theta = \pm\frac{\pi}{2}, \pm\frac{3\pi}{2}, \pm\frac{5\pi}{2}, \dots$$

so $\pi/4$ is the only critical point in the interval $0 \le \theta \le \pi/2$. The range at $\theta = \pi/4$ is $R = v_0{}^2/g$.

Now we must check the value of R at the endpoints $\theta = 0$ and $\theta = \pi/2$. Since $R = 0$ at each endpoint, the critical point $\theta = \pi/4$ gives both a local and a global maximum on $0 \le \theta \le \pi/2$. Therefore, the arrow goes farthest if shot at an angle of $\pi/4$, or $45°$.

A Graphical Example: Minimizing Gas Consumption

Next we look at an example in which a function is given graphically and the optimum values are read from a graph. You already know how to estimate the optimum values of $f(x)$ from a graph of $f(x)$—read off the highest and lowest values. In this example, we see how to estimate the optimum value of the quantity $f(x)/x$ from a graph of $f(x)$ against x.

The question we investigate is how to set driving speeds to maximize fuel efficiency.[2] We assume that gas consumption, g (in gallons/hour), as a function of velocity, v (in mph) is as shown in Figure 5.36. We want to minimize the gas consumption per *mile*, not the gas consumption per hour. Let $G = g/v$ represent the average gas consumption per mile. (The units of G are gallons/mile.)

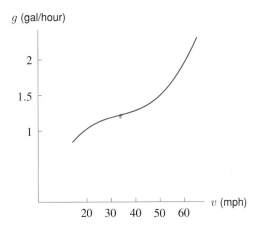

Figure 5.36: Gas consumption versus velocity

Example 3   Using Figure 5.36, estimate the velocity which minimizes G.

Solution   We want to find the minimum value of $G = g/v$ when g and v are related by the graph in Figure 5.36. We could use Figure 5.36 to sketch a graph of G against v and estimate a critical point. But there is an easier way.

---

[2]Adapted from Peter D. Taylor, *Calculus: The Analysis of Functions* (Toronto: Wall & Emerson, 1992).

Figure 5.37 shows that g/v is the slope of the line from the origin to the point P. Where on the curve should P be to make the slope a minimum? From the possible positions of the line shown in Figure 5.37, we see that the slope of the line is both a local and global minimum when the line is tangent to the curve. From Figure 5.38, we can see that the velocity at this point is about 50 mph. Thus to minimize gas consumption per mile, we should drive about 50 mph.

Figure 5.37: Graphical representation of gas consumption per mile, $G = \dfrac{g}{v}$

Figure 5.38: Velocity for maximum fuel efficiency

Finding Upper and Lower Bounds

A problem which is closely related to finding maxima and minima is finding the *bounds* of a function. In Example 1, the value of $f(x)$ on the interval $[-5, 12]$ ranges from -396 to 104. Thus

$$-396 \leq f(x) \leq 104$$

and we say that the function f is *bounded below* by -396 and *bounded above* by 104 on $[-5, 12]$. Of course, we could also say that

$$-400 \leq f(x) \leq 150,$$

so that f is also bounded below by -400 and above by 150 on $[-5, 12]$. However, we consider the -396 and 104 to be the *best possible bounds* because they describe more accurately how the function $f(x)$ behaves on $[-5, 12]$. (The bounds of a function are discussed in more detail in Appendix A.)

Example 4   Suppose an object on a spring oscillates about its equilibrium position at $y = 0$. Its distance from equilibrium is given as a function of time, t, by

$$y = e^{-t} \cos t.$$

Find the greatest distance the object goes above and below the equilibrium for $t \geq 0$.

Solution   We are looking for the bounds of the function. What does the graph of the function look like? We can think of it as a cosine curve with a decreasing amplitude of e^{-t}; in other words, it is a cosine curve squashed between the graphs of $y = e^{-t}$ and $y = -e^{-t}$, forming a wave with lower and lower crests and shallower and shallower troughs. (See Figure 5.39.)

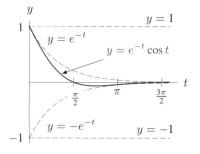

Figure 5.39: $f(t) = e^{-t} \cos t$ for $t \geq 0$

From the graph we can see that for $t \geq 0$, the graph lies between the horizontal lines $y = -1$ and $y = 1$. This means that -1 and 1 are bounds:

$$-1 \leq e^{-t} \cos t \leq 1.$$

The line $y = 1$ is the best possible upper bound because the graph does come up that high (at $t = 0$). However, we can find a better lower bound if we find the global minimum value of f for $t \geq 0$; this minimum occurs in the first trough between $t = \pi/2$ and $t = 3\pi/2$ because later troughs are squashed closer to the t-axis. At the minimum, $dy/dt = 0$. The product rule gives

$$\frac{dy}{dt} = (-e^{-t}) \cos t + e^{-t}(-\sin t) = -e^{-t}(\cos t + \sin t) = 0.$$

Since e^{-t} is never 0, we must have

$$\cos t + \sin t = 0, \quad \text{so} \quad \frac{\sin t}{\cos t} = -1.$$

Hence

$$\tan t = -1, \quad \text{giving} \quad t = \frac{3\pi}{4}.$$

Thus, the global minimum we see on the graph occurs at $t = 3\pi/4$. The value of y at that minimum is

$$y = e^{-3\pi/4} \cos\left(\frac{3\pi}{4}\right) \approx -0.067.$$

Rounding down so that the inequalities still hold for all $t \geq 0$ gives

$$-0.07 < e^{-t} \cos t \leq 1.$$

Notice how much smaller in magnitude the lower bound is than the upper. This is a reflection of how quickly the factor e^{-t} causes the oscillation to die out.

Problems for Section 5.3

For Problems 1–2, indicate all critical points on the given graphs. Determine which correspond to local minima, local maxima, global minima, global maxima, or none of these. (Note that the graphs are on closed intervals.)

1.

Figure 5.40

2.

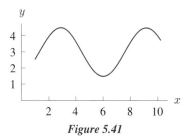

Figure 5.41

3. For $y = f(x) = x^{10} - 10x$, and $0 \leq x \leq 2$, find the value(s) of x for which:

   (a) $f(x)$ has a local maximum or local minimum. Indicate which ones are maxima and which are minima.

   (b) $f(x)$ has a global maximum or global minimum.

4. For $f(x) = x - \ln x$, and $0.1 \leq x \leq 2$, find the value(s) of x for which:

   (a) $f(x)$ has a local maximum or local minimum. Indicate which ones are maxima and which are minima.

   (b) $f(x)$ has a global maximum or global minimum.

5. For $f(x) = \sin^2 x - \cos x$, and $0 \leq x \leq \pi$, find, to two decimal places, the value(s) of x for which:

   (a) $f(x)$ has a local maximum or local minimum. Indicate which ones are maxima and which are minima.

   (b) $f(x)$ has a global maximum or global minimum.

6. The function $y = t(x)$ is positive and continuous with a global maximum at the point $(3, 3)$. Sketch a possible graph of $t(x)$ if $t'(x)$ and $t''(x)$ have the same sign for $x < 3$, but opposite signs for $x > 3$.

7. A function $y = g(x)$ has a derivative that is given in Figure 5.42 for $-2 \leq x \leq 2$.

    (a) Write a few sentences describing the behavior of $g(x)$ on this interval.
    (b) Does the graph of $g(x)$ have any inflection points? If so, give the approximate x-coordinates of their locations. Explain your reasoning.
    (c) What are the global maxima and minima of g on $[-2, 2]$?
    (d) If $g(-2) = 5$, what do you know about $g(0)$ and $g(2)$? Explain.

Figure 5.42   **Figure 5.43**

8. A function $y = h(x)$ has a second derivative shown in Figure 5.43 for $-2 \leq x \leq 1$. If $h'(-1) = 0$ and $h(-1) = 2$,

    (a) Explain why $h'(x)$ is never negative on this interval.
    (b) Explain why $h(x)$ must have a global maximum at $x = 1$.
    (c) Sketch a possible graph of $h(x)$ for $-2 \leq x \leq 1$.

9. A grapefruit is tossed straight up with an initial velocity of 50 ft/sec. The grapefruit is 5 feet above the ground when it is released. Its height at time t is given by

    $$y = -16t^2 + 50t + 5.$$

    How high does it go before returning to the ground?

10. For some positive constant C, the temperature change, T, in a patient generated by a dose, D, of a drug is given by
    $$T = \left(\frac{C}{2} - \frac{D}{3}\right)D^2.$$

    (a) What dosage maximizes the temperature change?
    (b) The sensitivity of the body, at dosage D, to the drug is defined as dT/dD. What dosage maximizes sensitivity?

11. When you cough, your windpipe contracts. The speed, v, with which air comes out depends on the radius, r, of your windpipe. If R is the normal (rest) radius of your windpipe, then for $r \leq R$, the speed is given by:
    $$v = a(R - r)r^2$$
    where a is a positive constant. What value of r maximizes the speed?

12. The bending moment M of a beam, supported at one end, at a distance x from the support is given by
    $$M = \tfrac{1}{2}wLx - \tfrac{1}{2}wx^2,$$
    where L is the length of the beam, and w is the uniform load per unit length. Find the point on the beam where the moment is greatest.

13. The efficiency of a screw, E, is given by
    $$E = \frac{(\theta - \mu\theta^2)}{\mu + \theta}, \quad \theta > 0,$$
    where θ is the angle of pitch of the thread and μ is the coefficient of friction of the material, a (positive) constant. What value of θ maximizes E?

14. A woman pulls a sled which, together with its load, has a mass of m kg. If her arm makes an angle of θ with her body (assumed vertical) and the coefficient of friction (a positive constant) is μ, the least force, F, she must exert to move the sled is given by

$$F = \frac{mg\mu}{\sin\theta + \mu\cos\theta}.$$

If $\mu = 0.15$, find the maximum and minimum values of F for $0 \le \theta \le \pi/2$. Give your answers as multiples of mg.

15. A circular ring of wire of radius r_0 lies in a plane perpendicular to the x-axis and is centered at the origin. The ring has a positive electric charge spread uniformly over it. The electric field in the x-direction, E, at the point x on the axis is given by

$$E = \frac{kx}{\left(x^2 + r_0^2\right)^{3/2}} \quad \text{for} \quad k > 0.$$

At what point on the x-axis is the field greatest? Least?

16. An electric current, I, in amps, is given by

$$I = \cos(wt) + \sqrt{3}\sin(wt),$$

where $w \ne 0$ is a constant. What are the maximum and minimum values of I?

17. (a) Show that $x > 2\ln x$ for all $x > 0$. [Hint: Find the minimum of $f(x) = x - 2\ln x$.]
    (b) Use the above result to show that $e^x > x^2$ for all positive x.
    (c) Is $x > 3\ln x$ for all positive x?

Find the best possible bounds for each of the functions in Problems 18–22.

18. e^{-x^2}, for $|x| \le 0.3$

19. $\ln(1 + x)$, for $x \ge 0$

20. $\ln(1 + x^2)$, for $-1 \le x \le 2$

21. $x^3 - 4x^2 + 4x$, for $0 \le x \le 4$

22. $x + \sin x$, for $0 \le x \le 2\pi$

23. When birds lay eggs, they do so in clutches of several at a time. When the eggs hatch, each clutch gives rise to a brood of baby birds. We want to determine the clutch size which maximizes the number of birds surviving to adulthood per brood. If the clutch is small, there are few baby birds in the brood; if the clutch is large, there are so many baby birds to feed that most die of starvation. The number of surviving birds per brood as a function of clutch size is shown by the benefit curve in Figure 5.44.[3]

    (a) Estimate the clutch size which maximizes the number of survivors per brood.
    (b) Suppose also that there is a biological cost to having a larger clutch: the female survival rate is reduced by large clutches. This cost is represented by the dotted line in Figure 5.44. If we take cost into account by assuming that the optimal clutch size in fact maximizes the vertical distance between the curves, what is the new optimal clutch size?

Figure 5.44

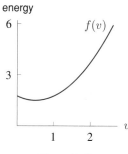

Figure 5.45

[3] Data from C. M. Perrins and D. Lack reported by J. R. Krebs and N. B. Davies in *An Introduction to Behavioural Ecology* (Oxford: Blackwell, 1987).

24. Let $f(v)$ be the amount of energy consumed by a flying bird, measured in joules per second (a joule is a unit of energy), as a function of its speed v (in meters/sec). See Figure 5.45.

    (a) Suggest a reason for the shape of this graph (in terms of the way birds fly).

    Now let $a(v)$ be the amount of energy consumed by the same bird, measured in joules *per meter*.

    (b) What is the relationship between $f(v)$ and $a(v)$?
    (c) Where is $a(v)$ a minimum?
    (d) Should the bird try to minimize $f(v)$ or $a(v)$ when it is flying? Why?

25. The forward motion of an aircraft in level flight is reduced by two kinds of forces, known as *induced drag* and *parasite drag*. Induced drag is a consequence of the downward deflection of air as the wings produce lift. Parasite drag results from friction between the air and the entire surface of the aircraft. Induced drag is inversely proportional to the square of speed and parasite drag is directly proportional to the square of speed. The sum of induced drag and parasite drag is called total drag. The graph in Figure 5.46 shows a certain aircraft's induced drag and parasite drag functions.

    (a) Sketch a graph of the total drag as a function of airspeed.
    (b) Estimate two different airspeeds which each result in a total drag of 1000 pounds. Does the total drag function have an inverse? What about the induced and parasite drag functions?
    (c) Fuel consumption (in gallons per hour) is roughly proportional to total drag. Suppose you are low on fuel and the control tower has just instructed you to enter a circular holding pattern of indefinite duration to await the passage of a storm at your landing field. At what airspeed should you fly the holding pattern? Why?

Figure 5.46

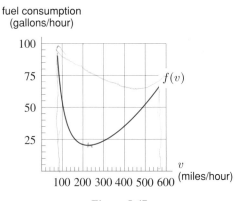

Figure 5.47

26. Let $f(v)$ be the fuel consumption, measured in gallons per hour, of a certain aircraft as a function of its airspeed, v, given in miles per hour. A graph of $f(v)$ is given in Figure 5.47.

    (a) Let $g(v)$ be the fuel consumption of the same aircraft, but measured in gallons per mile instead of gallons per hour. What is the relationship between $f(v)$ and $g(v)$?
    (b) For what value of v is $f(v)$ minimized?
    (c) For what value of v is $g(v)$ minimized?
    (d) Should a pilot try to minimize $f(v)$ or $g(v)$?

5.4 APPLICATIONS TO MARGINALITY

Management decisions within a particular business usually aim at maximizing profit for the company. In this section we will see how the derivative can be used to maximize profit. Profit depends on both production cost and revenue (or income) from sales. We begin by looking at the cost and revenue functions.

> The **cost function**, $C(q)$, gives the total cost of producing a quantity q of some good.

What sort of function do we expect C to be? The more goods that are made, the higher the total cost, so C is an increasing function. In fact, cost functions usually have the general shape shown in Figure 5.48. The intercept on the C-axis represents the *fixed costs*, which are incurred even if nothing is produced. (This includes, for instance, the machinery needed to begin production.) The cost function increases quickly at first and then more slowly because producing larger quantities of a good is usually more efficient than producing smaller quantities—this is called *economy of scale*. At still higher production levels, the cost function starts to increase faster again as resources become scarce, and sharp increases may occur when new factories have to be built. Thus, the graph of $C(q)$ may start out concave down and become concave up later on.

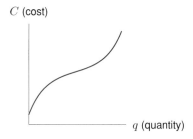

Figure 5.48: Cost as a function of quantity

> The **revenue function**, $R(q)$, gives the total revenue received by a firm from selling a quantity q of some good.

Revenue is income obtained from sales. If the price per item is p, and the quantity sold is q, then

$$\text{Revenue} = \text{Price} \times \text{Quantity}, \quad \text{so} \quad R = pq.$$

If the price per item does not depend on the quantity sold, then the graph of $R(q)$ is a straight line through the origin with slope equal to the price p. See Figure 5.49. In practice, for large values of q, the market may become glutted, causing the price to drop, giving $R(q)$ the shape in Figure 5.50.

Figure 5.49: Revenue: Constant price

Figure 5.50: Revenue: Decreasing price

The profit is usually written as π (to distinguish it from the price, p; this π has nothing to do with the area of a circle, and merely stands for the Greek equivalent of the letter "p"). The profit resulting from producing and selling q items is defined by

> $$\text{Profit} = \text{Revenue} - \text{Cost}, \quad \text{so} \quad \pi(q) = R(q) - C(q).$$

Example 1 If cost, C, and revenue, R, are given by the graph in Figure 5.51, for what production quantities, q, does the firm make a profit? Approximately what production level maximizes profit?

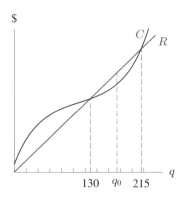

$Figure\ 5.51$: Costs and revenues for Example 1

Solution The firm makes a profit whenever revenues are greater than costs, that is, when $R > C$. The graph of R is above the graph of C approximately when $130 < q < 215$. Production between $q = 130$ units and $q = 215$ units will generate a profit. The vertical distance between the cost and revenue curves is largest at q_0, so q_0 units gives maximum profit.

Marginal Analysis

Many economic decisions are based on an analysis of the costs and revenues "at the margin." Let's look at this idea through an example.

Suppose we are running an airline and we are trying to decide whether to offer an additional flight. How should we decide? We'll assume that the decision is to be made purely on financial grounds: if the flight will make money for the company, it should be added. Obviously we need to consider the costs and revenues involved. Since the choice is between adding this flight and leaving things the way they are, the crucial question is whether the *additional costs* incurred are greater or smaller than the *additional revenues* generated by the flight. These additional costs and revenues are called the *marginal costs* and *marginal revenues*.

Suppose $C(q)$ is the function giving the total cost of running q flights. If the airline had originally planned to run 100 flights, its costs would be $C(100)$. With the additional flight, its costs would be $C(101)$. Therefore,

$$\text{Marginal cost} = C(101) - C(100).$$

Now

$$C(101) - C(100) = \frac{C(101) - C(100)}{101 - 100},$$

and this quantity is the average rate of change of cost between 100 and 101 flights. In Figure 5.52 the average rate of change is the slope of the line joining the $C(100)$ and $C(101)$ points on the graph. If the graph of the cost function is not curving too fast near the point, the slope of this line is close to the slope of the tangent line there. Therefore, the average rate of change is close to the instantaneous rate of change. Since these rates of change are not very different, many economists choose to define marginal cost, MC, as the instantaneous rate of change of cost with respect to quantity:

$$\boxed{\text{Marginal cost} = MC = C'(q).}$$

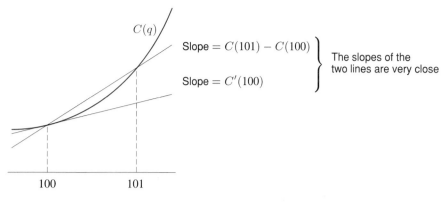

Figure 5.52: Marginal cost: Slope of one of these lines

Similarly if the revenue generated by q flights is $R(q)$, the additional revenue generated by increasing the number of flights from 100 to 101 is

$$\text{Marginal revenue} = R(101) - R(100).$$

Now $R(101) - R(100)$ is the average rate of change of revenue between 100 and 101 flights. As before, the average rate of change is usually almost equal to the instantaneous rate of change, so economists often define:

$$\boxed{\text{Marginal revenue} = MR = R'(q).}$$

We often refer to total cost and total revenue to distinguish them from marginal cost and marginal revenue. If the words cost and revenue are used alone, they are understood to mean total cost and total revenue.

Example 2   If $C(q)$ and $R(q)$ for the airline are given in Figure 5.53, should the company add the 101st flight?

Solution   The marginal revenue is the slope of the revenue curve, and the marginal cost is the slope of the cost curve at the point 100. From Figure 5.53, you can see that the slope at the point A is smaller than the slope at B, so $MC < MR$. This means that the airline will make more in extra revenue than it will spend in extra costs if it runs another flight, so it should go ahead and run the 101st flight.

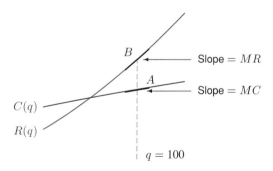

Figure 5.53: Cost and revenue for Example 2

Since MC and MR are derivative functions, they can be estimated from the graphs of total cost and total revenue.

Example 3   If R and C are given by the graphs in Figure 5.54, sketch graphs of $MR = R'(q)$ and $MC = C'(q)$.

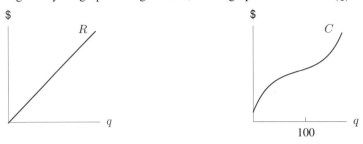

Figure 5.54: Total revenue and total cost for Example 3

Solution   The revenue graph is a line through the origin, with equation

$$R = pq$$

where p is the price, which is a constant. The slope is p and

$$MR = R'(q) = p.$$

The total cost is increasing, so the marginal cost is always positive (above the q-axis). For small q values, the total cost curve is concave down, so the marginal cost is decreasing. For larger q, say $q > 100$, the total cost curve is concave up and the marginal cost is increasing. Thus the marginal cost has a minimum at about $q = 100$. (See Figure 5.55.)

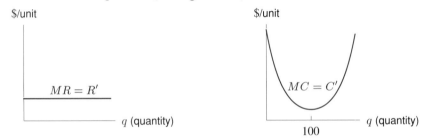

Figure 5.55: Marginal revenue and costs for Example 3

Maximizing Profit

Now let's look at how to maximize total profit, given functions for total revenue and total cost.

Example 4   Find the maximum profit if the total revenue and total cost are given, for $0 \leq q \leq 200$, by the curves R and C in Figure 5.56.

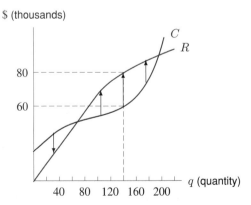

Figure 5.56: Maximum profit at $q = 140$

Solution   The profit is represented by the vertical difference between the curves and is marked by the vertical arrows on the graph. When revenue is below cost, the company is taking a loss; when revenue is above cost, the company is making a profit. We can see that the profit is largest at about $q = 140$, so this is the production level we're looking for. To be sure that the local maximum is a global maximum, we need to check the endpoints. At $q = 0$ and $q = 200$, the profit is negative, so the global maximum is indeed at $q = 140$.

To find the actual maximum profit, we estimate the vertical distance between the curves at $q = 140$. This gives a maximum profit of $80,000 - $60,000 = $20,000.

Suppose we wanted to find the minimum profit. In this example, we must look at the endpoints, when $q = 0$ or $q = 200$. We see the minimum profit is negative (a loss), and it occurs at $q = 0$.

Maximum Profit Occurs Where *MR = MC*

In Example 4, observe that at $q = 140$ the slopes of the two curves in Figure 5.56 are equal. To the left of $q = 140$, the revenue curve has a larger slope than the cost curve, and the profit increases as q increases. The company will make more money by producing more units, so production should increase toward $q = 140$. To the right of $q = 140$, the slope of the revenue curve is less than the slope of the cost curve, and the profit is decreasing. The company will make more money by producing fewer units so production should decrease toward $q = 140$. At the point where the slopes are equal, the profit has a local maximum; otherwise the profit could be increased by moving toward that point. Since the slopes are equal at $q = 140$, we have $MR = MC$ there.

Now let's look at the general situation. To maximize or minimize profit over an interval, we optimize the profit, π, where

$$\pi(q) = R(q) - C(q).$$

We know that global maxima and minima can only occur at critical points or at endpoints of an interval. To find critical points of π, look for zeros of the derivative:

$$\pi'(q) = R'(q) - C'(q) = 0.$$

So

$$R'(q) = C'(q),$$

that is, the slopes of the revenue and cost curves are equal. This is the same observation that we made in the previous example. In economic language,

> The maximum (or minimum) profit can occur where
>
> Marginal cost = Marginal revenue.

Of course, maximal or minimal profit does not *have* to occur where $MR = MC$; there are also the endpoints to consider.

Example 5   Find the quantity q which maximizes profit if the total revenue and total cost are given by

$$R(q) = 5q - 0.003q^2$$
$$C(q) = 300 + 1.1q,$$

where $0 \leq q \leq 800$ units and $R(q)$ and $C(q)$ are in dollars. What production level gives the minimum profit?

Solution We look for production levels that give marginal revenue = marginal cost:

$$MR = R'(q) = 5 - 0.006q$$
$$MC = C'(q) = 1.1.$$

So $5 - 0.006q = 1.1$, or

$$q = 3.9/0.006 = 650 \text{ units.}$$

Does this value of q represent a local maximum or minimum of π? We can tell by looking at production levels of 649 units and 651 units. When $q = 649$ we have $MR = \$1.106$, which is greater than the (constant) marginal cost of \$1.1. This means that producing one more unit will bring in more revenue than its cost, so profit will increase. When $q = 651$, $MR = \$1.094$, which is *less* than MC, so it is not profitable to produce the 651st unit. We conclude that $q = 650$ is a local maximum for the profit function π. The profit earned by producing and selling this quantity is $\pi(650) = R(650) - C(650) = \$967.50$.

To check for global maxima we need to look at the endpoints. If $q = 0$, the only cost is \$300 (the fixed costs) and there is no revenue, so $\pi(0) = -300$. At the upper limit of $q = 800$, we have $\pi(800) = \$900$. Therefore, the maximum profit is at the production level of 650 units, where $MR = MC$. The minimum profit (a loss) occurs when $q = 0$ and there is no production at all.

Problems for Section 5.4

1. A manufacturer reports the cost and revenue functions shown in Figure 5.57. Sketch graphs, as a function of quantity, of
    (a) Total profit   (b) Marginal cost   (c) Marginal revenue
    Label the points q_1 and q_2 on your graphs.

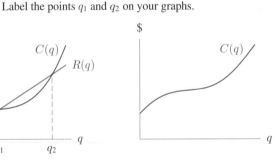

| Figure 5.57 | Figure 5.58 | Figure 5.59 |

2. Assume a manufacturer's cost of producing a good is given by the graph of $C(q)$ in Figure 5.57. Assume also that the manufacturer can sell the product for a price p each (regardless of the quantity sold), so that the total revenue from selling a quantity q is $R(q) = pq$.

    (a) The difference $\pi(q) = R(q) - C(q)$ is the total profit. For which quantity q_0 is the profit a maximum? Mark your answer on a sketch of the graph.
    (b) What is the relationship between p and $C'(q_0)$? Explain your result both graphically and analytically. What does this mean in terms of economics? (Note that p is the slope of the line $R(q) = pq$. Note also that $\pi(q)$ has a maximum at $q = q_0$, so $\pi'(q_0) = 0$.)
    (c) Graph $C'(q)$ and p (as a horizontal line) on the same axes. Mark q_0 on the q-axis.

3. Let $C(q)$ be the total cost of producing a quantity q of a certain good. (See Figure 5.58.)

    (a) What is the meaning of $C(0)$?
    (b) Describe in words how the marginal cost changes as the quantity produced increases.
    (c) Explain the concavity of the graph (in terms of economics).
    (d) Explain the economic significance (in terms of marginal cost) of the point at which the concavity changes.
    (e) Do you expect the graph of $C(q)$ to look like this for all types of goods?

4. The marginal revenue and marginal cost for a certain item are graphed in Figure 5.59. Do the following quantities maximize revenue for the company? Explain your answer.

   (a) $q = a$ (b) $q = b$

5. Suppose the total cost $C(q)$ of producing q goods is given by:

$$C(q) = 0.01q^3 - 0.6q^2 + 13q.$$

   (a) What is the fixed cost?
   (b) What is the maximum profit if each item is sold for \$7? (Assume you sell everything you produce.)
   (c) Suppose exactly 34 goods are produced. They all sell when the price is \$7 each, but for each \$1 increase in price, 2 fewer goods are sold. Should the price be raised, and if so by how much?

6. Suppose a company manufactures only one product. The quantity, q, of this product produced per month depends on the amount of capital, K, invested (i.e., the number of machines the company owns, the size of its building, and so on) and the amount of labor, L, available each month. It is often assumed that q can be expressed as a function of K and L by a *Cobb-Douglas production function*:

$$q = cK^\alpha L^\beta$$

   where c, α, β are positive constants, with $0 < \alpha < 1$ and $0 < \beta < 1$.

   In this problem we will see how the Russian government could use a Cobb-Douglas function to estimate how many people a newly privatized industry might employ. A company in such an industry will have only a small amount of capital available to it, and will need to use all of it; K is therefore fixed. Suppose L is measured in man-hours per month, and that each man-hour costs the company w rubles (a ruble is the unit of Russian currency). Suppose that the company has no other costs besides labor, and that each unit of the good can be sold for a fixed price of p rubles. How many man-hours of labor per month should the company use in order to maximize its profit?

7. An agricultural worker in Uganda is interested in planting clover to increase the number of bees making their home in the region. There are 100 bees in the region naturally, and for every acre put under clover, 20 more bees are found in the region.

   (a) Draw a graph of the total number, $N(x)$, of bees as a function of x, the number of acres devoted to clover.
   (b) Explain, both geometrically and algebraically, the shape of the graph of:

       (i) The marginal rate of increase of the number of bees with acres of clover, $N'(x)$.

       (ii) The average number of bees per acre of clover, $N(x)/x$.

8. If you invest x dollars in a certain project, your return is $R(x)$. Suppose you want to choose x to maximize your return per dollar invested,[4] which is

$$r(x) = \frac{R(x)}{x}.$$

   (a) Suppose the graph of $R(x)$ has the form in Figure 5.60 on page 270, with $R(0) = 0$. Illustrate on a copy of this graph that the maximum value of $r(x)$ is obtained at a point on the graph of $R(x)$ at which the line from the origin to the point is tangent to the graph.
   (b) It is also true that the maximum of $r(x)$ should occur at a point at which the slope of the graph of $r(x)$ is zero. On the same set of axes as part (a), draw a rough version of the graph of $r(x)$, corresponding to your R graph, and illustrate that the maximum occurs where the slope is 0.
   (c) Show, by taking the derivative of the preceding formula for $r(x)$, that the conditions in part (a) and (b) are equivalent: the point where the line from the origin is tangent to the graph of R is the same point where the graph of r has zero slope.

---

[4]From Peter D. Taylor, *Calculus: The Analysis of Functions* (Toronto: Wall & Emerson, 1992).

Figure 5.60

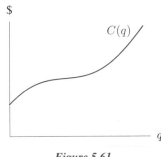

Figure 5.61

Problems 9–11 involve the *average cost* of manufacturing a quantity q of a good, which is defined to be

$$a(q) = \frac{C(q)}{q}.$$

9. Figure 5.61 shows the cost of production, $C(q)$, as a function of quantity produced, q.

   (a) For some q_0, sketch a line whose slope is the marginal cost, MC, at that point.
   (b) For the same q_0, explain why the average cost $a(q_0)$ can be represented by the slope of the line from that point on the curve to the origin.
   (c) Use the method of Example 3 on page 257 to explain why at the value of q which minimizes $a(q)$, the average and marginal costs are equal.

10. Suppose a firm produces a quantity q of some good and that the average cost per item is given by

$$a(q) = 0.01q^2 - 0.6q + 13, \quad \text{for} \quad q > 0.$$

   (a) What is the total cost, $C(q)$, of producing q goods?
   (b) What is the minimum marginal cost? What is the practical interpretation of this result?
   (c) At what production level is the average cost a minimum? What is the lowest average cost?
   (d) Compute the marginal cost at $q = 30$. How does this relate to your answer to part (c)? Explain this relationship both analytically and in words.

11. A reasonably realistic model of a firm's costs is given by the *short-run Cobb-Douglas cost curve*

$$C(q) = Kq^{1/a} + F,$$

   where a is a positive constant, F is the fixed cost, and K measures the technology available to the firm.

   (a) Show that C is concave down if $a > 1$.
   (b) Assuming that $a < 1$ and that average cost is minimized when average cost equals marginal cost, find what value of q minimizes the average cost.

5.5 MORE OPTIMIZATION: INTRODUCTION TO MODELING

Finding global maxima and minima is made much easier by having a formula for the function to be maximized or minimized. The process of translating a problem into a function whose formula we know is called *mathematical modeling*. By working through the examples that follow, you will get the flavor of some kinds of modeling.

Example 1 What are the dimensions of an aluminum can that holds 40 in³ of juice and that uses the least material (i.e., aluminum)? Assume that the can is cylindrical, and is capped on both ends.

Solution   It is often a good idea to think about a problem in general terms before trying to solve it. Since we're trying to use as little material as possible, why not make the can very small, say, the size of a peanut? We can't, since the can must hold 40 in^3. If we make the can short, to try to use less material in the sides, we'll have to make it fat as well, so that it can hold 40 in^3. Saving aluminum in the sides might actually use more aluminum in the top and bottom in a short, fat can. See Figure 5.62(a).

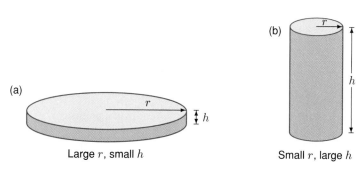

TABLE 5.1   *Amount of material, M, used in the can for various choices of radius, r, and height, h*

r (in)	h (in)	M (in^2)
0.2	318.31	400.25
1.0	12.73	86.27
2.0	3.18	65.09
3.0	1.41	83.13
4.0	0.80	120.64
10.0	0.13	636.49

Figure 5.62: Various cylindrical-shaped cans

If we try to save material by making the top and bottom small, the can has to be tall to accommodate the 40 in^3 of juice. So any savings we get by using a small top and bottom might be outweighed by the height of the sides. See Figure 5.62(b). This is, in fact, true, as Table 5.1 shows.

The table gives the amount of material used in the can for some choices of the radius, r, and the height, h. You can see that r and h change in opposite directions, and that more material is used at the extremes (very large or very small r and h) than in the middle. From the table it appears that the optimal radius for the can lies somewhere in $1.0 \leq r \leq 3.0$. If we consider the material used, M, as a function of the radius, r, a graph of this function looks like Figure 5.63. The graph shows that the global minimum we want is at a critical point.

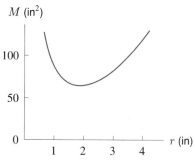

Figure 5.63: Total material used in can, M, as a function of radius, r

Both the table and the graph were obtained from a mathematical model, which in this case is a formula for the material used in making the can. Finding such a formula depends on knowing the geometry of a cylinder, in particular its area and volume. We have

$$M = \text{Material used in the can} = \text{Material in ends} + \text{Material in the side}$$

where

$$\text{Material in ends} = 2 \cdot \text{Area of a circle with radius } r = 2 \cdot \pi r^2,$$
$$\text{Material in the side} = \text{Area of cylinder with height } h \text{ and radius } r = 2\pi r h.$$

However, h is not independent of r: if r grows, h shrinks, and vice-versa. To find the relationship, we use the fact that the volume of the cylinder, $\pi r^2 h$, is equal to the constant 40 in^3:

$$\pi r^2 h = 40, \quad \text{giving} \quad h = \frac{40}{\pi r^2}.$$

This means

$$\text{Material in the side} = 2\pi r h = 2\pi r \frac{40}{\pi r^2} = \frac{80}{r}.$$

Thus we obtain the formula for the total material used in a can of radius r:

$$M(r) = 2\pi r^2 + \frac{80}{r}.$$

The domain of this function is all $r > 0$.

Now we use calculus to find the minimum of M. We look for critical points:

$$\frac{dM}{dr} = 4\pi r - \frac{80}{r^2} = 0 \quad \text{at a critical point,} \quad \text{so} \quad 4\pi r = \frac{80}{r^2}.$$

Therefore,

$$\pi r^3 = 20, \quad \text{so} \quad r = \left(\frac{20}{\pi}\right)^{1/3} \approx 1.85 \text{ inches,}$$

which agrees with our graph. We also have

$$h = \frac{40}{\pi r^2} \approx \frac{40}{\pi(1.85)^2} \approx 3.7 \text{ inches.}$$

Thus, the material used, $M(1.85)$, is about 64.7 in^2.

Practical Tips for Modeling Optimization Problems

1. Make sure that you know what quantity or function is to be optimized.

2. If possible, make several sketches showing how the elements that vary are related. Label your sketches clearly by assigning variables to quantities which change.

3. Try to obtain a formula for the function to be optimized in terms of the variables that you identified in the previous step. If necessary, eliminate from this formula all but one variable. Identify the domain over which this variable varies.

4. Find the critical points and evaluate the function at these points and the endpoints to find the global maxima and minima.

The next example, another problem in geometry, illustrates this approach.

Example 2   Alaina wants to get to the bus stop as quickly as possible. The bus stop is across a grassy park, 2000 feet west and 600 feet north of her starting position. Alaina can walk west along the edge of the park on the sidewalk at a speed of 6 ft/sec. She can also travel through the grass in the park, but only at a rate of 4 ft/sec (the park is a favorite place to walk dogs, so she must move with care). What path will get her to the bus stop the fastest?

Solution

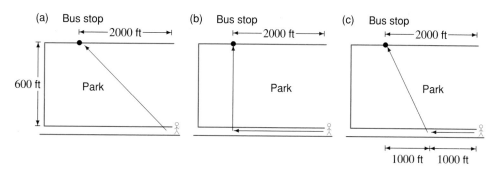

Figure 5.64: Three possible paths to the bus stop

We might first think that she should take a path that is the shortest distance. Unfortunately, the path that follows the shortest distance to the bus stop is entirely in the park, where her speed is slow. (See Figure 5.64(a).) That distance is $\sqrt{2000^2 + 600^2} \approx 2100$ feet, which will take her about 525 seconds to traverse. She could instead walk quickly the entire 2000 feet along the sidewalk, which would leave her just the 600-foot northward journey through the park. (See Figure 5.64(b).) This method would take $2000/6 + 600/4 \approx 483$ seconds total walking time.

But can she do even better? Perhaps another combination of sidewalk and park will give a shorter travel time. For example, what is the travel time if she walks 1000 feet west along the sidewalk and the rest of the way through the park? (See Figure 5.64(c).) The answer is about 458 seconds.

We make a model for this problem. We label the distance that Alaina walks west along the sidewalk x and the distance she walks through the park y, as in Figure 5.65. Then the total time, t, will be

$$t = t_{\text{sidewalk}} + t_{\text{park}}.$$

Since

$$\text{Time} = \text{Distance}/\text{Speed},$$

and she can walk 6 ft/sec on the sidewalk and 4 ft/sec in the park, we have that

$$t = \frac{x}{6} + \frac{y}{4}.$$

Now, by the Pythagorean Theorem, $y = \sqrt{(2000 - x)^2 + 600^2}$. Therefore

$$t = \frac{x}{6} + \frac{\sqrt{(2000 - x)^2 + 600^2}}{4}.$$

We can find the critical points of this function algebraically. (See Problem 1 on page 275.) Alternatively, we can graph the function on a calculator and estimate the critical point, which is $x \approx 1463$ feet. This gives a minimum total time of about 445 seconds.

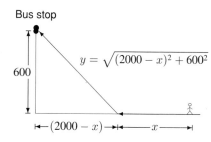

Figure 5.65: Modeling time to bus stop

Example 3   One hallway which is 4 feet wide meets another hallway which is 8 feet wide in a right angle. (See Figure 5.66.) What is the length of the longest ladder which can be carried horizontally around the corner?

Solution   We imagine the ladder being carried on its side and ignore its width. To allow the longest ladder possible we carry the ladder around the corner so that it just touches both walls (at A and C) and just touches the corner at B. Let's draw some lines that do this. (See Figure 5.66.) The length of the line $\overline{ABC}$ decreases as the corner is turned, then increases again. The minimum such length would be the length of the longest ladder that could make it around the corner. Certainly a smaller ladder would work (it wouldn't touch A, B, and C simultaneously), but a larger one would not fit.

Figure 5.66: Various ladders that touch both walls and corner   *Figure 5.67:* Ladder and hallway

So we want the *smallest* length of the line $\overline{ABC}$. We will express the length, l, as a function of θ, the angle between the line and the wall of the narrow hall. (See Figure 5.67.)

We have
$$l = \overline{AB} + \overline{BC}.$$
Now $\overline{AB} = 4/\sin\theta$ and $\overline{BC} = 8/\cos\theta$, so
$$l = \frac{4}{\sin\theta} + \frac{8}{\cos\theta}.$$
Here the domain of θ is $0 < \theta < \pi/2$.

Now we differentiate:
$$\frac{dl}{d\theta} = -\frac{4}{(\sin\theta)^2}(\cos\theta) - \frac{8}{(\cos\theta)^2}(-\sin\theta).$$

To minimize l, we solve $dl/d\theta = 0$:
$$-4\frac{\cos\theta}{(\sin\theta)^2} + 8\frac{\sin\theta}{(\cos\theta)^2} = 0,$$
so
$$2(\sin\theta)^3 = (\cos\theta)^3, \quad \text{or} \quad \frac{(\sin\theta)^3}{(\cos\theta)^3} = \frac{1}{2}.$$

This means
$$\tan\theta = \sqrt[3]{0.5} \approx 0.79, \quad \text{so} \quad \theta \approx 0.67 \text{ radians}.$$

Thus $\theta \approx 0.67$ is a critical point, and an investigation of what happens for θ near 0 and θ near $\pi/2$ shows that l has a global minimum at $\theta = 0.67$. The minimum value of l is therefore
$$l = \frac{4}{\sin(0.67)} + \frac{8}{\cos(0.67)} \approx 16.65 \text{ feet},$$
so the ladder can be at most 16.65 feet long and still make it around the corner.

Problems for Section 5.5

1. Find analytically the exact critical point of the function which represents the time, t, to walk to the bus stop in Example 2. Recall that t is given by

$$t = \frac{x}{6} + \frac{\sqrt{(2000 - x)^2 + 600^2}}{4}.$$

2. A smokestack deposits soot on the ground with a concentration inversely proportional to the square of the distance from the stack. With two smokestacks 20 miles apart, the concentration of the combined deposits on the line joining them, at a distance x from one stack, is given by

$$S = \frac{k_1}{x^2} + \frac{k_2}{(20 - x)^2}$$

   where k_1 and k_2 are positive constants which depend on the quantity of smoke each stack is emitting. If $k_1 = 7k_2$, find the point on the line joining the stacks where the concentration of the deposit is a minimum.

3. A wave of wavelength λ traveling in deep water has speed, v, given by

$$v = k\sqrt{\frac{\lambda}{c} + \frac{c}{\lambda}},$$

   where c and k are positive constants. As λ varies, does such a wave have a maximum or minimum velocity? If so, what is it? Explain.

4. If you have 100 feet of fencing and want to enclose a rectangular area up against a long, straight wall, what is the largest area you can enclose?

5. A closed box has a fixed surface area A and a square base with side x.

   (a) Find a formula for its volume, V, as a function of x.
   (b) Sketch a graph of V against x.
   (c) Find the maximum value of V.

6. A landscape architect plans to enclose a 3000 square foot rectangular region in a botanical garden. She will use shrubs costing \$25 per foot along three sides and fencing costing \$10 per foot along the fourth side. Find the minimum total cost.

7. A rectangular swimming pool is to be built with an area of 1800 square feet. The owner wants 5-foot wide decks along either side and 10-foot wide decks at the two ends. Find the dimensions of the smallest piece of property on which the pool can be built satisfying these conditions.

8. Suppose a rectangular beam is cut from a cylindrical log of radius 30 cm. The strength of a beam of width w and height h is proportional to wh^2. (See Figure 5.68.) Find the width and height of the beam of maximum strength.

9. A light is suspended at a height h above the floor. (See Figure 5.69.) The illumination at the point P is inversely proportional to the square of the distance from the point P to the light and directly proportional to the cosine of the angle θ. How far from the floor should the light be to maximize the illumination at the point P?

Figure 5.68

Figure 5.69

Figure 5.70

10. The cross-section of a tunnel is a rectangle of height h surmounted by a semicircular roof section of radius r (See Figure 5.70 on page 275). If the cross-sectional area is A, determine the dimensions of the cross section which minimize the perimeter.

11. Which point on the parabola $y = x^2$ is nearest to $(1, 0)$? Find the coordinates to two decimals. [Hint: Minimize the square of the distance—this avoids square roots.]

12. Find the coordinates of the point on the parabola $y = x^2$ which is closest to the point $(3,0)$.

13. Of all rectangles with given area, A, which has the shortest diagonals?

14. Suppose you run a small independent furniture business. Your assistant signs a deal with a customer to deliver up to 400 chairs, the exact number to be determined by the customer later. The price will be $90 per chair up to 300 chairs, and above 300, the price will be reduced by $0.25 per chair (on the whole order) for every additional chair over 300 ordered. What are the largest and smallest revenues your company can make under this deal?

15. The cost of fuel to propel a boat through the water (in dollars per hour) is proportional to the cube of the speed. A certain ferry boat uses $100 worth of fuel per hour when cruising at 10 miles per hour. Apart from fuel, the cost of running this ferry (labor, maintenance, and so on) is $675 per hour. At what speed should it travel so as to minimize the cost *per mile* traveled?

16. (a) For which positive number x is $x^{1/x}$ largest? Justify your answer.
    [Hint: You may want to write $x^{1/x} = e^{\ln(x^{1/x})}$.]
    (b) For which positive integer n is $n^{1/n}$ largest? Justify your answer.
    (c) Use your answer to parts (a) and (b) to decide which is larger: $3^{1/3}$ or $\pi^{1/\pi}$.

17. The *arithmetic mean* of two numbers a and b is defined as $(a+b)/2$; the *geometric mean* of two positive numbers a and b is defined as $\sqrt{ab}$.

    (a) For two positive numbers, which of the two means is larger? Justify your answer.
    [Hint: Define $f(x) = (a+x)/2 - \sqrt{ax}$ for fixed a.]
    (b) For three positive numbers a, b, c, the arithmetic and geometric mean are $(a+b+c)/3$ and $\sqrt[3]{abc}$, respectively. Which of the two means of three numbers is larger?
    [Hint: Redefine $f(x)$ for fixed a *and* b.]

18. On the same side of a straight river are two towns, and the townspeople want to build a pumping station, S, that supplies water to them. The pumping station is to be at the river's edge with pipes extending straight to the two towns. The distances are shown in Figure 5.71. Where should the pumping station be located to minimize the total length of pipe?

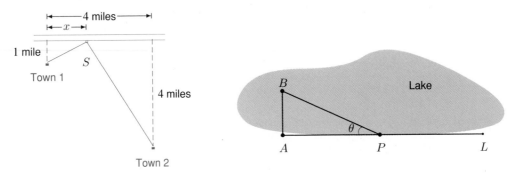

Figure 5.71   *Figure 5.72*

19. Assume that a pigeon is released from a boat (point B in Figure 5.72) floating on a lake. Because of falling air over the cool water, the energy required to fly one meter over the lake is twice the corresponding energy e required for flying over the bank ($e = 3$ joule/meter). To minimize the energy required to fly from B to the loft, L, the pigeon heads to a point P on the bank and then flies along the bank to L. The distance $\overline{AL}$ is 2000 m, and $\overline{AB}$ is 500 m.

    (a) Express the energy required to fly from B to L via P as a function of the angle θ (the angle BPA).
    (b) What is the optimal angle θ?
    (c) Does your answer change if $\overline{AL}$, $\overline{AB}$, and e have different numerical values?

20. To get the best view of the Statue of Liberty in Figure 5.73, you should be at the position where θ is a maximum. If the statue stands 92 meters high, including the pedestal, which is 46 meters high, how far from the base should you be? [Hint: Find a formula for θ in terms of your distance from the base. Use this function to maximize θ, noting that $0 \leq \theta \leq \pi/2$.]

Figure 5.73

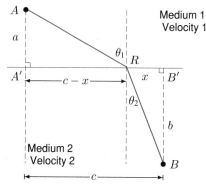

Figure 5.74

21. When a ray of light travels from one medium to another (for example, from air to water), it changes direction. This phenomenon is known as *refraction*. In Figure 5.74, light is traveling from A to B. The amount of refraction depends on the velocities, v_1 and v_2, of light in the two media and on *Fermat's Principle* which states that the light's time of travel, T, from A to B, is a minimum.

   (a) Find an expression for T in terms of x and the constants a, b, v_1, v_2, and c.
   (b) Show that if R is chosen so that the time of travel is minimized, then

$$\frac{\sin \theta_1}{\sin \theta_2} = \frac{v_1}{v_2}.$$

   This result is known as *Snell's Law* and the ratio v_1/v_2 is called the *index of refraction* of the second medium with respect to the first medium.

22. Show that when the value of x in Figure 5.74 is chosen according to Snell's Law, the time taken by the light ray is a minimum.

5.6 HYPERBOLIC FUNCTIONS

There are two combinations of e^x and e^{-x} which are used so often in engineering that they are given their own name. They are the *hyperbolic sine*, abbreviated sinh, and the *hyperbolic cosine*, abbreviated cosh. They are defined as follows:

Hyperbolic Functions

$$\cosh x = \frac{e^x + e^{-x}}{2} \qquad \sinh x = \frac{e^x - e^{-x}}{2}$$

Properties of Hyperbolic Functions

The graphs of the $\cosh x$ and $\sinh x$ are given in Figures 5.75 and 5.76 together with the graphs of multiples of e^x and e^{-x}. The graph of $\cosh x$ is called a *catenary*; it is the shape of a hanging cable.

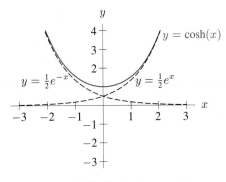

Figure 5.75: Graph of $y = \cosh x$ *Figure 5.76:* Graph of $y = \sinh x$

The graphs suggest that the following results hold:

$$\cosh 0 = 1 \qquad \sinh 0 = 0$$

$$\cosh(-x) = \cosh x \qquad \sinh(-x) = -\sinh x$$

To show that the hyperbolic functions really do have these properties, we use their formulas.

Example 1   Show that (a) $\cosh(0) = 1$ (b) $\cosh(-x) = \cosh x$

Solution   (a)   Substituting $x = 0$ into the formula for $\cosh x$ gives

$$\cosh 0 = \frac{e^0 + e^{-0}}{2} = \frac{1 + 1}{2} = 1.$$

(b)   Substituting $-x$ for x gives

$$\cosh(-x) = \frac{e^{-x} + e^{-(-x)}}{2} = \frac{e^{-x} + e^x}{2} = \cosh x.$$

Thus, we know that $\cosh x$ is an even function.

Example 2   Describe and explain the behavior of $\cosh x$ as $x \to \infty$ and $x \to -\infty$.

Solution   From Figure 5.75, it appears that as $x \to \infty$, the graph of $\cosh x$ resembles the graph of $\frac{1}{2}e^x$. Similarly, as $x \to -\infty$, the graph of $\cosh x$ resembles the graph of $\frac{1}{2}e^{-x}$. This behavior is explained by using the formula for $\cosh x$ and the facts that $e^{-x} \to 0$ as $x \to \infty$ and $e^x \to 0$ as $x \to -\infty$:

$$\text{As } x \to \infty, \qquad \cosh x = \frac{e^x + e^{-x}}{2} \to \frac{1}{2}e^x.$$

$$\text{As } x \to -\infty, \qquad \cosh x = \frac{e^x + e^{-x}}{2} \to \frac{1}{2}e^{-x}.$$

Identities Involving cosh x and sinh x

The reason the hyperbolic functions have names that remind us of the trigonometric functions is that they share some similar properties. A familiar identity for trigonometric functions is

$$(\cos x)^2 + (\sin x)^2 = 1.$$

To discover an analogous identity relating $(\cosh x)^2$ and $(\sinh x)^2$, we first calculate

$$(\cosh x)^2 = \left(\frac{e^x + e^{-x}}{2}\right)^2 = \frac{e^{2x} + 2e^x e^{-x} + e^{-2x}}{4} = \frac{e^{2x} + 2 + e^{-2x}}{4}$$

$$(\sinh x)^2 = \left(\frac{e^x - e^{-x}}{2}\right)^2 = \frac{e^{2x} - 2e^x e^{-x} + e^{-2x}}{4} = \frac{e^{2x} - 2 + e^{-2x}}{4}.$$

If we add these expressions, the resulting right-hand side contains terms involving both e^{2x} and e^{-2x}. If, however, we subtract the expressions for $(\cosh x)^2$ and $(\sinh x)^2$, we obtain a simple result:

$$(\cosh x)^2 - (\sinh x)^2 = \frac{e^{2x} + 2 + e^{-2x}}{4} - \frac{e^{2x} - 2 + e^{-2x}}{4} = \frac{4}{4} = 1.$$

Thus, writing $\cosh^2 x$ for $(\cosh x)^2$ and $\sinh^2 x$ for $(\sinh x)^2$, we have the identity

$$\boxed{\cosh^2 x - \sinh^2 x = 1}$$

The Hyperbolic Tangent

Extending the analogy to the trigonometric functions, we define

$$\boxed{\tanh x = \frac{\sinh x}{\cosh x}}$$

Derivatives of Hyperbolic Functions

We calculate the derivatives using the fact that $\frac{d}{dx}(e^x) = e^x$. The results are again reminiscent of the trigonometric functions. For example,

$$\frac{d}{dx}(\cosh x) = \frac{d}{dx}\left(\frac{e^x + e^{-x}}{2}\right) = \frac{e^x - e^{-x}}{2} = \sinh x.$$

We find $\frac{d}{dx}(\sinh x)$ similarly, giving the following results:

$$\boxed{\frac{d}{dx}(\cosh x) = \sinh x \qquad \frac{d}{dx}(\sinh x) = \cosh x}$$

Example 3   Compute the derivative of $\tanh x$.

Solution   Using the quotient rule gives

$$\frac{d}{dx}(\tanh x) = \frac{d}{dx}\left(\frac{\sinh x}{\cosh x}\right) = \frac{(\cosh x)^2 - (\sinh x)^2}{(\cosh x)^2} = \frac{1}{\cosh^2 x}.$$

Problems for Section 5.6

1. Show that $\sinh 0 = 0$.

2. Show that $\sinh(-x) = -\sinh(x)$.

3. Describe and explain the behavior of $\sinh x$ as $x \to \infty$ and as $x \to -\infty$.

4. Is there an identity analogous to $\sin(2x) = 2 \sin x \cos x$ for the hyperbolic functions? Explain.

5. Is there an identity analogous to $\cos(2x) = \cos^2 x - \sin^2 x$ for the hyperbolic functions? Explain.

6. Show that $\dfrac{d}{dx}(\sinh x) = \cosh x$.

Find the derivatives of the functions in Problems 7–11.

7. $y = \cosh(2x)$

8. $y = \sinh(3z + 5)$

9. $f(t) = \cosh(e^{t^2})$

10. $f(y) = \sinh\left(\sinh(3y)\right)$

11. $g(\theta) = \ln\left(\cosh(1 + \theta)\right)$

12. Consider the family of functions $y = a \cosh(x/a)$ for $a > 0$. Sketch graphs for $a = 1, 2, 3$. Describe in words the effect of increasing a.

13. (a) Using a calculator or computer, sketch the graph of $y = 2e^x + 5e^{-x}$ for $-3 \le x \le 3, 0 \le y \le 20$. Observe that it looks like the graph of $y = \cosh x$. Approximately where is its minimum?

    (b) Show algebraically that $y = 2e^x + 5e^{-x}$ can be written in the form $y = A \cosh(x - c)$. Calculate the values of A and c. Explain what this tells you about the graph in part (a).

14. The following problem is a generalization of Problem 13. Show that any function of the form

    $$y = Ae^x + Be^{-x}, \quad A > 0,\ B > 0,$$

    can be written, for some K and c, in the form

    $$y = K \cosh(x - c).$$

    What does this tell you about the graph of $y = Ae^x + Be^{-x}$?

15. Consider the family of functions of the form $y = Ae^x + Be^{-x}$, for any constants A, B.

    (a) Sketch the function for

(i) $A = 1,\ B = 1$	(ii) $A = 1,\ B = -1$	(iii) $A = 2,\ B = 1$
(iv) $A = 2,\ B = -1$	(v) $A = -2,\ B = -1$	(vi) $A = -2,\ B = 1$

    (b) Describe in words the general shape of the graph if A and B have the same sign. What effect does the sign of A have on the graph?

    (c) Describe in words the general shape of the graph if A and B have different signs. What effect does the sign of A have on the graph?

    (d) For what values of A and B does the function have a local maximum? A local minimum? Justify your answer using derivatives.

16. The cable between the two towers of a suspension bridge hangs in the shape of the curve

    $$y = \frac{T}{w} \cosh\left(\frac{wx}{T}\right),$$

    where T is the tension in the cable at its lowest point and w is the weight of the cable per unit length. This curve is called a *catenary*.

    (a) Suppose the cable stretches between the points $x = -T/w$ and $x = T/w$. Find an expression for the "sag" in the cable. (That is, find the difference between the height of the cable at the highest and lowest points.)

    (b) Show that the shape of the cable satisfies the differential equation

    $$\frac{d^2y}{dx^2} = \frac{w}{T}\sqrt{1 + \left(\frac{dy}{dx}\right)^2}.$$

17.   The Saint Louis arch can be approximated by using a function of the form $y = b - a \cosh(x/a)$. Putting the origin on the ground in the center of the arch and the y-axis upward, find an approximate equation for the arch given the dimensions shown in Figure 5.77. (In other words, find a and b.)

Figure 5.77

CHAPTER SUMMARY

- **Local extrema**
    Maximum, minimum, critical point, tests for local maxima/minima
- **Using the second derivative**
    Concavity, inflection point
- **Families of curves**
- **Optimization**
    Global extremum, modeling problems, graphical optimization, upper and lower bounds
- **Marginality**
    Cost/revenue functions, marginal cost/marginal revenue functions
- **Hyperbolic functions**

REVIEW PROBLEMS FOR CHAPTER FIVE

For Problems 1–2, indicate all critical points on the given graphs. Determine which correspond to local minima, local maxima, global maxima, global minima, or none of these. (Note that the graphs are on closed intervals.)

1.

Figure 5.78

2.

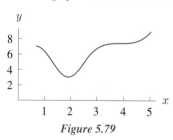

Figure 5.79

For the graphs of f' in Problems 3–6 decide:
(a)   Over what intervals is f increasing? Decreasing?
(b)   Does f have maxima or minima? If so, which, and where?

3.

4.

5.

6.

For each of the functions in Problems 7–10, do the following:

(a)   Find f' and f''.

(b)   Find the critical points of f.

(c)   Find any inflection points.

(d)   Evaluate f at the critical points and the endpoints. Identify the global maxima and minima of f.

(e)   Sketch f. Indicate clearly where f is increasing or decreasing, and its concavity.

7. $f(x) = x^3 - 3x^2$ $(-1 \leq x \leq 3)$

8. $f(x) = x + \sin x$ $(0 \leq x \leq 2\pi)$

9. $f(x) = e^{-x} \sin x$ $(0 \leq x \leq 2\pi)$

10. $f(x) = x^{-2/3} + x^{1/3}$ $(1.2 \leq x \leq 3.5)$

For each of the functions in Problems 11–13, find the limits as x tends to $+\infty$ and $-\infty$, and then proceed as in Problems 7–10 (That is, find f', etc.).

11. $f(x) = 2x^3 - 9x^2 + 12x + 1$

12. $f(x) = \dfrac{4x^2}{x^2 + 1}$

13. $f(x) = xe^{-x}$

14.   Determine local maxima and minima and points of inflection of $e^{-x^2/2}$. Sketch a graph.

For each of the functions in Problems 15–20, use derivatives to identify local maxima and minima and points of inflection. Sketch a graph of the function. Confirm your answers using a calculator or computer.

15. $f(x) = x^3 + 3x^2 - 9x - 15$ **16.** $f(x) = x^5 - 15x^3 + 10$ **17.** $f(x) = x - 2 \ln x$ for $x > 0$

18. $f(x) = x^2 e^{5x}$

19. $f(x) = e^{-x^2}$

20. $f(x) = \dfrac{x^2}{x^2 + 1}$

21.   On the graph of the derivative function f' in Figure 5.80, indicate the x-values that are critical points of the function f itself. Are they local maxima, local minima, or neither?

Figure 5.80: Graph of f', not f

Figure 5.81

22.   For the function, f, given in the graph in Figure 5.81:

(a)   Sketch $f'(x)$.

(b)   Where does $f'(x)$ change its sign?

(c)   Where does $f'(x)$ have local maxima or minima?

23.   Using your answer to Problem 22 as a guide, write a short paragraph (using complete sentences) which describes the relationships between the following features of a function f:

(a)   The local maxima and minima of f.

(b)   The points at which the graph of f changes concavity.

(c)   The sign changes of f'.

(d)   The local maxima and minima of f'.

24. A square-bottomed box with a top has a fixed volume, V. What dimensions minimize the surface area?

25. A square-bottomed box with no top has a fixed volume, V. What dimensions minimize the surface area?

Find the best possible bounds for the functions in Problems 26–27.

26. $e^{-x}\sin x$, for $x \geq 0$

27. $x\sin x$, for $0 \leq x \leq 2\pi$

28. Sketch several members of the family $y = x^3 - ax^2$ on the same axes. Show that the critical points lie on the curve $y = -\frac{1}{2}x^3$.

29. Suppose $g(t) = (\ln t)/t$ for $t > 0$.

(a) Does g have either a global maximum or a global minimum on $0 < t < \infty$? If so, where, and what are their values?

(b) What does your answer to part (a) tell you about the number of solutions to the equation

$$\frac{\ln x}{x} = \frac{\ln 5}{5}?$$

(Note: There are many ways to investigate the number of solutions to this equation. We are asking you to draw a conclusion from your answer to part (a).)

(c) Estimate the solution(s).

30. Populations with a growth limit have been modeled with the logistic family

$$y = \frac{A}{1 + Be^{-Cx}} \quad \text{for} \quad -\infty < x < \infty \quad \text{and} \quad A, B, C > 0.$$

(a) Sketch a graph of $g(x) = A/\left(1 + e^{-Cx}\right)$. What is the significance of the parameter A?

(b) Show that $g(-x) + g(x) = A$. What does this result mean about the graphs of $g(x)$ and $g(-x)$?

(c) What happens to the graph of g if A is kept constant and C is increased?

(d) Show that the curve $y = A/\left(1 + Be^{-Cx}\right)$ is a horizontal shift of the graph of g.

31. For $a > 0$, the line

$$a(a^2 + 1)y = a - x$$

forms a triangle in the first quadrant with the x- and y-axes.

(a) Find, in terms of a, the x- and y-intercepts of the line.

(b) Find the area of the triangle, as a function of a.

(c) Find the value of a making the area a maximum.

(d) What is this greatest area?

(e) If you want the triangle to have area $1/5$, what choices do you have for a?

32. A ship is steaming due north at 12 knots (1 knot = 1.85 kilometers/hour) and sights a large tanker 3 kilometers away northwest steaming at 15 knots due east. For reasons of safety, the ships want to maintain a distance of at least 100 meters between them. Use a calculator or computer to determine the shortest distance between them if they remain on their current headings, and hence decide if they need to change course.

33. Consider the vase in Figure 5.82 on page 284. Assume the vase is filled with water at a constant rate (i.e., constant volume per unit time).

(a) Graph $y = f(t)$, the depth of the water, against time, t. Show on your graph the points at which the concavity changes.

(b) At what depth is $y = f(t)$ growing most quickly? Most slowly? Estimate the ratio between the growth rates at these two depths.

Figure 5.82

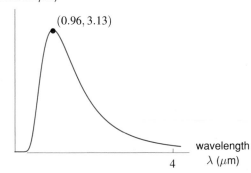

intensity of radiation
$r(\lambda)$ (MW/m²/μm)

$(0.96, 3.13)$

wavelength
λ (μm)

Figure 5.83

34. Any body radiates energy at various wavelengths. The power of the radiation (per meter² of surface) and the distribution of the radiation among the wavelengths varies with temperature.

The function graphed in Figure 5.83 represents the intensity of the radiation of a black body at a temperature $T = 3000°$ Kelvin, as a function of the wavelength. The intensity of the radiation is highest in the infrared range, that is, at wavelengths longer than that of visible light (0.4–0.7 μm). Max Planck's radiation law, announced to the Berlin Physical Society on October 19, 1900, states that

$$r(\lambda) = \frac{a}{\lambda^5(e^{b/\lambda} - 1)},$$

where a and b are empirical constants chosen to best fit the experimental data. Find a and b so that the formula fits the graph.

(Later in 1900 Planck was able to derive his radiation law entirely from theory. He found that $a = 2\pi c^2 h$ and $b = \frac{hc}{Tk}$ where c = speed of light, h = Planck's constant, and k = Boltzmann's constant.)

PROJECTS

1. **Resonance**

The study of resonance leads to the family

$$y = \frac{1}{(1 - x^2)^2 + 2ax^2}, \qquad x \geq 0, \quad a > 0.$$

(a) Find and classify the critical points, and then explain why the family is most interesting for $0 < a < 1$.

(b) Show that for a very near 0, there is a critical point at approximately $(1, 1/2a)$.

(c) Show that $y < 1/(1 - x^2)^2$ for $x > 0$.

(d) Sketch the curves for $a = 0.05, 0.10, 1.0,$ and 3.0 on one set of axes together with $y = 1/(1 - x^2)^2$, which is the curve with $a = 0$.

(e) What is the geometric significance of the parameter a?

2. **Building a Greenhouse**[5]

Your parents are going to knock out the bottom of the entire length of the south wall of their house and turn it into a greenhouse by replacing the bottom portion of the wall with a huge sloped piece of glass (which is expensive). They have already decided they are going to spend a certain fixed amount. The triangular ends of the greenhouse will be made of various materials they already have lying around.

The floor space in the greenhouse is only considered usable if they can both stand up in it, so part of it will be unusable. They want to choose the dimensions of the greenhouse to get the most usable floor space. What should the dimensions of the greenhouse be and how much usable space will your parents get?

3. **Average and Marginal Costs**

Suppose $C(q)$ is the total cost of producing a quantity q. The average cost $a(q)$ is given in Figure 5.84. The following rule is used by economists to determine the marginal cost $C'(q_0)$, for any q_0:

- Construct the tangent t_1 to $a(q)$ at q_0.
- Let t_2 be the line with the same vertical intercept as t_1 but with twice the slope of t_1.

Then $C'(q_0)$ is as shown in Figure 5.84. Explain why this rule works.

Figure 5.84

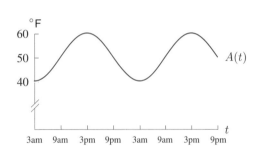

Figure 5.85

4. **Air and Water Temperature**

Consider a large tank of water, with temperature $W(t)$. The ambient temperature $A(t)$ (that is, the temperature of the surrounding air), is given by the graph in Figure 5.85. The temperature of the water is affected by the temperature of the surrounding air.

(a) How does the temperature of the water change if the water is colder than the surrounding air? What if the water is warmer?

(b) Assume that the temperature of the water is 50°F at 3 am. Using your answer in part (a), sketch a possible graph for $W(t)$ on the same axes as $A(t)$.

(c) Explain the relationship between the maxima and minima of $W(t)$ and the points where the two graphs intersect.

(d) What is the relationship between the rate at which the temperature of the water changes and the difference $A(t) - W(t)$?

(e) What is the relationship between the inflection points of $W(t)$ and the points where $A(t) - W(t)$ has a maxima or minima?

(f) Assume the tank is filled with cold water (35°F) at 3 am. Sketch a possible graph for $W(t)$. Pay attention to the concavity.

---

[5]Adapted from M. Cohen, E. Gaughan, A. Knoebel, D. Kurtz, D. Pengelley, *Student Research Projects in Calculus* (Washington DC: Mathematical Association of America, 1992).

FOCUS ON THEORY

THEOREMS ABOUT CONTINUOUS AND DIFFERENTIABLE FUNCTIONS

In this chapter we have used some basic facts without proof: for example, that a continuous function has a maximum on a bounded, closed interval, or that a function whose derivative is positive on an interval is increasing on that interval.

From a geometric point of view, these facts seem obvious. If we draw the graph of a continuous function, starting at one end of a bounded, closed interval and going to the other, it seems obvious that we must pass a highest point on the way. And if the derivative of a function is positive, then its graph must be sloping up, so the function has to be increasing.

However, this sort of graphical reasoning is not a rigorous proof, for two reasons. First, no matter how many pictures we imagine, we can't be sure we have covered all possibilities. Second, our pictures often depend on the theorems we are trying to prove.

A Continuous Function on a Closed Interval Has a Maximum

> **The Extreme Value Theorem**
> If f is continuous on the interval $[a, b]$, then f has a global maximum and a global minimum on that interval.

Our proof has two parts: The first is to show that f has an upper bound on $[a, b]$, the second is to show that if f has an upper bound then it has a global maximum on the interval. Here we prove the second part; the first part is proved in Problems 16 and 17. Then in Problem 5 we extend the result from maxima to minima.

Proof   We assume that f is continuous and has an upper bound on the interval $[a, b]$. This means f has a least upper bound M on $[a, b]$. Divide $[a, b]$ into two halves. Then, on one of the halves, the least upper bound for f is M, for if it were less than M on both halves, it would be less than M on the whole. Choose a half on which the least upper bound is equal to M. Continue bisecting and at each stage choose the half-interval where the least upper bound for f is M. See Figure 5.86. This results in a sequence of nested intervals. By the Nested Interval Theorem on page 84, there is a number c in $[a, b]$ which is contained in all these intervals.

Since M is the least upper bound for f, we have $f(c) \leq M$. It is not possible that $f(c) < M$. For if $f(c) < M$, then $f(c) < M_0$ for some number $M_0 < M$. (For example, we could take M_0 to be half-way between M and $f(c)$.) But then, since f is continuous, there would be a $\delta > 0$ such that $f(x) < M_0$ for all x in $[a, b]$ with $c - \delta < x < c + \delta$. (See Problem 15.) Since the nested

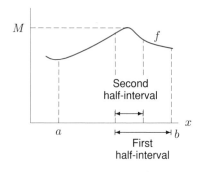

Figure 5.86: Successively choosing the half-interval where the least upper bound of f is M

intervals we constructed above have width tending to zero, one of them would be contained in the interval $c - \delta < x < c + \delta$. Therefore, f would be bounded above by M_0 on one of the nested intervals. However, we chose each nested interval so that the least upper bound for f is M. This is a contradiction of $M_0 < M$.

So it is not possible that $f(c) < M$; we must have $f(c) = M$. Thus, M is the global maximum of f on $[a, b]$, which is what we wanted to show.

The Extreme Value Theorem guarantees the existence of global maxima (and minima) on an interval. To actually find the global maxima, we look at all the local maxima. The following theorem tells us that inside an interval, local maxima only occur at critical points, where the derivative is either zero or undefined.

> **Theorem: Local Extrema and Critical Points**
> Suppose f is defined on an interval and has a local maximum or minimum at the point $x = a$, which is not an endpoint of the interval. If f is differentiable at $x = a$, then $f'(a) = 0$.

Proof We start with the definition of the derivative:

$$f'(a) = \lim_{h \to 0} \frac{f(a + h) - f(a)}{h}.$$

Remember that this is a two-sided limit:

$$f'(a) = \lim_{h \to 0^-} \frac{f(a + h) - f(a)}{h} = \lim_{h \to 0^+} \frac{f(a + h) - f(a)}{h}.$$

Suppose that f has a local maximum at $x = a$. By the definition of local maximum, $f(a + h) \leq f(a)$ for all sufficiently small h. Thus $f(a + h) - f(a) \leq 0$ for sufficiently small h. The denominator, h, is positive when we take the limit from the right and negative when we take the limit from the left. Thus

$$\lim_{h \to 0^-} \frac{f(a + h) - f(a)}{h} \geq 0 \quad \text{and} \quad \lim_{h \to 0^+} \frac{f(a + h) - f(a)}{h} \leq 0.$$

Since both these limits are equal to $f'(a)$, we have $f'(a) \geq 0$ and $f'(a) \leq 0$, so we must have $f'(a) = 0$.

A Relationship Between Local and Global: The Mean Value Theorem

We often want to infer a global conclusion (for example, f is increasing on an interval) from local information (f' is positive.) The following theorem relates the average rate of change of a function on an interval (global information) to the instantaneous rate of change at a point in the interval (local information).

> **The Mean Value Theorem**
> If f is continuous on $[a, b]$ and differentiable on (a, b), then there exists a number c, with $a < c < b$, such that
> $$f'(c) = \frac{f(b) - f(a)}{b - a}.$$
> In other words, $f(b) - f(a) = f'(c)(b - a)$.

To understand what this theorem is saying geometrically, consider the graph in Figure 5.87. Join the points on the curve where $x = a$ and $x = b$ with a line and observe that the slope of this secant line AB is given by

$$m = \frac{f(b) - f(a)}{b - a}.$$

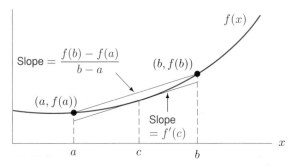

Figure 5.87: The point c with $f'(c) = \frac{f(b) - f(a)}{b - a}$

Now consider the tangent line drawn to the curve at each point between $x = a$ and $x = b$. In general, these lines will have different slopes. For the curve shown in Figure 5.87, the tangent line at $x = a$ is flatter than the secant line from A to B. Similarly, the tangent line at $x = b$ is steeper than the secant line. However, there is at least one point between a and b where the slope of the tangent line to the curve is precisely the same as the slope of the secant line. Suppose this occurs at $x = c$. Then

$$f'(c) = m = \frac{f(b) - f(a)}{b - a}.$$

The Mean Value Theorem tells us that the point $x = c$ exists, but it does not tell us how to find c.

Problems 18 and 19 show how the Mean Value Theorem can be deduced from the Extreme Value Theorem.

The Increasing Function Theorem

We say that a function f is *increasing* on an interval if, for any two numbers x_1 and x_2 in the interval such that $x_1 < x_2$, we have $f(x_1) < f(x_2)$. If instead we have $f(x_1) \leq f(x_2)$, we say f is *nondecreasing*.

> **The Increasing Function Theorem**
> Suppose that f is continuous on $[a, b]$ and differentiable on (a, b).
> - If $f'(x) > 0$ on (a, b), then f is increasing on $[a, b]$.
> - If $f'(x) \geq 0$ on (a, b), then f is nondecreasing on $[a, b]$.

Proof   Suppose $a \leq x_1 < x_2 \leq b$. By the Mean Value Theorem, there is a number c, with $x_1 < c < x_2$, such that

$$f(x_2) - f(x_1) = f'(c)(x_2 - x_1).$$

If $f'(c) > 0$, this says $f(x_2) - f(x_1) > 0$, which means f is increasing. If $f'(c) \geq 0$, this says $f(x_2) - f(x_1) \geq 0$, which means f is nondecreasing.

It may seem that something as simple as the Increasing Function Theorem should follow immediately from the definition of the derivative, and that the use of the Mean Value Theorem (which in turn depends on the Extreme Value Theorem) is surprising. A more direct, self-contained proof is given in Problem 11, but that proof also has some subtleties.

The Constant Function Theorem

If f is constant on an interval, then we know that $f'(x) = 0$ on the interval. The following theorem is the converse.

> **The Constant Function Theorem**
> Suppose that f is continuous on $[a, b]$ and differentiable on (a, b). If $f'(x) = 0$ on (a, b), then f is constant on $[a, b]$.

Proof   The proof is the same as for the Increasing Function Theorem, only in this case $f'(c) = 0$ so $f(x_2) - f(x_1) = 0$. Thus $f(x_2) = f(x_1)$ for $a \leq x_1 < x_2 \leq b$, so f is constant.

A proof of the Constant Function Theorem using the Increasing Function Theorem is given in Problems 6 and 8.

The Racetrack Principle

> **The Racetrack Principle[6]**
>
> Suppose that g and h are continuous on $[a, b]$ and differentiable on (a, b), and that $g'(x) \leq h'(x)$ for $a < x < b$.
> - If $g(a) = h(a)$, then $g(x) \leq h(x)$ for $a \leq x \leq b$.
> - If $g(b) = h(b)$, then $g(x) \geq h(x)$ for $a \leq x \leq b$.

The Racetrack Principle has the following interpretation. We can think of $g(x)$ and $h(x)$ as the positions of two racehorses at time x, with horse h always moving faster than horse g. If they start together, horse h is ahead during the whole race. If they finish together, horse g was ahead during the whole race.

Proof   Consider the function $f(x) = h(x) - g(x)$. Since $f'(x) = h'(x) - g'(x) \geq 0$, we know that f is nondecreasing by the Increasing Function Theorem. So $f(x) \geq f(a) = h(a) - g(a) = 0$. Thus $g(x) \leq h(x)$ for $a \leq x \leq b$. This proves the first part of the Racetrack Principle. Problem 7 asks for a proof of the second part.

Example 1   Explain graphically why $e^x \geq 1 + x$ for all values of x. Then use the Racetrack Principle to prove the inequality.

Solution   The graph of the function $f(x) = e^x$ is concave up everywhere and the equation of its tangent line at the point $(0, 1)$ is $y = x + 1$. (See Figure 5.88.) Since the graph always lies above its tangent, we have the inequality
$$e^x \geq 1 + x.$$

Now we prove the inequality using the Racetrack Principle. Let $g(x) = 1 + x$ and $h(x) = e^x$. Then $g(0) = h(0) = 1$. Furthermore, $g'(x) = 1$ and $h'(x) = e^x$. Hence $g'(x) \leq h'(x)$ for $x \geq 0$. So by the Racetrack Principle, with $a = 0$, we have $g(x) \leq h(x)$, that is, $1 + x \leq e^x$.

For $x \leq 0$ we have $h'(x) \leq g'(x)$. So by the Racetrack Principle, with $b = 0$, we have $g(x) \leq h(x)$, that is, $1 + x \leq e^x$.

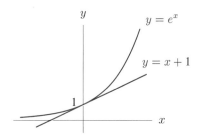

Figure 5.88: Graph showing that $e^x \geq 1 + x$

---

[6]Based on the Racetrack Principle in *Calculus&Mathematica,* by W li am Davis, Horacio Porta, Jerry Uhl (Reading: Addison Wesley, 1994).

Problems on the Theorems about Continuous and Differentiable Functions ▬▬▬

1. Use the Racetrack Principle and the fact that $\sin 0 = 0$ to show that $\sin x \leq x$ for all $x \geq 0$.

2. Use the Racetrack Principle to show that $\ln x \leq x - 1$.

3. Use the fact that $\ln x$ and e^x are inverse functions to show that the inequalities $e^x \geq 1 + x$ and $\ln x \leq x - 1$ are equivalent for $x > 0$.

4. Suppose that the position of a particle moving along the x-axis is given by $s = f(t)$, and that the initial position and velocity of the particle are $f(0) = 3$ and $f'(0) = 4$. Suppose that the acceleration is bounded by $5 \leq f''(t) \leq 7$ for $0 \leq t \leq 2$. What can we say about the position $f(2)$ of the particle at $t = 2$?

5. Show that if every continuous function on an interval $[a, b]$ has a global maximum, then every continuous function has a global minimum as well. [Hint: Consider $-f$.]

6. State a Decreasing Function Theorem, analogous to the Increasing Function Theorem. Deduce your theorem from the Increasing Function Theorem. [Hint: Apply the Increasing Function Theorem to $-f$.]

7. Suppose that g and h are continuous on $[a, b]$ and differentiable on (a, b). Prove that if $g'(x) \leq h'(x)$ for $a < x < b$ and $g(b) = h(b)$, then $h(x) \leq g(x)$ for $a \leq x \leq b$.

8. Deduce the Constant Function Theorem from the Increasing Function Theorem and the Decreasing Function Theorem (see problem 6).

9. Prove that if $f'(x) = g'(x)$ for all x in (a, b), then there is a constant C such that $f(x) = g(x) + C$ on (a, b). [Hint: Apply the Constant Function Theorem to $h(x) = f(x) - g(x)$.]

10. Suppose that $f'(x) = f(x)$ for all x. Prove that $f(x) = Ce^x$ for some constant C. [Hint: Consider the function $f(x)/e^x$.]

11. In this problem we give a proof of the Increasing Function Theorem that does not use the Mean Value Theorem. Suppose that f is continuous on $[a, b]$ and differentiable on (a, b), and that $f'(x) \geq 0$ for all x in (a, b).

    (a) We write Slope(c, d) to denote the slope between $(c, f(c))$ and $(d, f(d))$, so

    $$\text{Slope}(c, d) = \frac{f(d) - f(c)}{d - c}.$$

    Show that if $c < e < d$, then

    $$\text{Slope}(c, d) = \left(\frac{e - c}{d - c}\right) \cdot \text{Slope}(c, e) + \left(\frac{d - e}{d - c}\right) \cdot \text{Slope}(e, d).$$

    Use this to show that Slope(c, d) lies between Slope(c, e) and Slope(e, d).

    (b) First we prove f is nondecreasing on (a, b), that is, if $a < a_1 < b_1 < b$, then $f(a_1) \leq f(b_1)$. Suppose, contrary to what we want, that $f(a_1) > f(b_1)$, so that Slope(a_1, b_1) is negative. Show that there is a sequence of nested intervals $[a_n, b_n]$, each of which is one half of the previous one, such that Slope$(a_n, b_n) \leq$ Slope(a_1, b_1) for all n.

    (c) By the Nested Interval Theorem, there is a number c in (a, b) which is contained in all the intervals $[a_n, b_n]$. Show that, for all n, either Slope(a_n, c) or Slope(c, b_n) is less than or equal to Slope(a_1, b_1).

    (d) Using $\lim_{x \to c}$ Slope$(x, c) = \lim_{x \to c}$ Slope$(c, x) = f'(c) \geq 0$, show that for large enough n, both Slope(a_n, c) and Slope(c, b_n) are greater than Slope(a_1, b_1), contradicting part (c).

    (e) The contradiction in part (d) proves that our assumption $f(a_1) > f(b_1)$ in part (b) must be false. Therefore, $f(a_1) \leq f(b_1)$. Since this is true for all a_1 and b_1 such that $a < a_1 < b_1 < b$, we have shown that f is nondecreasing on (a, b). Use the continuity of f to deduce that f is nondecreasing on $[a, b]$.

    (f) Now suppose that $f'(x) > 0$ on (a, b). Show that if $a \leq a_1 < b_1 \leq b$, and $f(a_1) = f(b_1)$, then f is constant on $[a_1, b_1]$ so $f'(x) = 0$ on $[a_1, b_1]$. This contradicts $f'(x) > 0$, so it is not possible that $f(a_1) = f(b_1)$, and we must have $f(a_1) < f(b_1)$. Therefore, f is increasing on $[a, b]$.

12. Suppose that f is continuous on $[a, b]$ and differentiable on (a, b) and that $m \leq f'(x) \leq M$ on (a, b). Use the Racetrack Principle to prove that $f(x) - f(a) \leq M(x - a)$ for all x in $[a, b]$, and that $m(x - a) \leq f(x) - f(a)$ for all x in $[a, b]$. Conclude that $m \leq (f(b) - f(a))/(b - a) \leq M$. This is called the Mean Value Inequality. In words: If the instantaneous rate of change of f is between m and M on an interval, so is the average rate of change of f over the interval.

13. Suppose that $f''(x) \geq 0$ for all x in (a, b). We will show the graph of f lies above the tangent line at $(c, f(c))$ for any c with $a < c < b$.

    (a) Use the Increasing Function Theorem to prove that $f'(c) \leq f'(x)$ for $c \leq x < b$ and that $f'(x) \leq f'(c)$ for $a < x \leq c$.
    (b) Use (a) and the Racetrack Principle to conclude that $f(c) + f'(c)(x - c) \leq f(x)$, for $a < x < b$.

14. In this problem we use the Mean Value Theorem to give a proof of the Fundamental Theorem of Calculus. Let f be continuous, with antiderivative F.

    (a) Let $[a, b]$ be an interval contained in the domain of f, and let

    $$a = x_0 < x_1 < \cdots < x_{n-1} < x_n = b$$

    be a subdivision of $[a, b]$. Show that there is a Riemann sum for f using this subdivision which is equal to $F(b) - F(a)$. [Hint: Apply the Mean Value Theorem to

    $$F(b) - F(a) = ((F(b) - F(x_{n-1})) + (F(x_{n-1}) - F(x_{n-2})) + \cdots + (F(x_1) - F(a)).]$$

    (b) Deduce that $F(b) - F(a) = \int_a^b f(x)\, dx$.

15. Suppose that f is continuous on $[a, b]$, and let c be in $[a, b]$. Show that if $f(c) < M$, then there is a δ such that $f(x) < M$ for all x in $[a, b]$ such that $c - \delta < x < c + \delta$. [Hint: Let $\epsilon = M - f(c)$, and choose δ such that $|f(x) - f(c)| < \epsilon$ if $|x - c| < \delta$.]

On page 286 we proved that a continuous function f has a global maximum on the interval $[a, b]$ under the assumption that f has an upper bound on $[a, b]$. In Problems 16–17 we prove this claim.

16. (a) Suppose that f has no upper bound on $[a, b]$. Bisect $[a, b]$ into two halves. Deduce that f has no upper bound on at least one of the halves. Call that half $[a_1, b_1]$.
    (b) Continue bisecting so that at the nth stage you obtain an interval $[a_n, b_n]$ on which f has no upper bound. By the Nested Interval Theorem on page 84, there is a point c in all the intervals $[a_n, b_n]$.
    (c) Use continuity of f at c to deduce that f has an upper bound on $[a_n, b_n]$ for n sufficiently large. This contradicts the original supposition, so f must have an upper bound on $[a, b]$.

17. (a) Show that if $y \geq 0$, then $y/(1 + y) < 1$.
    (b) Suppose that f is continuous on $[a, b]$ and that $f(x) \geq 0$ on $[a, b]$. Define a function g by $g(x) = f(x)/(1 + f(x))$. Show that g is continuous and bounded on $[a, b]$. It follows from the partial proof of the Extreme Value Theorem on page 286 that g has a global maximum on $[a, b]$ at some point $x = c$.
    (c) Suppose that $y_1 \geq 0$ and $y_2 \geq 0$, and that $y_1/(1 + y_1) \leq y_2/(1 + y_2)$. Show that $y_1 \leq y_2$.
    (d) Use parts (c) and (d) to show that f has a global maximum at $x = c$.
    (e) We have shown that if f is continuous and non-negative on $[a, b]$, then it is bounded above on $[a, b]$. Now suppose that f is continuous, but not necessarily non-negative. By applying the argument to $|f|$, deduce that f is also bounded above.

18. In this problem we prove a special case of the Mean Value Theorem where $f(a) = f(b) = 0$. This special case is called Rolle's Theorem: If f is continuous on $[a, b]$ and differentiable on (a, b), and if $f(a) = f(b) = 0$, then there is a number c, with $a < c < b$, such that

$$f'(c) = 0.$$

By the Extreme Value Theorem, f has a global maximum and a global minimum on $[a, b]$.

    (a) Prove Rolle's theorem in the case that both the global maximum and the global minimum are at endpoints of $[a, b]$. [Hint: $f(x)$ must be a very simple function in this case.]
    (b) Prove Rolle's theorem in the case that either the global maximum or the global minimum is not at an endpoint. [Hint: Think about local maxima and minima.]

19. Use Rolle's Theorem to prove the Mean Value Theorem. Suppose that $f(x)$ is continuous on $[a, b]$ and differentiable on (a, b).

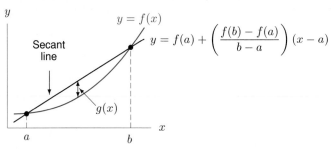

Figure 5.89: $g(x)$ is the difference between the secant line and the
graph of $f(x)$

(a) Let $g(x)$ be the difference between $f(x)$ and the y-value on the secant line joining $(a, f(a))$ to $(b, f(b))$. See Figure 5.89. Show that

$$g(x) = f(x) - f(a) - \frac{f(b) - f(a)}{b - a}(x - a).$$

(b) Use Rolle's Theorem to show that there must be a point c in (a, b) such that $g'(c) = 0$.

(c) Show that if c is the point in part (b), then

$$f'(c) = \frac{f(b) - f(a)}{b - a}.$$

CHAPTER SIX

CONSTRUCTING ANTIDERIVATIVES

In Chapter 2, we saw how to calculate velocity given position, and in Chapter 3, we saw how to reconstruct distance from velocity. In this chapter we will look in more detail at how to reconstruct a function from its derivative.

We already know how to get a rough idea of the graph of f from the graph of f' and how to use f' to estimate values of f numerically. We extend these ideas and we start to see how to go from f' to f analytically.

6.1 ANTIDERIVATIVES GRAPHICALLY AND NUMERICALLY

The Family of Antiderivatives

If the derivative of F is f, we call F an *antiderivative* of f. For example, since the derivative of x^2 is $2x$, we say that

$$x^2 \text{ is an antiderivative of } 2x.$$

Notice that $2x$ has many antiderivatives, since $x^2 + 1$, $x^2 + 2$, and $x^2 + 3$, all have derivative $2x$. In fact, if C is any constant, we have

$$\frac{d}{dx}(x^2 + C) = 2x + 0 = 2x,$$

and so any function of the form $x^2 + C$ is an antiderivative of $2x$. The function $f(x) = 2x$ has a *family of antiderivatives*.

Let us look at another example. If v is the velocity of a car and s is its position, then $v = ds/dt$ and s is an antiderivative of v. As before, $s + C$ is an antiderivative of v for any constant C. In terms of the car, adding C to s is equivalent to adding C to the odometer reading. Adding a fixed distance to the odometer reading simply means measuring distance from a different point, which doesn't alter the car's velocity.

Visualizing Antiderivatives Using Slopes

Suppose we have the graph of f', and we want to sketch an approximate graph of f. We are looking for the graph of f whose slope at any point is equal to the value of f' there. Where f' is above the x-axis, f is increasing; where f' is below the x-axis, f is decreasing. If f' is increasing, f is concave up; if f' is decreasing, f is concave down.

Example 1   The graph of f' is given in Figure 6.1. Sketch a graph of f in the cases when $f(0) = 0$ and $f(0) = 1$.

Figure 6.1: Graph of f'

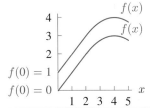

Figure 6.2: Two different f's which have the same derivative f'

Solution   For $0 \leq x \leq 2$, f has a constant slope of 1 and so the graph of f is a straight line. For $2 \leq x \leq 4$, f is increasing but more and more slowly; it has a maximum at $x = 4$ and decreases thereafter. (See Figure 6.2.)

Notice that the solutions with $f(0) = 0$ and $f(0) = 1$ start at different points on the vertical axis but have the same shape.

Example 2   Sketch a graph of the antiderivative F of $f(x) = e^{-x^2}$ satisfying $F(0) = 0$.

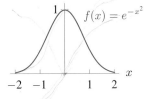

Figure 6.3: Graph of $f(x) = e^{-x^2}$

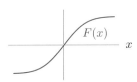

Figure 6.4: An antiderivative $F(x)$ of $f(x) = e^{-x^2}$

Solution   The graph of $f(x) = e^{-x^2}$ is shown in Figure 6.3. The slope of the antiderivative $F(x)$ is given by $f(x)$. Since $f(x)$ is always positive, the antiderivative $F(x)$ is always increasing. Since $f(x)$ is increasing for negative x, we know that $F(x)$ is concave up for negative x. Since $f(x)$ is decreasing for positive x, we know that $F(x)$ is concave down for positive x. Since $f(x) \to 0$ as $x \to \pm\infty$, $f(x) \to 0$ and so the graph of $F(x)$ levels off at both ends. A graph of $F(x)$ with $F(0) = 0$ is shown in Figure 6.4.

Example 3   For the function f' given in Figure 6.5, sketch a graph of three antiderivative functions f, where
(a) $f(0) = 0$,   (b) $f(0) = 1$,   (c) $f(0) = 2$.

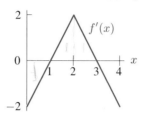

Figure 6.5: Slope function, f'

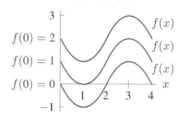

Figure 6.6: Antiderivatives f

Solution   To draw a graph of f, start at the point on the vertical axis specified by the initial condition and move with slope given by the value of f'. (See Figure 6.6.) Different initial conditions lead to different graphs for f, but for a given x-value they all have the same slope (because the value of f' is the same for each). Thus, the different f curves are obtained from one another by a vertical shift.

- Where f' is positive ($1 < x < 3$), we see f is increasing; where f' is negative ($0 < x < 1$ or $3 < x < 4$), we see f is decreasing.
- Where f' is increasing ($0 < x < 2$), we see f is concave up; where f' is decreasing ($2 < x < 4$), we see f is concave down.
- Where $f' = 0$, we see f has a local maximum ($x = 3$) or local minimum ($x = 1$).
- Where f' has a maximum ($x = 2$), we see f has a point of inflection.

Computing Values of an Antiderivative Using the Fundamental Theorem

A graph of f' shows where f is increasing and where f is decreasing. We can find the actual value of the function f by using the Fundamental Theorem:

Fundamental Theorem of Calculus

If F is an antiderivative of a continuous function, f, then

$$\int_a^b f(x)\,dx = F(b) - F(a).$$

Example 4   Figure 6.7 is the graph of the derivative $f'(x)$ of a function $f(x)$. It is given that $f(0) = 100$. Sketch the graph of $f(x)$, showing all critical points and inflection points of f and giving their coordinates.

Figure 6.7: Graph of derivative

Solution   The critical points of f occur at $x = 0$, $x = 20$, and $x = 30$, where $f'(x) = 0$. The inflection points of f occur at $x = 10$ and $x = 25$, where $f'(x)$ has a maximum or minimum. To find the coordinates of the critical points and inflection points of f, we evaluate $f(x)$ for $x = 0, 10, 20, 25, 30$. We use the Fundamental Theorem to express the values of $f(x)$ in terms of definite integrals. We evaluate the definite integrals using the areas of triangular regions under the graph of $f'(x)$, remembering that areas below the x-axis are subtracted. (See Figure 6.8.)

Figure 6.8: Finding $f(10) = f(0) + \int_0^{10} f'(x)\, dx$ *Figure 6.9:* Graph of $f(x)$

Since $f(0) = 100$, the Fundamental Theorem says

$$f(10) = f(0) + \int_0^{10} f'(x)\, dx = 100 + (\text{Shaded area in Figure 6.8}) = 100 + \frac{1}{2}(10)(20) = 200,$$

$$f(20) = f(10) + \int_{10}^{20} f'(x)\, dx = 200 + \frac{1}{2}(10)(20) = 300,$$

$$f(25) = f(20) + \int_{20}^{25} f'(x)\, dx = 300 - \frac{1}{2}(5)(10) = 275,$$

$$f(30) = f(25) + \int_{25}^{30} f'(x)\, dx = 275 - \frac{1}{2}(5)(10) = 250.$$

The graph of f is sketched in Figure 6.9.

Example 5   Suppose you are given that $F'(t) = t \cos t$ and $F(0) = 2$. Find the values of $F(b)$ at the points $b = 0, 0.1, 0.2, \ldots, 1.0$.

Solution   We apply the Fundamental Theorem with $f(t) = t \cos t$ and $a = 0$ to get values for $F(b)$. Since

$$F(b) - F(0) = \int_0^b F'(t)\, dt = \int_0^b t \cos t \, dt$$

and $F(0) = 2$, we have

$$F(b) = 2 + \int_0^b t \cos t \, dt.$$

Using numerical methods to estimate the definite integral $\int_0^b t \cos t \, dt$ for each of the values $b = 0, 0.1, 0.2, \ldots, 1.0$ gives the approximate values for F in Table 6.1:

TABLE 6.1   *Approximate values for F*

b	0	0.1	0.2	0.3	0.4	0.5	0.6	0.7	0.8	0.9	1.0
$F(b)$	2.000	2.005	2.020	2.044	2.077	2.117	2.164	2.216	2.271	2.327	2.382

Notice from the table that the function $F(b)$ appears to be increasing between $b = 0$ and $b = 1$. This is in fact the case, and could have been predicted without the use of the Fundamental Theorem from the fact that $t \cos t$, the derivative of $F(t)$, is positive for t between 0 and 1.

Problems for Section 6.1

For each function in Problems 1–4, sketch two functions F such that $F' = f$. In one case let $F(0) = 0$ and in the other, let $F(0) = 1$.

1.

2.

3.

4.

For each function $f(x)$ in Problems 5–7, assume $F(x)$ is such that $F'(x) = f(x)$.
(a) What are the critical points of $F(x)$?
(b) Which critical points are local maxima, local minima, or neither?
(c) Sketch a possible graph of $F(x)$.

5.

6.

7.

For each function in Problems 8–10, sketch two functions F where $F'(x) = f(x)$. In one case, let $F(0) = 0$, and in the other, let $F(0) = 1$. In each case, mark x_1, x_2, and x_3 on the x-axis of your graph. Identify local maxima, minima, and inflection points of $F(x)$.

8.

9.

10.

11. Assume f' is given by the graph in Figure 6.10. Suppose f is continuous and that $f(0) = 0$.

(a) Find $f(3)$ and $f(7)$.
(b) Find all x with $f(x) = 0$.
(c) Sketch a graph of f over the interval $0 \le x \le 7$.

12. Assume f' is given by the graph in Figure 6.10. Suppose f is continuous and that $f(3) = 0$.

(a) Sketch a graph of f.
(b) Find $f(0)$ and $f(7)$.
(c) Find $\int_0^7 f'(x)\,dx$ in two different ways.

13. Figure 6.11 shows the graph of the derivative $g'(x)$ of a function $g(x)$. It is given that $g(0) = 50$. Sketch the graph of $g(x)$, showing all critical points and inflection points of g and giving their coordinates.

Figure 6.10

Figure 6.11

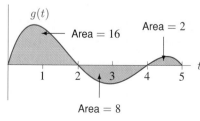

Figure 6.12

14.   A graph of $g(t)$ is given in Figure 6.12 on page 297. Sketch a graph of an antiderivative $G(t)$ of $g(t)$ satisfying $G(0) = 5$. Label each critical point of $G(t)$ with its coordinates.

15.   The vertical velocity of a cork bobbing up and down on the waves in the sea is given by Figure 6.13. Upward is considered positive. Describe the motion of the cork at each of the labeled points. At which point(s), if any, is the acceleration zero? Sketch a graph of the height of the cork above the sea floor as a function of time.

Figure 6.13

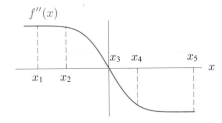

Figure 6.14

16.   The graph of f'' is given in Figure 6.14. Draw graphs of f and f', assuming both go through the origin, and use them to decide at which of the labeled x-values:

    (a) $f(x)$ is greatest.   (c) $f'(x)$ is greatest.   (e) $f''(x)$ is greatest.

    (b) $f(x)$ is least.   (d) $f'(x)$ is least.   (f) $f''(x)$ is least.

17.   Two functions, $f(x)$ and $g(x)$, are shown in Figure 6.15. Let $F(x)$ and $G(x)$ be antiderivatives, respectively, of these two functions. On the same axes, sketch graphs of the antiderivatives $F(x)$ and $G(x)$ satisfying $F(0) = 0$ and $G(0) = 0$. Compare the two antiderivatives, including a discussion of zeros and x- and y-coordinates of critical points.

18.   The Quabbin Reservoir in the western part of Massachusetts provides most of Boston's water. The graph in Figure 6.16 represents the flow of water in and out of the Quabbin Reservoir throughout 1993.

    (a)   Sketch a possible graph for the quantity of water in the reservoir, as a function of time.

    (b)   When, in the course of 1993, was the quantity of water in the reservoir largest? Smallest? Mark and label these points on the graph you drew in part (a).

    (c)   When was the quantity of water increasing most rapidly? Decreasing most rapidly? Mark and label these times on both graphs.

    (d)   By July 1994 the quantity of water in the reservoir was about the same as in January 1993. Draw plausible graphs for the flow into and the flow out of the reservoir for the first half of 1994. Explain your graphs.

Figure 6.15

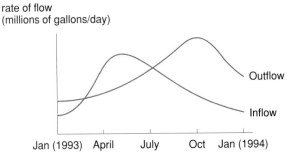

Figure 6.16

6.2   CONSTRUCTING ANTIDERIVATIVES ANALYTICALLY

What Is an Antiderivative of $f(x) = 0$?

A function whose derivative is zero everywhere on an interval must have a horizontal tangent line at every point of its graph, and the only way this can happen is if the function is constant. Alternatively, if we think of the derivative as a velocity, and if the velocity is always zero, then the object is standing still; the position function is constant. A rigorous proof of this result using the definition of the derivative is surprisingly subtle. (See the Constant Function Theorem on page 288.)

> If the derivative of a function is zero everywhere on an interval, then the function is constant on the interval:
>
> $$\text{If } F'(x) = 0 \text{ on an interval, then } F(x) = C \text{ on this interval.}$$

What Is the Most General Antiderivative of f?

We know that if a function f has an antiderivative F, then it has a family of antiderivatives of the form $F(x) + C$, where C is any constant. You might wonder if there are any others. To decide, suppose that we have two functions F and G with $F' = f$ and $G' = f$: that is, F and G are both antiderivatives of the same function f. Since $F' = G'$ we have $(F - G)' = 0$. But this means that we must have $F - G = C$, so $F(x) = G(x) + C$, where C is a constant. Thus, any two antiderivatives of the same function may only differ by a constant.

> If F and G are both antiderivatives of f, then $F(x) = G(x) + C$ for some constant C.

The Indefinite Integral

All antiderivatives of $f(x)$ are of the form $F(x) + C$. We introduce a notation for the general antiderivative that looks like the definite integral without the limits and is called the *indefinite integral*:

$$\int f(x)\, dx = F(x) + C.$$

It is important to understand the difference between

$$\int_a^b f(x)\, dx \qquad \text{and} \qquad \int f(x)\, dx.$$

The first is a number and the second is a family of *functions*. Because the notation is similar, the word "integration" is frequently used for the process of finding the antiderivative as well as of finding the definite integral. The context usually makes clear which is intended.

What is an Antiderivative of $f(x) = k$?

If k is a constant, the derivative of kx is k, so we have

$$\text{An antiderivative of } k \text{ is } kx.$$

Using the indefinite integral notation, we have

> If k is constant
>
> $$\int k\, dx = kx + C.$$

Finding Antiderivatives

Finding antiderivatives of functions is like taking square roots of numbers: if we pick a number at random, such as 7 or 493, we may have trouble saying what its square root is without a calculator. But if we happen to pick a number such as 25 or 64, which we know is a perfect square, then we can find its square root exactly. Similarly, if we pick a function which we recognize as a derivative, then we can find its antiderivative easily.

For example, to find an antiderivative of $f(x) = x$, notice that $2x$ is the derivative of x^2; this tells us that x^2 is an antiderivative of $2x$. If we divide by 2, then we guess that

$$\text{An antiderivative of } x \text{ is } \frac{x^2}{2}.$$

To check this statement, take the derivative of $x^2/2$:

$$\frac{d}{dx}\left(\frac{x^2}{2}\right) = \frac{1}{2} \cdot \frac{d}{dx}x^2 = \frac{1}{2} \cdot 2x = x.$$

What about an antiderivative of x^2? The derivative of x^3 is $3x^2$, so the derivative of $x^3/3$ is $3x^2/3 = x^2$. Thus,

$$\text{An antiderivative of } x^2 \text{ is } \frac{x^3}{3}.$$

Can you see the pattern? It looks like

$$\text{An antiderivative of } x^n \text{ is } \frac{x^{n+1}}{n+1}.$$

(We assume $n \neq -1$, or we would have $x^0/0$, which doesn't make sense.) It is easy to check this formula by differentiation:

$$\frac{d}{dx}\left(\frac{x^{n+1}}{n+1}\right) = \frac{(n+1)x^n}{n+1} = x^n.$$

Recall that the indefinite integral of a function is the family of all its antiderivatives. Thus, in indefinite integral notation, we have shown that

$$\int x^n \, dx = \frac{x^{n+1}}{n+1} + C, \quad n \neq -1.$$

So what about when $n = -1$? In other words, what is an antiderivative of $1/x$? Fortunately, we know a function whose derivative is $1/x$, namely, the natural logarithm. Thus, since

$$\frac{d}{dx}(\ln x) = \frac{1}{x},$$

we know that

$$\int \frac{1}{x} \, dx = \ln x + C, \quad \text{for } x > 0.$$

If $x < 0$, then $\ln x$ is not defined, so it can't be an antiderivative of $1/x$. In this case, we can try $\ln(-x)$:

$$\frac{d}{dx}\ln(-x) = (-1)\frac{1}{-x} = \frac{1}{x}$$

so

$$\int \frac{1}{x} \, dx = \ln(-x) + C, \quad \text{for } x < 0.$$

This means $\ln x$ is an antiderivative of $1/x$ if $x > 0$, and $\ln(-x)$ is an antiderivative of $1/x$ if $x < 0$. Since $|x| = x$ when $x > 0$ and $|x| = -x$ when $x < 0$ we can collapse these two formulas into:

$$\text{An antiderivative of } \frac{1}{x} \text{ is } \ln|x|.$$

Therefore

$$\int \frac{1}{x} \, dx = \ln|x| + C.$$

Since the exponential function is its own derivative, it is also its own antiderivative; thus

$$\int e^x \, dx = e^x + C.$$

Also, antiderivatives of the sine and cosine are easy to guess. Since

$$\frac{d}{dx} \sin x = \cos x \qquad \text{and} \qquad \frac{d}{dx} \cos x = -\sin x,$$

we get

$$\int \cos x \, dx = \sin x + C \qquad \text{and} \qquad \int \sin x \, dx = -\cos x + C.$$

Example 1   Find $\int (3x + x^2) \, dx$.

Solution   We know that $x^2/2$ is an antiderivative of x and that $x^3/3$ is an antiderivative of x^2, so we expect

$$\int (3x + x^2) \, dx = 3\left(\frac{x^2}{2}\right) + \frac{x^3}{3} + C.$$

You should always check your antiderivatives by differentiation—it's easy to do. Here

$$\frac{d}{dx}\left(\frac{3}{2}x^2 + \frac{x^3}{3} + C\right) = \frac{3}{2} \cdot 2x + \frac{3x^2}{3} = 3x + x^2.$$

The preceding example illustrates that the sum and constant multiplication rules of differentiation work in reverse:

Properties of Antiderivatives: Sums and Constant Multiples

In indefinite integral notation,

1. $\int \left[f(x) \pm g(x) \right] \, dx = \int f(x) \, dx \pm \int g(x) \, dx$

2. $\int cf(x) \, dx = c \int f(x) \, dx.$

In words,

1.   An antiderivative of the sum (or difference) of two functions is the sum (or difference) of their antiderivatives.

2.   An antiderivative of a constant times a function is the constant times an antiderivative of the function.

Example 2   Find $\int (\sin x + 3 \cos x) \, dx$.

Solution   We break the antiderivative into two terms:

$$\int (\sin x + 3 \cos x) \, dx = \int \sin x \, dx + 3 \int \cos x \, dx = -\cos x + 3 \sin x + C.$$

Check by differentiating:

$$\frac{d}{dx}(-\cos x + 3 \sin x + C) = \sin x + 3 \cos x.$$

Using Antiderivatives to Compute Definite Integrals

The Fundamental Theorem of Calculus, which tells us that if $F' = f$, then

$$\int_a^b f(x)\,dx = F(b) - F(a),$$

gives us a new way of calculating definite integrals. To find $\int_a^b f(x)\,dx$, we first try to find F, and then calculate $F(b) - F(a)$. This method of computing definite integrals has an important advantage over using left- and right-hand sums: it gives an exact answer quickly. However, the method only works in situations where you can find the antiderivative $F(x)$. This is not always easy; for example, none of the functions we have encountered so far is an antiderivative of $\sin(x^2)$.

Example 3   Compute $\displaystyle\int_1^3 2x\,dx$ using the Fundamental Theorem.

Solution   Since $F(x) = x^2$ is an antiderivative of $f(x) = 2x$, we have

$$\int_1^3 2x\,dx = F(3) - F(1) = 3^2 - 1^2 = 8.$$

Notice in this example we used the antiderivative x^2, but $x^2 + C$ works just as well for any constant C. It makes no difference which antiderivative we use to compute a definite integral because the constant cancels out when we subtract $F(a)$ from $F(b)$. For example, if we had used $x^2 + C$, we would have arrived at the same answer:

$$\int_1^3 2x\,dx = F(3) - F(1) = (3^2 + C) - (1^2 + C) = 8.$$

It is helpful to introduce a shorthand notation for $F(b) - F(a)$: we write it as

$$F(x)\Big|_a^b$$

For example:

$$\int_1^3 2x\,dx = x^2\Big|_1^3 = 3^2 - 1^2 = 8.$$

Problems for Section 6.2

For each of the functions in Problems 1–21, find an antiderivative.

1. $f(x) = 5$

2. $f(x) = 5x$

3. $f(x) = x^2$

4. $g(t) = t^2 + t$

5. $h(t) = \cos t$

6. $g(z) = \sqrt{z}$

7. $h(z) = \dfrac{1}{z}$

8. $r(t) = \dfrac{1}{t^2}$

9. $g(z) = \dfrac{1}{z^3}$

10. $f(z) = e^z$

11. $g(t) = \sin t$

12. $f(t) = 2t^2 + 3t^3 + 4t^4$

13. $p(t) = t^3 - \dfrac{t^2}{2} - t$

14. $q(y) = y^4 + \dfrac{1}{y}$

15. $f(x) = 5x - \sqrt{x}$

16. $f(t) = \dfrac{t^2 + 1}{t}$

17. $p(\theta) = 2\sin(2\theta)$

18. $r(t) = e^t + 5e^{5t}$

19. $q(t) = (t + 1)^2$

20. $f(x) = 5^x$

21. $p(t) = \cos t + \dfrac{1}{\cos^2 t}$

For Problems 22–29, find an antiderivative $F(x)$ with $F'(x) = f(x)$ and $F(0) = 0$. Is there only one possible solution in each case?

22. $f(x) = 3$

23. $f(x) = 2x$

24. $f(x) = -7x$

25. $f(x) = \frac{1}{4}x$

26. $f(x) = x^2$

27. $f(x) = \sqrt{x}$

28. $f(x) = 2 + 4x + 5x^2$

29. $f(x) = \sin x$

Find the indefinite integrals in Problems 30–45.

30. $\int 3x \, dx$

31. $\int (4t + 7) \, dt$

32. $\int \cos \theta \, d\theta$

33. $\int 5e^z \, dz$

34. $\int \left(x + \frac{1}{\sqrt{x}} \right) dx$

35. $\int \sin t \, dt$

36. $\int (\pi + x^{11}) \, dx$

37. $\int \left(t\sqrt{t} + \frac{1}{t\sqrt{t}} \right) dt$

38. $\int \cos(x + 1) \, dx$

39. $\int e^{2r} \, dr$

40. $\int \frac{1}{e^z} \, dz$

41. $\int \left(\frac{y^2 - 1}{y} \right)^2 dy$

42. $\int \frac{1}{x + 1} \, dx$ [Hint: $\frac{d}{dx} \left(\ln(x + 1) \right) = ?$]

43. $\int \frac{1}{2x - 1} \, dx$ [Hint: $\frac{d}{dx} \left(\ln(2x - 1) \right) = ?$]

44. $\int (e^{5+x} + e^{5x}) \, dx$ [Hint: $\frac{d}{dx} (e^{5x}) = ?$]

45. $\int (\cos 2x - 2 \sin x) \, dx$ [Hint: $\frac{d}{dx} (\sin 2x) = ?$]

Using the Fundamental Theorem, evaluate the definite integrals in Problems 46–55 exactly [as in $\ln(3\pi)$] and numerically [$\ln(3\pi) \approx 2.243$]:

46. $\int_2^5 (x^3 - \pi x^2) \, dx$

47. $\int_0^1 \sin \theta \, d\theta$

48. $\int_1^2 \frac{1 + y^2}{y} \, dy$

49. $\int_0^2 \left(\frac{x^3}{3} + 2x \right) dx$

50. $\int_0^{\pi/4} (\sin t + \cos t) \, dt$

51. $\int_{-3}^{-1} \frac{2}{r^3} \, dr$

52. $\int_0^1 2e^x \, dx$

53. $\int_0^{\pi/4} \frac{1}{\cos^2 x} \, dx$

54. $\int_{-1}^1 2^x \, dx$

55. $\int_0^{\pi/6} (\sin x + \cos 2x) \, dx$ [Hint: What is $\frac{d}{dx} (\sin 2x)$?]

56. Calculate the exact area between the graph of $y = 7 - 8x + x^2$ and the x-axis.

57. Calculate the exact area above the graph of $y = \sin \theta$ and below the graph of $y = \cos \theta$ for $0 \le \theta \le \pi/4$.

58. Find the exact value of the area between the graphs of $y = \cos x$ and $y = e^x$ for $0 \le x \le 1$.

59. Find the exact positive value of c making the area between the graph of $y = x^2 - c^2$ and the x-axis equal to 36.

60. The average value of the function $v(x) = 6/x^2$ on the interval $[1, c]$ is equal to 1. Find the value of c.

61. (a)   What is the average value of $f(t) = \sin t$ over $0 \le t \le 2\pi$? Why is this a reasonable answer?
     (b)   Find the average value of $f(t) = \sin t$ over $0 \le t \le \pi$.

62. Consider the costs of drilling an oil well. There are two types of costs: *fixed costs* (independent of the depth of the well) and *marginal cost* (the increment in cost that comes from drilling one more meter). By using both pieces of information, you can determine the total cost, C. The marginal costs depend on the depth at which you are drilling; drilling becomes more expensive, per meter, as you dig deeper into the earth. Suppose the fixed costs are 1,000,000 riyals (the riyal is the unit of currency of Saudi Arabia), and the marginal costs are

$$C'(x) = 4000 + 10x$$

in riyals/meter, where x is the depth in meters. Find the total cost of drilling a well x meters deep.

63. One of the earliest pollution problems brought to the attention of the Environmental Protection Agency (EPA) was the case of the Sioux Lake in eastern South Dakota. For years a small paper plant located nearby had been discharging waste containing carbon tetrachloride (CCl_4) into the waters of the lake. At the time the EPA learned of the situation, the chemical was entering at a rate of 16 cubic yards/year.

  The agency ordered the installation of filters designed to slow (and eventually stop) the flow of CCl_4 from the mill. Implementation of this program took exactly three years, during which the flow of pollutant was steady at 16 cubic yards/year. Once the filters were installed, the flow declined. If t is time measured in years since the EPA learned of the situation, the rate of flow was well approximated by

$$\text{Rate (in cubic yards/year)} = t^2 - 14t + 49,$$

between the time the filters were installed and the time the flow stopped.

(a) Draw a graph showing the rate of CCl_4 flow into the lake as a function of time, beginning at the time the EPA first learned of the situation.

(b) How many years elapsed between the time the EPA learned of the situation and the time the pollution flow stopped entirely?

(c) How much CCl_4 entered the waters during the time shown in the graph in part (a)?

6.3 DIFFERENTIAL EQUATIONS

In Chapter 2 we saw that velocity is the derivative of distance and that acceleration is the derivative of velocity. In this section we analyze the motion of an object falling freely under the influence of gravity. This involves going "backward" from acceleration to velocity to position.

Motion With Constant Velocity

Let's briefly consider a familiar problem: An object moving in a straight line with constant velocity. Imagine a car moving at 50 mph. How far does it go in a given time? The answer is given by

$$\text{Distance} = \text{Rate} \times \text{Time}$$

or

$$s = 50t$$

where s is the distance of the car (in miles) from a fixed reference point and t is the time in hours. Alternatively, we can describe the motion by writing the equation

$$\frac{ds}{dt} = 50.$$

This is called a *differential equation* for the *function* s. The solution to this equation is the general antiderivative

$$s = 50t + C.$$

The equation $s = 50t + C$ tells us that $s = C$ when $t = 0$. Thus, the constant C represents the initial distance, s_0, of the car from the reference point.

Uniformly Accelerated Motion

Now we consider an object moving with *constant acceleration* along a straight line, or *uniformly accelerated motion*. It has been known since Galileo's time that this is the motion of an object moving under the influence of gravity (ignoring air resistance). Thus, if v is the upward velocity and t is the time,

$$\frac{dv}{dt} = -g$$

where g is the constant *acceleration due to gravity*, whose value is determined by experiment. In the most frequently used units, its value is approximately

$$g = 9.81 \text{ m/sec}^2, \quad \text{or} \quad g = 32 \text{ ft/sec}^2.$$

The negative sign comes from the fact that velocity is measured upward, whereas gravity acts downward.

Example 1 A stone is dropped from a 100-foot-high building. Find, as a function of time, its position and velocity. When does it hit the ground, and how fast is it going at that time?

Solution Suppose t is measured in seconds from when the stone was dropped. If we measure distance, s, in feet above the ground, then the velocity, v, is in ft/sec upward, and the acceleration due to gravity is 32 ft/sec^2 downward, so

$$\frac{dv}{dt} = -32.$$

From what we know about antiderivatives, we must have

$$v = -32t + C$$

where C is some constant. Since $v = C$ when $t = 0$, the constant C represents the initial velocity, v_0. The fact that the stone is dropped rather than thrown off the top of the building tells us that the initial velocity is zero, so $v_0 = 0$. Substituting gives

$$0 = -32(0) + C \quad \text{so} \quad C = 0.$$

Thus,

$$v = -32t.$$

But now we can write

$$v = \frac{ds}{dt} = -32t.$$

The general antiderivative of $-32t$ is

$$s = -16t^2 + K,$$

where K is another constant.

Since the stone starts at the top of the building, $s = 100$ when $t = 0$. Substituting into $s = -16t^2 + K$ gives

$$100 = -16(0^2) + K,$$

so

$$K = 100,$$

and therefore

$$s = -16t^2 + 100.$$

Thus, we have found both v and s as functions of time.

The stone hits the ground when $s = 0$, so we must solve

$$0 = -16t^2 + 100$$

giving $t^2 = 100/16$ or $t = \pm 10/4 = \pm 2.5$ sec. Since t must be positive, $t = 2.5$ sec. At that time, $v = -32(2.5) = -80$ ft/sec. (Note that the velocity is negative because we are considering up as the positive direction and down as the negative direction.)

Example 2 An object is thrown vertically upward with a speed of 10 m/sec from a height of 2 meters. Find the highest point it reaches and when it hits the ground.

Solution We must find the position as a function of time. In this example, the velocity is in m/sec, so we use $g = 9.8$ m/sec^2. Measuring distance in meters upward from the ground, we have

$$\frac{dv}{dt} = -9.8.$$

As before, v is a function whose derivative is constant, so

$$v = -9.8t + C.$$

Since the initial velocity is 10 m/sec upward, we know that $v = 10$ when $t = 0$. Substituting gives

$$10 = -9.8(0) + C \quad \text{so} \quad C = 10.$$

Thus,
$$v = -9.8t + 10.$$

To find s, we use
$$v = \frac{ds}{dt} = -9.8t + 10$$

and look for a function that has $-9.8t + 10$ as its derivative. The general antiderivative of $-9.8t + 10$ is
$$s = -4.9t^2 + 10t + K,$$

where K is any constant. To find K, we use the fact that the object starts at a height of 2 meters, so $s = 2$ when $t = 0$. Substituting gives
$$2 = -4.9(0)^2 + 10(0) + K, \quad \text{so} \quad K = 2,$$

and therefore
$$s = -4.9t^2 + 10t + 2.$$

The object reaches its highest point when the velocity is 0, so at that time
$$v = -9.8t + 10 = 0.$$

This occurs when
$$t = \frac{10}{9.8} \approx 1.02 \text{ sec.}$$

When $t = 1.02$ seconds,
$$s = -4.9(1.02)^2 + 10(1.02) + 2 \approx 7.10 \text{ m.}$$

So the maximum height reached is 7.10 meters. The object reaches the ground when $s = 0$:
$$0 = -4.9t^2 + 10t + 2.$$

Solving this using the quadratic formula gives
$$t \approx -0.18 \quad \text{and} \quad t \approx 2.22 \text{ sec.}$$

Since the time at which the object hits the ground must be positive, $t \approx 2.22$ seconds.

Antiderivatives and Differential Equations

We solved the problem of uniformly accelerated motion by working backward from the derivative of a function to the function itself. If f is a known function, finding the *general solution* to the differential equation
$$\frac{dy}{dx} = f(x)$$
means finding the general antiderivative $y = F(x) + C$ with $F'(x) = f(x)$.

Example 3   Find and graph the general solution of the differential equation
$$\frac{dy}{dx} = \sin x + 2.$$

Solution   We are asking for a function whose derivative is $\sin x + 2$. An antiderivative of $\sin x + 2$ is
$$y = -\cos x + 2x.$$

The general solution is therefore
$$y = -\cos x + 2x + C,$$

where C is any constant. Figure 6.17 shows several curves in this family.

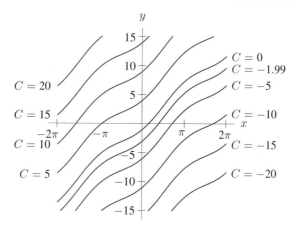

Figure 6.17: Solution curves of $\frac{dy}{dx} = \sin x + 2$

How Can We Pick One Solution to the Differential Equation $\dfrac{dy}{dx} = f(x)$?

Picking one antiderivative is equivalent to selecting a value of C. This requires knowing an extra piece of information, such as the initial velocity or initial position. In general, we pick a particular curve in the family of solutions by specifying that the curve should pass through a given point (x_0, y_0). The differential equation plus an extra condition

$$\frac{dy}{dx} = f(x), \qquad y(x_0) = y_0$$

is called an *initial value problem*. An initial value problem usually has a unique solution. Note that $y(x_0) = y_0$ is shorthand for $y = y_0$ when $x = x_0$.

Example 4   Find the solution of the initial value problem

$$\frac{dy}{dx} = \sin x + 2, \quad y(3) = 5.$$

Solution   We have already seen that the general solution to the differential equation is $y = -\cos x + 2x + C$. The initial condition allows us to determine the constant C. Substituting $y(3) = 5$ gives

$$5 = y(3) = -\cos 3 + 2 \cdot 3 + C,$$

so C is given by

$$C = 5 + \cos 3 - 6 \approx -1.99.$$

Thus, the (unique) solution is

$$y = -\cos x + 2x - 1.99.$$

Figure 6.17 shows this particular solution (marked with $C = -1.99$).

Problems for Section 6.3

Find the general solution of the differential equations in Problems 1–4.

1. $\dfrac{dy}{dx} = x^3 + 5$ 2. $\dfrac{dy}{dx} = 8x + \dfrac{1}{x}$ 3. $\dfrac{dW}{dt} = 4\sqrt{t}$ 4. $\dfrac{dr}{dp} = 3 \sin p$

In Problems 5–7, find the solution of the initial value problem.

5. $\dfrac{dy}{dx} = 6x^2 + 4x, \quad y(2) = 10$ 6. $\dfrac{dP}{dt} = 10e^t, \quad P(0) = 25$

7. $\dfrac{ds}{dt} = -32t + 100, \quad s = 50$ when $t = 0$

8. Show that $y = x + \sin x - \pi$ satisfies the initial value problem

$$\frac{dy}{dx} = 1 + \cos x, \quad y(\pi) = 0.$$

9. Show that $y = xe^{-x} + 2$ is a solution of the initial value problem

$$\frac{dy}{dx} = (1 - x)e^{-x}, \quad y(0) = 2.$$

10. (a) Find the general solution of the differential equation $\dfrac{dy}{dx} = 2x + 1$.
    (b) Sketch a graph of at least three solutions.
    (c) Find the solution satisfying $y(1) = 5$. Graph this solution with the others from part (b).

11. Ice is forming on a pond at a rate given by

$$\frac{dy}{dt} = k\sqrt{t},$$

where y is the thickness of the ice in inches at time t measured in hours since the ice started forming, and k is a positive constant. Find y as a function of t.

12. If a car goes from 0 to 80 mph in six seconds with constant acceleration, what is that acceleration?

13. A car going 80 ft/sec (about 55 mph) brakes to a stop in five seconds. Assume the deceleration is constant.

    (a) Graph the velocity against time, t, for $0 \le t \le 5$ seconds.
    (b) Represent, as an area on the graph, the total distance traveled from the time the brakes are applied until the car comes to a stop.
    (c) Find this area and hence the distance traveled.
    (d) Now find the total distance traveled using antidifferentiation.

14. A 727 jet needs to be flying 200 mph to take off. If it can accelerate from 0 to 200 mph in 30 seconds, how long must the runway be? (Assume constant acceleration.)

15. Suppose a car going at 30 ft/sec decelerates at a constant rate of 5 ft/sec^2.

    (a) Draw up a table showing the velocity of the car every half second. When does the car come to rest?
    (b) Using your table, find left and right sums which estimate the total distance traveled before the car comes to rest. Which is an overestimate, and which is an underestimate?
    (c) Sketch a graph of velocity against time. On the graph, show an area representing the distance traveled before the car comes to rest. Use the graph to calculate this distance.
    (d) Now find a formula for the velocity of the car as a function of time, and then find the total distance traveled by antidifferentiation. What is the relationship between your answer to parts (c) and (d) and your estimates in part (b)?

16. A water balloon launcher has been installed on the roof of a building. The vertical velocity (in feet/sec) of the water balloon at time t seconds after launch is given by

$$v(t) = -32t + 40,$$

with $v > 0$ corresponding to upward motion.

    (a) If the roof of the building is 30 feet above the ground, find an expression for the height of the water balloon above the ground at any time t.
    (b) What is the average velocity of the balloon between $t = 1.5$ and $t = 3$ seconds?
    (c) A 6-foot person is standing on the ground. How fast is the water balloon falling when it strikes the person on the top of the head?

17. An object is shot vertically upward from the ground with a velocity of 160 ft/sec.

    (a) Sketch a graph of the velocity of the object (with upward as positive) against time.
    (b) Mark on the graph the points when the object reaches its highest point and when it lands.
    (c) Find the maximum height reached by the object by considering an area on the graph.
    (d) Now express velocity as a function of time, and find the greatest height by antidifferentiation.

18. A stone thrown upwards from the top of a 320-foot cliff at 128 ft/sec eventually falls to the beach below.

    (a) How long does the stone take to reach its highest point?
    (b) What is its maximum height?
    (c) How long before the stone hits the beach?
    (d) What is the velocity of the stone on impact?

19. On the moon, the acceleration due to gravity is about 1.6 m/sec^2 (compared to $g \approx 9.8$ m/sec^2 on earth). If you drop a rock on the moon (with initial velocity 0), find formulas for:

    (a) Its velocity, $v(t)$, at time t.
    (b) The distance, $s(t)$, it falls in time t.

20. (a) Imagine throwing a rock straight up in the air. What should its initial velocity be if the rock reaches a maximum height of 100 feet above its starting point?
    (b) Now imagine being transplanted to the moon and throwing a moon rock vertically upward with the same velocity as in part (a). How high will it go? (On the moon, $g = 5$ ft/sec^2.)

21. A cat, walking along the window ledge of a New York apartment, knocks off a flower pot, which falls to the street 200 feet below. How fast is the flower pot traveling when it hits the street? (Give your answer in ft/sec and in mph, given that 1 ft/sec = 15/22 mph.)

22. Of all the cars surveyed in the April 1991 issue of *Car and Driver*, the Acura NSX, going at 70 mph, stops in the shortest distance: 157 feet. Find the acceleration, assuming it is constant.

6.4 SECOND FUNDAMENTAL THEOREM OF CALCULUS

Suppose f is an elementary function, that is, a combination of constants, powers of x, $\sin x$, $\cos x$, e^x, and $\ln x$. Then we have to be lucky to find an antiderivative F which is also an elementary function. But if we can't find F as an elementary function, how can we be sure that F exists at all? In this section we learn to use the definite integral to construct antiderivatives.

Construction of Antiderivatives Using the Definite Integral

Consider the function $f(x) = e^{-x^2}$. We would like to find a way of calculating values of its antiderivative. If the antiderivative, F, were an elementary function, we could evaluate it using a calculator, but it isn't. However, we know from the Fundamental Theorem of Calculus that

$$F(b) - F(a) = \int_a^b e^{-t^2}\, dt$$

so

$$F(x) - F(0) = \int_0^x e^{-t^2}\, dt.$$

All we have done here is to set $a = 0$ and replace the usual b in the upper limit with an x. Suppose we want the antiderivative that satisfies $F(0) = 0$. Then we get

$$F(x) = \int_0^x e^{-t^2}\, dt.$$

This is a formula for F. For any value of x, there is a unique value for $F(x)$, so F is a function. For any fixed x, we can calculate $F(x)$ numerically to any accuracy we like. For example, we can calculate

$$F(2) = \int_0^2 e^{-t^2}\, dt = 0.88208\ldots.$$

Notice that our expression for F is not an elementary function; we have *created* a new function using the definite integral. The next theorem says that this method of constructing antiderivatives works in general. This means that if we define F by

$$F(x) = \int_a^x f(t)\, dt$$

then F must be an antiderivative of f.

> **Construction Theorem for Antiderivatives (Second Fundamental Theorem of Calculus)**
> If f is a continuous function on an interval, and if a is any number in that interval, then the function F defined by
> $$F(x) = \int_a^x f(t)\, dt$$
> is an antiderivative of f.

Notice that if we already know an antiderivative of f, which we may call G, then the Fundamental Theorem of Calculus tells us that

$$G(x) - G(a) = \int_a^x f(t)\, dt.$$

By the definition of F, we have

$$F(x) = \int_a^x f(t)\, dt.$$

Therefore,

$$G(x) - G(a) = F(x),$$

so F and G differ by a constant (namely $G(a)$), and so F and G are *both* antiderivatives of f.

Justification of the Construction Theorem

Our task is to show that F, defined by this integral, is an antiderivative of f. We want to show that $F'(x) = f(x)$. By the definition of the derivative,

$$F'(x) = \lim_{h \to 0} \frac{F(x + h) - F(x)}{h}.$$

To gain some geometric insight, let's suppose f is positive and h is positive. Then we can visualize

$$F(x) = \int_a^x f(t)\, dt$$

and

$$F(x + h) = \int_a^{x+h} f(t)\, dt,$$

as areas, which leads to representing

$$F(x + h) - F(x) = \int_x^{x+h} f(t)\, dt$$

as a difference of two areas. From Figure 6.18, we see that $F(x + h) - F(x)$ is roughly the area of a rectangle of height $f(x)$ and width h (shaded darker in Figure 6.18), so we have

$$F(x + h) - F(x) \approx f(x)h,$$

hence

$$\frac{F(x + h) - F(x)}{h} \approx f(x).$$

More precisely, we can use the theorem on comparing integrals on page 172 to conclude that

$$mh \le \int_x^{x+h} f(t)\, dt \le Mh,$$

where m is the greatest lower bound for f on the interval from x to $x + h$ and M is the least upper

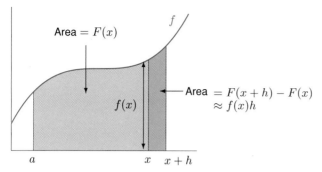

Figure 6.18: $F(x + h) - F(x)$ is area of roughly rectangular region

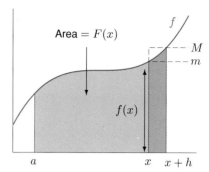

Figure 6.19: Upper and lower bounds for $F(x + h) - F(x)$

bound on that interval. (See Figure 6.19.) Hence

$$mh \leq F(x + h) - F(x) \leq Mh,$$

so

$$m \leq \frac{F(x + h) - F(x)}{h} \leq M.$$

Since f is continuous, both m and M approach $f(x)$ as h approaches zero. Thus

$$f(x) \leq \lim_{h \to 0} \frac{F(x + h) - F(x)}{h} \leq f(x).$$

Thus both inequalities must actually be equalities, so we have the result we want:

$$f(x) = \lim_{h \to 0} \frac{F(x + h) - F(x)}{h} = F'(x).$$

Using the Construction Theorem for Antiderivatives

The construction theorem enables us to write down antiderivatives of functions that do not have elementary antiderivatives. For example, an antiderivative of $(\sin x)/x$ is

$$F(x) = \int_0^x \frac{\sin t}{t} \, dt.$$

Notice that F is a function; we can calculate its values to any degree of accuracy. This function already has a name: it is called the *sine-integral*, and it is denoted $\text{Si}(x)$.

Example 1   Construct a table of values of $\text{Si}(x)$ for $x = 0, 1, 2, 3$.

Solution   Using numerical methods, we calculate the values of $\text{Si}(x) = \int_0^x \frac{\sin t}{t} \, dt$ given in Table 6.2. Since the integrand is undefined at $t = 0$, we took the lower limit as 0.00001 instead of 0.

TABLE 6.2   *A table of values of Si(x)*

x	0	1	2	3
$\text{Si}(x)$	0	0.95	1.61	1.85

The reason the sine-integral has a name is that scientists and engineers have found a use for it (for example, in optics). For people who use it all the time, it is just another common function like

sine or cosine. Its derivative,

$$\frac{d}{dx}\,\mathrm{Si}(x) = \frac{\sin x}{x},$$

can be used with the other differentiation rules, as in the next example.

Example 2   Find the derivative of $x\,\mathrm{Si}(x)$.

Solution   Using the product rule,

$$\frac{d}{dx}\,(x\,\mathrm{Si}(x)) = \left(\frac{d}{dx}\,x\right)\mathrm{Si}(x) + x\left(\frac{d}{dx}\,\mathrm{Si}(x)\right)$$

$$= 1\cdot\mathrm{Si}(x) + x\,\frac{\sin x}{x}$$

$$= \mathrm{Si}(x) + \sin x.$$

Problems for Section 6.4

1. The graph of the derivative F' of some function F is given in Figure 6.20. If $F(20) = 150$, estimate the maximum value attained by F.

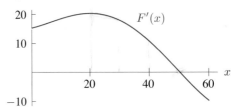

Figure 6.20

2. Assume that $F'(t) = \sin t \cos t$ and $F(0) = 1$. Find $F(b)$ for $b = 0,\ 0.5,\ 1,\ 1.5,\ 2,\ 2.5,$ and 3.

For the functions in Problems 3–5, let $F(x) = \int_0^x f(t)\,dt$. Graph $F(x)$ as a function of x.

3.   4.   5.

6. Make a table of values for the following function for $x = 0,\ 0.5,\ 1.0,\ 1.5,$ and 2.0.

$$I(x) = \int_0^x \sqrt{t^4 + 1}\,dt$$

7. (a) Continue the table of values for $\mathrm{Si}(x) = \displaystyle\int_0^x \frac{\sin t}{t}\,dt$, in Example 1, page 311 for $x = 4$ and $x = 5$.

   (b) Why is $\mathrm{Si}(x)$ decreasing between $x = 4$ and $x = 5$?

Find the derivatives in Problems 8–13.

8. $\dfrac{d}{dx}\displaystyle\int_0^x \sqrt{3 + \cos(t^2)}\,dt$ 9. $\dfrac{d}{dx}\displaystyle\int_1^x (1 + t)^{200}\,dt$ 10. $\dfrac{d}{dx}\displaystyle\int_{0.5}^x \arctan(t^2)\,dt$

11. $\dfrac{d}{dt}\displaystyle\int_t^\pi \cos(z^3)\,dz$ 12. $\dfrac{d}{dx}\displaystyle\int_x^1 \ln t\,dt$ 13. $\dfrac{d}{dx}\left[\mathrm{Si}(x^2)\right]$

14. Let $F(x) = \int_0^x \sin(2t)\, dt$.

    (a) Evaluate $F(\pi)$.

    (b) Draw a sketch to explain geometrically why the answer to part (a) is correct.

    (c) For what values of x is $F(x)$ positive? negative?

15. Let $F(x) = \int_2^x \frac{1}{\ln t}\, dt$ for $x \geq 2$.

    (a) Find $F'(x)$.

    (b) Is F increasing or decreasing? What can you say about the concavity of its graph?

    (c) Sketch a graph of $F(x)$.

Problems 16 and 17 concern the *error function*, erf(x), defined by

$$\text{erf}(x) = \frac{2}{\sqrt{\pi}} \int_0^x e^{-t^2}\, dt.$$

In Problems 16–19, find each of the given quantities.

16. $\dfrac{d}{dx}\left(x\,\text{erf}(x)\right)$ 17. $\dfrac{d}{dx}\left(\text{erf}(\sqrt{x})\right)$ 18. $\dfrac{d}{dx}\left(\int_0^{x^3} e^{-t^2}\, dt\right)$ 19. $\dfrac{d}{dx}\left(\int_x^{x^3} e^{-t^2}\, dt\right)$

CHAPTER SUMMARY

- **Constructing antiderivatives**
  Graphically, numerically, analytically

- **The family of antiderivatives**
  The indefinite integral

- **Differential equations**
  Initial value problems, uniform motion

- **Construction theorem (Second Fundamental Theorem of Calculus)**
  Constructing antiderivatives using definite integrals

REVIEW PROBLEMS FOR CHAPTER SIX

Find the indefinite integrals in Problems 1–15.

1. $\displaystyle\int (5x + 7)\, dx$ 2. $\displaystyle\int \left(\frac{3}{t} - \frac{2}{t^2}\right) dt$ 3. $\displaystyle\int (e^x + 5)\, dx$

4. $\displaystyle\int \left(x^{3/2} + \frac{\sqrt{x}}{5} - \frac{2}{x}\right) dx$ 5. $\displaystyle\int \frac{1}{\cos^2 x}\, dx$ 6. $\displaystyle\int 2^x\, dx$

7. $\displaystyle\int (x + 1)^2\, dx$ 8. $\displaystyle\int (x + 1)^3\, dx$ 9. $\displaystyle\int (x + 1)^9\, dx$

10. $\displaystyle\int \left(\frac{x + 1}{x}\right) dx$ 11. $\displaystyle\int \left(\frac{x^2 + x + 1}{x}\right) dx$ 12. $\displaystyle\int \left(3\cos\psi + 3\sqrt{\psi}\right) d\psi$

13. $\displaystyle\int (3\cos x - 7\sin x)\, dx$ 14. $\displaystyle\int \left(x + \frac{2}{x} + \pi\sin x\right) dx$ 15. $\displaystyle\int (2e^x - 8\cos x)\, dx$

Find antiderivatives for the functions in Problems 16–38. Check by differentiation.

16. $f(x) = x + x^5 + x^{-5}$

17. $g(x) = \dfrac{1}{x} + \dfrac{1}{x^2} + \dfrac{1}{x^3}$

18. $f(x) = x^6 - \dfrac{1}{7x^6}$

19. $g(t) = 5 + \cos t$

20. $g(\alpha) = 3\sin(3\alpha)$

21. $h(r) = 2\sqrt{r} + \dfrac{1}{2\sqrt{r}}$

22. $g(x) = (x+1)^3$

23. $h(t) = \dfrac{(t-1)^2}{t^2}$

24. $p(y) = \dfrac{1}{y} + y + 1$

25. $f(z) = e^z + 3$

26. $f(x) = 2xe^{x^2}$

27. $g(x) = e^x + e^{1+x}$

28. $g(\theta) = \sin\theta - 2\cos\theta$

29. $p(r) = 2\pi r$

30. $f(x) = 2^x + \dfrac{1}{2^x}$

31. $h(t) = \cos t(1 + \sin t)^{29}$

32. $h(t) = 2t\cos(t^2)$

33. $f(t) = t\cos(t^2)$

34. $r(x) = 3x^2\cos(x^3 + 7)$

35. $g(y) = e^2 + 2^y$

36. $g(\theta) = \sin\theta + \dfrac{1}{1+\theta^2}$

37. $f(x) = xe^{x^2}$

38. $f(x) = 6\sqrt{x} - \dfrac{1}{x^2} + \dfrac{10}{x}$

39. Calculate the exact area above the graph of $y = \frac{1}{2}\left(\frac{3}{\pi}x\right)^2$ and below the graph of $y = \cos x$. The curves intersect at $x = \pm\pi/3$.

40. Find the exact area of the region bounded by the x-axis and the graph of $y = x^3 - x$.

41. Find the exact positive value of c which makes the area under the graph of $y = c(1 - x^2)$ and above the x-axis equal to 1.

42. Sketch the parabola $y = x(x - \pi)$ and the curve $y = \sin x$, showing their points of intersection. Find the exact area between the two graphs.

For Problems 43–44 the graph of $f'(x)$ is given. Sketch a possible graph for $f(x)$. Mark the points $x_1 \ldots x_4$ on your graph and label local maxima, local minima and points of inflection.

43.

44.

45. If the graph of f is as shown in Figure 6.21, sketch a graph of the function $F(x) = \displaystyle\int_0^x f(t)\,dt$.

46. An object is attached to a coiled spring which is suspended from the ceiling of a room. The function $h(t)$ gives the height of the object above the floor of the room at time, t. The graph of $h'(t)$ is given in Figure 6.22. Sketch a possible graph of $h(t)$.

Figure 6.21

Figure 6.22

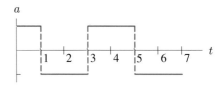

Figure 6.23

47. The acceleration, a, of a particle as a function of time, t, is shown in Figure 6.23. Sketch graphs of velocity and position against time. Assume the particle starts at rest at the origin.

48. An object thrown in the air on a planet in a distant galaxy is at height $s = -25t^2 + 72t + 40$ feet at time t seconds after it is thrown. What is the acceleration due to gravity on this planet? With what velocity was the object thrown? From what height?

49. The angular speed of a car engine increases from 1100 revs/min to 2500 revs/min in 6 sec. Assuming that the angular acceleration was constant,

(a) Find the angular acceleration in revs/min^2.

(b) How many revolutions does the engine make in this time?

50. A helicopter rotor slows down at a constant rate from 350 revs/min to 260 revs/min in 1.5 minutes.

(a) Find the angular acceleration during this time interval. What are the units of this acceleration?

(b) Assuming the angular acceleration remains constant, how long does it take for the rotor to stop? (Measure time from the moment when speed was 350 revs/min.)

(c) How many revolutions does the rotor make between the time the angular speed was 350 revs/min and stopping?

51. An object is thrown vertically upward with a velocity of 80 ft/sec.

(a) Make a table showing its velocity every second.

(b) When does it reach its highest point? When does it hit the ground?

(c) Using your table, write left and right sums which under- and overestimate the height the object attains.

(d) Use antidifferentiation to find the greatest height it reaches.

52. A car, initially moving at 60 mph, has a constant deceleration and stops in a distance of 200 feet. What is its deceleration? (Give your answer in ft/sec^2. Note that 1 mph = 22/15 ft/sec.)

53. Consider a bacteria population whose birth rate, B, is given by the curve in Figure 6.24 as a function of time in hours. The birth rate is in births per hour. The curve marked D gives the death rate (in deaths per hour) of the same population.

(a) Explain what the shape of each of these graphs tells you about the population.

(b) Use the graphs to find the time at which the net rate of increase of the population is at a maximum.

(c) Suppose at time $t = 0$ the population has size N. Sketch the graph of the total number born by time t. Also sketch the graph of the number alive at time t. Use the given graphs to find the time at which the population size is a maximum.

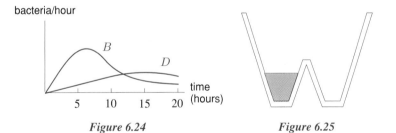

Figure 6.24   Figure 6.25   Figure 6.26

54. Water flows at a constant rate into the left side of the W-shaped container shown in Figure 6.25. Sketch a graph of the height, H, of the water in the left side of the container as a function of time, t. Suppose the container starts out empty.

55. Suppose a particle of mass, m, acted on by a force, F, moves in a straight line. Its acceleration, a, is given by Newton's Law:

$$F = ma.$$

The work, W, done by a *constant* force when the particle moves through a displacement, d, is

$$W = Fd.$$

Suppose the velocity, v, of the particle as a function of time, t, is given in Figure 6.26. What is the sign of the work done during each of the time intervals: $[0, t_1]$, $[t_1, t_2]$, $[t_2, t_3]$, $[t_3, t_4]$,$[t_2, t_4]$?

56. Let $F(x) = \displaystyle\int_{\pi/2}^{x} \frac{\sin t}{t} \, dt$. Find the value(s) of x between $\pi/2$ and $3\pi/2$ for which $F(x)$ has a global maximum and global minimum. Show your reasoning.

PROJECTS

1. **Distribution of Resources**

   Whether a resource is distributed evenly among members of a population is often an important political or economic question. How can we measure this? How can we decide if the distribution of wealth in this country is becoming more or less equitable over time? How can we measure which country has the most equitable income distribution? This problem describes a way of making such measurements.

   Suppose the resource is distributed evenly. Then any 20% of the population will have 20% of the resource. Similarly, any 30% will have 30% of the resource and so on. If, however, the resource is not distributed evenly, the poorest p% of the population (in terms of this resource) will not have p% of the goods. Suppose $F(x)$ represents the fraction of the resources owned by the poorest fraction x of the population. Thus $F(0.4) = 0.1$ means that the poorest 40% of the population owns 10% of the resource.

   (a) What would F be if the resource were distributed evenly?

   (b) What must be true of any such F? What must $F(0)$ and $F(1)$ equal? Is F increasing or decreasing? Is the graph of F concave up or concave down?

   (c) Gini's index of inequality, G, is one way to measure how evenly the resource is distributed. It is defined by

   $$G = 2 \int_0^1 [x - F(x)] \, dx.$$

   Show graphically what G represents.

2. **Yield from an Apple Orchard**

   Figure 6.27 is a graph of the annual yield, $y(t)$ (in bushels per year), from an orchard t years after planting. The trees take about 10 years to get established, but for the next 20 years they give a substantial yield. After about 30 years, however, age and disease start to take their toll, and the annual yield falls off.[1]

   (a) Represent on a sketch of Figure 6.27 the total yield, $Y(T)$, until time T. Write an expression for $Y(T)$ in terms of $y(t)$.

   (b) Sketch a graph of $Y(T)$ against T.

   (c) Write an expression for the average annual yield, $a(T)$, up until time T.

   (d) The important question is: When should the orchard be cut down and replanted? Assume that you want to maximize your average revenue per year, and that fruit prices remain constant, so that this is achieved by maximizing your average annual yield.

   (i) What condition on $Y(T)$ maximizes the average annual yield?

   (ii) What condition on $y(T)$ maximizes the average annual yield? Estimate the T value which yields the maximum average annual yield.

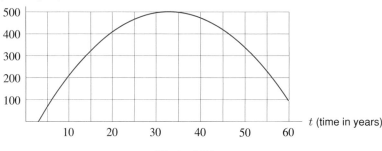

Figure 6.27

---

[1]From Peter D. Taylor, *Calculus: The Analysis of Functions*, (Toronto: Wall & Emerson, Inc., 1992).

FOCUS ON MODELING

THE EQUATIONS OF MOTION

The problem of a body moving freely under the influence of gravity near the surface of the earth intrigued mathematicians and philosophers from Greek times onward and was finally solved by Galileo and Newton. The question to be answered was: How do the velocity and the position of the body vary with time? We define s to be the position, or height, of the body above a fixed point (often ground level); v then is the velocity of the body measured upward. We assume that the acceleration of the body is a constant, $-g$ (the negative sign means that the acceleration is downward), so

$$\text{Acceleration } = \frac{dv}{dt} = -g.$$

Thus, velocity is the antiderivative of $-g$:

$$v = -gt + C.$$

If the initial velocity is v_0, then $C = v_0$, so

$$v = -gt + v_0.$$

How about the position? We know that

$$\frac{ds}{dt} = v = -gt + v_0.$$

Therefore, we can find s by antidifferentiating again, giving:

$$s = -\frac{gt^2}{2} + v_0 t + C.$$

If the initial position is s_0, then we must have

$$s = -\frac{gt^2}{2} + v_0 t + s_0.$$

Our derivation of the formulas for the velocity and the position of the body took little effort. It hides an almost 2000-year struggle to understand the mechanics of falling bodies, from Aristotle's *Physics* to Galileo's *Dialogues Concerning Two New Sciences*. Before sketching Galileo's solution of the problem, we make a few more observations on uniform and uniformly accelerated motion.

Though it is an oversimplification of his ideas, we can say that Aristotle's conception of motion was primarily in terms of *change of position*. This seems entirely reasonable; it is what we commonly observe, and this view dominated discussions of motion for centuries. But it misses a subtlety that was brought to light by Descartes, Galileo, and, with a different emphasis, by Newton. That subtlety is now usually referred to as the *principle of inertia*.

This principle holds that a body traveling undisturbed at constant velocity in a straight line will continue in this motion indefinitely. Stated another way, it says that one cannot distinguish in any absolute sense (that is, by performing an experiment), between being at rest and moving with constant velocity in a straight line. If you are reading this book in a closed room and have no external reference points, there is no experiment that will tell you, one way or the other, whether you are at rest or whether you, the room, and everything in it are moving with constant velocity in a straight line. Therefore, as Newton saw, an understanding of motion should be based on *change of velocity* rather than change of position. Since acceleration is the rate of change of velocity, it is acceleration that must play a central role in the description of motion.

How does acceleration come about? How does the velocity change? Through the action of *forces*. Newton placed a new emphasis on the importance of forces. Newton's laws of motion do not say what a force *is*, they say how it *acts*. His first law is the principle of inertia, which says what happens in the *absence* of a force—there is no change in velocity. His second law says that a force acts to produce a change in velocity, that is, an acceleration. It states that $F = ma$, where m is the mass of the object, F is the net force, and a is the acceleration produced by this force.

Let's return to Galileo for a moment. He demonstrated that a body falling under the influence of gravity does so with constant acceleration. Furthermore, assuming we can neglect air resistance, this constant acceleration is independent of the mass of the body. This last fact was the outcome of Galileo's famous observation that a heavy ball and a light ball dropped off the Leaning Tower of Pisa hit the ground at the same time. Whether or not he actually performed this experiment, Galileo presented a very clear thought experiment in the *Dialogues* to prove the same point. (This point was counter to Aristotle's more common sense notion that the heavier ball would reach the ground first.) Galileo showed that the mass of the object need not appear as a variable in the equation of motion. Thus, the same constant acceleration equation applies to all bodies falling under the influence of gravity.

Nearly a hundred years after Galileo's experiment, Newton formulated his laws of motion and gravity, which led to a differential equation describing the motion of a falling body. According to Newton, acceleration is caused by force, and in the case of falling bodies, that force is the force of gravity. Newton's law of gravity says that the gravitational force between two bodies is attractive and given by

$$F = \frac{GMm}{r^2},$$

where G is the gravitational constant, m and M are the masses of the two bodies, and r is the distance between them. This is the famous *inverse square law*. For a falling body, we take M to be the mass of the earth and r to be the distance from the body to the center of the earth. So, actually, r changes as the body falls, but for anything we can easily observe (say, a ball dropped from the Tower of Pisa), it won't change significantly over the course of the motion. Hence, as an approximation, it is reasonable to assume that the force is constant. According to Newton's second law,

$$\text{Force} = \text{Mass} \times \text{Acceleration}.$$

Since the gravitational force is acting downwards

$$-\frac{GMm}{r^2} = m\frac{d^2 s}{dt^2}.$$

Hence,

$$\frac{d^2 s}{dt^2} = -\frac{GM}{r^2} = \text{constant}.$$

If we define $g = GM/r^2$, then

$$\frac{d^2 s}{dt^2} = -g.$$

The fact that the mass cancels out of Newton's equations of motion reflects Galileo's experimental observation that the acceleration due to gravity is independent of the mass of the body.

Problems on the Equations of Motion

1. An object is dropped from a 400-foot tower. When does it hit the ground and how fast is it going at the time of the impact?

2. Consider the object in Problem 1, and suppose it falls off the same 400-foot tower. What would the acceleration due to gravity have to be to make it reach the ground in half the time?

3. A ball that is dropped from a window hits the ground in five seconds. How high is the window? (Give your answer in feet.)

4. On the moon the acceleration due to gravity is 5 ft/sec^2. An astronaut jumps into the air with an initial upward velocity of 10 ft/sec. How high does he go? How long is the astronaut off the ground?

5. Galileo was the first person to show that the distance traveled by a body falling from rest is proportional to the square of the time it has traveled, and independent of the mass of the body. Derive this result from the fact that the acceleration due to gravity is a constant.

6. While attempting to understand the motion of bodies under gravity, Galileo stated that:

   The time in which any space is traversed by a body starting at rest and uniformly accelerated is equal to the time in which that same space would be traversed by the same body moving at a uniform speed whose value is the mean of the highest velocity and the velocity just before acceleration began.

   (a) Write Galileo's statement in symbols, defining all the symbols you use.
   (b) Verify Galileo's statement for a body dropped off a 100-foot building accelerating from rest under gravity until it hits the ground.
   (c) Show why Galileo's statement is true in general.

7. In his *Dialogues Concerning Two New Sciences*, Galileo wrote:

   The distances traversed during equal intervals of time by a body falling from rest stand to one another in the same ratio as the odd numbers beginning with unity.

   Assume, as is now believed, that $s = -(gt^2)/2$, where s is the total distance traveled in time t, and g is the acceleration due to gravity.

   (a) How far does a falling body travel in the first second (between $t = 0$ and $t = 1$)? During the second second (between $t = 1$ and $t = 2$)? The third second? The fourth second?
   (b) What do your answers tell you about the truth of Galileo's statement?

8. The acceleration due to gravity 2 meters from the ground is 9.8 m/sec^2. What is the acceleration due to gravity 100 meters from the ground? At 100,000 meters? (The radius of the earth is 6.4×10^6 meters.)

CHAPTER SEVEN

INTEGRATION

In Chapter 3, we introduced the definite integral, and in Chapter 6, we introduced antiderivatives. In this chapter, we focus on methods of finding antiderivatives and definite integrals. We first learn additional ways to evaluate antiderivatives and definite integrals exactly, and we then develop more efficient numerical methods for approximating definite integrals. Finally, we introduce improper integrals and discuss methods of evaluating them.

7.1 INTEGRATION BY SUBSTITUTION: PART I

In Chapter 4, we learned rules which enabled us to differentiate any function obtained by combining constants, powers of x, $\sin x$, $\cos x$, e^x, and $\ln x$, using addition, multiplication, division, or composition of functions. These are called *elementary* functions.

Are there similar rules for antidifferentiation? For example, is there a product rule for antiderivatives? Or a chain rule? In the next few sections, we introduce two methods of antidifferentiation: substitution and integration by parts, which reverse the chain and product rules, respectively.

There is a great difference between looking for derivatives and looking for antiderivatives. Every elementary function has elementary derivatives. However, many elementary functions do not have elementary antiderivatives. Some examples are $\sqrt{x^3 + 1}$, $(\sin x)/x$, and e^{-x^2}. These are not exotic functions, but ordinary functions that arise naturally.

The Guess-and-Check Method

A good strategy for finding simple antiderivatives is to *guess* an answer (using knowledge of differentiation rules) and then *check* the answer by differentiating it. If we get the expected result, then we're done; otherwise, we revise the guess and check again.

The method of guess-and-check is most useful in reversing the chain rule. According to the chain rule,

$$\frac{d}{dx}(f(g(x))) = \underbrace{f'}_{\text{Derivative of outside}} \overbrace{(g(x))}^{\text{Inside}} \cdot \underbrace{g'(x)}_{\text{Derivative of inside}}.$$

Thus, any function which is the result of applying the chain rule will be the product of two factors: the "derivative of the outside" and the "derivative of the inside." If a function has this form, its antiderivative is $f(g(x))$.

Example 1 Find $\displaystyle\int 3x^2 \cos(x^3)\, dx$.

Solution The function $3x^2 \cos(x^3)$ looks like the result of applying the chain rule: there is an "inside" function x^3 and its derivative $3x^2$ appearing as a factor. We guess $\sin(x^3)$ for the antiderivative. Differentiating to check gives

$$\frac{d}{dx}(\sin(x^3)) = \cos(x^3) \cdot (3x^2).$$

Since this is what we began with, we have

$$\int 3x^2 \cos(x^3)\, dx = \sin(x^3) + C.$$

The basic idea of this method is to try to find an inside function whose derivative appears as a factor. This works even when the derivative is missing a constant factor, as in the next example.

Example 2 Find $\displaystyle\int te^{(t^2+1)}\, dt$.

Solution It looks like $t^2 + 1$ is an inside function. So we guess $e^{(t^2+1)}$ for the antiderivative, since taking the derivative of an exponential results in the reappearance of the exponential together with other terms from the chain rule. Now we check:

$$\frac{d}{dt}\left(e^{(t^2+1)}\right) = \left(e^{(t^2+1)}\right) \cdot 2t.$$

The original guess was too large by a factor of 2. We change the guess to $\frac{1}{2}e^{(t^2+1)}$ and check again:

$$\frac{d}{dt}\left(\frac{1}{2}e^{(t^2+1)}\right) = \frac{1}{2}e^{(t^2+1)} \cdot 2t = e^{(t^2+1)} \cdot t.$$

Thus,

$$\int te^{(t^2+1)}\, dt = \frac{1}{2}e^{(t^2+1)} + C.$$

Example 3   Find $\displaystyle\int x^3\sqrt{x^4+5}\, dx.$

Solution   Here the inside function is x^4+5, and its derivative appears as a factor, with the exception of a missing 4. Thus, the integrand we have is more or less of the form

$$g'(x)\sqrt{g(x)},$$

with $g(x) = x^4 + 5$. We might guess that an antiderivative is

$$\frac{(g(x))^{3/2}}{3/2} = \frac{(x^4+5)^{3/2}}{3/2}.$$

Let's check and see:

$$\frac{d}{dx}\left(\frac{(x^4+5)^{3/2}}{3/2}\right) = \frac{3}{2}\frac{(x^4+5)^{1/2}}{3/2} \cdot 4x^3 = 4x^3(x^4+5)^{1/2},$$

so $\dfrac{(x^4+5)^{3/2}}{3/2}$ is too big by a factor of 4. The correct antiderivative is

$$\frac{1}{4}\frac{(x^4+5)^{3/2}}{3/2} = \frac{1}{6}(x^4+5)^{3/2}.$$

Thus

$$\int x^3\sqrt{x^4+5}\, dx = \frac{1}{6}(x^4+5)^{3/2} + C.$$

As a final check:

$$\frac{d}{dx}\left(\frac{1}{6}(x^4+5)^{3/2}\right) = \frac{1}{6} \cdot \frac{3}{2}(x^4+5)^{1/2} \cdot 4x^3 = x^3(x^4+5)^{1/2}.$$

As we have seen in the preceding examples, antidifferentiating a function often involves "correcting for" constant factors: if differentiation produces an extra factor of 4, antidifferentiation will require a factor of $\frac{1}{4}$.

The Method of Substitution

When the integrand is complicated, it helps to formalize this guess-and-check method as follows:

> **To Make a Substitution**
>
> Let w be the "inside function" and $dw = w'(x)\, dx = \dfrac{dw}{dx}dx.$

Let's redo the first example that we just did by guess-and-check.

Example 4   Find $\displaystyle\int 3x^2\cos(x^3)\, dx.$

Solution As before, we look for an inside function whose derivative appears—in this case x^3. We let $w = x^3$. Then $dw = w'(x)\, dx = 3x^2\, dx$. The original integrand can now be completely rewritten in terms of the new variable w:

$$\int 3x^2 \cos(x^3)\, dx = \int \cos\underbrace{(x^3)}_{w} \cdot \underbrace{3x^2\, dx}_{dw} = \int \cos w\, dw = \sin w + C = \sin(x^3) + C.$$

By changing the variable to w, we can simplify the integrand. We now have $\cos w$, which can be antidifferentiated more easily. The final step, after antidifferentiating, is to convert back to the original variable, x.

Why Does Substitution Work?

The substitution method makes it look as if we can treat dw and dx as separate entities, even canceling them in the equation $dw = (dw/dx)dx$. Let's see why this works. Suppose we have an integral of the form $\int f(g(x))g'(x)\, dx$, where $g(x)$ is the inside function and $f(x)$ is the outside function. If F is an antiderivative of f, then $F' = f$, and by the chain rule $\frac{d}{dx}(F(g(x))) = f(g(x))g'(x)$. Therefore,

$$\int f(g(x))g'(x)\, dx = F(g(x)) + C.$$

Now write $w = g(x)$ and $dw/dx = g'(x)$ on both sides of this equation:

$$\int f(w)\frac{dw}{dx}\, dx = F(w) + C.$$

On the other hand, knowing that $F' = f$ tells us that

$$\int f(w)\, dw = F(w) + C.$$

Thus, the following two integrals are equal:

$$\int f(w)\frac{dw}{dx}\, dx = \int f(w)\, dw.$$

Substituting w for the inside function and writing $dw = w'(x)dx$ leaves the indefinite integral unchanged.

Let's revisit the second example that we did by guess-and-check.

Example 5 Find $\int te^{(t^2+1)}\, dt$.

Solution Here the inside function is $t^2 + 1$, with derivative $2t$. Since there is a factor of t in the integrand, we try

$$w = t^2 + 1.$$

Then

$$dw = w'(t)\, dt = 2t\, dt.$$

Notice, however, that the original integrand has only $t\, dt$, not $2t\, dt$. We therefore write

$$\frac{1}{2}\, dw = t\, dt$$

and then substitute:

$$\int te^{(t^2+1)}\, dt = \int e^{\overbrace{(t^2+1)}^{w}} \cdot \underbrace{t\, dt}_{\frac{1}{2}dw} = \int e^w \frac{1}{2}\, dw = \frac{1}{2}\int e^w\, dw = \frac{1}{2}e^w + C = \frac{1}{2}e^{(t^2+1)} + C.$$

This gives the same answer as we found using guess-and-check.

Now let's redo the third example that we solved previously by guess-and-check.

Example 6   Find $\int x^3 \sqrt{x^4 + 5} \, dx$.

Solution   The inside function is $x^4 + 5$, with derivative $4x^3$. The integrand has a factor of x^3, and since the only thing missing is a constant factor, we try

$$w = x^4 + 5.$$

Then

$$dw = w'(x) \, dx = 4x^3 \, dx,$$

giving

$$\frac{1}{4} dw = x^3 \, dx.$$

Thus,

$$\int x^3 \sqrt{x^4 + 5} \, dx = \int \sqrt{w} \, \frac{1}{4} dw = \frac{1}{4} \int w^{1/2} \, dw = \frac{1}{4} \cdot \frac{w^{3/2}}{\frac{3}{2}} + C = \frac{1}{6}(x^4 + 5)^{3/2} + C.$$

Once again, we get the same result as with guess-and-check.

Warning

We saw in the preceding examples that we can apply the substitution method when a *constant* factor is missing from the derivative of the inside function. However, we may not be able to use substitution if anything other than a constant factor is missing. For example, setting $w = x^4 + 5$ to find

$$\int x^2 \sqrt{x^4 + 5} \, dx$$

does us no good because we can't easily get $x^2 \, dx$ in terms of w. In order to use substitution, the integrand must contain the derivative of the inside function, *to within a constant factor*.

Example 7   Find $\int e^{\cos \theta} \sin \theta \, d\theta$.

Solution   We let $w = \cos \theta$ since its derivative is $-\sin \theta$ and there is a factor of $\sin \theta$ in the integrand. This gives

$$dw = w'(\theta) \, d\theta = -\sin \theta \, d\theta,$$

and so

$$-dw = \sin \theta \, d\theta.$$

Thus

$$\int e^{\cos \theta} \sin \theta \, d\theta = \int e^w \, (-dw) = (-1) \int e^w \, dw = -e^w + C = -e^{\cos \theta} + C.$$

Why didn't we put $(-1) \int e^w \, dw = -e^w - C$ in the preceding example? Since the constant C is arbitrary, it doesn't really matter whether we add or subtract it. However, the convention is always to add it on to whatever antiderivative we have calculated. If we think of $-e^{\cos \theta}$ as one of the antiderivatives of $e^{\cos \theta} \sin \theta$, then we write

$$-e^{\cos \theta} + C$$

to remind ourselves that there are lots of other antiderivatives.

Example 8 Find $\int \dfrac{e^t}{1+e^t}\, dt$.

Solution Observing that the derivative of $1+e^t$ is e^t, we see $w = 1 + e^t$ is a good choice. Then $dw = e^t\, dt$, so that

$$\int \frac{e^t}{1+e^t}\, dt = \int \frac{1}{1+e^t}\, e^t\, dt = \int \frac{1}{w}\, dw = \ln |w| + C$$

$$= \ln |1 + e^t| + C$$

$$= \ln(1 + e^t) + C \qquad \text{(Since } (1 + e^t) \text{ is always positive.)}$$

Since the numerator is $e^t\, dt$, we might also have tried $w = e^t$. This substitution leads to the integral $\int (1/(1+w))dw$, which is better than the original integral but requires another substitution, $u = 1 + w$, to finish. There are often several different ways of doing an integral by substitution.

Notice the pattern in the previous example: having a function in the denominator and its derivative in the numerator leads to a natural logarithm. The next example follows the same pattern.

Example 9 Find $\int \tan \theta\, d\theta$.

Solution Recall that $\tan \theta = (\sin \theta)/(\cos \theta)$. If $w = \cos \theta$, then $dw = -\sin \theta\, d\theta$, so

$$\int \tan \theta\, d\theta = \int \frac{\sin \theta}{\cos \theta}\, d\theta = \int \frac{-dw}{w} = -\ln |w| + C = -\ln |\cos \theta| + C.$$

Some people prefer the substitution method over guess-and-check since it is more systematic, but both methods achieve the same result. For simple problems, guess-and-check can be faster.

Problems for Section 7.1

1. (a) Find the derivatives of $\sin(x^2 + 1)$ and $\sin(x^3 + 1)$.
   (b) Use your answer to part (a) to find antiderivatives of:
      (i) $x \cos(x^2 + 1)$ (ii) $x^2 \cos(x^3 + 1)$
   (c) Find the antiderivatives of:
      (i) $x \sin(x^2 + 1)$ (ii) $x^2 \sin(x^3 + 1)$

For each of the functions in Problems 2–10, find the general antiderivative.

2. $p(t) = \pi t^3 + 4t$

3. $f(x) = \sin 3x$

4. $f(x) = 2x \cos(x^2)$

5. $r(t) = 12t^2 \cos(t^3)$

6. $f(x) = \sin(2 - 5x)$

7. $f(x) = e^{\sin x} \cos x$

8. $g(t) = te^{t^2}$

9. $f(x) = \dfrac{x}{x^2 + 1}$

10. $f(x) = \dfrac{1}{3\cos^2(2x)}$

Find the general antiderivatives in Problems 11–41. Remember, you can check your answers.

11. $\displaystyle\int y(y^2 + 5)^8\, dy$

12. $\displaystyle\int t^2 (t^3 - 3)^{10}\, dt$

13. $\displaystyle\int x(x^2 - 4)^{7/2}\, dx$

14. $\displaystyle\int \frac{1}{\sqrt{4 - x}}\, dx$

15. $\displaystyle\int x(x^2 + 3)^2\, dx$

16. $\displaystyle\int \frac{dy}{y + 5}$

17. $\displaystyle\int (2t - 7)^{73}\, dt$

18. $\displaystyle\int (x^2 + 3)^2\, dx$

19. $\displaystyle\int \sin \theta(\cos \theta + 5)^7\, d\theta$

20. $\int \sqrt{\cos 3t} \sin 3t \, dt$

21. $\int x e^{-x^2} \, dx$

22. $\int \sin^6 \theta \cos \theta \, d\theta$

23. $\int \sin^6(5\theta) \cos(5\theta) \, d\theta$

24. $\int x^2 e^{x^3+1} \, dx$

25. $\int \sin^3 \alpha \cos \alpha \, d\alpha$

26. $\int \frac{(\ln z)^2}{z} \, dz$

27. $\int \frac{e^t + 1}{e^t + t} \, dt$

28. $\int \frac{y}{y^2 + 4} \, dy$

29. $\int \tan(2x) \, dx$

30. $\int \frac{\cos \sqrt{x}}{\sqrt{x}} \, dx$

31. $\int \frac{e^{\sqrt{y}}}{\sqrt{y}} \, dy$

32. $\int \frac{1 + e^x}{\sqrt{x + e^x}} \, dx$

33. $\int \frac{e^x}{2 + e^x} \, dx$

34. $\int \frac{x + 1}{x^2 + 2x + 19} \, dx$

35. $\int y^2(1 + y)^2 \, dy$

36. $\int x^2(1 + 2x^3)^2 \, dx$

37. $\int \frac{t}{1 + 3t^2} \, dt$

38. $\int \frac{e^x - e^{-x}}{e^x + e^{-x}} \, dx$

39. $\int \frac{(t + 1)^2}{t^2} \, dt$

40. $\int \frac{x \cos(x^2)}{\sqrt{\sin(x^2)}} \, dx$

41. $\int (2x + 1)e^{x^2} e^x \, dx$ [Hint: Rewrite $e^{x^2}e^x = e^?$.]

42. Use the identity $\cos^2 x + \sin^2 x = 1$ to find $\int \sin^3 x \, dx$.

43. Find $\int 4x(x^2 + 1) \, dx$ using two methods:

    (a) Do the multiplication first, and then antidifferentiate.
    (b) Now do the integral using the substitution $w = x^2 + 1$.
    (c) Explain why the two expressions formed in parts (a) and (b) are different. Are they both correct?

44. Throughout much of this century, the yearly consumption of electricity in the US has been increasing exponentially at a continuous rate of 7% per year. Assuming this trend continues and that the electrical energy consumed in 1900 was 1.4 million megawatt-hours (a megawatt-hour is a measure of electrical energy),

    (a) Write an expression for yearly electricity consumption as a function of time, t, measured in years since 1900.
    (b) Find the average yearly electrical consumption throughout this century.
    (c) During what year was electrical consumption closest to the average for the century?
    (d) Without doing the calculation for part (c), how could you have predicted which half of the century the answer would be in?

45. If we assume that wind resistance is proportional to velocity, then the velocity, v, of a body of mass m falling vertically is given by

$$v = \frac{mg}{k} \left(1 - e^{-kt/m}\right),$$

where g is the acceleration due to gravity and k is a constant. Find the height, h, above the surface of the earth as a function of time. Assume the body starts at height h_0.

46. (a) During the 1970s, ACME Widgets sold at a continuous rate of $R = R_0 e^{0.15t}$ widgets per year, where t is time in years since January 1, 1970. Suppose they were selling widgets at a rate of 1000 per year on the first day of the decade. How many widgets did they sell during the decade? How many did they sell if the rate on January 1, 1970 was 150,000,000 widgets per year?

    (b) In the first case above (1000 widgets per year on January 1, 1970), how long did it take for half the widgets in the 1970s to be sold? In the second case (150,000,000 widgets per year on January 1, 1970), when had half the widgets in the 1970s been sold?

    (c) In 1980 ACME began an advertising campaign claiming that half the widgets it had sold in the previous ten years were still in use. Based on your answer to part (b), about how long must a widget last in order to justify this claim?

7.2   INTEGRATION BY SUBSTITUTION: PART II

In this section we use substitution to evaluate *definite* integrals. We then look at some examples where it is harder to see which substitution to use, but where the method works nonetheless.

Definite Integrals by Substitution

Example 1   Compute $\displaystyle\int_0^2 xe^{x^2}\,dx$.

Solution   To evaluate this definite integral using the Fundamental Theorem of Calculus, we first need to find an antiderivative of $f(x) = xe^{x^2}$. The inside function is x^2, so let $w = x^2$. Then $dw = 2x\,dx$, so $\frac{1}{2}\,dw = x\,dx$. Thus,

$$\int xe^{x^2}\,dx = \int e^w \frac{1}{2}\,dw = \frac{1}{2}e^w + C = \frac{1}{2}e^{x^2} + C.$$

Now we find the definite integral

$$\int_0^2 xe^{x^2}\,dx = \frac{1}{2}e^{x^2}\Big|_0^2 = \frac{1}{2}(e^4 - e^0) = \frac{1}{2}(e^4 - 1).$$

There is another way to look at the same problem. After we established that

$$\int xe^{x^2}\,dx = \frac{1}{2}e^w + C,$$

our next two steps were to replace w by x^2, and then x by 2 and 0. We could have directly replaced the original limits of integration, $x = 0$ and $x = 2$, by the corresponding w limits. Since $w = x^2$, the w limits are $w = 0^2 = 0$ (when $x = 0$) and $w = 2^2 = 4$ (when $x = 2$), so we get

$$\int_{x=0}^{x=2} xe^{x^2}\,dx = \frac{1}{2}\int_{w=0}^{w=4} e^w\,dw = \frac{1}{2}e^w\Big|_0^4 = \frac{1}{2}\left(e^4 - e^0\right) = \frac{1}{2}(e^4 - 1).$$

As we would expect, both methods give the same answer.

To Use Substitution to Find Definite Integrals

Either
- Compute the indefinite integral, expressing an antiderivative in terms of the original variable, and then evaluate the result at the original limits,

or
- Convert the original limits to new limits in terms of the new variable and do not convert the antiderivative back to the original variable.

Example 2   Evaluate $\displaystyle\int_0^{\pi/4} \frac{\tan^3 \theta}{\cos^2 \theta}\,d\theta$.

Solution   To use substitution, we must decide what w should be. There are two possible inside functions, $\tan \theta$ and $\cos \theta$. Now

$$\frac{d}{d\theta}(\tan \theta) = \frac{1}{\cos^2 \theta} \quad \text{and} \quad \frac{d}{d\theta}(\cos \theta) = -\sin \theta,$$

and since the integral contains a factor of $1/\cos^2 \theta$ but not of $\sin \theta$, we let $w = \tan \theta$. Then $dw = (1/\cos^2 \theta)d\theta$. When $\theta = 0$, $w = \tan 0 = 0$, and when $\theta = \pi/4$, $w = \tan(\pi/4) = 1$, so

$$\int_0^{\pi/4} \frac{\tan^3 \theta}{\cos^2 \theta}\, d\theta = \int_0^{\pi/4} (\tan \theta)^3 \cdot \frac{1}{\cos^2 \theta}\, d\theta = \int_0^1 w^3\, dw = \frac{1}{4}w^4 \Big|_0^1 = \frac{1}{4}.$$

Example 3   Evaluate $\displaystyle \int_1^3 \frac{dx}{5-x}$.

Solution   Let $w = 5 - x$, so $dw = -dx$. When $x = 1$, $w = 4$, and when $x = 3$, $w = 2$, so

$$\int_1^3 \frac{dx}{5-x} = \int_4^2 \frac{-dw}{w} = -\ln|w| \Big|_4^2 = -(\ln 2 - \ln 4) = \ln\left(\frac{4}{2}\right) = \ln 2 \approx 0.69.$$

Notice that we write the limit $w = 4$ at the bottom, even though it is larger than $w = 2$, because $w = 4$ corresponds to the lower limit $x = 1$.

More Complex Substitutions

In the examples of substitution presented so far, we had to guess an expression for w and then we had to be lucky enough to have dw (or some constant multiple of it) lying around in the integrand. What if we are not so lucky? It turns out that we can often get somewhere by letting w be some messy expression contained inside, say, a cosine or under a root, even if we cannot see immediately how such a substitution helps.

Example 4   Find $\displaystyle \int (x + 7)\sqrt[3]{3 - 2x}\, dx$.

Solution   Here, instead of the derivative of the inside function (which is -2), we have the factor $(x + 7)$. However, substituting $w = 3 - 2x$ turns out to help. Then $dw = -2\, dx$, so $(-1/2)\, dw = dx$. Now we must convert everything to w, including $x + 7$. If $w = 3 - 2x$, then $2x = 3 - w$, so $x = 3/2 - w/2$, and therefore we can write $x + 7$ in terms of w. Thus

$$\int (x + 7)\sqrt[3]{3 - 2x}\, dx = \int \left(\frac{3}{2} - \frac{w}{2} + 7\right)\sqrt[3]{w}\left(-\frac{1}{2}\right) dw$$

$$= -\frac{1}{2} \int \left(\frac{17}{2} - \frac{w}{2}\right) w^{1/3}\, dw$$

$$= -\frac{1}{4} \int (17 - w)w^{1/3}\, dw$$

$$= -\frac{1}{4} \int (17w^{1/3} - w^{4/3})\, dw$$

$$= -\frac{1}{4}\left(17\frac{w^{4/3}}{4/3} - \frac{w^{7/3}}{7/3}\right) + C$$

$$= -\frac{1}{4}\left(\frac{51}{4}(3 - 2x)^{4/3} - \frac{3}{7}(3 - 2x)^{7/3}\right) + C.$$

Looking back over the solution, the reason this substitution works is that it converts $\sqrt[3]{3 - 2x}$, the messiest part of the integrand, to $\sqrt[3]{w}$, which can be combined with the other term and then integrated.

Identities can sometimes help us simplify an integral, as the next example illustrates.

Example 5   Find $\int \cos^3 \theta \, d\theta$.

Solution   This integral can be simplified by using the Pythagorean identity $\cos^2 \theta + \sin^2 \theta = 1$ to rewrite the integrand:

$$\int \cos^3 \theta = \int \cos^2 \theta \cdot \cos \theta \, d\theta = \int (1 - \sin^2 \theta) \cos \theta \, d\theta$$

$$= \int \cos \theta \, d\theta - \int \sin^2 \theta \cos \theta \, d\theta.$$

The first integral is straightforward; for the second, we use substitution, letting $w = \sin \theta$ and $dw = \cos \theta \, d\theta$. We have

$$\int \sin^2 \theta \cos \theta \, d\theta = \int w^2 \, dw = \frac{1}{3}w^3 + C = \frac{1}{3}\sin^3 \theta + C.$$

Combining the constants from both integrals into C, we have

$$\int \cos^3 \theta \, d\theta = \int \cos \theta \, d\theta - \int \sin^2 \theta \cos \theta \, d\theta = \sin \theta - \frac{1}{3}\sin^3 \theta + C.$$

Example 6   Find $\int \sqrt{1 + \sqrt{x}} \, dx$.

Solution   This time, the derivative of the inside function is nowhere to be seen. Nevertheless, we try $w = 1 + \sqrt{x}$. Then $w - 1 = \sqrt{x}$, so $(w - 1)^2 = x$. Therefore $2(w - 1) \, dw = dx$. We then have

$$\int \sqrt{1 + \sqrt{x}} \, dx = \int \sqrt{w} \, 2(w - 1) \, dw = 2 \int w^{1/2}(w - 1) \, dw$$

$$= 2 \int (w^{3/2} - w^{1/2}) \, dw = 2 \left(\frac{2}{5}w^{5/2} - \frac{2}{3}w^{3/2} \right) + C$$

$$= 2 \left(\frac{2}{5}(1 + \sqrt{x})^{5/2} - \frac{2}{3}(1 + \sqrt{x})^{3/2} \right) + C.$$

Notice that the substitution in the preceding example again converts the inside of the messiest function into something simple. In addition, since the derivative of the inside function is not waiting for us, we have to solve for x so that we can get dx entirely in terms of w and dw.

Substitution can be used to express one definite integral in terms of another, even in situations where we don't know what the indefinite integral is.

Example 7   Derive a formula for the area of the ellipse

$$\frac{x^2}{a^2} + \frac{y^2}{b^2} = 1$$

assuming $a, b > 0$.

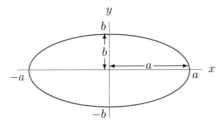

Figure 7.1: The ellipse: $\frac{x^2}{a^2} + \frac{y^2}{b^2} = 1$

Solution   Solving for y gives

$$y = \pm b\sqrt{1 - \frac{x^2}{a^2}},$$

where the positive square root gives the upper half of the ellipse, and the negative square root gives the lower. (See Figure 7.1.) Thus, the area of the ellipse is given by

$$\text{Area} = 2 \int_{-a}^{a} b\sqrt{1 - \frac{x^2}{a^2}}\, dx$$

Substitution allows us to convert this integral to one whose value is known. Let $w = x/a$, so $dx = a\, dw$. When $x = -a$, $w = -1$, and when $x = a$, $w = 1$. Thus

$$2 \int_{-a}^{a} b\sqrt{1 - \frac{x^2}{a^2}}\, dx = 2 \int_{-1}^{1} b\sqrt{1 - w^2}\, a\, dw = 2ab \int_{-1}^{1} \sqrt{1 - w^2}\, dw.$$

Now the integral $\int_{-1}^{1} \sqrt{1 - w^2}\, dw$ is the area of half a circle of radius 1 (because $y = \sqrt{1 - w^2}$ is the top half of the circle $w^2 + y^2 = 1$). Thus,

$$\int_{-1}^{1} \sqrt{1 - w^2}\, dw = \frac{\pi}{2}$$

so

$$\text{Area of ellipse} = 2ab \int_{-1}^{1} \sqrt{1 - w^2}\, dw = \pi ab.$$

Problems for Section 7.2

1. Use substitution to express each of the following integrals as a multiple of $\int_a^b \frac{1}{w}\, dw$ for some a and b. Then evaluate the integrals.

(a) $\displaystyle\int_0^1 \frac{x}{1 + x^2}\, dx$ (b) $\displaystyle\int_0^{\pi/4} \frac{\sin x}{\cos x}\, dx$

For Problems 2–10, use the Fundamental Theorem to calculate the definite integrals. Show your work.

2. $\displaystyle\int_0^{\pi} \cos(x + \pi)\, dx$ 3. $\displaystyle\int_0^{1/2} \cos(\pi x)\, dx$ 4. $\displaystyle\int_0^{\pi/2} e^{-\cos\theta} \sin\theta\, d\theta$

5. $\displaystyle\int_1^2 2x e^{x^2}\, dx$ 6. $\displaystyle\int_1^8 \frac{e^{\sqrt[3]{x}}}{\sqrt[3]{x^2}}\, dx$ 7. $\displaystyle\int_{-1}^{e-2} \frac{1}{t + 2}\, dt$

8. $\displaystyle\int_1^4 \frac{\cos\sqrt{x}}{\sqrt{x}}\, dx$ 9. $\displaystyle\int_0^2 \frac{x}{(1 + x^2)^2}\, dx$ 10. $\displaystyle\int_{-\pi/4}^{\pi/4} \cos^2\theta \sin^5\theta\, d\theta$

[Hint: $\cos^2\theta + \sin^2\theta = 1$.]

For Problems 11–16, evaluate the definite integrals. Whenever possible, use the Fundamental Theorem of Calculus, perhaps after a substitution. Otherwise, use numerical methods.

11. $\displaystyle\int_{-1}^3 (x^3 + 5x)\, dx$ 12. $\displaystyle\int_{-1}^1 \frac{1}{1 + y^2}\, dy$ 13. $\displaystyle\int_1^3 \frac{1}{x}\, dx$

14. $\displaystyle\int_1^3 \frac{dt}{(t + 7)^2}$ 15. $\displaystyle\int_{-1}^2 \sqrt{x + 2}\, dx$ 16. $\displaystyle\int_1^2 \frac{\sin t}{t}\, dt$

For Problems 17–27, evaluate the definite integrals. Whenever possible, use the Fundamental Theorem of Calculus, perhaps after a substitution. Otherwise, use numerical methods.

17. $\displaystyle\int_0^1 x(1+x^2)^{20}\,dx$

18. $\displaystyle\int_0^\pi \sin\theta(\cos\theta+5)^7\,d\theta$

19. $\displaystyle\int_0^{\pi/12} \sin(3\alpha)\,d\alpha$

20. $\displaystyle\int_0^1 \frac{x}{1+5x^2}\,dx$

21. $\displaystyle\int_1^2 \frac{x^2+1}{x}\,dx$

22. $\displaystyle\int_0^1 \frac{1}{x^2+2x+1}\,dx$

23. $\displaystyle\int_0^1 \frac{(x+2)}{(x+2)^2+1}\,dx$

24. $\displaystyle\int_4^1 x\sqrt{x^2+4}\,dx$

25. $\displaystyle\int_0^{1/\sqrt{2}} \frac{x}{\sqrt{1-x^4}}\,dx$

26. $\displaystyle\int_{1/4}^1 \sin\frac{1}{t}\,dt$

27. $\displaystyle\int_{-2}^0 \frac{2x+4}{x^2+4x+5}\,dx$

In Problems 28–32, explain why the two antiderivatives are really, despite their apparent dissimilarity, different expressions of the same problem. You do not need to evaluate the integrals.

28. $\displaystyle\int \frac{e^x\,dx}{1+e^{2x}}$ and $\displaystyle\int \frac{\cos x\,dx}{1+\sin^2 x}$

29. $\displaystyle\int \frac{\ln x}{x}\,dx$ and $\displaystyle\int x\,dx$

30. $\displaystyle\int e^{\sin x}\cos x\,dx$ and $\displaystyle\int \frac{e^{\arcsin x}}{\sqrt{1-x^2}}\,dx$

31. $\displaystyle\int \sqrt{x+1}\,dx$ and $\displaystyle\int \frac{\sqrt{1+\sqrt{x}}}{\sqrt{x}}\,dx$

32. $\displaystyle\int (\sin x)^3\cos x\,dx$ and $\displaystyle\int (x^3+1)^3 x^2\,dx$

33. Find $\displaystyle\int \frac{dx}{x^2+4x+5}$ by completing the square and using the substitution $x+2=\tan\theta$.

34. Integrate:

(a) $\displaystyle\int \frac{1}{\sqrt{x}}\,dx$

(b) $\displaystyle\int \frac{1}{\sqrt{x+1}}\,dx$

(c) $\displaystyle\int \frac{1}{\sqrt{x}+1}\,dx$

35. Find the exact area under the graph of $f(x)=xe^{x^2}$ between $x=0$ and $x=2$.

36. Find the exact value of the area under one arch of the curve $V(t)=V_0\sin(\omega t)$.

37. (a) Calculate exactly: $\displaystyle\int_{-\pi}^\pi \cos^2\theta\sin\theta\,d\theta$.

(b) Calculate the exact area under the curve $y=\cos^2\theta\sin\theta$ between $\theta=0$ and $\theta=\pi$.

38. Suppose $\displaystyle\int_0^1 f(t)\,dt=3$. Calculate each of the following integrals:

(a) $\displaystyle\int_0^{0.5} f(2t)\,dt$

(b) $\displaystyle\int_0^1 f(1-t)\,dt$

(c) $\displaystyle\int_1^{1.5} f(3-2t)\,dt$

39. Find the average value of $f(x)=\dfrac{1}{x+1}$ on the interval $x=0$ to $x=2$. Sketch a graph showing the function and the average value.

40. (a) Find $\displaystyle\int \sin\theta\cos\theta\,d\theta$.

(b) You probably solved part (a) by making the substitution $w=\sin\theta$ or $w=\cos\theta$. (If not, go back and do it that way.) Now find $\int \sin\theta\cos\theta\,d\theta$ by making the *other* substitution.

(c) There is yet another way of finding this integral which involves the trigonometric identities

$$\sin(2\theta)=2\sin\theta\cos\theta \quad\text{and}\quad \cos(2\theta)=\cos^2\theta-\sin^2\theta.$$

Find $\int \sin\theta\cos\theta\,d\theta$ using one of these identities and then the substitution $w=2\theta$.

(d) You should now have three different expressions for the indefinite integral $\int \sin\theta\cos\theta\,d\theta$. Are they really different? Are they all correct? Explain the reasons for your answer.

41. It is said that Charles Dickens was paid a penny per word for his novels. We can express the payment for a novel containing L words as $\int_0^L 1\,dw$. Instead of the constant 1 penny per word, suppose that for each new manuscript he produced he would be paid $\int_0^L f(w)\,dw$, where L is the total length of the novel (in words), and f is a function depending on w, the number of words, which represents the pay rate, given by:

$$f(w) = \begin{cases} (1/2) + w/2000 & 0 \le w \le 2000 \\ 3/2 & 2000 \le w \le 20{,}000 \\ (3/2)e^{20 - w/1000} & w \ge 20{,}000. \end{cases}$$

(a) How much would Dickens be paid for a 40,000-word novel?

(b) Would he be paid more for writing one 30,000-word novel or for writing two 15,000-word novels? Try to answer part (b) without computing any integrals. [Hint: Does the exponentially decaying part of the definition of $f(w)$ encourage or discourage long novels?]

42. The rate at which water is flowing into a tank is $r(t)$ gallons/minute, with t in minutes.

(a) Write an expression approximating the amount of water entering the tank during the interval from time t to time $t + \Delta t$, where Δt is small.

(b) Write a Riemann sum approximating the total amount of water entering the tank between $t = 0$ and $t = 5$. Write an exact expression for this amount.

(c) By how much has the amount of water in the tank changed between $t = 0$ and $t = 5$ if $r(t) = 20e^{0.02t}$?

(d) If $r(t)$ is as given in part (c), and if the tank contains 3000 gallons initially, find a formula for $Q(t)$, the amount of water in the tank at time t.

43. If we assume that wind resistance is proportional to the square of velocity, then the downward velocity, v, of a falling body is given by

$$v = \sqrt{\frac{g}{k}} \left(\frac{e^{t\sqrt{gk}} - e^{-t\sqrt{gk}}}{e^{t\sqrt{gk}} + e^{-t\sqrt{gk}}} \right).$$

Use the substitution $w = e^{t\sqrt{gk}} + e^{-t\sqrt{gk}}$ to find the height, h, of the body above the surface of the earth as a function of time. Assume the body starts at a height h_0.

7.3 INTEGRATION BY PARTS

We have seen how the method of substitution reverses the chain rule. Now we will introduce a method called *integration by parts* that is based on the product rule.

Example 1 Find $\int xe^x\,dx$.

Solution We are looking for a function whose derivative is xe^x. If we think about how the product rule works, we might be led to guess xe^x, because we know that when we take the derivative we will get two terms, one of which will be xe^x:

$$\frac{d}{dx}(xe^x) = \frac{d}{dx}(x)e^x + x\frac{d}{dx}(e^x) = e^x + xe^x.$$

Of course, our guess is wrong because of the extra e^x. Maybe we can adjust our guess by subtracting from xe^x something to cancel out the extra e^x. This leads us to try $xe^x - e^x$. Let's check it:

$$\frac{d}{dx}(xe^x - e^x) = \frac{d}{dx}(xe^x) - \frac{d}{dx}(e^x) = e^x + xe^x - e^x = xe^x.$$

It works, so $\int xe^x\,dx = xe^x - e^x + C.$

Example 2 Find $\int \theta \cos\theta\,d\theta$.

Solution   We guess the antiderivative is $\theta \sin \theta$ and use the product rule to check:

$$\frac{d}{d\theta}(\theta \sin \theta) = \frac{d(\theta)}{d\theta} \sin \theta + \theta \frac{d}{d\theta}(\sin \theta) = \sin \theta + \theta \cos \theta.$$

To correct for the extra $\sin \theta$ term, we must subtract from our original guess something whose derivative is $\sin \theta$. Since $\frac{d}{d\theta}(\cos \theta) = -\sin \theta$, we'll try:

$$\frac{d}{d\theta}(\theta \sin \theta + \cos \theta) = \frac{d}{d\theta}(\theta \sin \theta) + \frac{d}{d\theta}(\cos \theta) = \sin \theta + \theta \cos \theta - \sin \theta = \theta \cos \theta.$$

Thus, $\displaystyle \int \theta \cos \theta \, d\theta = \theta \sin \theta + \cos \theta + C.$

The General Formula for Integration by Parts

We can formalize the process illustrated in the last two examples in the following way. We begin with the product rule:

$$\frac{d}{dx}(uv) = u'v + uv'$$

where u and v are functions of x with derivatives u' and v', respectively. We rewrite this as:

$$uv' = \frac{d}{dx}(uv) - u'v$$

and then integrate both sides:

$$\int uv' \, dx = \int \frac{d}{dx}(uv) \, dx - \int u'v \, dx.$$

Since an antiderivative of $\dfrac{d}{dx}(uv)$ is just uv, we get the following formula:

Integration by Parts

$$\int uv' \, dx = uv - \int u'v \, dx.$$

This formula is useful when the integrand can be viewed as a product and when the integral on the right-hand side is simpler than that on the left. In effect, we were using integration by parts in the previous two examples. In Example 1, we let $xe^x = (x) \cdot (e^x) = uv'$, and choose $u = x$ and $v' = e^x$. Thus, $u' = 1$ and $v = e^x$, so

$$\int \underbrace{(x)}_{u} \underbrace{(e^x)}_{v'} \, dx = \underbrace{(x)}_{u} \underbrace{(e^x)}_{v} - \int \underbrace{(1)}_{u'} \underbrace{(e^x)}_{v} \, dx = xe^x - e^x + C.$$

So uv represents our first guess, and $\int u'v \, dx$ represents the correction to our guess.

Notice what would have happened in Example 1 if we took $u = x$ and $v = e^x + C_1$. Then

$$\int xe^x \, dx = x(e^x + C_1) - \int (e^x + C_1) \, dx$$
$$= xe^x + C_1 x - e^x - C_1 x + C$$
$$= xe^x - e^x + C,$$

as before. It is not necessary to include an arbitrary constant in the antiderivative for v; any antiderivative will do.

What would have happened in Example 1 if we had picked u and v' the other way around? If $u = e^x$ and $v' = x$, then $u' = e^x$ and $v = x^2/2$. The formula for integration by parts then gives

$$\int xe^x \, dx = \frac{x^2}{2} e^x - \int \frac{x^2}{2} \cdot e^x \, dx,$$

which is true but not helpful since the integral on the right is worse than the one on the left. To use this method, we must choose u and v' to make the integral on the right easier to find than the integral on the left.

Advice on How to Choose u and v'

- Whatever you let v' be, you need to be able to find v.
- It helps if u' is simpler than u (or at least no more complicated than u).
- It helps if v is simpler than v' (or at least no more complicated than v').

If we pick $v' = x$ in Example 1, then $v = x^2/2$, which is certainly "worse" than v'.

There are some examples which don't look like good candidates for integration by parts because they don't appear to involve products, but for which the process works well. Such examples often involve $\ln x$ or the inverse trigonometric functions. Here is one:

Example 3   Find $\displaystyle\int_2^3 \ln x \, dx$.

Solution   This doesn't look like a product unless we write $\ln x = (1)(\ln x)$. Then we might say $u = 1$ so $u' = 0$, which certainly makes things simpler. But if $v' = \ln x$, what is v? If we knew, we wouldn't need integration by parts. Let's try the other way: if $u = \ln x$, $u' = 1/x$ and if $v' = 1$, $v = x$, so

$$\int_2^3 \underbrace{(\ln x)}_{u}\,\underbrace{(1)}_{v'} \, dx = \underbrace{(\ln x)}_{u}\,\underbrace{(x)}_{v}\Big|_2^3 - \int_2^3 \underbrace{\left(\frac{1}{x}\right)}_{u'} \cdot \underbrace{(x)}_{v} \, dx$$

$$= x\ln x\Big|_2^3 - \int_2^3 1\, dx = (x\ln x - x)\Big|_2^3$$

$$= 3\ln 3 - 3 - 2\ln 2 + 2 = 3\ln 3 - 2\ln 2 - 1.$$

Notice that when doing a definite integral by parts, we must remember to put the limits of integration (here 2 and 3) on the uv term (in this case $x\ln x$) as well as on the integral $\int u'v\, dx$.

Example 4   Find $\displaystyle\int x^6 \ln x \, dx$.

Solution   View $x^6 \ln x$ as uv' where $u = \ln x$ and $v' = x^6$. Then $v = \frac{1}{7}x^7$ and $u' = 1/x$, so integration by parts gives us:

$$\int x^6 \ln x\, dx = \int (\ln x)x^6\, dx = (\ln x)\left(\frac{1}{7}x^7\right) - \int \frac{1}{7}x^7 \cdot \frac{1}{x}\, dx$$

$$= \frac{1}{7}x^7 \ln x - \frac{1}{7}\int x^6\, dx$$

$$= \frac{1}{7}x^7 \ln x - \frac{1}{49}x^7 + C.$$

In Example 4 we did not choose $v' = \ln x$, because it is not immediately clear what v would be. In fact, we used integration by parts in Example 3 to find out what the antiderivative of $\ln x$ was. Also, using $u = \ln x$, as we have done, gives $u' = 1/x$, which most people would consider simpler than $u = \ln x$. This shows that u does not have to be the first factor in the integrand (here x^6).

Example 5   Find $\int x^2 \sin 4x \, dx$.

Solution   If we let $v' = \sin 4x$, then $v = -\frac{1}{4} \cos 4x$, which is no worse than v'. Also letting $u = x^2$, we get $u' = 2x$, which is simpler than $u = x^2$. Using integration by parts:

$$\int x^2 \sin 4x \, dx = x^2 \left(-\frac{1}{4} \cos 4x \right) - \int 2x \left(-\frac{1}{4} \cos 4x \right) \, dx$$

$$= -\frac{1}{4} x^2 \cos 4x + \frac{1}{2} \int x \cos 4x \, dx.$$

The trouble is we still have to grapple with $\int x \cos 4x \, dx$. This can be done by using integration by parts again with a new u and v, namely $u = x$ and $v' = \cos 4x$:

$$\int x \cos 4x \, dx = x \left(\frac{1}{4} \sin 4x \right) - \int 1 \cdot \frac{1}{4} \sin 4x \, dx$$

$$= \frac{1}{4} x \sin 4x - \frac{1}{4} \cdot \left(-\frac{1}{4} \cos 4x \right) + C$$

$$= \frac{1}{4} x \sin 4x + \frac{1}{16} \cos 4x + C.$$

Thus,

$$\int x^2 \sin 4x \, dx = -\frac{1}{4} x^2 \cos 4x + \frac{1}{2} \int x \cos 4x \, dx$$

$$= -\frac{1}{4} x^2 \cos 4x + \frac{1}{2} \left(\frac{1}{4} x \sin 4x + \frac{1}{16} \cos 4x + C \right)$$

$$= -\frac{1}{4} x^2 \cos 4x + \frac{1}{8} x \sin 4x + \frac{1}{32} \cos 4x + C_1$$

where C_1 is the arbitrary constant $\frac{1}{2} C$. Notice that, in this example, each time we used integration by parts, the exponent of x went down by 1.

Example 6   Find $\int \cos^2 \theta \, d\theta$.

Solution   Using integration by parts with $u = \cos \theta$, $v' = \cos \theta$ gives $u' = -\sin \theta$, $v = \sin \theta$, so we get

$$\int \cos^2 \theta \, d\theta = \cos \theta \sin \theta + \int \sin^2 \theta \, d\theta$$

Substituting $\sin^2 \theta = 1 - \cos^2 \theta$ leads to

$$\int \cos^2 \theta \, d\theta = \cos \theta \sin \theta + \int (1 - \cos^2 \theta) \, d\theta$$

$$= \cos \theta \sin \theta + \int 1 \, d\theta - \int \cos^2 \theta \, d\theta.$$

Looking at the right side, we see that the original integral has reappeared. If we move it to the left, we get

$$2 \int \cos^2 \theta \, d\theta = \cos \theta \sin \theta + \int 1 \, d\theta = \cos \theta \sin \theta + \theta + C.$$

Dividing by 2 gives

$$\int \cos^2 \theta \, d\theta = \frac{1}{2} \cos \theta \sin \theta + \frac{1}{2} \theta + C_1$$

where C_1 is the arbitrary constant $\frac{1}{2} C$. Problem 38 does this integral by another method.

You can check these results by differentiation.

Problems for Section 7.3

1. By writing $\arctan x = (1) \cdot (\arctan x)$, find $\int \arctan x \, dx$.

Find the indefinite integrals in Problems 2–27.

2. $\int te^{5t} \, dt$

3. $\int t^2 e^{5t} \, dt$

4. $\int pe^{-0.1p} \, dp$

5. $\int t \sin t \, dt$

6. $\int y \ln y \, dy$

7. $\int x^3 \ln x \, dx$

8. $\int (z+1)e^{2z} \, dz$

9. $\int \frac{z}{e^z} \, dz$

10. $\int t^2 \sin t \, dt$

11. $\int \theta^2 \cos 3\theta \, d\theta$

12. $\int \sin^2 \theta \, d\theta$

13. $\int (\theta+1) \sin(\theta+1) \, d\theta$

14. $\int \cos^2(3\alpha+1) \, d\alpha$

15. $\int \frac{\ln x}{x^2} \, dx$

16. $\int q^5 \ln 5q \, dq$

17. $\int y \sqrt{y+3} \, dy$

18. $\int (t+2)\sqrt{2+3t} \, dt$

19. $\int \frac{y}{\sqrt{5-y}} \, dy$

20. $\int \frac{t+7}{\sqrt{5-t}} \, dt$

21. $\int (\ln t)^2 \, dt$

22. $\int x(\ln x)^4 \, dx$

23. $\int \arcsin w \, dw$

24. $\int \arctan 7z \, dz$

25. $\int x \arctan x^2 \, dx$

26. $\int x^3 e^{x^2} \, dx$

27. $\int x^5 \cos x^3 \, dx$

Evaluate the integrals in Problems 28–36 both exactly [e.g., $\ln(3\pi)$] and numerically [e.g. $\ln(3\pi) \approx 2.243$]:

28. $\int_1^5 \ln t \, dt$

29. $\int_3^5 x \cos x \, dx$

30. $\int_0^{10} ze^{-z} \, dz$

31. $\int_1^3 t \ln t \, dt$

32. $\int_0^1 \arctan y \, dy$

33. $\int_0^5 \ln(1+t) \, dt$

34. $\int_0^1 x \arctan x^2 \, dx$

35. $\int_0^1 \arcsin z \, dz$

36. $\int_0^1 u \arcsin u^2 \, du$

37. In Problem 12, you evaluated $\int \sin^2 \theta \, d\theta$ using integration by parts. (If you didn't do it by parts, do so now!) Redo this integral by changing the form of the integrand using the identity $\sin^2 \theta = (1-\cos 2\theta)/2$. Explain any differences in the form of the answer obtained by the two methods.

38. Compute $\int \cos^2 \theta \, d\theta$ in two different ways and explain any differences in the form of your answers. (The identity $\cos^2 \theta = (1 + \cos 2\theta)/2$ may be useful.)

39. Use integration by parts twice to find $\int e^x \sin x \, dx$.

40. Use integration by parts twice to find $\int e^\theta \cos \theta \, d\theta$.

41. Use the results from Problems 39 and 40 and integration by parts to find $\int xe^x \sin x \, dx$.

42. Use the results from Problems 39 and 40 and integration by parts to find $\int \theta e^\theta \cos \theta \, d\theta$.

43. Show that $\int x^n e^x \, dx = x^n e^x - n \int x^{n-1} e^x \, dx$.

44. Show that $\int x^n \cos ax \, dx = \frac{1}{a} x^n \sin ax - \frac{n}{a} \int x^{n-1} \sin ax \, dx$.

45. Show that $\int x^n \sin ax \, dx = -\frac{1}{a} x^n \cos ax + \frac{n}{a} \int x^{n-1} \cos ax \, dx$.

46. Show that $\int \cos^n x \, dx = \frac{1}{n} \cos^{n-1} x \sin x + \frac{n-1}{n} \int \cos^{n-2} x \, dx$.

47. Find the exact value of the area under the first arch of $f(x) = x \sin x$.

48. Let f be a twice differentiable function such that $f(0) = 6$, $f(1) = 5$, and $f'(1) = 2$. Evaluate the integral $\int_0^1 x f''(x) dx$.

49. Suppose $F(a)$ represents the area under the graph of $y = x^2 e^{-x}$ between $x = 0$ and $x = a$. (Assume $a > 0$.)

    (a) Find a formula for $F(a)$.
    (b) Is F an increasing or decreasing function?
    (c) Is F concave up or concave down for $0 < a < 2$?

50. The voltage, V, in an electric circuit is given as a function of time, t, by

$$V = V_0 \cos(\omega t + \phi).$$

Suppose each of the positive constants, V_0, ω, ϕ is increased (while the other two are held constant). What is the effect of each increase on each of the following quantities:

    (a) The maximum value of V?
    (b) The maximum value of dV/dt?
    (c) The average value of V^2 over one period?

51. Integrating $e^{ax} \sin bx$ by parts twice yields a result of the form

$$\int e^{ax} \sin bx \, dx = e^{ax} (A \sin bx + B \cos bx) + C.$$

    (a) Find the constants A and B in terms of a and b. [Hint: Don't actually perform the integration by parts.]
    (b) Evaluate $\int e^{ax} \cos bx \, dx$ by modifying the result in part (a). [Again, it is not necessary to perform integration by parts, as the result is of the same form as that in part (a).]

52. During a surge in the demand for electricity, the rate, r, at which energy is used can be approximated by

$$r = te^{-at},$$

where t is the time in hours and a is a positive constant.

    (a) Find the total energy, E, used in the first T hours. Give your answer as a function of a.
    (b) What happens to E as $T \to \infty$?

53. In describing the behavior of an electron, we use wave functions $\Psi_1, \Psi_2, \Psi_3, \ldots$ of the form

$$\Psi_n(x) = C_n \sin(n\pi x) \qquad \text{for } n = 1, 2, 3, \ldots$$

where x is the distance from a fixed point and C_n is a positive constant.

    (a) Find C_1 so that Ψ_1 satisfies

$$\int_0^1 \left(\Psi_1(x) \right)^2 \, dx = 1.$$

    This process is called normalizing the wave function.
    (b) For any integer n, find C_n so that Ψ_n is normalized.

7.4 TABLES OF INTEGRALS

Since so few functions have elementary antiderivatives, they have been compiled in a list called a table of integrals. Such integral tables are available[1] so that if you need an antiderivative, you can simply look it up. A short table of indefinite integrals is given inside the back cover. The key to using these tables is being able to recognize the general class of function that you are trying to integrate, so you can know in what section of the table to look.

Warning: This section involves long division of polynomials and completing the square. You may want to review these topics!

Using the Table of Integrals

Part I of the table inside the back cover gives the antiderivatives of the basic functions x^n, a^x, $\ln x$, $\sin x$, $\cos x$, and $\tan x$. (The antiderivative for $\ln x$ is found using integration by parts and is a special case of the more general formula III-13.) Most of these are already familiar.

Part II of the table contains antiderivatives of functions involving products of e^x, $\sin x$, and $\cos x$. All of these antiderivatives were obtained using integration by parts.

Example 1   Find $\int \sin 7z \sin 3z \, dz$.

Solution   Since the integrand is the product of two sines, we should use II-10 in the table,

$$\int \sin 7z \sin 3z \, dz = -\frac{1}{40}(7 \cos 7z \sin 3z - 3 \cos 3z \sin 7z) + C.$$

Part III of the table contains antiderivatives for products of a polynomial and e^x, $\sin x$, or $\cos x$. It also has an antiderivative for $x^n \ln x$, which can easily be used to find the antiderivatives of the product of a general polynomial and $\ln x$. Each *reduction formula* is used repeatedly to reduce the degree of the polynomial until a zero degree polynomial is obtained.

Example 2   Find $\int (x^5 + 2x^3 - 8)e^{3x} \, dx$.

Solution   Since $p(x) = x^5 + 2x^3 - 8$ is a polynomial multiplied by e^{3x}, this is of the form in III-14. Now $p'(x) = 5x^4 + 6x^2$ and $p''(x) = 20x^3 + 12x$, and so on, giving

$$\int (x^5 + 2x^3 - 8)e^{3x} \, dx = e^{3x}\left[\frac{1}{3}(x^5 + 2x^3 - 8) - \frac{1}{9}(5x^4 + 6x^2) + \frac{1}{27}(20x^3 + 12x)\right.$$
$$\left. -\frac{1}{81}(60x^2 + 12) + \frac{1}{243}(120x) - \frac{1}{729} \cdot 120\right] + C.$$

Here we have the successive derivatives of the original polynomial $x^5 + 2x^3 - 8$, occurring with alternating signs and multiplied by successive powers of 1/3.

Part IV of the table contains reduction formulas for the antiderivatives of $\cos^n x$ and $\sin^n x$, which can be obtained by integration by parts. When n is a positive integer, formulas IV-17 and IV-18 can be used repeatedly to reduce the power n until it is 0 or 1.

---

[1] See, for example, *CRC Standard Mathematical Tables* (Boca Raton, Fl: CRC Press). There are now computer programs and calculators that can compute antiderivatives as well.

Example 3   Find $\int \sin^6 \theta \, d\theta$.

Solution   Use IV-17 repeatedly:

$$\int \sin^6 \theta \, d\theta = -\frac{1}{6} \sin^5 \theta \cos \theta + \frac{5}{6} \int \sin^4 \theta \, d\theta$$

$$\int \sin^4 \theta \, d\theta = -\frac{1}{4} \sin^3 \theta \cos \theta + \frac{3}{4} \int \sin^2 \theta \, d\theta$$

$$\int \sin^2 \theta \, d\theta = -\frac{1}{2} \sin \theta \cos \theta + \frac{1}{2} \int 1 \, d\theta.$$

Calculate $\int \sin^2 \theta \, d\theta$ first, and use this to find $\int \sin^4 \theta \, d\theta$; then calculate $\int \sin^6 \theta \, d\theta$. Putting this all together, we get

$$\int \sin^6 \theta \, d\theta = -\frac{1}{6} \sin^5 \theta \cos \theta - \frac{5}{24} \sin^3 \theta \cos \theta - \frac{15}{48} \sin \theta \cos \theta + \frac{15}{48}\theta + C.$$

The last item in **Part IV** of the table is not a formula at all: it is, instead, advice on how to antidifferentiate products of integer powers of $\sin x$ and $\cos x$. There are various techniques to choose from, depending on the nature (odd or even, positive or negative) of the exponents.

Example 4   Find $\int \cos^3 t \sin^4 t \, dt$.

Solution   Here the exponent of $\cos t$ is odd, so IV-23 recommends making the substitution $w = \sin t$. Then $dw = \cos t \, dt$. To make this work, we'll have to separate off one of the cosines to be part of dw. Also, the remaining even power of $\cos t$ can be rewritten in terms of $\sin t$ by using $\cos^2 t = 1 - \sin^2 t = 1 - w^2$, so that

$$\int \cos^3 t \sin^4 t \, dt = \int \cos^2 t \sin^4 t \cos t \, dt$$

$$= \int (1 - w^2)w^4 \, dw = \int (w^4 - w^6) \, dw$$

$$= \frac{1}{5}w^5 - \frac{1}{7}w^7 + C = \frac{1}{5} \sin^5 t - \frac{1}{7} \sin^7 t + C.$$

Example 5   Find $\int \cos^2 x \sin^4 x \, dx$.

Solution   In this example, both exponents are even. The advice given in IV-23 is to convert to all sines or all cosines. We'll convert to all sines by substituting $\cos^2 x = 1 - \sin^2 x$, and then we'll multiply out the integrand:

$$\int \cos^2 x \sin^4 x \, dx = \int (1 - \sin^2 x) \sin^4 x \, dx = \int \sin^4 x \, dx - \int \sin^6 x \, dx.$$

In Example 3 we found $\int \sin^4 x \, dx$ and $\int \sin^6 x \, dx$. Put them together to get

$$\int \cos^2 x \sin^4 x \, dx = -\frac{1}{4} \sin^3 x \cos x - \frac{3}{8} \sin x \cos x + \frac{3}{8}x$$

$$- \left(-\frac{1}{6} \sin^5 x \cos x - \frac{5}{24} \sin^3 x \cos x - \frac{15}{48} \sin x \cos x + \frac{15}{48}x \right) + C$$

$$= \frac{1}{6} \sin^5 x \cos x - \frac{1}{24} \sin^3 x \cos x - \frac{3}{48} \sin x \cos x + \frac{3}{48}x + C.$$

The last two parts of the table are concerned with quadratic functions: **Part V** has expressions with quadratic denominators; **Part VI** contains square roots of quadratics. The quadratics that appear in these formulas are of the form $x^2 \pm a^2$ or $a^2 - x^2$, or in factored form $(x - a)(x - b)$, where a and b are constants. Many integrands with quadratics in them are either in one of these forms or can be transformed into one of them by completing the square or factoring.

Using Factoring

Example 6   Find $\int \dfrac{3x + 7}{x^2 + 6x + 8} \, dx$.

Solution   In this case the denominator factors

$$x^2 + 6x + 8 = (x + 2)(x + 4).$$

Now in V-27 we let $a = -2$, $b = -4$, $c = 3$, and $d = 7$, to obtain

$$\int \frac{3x + 7}{x^2 + 6x + 8} \, dx = \frac{1}{2}(\ln|x + 2| - (-5)\ln|x + 4|) + C.$$

Completing the Square to Rewrite the Quadratic in the Form $w^2 + a^2$

Example 7   Find $\int \dfrac{1}{x^2 + 6x + 14} \, dx$.

Solution   By completing the square, we get

$$x^2 + 6x + 14 = (x^2 + 6x + 9) - 9 + 14$$
$$= (x + 3)^2 + 5.$$

Let $w = x + 3$. Then $dw = dx$ and so the substitution gives

$$\int \frac{1}{x^2 + 6x + 14} \, dx = \int \frac{1}{w^2 + 5} \, dw = \frac{1}{\sqrt{5}} \arctan \frac{w}{\sqrt{5}} + C = \frac{1}{\sqrt{5}} \arctan \frac{x + 3}{\sqrt{5}} + C,$$

where the antidifferentiation uses V-24 with $a^2 = 5$.

Preparing to Use the Table: Transforming the Integrand

Some integration problems will arise in forms that look nearly the same as the formulas in our table, but many will not. In order to use the table, we often need to manipulate or reshape integrands to fit entries in the table. The kinds of manipulation that tend to be useful are expansion, factoring, long division, completing the square, and substitution.

Example 8   Find $\int \dfrac{x^2}{x^2 + 4} \, dx$.

Solution   A good rule of thumb when integrating a rational function whose numerator has a degree greater than or equal to that of the denominator is to start by doing *long division*. This results in a polynomial plus a simpler rational function as a remainder. Performing long division here, we obtain:

$$\frac{x^2}{x^2 + 4} = 1 - \frac{4}{x^2 + 4}.$$

Then, by V-24 with $a = 2$, we obtain:

$$\int \frac{x^2}{x^2 + 4} \, dx = \int 1 \, dx - 4 \int \frac{1}{x^2 + 4} \, dx = x - 4 \cdot \frac{1}{2} \arctan \frac{x}{2} + C = x - 2 \arctan \frac{x}{2} + C.$$

Example 9   Find $\int e^t \sin(5t + 7)\, dt$.

Solution   This looks pretty similar to II-8. To make the correspondence more complete, let's try the substitution $w = 5t + 7$. Then $dw = 5\, dt$, so $dt = \frac{1}{5}\, dw$. Also, $t = (w - 7)/5$. Then the integral becomes

$$\int e^t \sin(5t + 7)\, dt = \int e^{(w-7)/5} \sin w \,\frac{dw}{5}$$

$$= \frac{e^{-7/5}}{5} \int e^{w/5} \sin w \, dw \quad \text{(Since } e^{(w-7)/5} = e^{w/5} e^{-7/5} \text{ and } e^{-7/5} \text{ is a constant)}$$

Now we can use II-8 with $a = \frac{1}{5}$ and $b = 1$ to write

$$\int e^{w/5} \sin w \, dw = \frac{1}{(\frac{1}{5})^2 + 1^2} e^{w/5} \left(\frac{\sin w}{5} - \cos w \right) + C,$$

so

$$\int e^t \sin(5t + 7)\, dt = \frac{e^{-7/5}}{5} \left(\frac{25}{26} e^{(5t+7)/5} \left(\frac{\sin(5t + 7)}{5} - \cos(5t + 7) \right) \right) + C$$

$$= \frac{5e^t}{26} \left(\frac{\sin(5t + 7)}{5} - \cos(5t + 7) \right) + C.$$

Method of Partial Fractions

The formulas in the table involving rational functions of the type in V-26 and V-27 were obtained by splitting the integrand into *partial fractions*. For example, to find

$$\int \frac{1}{(x - 2)(x - 5)}\, dx,$$

the integrand is split into partial fractions with denominators $(x - 2)$ and $(x - 5)$. We write

$$\frac{1}{(x - 2)(x - 5)} = \frac{A}{x - 2} + \frac{B}{x - 5}.$$

Multiplying by $(x - 2)(x - 5)$ gives

$$1 = A(x - 5) + B(x - 2)$$

so,

$$1 = (A + B)x - 5A - 2B.$$

Since this equation holds for all x, the constant terms on both sides must be equal. Similarly, the coefficient of x on both sides must be equal. So

$$-5A - 2B = 1 \quad \text{and} \quad A + B = 0.$$

Solving these equations gives $A = -1/3$, $B = 1/3$.

Example 10   Use partial fractions to rewrite the integrand in such a way that $\int \dfrac{1}{(x - 2)(x - 5)}\, dx$ can be integrated.

Solution   The calculation we have just done gives

$$\int \frac{1}{(x - 2)(x - 5)}\, dx = \int \left(\frac{-1/3}{x - 2} + \frac{1/3}{x - 5} \right) dx = -\frac{1}{3} \ln|x - 2| + \frac{1}{3} \ln|x - 5| + C.$$

You can check that using V-26 gives the same result.

A generalization of this method can be used to find an antiderivative of any rational function.

Example 11 Find $\int \dfrac{x+2}{x^2+x}\,dx$.

Solution We factor the denominator and split the integrand into partial fractions:

$$\frac{x+2}{x^2+x} = \frac{x+2}{x(x+1)} = \frac{A}{x} + \frac{B}{x+1}.$$

Multiplying by $x(x+1)$ gives

$$x+2 = A(x+1) + Bx$$
$$= (A+B)x + A.$$

Equating constant terms and coefficients of x, we must have $A=2$ and $A+B=1$, so $B=-1$. Then

$$\int \frac{x+2}{x^2+x}\,dx = \int \left(\frac{2}{x} - \frac{1}{x+1}\right)\,dx = 2\ln|x| - \ln|x+1| + C.$$

Problems for Section 7.4

For Problems 1–30, antidifferentiate, using the table of integrals and a substitution if necessary.

1. $\displaystyle\int e^{-3\theta}\cos\theta\,d\theta$

2. $\displaystyle\int x^5 \ln x\,dx$

3. $\displaystyle\int \sin w \cos^4 w\,dw$

4. $\displaystyle\int \sin^4 x\,dx$

5. $\displaystyle\int \frac{1}{3+y^2}\,dy$

6. $\displaystyle\int \frac{1}{\cos^3 x}\,dx$

7. $\displaystyle\int x^3 e^{2x}\,dx$

8. $\displaystyle\int \sin 3\theta \cos 5\theta\,d\theta$

9. $\displaystyle\int \sin 3\theta \sin 5\theta\,d\theta$

10. $\displaystyle\int x^2 e^{3x}\,dx$

11. $\displaystyle\int x^2 e^{x^3}\,dx$

12. $\displaystyle\int x^4 e^{3x}\,dx$

13. $\displaystyle\int u^5 \ln(5u)\,du$

14. $\displaystyle\int \frac{t^2+1}{t^2-1}\,dt$

15. $\displaystyle\int x^3 \sin x^2\,dx$

16. $\displaystyle\int \cos 2y \cos 7y\,dy$

17. $\displaystyle\int y^2 \sin 2y\,dy$

18. $\displaystyle\int e^{5x}\sin 3x\,dx$

19. $\displaystyle\int \frac{1}{\sin^2 2\theta}\,d\theta$

20. $\displaystyle\int \frac{1}{\sin^3 3\theta}\,d\theta$

21. $\displaystyle\int \frac{1}{\cos^4 7x}\,dx$

22. $\displaystyle\int \frac{1}{x^2+4x+3}\,dx$

23. $\displaystyle\int \tan^4 x\,dx$

24. $\displaystyle\int \frac{dz}{z(z-3)}$

25. $\displaystyle\int \frac{dy}{4-y^2}$

26. $\displaystyle\int \frac{1}{1+(z+2)^2}\,dz$

27. $\displaystyle\int \frac{1}{y^2+4y+5}\,dy$

28. $\displaystyle\int \frac{1}{x^2+4x+4}\,dx$

29. $\displaystyle\int \sin^3 3\theta \cos^2 3\theta\,d\theta$

30. $\displaystyle\int z e^{2z^2}\cos(2z^2)\,dz$

31. Show that for all integers m and n, with $m \neq \pm n$, $\displaystyle\int_{-\pi}^{\pi} \sin m\theta \sin n\theta\,d\theta = 0$.

32. Show that for all integers m and n, with $m \neq \pm n$, $\displaystyle\int_{-\pi}^{\pi} \cos m\theta \cos n\theta\,d\theta = 0$.

33. (a) Show how combining terms in the expression $\dfrac{2}{x} + \dfrac{1}{x+3}$ yields $\dfrac{3x+6}{x^2+3x}$, and use this to evaluate $\displaystyle\int \frac{3x+6}{x^2+3x}\,dx$.

    (b) Show that your answer to part (a) agrees with the answer you get by using the integral tables.

34. Rewrite $\dfrac{1}{x^2 - 1}$ using partial fractions and find $\displaystyle\int \dfrac{1}{x^2 - 1}\,dx$.

35. (a) Use partial fractions to find $\displaystyle\int \dfrac{1}{x^2 - x}\,dx$.
    (b) Show that your answer to part (a) agrees with the answer you get by using the integral tables.

36. Use partial fractions to find $\displaystyle\int \dfrac{1}{x(L - x)}\,dx$, where L is constant.

37. Use the method of partial fractions to evaluate $\displaystyle\int \dfrac{dP}{3P - 3P^2}$.

38. Use the method of partial fractions to find $\displaystyle\int \dfrac{3x + 1}{x^2 - 3x + 2}\,dx$.

39. The voltage, V, in an electrical outlet is given as a function of time, t, by the function $V = V_0 \cos\left(120\pi t\right)$, where V is in volts and t is in seconds, and V_0 is a constant representing the maximum voltage.

    (a) What is the average value of the voltage over 1 second?
    (b) Engineers do not use the average voltage. They use the *root mean square* voltage defined by $\overline{V} = \sqrt{\text{average of } (V^2)}$. Find $\overline{V}$ in terms of V_0. (Take the average over 1 second.)
    (c) The standard voltage in an American house is 110 volts, meaning that $\overline{V} = 110$. What is V_0?

40. An economist studying the rate of production, $R(t)$, of oil in a new oil well has proposed the following model:
$$R(t) = A + Be^{-t} \sin(2\pi t)$$
    where t is the time in years, A is the equilibrium rate (a constant), and B is the "variability" coefficient (a constant).

    (a) Find the total amount of oil produced in the first N years of operation. (Take N to be an integer.)
    (b) Find the average amount of oil produced per year over the first N years (where N is an integer).
    (c) From your answer to part (b), find the average amount of oil produced per year as $N \to \infty$.
    (d) Looking at the function $R(t)$, explain how you might have predicted your answer to part (c) without doing any calculations.
    (e) Do you think it is reasonable to expect this model to hold over a very long period? Why or why not?

41. Suppose n is a positive integer and $\Psi_n = C_n \sin(n\pi x)$ is the wave function used in describing the behavior of an electron. If n and m are different integers, find
$$\int_0^1 \Psi_n(x) \cdot \Psi_m(x)\,dx.$$

7.5 APPROXIMATING DEFINITE INTEGRALS

The methods of the last few sections allow us to get exact answers for definite integrals in a variety of special cases. However, many functions do not have elementary antiderivatives. To evaluate the definite integrals of such functions, we cannot use the Fundamental Theorem and so we must use numerical methods. In addition, many real-world applications of calculus do not require exact answers.

We already know how to approximate a definite integral numerically using left- and right-hand Riemann sums. In the next two sections we introduce better methods for approximating definite integrals—better in the sense that they give more accurate results with less work than that required to find the left- and right-hand sums.

The Midpoint Rule

In the left- and right-hand Riemann sums, the heights of the rectangles are found using the left-hand or right-hand endpoints, respectively, of the subintervals. For the *midpoint rule*, we use the midpoint of each of the subintervals.

For example, in approximating $\int_1^2 f(x)\,dx$ by a Riemann sum with two subdivisions, we first divide the interval $1 \le x \le 2$ into two pieces. The midpoint of the first subinterval is 1.25 and the midpoint of the second is 1.75. The heights of the two rectangles are $f(1.25)$ and $f(1.75)$, respectively. (See Figure 7.2.) The Riemann sum is

$$f(1.25)0.5 + f(1.75)0.5.$$

Figure 7.2 shows that evaluating f at the midpoint of each subdivision usually gives a better approximation to the area under the curve than evaluating f at either end. For this particular f, it appears that each rectangle is partly above and partly below the graph on each subinterval. Furthermore, the area under the curve which is not under the rectangle appears to be nearly equal to the area under the rectangle which is above the curve. In fact, this new midpoint Riemann sum is generally a better approximation to the definite integral than the left- or right-hand sum with the same number of subdivisions, n.

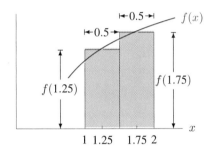

Figure 7.2: Midpoint rule with two subdivisions

So far, we have three ways of estimating an integral using a Riemann sum:

1. The **left rule** uses the left endpoint of each subinterval.

2. The **right rule** uses the right endpoint of each subinterval.

3. The **midpoint rule** use the midpoint of each subinterval.

We write LEFT(n), RIGHT(n), and MID(n) to denote the results obtained by using these rules with n subdivisions.

Example 1   For $\int_1^2 \dfrac{1}{x}\,dx$, compute LEFT(2), RIGHT(2) and MID(2), and compare your answers with the exact value of the integral.

Solution   For $n = 2$ subdivisions of the interval $[1, 2]$, we use $\Delta x = 0.5$. Then

$$\text{LEFT}(2) = f(1)(0.5) + f(1.5)(0.5) = \frac{1}{1}(0.5) + \frac{1}{1.5}(0.5) = 0.8333\ldots$$

$$\text{RIGHT}(2) = f(1.5)(0.5) + f(2)(0.5) = \frac{1}{1.5}(0.5) + \frac{1}{2}(0.5) = 0.5833\ldots$$

$$\text{MID}(2) = f(1.25)(0.5) + f(1.75)(0.5) = \frac{1}{1.25}(0.5) + \frac{1}{1.75}(0.5) = 0.6857\ldots$$

All three Riemann sums in this example are approximating

$$\int_1^2 \frac{1}{x}\,dx = \ln x \Big|_1^2 = \ln 2 - \ln 1 = \ln 2 = 0.6931\ldots.$$

With only two subdivisions, the left and right rules give quite poor approximations but the midpoint rule is already fairly close to the exact answer.

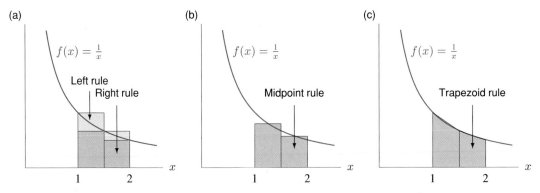

Figure 7.3: Left, right, midpoint, and trapezoid approximations to $\int_1^2 \frac{1}{x}\,dx$

Figure 7.3(a) illustrates why the left and right rules are so inaccurate. Since $f(x) = 1/x$ is decreasing from 1 to 2, the left rule overestimates on each subdivision while the right rule underestimates. However, the midpoint rule approximates with rectangles on each subdivision that are each partly above and partly below the graph, so the errors tend to balance out. (See Figure 7.3(b).)

The Trapezoid Rule

We have just seen how the midpoint rule can have the effect of balancing out the errors of the left and right rules. There is another way of balancing these errors: we average the results from the left and right rules. This approximation is called the *trapezoid rule*:

$$\text{TRAP}(n) = \frac{\text{LEFT}(n) + \text{RIGHT}(n)}{2}.$$

The trapezoid rule averages the values of f at the left and right endpoints of each subinterval and multiplies by Δx. This is the same as approximating the area under the graph of f in each subinterval by a trapezoid (see Figure 7.4).

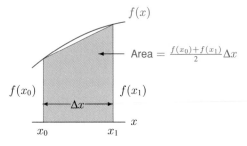

Figure 7.4: Area used in the trapezoid rule

Example 2 For $\displaystyle\int_1^2 \frac{1}{x}\,dx$, compare the trapezoid rule with two subdivisions with the left, right, and midpoint rules.

Solution In the previous example we got $\text{LEFT}(2) = 0.8333\ldots$ and $\text{RIGHT}(2) = 0.5833\ldots$. The trapezoid rule is the average of these, so $\text{TRAP}(2) = 0.7083\ldots$. (See Figure 7.3(c).) The exact value of the integral is $0.6931\ldots$, so the trapezoid rule is better than the left or right rules. The midpoint rule is still the best, however, since $\text{MID}(2) = 0.6857\ldots$.

Is the Approximation an Over- or Underestimate?

It is useful to know when a rule is producing an overestimate and when it is producing an underestimate. In Chapter 3 we saw that the following relationship holds.

If f is increasing on $[a, b]$, then

$$\text{LEFT}(n) \leq \int_a^b f(x)\,dx \leq \text{RIGHT}(n).$$

If f is decreasing on $[a, b]$, then

$$\text{RIGHT}(n) \leq \int_a^b f(x)\,dx \leq \text{LEFT}(n).$$

The Trapezoid Rule

If the graph of the function is concave down on $[a, b]$, then each trapezoid lies below the graph and the trapezoid rule underestimates. If the graph is concave up on $[a, b]$, the trapezoid rule overestimates. (See Figure 7.5.)

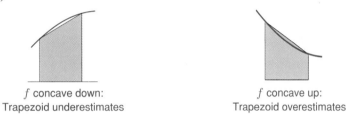

f concave down:
Trapezoid underestimates

f concave up:
Trapezoid overestimates

Figure 7.5: Error in the trapezoid rule

The Midpoint Rule

To understand the relationship between the midpoint rule and concavity, take a rectangle whose top intersects the curve at the midpoint of a subinterval. Draw a tangent to the curve at the midpoint; this gives a trapezoid. See Figure 7.6. (This is *not* the same trapezoid as in the trapezoid rule.) The midpoint rectangle and the new trapezoid have the same area, because the shaded triangles in Figure 7.6 are congruent. Hence, if the graph of the function is concave down, the midpoint rule overestimates; if the graph is concave up, the midpoint rule underestimates. (See Figure 7.7.)

If the graph of f is concave down on $[a, b]$, then

$$\text{TRAP}(n) \leq \int_a^b f(x)\,dx \leq \text{MID}(n).$$

If the graph of f is concave up on $[a, b]$, then

$$\text{MID}(n) \leq \int_a^b f(x)\,dx \leq \text{TRAP}(n).$$

f concave down:
Midpoint overestimates

f concave up:
Midpoint underestimates

Figure 7.6: Midpoint rectangle and trapezoid with same area

Figure 7.7: Error in the midpoint rule

Problems for Section 7.5

1. (a) By hand, find LEFT(2) and RIGHT(2) for $\int_0^4 (x^2 + 1)\, dx$. Show your work.
   (b) Illustrate your answers to part (a) graphically. Is each estimate an underestimate or overestimate?

2. (a) By hand, find MID(2) and TRAP(2) for $\int_0^4 (x^2 + 1)\, dx$. Show your work.
   (b) Illustrate your answers to part (a) graphically. Is each estimate an underestimate or overestimate?

3. The following table contains approximations to $\int_0^4 \sqrt{100 + x^3}\, dx$. Fill in the rest of the table with values rounded to four decimal places.

	$n = 1$	$n = 2$	$n = 4$
LEFT			
RIGHT	51.2250		
TRAP		43.5909	
MID			

4. (a) Estimate $\int_0^1 \dfrac{1}{1 + x^2}\, dx$ by subdividing the interval into eight parts using:
   - (i) The left-hand Riemann sum.
   - (ii) The right-hand Riemann sum.
   - (iii) The trapezoidal rule.
   (b) Since the exact value of the integral is $\pi/4$, you can estimate the value of π. Explain why your first estimate is too large and your second estimate too small.

5. Consider the following integrals:

   $$\text{(i)} \quad \int_1^{10} \ln x\, dx \qquad \text{(ii)} \quad \int_0^4 e^x\, dx$$

   (a) For each integral, find LEFT(32), RIGHT(32), and TRAP(32). Also find the exact value of each integral.
   (b) For each integral, arrange LEFT(32), RIGHT(32), TRAP(32), and the true value in ascending order. Explain, using diagrams, how you could predict this ordering without doing the calculations in part (a).

6. Given the velocity-time data in the table, estimate the total distance traveled from time $t = 0$ to time $t = 6$ using LEFT, RIGHT, and TRAP.

t	0	1	2	3	4	5	6
v	3	4	5	4	7	8	11

7. The graph of g is shown in Figure 7.8. The results from the left, right, trapezoid, and midpoint rules used to approximate $\int_0^1 g(t)\, dt$, with the same number of subdivisions for each rule, are as follows:
   0.601, 0.632, 0.633, 0.664.

   (a) Match each rule with its approximation.
   (b) Between which two consecutive approximations does the true value of the integral lie?

Figure 7.8   *Figure 7.9*

8. We approximate the integrals of the functions in Figure 7.9 on the interval shown, using a fixed number of subdivisions.

   (a) For which function, f or g, is LEFT more accurate? RIGHT? Explain.
   (b) For which function, f or g, is TRAP more accurate? MID? Explain.

9. Given the values for $y = f(x)$ in the table:

   (a) Which of the four approximation methods in this section is most likely to give the best estimate of $\int_0^{12} f(x)\,dx$? Estimate the integral using this method.
   (b) Assume $f(x)$ is continuous with no critical points or points of inflection on the interval $0 \leq x \leq 12$. Is the estimate found in part (a) an over- or underestimate? Explain.

x	0	3	6	9	12
$f(x)$	100	97	90	78	55

Suppose you want to approximate $\int_0^5 f(x)\,dx$ for the functions graphed in Problems 10–13. In each case, pick which of the approximations — left, right, trapezoid, or midpoint — is guaranteed to give an overestimate for the integral, and which is guaranteed to give an underestimate. (There may be more than one answer.)

10.

11.

12.

13.

14. (a) Find the exact value of $\displaystyle\int_0^{2\pi} \sin\theta\,d\theta$.

    (b) Explain, using pictures, why the MID(1) and MID(2) approximations to this integral give the exact value.

    (c) Does MID(3) give the exact value of this integral? How about MID(n)? Explain.

15. (a) Show that $\displaystyle\int_0^1 \sqrt{2 - x^2}\,dx = \frac{\pi}{4} + \frac{1}{2}$. [Hint: Break up the area under $y = \sqrt{2 - x^2}$ from $x = 0$ to $x = 1$ into two pieces: a sector of a circle and a right triangle.]

    (b) Approximate $\int_0^1 \sqrt{2 - x^2}\,dx$ for $n = 5$ using the left, right, trapezoid, and midpoint rules. Compute the error in each case using the answer to part (a), and compare the errors.

16. Sketch the graph of a function for which LEFT(2) is more accurate than MID(2). (You don't need to give a formula for the function, just a graph.)

17. The width, in feet, at various points along the fairway of a hole on a golf course is given in Figure 7.10. If one pound of fertilizer covers 200 square feet, estimate the amount of fertilizer needed to fertilize the fairway.

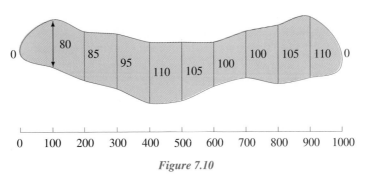

Figure 7.10

18. (a)   Explain why the area of the trapezoid in Figure 7.11 is $h \cdot (l_1 + l_2)/2$.
     (b)   On a graph similar to Figure 7.12, sketch areas representing each of the following quantities:

$$E = h \cdot f(0), \quad F = h \cdot f(h), \quad R = h \cdot f\left(\frac{h}{2}\right), \quad C = h \cdot \frac{f(0) + f(h)}{2} = \frac{E + F}{2}$$

$$N = \frac{h}{2} \cdot \frac{f(0) + f(\frac{h}{2})}{2} + \frac{h}{2} \cdot \frac{f(\frac{h}{2}) + f(h)}{2} = \frac{R + C}{2}.$$

     (c)   Let A be the area under the function shown in Figure 7.12. Write the values A, E, F, R, C, and N in increasing order.
     (d)   Which is the better approximation to A: E or F?
     (e)   Which is the better approximation to A: R or C?

Figure 7.11

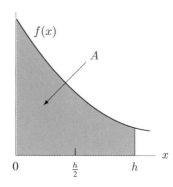

Figure 7.12

Problems 19–23 involve approximating $\int_a^b f(x)\,dx$.

19. Show that $\text{RIGHT}(n) = \text{LEFT}(n) + f(b)\Delta x - f(a)\Delta x$.

20. Show that $\text{TRAP}(n) = \text{LEFT}(n) + \frac{1}{2}\left(f(b) - f(a)\right)\Delta x$.

21. Show that $\text{LEFT}(2n) = \frac{1}{2}\left(\text{LEFT}(n) + \text{MID}(n)\right)$.

22. Using a computer or calculator, verify that the equations given in Problems 19 and 20 hold for $\int_1^2 (1/x)\,dx$, when $n = 10$.

23. Suppose that $a = 2$, $b = 5$, $f(2) = 13$, $f(5) = 21$ and that $\text{LEFT}(10) = 3.156$ and $\text{MID}(10) = 3.242$. Use the equations given in Problems 19–21 to compute $\text{RIGHT}(10)$, $\text{TRAP}(10)$, $\text{LEFT}(20)$, $\text{RIGHT}(20)$, and $\text{TRAP}(20)$.

7.6   APPROXIMATION ERRORS AND SIMPSON'S RULE

When we compute an approximation, we are always concerned about the error, namely the difference between the exact answer and the approximation. We usually do not know the exact error; if we did, we would also know the exact answer. Often the best we can get is an upper bound on the error and some idea of how much work is involved in making the error smaller. The study of numerical approximations is really the study of errors. The errors for some methods are much smaller than those for others. The errors for the midpoint and trapezoid rules are related to each other in a way that suggests an even better method, called Simpson's rule. We will work with the example $\int_1^2 \frac{1}{x}\,dx$ because we know the exact value of this integral ($\ln 2$) and so we can investigate the behavior of the errors.

Error in Left and Right Rules

For any approximation, we take

$$\text{Error} = \text{Actual value} - \text{Approximate value}.$$

TABLE 7.1 *Errors for the left and right rule approximation to* $\int_1^2 \frac{1}{x}\,dx = \ln 2 \approx 0.6931471806$

n	Error in left rule	Error in right rule
2	−0.1402	0.1098
10	−0.0256	0.0244
50	−0.0050	0.0050
250	−0.0010	0.0010

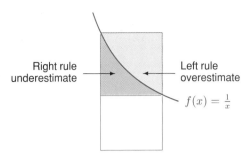

Figure 7.13: Errors in left and right sums

Let us see what happens to the error in the left and right rules as we increase n. We increase n each time by a factor of 5 starting at $n = 2$. The results are in Table 7.1. A positive error indicates that the Riemann sum is less than the exact value, ln 2. Notice that the errors for the left and right rules have opposite signs but are approximately equal in magnitude. (See Figure 7.13.) The best way to try to get the errors to cancel is to average the left and right rules; this average is the trapezoid rule. If we had not already thought of the trapezoid rule, we might have been led to invent it by this observation.

There is another pattern to the errors in Table 7.1. If we compute the *ratio* of the errors in Table 7.2, we see that the error[2] in both the left and right rules decreases by a factor of about 5 as n increases by a factor of 5.

There is nothing special about the number 5; the same holds for any factor. To get one extra digit of accuracy in any calculation, we must make the error $1/10$ as big, so we must increase n by a factor of 10. In fact, *for the left or right rules, each extra digit of accuracy requires about 10 times the work.* The calculator used to produce these tables took about half a second to compute the left rule approximation for $n = 50$, and this yields ln 2 to two digits. To get three correct digits, n would need to be around 500 and the time would be about 5 seconds. Four digits requires $n = 5000$ and 50 seconds. Ten digits requires $n = 5 \times 10^9$ and 5×10^7 seconds, which is more than a year! Clearly, the errors for the left and right rules do not decrease fast enough as n increases for practical use.

TABLE 7.2 *Ratio of the errors as n increases for $\int_1^2 \frac{1}{x}\,dx$*

	Ratio of errors in left rule	Ratio of errors in right rule
Error(2)/Error(10)	5.47	4.51
Error(10)/Error(50)	5.10	4.90
Error(50)/Error(250)	5.02	4.98

Error in Trapezoid and Midpoint Rules

Table 7.3 shows that the trapezoid and midpoint rules produce much better approximations to $\int_1^2 \frac{1}{x}\,dx$ than the left and right rules.

Again there is a pattern to the errors. For each n, the midpoint rule is noticeably better than the trapezoid rule; the error for the midpoint rule, in absolute value, seems to be about half the error of the trapezoid rule. To see why, compare the shaded areas in Figure 7.14. Also, notice in Table 7.3 that the errors for the two rules have opposite signs; this is due to concavity.

We are interested in how the errors behave as n increases. Table 7.4 gives the ratios of the errors for each rule. For each rule, we see that as n increases by a factor of 5, the error decreases by a factor of about $25 = 5^2$. In fact, it can be shown that this squaring relationship holds for any factor, so increasing n by a factor of 10 will decrease the error by a factor of about $100 = 10^2$. Reducing the error by a factor of 100 is equivalent to adding two more decimal places of accuracy to the result.

---

[2]The values in Table 7.1 are rounded to 4 decimal places; those in Table 7.2 were computed using more decimal places and then rounded.

TABLE 7.3   *The errors for the trapezoid and midpoint rules for $\int_1^2 \frac{1}{x}\,dx$*

n	Error in trapezoid rule	Error in midpoint rule
2	−0.0152	0.0074
10	−0.00062	0.00031
50	−0.0000250	0.0000125
250	−0.0000010	0.0000005

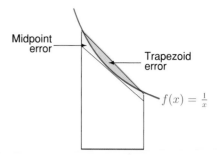

Figure 7.14: Errors in the midpoint and trapezoid rules

In other words: *In the trapezoid or midpoint rules, each extra 2 digits of accuracy requires about 10 times the work.*

This result shows the advantage of the midpoint and trapezoid rules over the left and right rules: less additional work needs to be done to get another decimal place of accuracy. The calculator used to produce these tables again took about half a second to compute the midpoint rule for $\int_1^2 \frac{1}{x}\,dx$ with $n = 50$, and this gets 4 digits correct. Thus to get 6 digits would take $n = 500$ and 5 seconds, to get 8 digits would take 50 seconds, and to get 10 digits would take 500 seconds, or about 10 minutes. That is still not great, but it is certainly better than the 1 year required by the left or right rule.

TABLE 7.4   *Ratios of the errors as n increases for $\int_1^2 \frac{1}{x}\,dx$*

	Ratio of errors in trapezoid rule	Ratio of errors in midpoint rule
Error(2)/Error(10)	24.33	23.84
Error(10)/Error(50)	24.97	24.95
Error(50)/Error(250)	25.00	25.00

Simpson's Rule

Still more improvement is possible. Observing that the trapezoid error has the opposite sign and about twice the magnitude of the midpoint error, we may guess that a weighted average of the two rules, with the midpoint rule weighted twice the trapezoid rule, will have a much smaller error. This approximation is called *Simpson's rule*[3]:

$$\text{SIMP}(n) = \frac{2 \cdot \text{MID}(n) + \text{TRAP}(n)}{3}.$$

Table 7.5 gives the errors for Simpson's rule. Notice how much smaller the errors are than the previous errors. Of course, it is a little unfair to compare Simpson's rule at $n = 50$, say, with the previous rules, because Simpson's rule must compute the value of f at both the midpoint and the endpoints of each subinterval and hence involves evaluating the function at twice as many points. We know by our previous analysis, however, that even if we did compute the other rules at $n = 100$ to compare with Simpson's rule at $n = 50$, the other errors would only decrease by a factor of 2 for the left and right rules and by a factor of 4 for the trapezoid and midpoint rules.

We see in Table 7.5 that as n increases by a factor of 5, the errors decrease by a factor of about 600, or about 5^4. Again this behavior holds for any factor, so increasing n by a factor of 10 decreases the error by a factor of about 10^4. In other words: *In Simpson's rule, each extra 4 digits of accuracy requires about 10 times the work.*

---

[3]Some books and computer programs use slightly different terminology for Simpson's rule; what we call $n = 50$, they call $n = 100$.

TABLE 7.5 *The errors for Simpson's rule and the ratios of the errors*

n	Error	Ratio
2	-0.0001067877	550.15
10	-0.0000001940	632.27
50	-0.0000000003	

This is a great improvement over either the midpoint or trapezoid rules, which only give two extra digits of accuracy when n is increased by a factor of 10. Simpson's rule is so efficient that we get 9 digits correct with $n = 50$ in about 1 second on our calculator. Doubling n will decrease the error by a factor of about $2^4 = 16$ and hence will give the tenth digit. The total time is 2 seconds, which is pretty good.

In general, Simpson's rule achieves a reasonable degree of accuracy when using relatively small values of n, and is a good choice for an all-purpose method for estimating definite integrals.

Analytical View of the Trapezoid and Simpson's Rules

Our approach to approximating $\int_a^b f(x)\, dx$ numerically has been empirical: try a method, see how the error behaves, and then try to improve it. We can also develop the various rules for numerical integration by making better and better approximations to the integrand, f. The left, right, and midpoint rules are all examples of approximating f by a constant (flat) function on each subinterval. The trapezoid rule is obtained by approximating f by a linear function on each subinterval. Simpson's rule can, in the same spirit, be obtained by approximating f by quadratic functions. The details are given in Problems 20 and 21 on page 356.

How the Error Depends on the Integrand

Other factors besides the size of n affect the size of the error in each of the rules. Instead of looking at how the error behaves as we increase n, let's leave n fixed and imagine trying our approximation methods on different functions. We observe that the error in the left or right rule depends on how steeply the graph of f rises or falls. A steep curve makes the triangular regions missed by the left or right rectangles tall and hence large in area. This observation suggests that the error in the left or right rules depends on the size of the derivative of f (see Figure 7.15).

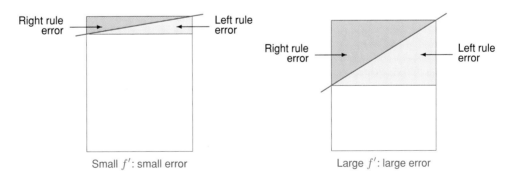

Figure 7.15: The error in the left and right rules depends on the steepness of the curve

From Figure 7.16 it appears that the errors in the trapezoid and midpoint rules depend on how much the curve is bent up or down. In other words, the concavity, and hence the size of the second derivative of f, has an effect on the errors of these two rules. Finally, it can be shown[4] that the error in Simpson's rule depends on the size of the *fourth* derivative of f, written $f^{(4)}$.

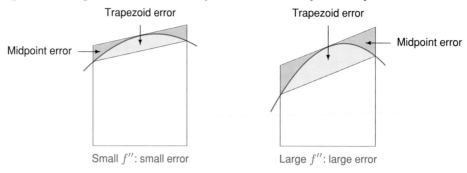

Small f'': small error   Large f'': large error

Figure 7.16: The error in the trapezoid and midpoint rules depends on how bent the curve is

Problems for Section 7.6

1. (a) Using the result of Problem 2 on page 348, compute SIMP(2) for $\int_0^4 (x^2 + 1)\, dx$.
   (b) Use the Fundamental Theorem of Calculus to find $\int_0^4 (x^2 + 1)\, dx$ exactly.
   (c) What is the error in SIMP(2) for this integral?

For Problems 2–7 use Simpson's rule with various values of n to evaluate the definite integrals with error less than 0.001. Explain why you think you have reached a large enough value of n.

2. $\displaystyle\int_0^1 \frac{1}{4 + x^2}\, dx$ 3. $\displaystyle\int_0^2 e^{\sin t}\, dt$ 4. $\displaystyle\int_0^3 \cos^2 \theta\, d\theta$

5. $\displaystyle\int_0^1 \sqrt{1 + x^4}\, dx$ 6. $\displaystyle\int_0^{10} \ln(z^2 + 1)\, dz$ 7. $\displaystyle\int_0^1 \cos(t^2)\, dt$

8. In this problem you will investigate the behavior of the errors in the approximation of the integral
$$\int_1^2 \frac{1}{x}\, dx \approx 0.6931471806\ldots.$$
   (a) For $n = 2, 4, 8, 16, 32, 64, 128$ subdivisions, compute the left and right approximations and the errors in each.
   (b) What are the signs of the errors in the left and right approximations? How do the errors change if n is doubled?
   (c) For the values of n in part (a), compute the midpoint and trapezoid approximations and the errors in each.
   (d) What are the signs of the errors in the midpoint and trapezoid approximations? How do the errors change if n is doubled?
   (e) For $n = 2, 4, 8, 16, 32$, compute Simpson's rule approximation and the error in each. How do these errors change as n doubles?

9. (a) What is the exact value of $\int_0^4 e^x\, dx$?
   (b) Find LEFT(2), RIGHT(2), TRAP(2), MID(2), and SIMP(2). Compute the error for each.
   (c) Repeat part (b) with $n = 4$ (instead of $n = 2$).
   (d) For each rule in part (b), as n goes from $n = 2$ to $n = 4$, does the error go down approximately as you would expect? Explain.

10. Consider Simpson's rule approximations to $\int_0^2 (x^3 + 3x^2)\, dx$.
    (a) What is the exact value of this integral?
    (b) Find SIMP(n) for $n = 2, 4$, and 100. What do you notice?

[4] See Kendall E. Atkinson, *An Introduction to Numerical Analysis* (New York: John Wiley and Sons, 1978).

11. Suppose that the approximation to a certain definite integral using $n = 10$ is 2.346 and that the exact value is 4.0. If the approximation was found using each of the following rules, use the same rule to estimate the integral with $n = 30$.

    (a) LEFT   (b) TRAP   (c) SIMP

12. (a) Suppose a certain computer takes two seconds to compute a certain definite integral accurate to 4 digits to the right of the decimal point, using the left rectangle rule. How long (in years) will it take to get 8 digits correct using the left rectangle rule? How about 12 digits? 20 digits?

    (b) Repeat part(a) but this time assume that the trapezoidal rule is being used throughout.

13. Suppose a certain computer takes 3 seconds to compute a certain definite integral accurate to 2 decimal places. How long will it take the computer to get 10 decimal places of accuracy if it is using the rule shown below? Give your answer in seconds and in appropriate time units (minutes, hours, days, or years).

    (a) LEFT   (b) MID   (c) SIMP

14. Suppose that for a certain definite integral, LEFT(10) = 0.38745 and LEFT(20) = 0.36517. Estimate the actual error for LEFT(10) (and thus the actual value of the integral) by assuming that the error is reduced by a factor of 2 in going from LEFT(10) to LEFT(20).

15. Suppose for a certain definite integral that MID(10) = 35.619 and MID(20) = 35.415. Estimate the actual error for MID(10) assuming that the error for MID(10) is reduced by a factor of 4 in going to MID(20).

16. Suppose that for a certain definite integral TRAP(10) = 12.676 and TRAP(30) = 10.420. Use this information to estimate the actual value of the integral. Justify your answer.

17. Suppose that for a certain definite integral, MID(10) = 5.364 and MID(20) = 4.926. Estimate as accurately as possible:

    (a) Actual value of the integral.   (b) Value of MID(60).

18. Some approximations for a certain definite integral are given in the following table. Use the fact that the exact value of the integral is 0.69315 to fill in the errors for $n = 2$. Then use what you know about errors to estimate the errors when $n = 20$.

	Approximation $n = 2$	Error $n = 2$	Error $n = 20$
LEFT	0.83333		
RIGHT	0.58333		
TRAP	0.70833		
MID	0.68571		
SIMP	0.69325		

19. Some approximations to an integral, using $n = 3$, are given in Table 7.6. The true value of the integral is 7.621372.

    (a) Does the integrand function appear to be increasing or decreasing? Concave up or concave down? Explain.

    (b) Find the error for each approximation, and fill in column (b) in Table 7.6.

    (c) Use what you know about errors and the errors for $n = 3$ to estimate errors for $n = 30$. Fill in column (c) in Table 7.6.

    TABLE 7.6

	Approximation $n = 3$	Error $n = 3$	Error $n = 30$
LEFT	5.416101		
RIGHT	9.307921		
TRAP	7.362011		
MID	7.742402		
SIMP	7.615605		

Problems 20–21 show how Simpson's rule can be obtained by approximating the integrand, f, by quadratic functions.

20. Suppose that $a < b$ and that m is the midpoint $m = (a+b)/2$. Let $h = b - a$. The purpose of this problem is to show that if f is a quadratic function, then

$$\int_a^b f(x)\,dx = \frac{h}{3}\left(\frac{f(a)}{2} + 2f(m) + \frac{f(b)}{2}\right).$$

(a) Show that this equation holds for the functions $f(x) = 1$, $f(x) = x$, and $f(x) = x^2$.

(b) Use part (a) and the "Properties about Sums and Constant Multiples of the Integrand" from Section 3.4, page 170, to show that the equation holds for any quadratic function,
$f(x) = Ax^2 + Bx + C$.

21. Consider the following method for approximating $\int_a^b f(x)dx$. Divide the interval $[a, b]$ into n equal subintervals. On each subinterval approximate f by a quadratic function that agrees with f at both endpoints and at the midpoint of the subinterval.

(a) Explain why the integral of f on the subinterval $[x_i, x_{i+1}]$ is approximately equal to the expression

$$\frac{h}{3}\left(\frac{f(x_i)}{2} + 2f(m_i) + \frac{f(x_{i+1})}{2}\right),$$

where m_i is the midpoint of the subinterval, $m_i = (x_i + x_{i+1})/2$. (See Problem 20.)

(b) Show that if we add up these approximations for each subinterval, we get Simpson's rule:

$$\int_a^b f(x)dx \approx \frac{2 \cdot \text{MID}(n) + \text{TRAP}(n)}{3}.$$

7.7 IMPROPER INTEGRALS

Our original discussion of the definite integral $\int_a^b f(x)\,dx$ assumed that the interval $a \leq x \leq b$ was of finite length and that f was continuous. Integrals that arise in applications don't necessarily have these nice properties. In this section we investigate a class of integrals, called *improper* integrals, in which one limit of integration is infinite or the integrand is unbounded. As an example, to estimate the mass of the earth's atmosphere, we might calculate an integral which sums the mass of the air up to different heights. In order to represent the fact that the atmosphere doesn't end at a specific height, we let the upper limit of integration get larger and larger, or tend to infinity.

We will usually consider only improper integrals with positive integrands since they are the most common.

One Type of Improper Integral: When the Limit of Integration Is Infinite

Here is an example of an improper integral:

$$\int_1^\infty \frac{1}{x^2}\,dx.$$

To evaluate this integral, we first compute the definite integral $\int_1^b \frac{1}{x^2}\,dx$:

$$\int_1^b \frac{1}{x^2}\,dx = -x^{-1}\bigg|_1^b = -\frac{1}{b} + \frac{1}{1}.$$

Now take the limit as $b \to \infty$. Since

$$\lim_{b \to \infty} \int_1^b \frac{1}{x^2} \, dx = \lim_{b \to \infty} \left(-\frac{1}{b} + 1 \right) = 1,$$

we say that the improper integral $\int_1^\infty \frac{1}{x^2} \, dx$ *converges* to 1.

If we think of this in terms of areas, it may seem strange that the region whose area is computed by $\int_1^\infty \frac{1}{x^2} \, dx$ extends from $x = 1$ infinitely far to the right. How could it have finite area? (See Figure 7.17(a).) What our limit computations are saying is that

$$\text{When } b = 10: \quad \int_1^{10} \frac{1}{x^2} \, dx = -\frac{1}{x} \Big|_1^{10} = -\frac{1}{10} + 1 = 0.9$$

$$\text{When } b = 100: \quad \int_1^{100} \frac{1}{x^2} \, dx = -\frac{1}{100} + 1 = 0.99$$

$$\text{When } b = 1000: \quad \int_1^{1000} \frac{1}{x^2} \, dx = -\frac{1}{1000} + 1 = 0.999$$

and so on. In other words, as b gets larger and larger, the area between $x = 1$ and $x = b$ tends to 1. See Figure 7.17(b). Thus, it does make sense to declare that $\int_1^\infty \frac{1}{x^2} \, dx = 1$.

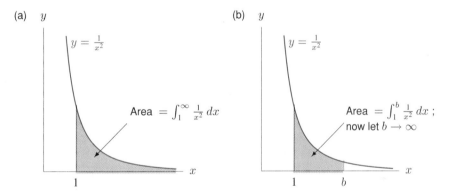

Figure 7.17: Area representation of improper integral

Of course, in another example, we might not get a finite limit as b gets larger and larger. In that case we say the improper integral *diverges*.

Suppose $f(x)$ is positive for $x \geq a$.

If $\lim_{b \to \infty} \int_a^b f(x) \, dx$ is a finite number, we say that $\int_a^\infty f(x) \, dx$ **converges** and define

$$\int_a^\infty f(x) \, dx = \lim_{b \to \infty} \int_a^b f(x) \, dx.$$

Otherwise, we say that $\int_a^\infty f(x) \, dx$ **diverges**. We define $\int_{-\infty}^b f(x) \, dx$ similarly.

Example 1   Does the improper integral $\int_1^\infty \frac{1}{\sqrt{x}}\,dx$ converge or diverge?

Solution   We consider

$$\int_1^b \frac{1}{\sqrt{x}}\,dx = \int_1^b x^{-1/2}\,dx = 2x^{1/2}\Big|_1^b = 2b^{1/2} - 2.$$

We see that $\int_1^b \frac{1}{\sqrt{x}}\,dx$ grows without bound as $b \to \infty$. Thus we say the integral $\int_1^\infty \frac{1}{\sqrt{x}}\,dx$ *diverges*. We have shown that the area under the curve in Figure 7.18 is not finite.

Notice that $f(x) \to 0$ as $x \to \infty$ does not guarantee convergence.

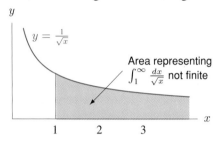

Figure 7.18: $\int_1^\infty \frac{1}{\sqrt{x}}\,dx$ diverges

What is the difference between the functions $1/x^2$ and $1/\sqrt{x}$ that makes the area under the graph of $1/x^2$ approach 1 as $x \to \infty$, whereas the area under $1/\sqrt{x}$ grows very large? Both functions approach 0 as x grows, so as b grows larger, smaller bits of area are being added to the definite integral. The difference between the functions is subtle: the values of the function $1/\sqrt{x}$ *don't shrink fast enough* for the integral to have a finite value. Of the two functions, $1/x^2$ drops to 0 much faster than $1/\sqrt{x}$, and this feature keeps the area under $1/x^2$ from growing beyond 1.

Example 2   Find $\int_0^\infty e^{-5x}\,dx$.

Solution   First we consider $\int_0^b e^{-5x}\,dx$:

$$\int_0^b e^{-5x}\,dx = -\frac{1}{5}e^{-5x}\Big|_0^b = -\frac{1}{5}e^{-5b} + \frac{1}{5}.$$

Since $e^{-5b} = \frac{1}{e^{5b}}$, this term tends to 0 as b approaches infinity, so $\int_0^\infty e^{-5x}\,dx$ converges. Its value is

$$\int_0^\infty e^{-5x}\,dx = \lim_{b \to \infty} \int_0^b e^{-5x}\,dx = \lim_{b \to \infty}\left(-\frac{1}{5}e^{-5b} + \frac{1}{5}\right) = 0 + \frac{1}{5} = \frac{1}{5}.$$

Since e^{5x} grows very rapidly, we expect that e^{-5x} will approach 0 rapidly. The fact that the area approaches $\frac{1}{5}$ instead of growing without bound is a consequence of the speed with which the integrand e^{-5x} approaches 0.

Example 3   Determine for which values of the exponent, p, the improper integral $\int_1^\infty \frac{1}{x^p}\,dx$ diverges.

Solution   For $p \neq 1$,

$$\int_1^b x^{-p}\,dx = \frac{1}{-p+1}x^{-p+1}\Big|_1^b = \left(\frac{1}{-p+1}b^{-p+1} - \frac{1}{-p+1}\right).$$

The important question is whether the exponent of b is positive or negative. If it is negative, then as b approaches infinity, b^{-p+1} approaches 0. If the exponent is positive, then b^{-p+1} grows without bound as b approaches infinity.

What happens if $p = 1$? In this case we get

$$\int_1^\infty \frac{1}{x}\, dx = \lim_{b \to \infty} \ln x \Big|_1^b = \lim_{b \to \infty} \ln b - \ln 1.$$

Since $\ln b$ becomes arbitrarily large as b approaches infinity, the integral grows without bound. We conclude that $\int_1^\infty \frac{1}{x^p}\, dx$ diverges precisely when $p \leq 1$. For $p > 1$ the integral has the value

$$\int_1^\infty \frac{1}{x^p}\, dx = \lim_{b \to \infty} \int_1^b \frac{1}{x^p}\, dx = \lim_{b \to \infty} \left(\frac{1}{-p+1} b^{-p+1} - \frac{1}{-p+1} \right) = -\left(\frac{1}{-p+1} \right) = \frac{1}{p-1}.$$

Application of Improper Integrals to Energy

The energy, E, required to separate two charged particles, originally a distance a apart, to a distance b, is given by the integral

$$E = \int_a^b \frac{kq_1 q_2}{r^2}\, dr$$

where q_1 and q_2 are the magnitudes of the charges and k is a constant. If q_1 and q_2 are in coulombs, a and b are in meters, and E is in joules, the value of the constant k is 9×10^9.

Example 4   A hydrogen atom consists of a proton and an electron, with opposite charges of magnitude 1.6×10^{-19} coulombs. Find the energy required to take a hydrogen atom apart (that is, to move the electron from its orbit to an infinite distance from the proton). Assume that the initial distance between the electron and the proton is the Bohr radius, $R_B = 5.3 \times 10^{-11}$ meter.

Solution   Since we are moving from an initial distance of R_B to a final distance of ∞, the energy is represented by the improper integral

$$E = \int_{R_B}^\infty k \frac{q_1 q_2}{r^2}\, dr = kq_1 q_2 \lim_{b \to \infty} \int_{R_B}^b \frac{1}{r^2}\, dr$$

$$= kq_1 q_2 \lim_{b \to \infty} -\frac{1}{r} \Big|_{R_B}^b = kq_1 q_2 \lim_{b \to \infty} \left(-\frac{1}{b} + \frac{1}{R_B} \right) = \frac{kq_1 q_2}{R_B}.$$

Substituting numerical values, we get

$$E = \frac{(9 \times 10^9)(1.6 \times 10^{-19})^2}{5.3 \times 10^{-11}} \approx 4.35 \times 10^{-18} \text{ joules}.$$

This is about the amount of energy needed to lift a speck of dust 0.000000025 inch off the ground. (In other words, not much!)

What happens if the limits of integration are $-\infty$ and ∞? In this case, we break the integral at any point and write the original integral as a sum of two new improper integrals.

We can use any (finite) number c to define

$$\int_{-\infty}^\infty f(x)\, dx = \int_{-\infty}^c f(x)\, dx + \int_c^\infty f(x)\, dx.$$

If *either* of the two new improper integrals diverges, we say the original integral diverges. Only if both of the new integrals have a finite value do we add the values to get a finite value for the original integral.

It is not hard to show that the preceding definition does not depend on the choice for c.

Another Type of Improper Integral: When the Integrand Becomes Infinite

There is another way for an integral to be improper. The interval may be finite but the function may be unbounded near some points in the interval. For example, consider $\int_0^1 \frac{1}{\sqrt{x}} \, dx$. Since the graph of $y = 1/\sqrt{x}$ has a vertical asymptote at $x = 0$, the region between the graph, the x-axis, and the lines $x = 0$ and $x = 1$ is unbounded. Instead of extending to infinity in the horizontal direction as in the previous improper integrals, this region extends to infinity in the vertical direction. See Figure 7.19(a). We handle this improper integral in a similar way as before: we compute $\int_a^1 \frac{1}{\sqrt{x}} \, dx$ for values of a slightly larger than 0 and look at what happens as a approaches 0 from the positive side. (This is written as $a \to 0^+$.)

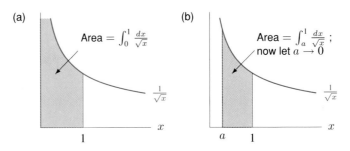

Figure 7.19: Area representation of improper integral

First we compute the integral:

$$\int_a^1 \frac{1}{\sqrt{x}} \, dx = 2x^{1/2} \Big|_a^1 = 2 - 2a^{1/2}.$$

Now we take the limit:

$$\lim_{a \to 0^+} \int_a^1 \frac{1}{\sqrt{x}} \, dx = \lim_{a \to 0^+} (2 - 2a^{1/2}) = 2.$$

Since the limit is finite, we say the improper integral converges, and that

$$\int_0^1 \frac{1}{\sqrt{x}} \, dx = 2.$$

Geometrically, what we have done is to calculate the finite area between $x = a$ and $x = 1$ and take the limit as a tends to 0 from the right. See Figure 7.19(b). Since the limit exists, the integral converges to 2. If the limit did not exist, we would say the improper integral diverges.

Example 5   Investigate the convergence of $\displaystyle\int_0^2 \frac{1}{(x-2)^2} \, dx$.

Solution   This is an improper integral since the integrand tends to infinity as x approaches 2, and is undefined at $x = 2$. Since the trouble is at the right endpoint, we replace the upper limit by b, and let b tend to 2 from the left. This is written $b \to 2^-$, with the "$-$" signifying that 2 is approached from below. See Figure 7.20.

$$\int_0^2 \frac{1}{(x-2)^2} \, dx = \lim_{b \to 2^-} \int_0^b \frac{1}{(x-2)^2} \, dx = \lim_{b \to 2^-} (-1)(x-2)^{-1} \Big|_0^b = \lim_{b \to 2^-} \left(-\frac{1}{(b-2)} - \frac{1}{2} \right).$$

Therefore, since $\displaystyle\lim_{b \to 2^-} \left(-\frac{1}{b-2} \right)$ does not exist, the integral diverges.

Figure 7.20: Shaded area represents $\int_0^2 \frac{1}{(x-2)^2}\, dx$

Figure 7.21: Shaded area represents $\int_{-1}^2 \frac{1}{x^4}\, dx$

Suppose $f(x)$ is positive and continuous on $a \le x < b$ and tends to infinity as $x \to b$. If $\displaystyle\lim_{c \to b^-} \int_a^c f(x)\, dx$ is a finite number, we say that $\displaystyle\int_a^b f(x)\, dx$ **converges** and define

$$\int_a^b f(x)\, dx = \lim_{c \to b^-} \int_a^c f(x)\, dx.$$

Otherwise, we say that $\displaystyle\int_a^b f(x)\, dx$ **diverges**.

When $f(x)$ tends to infinity as x approaches a, we define convergence in a similar way. In addition, an integral can be improper because the integrand tends to infinity *inside* the interval of integration rather than at an endpoint. In this case, we break the given integral into two (or more) improper integrals so that the integrand tends to infinity only at endpoints.

Suppose that $f(x)$ is positive and continuous on $[a, b]$ except at the point c. If $f(x)$ tends to infinity as $x \to c$, then we define

$$\int_a^b f(x)\, dx = \int_a^c f(x)\, dx + \int_c^b f(x)\, dx.$$

If *either* of the two new improper integrals diverges, we say the original integral diverges. Only if *both* of the new integrals have a finite value do we add the values to get a finite value for the original integral.

Example 6   Investigate the convergence of $\displaystyle\int_{-1}^2 \frac{1}{x^4}\, dx$.

Solution   See the graph in Figure 7.21. The trouble spot is $x = 0$, rather than $x = -1$ or $x = 2$. To handle this situation, we break the given improper integral into two other improper integrals each of which have $x = 0$ as one of the endpoints:

$$\int_{-1}^2 \frac{1}{x^4}\, dx = \int_{-1}^0 \frac{1}{x^4}\, dx + \int_0^2 \frac{1}{x^4}\, dx.$$

We can now use the previous technique to evaluate the new integrals, if they converge. Since

$$\int_0^2 \frac{1}{x^4}\, dx = \lim_{a \to 0^+} \left. -\frac{1}{3} x^{-3} \right|_a^2 = \lim_{a \to 0^+} \left(-\frac{1}{3}\right)\left(\frac{1}{8} - \frac{1}{a^3}\right)$$

the integral $\int_0^2 \frac{1}{x^4} dx$ does not exist. A similar computation shows that $\int_{-1}^0 \frac{1}{x^4} dx$ also diverges. Thus, the original integral diverges.

It is easy to miss an improper integral when the integrand tends to infinity inside the interval. For example, it is fundamentally incorrect to say that $\int_{-1}^2 \frac{1}{x^4} dx = -\frac{1}{3}x^{-3}\Big|_{-1}^2 = -\frac{1}{24} - \frac{1}{3} = -\frac{3}{8}.$

Example 7   Find $\displaystyle\int_0^6 \frac{1}{(x-4)^{2/3}} dx$.

Solution   Figure 7.22 shows that the trouble spot is at $x = 4$, so we break the integral at $x = 4$ and consider the separate parts.

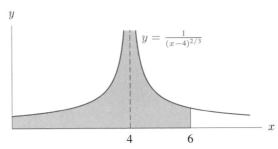

Figure 7.22: Shaded area represents $\int_0^6 \frac{1}{(x-4)^{2/3}} dx$

We have

$$\int_0^4 \frac{1}{(x-4)^{2/3}} dx = \lim_{b \to 4^-} 3(x-4)^{1/3}\Big|_0^b = \lim_{b \to 4^-} \left(3(b-4)^{1/3} - 3(-4)^{1/3}\right) = 3(4)^{1/3}.$$

Similarly,

$$\int_4^6 \frac{1}{(x-4)^{2/3}} dx = \lim_{a \to 4^+} 3(x-4)^{1/3}\Big|_a^6 = \lim_{a \to 4^+} \left(3 \cdot 2^{1/3} - 3(a-4)^{1/3}\right) = 3(2)^{1/3}.$$

Since both of these integrals converge, the original integral converges:

$$\int_0^6 \frac{1}{(x-4)^{2/3}} dx = 3(4)^{1/3} + 3(2)^{1/3} \approx 8.54.$$

Finally, there is a question of what to do when an integral is improper at both endpoints. In this case, we just break the integral at any interior point of the interval. The original integral diverges if either or both of the new integrals diverge.

Example 8   Investigate the convergence of $\displaystyle\int_0^\infty \frac{1}{x^2} dx$.

Solution   This integral is improper both because the upper limit is ∞ and because the function is undefined at $x = 0$. We break the integral into two parts at, say, $x = 1$. We know by Example 3 that $\int_1^\infty \frac{1}{x^2} dx$ has a finite value. However, the other part, $\int_0^1 \frac{1}{x^2} dx$, diverges since:

$$\int_0^1 \frac{1}{x^2} dx = \lim_{a \to 0^+} -x^{-1}\Big|_a^1 = \lim_{a \to 0^+} \left(\frac{1}{a} - 1\right).$$

Therefore $\int_0^\infty \frac{1}{x^2} dx$ diverges as well.

Problems for Section 7.7

Calculate the values of the integrals in Problems 1–24, if they converge.

1. $\displaystyle\int_1^\infty e^{-2x}\,dx$

2. $\displaystyle\int_0^\infty \frac{x}{e^x}\,dx$

3. $\displaystyle\int_1^\infty \frac{x}{4+x^2}\,dx$

4. $\displaystyle\int_{-\infty}^0 \frac{e^x}{1+e^x}\,dx$

5. $\displaystyle\int_{-\infty}^\infty \frac{dz}{z^2+25}$

6. $\displaystyle\int_0^4 \frac{dx}{\sqrt{16-x^2}}$

7. $\displaystyle\int_{\pi/4}^{\pi/2} \frac{\sin x}{\sqrt{\cos x}}\,dx$

8. $\displaystyle\int_{-1}^1 \frac{1}{v}\,dv$

9. $\displaystyle\int_0^1 \frac{x^4+1}{x}\,dx$

10. $\displaystyle\int_1^\infty \frac{1}{x^2+1}\,dx$

11. $\displaystyle\int_1^\infty \frac{1}{\sqrt{x^2+1}}\,dx$

12. $\displaystyle\int_0^4 \frac{1}{u^2-16}\,du$

13. $\displaystyle\int_1^\infty \frac{y}{y^4+1}\,dy$

14. $\displaystyle\int_2^\infty \frac{dx}{x\ln x}$

15. $\displaystyle\int_0^1 \frac{\ln x}{x}\,dx$

16. $\displaystyle\int_{16}^{20} \frac{1}{y^2-16}\,dy$

17. $\displaystyle\int_1^2 \frac{dx}{x\ln x}$

18. $\displaystyle\int_0^\pi \frac{1}{\sqrt{x}}e^{-\sqrt{x}}\,dx$

19. $\displaystyle\int_3^\infty \frac{dx}{x(\ln x)^2}$

20. $\displaystyle\int_0^2 \frac{1}{\sqrt{4-x^2}}\,dx$

21. $\displaystyle\int_4^\infty \frac{dx}{(x-1)^2}$

22. $\displaystyle\int_4^\infty \frac{dx}{x^2-1}$

23. $\displaystyle\int_7^\infty \frac{dy}{\sqrt{y-5}}$

24. $\displaystyle\int_\pi^\infty \sin y\,dy$

25. Find the area under the curve $y = xe^{-x}$ for $x \geq 0$.

26. Find the area under the curve $y = 1/\cos^2 t$ between $t = 0$ and $t = \pi/2$.

27. Suppose a function h is defined by $h(x) = \dfrac{1}{x\sqrt{x}} - \dfrac{1}{16},\, 0 < x \leq 4$, and $h(x) = \dfrac{1}{x^2},\, x > 4$;

    (a) Evaluate $\int_0^\infty h(x)\,dx$.
    (b) Is h differentiable at $x = 4$? If not, why not? If so, find $h'(4)$.

28. Given that $\displaystyle\int_{-\infty}^\infty e^{-x^2}\,dx = \sqrt{\pi}$, calculate the exact value of

$$\int_{-\infty}^\infty e^{-(x-a)^2/b}\,dx.$$

29. The gamma function is defined for all $x > 0$ by the rule

$$\Gamma(x) = \int_0^\infty t^{x-1}e^{-t}\,dt.$$

    (a) Find $\Gamma(1)$ and $\Gamma(2)$.
    (b) Integrate by parts with respect to t to show that, for positive n,

$$\Gamma(n+1) = n\Gamma(n).$$

    (c) Find a simple expression for $\Gamma(n)$ for positive integers n .

30. The rate, r, at which people get sick during an epidemic of the flu can be approximated by

$$r = 1000te^{-0.5t}$$

where r is measured in people/day and t is measured in days since the start of the epidemic.

(a) Sketch a graph of r as a function of t.   (b) When are people getting sick fastest?

(c) How many people get sick altogether?

31. Find the energy required to separate opposite electric charges of magnitude 1 coulomb. Assume the charges are initially 1 meter apart and that one is moved infinitely far away from the other. (The relevant definition of energy is on page 359.)

32. For what values of p does the following integral converge?

$$\int_e^\infty x^p \ln x \, dx$$

What is the value of the integral when it converges?

33. For what values of p does the following integral converge?

$$\int_0^e x^p \ln x \, dx$$

What is the value of the integral when it converges?

7.8 MORE ON IMPROPER INTEGRALS

Making Comparisons

Sometimes it is difficult to find the exact value of an improper integral by antidifferentiation, but it may be possible to determine whether an integral converges or diverges. The key is to *compare* the given integral to one whose behavior we already know. Let's look at an example.

Example 1   Determine whether $\displaystyle\int_1^\infty \frac{1}{\sqrt{x^3 + 5}} \, dx$ converges.

Solution   First, let's see what this integrand does as $x \to \infty$. For large x, the 5 becomes insignificant compared with the x^3, so

$$\frac{1}{\sqrt{x^3 + 5}} \approx \frac{1}{\sqrt{x^3}} = \frac{1}{x^{3/2}}.$$

Since

$$\int_1^\infty \frac{1}{\sqrt{x^3}} \, dx = \int_1^\infty \frac{1}{x^{3/2}} \, dx = \lim_{b\to\infty} \int_1^b \frac{1}{x^{3/2}} \, dx = \lim_{b\to\infty} \left. -2x^{-1/2} \right|_1^b = \lim_{b\to\infty} \left(2 - 2b^{-1/2} \right) = 2.$$

the integral $\int_1^\infty \frac{1}{x^{3/2}} \, dx$ converges. So we expect our integral to converge as well.

In order to confirm this, we observe that for $0 \le x^3 \le x^3 + 5$, we have

$$\frac{1}{\sqrt{x^3 + 5}} \le \frac{1}{\sqrt{x^3}}.$$

and so for $b \ge 1$,

$$\int_1^b \frac{1}{\sqrt{x^3 + 5}} \, dx \le \int_1^b \frac{1}{\sqrt{x^3}} \, dx.$$

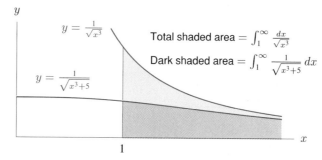

Figure 7.23: Graph showing $\int_1^\infty \frac{1}{\sqrt{x^3+5}}\, dx \leq \int_1^\infty \frac{dx}{\sqrt{x^3}}$

(See Figure 7.23.) Since $\int_1^b \frac{1}{\sqrt{x^3+5}}\, dx$ increases as b approaches infinity but is always smaller than $\int_1^b \frac{1}{x^{3/2}}\, dx < \int_1^\infty \frac{1}{x^{3/2}}\, dx = 2$, we know $\int_1^\infty \frac{1}{\sqrt{x^3+5}}\, dx$ must have a finite value less than 2. Thus,

$$\int_1^\infty \frac{dx}{\sqrt{x^3 + 5}} \quad \text{converges to a value less than 2.}$$

Notice that we first looked at the behavior of the integrand as $x \to \infty$. This is useful because the convergence or divergence of the integral is determined by what happens as $x \to \infty$.

The Comparison Test for $\displaystyle\int_a^\infty f(x)\, dx$

Assume $f(x)$ is positive. Making a comparison involves two stages:
1. Guess, by looking at the behavior of the integrand for large x, whether the integral converges or not. (This is the "behaves like" principle.)
2. Confirm the guess by comparison:
   - If $0 \leq f(x) \leq g(x)$ and $\int_a^\infty g(x)\, dx$ converges, then $\int_a^\infty f(x)\, dx$ converges
   - If $0 \leq g(x) \leq f(x)$ and $\int_a^\infty g(x)\, dx$ diverges, then $\int_a^\infty f(x)\, dx$ diverges.

Example 2   Decide whether $\displaystyle\int_4^\infty \frac{dt}{(\ln t) - 1}$ converges or diverges.

Solution   Since $\ln t$ grows without bound as $t \to \infty$, the -1 is eventually going to be insignificant in comparison to $\ln t$. Thus, as far as convergence is concerned,

$$\int_4^\infty \frac{1}{(\ln t) - 1}\, dt \quad \text{behaves like} \quad \int_4^\infty \frac{1}{\ln t}\, dt.$$

Does $\int_4^\infty \frac{1}{\ln t}\, dt$ converge or diverge? Since $\ln t$ grows very slowly, $1/\ln t$ goes to zero very slowly, and so the integral probably doesn't converge. We know that $(\ln t) - 1 < \ln t < t$ for all positive t. So, provided $t > e$, we take reciprocals:

$$\frac{1}{(\ln t) - 1} > \frac{1}{\ln t} > \frac{1}{t}.$$

Since $\int_4^\infty \frac{1}{t}\, dt$ diverges, we conclude that

$$\int_4^\infty \frac{1}{(\ln t) - 1}\, dt \quad \text{diverges.}$$

How Do We Know What To Compare With?

In Examples 1 and 2, we investigated the convergence of an integral by comparing it with an easier integral. How did we pick the easier integral? This is a matter of trial and error, guided by any information we get by looking at the original integrand as $x \to \infty$. We want the comparison integrand to be easy and, in particular, to have a simple antiderivative.

Useful Integrals for Comparison

- $\int_1^\infty \dfrac{1}{x^p} \, dx$ converges for $p > 1$ and diverges for $p \leq 1$.

- $\int_0^1 \dfrac{1}{x^p} \, dx$ converges for $p < 1$ and diverges for $p \geq 1$.

- $\int_0^\infty e^{-ax} \, dx$ converges for $a > 0$.

Of course, we can use any function for comparison, provided we can determine its behavior.

Example 3 Investigate the convergence of $\int_1^\infty \dfrac{(\sin x) + 3}{\sqrt{x}} \, dx$.

Solution Since it looks difficult to find an antiderivative of this function, we try comparison. What happens to this integrand as $x \to \infty$? Since $\sin x$ oscillates between -1 and 1,

$$\frac{2}{\sqrt{x}} = \frac{-1+3}{\sqrt{x}} \leq \frac{(\sin x) + 3}{\sqrt{x}} \leq \frac{1+3}{\sqrt{x}} = \frac{4}{\sqrt{x}},$$

the integrand oscillates between $2/\sqrt{x}$ and $4/\sqrt{x}$. (See Figure 7.24.)

What do $\int_1^\infty \frac{2}{\sqrt{x}} \, dx$ and $\int_1^\infty \frac{4}{\sqrt{x}} \, dx$ do? As far as convergence is concerned, they certainly do the same thing, and whatever that is, the original integral does it too. It is important to notice that $\sqrt{x}$ grows very slowly. This means that $1/\sqrt{x}$ gets small slowly, which means that convergence is unlikely. Since $\sqrt{x} = x^{1/2}$, the result in the preceding box (with $p = \frac{1}{2}$) tells us that $\int_1^\infty \frac{dx}{\sqrt{x}}$ diverges. So the comparison test tells us that the original integral diverges.

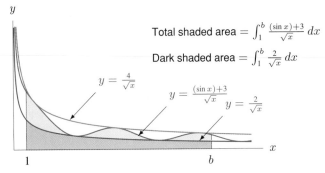

Figure 7.24: Graph showing $\int_1^b \frac{2}{\sqrt{x}} \, dx \leq \int_1^b \frac{(\sin x)+3}{\sqrt{x}} \, dx$, for $b \geq 1$

Notice that there are two possible comparisons we could have made in Example 3:

$$\frac{2}{\sqrt{x}} \leq \frac{(\sin x) + 3}{\sqrt{x}} \qquad \text{or} \qquad \frac{(\sin x) + 3}{\sqrt{x}} \leq \frac{4}{\sqrt{x}}.$$

Since both $\int_1^\infty \frac{2}{\sqrt{x}} \, dx$ and $\int_1^\infty \frac{4}{\sqrt{x}} \, dx$ diverge, only the first comparison is useful. Knowing that an integral is *smaller* than a divergent integral is of no help whatsoever!

The next example shows what to do if the comparison does not hold throughout the interval of integration.

Example 4   Show $\int_1^\infty e^{-x^2/2}\,dx$ converges to a finite value.

Solution   We know that $e^{-x^2/2}$ goes very rapidly to zero as $x \to \infty$, so we expect this integral to converge. Hence we look for some larger integrand which has a convergent integral. One possibility is $\int_1^\infty e^{-x}\,dx$, because e^{-x} has an elementary antiderivative and $\int_1^\infty e^{-x}\,dx$ converges. What is the relationship between $e^{-x^2/2}$ and e^{-x}? We know that for $x \geq 2$,

$$x \leq \frac{x^2}{2} \quad \text{so} \quad -\frac{x^2}{2} \leq -x,$$

and so, for $x \geq 2$

$$e^{-x^2/2} \leq e^{-x}.$$

Since this inequality holds only for $x \geq 2$, we split the interval of integration into two pieces:

$$\int_1^\infty e^{-x^2/2}\,dx = \int_1^2 e^{-x^2/2}\,dx + \int_2^\infty e^{-x^2/2}\,dx.$$

Now $\int_1^2 e^{-x^2/2}\,dx$ is finite (it is not improper) and $\int_2^\infty e^{-x^2/2}\,dx$ is finite by comparison with $\int_2^\infty e^{-x}\,dx$. Therefore, $\int_1^\infty e^{-x^2/2}\,dx$ is the sum of two finite pieces and therefore must be finite.

Problems for Section 7.8

1. The graphs of $y = 1/x, y = 1/x^2$ and the functions $f(x)$, $g(x)$, $h(x)$, and $k(x)$ are shown in Figure 7.25.

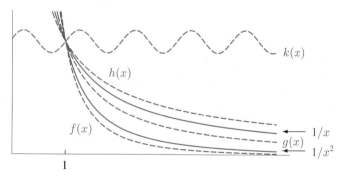

Figure 7.25

(a)   What is the area between $y = 1/x$ and $y = 1/x^2$ from $x = 1$ to $x = \infty$? Explain your answer.

(b)   Using the graph, decide whether the integral of each of the functions $f(x)$, $g(x)$, $h(x)$ and $k(x)$ from $x = 1$ to $x = \infty$ converges, diverges, or whether it is impossible to tell.

In Problems 2–16 decide if the improper integral converges or diverges. Explain your reasoning.

2. $\int_{50}^\infty \dfrac{dz}{z^3}$

3. $\int_{0.5}^1 \dfrac{1}{x^{(19/20)}}\,dx$

4. $\int_1^\infty \dfrac{dx}{1+x}$

5. $\int_1^\infty \dfrac{dx}{x^3+1}$

6. $\int_2^\infty \dfrac{d\theta}{\sqrt{\theta^3+1}}$

7. $\int_{-1}^5 \dfrac{dt}{(t+1)^2}$

8. $\int_0^\infty \dfrac{dy}{1+e^y}$

9. $\int_1^\infty \dfrac{2+\cos\phi}{\phi^2}\,d\phi$

10. $\int_{-\infty}^\infty \dfrac{du}{1+u^2}$

11. $\int_1^\infty \dfrac{du}{u+u^2}$

12. $\int_1^\infty \dfrac{d\theta}{\sqrt{\theta^2+1}}$

13. $\int_0^1 \dfrac{d\theta}{\sqrt{\theta^3+\theta}}$

14. $\int_0^\infty \dfrac{dz}{e^z+2^z}$

15. $\int_0^\pi \dfrac{2-\sin\phi}{\phi^2}\,d\phi$

16. $\int_4^\infty \dfrac{3+\sin\alpha}{\alpha}\,d\alpha$

Estimate the values of the integrals in Problems 17–18 correct to two decimal places by integrating the functions on your calculator or computer for large values of the upper bound of integration.

17. $\displaystyle\int_1^\infty e^{-x^2}\, dx$

18. $\displaystyle\int_0^\infty e^{-x^2}\cos^2 x\, dx$

19. The bell-shaped curve of statistics has the equation

$$f(x) = ae^{-x^2/2}.$$

Find the value of a (to three decimal places) that makes

$$\int_{-\infty}^\infty f(x)\, dx = 1.$$

20. Statisticians often use the function
$$g(x) = ae^{-(x-k)^2/2}.$$

   (a) To three decimal places, what value of a should be chosen to ensure that

$$\int_{-\infty}^\infty g(x)\, dx = 1?$$

   (b) Is your answer the same as or different from your answer to Problem 19? Why?

21. (a) Find an upper bound for

$$\int_3^\infty e^{-x^2}\, dx.$$

   [Hint: $e^{-x^2} \le e^{-3x}$ for $x \ge 3$.]

   (b) For any positive n, generalize the result of part (a) to find an upper bound for

$$\int_n^\infty e^{-x^2}\, dx$$

   by noting that $nx \le x^2$ for $x \ge n$.

22. For what values of p does the integral $\displaystyle\int_2^\infty \frac{dx}{x(\ln x)^p}$ converge?

23. For what values of p does the integral $\displaystyle\int_1^2 \frac{dx}{x(\ln x)^p}$ converge?

24. Do the following integrals converge? If so, give an upper bound for the value of the integral.

   (a) $\displaystyle\int_1^\infty \frac{2x^2 + 1}{4x^4 + 4x^2 - 2}\, dx$

   (b) $\displaystyle\int_1^\infty \left(\frac{2x^2 + 1}{4x^4 + 4x^2 - 2}\right)^{1/4} dx$

25. In Planck's Radiation Law, we encounter the integral

$$\int_1^\infty \frac{dx}{x^5(e^{1/x} - 1)}.$$

   (a) Explain why a graph of the tangent line to e^t at $t = 0$ tells us that for all t

$$1 + t \le e^t.$$

   (b) Substituting $t = 1/x$, show that for all x

$$e^{1/x} - 1 > \frac{1}{x}.$$

   (c) Use the comparison test to show that the original integral converges.

CHAPTER SUMMARY

- **Integration techniques**
  Substitution, parts, partial fractions, using tables
- **Improper integrals**
  Convergence/divergence, comparison test for integrals
- **Numerical approximations**
  Riemann sums (left, right, midpoint), trapezoid rule, Simpson's rule, approximation errors

REVIEW PROBLEMS FOR CHAPTER SEVEN

For each region in Problems 1–3, write a definite integral which represents its area. Evaluate the integral to derive a formula for the area.

1. A rectangle with base b and height h: **2.** A right triangle of base b and height h: **3.** A circle of radius r:

4. Suppose the function f is defined by $f(t) = t^2$ for $0 \le t \le 1$ and $f(t) = 2 - t$ for $1 < t \le 2$. Compute $\int_0^2 f(t)\,dt$.

5. (a) Find the average value of the following functions over one cycle:

     (i) $f(t) = \cos t$ (ii) $g(t) = |\cos t|$ (iii) $k(t) = (\cos t)^2$

    (b) Write the averages you have just found in ascending order. Explain clearly, using words and graphs, why the averages should be expected to come out in the order they did.

6. Integrate:

    (a) $\displaystyle\int \frac{e^x}{1 + e^x}\,dx$ (b) $\displaystyle\int \frac{e^x}{1 + e^{2x}}\,dx$ (c) $\displaystyle\int \frac{1}{1 + e^x}\,dx$

7. (a) Using a substitution, find $\displaystyle\int \frac{dx}{\sqrt{1 - 4x^2}}$ from the table of integrals.

    (b) Use your answer to part (a) to find $\displaystyle\int_0^{\pi/8} \frac{dx}{\sqrt{1 - 4x^2}}$.

    (c) Check your answer to part (b) by numerical integration.

8. (a) Explain why you can rewrite x^x as $x^x = e^{x \ln x}$ for $x > 0$.

    (b) Use your answer to part (a) to find $\dfrac{d}{dx}(x^x)$.

    (c) Find $\displaystyle\int x^x (1 + \ln x)\,dx$.

    (d) Find $\displaystyle\int_1^2 x^x (1 + \ln x)\,dx$ using part (c) and check your answer using numerical methods.

In Problems 9–12 explain why the following pairs of antiderivatives are really, despite their apparent dissimilarity, different expressions of the same problem. You do not need to evaluate the integrals.

9. $\displaystyle\int \frac{dx}{x^2 + 4x + 4}$ and $\displaystyle\int \frac{x}{(x^2 + 1)^2}\, dx$ **10.** $\displaystyle\int \frac{1}{\sqrt{1 - x^2}}\, dx$ and $\displaystyle\int \frac{x\, dx}{\sqrt{1 - x^4}}$

11. $\displaystyle\int \frac{x}{1 - x^2}\, dx$ and $\displaystyle\int \frac{1}{x \ln x}\, dx$ **12.** $\displaystyle\int \frac{x}{x + 1}\, dx$ and $\displaystyle\int \frac{1}{x + 1}\, dx$

For Problems 13–21 decide if the integral converges or diverges. If the integral converges, find its value.

13. $\displaystyle\int_{4}^{\infty} \frac{dt}{t^{3/2}}$ **14.** $\displaystyle\int_{10}^{\infty} \frac{dx}{x \ln x}$ **15.** $\displaystyle\int_{0}^{\infty} w e^{-w}\, dw$

16. $\displaystyle\int_{-1}^{1} \frac{1}{x^4}\, dx$ **17.** $\displaystyle\int_{-\pi/4}^{\pi/4} \tan \theta\, d\theta$ **18.** $\displaystyle\int_{2}^{\infty} \frac{1}{4 + z^2}\, dz$

19. $\displaystyle\int_{10}^{\infty} \frac{1}{z^2 - 4}\, dz$ **20.** $\displaystyle\int_{-5}^{10} \frac{dt}{\sqrt{t + 5}}$ **21.** $\displaystyle\int_{0}^{\pi/2} \frac{1}{\sin \phi}\, d\phi$

For Problems 22–26, decide if the integral converges or diverges. If the integral converges, find its value, or give a bound for the value.

22. $\displaystyle\int_{0}^{\pi/4} \tan 2\theta\, d\theta$ **23.** $\displaystyle\int_{1}^{\infty} \frac{x}{x + 1}\, dx$ **24.** $\displaystyle\int_{0}^{\infty} \frac{\sin^2 \theta}{\theta^2 + 1}\, d\theta$

25. $\displaystyle\int_{0}^{\pi} \tan^2 \theta\, d\theta$ **26.** $\displaystyle\int_{0}^{1} (\sin x)^{-3/2}\, dx$

27. Find the average (vertical) height of the shaded area in Figure 7.26.
28. Find the average (horizontal) width of the shaded area in Figure 7.26.

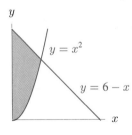

Figure 7.26

29. The curves $y = \sin x$ and $y = \cos x$ cross each other infinitely often. What is the area of the region bounded by these two curves between two consecutive crossings?

Are the statements in Problems 30–33 true or false? Why?

30. The midpoint rule approximation to $\displaystyle\int_{0}^{1} (y^2 - 1)\, dy$ is always smaller than the exact value of the integral.

31. The trapezoid rule approximation is never exact.

32. If $\mathrm{LEFT}(2) < \displaystyle\int_{a}^{b} f(x)\, dx$, then $\mathrm{LEFT}(4) < \displaystyle\int_{a}^{b} f(x)\, dx$.

33. If $0 < f' < g'$ everywhere, then the error in approximating $\displaystyle\int_{a}^{b} f(x)\, dx$ by $\mathrm{LEFT}(n)$ is less than the error in approximating $\displaystyle\int_{a}^{b} g(x)\, dx$ by $\mathrm{LEFT}(n)$.

34. Does the improper integral $\int_0^1 \frac{e^x}{x}\, dx$ converge or diverge?

35. Suppose you estimate $\int_0^{0.5} f(x)\, dx$ by the trapezoid and midpoint rules with 100 steps. Explain which of the two estimates overestimates, and which underestimates, the true value of the integral if

(a) $f(x) = 1 + e^{-x}$ (b) $f(x) = e^{-x^2}$ (c) $f(x)$ is a straight line.

36. Suppose for a certain definite integral that TRAP(10) = 4.6891 and TRAP(50) = 4.6966. Estimate the actual error for TRAP(10) (and thus the actual value of the integral) by assuming that the error is reduced by a factor of roughly 25 in going from TRAP(10) to TRAP(50).

37. Suppose for a definite integral that SIMP(5) = 7.41562 and SIMP(10) = 7.41738. Estimate the actual value of the integral by using the fact that the error is reduced by a factor of roughly 16 in going from SIMP(5) to SIMP(10).

38. In 1987 the average per capita income in the US was $26,000. Suppose that average per capita income is increasing continuously at a rate in dollars per year given by

$$r(t) = 480(1.024)^t,$$

where t is the number of years since 1987.

(a) Estimate the average per capita income in 1995.

(b) Find a formula for the average per capita income as a function of time after 1987.

39. In 1990 humans generated 1.4×10^{20} joules of energy through the combustion of petroleum. It is estimated that all of the earth's petroleum will generate approximately 10^{22} joules. Assuming the use of energy generated by petroleum combustion will increase by 2% each year, how long will it be before all of our petroleum resources are used up?

PROJECTS

1. **Taylor Polynomial Inequalities**

    (a) Use the fact that $e^x \geq 1 + x$ for all values of x and the formula

    $$e^x = 1 + \int_0^x e^t\, dt$$

    to show that

    $$e^x \geq 1 + x + \frac{x^2}{2}$$

    for all positive values of x. Generalize this idea to get inequalities involving higher-degree polynomials.

    (b) Use the fact that $\cos x \leq 1$ for all x and repeated integration to show that

    $$\cos x \leq 1 - \frac{x^2}{2!} + \frac{x^4}{4!}.$$

2. **Slope Fields**

    Suppose we want to sketch the antiderivative, F, of the function f. To get an accurate graph of F, we must be careful about making F have the right slope at every point. The slope of F at any point (x, y) on its graph should be $f(x)$, since $F'(x) = f(x)$. We arrange this as follows: at the point (x, y) in the plane, draw a small line segment with slope $f(x)$. Do this at many points. We call such a diagram a *slope field*. If $f(x) = x$, we get the slope field in Figure 7.27.

Figure 7.27: Slope field of
$f(x) = x$

Figure 7.28: Slope field of
$f(x) = e^{-x^2}$

Notice how the lines in Figure 7.27 seem to be arranged in a parabolic pattern. This is because the general antiderivative of x is $x^2/2 + C$, so the lines are all the tangent lines to the family of parabolas $y = x^2/2 + C$. This suggests a way of finding antiderivatives graphically even if we can't write down a formula for them: plot the slopes, and see if they suggest the graph of an antiderivative. For example, if you do this with $f(x) = e^{-x^2}$, which is one of the functions that does not have an elementary antiderivative, you get Figure 7.28.

You can see the ghost of the graph of a function lurking behind the slopes in Figure 7.28; in fact there is a whole stack of them. If you move across the plane in the direction suggested by the slope field at every point, you will trace out a curve. The slope field is tangent to the curve everywhere, so this is the graph of an antiderivative of e^{-x^2}.

(a) (i)   Sketch a graph of $f(t) = \dfrac{\sin t}{t}$.

    (ii) What does your graph tell you about the behavior of

$$\mathrm{Si}(x) = \int_0^x \frac{\sin(t)}{t} \, dt$$

    for $x > 0$? Is $\mathrm{Si}(x)$ always increasing or always decreasing? Does $\mathrm{Si}(x)$ cross the x-axis for $x > 0$?

    (iii) By drawing the slope field for $f(t) = \dfrac{\sin t}{t}$, decide whether $\lim\limits_{x \to \infty} \mathrm{Si}(x)$ exists.

(b) (i)   Use your calculator or computer to sketch a graph of $y = x^{\sin x}$ for $0 < x \leq 20$.

    (ii) Using your answer to part (a), sketch by hand a graph of the function F, where

$$F(x) = \int_0^x t^{\sin t} \, dt.$$

    (iii) Use a slope field program to check your answer to part (b).

(c) Let $F(x)$ be the antiderivative of $\sin(x^2)$ satisfying $F(0) = 0$.

    (i)   Describe any general features of the graph of F that you can deduce by looking at the graph of $\sin(x^2)$ in Figure 7.29.

    (ii) By drawing a slope field (using a calculator or computer), sketch a graph of F. Does F ever cross the x-axis in the region $x > 0$? Does $\lim\limits_{x \to \infty} F(x)$ exist?

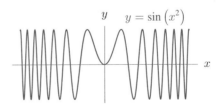

Figure 7.29

FOCUS ON PRACTICE

INTEGRATION

For Problems 1–113, evaluate the following integrals. Assume $a, b, c,$ and k are constants. Evaluate the definite integrals using the Fundamental Theorem of Calculus, if possible, and check your answers numerically. Problems 1– 75 can be done without an integral table, as can some of the later problems.

1. $\displaystyle\int \sin t \, dt$

2. $\displaystyle\int \cos 2t \, dt$

3. $\displaystyle\int e^{5z} \, dz$

4. $\displaystyle\int (3w + 7) \, dw$

5. $\displaystyle\int \sin 2\theta \, d\theta$

6. $\displaystyle\int (x^3 - 1)^4 x^2 \, dx$

7. $\displaystyle\int \left(x^{3/2} + x^{2/3}\right) \, dx$

8. $\displaystyle\int \left(e^x + 3^x\right) \, dx$

9. $\displaystyle\int (r + 1)^3 \, dr$

10. $\displaystyle\int \left(\frac{4}{x^2} - \frac{3}{x^3}\right) \, dx$

11. $\displaystyle\int \left(\frac{x^3 + x + 1}{x^2}\right) \, dx$

12. $\displaystyle\int \frac{(1 + \ln x)^2}{x} \, dx$

13. $\displaystyle\int t e^{t^2} \, dt$

14. $\displaystyle\int x \cos x \, dx$

15. $\displaystyle\int \sin x (\sqrt{2 + 3 \cos x}) dx$

16. $\displaystyle\int x^2 e^{2x} \, dx$

17. $\displaystyle\int x\sqrt{1 - x} \, dx$

18. $\displaystyle\int x \ln x \, dx$

19. $\displaystyle\int y \sin y \, dy$

20. $\displaystyle\int (\ln x)^2 \, dx$

21. $\displaystyle\int \ln(x^2) \, dx$

22. $\displaystyle\int e^{0.5 - 0.3t} \, dt$

23. $\displaystyle\int \sin^2 \theta \cos \theta \, d\theta$

24. $\displaystyle\int x\sqrt{4 - x^2} \, dx$

25. $\displaystyle\int \frac{(u + 1)^3}{u^2} \, du$

26. $\displaystyle\int \frac{\cos \sqrt{y}}{\sqrt{y}} \, dy$

27. $\displaystyle\int \frac{1}{\cos^2 z} \, dz$

28. $\displaystyle\int \cos^2 \theta \, d\theta$

29. $\displaystyle\int t^{10}(t - 10) \, dt$

30. $\displaystyle\int \tan(2x - 6) \, dx$

31. $\displaystyle\int_1^3 \ln(x^3) \, dx$

32. $\displaystyle\int_1^e (\ln x)^2 \, dx$

33. $\displaystyle\int_{-\pi}^{\pi} e^{2x} \sin 2x \, dx$

34. $\displaystyle\int_0^{10} z e^{-z} \, dz$

35. $\displaystyle\int_{-\pi/3}^{\pi/4} \sin^3 \theta \cos \theta \, d\theta$

36. $\displaystyle\int_{-\pi/4}^{\pi/4} x^3 \cos x^2 \, dx$

37. $\displaystyle\int_1^4 \frac{e^{\sqrt{x}}}{\sqrt{x}} \, dx$

38. $\displaystyle\int_0^1 \frac{dx}{x^2 + 1}$

39. $\displaystyle\int \frac{(\ln x)^2}{x} \, dx$

40. $\displaystyle\int \frac{(t + 2)^2}{t^3} \, dt$

41. $\displaystyle\int \left(x^2 + 2x + \frac{1}{x}\right) \, dx$

42. $\displaystyle\int \frac{t + 1}{t^2} \, dt$

43. $\displaystyle\int t e^{t^2 + 1} \, dt$

44. $\displaystyle\int \tan \theta \, d\theta$

45. $\displaystyle\int \sin(5\theta) \cos(5\theta) \, d\theta$

46. $\displaystyle\int \frac{x}{x^2 + 1} \, dx$

47. $\displaystyle\int \frac{dz}{1 + z^2}$

48. $\displaystyle\int \frac{dz}{1 + 4z^2}$

49. $\displaystyle\int \cos^3 2\theta \sin 2\theta \, d\theta$

50. $\displaystyle\int \sin 5\theta \cos^3 5\theta \, d\theta$

51. $\displaystyle\int \sin^3 z \cos^3 z \, dz$

52. $\displaystyle\int t(t-10)^{10}\,dt$

53. $\displaystyle\int \cos\theta\sqrt{1+\sin\theta}\,d\theta$

54. $\displaystyle\int xe^x\,dx$

55. $\displaystyle\int t^3 e^t\,dt$

56. $\displaystyle\int_1^3 x(x^2+1)^{70}\,dx$

57. $\displaystyle\int (3z+5)^3\,dz$

58. $\displaystyle\int \frac{du}{9+u^2}$

59. $\displaystyle\int \frac{\cos w}{1+\sin^2 w}\,dw$

60. $\displaystyle\int \frac{1}{x}\tan(\ln x)\,dx$

61. $\displaystyle\int \frac{1}{x}\sin(\ln x)\,dx$

62. $\displaystyle\int \frac{dx}{\sqrt{1-4x^2}}$

63. $\displaystyle\int \frac{w\,dw}{\sqrt{16-w^2}}$

64. $\displaystyle\int \frac{e^{2y}+1}{e^{2y}}\,dy$

65. $\displaystyle\int \frac{\sin w\,dw}{\sqrt{1-\cos w}}$

66. $\displaystyle\int \frac{dx}{x\ln x}$

67. $\displaystyle\int \frac{du}{3u+8}$

68. $\displaystyle\int \frac{x\cos\sqrt{x^2+1}}{\sqrt{x^2+1}}\,dx$

69. $\displaystyle\int \frac{t^3}{\sqrt{1+t^2}}\,dt$

70. $\displaystyle\int ue^{ku}\,du$

71. $\displaystyle\int (w+5)^4 w\,dw$

72. $\displaystyle\int e^{\sqrt{2x+3}}\,dx$

73. $\displaystyle\int r(\ln r)^2\,dr$

74. $\displaystyle\int (e^x+x)^2\,dx$

75. $\displaystyle\int u^2\ln u\,du$

76. $\displaystyle\int \frac{5x+6}{x^2+4}\,dx$

77. $\displaystyle\int \frac{1}{\sin^3(2x)}\,dx$

78. $\displaystyle\int \frac{dr}{r^2-100}$

79. $\displaystyle\int y^2\sin(cy)\,dy$

80. $\displaystyle\int e^{-ct}\sin kt\,dt$

81. $\displaystyle\int e^{5x}\cos(3x)\,dx$

82. $\displaystyle\int \left(x^{\sqrt{k}}+\sqrt{k}^{\,x}\right)\,dx$

83. $\displaystyle\int \sqrt{3+12x^2}\,dx.$

84. $\displaystyle\int (x^2-3x+2)e^{-4x}\,dx$

85. $\displaystyle\int \frac{dx}{x^2+5x+4}$

86. $\displaystyle\int \frac{1}{\sqrt{x^2-3x+2}}\,dx$

87. $\displaystyle\int \frac{x^3}{x^2+3x+2}\,dx$

88. $\displaystyle\int \frac{x^2+1}{x^2-3x+2}\,dx$

89. $\displaystyle\int \frac{dx}{ax^2+bx}$

90. $\displaystyle\int \frac{ax+b}{ax^2+2bx+c}\,dx$

91. $\displaystyle\int \frac{dz}{z^2+z}$

92. $\displaystyle\int \left(\frac{x}{3}+\frac{3}{x}\right)^2\,dx$

93. $\displaystyle\int \frac{2^t}{2^t+1}\,dt$

94. $\displaystyle\int 10^{1-x}\,dx$

95. $\displaystyle\int (x^2+5)^3\,dx$

96. $\displaystyle\int v\arcsin v\,dv$

97. $\displaystyle\int \sin^2(2\theta)\cos^3(2\theta)\,d\theta$

98. $\displaystyle\int \cos(2\sin x)\cos x\,dx$

99. $\displaystyle\int \sqrt{4-x^2}\,dx$

100. $\displaystyle\int \frac{z^3}{z-5}\,dz$

101. $\displaystyle\int \frac{\sin w\cos w}{1+\cos^2 w}\,dw$

102. $\displaystyle\int \frac{1}{\tan(3\theta)}\,d\theta$

103. $\displaystyle\int \frac{x}{\cos^2 x}\,dx$

104. $\displaystyle\int \frac{x+1}{\sqrt{x}}\,dx$

105. $\displaystyle\int \frac{x}{\sqrt{x}+1}\,dx$

106. $\displaystyle\int \frac{\sqrt{\sqrt{x}+1}}{\sqrt{x}}\,dx$

107. $\displaystyle\int \frac{e^{2y}}{e^{2y}+1}\,dy$

108. $\displaystyle\int \frac{z}{(z^2-5)^3}\,dz$

109. $\displaystyle\int \frac{z}{(z-5)^3}\,dz$

110. $\displaystyle\int \frac{(1+\tan x)^3}{\cos^2 x}\,dx$

111. $\displaystyle\int \frac{(2x-1)e^{x^2}}{e^x}\,dx$

112. $\displaystyle\int (x+\sin x)^3(1+\cos x)\,dx$

113. $\displaystyle\int \left(2x^3+3x+4\right)\cos(2x)\,dx$

CHAPTER EIGHT

USING THE DEFINITE INTEGRAL

In Chapter 3, we saw how a definite integral can be used to compute an area, an average value, or a total change. In Chapter 7, we focused on different ways of evaluating integrals. In this chapter we will use definite integrals to solve problems in geometry, physics, economics, and probability. In each section, we use the same method to represent a quantity as a definite integral: we chop up the quantity and approximate it by a Riemann sum.

8.1 APPLICATIONS TO GEOMETRY

Setting Up Riemann Sums

There are many situations where a quantity can be expressed as a definite integral. Typically this will happen when the quantity can be approximated by dividing it into little pieces. We then solve the problem approximately for each piece and add the resulting contributions.

For example, in Section 3.1, the way we learned that distance traveled is the definite integral of velocity was by adding up the distances traveled over small time intervals. For a particle moving with velocity $v = f(t)$ at time t, we find the net distance traveled between $t = a$ and $t = b$ by cutting the time interval into n small subintervals each of length Δt. We approximate the distance traveled during each subinterval Δt by assuming that the velocity is constant over the subinterval. Let's suppose the velocity throughout the ith interval is $v = f(t_i)$, the velocity at the start of the interval. The distance moved during that time is approximately

$$\text{Velocity} \times \text{Time} = f(t_i)\,\Delta t.$$

Therefore, the net change in position is approximated by the sum of all of these small distances, so

$$\text{Change in position} \approx \sum_{i=0}^{n-1} f(t_i)\,\Delta t.$$

In the limit as Δt tends to zero, this sum becomes an integral:

$$\text{Change in position} = \lim_{\Delta t \to 0} \sum_{i=0}^{n-1} f(t_i)\,\Delta t = \int_a^b f(t)\,dt.$$

Although the dt in the definite integral is not a number, it behaves as if it were a very small version of Δt. Informally, $f(t)\,dt$ may be thought of as the distance traveled by the particle over a very small time interval, since it is the velocity, $f(t)$, multiplied by the time, dt. The integral may be thought of as the sum of all these small distances.

In this chapter we will see how the same strategy of subdividing and summing can be used to calculate many other quantities. In this section we show how the definite integral can be used to compute the volume of a solid and the length of a curve .

To Compute a Volume or Length Using an Integral

- Divide the solid (or curve) into small pieces whose volume (or length) we can easily approximate;
- Add the contributions of all the pieces, obtaining a Riemann sum that approximates the total volume (or length);
- Take the limit as the number of terms in the sum tends to infinity, giving a definite integral for the total volume (or total length).

Finding Volumes by Slicing

When calculating the volume of a solid using Riemann sums, we chop the solid into smaller pieces whose volumes we can estimate.

Figure 8.1: Cone cut into
vertical slices

Figure 8.2: Cone cut into
horizontal slices

How to Slice a Cone

Let's see how we might slice a cone standing with the vertex uppermost. We could divide the cone into parallel vertical slices. These slices will be arch-shaped (they are actually hyperbolas); see Figure 8.1. We could also divide the cone horizontally, giving coin-shaped slices; see Figure 8.2.

To construct a Riemann sum, we prefer to cut the cone into the horizontal (circular) slices because it would be difficult to estimate the volumes of the arch-shaped slices.

Example 1 Compute the volume, in cubic feet, of the Great Pyramid of Egypt, whose base is a square 755 feet by 755 feet and whose height is 410 feet.

Solution You may already know the formula $V = \frac{1}{3}b^2 \cdot h$ for the volume V of a pyramid with height h and base length b. We will not use the formula, but our approach can be used to prove that $V = \frac{1}{3}b^2h$.

We view the pyramid as built up in layers starting from the base. Each layer is a square with thickness Δh. The bottom layer is a square slab 755 feet by 755 feet and volume about $(755)^2\Delta h$ ft^3. As we move up the pyramid, the layers have shorter side lengths. We pick points, $h_0, h_1, \ldots, h_n$, which divide the height into n subintervals of length Δh, with $h_0 = 0$ and $h_n = 410$. If we let s_i denote the side length of the ith layer, then the volume of this layer is approximately $s_i^2\, \Delta h$ ft^3. (The volume is only approximately $s_i^2\, \Delta h$ because the sides of the slab are not vertical.) See Figure 8.3.

Volume of slice
$\approx s_i^2\Delta h$

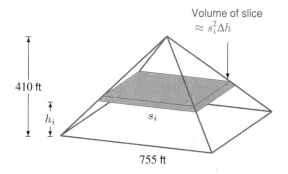

410 ft

h_i

s_i

755 ft

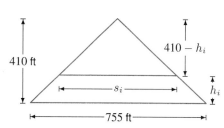

410 ft

$410 - h_i$

s_i

h_i

755 ft

Figure 8.3: The Great Pyramid

Figure 8.4: Cross-section relating s_i and h_i

The total volume of the pyramid is the sum of the volumes, $s_i^2\, \Delta h$, of each layer. We express s_i as a function of h_i using the triangular cross-section in Figure 8.4. By similar triangles, we get $s_i/755 = (410 - h_i)/410$. Thus, $s_i = (755/410)(410 - h_i)$, and the total volume, V, is approximated by adding the volumes of the n layers:

$$V \approx \sum_{i=0}^{n-1} s_i^2\, \Delta h = \sum_{i=0}^{n-1} \left[\left(\frac{755}{410}\right)(410 - h_i)\right]^2 \Delta h \text{ ft}^3.$$

As the thickness of each slice tends to zero, the sum becomes a definite integral. Finally, since h varies from 0 to 410, the height of the pyramid, we have

$$V = \int_{h=0}^{h=410} \left[\left(\frac{755}{410} \right) (410 - h) \right]^2 dh = \left(\frac{755}{410} \right)^2 \int_0^{410} (410 - h)^2 \, dh$$

$$= \left(\frac{755}{410} \right)^2 \left[-\frac{(410 - h)^3}{3} \right] \Big|_0^{410} = \left(\frac{755}{410} \right)^2 \frac{(410)^3}{3} = \frac{1}{3} (755)^2 (410) \approx 78 \text{ million ft}^3.$$

Note that $V = \frac{1}{3}(755)^2(410) = \frac{1}{3}b^2 \cdot h$, as expected.

Notice that the limits in the definite integral are the limits for the variable h. Once we decided to slice the pyramid horizontally, we knew that a typical slice has thickness Δh, so h is the variable in our definite integral, and the limits must be values of h.

Example 2   Find the volume of a hemisphere of radius 7 inches.

Solution   Again, you probably know the formula $\frac{4}{3}\pi r^3$ for the volume of a sphere; we will not use it here.

Imagine the hemisphere sitting flat on a plane and divide it into horizontal layers of thickness Δh inches. (See Figure 8.5.) Each layer is a circular slab with radius r, where r varies from 7 inches for the bottom layer to 0 inches for the top layer. The volume of each layer is approximately $\pi r^2 \, \Delta h \text{ in}^3$.

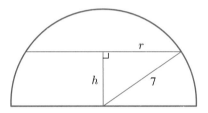

Figure 8.5: Slicing to find the volume of a hemisphere

Figure 8.6: Vertical slice through center of hemisphere showing relation between r and h

Our problem is to express r in terms of h so that we can write a Riemann sum (whose terms depend only on h) representing the sum of the volumes of the layers. From Figure 8.6, showing a vertical slice through the center of the hemisphere, we find that $r^2 = 7^2 - h^2$. So at height h,

$$\text{Volume of slice} \approx \pi r^2 \, \Delta h = \pi (7^2 - h^2) \, \Delta h \text{ in}^3.$$

Summing the volumes of all slices, we have

$$\text{Volume} = V \approx \sum \pi r^2 \, \Delta h = \sum \pi (7^2 - h^2) \, \Delta h \text{ in}^3.$$

As the thickness of each slab tends to zero, the sum becomes a definite integral. Since h varies from 0 to 7, the radius of the hemisphere, we have

$$V = \int_{h=0}^{h=7} \pi r^2 \, dh = \pi \int_0^7 (7^2 - h^2) \, dh = \pi \left(7^2 h - \frac{1}{3} h^3 \right) \Big|_0^7 = \pi \left(7^3 - \frac{1}{3} \cdot 7^3 \right) = \frac{2}{3} \pi 7^3 \text{ in}^3.$$

Notice that V is half of $\frac{4}{3}\pi 7^3 \text{ in}^3$, as we expected.

Note that the sum represented by the $\sum$ sign is over all the layers. In this example we did not bother to write down the points $h_0, h_1 \ldots$, that are needed to write the Riemann sum precisely. This is unnecessary if all we want is the final expression for the definite integral.

In Examples 1 and 2 all cross-sections of the solid perpendicular to a certain direction were the same geometric figure—a square in the case of the pyramid and a circle in the case of the hemisphere. The volumes of other solids with known cross-sections can be calculated by a similar method.

Example 3   Find the volume of the solid whose base is the region in the xy-plane bounded by the curves $y = x^2$ and $y = 8 - x^2$ and whose cross-sections perpendicular to the x-axis are squares with one side in the xy-plane. (See Figure 8.7.)

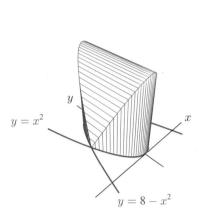

Figure 8.7: The solid for Example 3   *Figure 8.8:* A slice of the solid for Example 3

Solution   We view the solid as a loaf of bread sitting on the xy-plane and made up of square slices. These slices are analogous to the square layers making up the pyramid in our first example. A typical slice is shown in Figure 8.8. The thickness of each slice is Δx. The side length, s, of the square face is the distance (in the y direction) between the two curves. Therefore, in terms of x we have $s = (8 - x^2) - x^2 = 8 - 2x^2$.

Notice that this side length grows from $s = 0$ at $x = -2$ to $s = 8$ at $x = 0$, and then it shrinks back to $s = 0$ at $x = 2$. The volume of the solid is obtained by summing up the volumes of all the slices. Each slice has volume $s^2 \Delta x = (8 - 2x^2)^2 \Delta x$, so the total volume is

$$V \approx \sum s^2 \Delta x = \sum (8 - 2x^2)^2 \Delta x.$$

As the thickness Δx of each slice tends to zero, the sum becomes a definite integral. Since x varies between -2 and 2, we have

$$V = \int_{-2}^{2} (8 - 2x^2)^2 \, dx$$

$$= \int_{-2}^{2} (64 - 32x^2 + 4x^4) \, dx$$

$$= 64x - \frac{32}{3}x^3 + 4\frac{x^5}{5} \Big|_{-2}^{2}$$

$$= \frac{2048}{15} \approx 136.5.$$

We can check the plausibility of this answer by observing that the given solid is somewhat larger than the solid obtained by putting together two pyramids with 8×8 bases at $x = 0$ and apexes (highest points) at $x = 2$ and $x = -2$. The total volume of these two pyramids is $2(\frac{1}{3})b^2 h = 2(\frac{1}{3})8^2 \cdot 2 \approx 85.3$, which is, as expected, less than our answer, 136.5.

Volumes of Revolution

Another way to create a solid having known cross-sections is to revolve a region in the plane around some line in space, giving a *solid of revolution,* as in the following examples.

Example 4   The region bounded by the curve $y = e^{-x}$ and the x-axis between $x = 0$ and $x = 1$ is revolved around the x-axis. Find the volume of this solid of revolution.

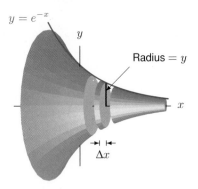

Figure 8.9: A thin strip rotated around the x-axis to form a circular slice

Solution   Imagine the plane region divided into thin strips perpendicular to the x-axis. Each strip has width Δx. As we rotate the whole region about the x-axis, each strip sweeps out a circular disk of thickness Δx. (See Figure 8.9.) The radius of the slice, as can be seen in Figure 8.9, is the distance from the x-axis to the curve, which is just the value of $y = e^{-x}$ on the slice. Thus the volume of the slice is $\pi y^2 \Delta x = \pi (e^{-x})^2 \Delta x$, and the total volume, V, of the solid is given by

$$V \approx \sum \pi y^2 \Delta x = \sum \pi \left(e^{-x}\right)^2 \Delta x.$$

As the thickness of each slice tends to zero, we get

$$V = \int_0^1 \pi (e^{-x})^2 \, dx = \pi \int_0^1 e^{-2x} \, dx = \pi \left(-\frac{1}{2}\right) e^{-2x} \Big|_0^1$$

$$= \pi \left(-\frac{1}{2}\right) (e^{-2} - e^0) = \frac{\pi}{2}(1 - e^{-2}) \approx 1.36.$$

Example 5   The region bounded by the curves $y = x$ and $y = x^2$ is rotated about the line $y = 3$. Compute the volume of the resulting solid.

Solution   Again imagine the region divided into thin vertical strips of thickness Δx, as in Figure 8.10.

Figure 8.10: The region for Example 5

As each strip is rotated around the line $y = 3$, it sweeps out a slice of the total solid shaped like a disk with a hole in it (see Figures 8.11 and 8.12). This disk-with-a-hole has two radii: an inner radius, r_{in}, which is the distance from the line $y = 3$ to the nearer curve $y = x$, and an outer radius, r_{out}, which is the distance from the line $y = 3$ to the more distant curve, $y = x^2$. So the slice at x has inner radius $r_{\text{in}} = 3 - x$ and outer radius $r_{\text{out}} = 3 - x^2$. Think of the slice as a circular disk of radius r_{out} from which has been removed a smaller circular disk of radius r_{in}. Thus, the volume of the slice is

$$\pi r_{\text{out}}^2 \Delta x - \pi r_{\text{in}}^2 \Delta x = \pi (3 - x^2)^2 \Delta x - \pi (3 - x)^2 \Delta x.$$

Adding up the volumes of all the slices, we have

$$V \approx \sum \left(\pi r_{\text{out}}^2 - \pi r_{\text{in}}^2\right) \Delta x = \sum \left[\pi(3 - x^2)^2 - \pi(3 - x)^2\right] \Delta x.$$

We let Δx, the thickness of each slice, tend to zero to obtain the definite integral. Since the curves $y = x$ and $y = x^2$ intersect at $x = 0$ and $x = 1$, we have

$$V = \int_0^1 \left[\pi(3 - x^2)^2 - \pi(3 - x)^2\right] dx = \pi \int_0^1 \left[(9 - 6x^2 + x^4) - (9 - 6x + x^2)\right] dx$$

$$= \pi \int_0^1 (6x - 7x^2 + x^4)\, dx = \pi \left(3x^2 - \frac{7x^3}{3} + \frac{x^5}{5}\right)\Bigg|_0^1 \approx 2.72.$$

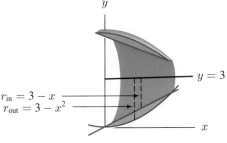

Figure 8.11: Cutaway view of volume showing inner and outer radii

Figure 8.12: One slice (a disk-with-a-hole)

Arc Length

The definite integral can also be used to compute the lengths of various curves. Suppose we wish to compute the *arc length*, as it is usually called, of a curve $y = f(x)$ from $x = a$ to $x = b$, where $a < b$. As usual, we divide the curve into many little pieces; each one is approximately straight.

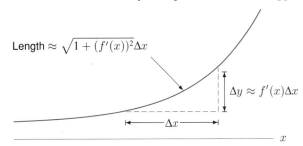

Figure 8.13: Length of a small piece of curve approximated using Pythagoras' theorem

Figure 8.13 shows that a small change Δx in the x coordinate produces a corresponding small change in the y coordinate of approximately $\Delta y \approx f'(x)\,\Delta x$. The length of the little piece of the curve is then approximately

$$\Delta L \approx \sqrt{(\Delta x)^2 + (\Delta y)^2} \approx \sqrt{(\Delta x)^2 + \left(f'(x)\,\Delta x\right)^2} = \sqrt{1 + (f'(x))^2}\, \Delta x.$$

Thus, for $a < b$, the arc length of the curve $y = f(x)$ from $x = a$ to $x = b$ is approximated by a Riemann sum:

$$\text{Arc length} = L \approx \sum \sqrt{1 + \left(f'(x)\right)^2}\, \Delta x.$$

As we let Δx tend to zero, the sum becomes a definite integral. Since x varies between a and b, we have the following expression for the arc length:

$$\boxed{\text{Arc length} = L = \int_a^b \sqrt{1 + (f'(x))^2}\, dx.}$$

Example 6   Set up and evaluate an integral to compute the length of the curve $y = x^3$ from $x = 0$ to $x = 5$.

Solution   If $f(x) = x^3$, then $f'(x) = 3x^2$, so

$$L = \int_0^5 \sqrt{1 + (3x^2)^2} \, dx.$$

Although the formula for the arc length of a curve is easy to apply, the integrands it generates often do not have elementary antiderivatives. Using numerical methods, we get $L = 125.68$. The curve starts at $(0, 0)$ and goes to $(5, 125)$, so its length must be at least the length of a straight line between these points, or $\sqrt{5^2 + 125^2} = 125.10$. (See Figure 8.14.)

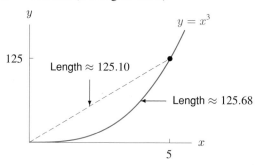

Figure 8.14: Arc length of $y = x^3$ (Note: the picture is distorted because the scales on the two axes are quite different)

Problems for Section 8.1

1. Imagine a hard-boiled egg lying on its side cut into thin slices. First think about vertical slices and then horizontal ones. What would these slices look like? Sketch them.

2. Find, by slicing, the volume of a cone whose height is 3 cm and whose base radius is 1 cm. Slice the cone as shown in Figure 8.2 on page 377.

3. Find, by slicing, a formula for the volume of a cone of height h and base radius r.

4. Find the volume of a sphere of radius r by slicing up the sphere, setting up a Riemann sum that approximates the volume, and then obtaining a definite integral by letting the thickness of each slice tend to zero.

5. When the ellipse $x^2/a^2 + y^2/b^2 = 1$ is rotated about the x-axis, it generates an *ellipsoid*. Compute its volume.

For Problems 6–8, sketch the solid obtained by rotating each region around the indicated axis. Using the sketch, show how to approximate the volume of the solid by a Riemann sum, and hence find the volume.

6. Region bounded by $y = x^3$, $x = 1$, $y = -1$. Axis: $y = -1$.

7. Region bounded by $y = \sqrt{x}$, $x = 1$, $y = 0$. Axis: $x = 1$.

8. Region bounded by the first arch of $y = \sin x$, $y = 0$. Axis: x axis.

For Problems 9–13 consider the region bounded by $y = e^x$, the x-axis, and the lines $x = 0$ and $x = 1$. Find the volume of the following solids.

9. The solid obtained by rotating the region about the x-axis.

10. The solid obtained by rotating the region about the horizontal line $y = -3$.

11. The solid obtained by rotating the region about the horizontal line $y = 7$.

12. The solid whose base is the given region and whose cross-sections perpendicular to the x-axis are squares.

13. The solid whose base is the given region and whose cross-sections perpendicular to the x-axis are semicircles.

14. The design of boats is based on Archimedes' Principle, which states that the buoyant force on an object in water is equal to the weight of the water displaced. Suppose you want to build a sailboat whose hull is parabolic with cross section $y = ax^2$, where a is a constant. Your boat will have length L and its maximum draft (the maximum vertical depth of any point of the boat beneath the water line) will be H. See Figure 8.15. Every cubic meter of water weighs 10,000 newtons. What is the maximum possible weight for your boat and cargo?

Figure 8.15   *Figure 8.16*

15. (a) A pie dish is 9 inches across the top, 7 inches across the bottom, and 3 inches deep. See Figure 8.16. Compute the volume of this dish.
    (b) Make a rough estimate of the volume in cubic inches of a single cut-up apple, and estimate the number of apples that is needed to make an apple pie that fills this dish.

16. Many communities expect to run out of space for their garbage soon. On Staten Island, solid garbage is packed into a pyramid-shaped dump with a square base of length 100 yards. One yard vertically above the base, the length of the side parallel to the base is 99 yards; the dump can be built up to a vertical height of 20 yards. (The top of the pyramid is never reached.) If 65 cubic yards of garbage arrive at the dump every day, how long will it be before the dump is full?

17. The circumference of a tree at different heights above the ground is given in the table below. Assume that all horizontal cross-sections of the tree are circles. Estimate the volume of the tree.

Height (inches)	0	20	40	60	80	100	120
Circumference (inches)	31	28	21	17	12	8	2

18. The hull of a certain boat has widths given by the following table. Reading across a row of the table gives widths at points 0, 10, ..., 60 feet from the front to the back at a certain level below waterline. Reading down a column of the table gives widths at levels 0, 2, 4, 6, 8 feet below waterline at a certain distance from the front. Use the trapezoidal rule to estimate the volume of the hull below waterline.

		Front of boat $\longrightarrow$ Back of boat						
		0	10	20	30	40	50	60
	0	2	8	13	16	17	16	10
Depth	2	1	4	8	10	11	10	8
below	4	0	3	4	6	7	6	4
waterline	6	0	1	2	3	4	3	2
(in feet)	8	0	0	1	1	1	1	1

For Problems 19–20, find the arc length of the given function from $x = 0$ to $x = 2$.

19. $f(x) = \sqrt{4 - x^2}$ 20. $f(x) = \sqrt{x^3}$

21. (a) Write an integral which represents the circumference of a circle of radius r.
    (b) Evaluate the integral, and show that you get the answer you expect.

22. Compute, with error at most 0.001, the perimeter of the region used for the base of the solids in Problems 9–13.

23. The graph of the function $y = \cosh x = \frac{1}{2}(e^x + e^{-x})$ is called a catenary and represents the shape of a hanging cable. Find the length of this catenary between $x = -1$ and $x = 1$.

24. Set up an integral for the circumference of an ellipse with semi-major axis $a = 2$ and semi-minor axis $b = 1$. Evaluating this integral numerically (i.e., using a calculator or computer) poses a problem. Describe this problem. Formulate a way to solve it.

25. There are very few elementary functions $y = f(x)$ for which arc length can be computed in elementary terms using the formula

$$\int_a^b \sqrt{1 + \left(\frac{dy}{dx}\right)^2}\, dx.$$

You have seen some such functions f in Problems 19, 20, and 23, namely, $f(x) = \sqrt{4 - x^2}$, $f(x) = \sqrt{x^3}$, and $f(x) = \frac{1}{2}(e^x + e^{-x})$. Try to find some other function that "works," that is, a function whose arc length you can find using this formula and antidifferentiation.

26. After doing Problem 25, you may wonder what sort of functions can represent arc length. If $g(0) = 0$ and g is differentiable and increasing, then can $g(x)$, $x \geq 0$, represent arc length? That is, can we find a function $f(t)$ such that

$$\int_0^x \sqrt{1 + (f'(t))^2}\, dt = g(x)?$$

(a) Show that $f(x) = \int_0^x \sqrt{(g'(t))^2 - 1}\, dt$ works as long as $g'(x) \geq 1$. In other words, show that the arc length of the graph of f from 0 to x is $g(x)$.

(b) Show that if $g'(x) < 1$ for some x, then $g(x)$ cannot represent the arc length of the graph of any function.

(c) Find a function f whose arc length from 0 to x is $2x$.

8.2 DENSITY AND CENTER OF MASS

Density and How to Slice a Region

The examples in this section involve the idea of *density*. For example,

- A population density is measured in, say, people per mile (along the edge of a road), or people per unit area (in a city), or bacteria per cubic centimeter (in a test tube).

- The density of a substance (e.g. air, wood, or metal), is the mass of a unit volume of the substance and is measured in, say, grams per cubic centimeter.

Suppose we want to calculate the total mass or total population, but the density is not constant over a region. Then to find the total quantity:

> Divide the region into small pieces in such a way that the density is approximately constant on each piece, and add up the contributions of all the pieces.

Example 1 The Massachusetts Turnpike ("the Pike") starts in the middle of Boston and heads west. The number of people living next to it varies as it gets further and further from the city. Suppose that, x miles out of town, the population density adjacent to the Pike is $P = f(x)$ people/mile. Express the total population living next to the Pike within 5 miles of Boston as a definite integral.

Solution Think about how we could estimate the population. Divide the Pike up into segments of length Δx. The population density at the center of Boston is $f(0)$; let's use that density for the first segment. This gives an estimate of

People living in first segment $\approx f(0)$ people/ mile $\cdot \Delta x$ mile $= f(0)\Delta x$ people.

Figure 8.17: Population along the Massachusetts Turnpike

Similarly, the population in a typical segment is the population density times the length of the interval, or roughly $f(x)\,\Delta x$. (See Figure 8.17.) The sum of all these estimates gives the estimate

$$\text{Total population} \approx \sum f(x)\,\Delta x.$$

Letting $\Delta x \to 0$ shows that

$$\text{Total population} = \lim_{\Delta x \to 0} \sum f(x)\,\Delta x = \int_0^5 f(x)\,dx.$$

The 5 and 0 in the limits of the integral are the upper and lower limits of the interval over which we are integrating.

Example 2 The air density (in kg/m³) h meters above the earth's surface is $P = f(h)$. Find the mass of a cylindrical column of air 2 meters in diameter and 25 kilometers high.

Solution The column of air is a circular cylinder 2 meters in diameter and 25 kilometers, or 25,000 meters, high. First we must decide how we are going to slice this column. Since the air density varies with altitude but remains constant horizontally, we take horizontal slices of air. That way, the density will be more or less constant over the whole slice, being close to its value at the bottom of the slice. (See Figure 8.18.)

Figure 8.18: Slicing a column of air horizontally

A slice is a cylinder of height Δh and diameter 2 m, so its radius will be 1 m. We find the approximate mass of the slice by multiplying its volume and its density together. If the thickness of the slice is Δh, then its volume is $\pi r^2 \cdot \Delta h = \pi 1^2 \cdot \Delta h = \pi \Delta h$ m³ (where m³ = cubic meter). The density of the slice is roughly $f(h)$. Thus,

$$\text{Mass of slice} \approx \text{Volume} \cdot \text{Approximate density} = (\pi \Delta h \text{ m}^3)(f(h) \text{ kg/m}^3) = \pi \Delta h \cdot f(h) \text{ kg}.$$

Adding these slices up yields a Riemann sum:

$$\text{Total mass} \approx \sum \pi f(h)\,\Delta h \text{ kg}.$$

As $\Delta h \to 0$, this sum approximates the definite integral:

$$\text{Total mass} = \int_0^{25,000} \pi f(h)\,dh \text{ kg}.$$

In order to get a numerical value for the mass of air, we need an explicit formula for the density as a function of height, as in the next example.

Example 3   Find the mass of the column of air in Example 2 if the density of air at height h is given by

$$P = f(h) = 1.28e^{-0.000124h} \text{ kg/m}^3.$$

Solution   Using the result of the previous example, we have

$$\text{Mass} = \int_0^{25,000} \pi 1.28 e^{-0.000124h} \, dh = \frac{-1.28\pi}{0.000124} \left(e^{-0.000124h} \Big|_0^{25,000} \right)$$

$$\approx 32{,}429(e^0 - e^{-0.000124(25{,}000)}) \approx 31{,}000 \text{ kg}.$$

Sometimes it requires some thought to figure out how to slice a quantity up. The key point to remember is that you want the density to be nearly constant within each piece.

Example 4   The population density in Ringsburg is a function of the distance from the city center. At r miles from the center, the density is $P = f(r)$ people per square mile. Ringsburg has a radius of 5 miles. Write a definite integral that expresses the total population of Ringsburg.

Solution   We want to slice Ringsburg up and estimate the population on each slice. If we were to take straight-line slices, the population density would vary on each slice, since it depends on the distance from the city center. We want the population density to be pretty close to constant on each slice. We therefore take slices that are thin rings around the center. (See Figure 8.19.) Since the ring is very thin, we can approximate its area by straightening it into a thin rectangle. (See Figure 8.20.) The width of the rectangle is Δr miles, and its length is approximately equal to the circumference of the ring, $2\pi r$ miles, so its area is about $2\pi r \, \Delta r \text{ mi}^2$. Since

$$\text{Population on ring} \approx \text{Density} \cdot \text{Area},$$

we get

$$\text{Population on ring} \approx (f(r) \text{ people/mi}^2)(2\pi r\Delta r \text{ mi}^2) = f(r) \cdot 2\pi r \, \Delta r \text{ people}.$$

Adding the contributions from each ring, we get

$$\text{Total population} \approx \sum 2\pi r f(r) \, \Delta r \text{ people}.$$

So

$$\text{Total population} = \int_0^5 2\pi r f(r) \, dr \text{ people}.$$

Figure 8.19: Ringsburg

Figure 8.20: Ring from Ringsburg (straightened out)

Note: You may wonder what happens if we calculate the area of the ring by subtracting the area of the inner circle (πr^2) from the area of the outer circle $[\pi(r + \Delta r)^2]$, giving

$$\text{Area} = \pi(r + \Delta r)^2 - \pi r^2.$$

Multiplying out and subtracting, we get

$$\text{Area} = \pi[r^2 + 2r\,\Delta r + (\Delta r)^2] - \pi r^2$$
$$= 2\pi r\,\Delta r + \pi(\Delta r)^2.$$

This expression differs from the one we used before by the $\pi(\Delta r)^2$ term. However, as Δr becomes very small, $\pi(\Delta r)^2$ becomes much, much smaller. We say it is *second order* of smallness, since the power of the small factor, Δr, is 2. In the limit as $\Delta r \to 0$, we can ignore $\pi(\Delta r)^2$.

Center of Mass

The *center of mass* of a mechanical system can be thought of as the balancing point. For a system consisting of n discrete masses m_i along the x-axis, each located at coordinate x_i, the center of mass is defined by the point whose x-coordinate is $\bar{x}$ where

$$\bar{x} = \frac{\sum_{i=1}^{n} x_i m_i}{\sum_{i=1}^{n} m_i}.$$

(See Figure 8.21.) The terms $x_i m_i$ in the numerator are called the *moments* of the masses m_i; the denominator is the total mass of the system. If the mechanical system is not composed of discrete masses, but is instead an object lying on the x-axis between $x = a$ and $x = b$, with mass density $\delta(x)$, then the center of mass is given by

$$\bar{x} = \frac{\int_a^b x\delta(x)\,dx}{\int_a^b \delta(x)\,dx}.$$

Again, the denominator is the total mass of the system.

We can see how this last formula relates to the formula for discrete masses using Riemann sums. If we divide the object into n small pieces, each of length Δx, then the ith piece has mass $m_i \approx \delta(x_i)\Delta x$ (where x_i is a point in the ith piece). Then the numerator of the formula for $\bar{x}$ is $\sum_{i=1}^{n} x_i \delta(x_i)\Delta x$, which approaches $\int_a^b x\delta(x)\,dx$ as $n \to \infty$. Similarly, the denominator $\sum_{i=1}^{n} m_i \approx \sum_{i=1}^{n} \delta(x_i)\Delta x$ approaches $\int_a^b \delta(x)\,dx$ as $n \to \infty$.

Figure 8.21: Center of mass of discrete masses, m_i

Example 5   Find the center of mass of a 2-meter rod lying on the x-axis with its left end at the origin if:
(a)   The density is constant and the total mass is 5 kg.   (b)   The density is $\delta(x) = 15x^2$ kg/m.

Solution   (a)   We expect that the balancing point will be in the middle, that is, $\bar{x} = 1$. To check this, we compute the density, which is the total mass divided by the length, or $\delta(x) = 5/2$ kg/m. Then

$$\bar{x} = \frac{\int_0^2 x \cdot \frac{5}{2}\,dx}{5} = \frac{1}{5} \cdot \frac{5}{2} \cdot \left. \frac{x^2}{2} \right|_0^2 = 1 \text{ meter.}$$

(b)   Since the rod is heavier in the interval $[1, 2]$ than in the interval $[0, 1]$, we expect the center of mass to be between $x = 1$ and $x = 2$. Now we have

$$\text{Total mass} = \int_0^2 15x^2\,dx = \left. 5x^3 \right|_0^2 = 40 \text{ kg.}$$

Thus,

$$\bar{x} = \frac{\int_0^2 x \cdot 15x^2\,dx}{40} = \frac{15}{40} \cdot \left. \frac{x^4}{4} \right|_0^2 = \frac{3}{2} \text{ meter.}$$

The definition of $\bar{x}$ can be used to find the x-coordinate of the center of mass for objects that don't lie entirely along the x-axis. The key in these cases is to find an expression for the mass of a small slice of the object located at position x_i.

Example 6   An isosceles triangle of altitude 1 unit and base 1 unit has its base along the y-axis and the positive x-axis passing through its vertex. See Figure 8.22. If the mass of the triangle is m and the triangle has constant density, find the x-coordinate of its center of mass.

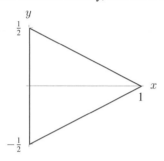

Figure 8.22: Find center of mass of this triangle

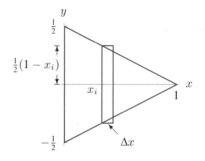

Figure 8.23: Sliced triangle

Solution   Because the mass of the triangle is symmetrically distributed with respect to the x-axis, the center of mass lies on the x-axis. We expect the center of mass to be closer to the origin than to the vertex at $x = 1$, since the triangle is wider near the origin.

The area of the triangle is $\frac{1}{2} \cdot 1 \cdot 1 = \frac{1}{2}$. Thus Density = Mass/Area = $2m$. If we slice the triangle into strips of width Δx, then the piece at position x_i has height $2 \cdot \frac{1}{2}(1 - x_i) = (1 - x_i)$. (See Figure 8.23.) So

$$\text{Area of strip} \approx (1 - x_i)\Delta x,$$

giving

$$\text{Mass} = \text{Area} \times \text{Density} \approx 2m \cdot (1 - x_i)\Delta x.$$

So the center of mass is

$$\bar{x} \approx \frac{\sum x_i 2m \cdot (1 - x_i)\Delta x}{m}.$$

As $n \to \infty$, we have

$$\bar{x} = \frac{\int_0^1 2mx(1 - x)\,dx}{m} = 2\left(\frac{x^2}{2} - \frac{x^3}{3}\right)\Bigg|_0^1 = \frac{1}{3}.$$

Similar ideas are used to find the coordinates of the center of mass for objects in the xy-plane.

Problems for Section 8.2

1. A rod has length 2 meters. At a distance x meters from its left end, the density of the rod is given by

    $$\rho(x) = 2 + 6x \text{ g/m.}$$

    (a) Write a Riemann sum approximating the total mass of the rod.
    (b) Find the exact mass by converting the sum into an integral.

2. The density of cars (in cars per mile) down a 20-mile stretch of the Pennsylvania Turnpike can be approximated by

    $$\rho(x) = 300\left(2 + \sin\left(4\sqrt{x + 0.15}\right)\right),$$

    where x is the distance in miles from the Breezewood toll plaza.

    (a) Sketch a graph of this function for $0 \leq x \leq 20$.
    (b) Write a sum that approximates the total number of cars on this 20-mile stretch.
    (c) Find the total number of cars on the 20-mile stretch.

3. Suppose you want to find the total mass of a 3×5 rectangular sheet, whose density per unit area at a distance x from one of the sides of length 5 is $1/(1 + x^4)$.

    (a) Find a Riemann sum which approximates the total mass.
    (b) Find the mass to one decimal place. How do you know your answer is accurate enough?

4. The density of oil in a circular oil slick on the surface of the ocean at a distance r meters from the center of the slick is given by $\rho(r) = 50/(1 + r)$ kg/m^2.

    (a) If the slick extends from $r = 0$ to $r = 10{,}000$ m, find a Riemann sum approximating the total mass of oil in the slick.
    (b) Find the exact value of the mass of oil in the slick by turning your sum into an integral and evaluating it.
    (c) Within what distance r is half the oil of the slick contained?

5. The soot produced by a garbage incinerator spreads out in a circular pattern. The depth, $H(r)$, in millimeters, of the soot deposited each month at a distance r kilometers from the incinerator is given by $H(r) = 0.115e^{-2r}$.

    (a) Write a definite integral giving the total volume of soot deposited within 5 kilometers of the incinerator each month.
    (b) Evaluate the integral you found in part (a), giving your answer in cubic meters.

6. If a rod lies along the x-axis between a and b, the moment of the rod is $\int_a^b x\rho(x)\,dx$, where $\rho(x)$ is its density at a position x. Find the moment of the rod in Problem 1.

7. A rod of length 3 with density $\delta(x) = 1 + x^2$ is positioned along the positive x-axis, with its left end at the origin. Find the total mass and the center of mass of the rod.

8. Consider a rod of length 1, with density $\delta(x) = 1 + kx^2$, where k is a positive constant. Suppose the rod is lying on the positive x-axis with one end at the origin.

    (a) Find the center of mass as a function of k.
    (b) Show that the center of mass of the rod satisfies $0.5 < \bar{x} < 0.75$.

9. A rod of length 2 meters and density $\delta(x) = 3 - e^{-x}$ kilograms per meter is placed on the x-axis with its ends at $x = \pm 1$.

    (a) Will the center of mass of the rod be on the left or right of the origin? Explain.
    (b) Find the coordinate of the center of mass.

10. One half of a uniform circular disk of radius 1 lies with its diameter along the y-axis and center at the origin. The mass of the half-disk is 1. Find the location of the center of mass $\bar{x}$.

11. Suppose an isosceles triangle with uniform density, altitude a, and base b is placed in the xy-plane as in Figure 8.24. Show that the center of mass is at $\bar{x} = a/3$. Hence show that the center of mass is independent of the triangle's base.

Figure 8.24 *Figure 8.25* *Figure 8.26*

12. A cardboard figure has the shape shown in Figure 8.25. The region is bounded on the left by the line $x = a$, on the right by the line $x = b$, above by $f(x)$, and below by $g(x)$. If the density $\rho(x)$ gm/cm^2 varies only with x, find an expression for the total mass of the figure, in terms of $f(x)$, $g(x)$, and $\rho(x)$.

13. The storage shed in Figure 8.26 is the shape of a half-cylinder of radius r and length l.

    (a) What is the volume of the shed?
    (b) Suppose the shed is filled with sawdust whose density at any point is proportional to the distance of that point from the floor. The constant of proportionality is k. Calculate the total mass of sawdust in the shed.

14.   The following table gives the density D (in gm/cm^3) of the earth at a depth x km below the earth's surface. The radius of the earth is about 6370 km. Find an upper and a lower bound for the earth's mass such that the upper bound is less than twice the lower bound. Explain your reasoning; in particular, what assumptions have you made about the density?

x	0	1000	2000	2900	3000	4000	5000	6000	6370
D	3.3	4.5	5.1	5.6	10.1	11.4	12.6	13.0	13.0

15.   An exponential model for the density of the earth's atmosphere says that if the temperature of the atmosphere were constant, then the density of the atmosphere as a function of height, h (in meters), above the surface of the earth would be given by

$$\rho(h) = 1.28e^{-0.000124h} \text{ kg/m}^3.$$

(a)   Write (but do not evaluate) a sum that approximates the mass of the portion of the atmosphere from $h = 0$ to $h = 100$ m (i.e., the first 100 meters above sea level). Assume the radius of the earth is 6370 km.

(b)   Find the exact answer by turning your sum in part (a) into an integral. Evaluate the integral.

16.   Water is flowing in a cylindrical pipe of radius 1 inch. Because water is viscous and sticks to the pipe, the rate of flow varies with the distance from the center. The speed of the water at a distance r inches from the center is $10(1 - r^2)$ inches per second. What is the rate (in cubic inches per second) at which water is flowing through the pipe?

8.3 APPLICATIONS TO PHYSICS

Although geometric problems of area, length, and volume were a driving force for the development of the calculus in the seventeenth century, it was Newton's spectacularly successful applications of the calculus to physics that most clearly demonstrated the power of this new mathematics.

Work

Our first application involves the concept of work. In physics the word "work" has a technical meaning which is different from its everyday meaning. Physicists say that if a constant force, F, is applied to some object to move it a distance, d, then the force has done work on the object. The force must be in the same direction as the motion. We make the following definition:

Work done = Force × Distance
$$W = F \cdot d$$

Notice that if we walk across a room holding a book, we do not accomplish any work on the book, since the only force we exert on the book is vertical, but the motion of the book is horizontal. On the other hand, if we lift the book from the floor to a height of 3 feet above the floor, we accomplish work.

Note on Units: Mass versus Weight

When we talk about how much something weighs, we're talking about the force that gravity exerts on that object. A lump of iron, for example, might weigh 10 pounds on the surface of the earth. This is because the earth's gravitational field is exerting 10 pounds of force on that lump of iron. If the same lump, however, is adrift in interstellar space, it is weightless because there is no gravitational field in its vicinity to exert a force on it. Of course, the quantity of iron present hasn't changed; there's no less iron there than there was before. So we make a distinction between the *mass* of the lump (the quantity of matter it comprises) and the *weight* of that mass (the force exerted on it due to gravity).

In the British system of units, a *pound* is a unit of weight. In the metric system, however, a *kilogram* is a unit of mass, not weight. Thus knowing we have 1 kilogram of iron means that we know the mass of iron present. To figure out how much our iron weighs, we have to figure out how much force is being exerted on it by gravity. Since Newton's second law tells us that Force = Mass × Acceleration, we see that we have to multiply the iron's mass by the acceleration due to gravity, which (on the surface of the earth) is $9.8 \text{ m}/\sec^2$. Thus the kilogram of iron weighs $1 \text{ kg} \cdot 9.8 \text{ m}/\sec^2 = 9.8 \text{ kg m}/\sec^2$, which is almost always written 9.8 newtons. (A newton = $1 \text{ kg} \cdot \text{ m}/\sec^2$ is a unit of force or weight. It is abbreviated nt.)

Example 1 How much work is done in lifting

(a) A 5-pound book 3 feet off the floor?   (b) A 1.5-kilogram book 2 meters off the floor?

Solution (a) $W = F \cdot d = (5 \text{ lb})(3 \text{ ft}) = 15$ foot-pounds.

(b) In this case we have not done $(1.5)(2) = 3$ units of work. We need to use the book's weight (not its mass), namely $(1.5 \text{ kg})(g \text{ m/sec}^2) = (1.5)(9.8)$ newtons. Thus, the work done is

$$W = F \cdot d = [(1.5 \text{ kg})(9.8 \text{ m/sec}^2)] \cdot (2 \text{ m}) = 29.4 \text{ newton-meters.}$$

A newton-meter is usually called a *joule*, so we did 29.4 joules of work in the metric system of units. In the British system the unit of work is the foot-pound, the work done when a 1-pound force moves an object a distance of 1 foot.

Calculating the Work Done

In the following examples, we calculate the work done where both the force and distance moved vary. To find the work done on a complicated object, *we slice the object up in such a way that we can find the work done on each piece.* We calculate the work for each piece using $W = Fd$, and we sum these pieces to approximate the total work as a Riemann sum. Letting the size of each piece tend to zero, we obtain a definite integral that represents the total work.

Example 2 A 28-meter uniform chain with a mass of 20 kilograms is dangling from the roof of a building. How much work is needed to pull the chain up to the top of the building?

Solution Since a 20-kilogram chain weighs $(20 \text{ kg})(9.8 \text{ m}/\sec^2) = 196$ newtons, it might seem that the answer should be $(196 \text{ nt})(28 \text{ m}) = 5488$ joules. But remember that not all of the chain has to move 28 meters—links near the top move less.

Let's divide the chain into small sections of length Δy, each weighing $7 \Delta y$ newtons. (A length of 28 meters weighs 196 newtons, so 1 meter weighs 7 newtons.) See Figure 8.27 on page 392. If Δy is small, all of this piece is hauled up approximately the same distance, namely y, against the force due to gravity of $7 \Delta y$ newtons. Thus,

Work done on the small piece $\approx (7 \Delta y \text{ newtons})(y \text{ meters}) = 7y \Delta y \text{ joules.}$

The work done on the entire chain is given by the total of the work done on each piece:

$$W \approx \sum 7y \, \Delta y \text{ joules.}$$

As $\Delta y \to 0$, we obtain a definite integral. Since y varies from 0 to 28 meters, the total work is

$$W = \int_0^{28} (7y) \, dy = \frac{7}{2} y^2 \Big|_0^{28} = 2744 \text{ joules.}$$

Figure 8.27: Chain for Example 2   *Figure 8.28:* Pyramid for Example 3

Example 3   It is reported that the Great Pyramid of Egypt was built in 20 years. If the stone making up the pyramid has density 200 pounds per cubic foot, find the total amount of work done in building the pyramid. Assume that the pyramid is 410 feet high and has a square base 755 feet by 755 feet. Estimate how many workers were needed to build the pyramid.

Solution   We assume that the stones were originally located at the approximate height of the construction site. Imagine the pyramid built in layers as we did in Example 1 on page 377.

We found by similar triangles that the layer at height h has a side length $s = \frac{755}{410}(410 - h)$ ft. The layer at height h has a volume of $s^2 \Delta h$ ft^3 and hence a weight of $200s^2 \Delta h$ lb. Constructing this layer involved lifting a weight of $200s^2 \Delta h$ through a height of h, so the work required is $(200s^2 \Delta h \text{ lb})(h \text{ ft}) = 200s^2 h \Delta h$ ft-lb. (See Figure 8.28.) The total work, W, of constructing all the layers is then

$$W \approx \sum 200s^2 \, h\Delta h = \sum 200 \left(\frac{755}{410}\right)^2 (410 - h)^2 h \, \Delta h \text{ ft-lb.}$$

As the thickness of each layer, Δh, tends to zero, we obtain a definite integral. Since h varies from 0 to 410, we have

$$W = \int_0^{410} 200s^2 h \, dh = 200 \int_0^{410} \left(\frac{755}{410}\right)^2 (410 - h)^2 h \, dh = 200 \left(\frac{755}{410}\right)^2 \int_0^{410} (410 - h)^2 h \, dh$$

$$= 200 \left(\frac{755}{410}\right)^2 \left(410^2 \frac{h^2}{2} - 2(410)\frac{h^3}{3} + \frac{h^4}{4}\right)\bigg|_0^{410} = 200 \left(\frac{755}{410}\right)^2 (410)^4 \left(\frac{1}{12}\right)$$

$$\approx 1{,}597{,}020{,}000{,}000 \approx 1.6 \times 10^{12} \text{ foot-pounds.}$$

We have calculated the total work done in building the pyramid; now we want to estimate the total number of workers needed. Let's assume every laborer worked 10 hours a day, 300 days a year, for 20 years. Assume that a typical worker lifted ten 50 pound blocks a distance of 4 feet every hour, thus performing 2000 foot-pounds of work per hour (this is a very rough estimate). Then each laborer performed $(10)(300)(20)(2000) = 1.2 \times 10^8$ foot-pounds of work over a twenty year period. Thus, the number of workers needed was about $(1.6 \times 10^{12})/(12 \times 10^7)$, or about 13,000.

Escape from the Earth's Gravitational Field

If you throw an object into the air, you expect gravity to bring it back to earth. However, the harder you throw it, the longer you expect to wait before it returns. You may wonder if it is possible to throw something hard enough so that it never returns to earth. We will calculate the total amount of work done to move an object infinitely far away from the earth. Amazingly enough, this total amount of work is finite, and it *is* possible to launch something hard enough that it never returns to earth.

Previously, when we analyzed the motion of a freely falling body, we assumed that the force of gravity is constant. Now, since we are thinking of a body moving into outer space, we have to take into account the fact that, as we move away from the earth, its gravitational force gets weaker. Newton's Law of Gravitation says that the force, F, exerted on a mass m at a distance, r, from the center of the earth is given by

$$F = \frac{GMm}{r^2}.$$

Here M is the mass of the earth, and G is called the gravitational constant, whose value is about 6.67×10^{-11} if mass is measured in kilograms, distance in meters, and force in newtons. We take $r > R$, the earth's radius.

Example 4   Find the total work done to move an object of mass m infinitely far from the earth.

Solution   The force pulling the object back to the earth is GMm/r^2. Thus,

$$\text{Work needed to move an object a distance } \Delta r \text{ further away} \approx \left(\frac{GMm}{r^2}\right)\Delta r.$$

Therefore the total work required to move the object from the surface of the earth $r = R$ to a point "infinitely far away" is approximated by

$$W \approx \sum \frac{GMm}{r^2}\Delta r$$

where the value of r runs from $r = R$ (at the earth's surface) to $r = \infty$ (very far away). As Δr tends to zero, we get the definite integral

$$W = \int_R^\infty \frac{GMm}{r^2}\,dr = \lim_{b\to\infty} -\frac{GMm}{r}\Big|_R^b = \frac{GMm}{R}.$$

Imagine an object shot upward with initial velocity v. The energy of the body by virtue of its motion is called its *kinetic energy* and is equal to $\frac{1}{2}mv^2$. When the object moves upward, it slows down, losing kinetic energy as work is done against gravity. The body moves upward until all the kinetic energy is expended; it will stop at the height at which

$$\text{Initial kinetic energy} = \text{Work done against gravity.}$$

Since the work done to move an object infinitely far from the earth is finite, a finite starting velocity can give the object enough kinetic energy to move infinitely far away. The *escape velocity* from the earth is defined to be the minimum vertical velocity that must be given to an object so that it is never drawn back to earth by gravity.

Example 5   Find the escape velocity of an object from the earth.

Solution   If v is the initial velocity, then setting the initial kinetic energy equal to the work done to escape infinitely far from the earth, we get

$$\frac{1}{2}mv^2 = \frac{GMm}{R}.$$

Solving for v gives

$$v = \sqrt{\frac{2GM}{R}}.$$

Since the mass of the earth is $M \approx 6 \times 10^{24}$ kg and its radius is $R \approx 6.4 \times 10^6$ meters, the escape velocity is about 11,000 meters/sec $\approx$ 25,000 mph, or about 30 times the speed of sound. Thus, the escape velocity is finite, though it is much higher than you can achieve by throwing or by shooting a gun. (A bullet's speed is slightly faster than sound.)

Force and Pressure

We can use the definite integral to compute the force exerted by a liquid on a surface, for example, the force of water on a dam. The idea is to get the force from the *pressure*. The pressure in a liquid is the force per unit area exerted by the liquid. Two things you need to know about pressure are:

- At any point, pressure is exerted equally in all directions—up, down, sideways.
- Pressure increases with depth. (That is one of the reasons why deep sea divers have to take much greater precautions than scuba divers.)

At a depth of h feet, the pressure, p, exerted by the liquid, measured in pounds per square foot, is given by computing the total weight of a column of liquid h feet high with a base of 1 square foot. The volume of such a column of liquid is just h cubic feet.

If the liquid has density ρ (mass per unit volume), then its weight per unit volume is ρg (where g is the acceleration due to gravity). The weight of the column of liquid is $\rho g h$, so

$$\boxed{\text{Pressure} = \text{Density} \times g \times \text{Depth} \quad \text{or} \quad p = \rho g h}$$

Sometimes we may be given the density of the liquid as a weight per unit volume, rather than a mass per unit volume. In that case, we do not need to multiply by g because it has already been done. Since pounds are units of weight, and not mass, knowing that water weighs 62.4 lb/ft^3 tells us that for water, $\rho g = 62.4$ lb/ft^3, and pressure at depth h is $62.4h$ lb/ft^2. See Figure 8.29.

Next we need to know the relation between force and pressure. Provided the pressure is constant over a given area, we have the following relation:

$$\boxed{\text{Force} = \text{Pressure} \times \text{Area}}$$

When the pressure isn't constant over a surface, we divide the surface into small pieces *in such a way that the pressure is nearly constant on each one*. Then we can use this formula on each piece and obtain a definite integral to get the force over an entire surface. Since the pressure varies with depth, we should divide the surface into horizontal strips, each of which is at an approximately constant depth.

Figure 8.29: Pressure exerted by a column of water

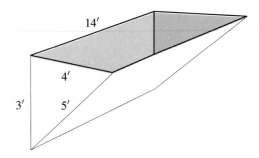

Figure 8.30: The trough for Example 6

Example 6   A trough 14 feet long has one rectangular vertical side 14 feet by 3 feet, one rectangular inclined side 14 feet by 5 feet, and two triangular ends with sides 3 feet, 4 feet, and 5 feet (see Figure 8.30). If the trough is filled to the top with water, compute the force on each end and each side.

Solution   We will do the long vertical side first. Divide the side into thin horizontal strips of width Δh. Since pressure depends only on the depth h and the strips are thin and horizontal, we can assume the pressure is the same at every point in the strip. The area of the strip is $14\,\Delta h$ ft^2, and the force on the

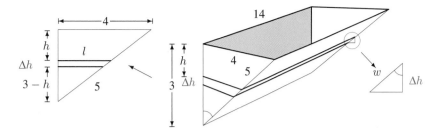

Figure 8.31: Calculating the widths of strips on the end and the inclined side

strip at depth h ft is the pressure at that depth times the area of the strip:

$$\text{Force on strip} = \text{Pressure} \times \text{Area} \approx (62.4h \text{ lb/ft}^2)(14 \, \Delta h \text{ ft}^2) = (62.4h)14 \, \Delta h \text{ lb.}$$

Summing the forces on these strips, we get an expression for the total force on the vertical side:

$$F_{\substack{\text{vertical} \\ \text{side}}} \approx \sum (62.4h)14 \, \Delta h \text{ lb.}$$

Letting $\Delta h \to 0$, we get a definite integral. Since h varies from 0 to 3 feet, we have

$$F_{\substack{\text{vertical} \\ \text{side}}} = \int_0^3 (62.4h)14 \, dh = (62.4)(14)\frac{h^2}{2}\Big|_0^3 = (62.4)(14)\left(\frac{9}{2}\right) \approx 3930 \text{ pounds.}$$

To compute the force on the inclined side, we use an inclined horizontal strip. Each strip has length 14, but its width is not Δh. By similar triangles (see Figure 8.31) the width w of the strip satisfies

$$\frac{\Delta h}{w} = \frac{3}{5}, \quad \text{so} \quad w = \frac{5}{3}\Delta h.$$

Therefore,

$$\text{Area of strip} = 14w = 14\left(\frac{5}{3}\right)\Delta h \text{ ft}^2,$$

so

$$\text{Force on strip} = (62.4h)(14)\left(\frac{5}{3}\right)\Delta h \text{ lb.}$$

The total force on the inclined side is obtained by adding the forces on the strips, giving

$$F_{\substack{\text{inclined} \\ \text{side}}} \approx \sum (62.4h)(14)\left(\frac{5}{3}\right)\Delta h \text{ lb.}$$

We let $\Delta h \to 0$ to arrive at a definite integral. Again, h varies between 0 and 3 feet, so

$$F_{\substack{\text{inclined} \\ \text{side}}} = \int_0^3 (62.4h)(14)\left(\frac{5}{3}\right) dh = \left(\frac{5}{3}\right)(62.4)(14)\left(\frac{9}{2}\right) \approx 6550 \text{ pounds.}$$

Notice that this is just $\frac{5}{3}$ the force on the vertical side.

Finally, to compute the force on each triangular end, we again divide the end into horizontal strips. The width of the strip is Δh as with the vertical side, but this time the length of the strip varies from 4 feet at the top of the trough to 0 feet at the bottom. By similar triangles (see Figure 8.31) the length l of a strip is related to the depth h by

$$\frac{l}{4} = \frac{3-h}{3}, \quad \text{so} \quad l = \frac{4}{3}(3-h).$$

Thus, the area of the strip is $l \, \Delta h = \frac{4}{3}(3-h)\Delta h$ ft^2. As before, this means the total force is

$$F_{\text{end}} \approx \sum (62.4h)\left(\frac{4}{3}(3-h)\right)\Delta h \text{ lb.}$$

Since h varies between 0 and 3, as $\Delta h \to 0$ we have

$$F_{\text{end}} = \int_0^3 (62.4h) \left(\frac{4}{3}(3 - h) \right) dh = (62.4) \left(\frac{4}{3} \right) \left(3\frac{h^2}{2} - \frac{h^3}{3} \right) \Big|_0^3$$

$$= (62.4) \left(\frac{4}{3} \right) \left(\frac{27}{6} \right) \approx 374 \text{ pounds.}$$

Problems for Section 8.3

1. Figure 8.32 shows thrust-time curves for two model rockets. The thrust or force, F, of the engine (in newtons) is plotted against time, t, (in seconds). The *total impulse* of the rocket's engine is defined as the definite integral of F with respect to t. The total impulse is a measure of the strength of the engine.

   (a) For approximately how many seconds is the thrust of rocket B greater than 10 newtons?

   (b) Estimate the total impulse for model rocket A.

   (c) What are the units for the total impulse calculated in part (b)?

   (d) Which rocket has the largest total impulse?

   (e) Which rocket has the largest maximum thrust?

Figure 8.32

2. A worker on a scaffolding 75 ft above the ground needs to lift a 500 lb bucket of cement from the ground to a point 30 ft above the ground by pulling on a rope weighing 5 lb/ft. How much work is required?

3. A 1000-lb weight is being lifted to a height 10 feet off the ground. It is lifted using a rope which weighs 4 lb per foot and which is being pulled up by construction workers standing on a roof 30 feet off the ground. Find the work done to lift the weight.

4. A rectangular water tank has length 20 ft, width 10 ft, and depth 15 ft. If the tank is full, how much work does it take to pump all the water out? (Note that 1 cubic foot of water weighs 62.4 lb.)

5. A water tank is in the form of a right circular cylinder with height 20 ft and radius 6 ft. If the tank is half full of water, find the work required to pump all of it over the top rim. (Note that 1 cubic foot of water weighs 62.4 lb.)

6. Suppose the tank in Problem 5 is full of water. Find the work required to pump all of it to a point 10 ft above the top of the tank.

7. A fuel oil tank is an upright cylinder, buried so that its circular top is 10 feet beneath ground level. The tank has a radius of 5 feet and is 15 feet high, although the current oil level is only 6 feet deep. Calculate the work required to pump all of the oil to the surface. Assume that the oil weighs 50 lb per cubic foot.

8. A gas station stores its gasoline in a tank under the ground. The tank is a cylinder lying horizontally on its side. (In other words, the tank is not standing vertically on one of its flat ends.) If the radius of the cylinder is 4 feet, its length is 12 feet, and its top is 10 feet under the ground, find the total amount of work needed to pump the gasoline out of the tank. (Gasoline weighs 42 lb/ft^3.)

9. How much work is required to lift a 1000-kg satellite to an altitude of 2×10^6 m above the surface of the earth? (The radius of the earth is 6.4×10^6 m, its mass is 6×10^{24} kg, and in these units the gravitational constant, G, is 6.67×10^{-11}.)

10. Show that the escape velocity needed by an object to escape the gravitational influence of a spherical planet of density ρ and radius R is proportional to R and to $\sqrt{\rho}$.

11. Calculate the escape velocity of an object from the moon. The acceleration due to gravity on the moon is 1.6 m/sec^2, whereas on earth it is 9.8 m/sec^2; the radius of the moon is approximately 1740 km. [Hint: If g is the acceleration due to gravity, the force on an object is $F = mg = GMm/R^2$. See page 393.]

12. A reservoir has a dam at one end. The dam is a rectangular wall, 1000 feet long and 50 feet high. The problem is to find the total force of the water acting on the dam.

     (a) Show how to approximate this force by a Riemann sum, explaining your reasoning clearly.
     (b) Write an integral which represents the force, and evaluate it.

13. Before any water was pumped out, what was the total force on the bottom and each side of the tank in Problem 4?

14. A lobster tank in a restaurant is 4 ft long by 3 ft wide by 2 ft deep. Find the water force on the bottom and on each of the four sides. (The density of water is 62.4 lb/ft^3.)

For Problems 15–16, find the kinetic energy of a rotating body, using the fact that the kinetic energy of a particle of mass m moving at a speed v is $\frac{1}{2}mv^2$. Slice the object into pieces in such a way that the velocity is approximately constant across each piece.

15. Find the kinetic energy of a rod of mass 10 kg and length 6 m rotating about an axis perpendicular to the rod at its midpoint, with an angular velocity of 2 radians per second. (Imagine a helicopter blade of uniform thickness.)

16. Find the kinetic energy of a phonograph record of uniform density, mass 50 gm and radius 10 cm rotating at $33\frac{1}{3}$ revolutions per minute.

17. We define the electric potential at a distance r from an electric charge q by q/r. The electric potential of a charge distribution is obtained by adding up the potential from each point. Suppose electric charge is sprayed (with constant density σ in units of charge/unit area) on to a circular disk of radius a. Consider the axis perpendicular to the disk and through its center. Find the electric potential at the point P on this axis at a distance R from the center. (See Figure 8.33.)

18. On a hot afternoon in Pétionville, Haiti, Lubonga is trying to cool off by sipping the glass of palm wine shown in Figure 8.34. How much work does she have to do to empty the glass? (The density of the palm wine is 1.2 gm/cm^3. Initially, the glass is filled to a vertical depth of 8 cm. Note that the force due to gravity acting on 1 gram is 980 dynes, and that in these units work is measured in ergs. The work done in lifting a small piece of fluid depends on the difference in height between the initial and final positions.)

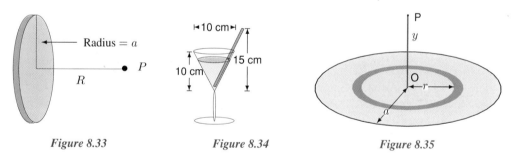

Figure 8.33   Figure 8.34   Figure 8.35

19. A uniform, thin, circular disk of radius a and mass M lies on a horizontal plane. The point P lies a distance y directly above O, the center of the disk. Calculate the gravitational force on a mass m at the point P. (See Figure 8.35.) Use the fact that the gravitational force exerted on the mass m by a thin horizontal ring of radius r, mass μ, and center O is toward O and given by

$$F = \frac{G\mu my}{(r^2 + y^2)^{3/2}}, \qquad \text{where } G \text{ is the gravitational constant.}$$

For Problems 20–21, find the gravitational force between two objects. Use the fact that the gravitational attraction between particles of mass m_1 and m_2 at a distance r apart is Gm_1m_2/r^2. For each problem, slice the objects into pieces, use this formula for the pieces, and sum using a definite integral.

20. What is the force of gravitational attraction between a thin uniform rod of mass M and length l and a particle of mass m lying in the same line as the rod at a distance a from one end?

21. Two long, thin, uniform rods of lengths l_1 and l_2 lie on a straight line with a gap between them of length a. Suppose their masses are M_1 and M_2, respectively, and the constant of the gravitation is G. What is the force of attraction between the rods? (Use the result of Problem 20.)

8.4 APPLICATIONS TO ECONOMICS

Present and Future Value

Many business deals involve payments in the future. If you buy a car or furniture, for example, you may buy it on credit and pay over a period of time. If you are going to accept payment in the future under such a deal, you obviously need to know how much you should be paid. Being paid $100 in the future is clearly worse than being paid $100 today for many reasons. If you are given the money today, you can do something else with it—for example, put it in the bank, invest it somewhere, or spend it. Thus, even without considering inflation, if you are to accept payment in the future, you would expect to be paid more to compensate for this loss of potential earnings. The question we will consider now is, how much more?

To simplify matters, we consider only what we would lose by not earning interest; we will not consider the effect of inflation. Let's look at some specific numbers. Suppose you deposit $100 in an account which earns 7% interest compounded annually, so that in a year's time you will have $107. Thus, $100 today will be worth $107 a year from now. We say that the $107 is the *future value* of the $100, and that the $100 is the *present value* of the $107. In general, we say the following:

- The **future value**, B, of a payment, P, is the amount to which the P would have grown if deposited in an interest bearing bank account.
- The **present value**, P, of a future payment, B, is the amount which would have to be deposited in a bank account today to produce exactly B in the account at the relevant time in the future.

As you might expect, the present value will always be smaller than the future value. In Appendix B, we see that with an interest rate of r, compounded annually, and a time period of t years, a deposit of P grows to a future balance of B, where

$$B = P(1 + r)^t, \quad \text{or equivalently,} \quad P = \frac{B}{(1 + r)^t}.$$

If the interest is compounded n times a year for t years at rate r, and if B is the *future value* of P after t years and P is the *present value* of B, then

$$B = P \left(1 + \frac{r}{n}\right)^{nt}, \quad \text{or equivalently,} \quad P = \frac{B}{\left(1 + r/n\right)^{nt}}.$$

Note that for a 7% interest rate, $r = 0.07$. In addition, as we see in Appendix B, when n tends towards infinity, we say that the compounding becomes continuous, and we get the following result:

$$B = Pe^{rt}, \quad \text{or equivalently,} \quad P = \frac{B}{e^{rt}} = Be^{-rt}.$$

Example 1   You win the lottery and are offered the choice between $1 million in four yearly installments of $250,000 each, starting now, and a lump-sum payment of $920,000 now. Assuming a 6% interest rate, compounded continuously, and ignoring taxes, which should you choose?

Solution   We will do the problem in two ways. First, we assume that you pick the option with the largest present value. The first of the four $250,000 payments is made now, so

Present value of first payment = $250,000.

The second payment is made one year from now and so

$$\text{Present value of second payment} = \$250{,}000e^{-0.06(1)}.$$

Calculating the present value of the third and fourth payments similarly, we find:

$$\begin{aligned}
\text{Total present value} &= \$250{,}000 + \$250{,}000e^{-0.06(1)} + \$250{,}000e^{-0.06(2)} + \$250{,}000e^{-0.06(3)} \\
&\approx \$250{,}000 + \$235{,}441 + \$221{,}730 + \$208{,}818 \\
&= \$915{,}989.
\end{aligned}$$

Since the present value of the four payments is less than $920,000, you are better off taking the $920,000 right now.

Alternatively, we can do the problem by comparing the future values of the two pay schemes. The scheme with the highest future value is the best from a purely financial point of view. We calculate the future value of both schemes three years from now, on the date of the last $250,000 payment. At that time,

$$\text{Future value of the lump sum payment} = \$920{,}000e^{0.06(3)} \approx \$1{,}101{,}440.$$

Now we calculate the future value of the first $250,000 payment:

$$\text{Future value of the first payment} = \$250{,}000e^{0.06(3)}.$$

Calculating the future value of the other payments similarly, we find:

$$\begin{aligned}
\text{Total future value} &= \$250{,}000e^{0.06(3)} + \$250{,}000e^{0.06(2)} + \$250{,}000e^{0.06(1)} + \$250{,}000 \\
&\approx \$299{,}304 + \$281{,}874 + \$265{,}459 + \$250{,}000 \\
&= \$1{,}096{,}637.
\end{aligned}$$

The future value of the $920,000 payment is greater, so you are better off taking the $920,000 right now. Of course, since the present value of the $920,000 payment is greater than the present value of the four separate payments, you would expect the future value of the $920,000 payment to be greater than the future value of the four separate payments.

(Note: If you read the fine print, you will find that many lotteries do not make their payments right away, but often spread them out, sometimes far into the future. This is to reduce the present value of the payments made, so that the value of the prizes is much less than it might first appear!)

Income Stream

When we consider payments made to or by an individual, we usually think of *discrete* payments, that is, payments made at specific moments in time. However, we may think of payments made by a company as being *continuous*. The revenues earned by a huge corporation, for example, come in essentially all the time, and therefore they can be represented by a continuous *income stream*. Since the rate at which revenue is earned may vary from time to time, the income stream is described by

$$P(t) \text{ dollars/year.}$$

Notice that $P(t)$ is a *rate* at which deposits are made (its units are dollars per year, for example) and that the rate depends on the time, t, usually measured in years from the present.

Present and Future Values of an Income Stream

Just as we can find the present and future values of a single payment, so we can find the present and future values of a stream of payments. As before, the future value represents the total amount of money obtained by depositing the income stream into a bank account and letting it gather interest. The present value represents the amount of money you would have to deposit today (in an interest-bearing bank account) in order to match what you would get from the income stream.

When we are working with a continuous income stream, we will assume that interest is compounded continuously. The reason for this is that the approximations we are going to make (of sums by integrals) are much simpler if both payments and interest are continuous.

Suppose that we want to calculate the present value of the income stream described by a rate of $P(t)$ dollars per year, and that we are interested in the period from now until T years in the future. In order to use what we know about single deposits to calculate the present values of an income stream, we must first divide the stream into many small deposits, each of which is made at approximately one instant. We divide the interval $0 \leq t \leq T$ into subintervals, each of length Δt:

Assuming Δt is small, the rate, $P(t)$, at which deposits are being made will not vary much within one subinterval. Thus, between t and $t + \Delta t$:

$$\text{Amount paid} \approx \text{Rate of deposits} \ \times \ \text{Time}$$
$$\approx (P(t) \text{ dollars/year})(\Delta t \text{ years})$$
$$= P(t)\Delta t \text{ dollars}.$$

Measured from the present, the deposit of $P(t)\Delta t$ is made t years in the future. Thus,

$$\begin{array}{l} \text{Present value of money} \\ \text{deposited in interval } t \text{ to } t + \Delta t \end{array} \approx P(t)\Delta t e^{-rt}.$$

Summing over all subintervals gives

$$\text{Total present value} \approx \sum P(t)e^{-rt}\Delta t \text{ dollars}.$$

In the limit as $\Delta t \to 0$, we get the following integral:

$$\boxed{\text{Present value} = \int_0^T P(t)e^{-rt}dt.}$$

In computing future value, the deposit of $P(t)\Delta t$ has a period of $(T - t)$ years to earn interest, and therefore

$$\text{Future value of money deposited in interval } t \text{ to } t + \Delta t \approx \left[P(t)\Delta t\right] e^{r(T-t)}.$$

Summing over all subintervals, we get:

$$\text{Total future value} \approx \sum P(t)\Delta t e^{r(T-t)} \text{ dollars}.$$

As the length of the subdivisions tends toward zero, the sum becomes an integral:

$$\boxed{\text{Future value} = \int_0^T P(t)e^{r(T-t)}dt.}$$

Example 2   Find the present and future values of a constant income stream of \$100 per year over a period of 20 years, assuming an interest rate of 10% compounded continuously.

Solution   Using $P(t) = 100$ and $r = 0.1$, we have

$$\text{Present value } = \int_0^{20} 100e^{-0.1t}dt = 100\left(-\frac{e^{-0.1t}}{0.1}\right)\Bigg|_0^{20} = 1000(1 - e^{-2}) \approx \$864.66.$$

$$\text{Future value } = \int_0^{20} 100e^{0.1(20-t)}dt = \int_0^{20} 100e^2e^{-0.1t}dt$$

$$= 100e^2\left(-\frac{e^{-0.1t}}{0.1}\right)\Bigg|_0^{20} = 1000e^2(1 - e^{-2}) \approx \$6389.06.$$

Example 3   What is the relationship between the present and future values in the previous example? Explain this relationship.

Solution   Since

$$\text{Present value } = 1000(1 - e^{-2}) \approx \$864.66$$
$$\text{Future value } = 1000e^2(1 - e^{-2}) \approx \$6389.06$$

you can see that

$$\text{Future value } = \text{(Present value)}e^2.$$

The reason for this is that the stream of payments is equivalent to a single payment of \$864.66 at time $t = 0$. With an interest rate of 10%, in 20 years that single payment will have grown to a future value of

$$864.66e^{0.1(20)} = 864.66e^2 \approx \$6389.02.$$

To within rounding error, this is the future value calculated previously.

Supply and Demand Curves

In a free market, the quantity of a certain item produced and sold can be described by the supply and demand curves of the item. The *supply curve* shows what quantity of the item the producers will supply at different price levels. It is usually assumed that as the price increases, the quantity supplied will increase. The consumers' behavior is reflected in the *demand curve*, which shows what quantity of goods are bought at various prices. An increase in price is usually assumed to cause a decrease in the quantity purchased. See Figure 8.36.

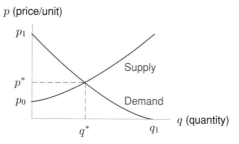

Figure 8.36: Supply and demand curves

It is assumed that the market will settle to the *equilibrium price* and *quantity*, p^* and q^*, where the graphs cross. This means that at the equilibrium point, a quantity q^* of an item will be produced and sold for a price of p^* each.

Consumer and Producer Surplus

Notice that at equilibrium, a number of consumers have bought the item at a lower price than they would have been willing to pay. (For example, there are some consumers who would have been willing to pay prices up to p_1.) Similarly, there are some suppliers who would have been willing to produce the item at a lower price (down to p_0, in fact). We define the following terms:

> • The **consumer surplus** measures the consumers' gain from trade. It is the total amount gained by consumers by buying the item at the current price rather than at the price they would have been willing to pay.
>
> • The **producer surplus** measures the suppliers' gain from trade. It is the total amount gained by producers by selling at the current price, rather than at the price they would have been willing to accept.
>
> In the absence of price controls, the current price is assumed to be the equilibrium price.

Both consumers and producers are richer for having traded. The consumer and producer surplus measure how much richer they are.

Suppose that all consumers buy the good at the maximum price they are willing to pay. Divide the interval from 0 to q^* into subintervals of length Δq. Figure 8.37 shows that a quantity Δq of items are sold at a price of about p_1, another Δq are sold for a slightly lower price of about p_2, the next Δq for a price of about p_3, and so on. Thus, the consumers' total expenditure is about

$$p_1\Delta q + p_2\Delta q + p_3\Delta q + \cdots = \sum p_i\Delta q.$$

If D is the demand function given by $p = D(q)$, and if all consumers who were willing to pay more than p^* paid as much as they were willing, then as $\Delta q \to 0$, we would have

$$\begin{array}{ll} \text{Consumer} \\ \text{expenditure} \end{array} = \int_0^{q^*} D(q)dq = \begin{array}{l} \text{Area under demand} \\ \text{curve from 0 to } q^*. \end{array}$$

Now if all goods are sold at the equilibrium price, the consumers' actual expenditure is p^*q^*, the area of the rectangle between the axes and the lines $q = q^*$ and $p = p^*$. Thus the consumer surplus may be calculated as follows:

$$\begin{array}{ll} \text{Consumer} \\ \text{surplus} \end{array} = \left(\int_0^{q^*} D(q)dq \right) - p^*q^* = \begin{array}{l} \text{Area under demand} \\ \text{curve above } p = p^*. \end{array}$$

Figure 8.37: Calculation of consumer surplus

Figure 8.38: Consumer and producer surplus

See Figure 8.38. Similarly, if the supply curve is given by the function $p = S(q)$, we define the producer surplus as follows:

$$\begin{array}{ll} \text{Producer} \\ \text{Surplus} \end{array} = p^*q^* - \left(\int_0^{q^*} S(q)dq \right) = \begin{array}{l} \text{Area between supply} \\ \text{curve and line } p = p^*. \end{array}$$

Problems for Section 8.4

1. Draw a graph, with time in years on the horizontal axis, of what an income stream might look like for a company that sells sunscreen in the northeast United States.

2. A business associate who owes you $3000 offers to pay you $2800 now, or else pay you three yearly installments of $1000 each, with the first installment paid now. If you use only financial reasons to make your decision, which option should you choose? Justify your answer, assuming a 6% market interest rate, compounded continuously.

3. Big Tree McGee is negotiating his rookie contract with a professional basketball team. They have agreed to a three-year deal which will pay Big Tree a fixed amount at the end of each of the three years, plus a signing bonus at the beginning of his first year. They are still haggling about the amounts and Big Tree must decide between a big signing bonus and fixed payments per year, or a smaller bonus with payments increasing each year. The two options are summarized in the table. All values are payments in millions of dollars.

	Signing bonus	Year 1	Year 2	Year 3
Option #1	6.0	2.0	2.0	2.0
Option #2	1.0	2.0	4.0	6.0

   (a) Big Tree decides to invest all income in stock funds which he expects to grow at a rate of 10% per year, compounded continuously. He would like to choose the contract option which gives him the greater future value at the end of the three years when the last payment is made. Which option should he choose? Justify your answer.

   (b) The local newspaper is running a story about Big Tree's contract. They wish to report the present value of each contract offer. Calculate these present values.

4. Find the present and future values of a constant income stream of $3000 per year over a period of 15 years, assuming a 6% annual interest rate compounded continuously.

5. (a) A bank account earns 10% interest compounded continuously. At what (constant, continuous) rate must a parent deposit money into such an account in order to save $100,000 in 10 years for a child's college expenses?

   (b) If the parent decides instead to deposit a lump sum now in order to attain the goal of $100,000 in 10 years, how much must be deposited now?

6. (a) If you deposit money continuously at a constant rate of $1000 per year into a bank account that earns 5% interest, how many years will it take for the balance to reach $10,000?

   (b) How many years would it take if the account had $2000 in it initially?

7. Suppose sales of Version 6.0 of a computer software package start out high and decrease exponentially. At time t, in years, the sales are $s(t) = 50e^{-t}$ thousands of dollars per year. After two years, Version 7.0 of the software is released and replaces Version 6.0. Assume that all income from software sales is immediately invested in government bonds which pay interest at a 6% rate compounded continuously, calculate the total value of sales of Version 6.0 over the two year period.

8. The value of good wine increases with age. Thus, if you are a wine dealer, you have the problem of deciding whether to sell your wine now, at a price of $\$P$ a bottle, or to sell it later at a higher price. Suppose you know that the amount a wine-drinker is willing to pay for a bottle of this wine t years from now is $\$P(1 + 20\sqrt{t})$. Assuming continuous compounding and a prevailing interest rate of 5% per year, when is the best time to sell your wine?

9. An oil company discovered an oil reserve of 100 million barrels. Suppose for time $t > 0$, in years, the company's extraction plan is a linear declining function of time as follows:

$$q(t) = a - bt,$$

where $q(t)$ is the rate of extraction of oil in millions of barrels per year at time t and $b = 0.1$ and $a = 10$.

   (a) How long does it take to exhaust the entire reserve?

   (b) If the oil price is a constant $20 per barrel, the extraction cost per barrel is a constant $10, and the market interest rate is 10%, what is the present value of the company's profit?

10. In 1980 West Germany made a loan of 20 billion Deutsche Marks to the Soviet Union, to be used for the construction of a natural gas pipeline connecting Siberia to Western Russia, and continuing to West Germany (Urengoi–Uschgorod–Berlin). Assume that the deal was as follows: In 1985, upon completion of the pipeline, the Soviet Union would deliver natural gas to West Germany, at a constant rate, for all future times. Assuming a constant price of natural gas of 0.10 Deutsche Mark per cubic meter, and assuming West Germany expects 10% annual interest on its investment (compounded continuously), at what rate does the Soviet Union have to deliver the gas, in billions of cubic meters per year? Keep in mind that delivery of gas could not begin until the pipeline was completed. Thus, West Germany received no return on its investment until after five years had passed. (Note: A more complex deal of this type was actually made between the two countries.)

11. In May 1991 *Car and Driver* described a Jaguar that sells for $980,000. At that price only 50 have been sold. It is estimated that 350 could have been sold if the price had been $560,000. Assuming that the demand curve is a straight line, and that $560,000 and 350 are the equilibrium price and quantity, find the consumer surplus.

12. Using Riemann sums, explain the economic significance of $\int_0^{q^*} S(q)\,dq$ to the producers.

13. Using Riemann sums, give an interpretation of producer surplus,

$$\int_0^{q^*} (p^* - S(q))\,dq$$

analogous to the interpretation of consumer surplus.

14. In Figure 8.38, page 402, find the regions with the following areas:

(a) $p^* q^*$ (b) $\int_0^{q^*} D(q)\,dq$ (c) $\int_0^{q^*} S(q)\,dq$

(d) $\left(\int_0^{q^*} D(q)\,dq\right) - p^* q^*$ (e) $p^* q^* - \int_0^{q^*} S(q)\,dq$ (f) $\int_0^{q^*} (D(q) - S(q))\,dq$

15. The dairy industry is an example of cartel pricing: the government has set milk prices artificially high. On a supply and demand graph, label p^+, a price above the equilibrium price. Using the graph, describe the effect of forcing the price up to p^+ on:

(a) The consumer surplus. (b) The producer surplus.

(c) The total gains from trade (Consumer surplus + Producer surplus).

16. Rent controls on apartments are an example of price controls on a commodity. They keep the price of the good artificially low (below the equilibrium price). Sketch a graph of supply and demand curves, and label on it a price p^- below the equilibrium price. What effect does forcing the price down to p^- have on:

(a) The producer surplus? (b) The consumer surplus?

(c) The total gains from trade (Consumer surplus + Producer surplus)?

CHAPTER SUMMARY

- **Geometry**
  Volume, arc length
- **Density**
  Finding total quantity from density, center of mass
- **Physics**
  Work, escape velocity, force and pressure
- **Economics**
  Present and future value of income stream, consumer and producer surplus

REVIEW PROBLEMS FOR CHAPTER EIGHT

1. (a) Sketch the solid obtained by rotating the region bounded by $y = \sqrt{x}$, $x = 1$, and $y = 0$ around the line $y = 0$.
   (b) Approximate its volume by Riemann sums, showing the volume represented by each term in your sum on the sketch.
   (c) Now find the volume of this solid using an integral.

2. Using the region of Problem 1, find the volume when it is rotated around
   (a) The line $y = 1$.   (b) The y-axis.

3. The reflector behind a car headlight is made in the shape of the parabola, $x = \frac{4}{9}y^2$, with a circular cross-section, as shown in Figure 8.39.

   (a) Find a Riemann sum approximating the volume contained by this headlight.
   (b) Find the volume exactly.

| *Figure 8.39* | *Figure 8.40* | *Figure 8.41* |

4. In this problem, you will derive the formula for the volume of a right circular cone with height l and base radius b by rotating the line $y = ax$ from $x = 0$ to $x = l$ around the x-axis. See Figure 8.40.

   (a) What value should you choose for a such that the cone will have height l and base radius b?
   (b) Given this value of a, find the volume of the cone.

5. Figure 8.41 shows a cross section through an apple. (Scale: One division = 1/2 inch.)

   (a) Give a rough estimate for the volume of this apple (in cubic inches).
   (b) The density of these apples is about 0.03 lb/in^3 (a little less than the density of water—as you might expect, since apples float). Estimate how much this apple would cost. (They go for 80 cents a pound.)

6. If the circle $x^2 + y^2 = 1$ is rotated about the line $y = 3$, it forms a torus (a doughnut-shaped figure). Find the volume of this torus.

For the curves described in Problems 7–8, write the integral that gives the exact length of the curve; you need not evaluate the integral.

7. One arch of the sine curve, from $x = 0$ to $x = \pi$.

8. The ellipse with equation $(x^2/a^2) + (y^2/b^2) = 1$.

9. A nuclear power plant produces strontium-90 at a rate of 1 kg/yr. How much of the strontium produced since 1971 (when the plant opened) was still around in 1992? (The half–life of strontium-90 is 28 years.)

10. A 200-lb weight is to be hoisted by a rope from the ground to a point on a flat roof 20 ft above the ground. If the rope weighs 2 lb/ft, find the work done in lifting the weight.

11. Water is to be raised from a well 40 ft deep by means of a bucket attached to a rope. When the bucket is full of water it weighs 30 lb, but the bucket has a leak that causes it to lose water at the constant rate of 1/4 lb for each foot that the bucket is raised. Neglecting the weight of the rope, find the work done in raising the bucket to the top.

12. A water tank is in the shape of a right circular cone with height 18 ft and radius 12 ft at the top. If it is filled with water to a depth of 15 ft, find the work done in pumping all of the water over the top of the tank. (The density of water is $\rho = 62.4$ lb/ft^3.)

13. The dam in Hannawa Falls, NY, on the Raquette River is approximately 60 feet across and 25 feet high. Find the water force on the dam. (The density of water is $\rho = 62.4$ lb/ft^3.)

14. (a) Find the present and future values of a constant income stream of $100 per year over a period of 20 years, assuming a 10% annual interest rate compounded continuously.
    (b) How many years will it take for the balance to reach $5000?

15. Suppose you are manufacturing a particular item, and that the rate at which you earn a profit on the item is decreasing with time according to the formula

    $$\text{Rate profit earned} = (2 - 0.1t) \text{ thousand dollars per year}$$

    after t years. (A negative profit represents a loss.) If interest is 10%, compounded continuously:

    (a) Write a Riemann sum approximating the present value of the total profit earned up to a time T years in the future.
    (b) Write an integral representing the present value in part (a). (You need not evaluate this integral.)
    (c) For what T is the present value of the stream of profits on this item maximized? What is the present value of the total profit earned up to that time?

16. A car moving at a speed of v mph achieves $25 + 0.1v$ mpg (miles per gallon) for v between 20 and 60 mph. Suppose your speed as a function of time, t, in hours, is given by

    $$v = 50\frac{t}{t+1} \text{ mph.}$$

    How many gallons do you consume between $t = 2$ and $t = 3$?

17. Mt. Shasta is a cone-like volcano whose radius at an elevation of h feet above sea level is approximately $(3.5 \cdot 10^5)/\sqrt{h + 600}$ feet. Its bottom is 400 feet above sea level, and its top is 14,400 feet above sea level. See Figure 8.42.

    (a) Give a Riemann sum approximating the volume of Mt. Shasta.
    (b) Find the volume in cubic feet. (Note: This data about Mt. Shasta is more or less accurate. Mt. Shasta lies in northern California, and for some time was thought to be the highest point in the US outside Alaska.)

Figure 8.42

Figure 8.43

Figure 8.44

18. Figure 8.43 shows an ancient Greek water clock called a clepsydra, which is designed so that the depth of the water decreases at a constant rate as the water runs out a hole in the bottom. This design allows the hours to be marked by a uniform scale. The tank of the clepsydra is a volume of revolution about a vertical axis. According to Torricelli's law, the exit speed of the water flowing through the hole is proportional to the square root of the depth of the water. Use this to find the formula $y = f(x)$ for this profile, assuming that $f(1) = 1$.

19. Find the volume of the snowman in Figure 8.44. If x and y are in meters, and the origin is on the ground, the x-axis is horizontal and the y-axis is vertical, then the body is approximated by rotating the curve $y = 1 - 4x^2$ about the y-axis. The neck is a cylinder of radius 0.1 meter and length 0.15 meter; the head is spherical with radius 0.2 meter.

20. A blood vessel is cylindrical with radius R and length l. The blood near the boundary moves slowly; blood at the center moves the fastest. Assume that the velocity, v, of the blood at a distance r from the

center of the artery is given by

$$v = \frac{P}{4\eta l}(R^2 - r^2)$$

where P is the pressure difference between the ends of the blood vessel and η is the viscosity of blood.

(a) Find the rate at which the blood is flowing down the blood vessel. (Give your answer as a volume per unit time.)

(b) Show that your result agrees with Poiseuille's Law which says that the rate at which blood is flowing down the blood vessel is proportional to the radius of the blood vessel to the fourth power.

21. Consider a bowl made by rotating the curve $y = ax^2$ around the y-axis (a is a constant).

(a) Assume the bowl is filled with water to depth h. What is the volume of water in the bowl? (Your answer will contain a and h.)

(b) What is the area of the surface of the water if the bowl is filled to depth h? (Your answer will contain a and h.)

(c) Suppose water is evaporating from the surface of the bowl at a rate proportional to the surface area, with proportionality constant k. Find a differential equation satisfied by h as a function of time, t. (That is, find an equation for dh/dt.)

(d) If the water starts at depth h_0, find the time taken for all the water to evaporate.

22. A cylindrical centrifuge of radius 1 m and height 2 m is filled with water to a depth of 1 meter (see Figure 8.45(I)). As the centrifuge accelerates, the water level rises along the wall and drops in the center; the cross-section will be a parabola. (See Figure 8.45(II).)

(a) Find the equation of the parabola in Figure 8.45(II) in terms of h, the depth of the water at its lowest point.

(b) As the centrifuge rotates faster and faster, either water will be spilled out the top, as in Figure 8.45(III), or the bottom of the centrifuge will be exposed, as in Figure 8.45(IV). Which happens first?

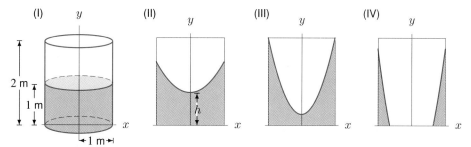

Figure 8.45

Problems 23–24 concern Figures 8.46 and 8.47. In each problem, you are given two objects which have the same mass M, the same radius R, and the same angular velocity about the indicated axes (say, one revolution per minute). For each problem, determine which object has the greater kinetic energy. (The kinetic energy of a particle of mass m with speed v is $\frac{1}{2}mv^2$.) Don't attempt to compute the kinetic energy of the objects to do this; just use reasoning.

23. **24.**

Figure 8.46   *Figure 8.47*

PROJECTS

1. **Volume Enclosed by Two Cylinders**

    Two cylinders are inscribed in a cube of side length 2, as shown in Figure 8.48. What is the volume of the solid that the two cylinders enclose? [Hint: Use horizontal slices.] Note: The solution was known to Archimedes. The Chinese mathematician Liu Hui (third century A.D.) tried to find this volume, but he failed; he wrote a poem about his efforts calling the enclosed volume a "box-lid":

    > Look inside the cube
    > And outside the box-lid;
    > Though the dimension increases,
    > It doesn't quite fit.
    > The marriage preparations are complete;
    > But square and circle wrangle,
    > Thick and thin are treacherous plots,
    > They are incompatible.
    > I wish to give my humble reflections,
    > But fear that I will miss the correct principle;
    > I dare to let the doubtful points stand,
    > Waiting
    > For one who can expound them.

    Figure 8.48

2. **Length of a Hanging Cable**

    The distance between the towers of the main span of the Golden Gate Bridge is about 1280 m; the sag of the cable halfway between the towers on a cold winter day is about 143 m. See Figure 8.49.

    Figure 8.49

    (a) How long is the cable, assuming it has an approximately parabolic shape? (Represent the cable as a parabola of the form $y = kx^2$ and determine k to at least 1 decimal place.)

    (b) On a hot summer day the cable is about 0.05% longer, due to thermal expansion. By how much does the sag increase? Assume no movement of the towers.

FOCUS ON MODELING

DISTRIBUTION FUNCTIONS

Understanding the distribution of various quantities through the population can be important to decision makers. For example, the income distribution gives useful information about the economic structure of a society. In this section we will look at the distribution of ages in the US. To allocate funding for education, health care, and social security, the government needs to know how many people are in each age group. We will see how to represent such information by a density function.

US Age Distribution

TABLE 8.1 *Distribution of ages in the US in 1995*

Age group	Percentage of total population
0–20	29%
20–40	31%
40–60	24%
60–80	13%
80–100	3%

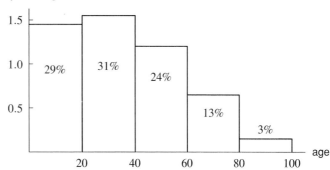

Figure 8.50: How ages were distributed in the US in 1995

Suppose we have the data in Table 8.1 showing how the ages of the US population were distributed in 1995. To represent this information graphically we use a type[1] of *histogram*, putting a vertical bar above each age group in such a way that the *area* of each bar represents the percentage in that age group. The total area of all the rectangles is 100% = 1. We only consider people who are less than 100 years old.[2] For the 0–20 age group, the base of the rectangle is 20, and we want the area to be 29%, so the height must be 29%/20 = 1.45%. We treat ages as though they were continuously distributed. The category 0–20, for example, contains people who are just one day short of their twentieth birthday. Notice that the vertical axis is measured in percent/year. (See Figure 8.50.)

Example 1   In 1995, estimate what percentage of the US population was:
  (a) Between 20 and 60 years old?   (b) Less than 10 years old?
  (c) Between 75 and 80 years old?   (d) Between 80 and 85 years old?

Solution   (a) We add the percentages, so 31% + 24% = 55%.
  (b) To find the percentage less than 10 years old, we could assume, for example, that the population was distributed evenly over the 0–20 group. (This means we are assuming that babies were born at a fairly constant rate over the last 20 years, which is probably reasonable.) If we make this assumption, then we can say that the population less than 10 years old was about half that in the 0–20 group, that is, 14.5%. Notice that we get the same result by computing the area of the rectangle from 0 to 10. (See Figure 8.51.)

---

[1]There are other types of histogram which have frequency on the vertical axis.
[2]In fact, 0.02% of the population is over 100, but this is too small to be visible on the histogram.

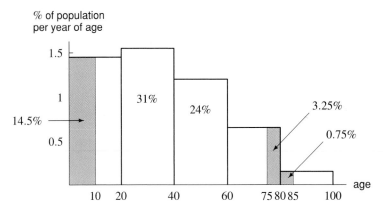

Figure 8.51: Ages in the US in 1995 — various subgroups (for Example 1)

(c) To find the population between 75 and 80 years old, since 13% of Americans in 1990 were in the 60-80 group, we might apply the same reasoning and say that $\frac{1}{4}(13\%) = 3.25\%$ of the population was in this age group. This result is represented as an area in Figure 8.51. The assumption that the population was evenly distributed is not a good one here; certainly there were more people between the ages of 60 and 65 than between 75 and 80. Thus, the estimate of 3.25% is certainly too high.

(d) Again using the (faulty) assumption that ages in each group were distributed uniformly, we would find that the percentage between 80 and 85 was $\frac{1}{4}(3\%) = 0.75\%$. (See Figure 8.51.) This estimate is also poor—there were certainly more people in the 80–85 group than, say, the 95–100 group, and so the 0.75% estimate is too low.

Smoothing Out the Histogram

We could get better estimates if we had smaller age groups (each age group in Figure 8.50 is 20 years, which is quite large) or if the histogram were smoother. Suppose we have the more detailed data in Table 8.2, which leads to the new histogram in Figure 8.52. As we get more detailed information, the upper silhouette of the histogram becomes smoother, but the area of any of the bars still represents the percentage of the population in that age group. Imagine, in the limit, replacing the upper silhouette of the histogram by a smooth curve in such a way that area under the curve above one age group is the same as the area in the corresponding rectangle. The total area under the whole curve is again 100% = 1. (See Figure 8.52.)

The Age Density Function

If t is age in years, we define $p(t)$, the age *density function*, to be a function which "smooths out" the age histogram. This function has the property that

$$\begin{array}{c} \text{Fraction of population} \\ \text{between ages } a \text{ and } b \end{array} = \begin{array}{c} \text{Area under} \\ \text{graph of } p \\ \text{between } a \text{ and } b \end{array} = \int_a^b p(t)\,dt.$$

If a and b are the smallest and largest possible ages (say, $a = 0$ and $b = 100$), so that the ages of all of the population are between a and b, then

$$\int_a^b p(t)\,dt = \int_0^{100} p(t)\,dt = 1.$$

What does the age density function p tell us? Notice that we have not talked about the meaning of $p(t)$ itself, but *only* of the integral $\int_a^b p(t)\,dt$. Let's look at this in a bit more detail. Suppose, for

TABLE 8.2 *Ages in the US in 1995 (more detailed)*

Age group	Percentage of total population
0–10	15%
10–20	14%
20–30	14%
30–40	17%
40–50	14%
50–60	10%
60–70	8%
70–80	5%
80–90	2%
90–100	1%

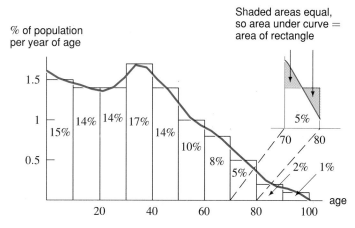

Figure 8.52: Smoothing out the age histogram

example, that $p(10) = 0.015 = 1.5\%$ per year. This is *not* telling us that 1.5% of the population is precisely 10 years old (where 10 years old means exactly 10, not $10\frac{1}{2}$, not $10\frac{1}{4}$, not 10.1). However, $p(10) = 0.015$ does tell us that for some small interval Δt around 10, the fraction of the population with ages in this interval is approximately $p(10)\,\Delta t = 0.015\,\Delta t$. Notice also that the units of $p(t)$ are *% per year*, so $p(t)$ must be multiplied by years to give a percentage of the population.

The Density Function

Suppose we are interested in how a certain characteristic, x, is distributed through a population. For example, x might be height or age if the population is people, or might be wattage for a population of light bulbs. Then we define a general density function with the following properties:

> The function, $p(x)$, is a **density function** if
>
> $$\begin{array}{ccc} \text{Fraction of population} & & \text{Area under} \\ \text{for which } x \text{ is} & = & \text{graph of } p & = \int_a^b p(x)\,dx. \\ \text{between } a \text{ and } b & & \text{between } a \text{ and } b \end{array}$$
>
> $$\int_{-\infty}^{\infty} p(x)\,dx = 1 \quad \text{and} \quad p(x) \geq 0 \quad \text{for all } x.$$

The density function must be nonnegative if its integral always gives a fraction of the population. Also, the fraction of the population with x between $-\infty$ and ∞ is 1 because the entire population has the characteristic x between $-\infty$ and ∞. The function p that was used to smooth out the age histogram satisfies this definition of a density function. We do not assign a meaning to the value $p(x)$ directly, but rather interpret $p(x)\,\Delta x$ as the fraction of the population with the characteristic in a short interval of length Δx around x.

The density function is often approximated by formulas, as in the next example.

Example 2   Find formulas to approximate the density function, p, for the US age distribution. Assume that p is continuous, that it is constant at 1.5% up to age 40 and then drops linearly.

Solution   We have $p(t) = 0.015$ for $0 \leq t < 40$. For $t \geq 40$, we need a linear function sloping downward. Because p is continuous, we have $p(40) = 1.5\% = 0.015$. Because p is a density function we have $\int_0^{100} p(t)dt = 1$. Suppose b is as in Figure 8.53 then

$$\int_0^{100} p(t)dt = \int_0^{40} p(t)dt + \int_{40}^{100} p(t)dt = 40(0.015) + \frac{1}{2}(0.015)b = 1,$$

where $\int_{40}^{100} p(t)dt$ is given by the area of the triangle. This gives

$$\frac{0.015}{2}b = 0.4, \quad \text{and so} \quad b \approx 53.3.$$

Thus the slope of the line is $-0.015/53.3 \approx -0.00028$, so for $40 \leq t \leq 40 + 53.3 = 93.3$, so

$$p(t) - 0.015 = -0.00028(t - 40),$$
$$p(t) = 0.0262 - 0.00028t.$$

According to this way of smoothing the data, there is no one over 93.3 years old, so $p(t) = 0$ for $t > 93.3$.

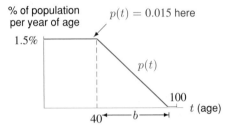

Figure 8.53: Age density function

Cumulative Distribution Function for Ages

Another way of showing how ages are distributed in the US is by using the *cumulative distribution function* $P(t)$, defined by

$$P(t) = \frac{\text{Fraction of population}}{\text{of age less than } t} = \int_0^t p(x)dx.$$

Thus, P is the antiderivative of p with $P(0) = 0$, and $P(t)$ gives the area under the density curve between 0 and t.

Notice that the cumulative distribution function is nonnegative and increasing (or at least nondecreasing), since the number of people younger than age t increases as t increases. Another way of seeing this is to notice that $P' = p$, and p is positive (or nonnegative). Thus the cumulative age distribution is a function which starts with $P(0) = 0$ and increases as t increases. $P(t) = 0$ for $t < 0$ because, when $t < 0$, there is no one whose age is less than t. The limiting value of P, as $t \to \infty$, is 1 since as t becomes very large (100 say), everyone will be younger than age t, so the fraction of people with age less than t tends towards 1. (See Figure 8.54.) Notice that for t less than 40, the graph of P is a straight line, because p is constant there. For $t > 40$, the graph of P levels off as p tends to 0.

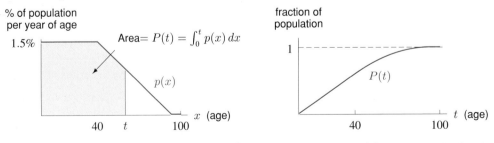

Figure 8.54: $P(t)$, the cumulative age distribution function, and its relation to $p(x)$, the age density function

Cumulative Distribution Function

A **cumulative distribution function**, $P(t)$, of a density function p, is defined by

$$P(t) = \int_{-\infty}^{t} p(x)\, dx = \begin{array}{l}\text{Fraction of population having} \\ \text{values of } x \text{ below } t.\end{array}$$

Thus, P is an antiderivative of p, that is, $P' = p$.

Any cumulative distribution has the following properties:

- P is increasing (or nondecreasing).
- $\lim\limits_{t \to \infty} P(t) = 1$ and $\lim\limits_{t \to -\infty} P(t) = 0.$
- $\begin{array}{l}\text{Fraction of population} \\ \text{having values of } x \\ \text{between } a \text{ and } b\end{array} = \int_{a}^{b} p(x)\, dx = P(b) - P(a).$

Problems on Distribution Functions

In Problems 1–3, sketch graphs of a density function and a cumulative distribution function which could represent the distribution of income through a population with the given characteristics.

1. A large middle class.

2. Small middle and upper classes and many poor people.

3. Small middle class, many poor and many rich people.

4. A large number of people take a standardized test, receiving scores described by the density function p graphed in Figure 8.55. Does the density function imply that most people receive a score near 50? Explain why or why not.

fraction of students per test score

Figure 8.55

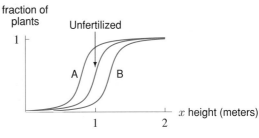

fraction of plants

Figure 8.56

5. An experiment is done to determine the effect of two new fertilizers A and B on the growth of a species of peas. The cumulative distribution functions of the heights of the mature peas without treatment and treated with each of A and B are graphed in Figure 8.56.

    (a) About what height are most of the unfertilized plants?
    (b) Explain in words the effect of the fertilizers A and B on the mature height of the plants.

6. Suppose $F(x)$ is the cumulative distribution function for heights (in meters) of trees in a forest.

    (a) Explain in terms of trees the meaning of the statement $F(7) = 0.6$.
    (b) Which is greater, $F(6)$ or $F(7)$? Justify your answer in terms of trees.

7. Suppose that $p(x)$ is the density function for heights of American men in inches. What is the meaning of the statement $p(68) = 0.2$?

8. Suppose $P(t)$ is the fraction of the US population of age less than t. Using Table 8.2 on page 411, make a table of values for $P(t)$.

9. Figure 8.57 shows a density function and the corresponding cumulative distribution function.[3]

    (a) Which curve represents the density distribution functions and which represents the cumulative distribution? Give a reason for your choice.

    (b) Put reasonable values on the tick marks on each of the axes.

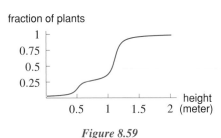

Figure 8.57

Figure 8.58

Figure 8.59

10. The heights of grass plants in a meadow were measured and the density function and cumulative distribution function are graphed in Figure 8.58 and Figure 8.59, respectively.

    (a) You know that there are two species of grass in the meadow, a short grass and a tall grass. Explain how the graph of the density function reflects this fact.

    (b) Explain how the graph of the cumulative distribution functions reflects the fact that there are two species of grass in the meadow.

    (c) About what percentage of the grasses in the meadow belong to the short grass species?

11. After measuring the duration of many telephone calls, the telephone company found their data was well-approximated by the density function $p(x) = 0.4e^{-0.4x}$, where x is the duration of a call, in minutes.

    (a) What percentage of calls last between 1 and 2 minutes?

    (b) What percentage of calls last 1 minute or less?

    (c) What percentage of calls last 3 minutes or more?

    (d) Find the cumulative distribution function.

12. Consider a population of individuals with a disease. Suppose that t is the number of years since the onset of the disease. The death density function, $f(t) = cte^{-kt}$, approximates the fraction of the sick individuals who die in the time interval $[t, t + \Delta t]$ as follows:

$$\text{Fraction who die} \approx f(t)\Delta t = cte^{-kt}\Delta t$$

    where c and k are positive constants whose values depend on the particular disease.

    (a) Find the value of c in terms of k.

    (b) If 40% of the population dies within 5 years, find c and k.

    (c) Find the cumulative death distribution function, $C(t)$. Give your answer in terms of k.

13. Suppose that the time to conduct a routine maintenance check has a cumulative distribution function, $P(t)$, where $P(t)$ gives the fraction of maintenance checks completed in less than or equal to t minutes. Values of $P(t)$ are in Table 8.3.

TABLE 8.3

t (minutes)	0	5	10	15	20	25	30
$P(t)$ (fraction completed)	0	0.03	0.08	0.21	0.38	0.80	0.98

    (a) What fraction of maintenance checks is completed in 15 minutes or less?

    (b) What fraction of maintenance checks takes longer than 30 minutes?

    (c) What fraction takes between 10 and 15 minutes?

    (d) Draw a histogram showing how times for routine maintenance checks are distributed.

    (e) In which of the given 5-minute intervals is the length of a maintenance check most likely to fall?

    (f) Give a rough sketch of the density function.

    (g) Sketch a graph of the cumulative distribution function.

[3] Adapted from *Calculus*, by David A. Smith and Lawerence C. Moore (Lexington: D.C. Heath, 1994).

14. Students at the University of California were surveyed and asked their grade point average . (The GPA ranges from 0 to 4, where 2 is just passing.) The distribution of GPAs is shown in Figure 8.60.[4]

    (a) Roughly what fraction of students are passing?
    (b) Roughly what fraction of the students have honor grades (GPAs above 3)?
    (c) Why do you think there is a peak around 2?
    (d) Sketch a graph of the cumulative distribution function.

fraction of students
per GPA

Figure 8.60

fraction of earth's surface
per mile of elevation

Figure 8.61

15. Figure 8.61[4] shows the distribution of elevation, in miles, across the earth's surface. Positive elevation denotes land above sea level; negative elevation shows land below sea level (i.e., the ocean floor).

    (a) Describe in words the elevation of most of the earth's surface.
    (b) Approximately what fraction of the earth's surface is below sea level?

16. A person who travels regularly on the 9:00 am bus from Oakland to San Francisco reports that the bus is almost always a few minutes late but rarely more than five minutes late. We are told further that the bus is never more than two minutes early, although it is on very rare occasions a little early.

    (a) Sketch a possible density function, $p(t)$, where t is the number of minutes that the bus is late. Shade the region under the graph between $t = 2$ minutes and $t = 4$ minutes. Explain what this region represents.
    (b) Now sketch the cumulative distribution function $P(t)$ for this situation. What measurement(s) on this graph correspond to the area shaded? What do the inflection point(s) on your graph of P correspond to on the graph of p? How can you interpret the inflection points on the graph of P without referring to the graph of p?

17. Suppose the density function for radii r (mm) of spherical raindrops during a storm is constant over the range $0 < r < 5$ and zero elsewhere.

    (a) Find the density function $f(r)$ for the radii.
    (b) Find the cumulative distribution function $F(r)$ for the radii.
    (c) Find the cumulative distribution function $G(v)$ for the volumes v (mm^3) of the raindrops.
    (d) Find the density function $g(v)$ for the volumes.

18. Consider a pendulum swinging through a small angle. The x-coordinate of the bob moves between $-a$ and a, as shown in Figure 8.62.

    (a) Draw the density function for the location of the x-coordinate of the pendulum bob (i.e., neglect up-and-down motion). In order to do this, imagine a camera taking pictures of the pendulum at random instants. Where is the bob most likely to be found? Least likely? [Hint: Consider the speed of the pendulum at different points in its path. Is the camera more likely to take a photograph of the bob at a point on its path where the bob is moving quickly, or where it is moving slowly?]

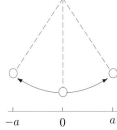

Figure 8.62

    (b) Now sketch the function

$$f(x) = \begin{cases} \dfrac{1}{\pi\sqrt{a^2 - x^2}} & -a < x < a; \\ 0 & |x| \geq a. \end{cases}$$

    How does this graph compare with the one you drew in part (a)?

---

[4]Adapted from *Statistics*, by Freedman, Pisani, Purves, and Adikhari (New York: Norton, 1991).

(c) Assuming that the function given in part (b) is the density function for the pendulum, what do you expect

$$\int_{-a}^{a} \frac{1}{\pi \sqrt{a^2 - x^2}}\, dx$$

to be? Check this by computing the integral.

(d) Does it seem reasonable, physically speaking, that $f(x)$ "blows up" at a and $-a$? Explain.

PROBABILITY AND MORE ON DISTRIBUTIONS

Probability

Suppose we pick a member of the US population at random and ask what is the probability that the person is between, say, the ages of 70 and 80. We saw in Table 8.2 on page 411 that 5% of the population is in this age group. We say that the probability, or chance, that the person is between 70 and 80 is 0.05. Using any age density function $p(t)$, we can define probabilities as follows:

$$\begin{array}{c} \text{Probability that} \\ \text{a person is between} \\ \text{ages } a \text{ and } b \end{array} = \begin{array}{c} \text{Fraction of population} \\ \text{between ages } a \text{ and } b \end{array} = \int_{a}^{b} p(t)\, dt.$$

Since the cumulative distribution function gives the fraction of the population younger than age t, the cumulative distribution can also be used to calculate the probability that a randomly selected person is in a given age group.

$$\begin{array}{c} \text{Probability that} \\ \text{a person is younger} \\ \text{than age } t \end{array} = \begin{array}{c} \text{Fraction of population} \\ \text{younger than age } t \end{array} = P(t) = \int_{0}^{t} p(x)\, dx.$$

In the next example, both a density function and a cumulative distribution function are used to describe the same situation.

Example 3   Suppose you want to analyze the fishing industry in a small town. Each day, the boats bring back at least 2 tons of fish, and never more than 8 tons.

(a) Using the density function describing the daily catch in Figure 8.63, find and graph the corresponding cumulative distribution function and explain its meaning.

(b) What is the probability that the catch will be between 5 and 7 tons?

Figure 8.63: Density function of daily catch

Solution   (a)   The cumulative distribution function $P(t)$ is equal to the fraction of days on which the catch is less than t tons of fish. Since the catch is never less than 2 tons, we have $P(t) = 0$ for $t \leq 2$. Since the catch is always less than 8 tons, we have $P(t) = 1$ for $t \geq 8$. For t in the range $2 < t < 8$, we must evaluate the integral

$$P(t) = \int_{-\infty}^{t} p(x)dx = \int_{2}^{t} p(x)dx.$$

This integral equals the area under the graph of $p(x)$ between $x = 2$ and $x = t$. It can be calculated by noting that $p(x)$ is given by the formula

$$p(x) = \begin{cases} 0.04x & \text{for } 2 \leq x \leq 6 \\ -0.06x + 0.6 & \text{for } 6 < x \leq 8 \end{cases}$$

and $p(x) = 0$ for $x < 2$ or $x > 8$. Thus, for $2 \leq t \leq 6$,

$$P(t) = \int_{2}^{t} 0.04x \, dx = 0.04 \frac{x^2}{2} \bigg|_{2}^{t} = 0.02t^2 - 0.08.$$

And for $6 \leq t \leq 8$,

$$P(t) = \int_{2}^{t} p(x) \, dx = \int_{2}^{6} p(x) \, dx + \int_{6}^{t} p(x) \, dx$$

$$= 0.64 + \int_{6}^{t} (-0.06x + 0.6) \, dx = 0.64 + \left(-0.06 \frac{x^2}{2} + 0.6x \right) \bigg|_{6}^{t}$$

$$= -0.03t^2 + 0.6t - 1.88.$$

Thus

$$P(t) = \begin{cases} 0.02t^2 - 0.08 & \text{for } 2 \leq t \leq 6 \\ -0.03t^2 + 0.6t - 1.88 & \text{for } 6 < t \leq 8. \end{cases}$$

In addition $P(t) = 0$ for $t < 2$ and $P(t) = 1$ for $8 < t$. (See Figure 8.64.)

fraction of days

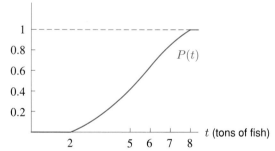

Figure 8.64: Cumulative distribution of daily catch

fraction of days per
ton of caught fish

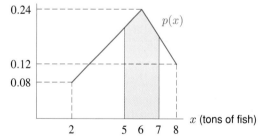

Figure 8.65: Shaded area represents the probability the catch is between 5 and 7 tons

(b)   The probability that the catch is between 5 and 7 tons can be found using either the density function, p, or the cumulative distribution function, P. If we use the density function, this probability can be represented by the shaded area in Figure 8.65, which is about 0.43.

$$\text{Probability catch is between 5 and 7 tons} = \int_{5}^{7} p(x) \, dx = 0.43.$$

The probability can be found from the cumulative distribution as follows:

$$\text{Probability catch is between 5 and 7 tons} = P(7) - P(5) = 0.85 - 0.42 = 0.43.$$

The Median and Mean

It is often useful to be able to give an "average" value for a distribution. Two measures that are in common use are the *median* and the *mean*.

The Median

A **median** of a quantity x distributed through a population is a value T such that half the population has values of x less than (or equal to) T, and half the population has values of x greater than (or equal to) T. Thus, a median T satisfies

$$\int_{-\infty}^{T} p(x)\, dx = 0.5,$$

where p is the density function. In other words, half the area under the graph of p lies to the left of T.

Example 4   Find the median age in the US in 1995, using the age density function given by

$$p(t) = \begin{cases} 0.015 & \text{for } 0 \le t \le 40 \\ 0.0262 - 0.00028t & \text{for } 40 < t \le 93.3. \end{cases}$$

Solution   We want to find the value of T such that

$$\int_{-\infty}^{T} p(t)\, dt = \int_{0}^{T} p(t)\, dt = 0.5.$$

Since $p(t) = 1.5\%$ up to age 40, we have

$$\text{Median} = T = \frac{50\%}{1.5\%} \approx 33 \text{ years.}$$

(See Figure 8.66.)

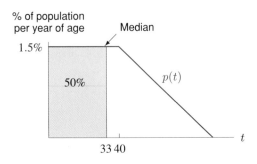

Figure 8.66: Median of age distribution

The Mean

Another commonly used average value is the *mean*. To find the mean of N numbers, you add the numbers and divide the sum by N. For example, the mean of the numbers 1, 2, 7, and 10 is $(1 + 2 + 7 + 10)/4 = 5$. The mean age of the entire US population is therefore defined as

$$\text{Mean age} = \frac{\sum \text{Ages of all people in the US}}{\text{Total number of people in the US}}.$$

Calculating the sum of all the ages directly would be an enormous task; we will approximate the sum by an integral. The idea is to "slice up" the age axis and consider the people whose age is between t and $t + \Delta t$. How many are there?

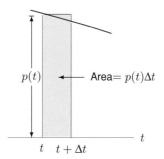

Figure 8.67: Shaded area is
percentage of population with age
between t and $t + \Delta t$

The fraction of the population between t and $t + \Delta t$ is the area under the graph of p between these points, which is well approximated by the area of the rectangle, $p(t)\Delta t$. (See Figure 8.67.) If the total number of people in the population is N, then

$$\begin{array}{c} \text{Number of people with age} \\ \text{between } t \text{ and } t + \Delta t \end{array} \approx p(t)\Delta t N.$$

The age of all of these people is approximately t:

$$\begin{array}{c} \text{Sum of ages of people} \\ \text{between age } t \text{ and } t + \Delta t \end{array} \approx tp(t)\Delta t N.$$

Therefore, adding and factoring out an N gives us

$$\text{Sum of ages of all people} \approx \left(\sum tp(t)\Delta t \right) N.$$

In the limit, as we allow Δt to shrink to 0, the sum becomes an integral, so

$$\text{Sum of ages of all people} = \left(\int_0^{100} tp(t)\,dt \right) N.$$

Therefore, with N equal to the total number of people in the US, and assuming no person is over 100 years old,

$$\text{Mean age} = \frac{\text{Sum of ages of all people in US}}{N} = \int_0^{100} tp(t)\,dt.$$

We can give the same argument for any[5] density function $p(x)$.

> If a quantity has density function $p(x)$,
>
> $$\textbf{Mean value} \text{ of the quantity} = \int_{-\infty}^{\infty} xp(x)\,dx.$$

It can be shown that the mean is the point on the horizontal axis where the region under the graph of the density function, if it were made out of cardboard, would balance.

---

[5]Provided all the relevant improper integrals converge.

Example 5   Find the mean age of the US population, using the density function of Example 4.

Solution   The formula for p is

$$p(t) = \begin{cases} 0.015 & \text{for } 0 \leq t \leq 40 \\ 0.0262 - 0.00028t & \text{for } 40 < t \leq 93.3. \end{cases}$$

Using these formulas, we compute

$$\text{Mean age} = \int_0^{100} t p(t) dt = \int_0^{40} t(0.015) dt + \int_{40}^{93.3} t(0.0262 - 0.00028t) dt$$

$$= 0.015 \frac{t^2}{2} \Big|_0^{40} + 0.0262 \frac{t^2}{2} \Big|_{40}^{93.3} - 0.00028 \frac{t^3}{3} \Big|_{40}^{93.3} \approx 35 \text{ years.}$$

The mean is shown is Figure 8.68.

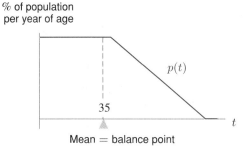

Figure 8.68: Mean of age distribution

Normal Distributions

How much rain do you expect will fall in your home town this year? If you live in Anchorage, Alaska, the answer is something close to 15 inches (including the snow). Of course, you don't expect exactly 15 inches. Some years there will be more than 15 inches, and some years there will be less. Most years, however, the amount of rainfall will be close to 15 inches; only rarely will it be well above or well below 15 inches. What does the density function for the rainfall look like? To answer this question, we look at rainfall data over many years. Records show that the distribution of rainfall is well-approximated by a *normal distribution*.Its graph is a bell-shaped curve which peaks at 15 inches and slopes downward approximately symmetrically on either side.

Normal distributions are frequently used to model real phenomena, from grades on an exam to the number of airline passengers on a particular flight. A normal distribution is characterized by its *mean*, μ, and its *standard deviation*, σ. The mean tells us the location of the central peak. The standard deviation tells us how closely the data is clustered around the mean. A small value of σ tells us that the data is close to the mean; a large σ tells us the data is spread out. The formula for a normal distribution is as follows. The factor of $1/(\sigma\sqrt{2\pi})$ makes the area under the graph equal to 1.

A **normal distribution** has a density function of the form

$$p(x) = \frac{1}{\sigma\sqrt{2\pi}} e^{-(x-\mu)^2/(2\sigma^2)},$$

where μ is the mean of the distribution and σ is the standard deviation, with $\sigma > 0$.

To model the rainfall in Anchorage, we use a normal distribution with $\mu = 15$. The standard deviation can be estimated by looking at the data; we will take it to be 1. (See Figure 8.69.)

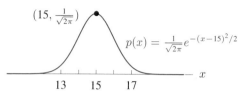

Figure 8.69: Normal distribution with $\mu = 15$ and $\sigma = 1$

Example 6 For Anchorage's rainfall, use the normal distribution with the density function with $\mu = 15$ and $\sigma = 1$ to compute the fraction of the years with rainfall between
(a) 14 and 16 inches,   (b) 13 and 17 inches,   (c) 12 and 18 inches.

Solution (a) The fraction of the years with annual rainfall between 14 and 16 inches is $\int_{14}^{16} \frac{1}{\sqrt{2\pi}} e^{-(x-15)^2/2}\,dx$. Since there is no elementary antiderivative for $e^{-(x-15)^2/2}$, we find the integral numerically. Its value is about 0.68.

$$\text{Fraction of years with rainfall between 14 and 16 inches} = \int_{14}^{16} \frac{1}{\sqrt{2\pi}} e^{-(x-15)^2/2}\,dx \approx 0.68.$$

(b) Finding the integral numerically again:

$$\text{Fraction of years with rainfall between 13 and 17 inches} = \int_{13}^{17} \frac{1}{\sqrt{2\pi}} e^{-(x-15)^2/2}\,dx \approx 0.95.$$

(c)

$$\text{Fraction of years with rainfall between 12 and 18 inches} = \int_{12}^{18} \frac{1}{\sqrt{2\pi}} e^{-(x-15)^2/2}\,dx \approx 0.997.$$

Since 0.95 is so close to 1, we expect that most of the time the rainfall will be between 13 and 17 inches a year.

Among the normal distributions, the one having $\mu = 0$, $\sigma = 1$ is called the *standard normal distribution*. Values of the corresponding cumulative distribution function are published in tables.

Problems on Probability and Distributions

1. Consider the fishing data given in Example 3 on page 416. Show that the area under the density function in Figure 8.63 is 1. Why is this to be expected?

2. Find the mean daily catch for the fishing data in Figure 8.63, page 416.

3. The probability of a transistor failing between $t = a$ months and $t = b$ months is given by $c\int_a^b e^{-ct}\,dt$ for some constant c.

   (a) If the probability of failure within the first six months is 10%, what is c?
   (b) Given the value of c in part (a), what is the probability the transistor will fail within the second six months?

4. Suppose that x measures the time (in hours) it takes for a student to complete an exam. Assume that all students are done within two hours and the density function for x is given by

$$p(x) = \begin{cases} x^3/4 & \text{if } 0 < x < 2 \\ 0 & \text{otherwise.} \end{cases}$$

   (a) What proportion of students take between 1.5 and 2.0 hours to finish the exam?
   (b) What is the mean time for students to complete the exam?
   (c) Compute the median of this distribution.

5. In 1950 an experiment was done observing the time gaps between successive cars on the Arroyo Seco Freeway.[6] The data show that the density function of these time gaps was given approximately by

$$p(x) = ae^{-0.122x}$$

   where x is the time in seconds and a is a constant.

   (a) Find a.
   (b) Find P, the cumulative distribution function.
   (c) Find the median and mean time gap.
   (d) Sketch rough graphs of p and P.

6. The distribution of IQ scores is often modeled by the normal distribution with mean 100 and standard deviation 15.

   (a) Write a formula for the density distribution of IQ scores.
   (b) Estimate the fraction of the population with IQ between 115 and 120.

7. For a normal population of mean 0, show that the fraction of the population within one standard deviation of the mean does not depend on the standard deviation.

   [Hint: Use the substitution $w = x/\sigma$.]

8. (a) Using a calculator or computer, sketch graphs of the density function of the normal distribution

$$p(x) = \frac{1}{\sigma\sqrt{2\pi}}e^{-(x-\mu)^2/(2\sigma^2)}.$$

   (i) For fixed μ (say, $\mu = 5$) and varying σ (say, $\sigma = 1, 2, 3$).
   (ii) For varying μ (say, $\mu = 4, 5, 6$) and fixed σ (say, $\sigma = 1$).

   (b) Explain how the graphs confirm that μ is the mean of the distribution and that σ is a measure of how closely the data is clustered around the mean.

9. Consider the normal distribution, $p(x)$.

   (a) Show that $p(x)$ is a maximum when $x = \mu$. What is that maximum value?
   (b) Show that $p(x)$ has points of inflection where $x = \mu + \sigma$ and $x = \mu - \sigma$.
   (c) Describe in your own words what μ and σ tell you about the distribution.

10. Consider a group of people who have received treatment for a disease such as cancer. Let t be the *survival time*, the number of years a person lives after receiving treatment. The density function giving the distribution of t is $p(t) = Ce^{-Ct}$ for some positive constant C.

    (a) What is the practical meaning for the cumulative distribution function $P(t) = \int_0^t p(x)\,dx$?
    (b) The survival function, $S(t)$, is the probability that a randomly selected person survives for at least t years. Find $S(t)$.
    (c) Suppose a patient has a 70% probability of surviving at least two years. Find C.

11. While taking a walk along the road where you live, you accidentally drop your glove. You don't know where you dropped it. Suppose the probability density $p(x)$ for having dropped the glove x kilometers from home (along the road) is

$$p(x) = 2e^{-2x} \quad \text{for } x \geq 0.$$

    (a) What is the probability that you dropped it within 1 kilometer of home?
    (b) At what distance y from home is the probability that you dropped it within y km of home equal to 0.95?

---

[6]Reported by Daniel Furlough and Frank Barnes.

12. Which of the following functions makes the most sense as a model for the probability density representing the time (in minutes, starting from $t = 0$) that the next customer walks into a store?

(a) $p(t) = \begin{cases} \cos t & 0 \le t \le 2\pi \\ e^{t-2\pi} & t \ge 2\pi \end{cases}$

(b) $p(t) = 3e^{-3t}$ for $t \ge 0$

(c) $p(t) = e^{-3t}$ for $t \ge 0$

(d) $p(t) = 1/4$ for $0 \le t \le 4$

13. Let $P(x)$ be the cumulative distribution function for the income distribution in the US in 1973 (income is measured in thousands of dollars). Some values of $P(x)$ are in the following table:

Income x (thousands)	1	4.4	7.8	12.6	20	50
$P(x)$ (%)	1	10	25	50	75	99

(a) What fraction of the population made between $20,000 and $50,000?

(b) What was the median income?

(c) Sketch a density function for this distribution. Where, approximately, does your density function have a maximum? What is the significance of this point, in terms of income distribution? How can you recognize this point on the graph of the density function and on the graph of the cumulative distribution?

14. Let v be the speed, in meters/second, of an oxygen molecule, and let $p(v)$ be the density function of the speed distribution of oxygen molecules at room temperature. Maxwell showed that

$$p(v) = av^2 e^{-mv^2/(2kT)},$$

where $k = 1.4 \times 10^{-23}$ is the Boltzmann constant, T is the temperature in Kelvin (at room temperature, $T = 293$), and $m = 5 \times 10^{-26}$ is the mass of the oxygen molecule in kilograms.

(a) Find the value of a.

(b) Estimate the median and the mean speed. Find the maximum of $p(v)$.

(c) How do your answers in part (b) for the mean and the maximum of $p(v)$ change as T changes?

15. If we think of an electron as a particle, the function

$$P(r) = 1 - (2r^2 + 2r + 1)e^{-2r}$$

is the cumulative distribution function of the distance, r, of the electron in a hydrogen atom from the center of the atom. The distance is measured in Bohr radii. (1 Bohr radius $= 5.29 \times 10^{-11}$ m. Niels Bohr (1885–1962) was a Danish physicist.)

For example, $P(1) = 1 - 5e^{-2} \approx 0.32$ means that the electron is within 1 Bohr radius from the center of the atom 32% of the time.

(a) Find a formula for the density function of this distribution. Sketch the density function and the cumulative distribution function.

(b) Find the median distance and the mean distance. Near what value of r is an electron most likely to be found?

(c) The Bohr radius is sometimes called the "radius of the hydrogen atom." Why?

CHAPTER NINE

APPROXIMATIONS AND SERIES

In this chapter we will see how to approximate functions by simpler functions. Taylor approximations use polynomials, which may be considered the simplest functions. They are easy to use because they can be evaluated by hand, unlike the transcendental functions, such as e^x and $\ln x$.

Fourier approximations use sines and cosines, the simplest periodic functions, instead of polynomials. Taylor approximations are generally good approximations to the function locally (that is, near a specific point), whereas Fourier approximations are generally good approximations over an interval.

9.1 TAYLOR POLYNOMIALS AND SERIES

Taylor Polynomial Approximations

We approximate a function $f(x)$ near the point $x = a$ by polynomials. To get more accuracy, we take higher-degree polynomials. The approximation is usually good near the point $x = a$, but often not so good farther away from that point.

Linear Approximations

We have already seen how to approximate a function using a degree 1 polynomial, namely the tangent line approximation:

$$f(x) \approx f(a) + f'(a)(x - a).$$

The tangent line is the best linear approximation to the function near $x = a$. The tangent line and the curve share the same slope at $x = a$. (See Figure 9.1.) If $a = 0$, this approximation is called a *Taylor polynomial of degree 1*.

Taylor Polynomial of Degree 1 Approximating $f(x)$ for x near 0

$$f(x) \approx P_1(x) = f(0) + f'(0)x$$

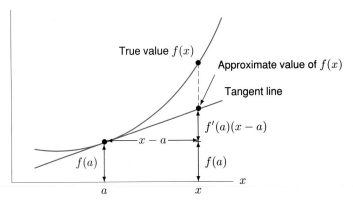

Figure 9.1: Tangent line approximation of $f(x)$ for x near a

---

Example 1   Approximate $f(x) = \cos x$, with x in radians, by its tangent line at $x = 0$.

Solution   The tangent line at $x = 0$ is just the horizontal line $y = 1$, as shown in Figure 9.2, so

$$f(x) = \cos x \approx 1, \quad \text{for } x \text{ near } 0.$$

If we take $x = 0.05$, then

$$f(0.05) = \cos(0.05) = 0.999\ldots,$$

which is quite close to the approximation $\cos x \approx 1$. Similarly, if $x = -0.1$, then

$$f(-0.1) = \cos(-0.1) = 0.995\ldots$$

is close to the approximation $\cos x \approx 1$. However, if $x = 0.4$, then

$$f(0.4) = \cos(0.4) = 0.921\ldots,$$

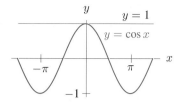

Figure 9.2: Graph of $\cos x$ and its tangent line at $x = 0$

so the approximation $\cos x \approx 1$ is less accurate. The graph suggests that the farther a point x is away from 0, the worse the approximation is likely to be.

Quadratic Approximations

Suppose we want a more accurate approximation to $f(x) = \cos x$ for x near 0. Instead of using a line, we use a quadratic function which not only has the same slope, but bends in the same way as the original curve. We require that at $x = 0$ the graphs of the original function f and the quadratic function have the same slope, $f'(0)$, and that they bend at the same rate—in other words, that they have the same second derivative, $f''(0)$.

Example 2   Find the quadratic approximation to $f(x) = \cos x$ for x near 0.

Solution   We take a quadratic polynomial

$$P_2(x) = C_0 + C_1 x + C_2 x^2,$$

and determine the values for C_0, C_1, and C_2. At $x = 0$, we want the two functions and their first and second derivatives to agree; that is, we want $P_2(0) = f(0)$, $P_2'(0) = f'(0)$, and $P_2''(0) = f''(0)$. Since

$$P_2(x) = C_0 + C_1 x + C_2 x^2 \qquad \text{and} \qquad f(x) = \cos x$$
$$P_2'(x) = C_1 + 2C_2 x \qquad\qquad\qquad\qquad f'(x) = -\sin x$$
$$P_2''(x) = 2C_2 \qquad\qquad\qquad\qquad\qquad f''(x) = -\cos x,$$

we have

$$C_0 = P_2(0) = f(0) = \cos 0 = 1 \qquad \text{so} \qquad C_0 = 1$$
$$C_1 = P_2'(0) = f'(0) = -\sin 0 = 0 \qquad\qquad C_1 = 0$$
$$2C_2 = P_2''(0) = f''(0) = -\cos 0 = -1, \qquad C_2 = -\tfrac{1}{2}.$$

Consequently, the quadratic approximation is

$$\cos x \approx P_2(x) = 1 + 0 \cdot x - \frac{1}{2}x^2 = 1 - \frac{x^2}{2}, \quad \text{for } x \text{ near } 0.$$

From Figure 9.3, we see that the quadratic approximation $\cos x \approx P_2(x)$ is better than the linear approximation $\cos x \approx P_1(x)$ for x near 0. Let's compare the accuracy of the two approximations numerically. At $x = 0.4$, $\cos(0.4) = 0.921\ldots$ and $P_2(0.4) = 0.920$, so the quadratic approximation is a significant improvement over the linear approximation. (The magnitude of the error is about 0.001 instead of 0.08.) In addition, the quadratic approximation is much better than the linear approximation for x-values near 0: $\cos(0.1) = 0.995004\ldots$ and $P_2(0.1) = 0.995$.

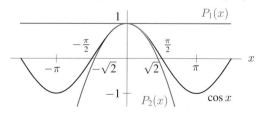

Figure 9.3: Graph of $\cos x$ and its linear, $P_1(x)$, and quadratic, $P_2(x)$, approximations for x near 0

We generalize the computations in Example 2 and define:

Taylor Polynomial of Degree 2 Approximating $f(x)$ for x near 0

$$f(x) \approx P_2(x) = f(0) + f'(0)x + \frac{f''(0)}{2}x^2$$

Higher-Degree Polynomials

As a rule, over a small interval around $x = 0$, the quadratic approximation to a function is a better approximation than the linear (tangent line) approximation. However, Figure 9.3 shows that even though we have matched up the function and the quadratic in terms of their values, their slopes, and their concavity at the point $x = 0$, the quadratic can still bend away for large x. We can fix this somewhat by using an approximating polynomial of higher degree. Suppose that we want to approximate a function $f(x)$ for x near 0 by a polynomial of degree n:

$$f(x) \approx P_n(x) = C_0 + C_1 x + C_2 x^2 + \cdots + C_{n-1} x^{n-1} + C_n x^n.$$

We need to find the values of the constants: $C_0,\ C_1,\ C_2, \ldots,\ C_n$. To do this, we require that the function $f(x)$ and each of its first n derivatives agree with those of the polynomial $P_n(x)$ at the point $x = 0$. Notice that the higher-order derivatives of a function contribute more subtle information about its graph than the first two derivatives do. (For instance, the third derivative measures how fast the concavity changes.) The more derivatives there are that agree at $x = 0$, the longer the function and the polynomial are likely to remain close to each other.

To see how to find the constants, let's take as an example

$$f(x) \approx P_3(x) = C_0 + C_1 x + C_2 x^2 + C_3 x^3.$$

Substituting $x = 0$ gives

$$f(0) = P_3(0) = C_0.$$

Differentiating $P_3(x)$ yields

$$P_3'(x) = C_1 + 2C_2 x + 3C_3 x^2,$$

so substituting $x = 0$ shows that

$$f'(0) = P_3'(0) = C_1.$$

Differentiating and substituting again, we get

$$P_3''(x) = 2 \cdot 1 C_2 + 3 \cdot 2 \cdot 1 C_3 x,$$

which gives

$$f''(0) = P_3''(0) = 2C_2,$$

so that

$$C_2 = \frac{f''(0)}{2}.$$

The third derivative, denoted by P_3''', is

$$P_3'''(x) = 3 \cdot 2 \cdot 1 C_3,$$

so

$$f'''(0) = P_3'''(0) = 3 \cdot 2 \cdot 1 C_3,$$

and then

$$C_3 = \frac{f'''(0)}{3 \cdot 2 \cdot 1}.$$

You can imagine a similar calculation starting with $P_4(x)$, using the fourth derivative $f^{(4)}$, which would give

$$C_4 = \frac{f^{(4)}(0)}{4 \cdot 3 \cdot 2 \cdot 1},$$

and so on. Using factorial notation,[1] we write these expressions as

$$C_3 = \frac{f'''(0)}{3!}, \quad C_4 = \frac{f^{(4)}(0)}{4!}.$$

In general, for any positive integer n,

$$C_n = \frac{f^{(n)}(0)}{n!},$$

where $f^{(n)}$ means the nth derivative of f. So we define:

Taylor Polynomial of Degree n Approximating $f(x)$ for x near 0

$$f(x) \approx P_n(x)$$
$$= f(0) + f'(0)x + \frac{f''(0)}{2!}x^2 + \frac{f'''(0)}{3!}x^3 + \frac{f^{(4)}(0)}{4!}x^4 + \cdots + \frac{f^{(n)}(0)}{n!}x^n$$

We call this a Taylor polynomial centered at $x = 0$, or a Taylor polynomial about $x = 0$.

Example 3   Construct the Taylor polynomial of degree 7 approximating the function $f(x) = \sin x$ for x near 0. Compare the value of the Taylor approximation with the true value of f at $x = \pi/3$.

Solution   We have

$$
\begin{aligned}
f(x) &= \sin x & \text{giving} \quad f(0) &= 0 \\
f'(x) &= \cos x & f'(0) &= 1 \\
f''(x) &= -\sin x & f''(0) &= 0 \\
f'''(x) &= -\cos x & f'''(0) &= -1 \\
f^{(4)}(x) &= \sin x & f^{(4)}(0) &= 0 \\
f^{(5)}(x) &= \cos x & f^{(5)}(0) &= 1 \\
f^{(6)}(x) &= -\sin x & f^{(6)}(0) &= 0 \\
f^{(7)}(x) &= -\cos x, & f^{(7)}(0) &= -1.
\end{aligned}
$$

Using these values, we see that the Taylor polynomial approximation of degree 7 is

$$\sin x \approx P_7(x) = 0 + x + 0 \cdot \frac{x^2}{2!} - \frac{x^3}{3!} + 0 \cdot \frac{x^4}{4!} + \frac{x^5}{5!} + 0 \cdot \frac{x^6}{6!} - \frac{x^7}{7!}$$
$$= x - \frac{x^3}{3!} + \frac{x^5}{5!} - \frac{x^7}{7!}, \quad \text{for } x \text{ near } 0.$$

In Figure 9.4 we show the graphs of the sine function and the approximating polynomial of degree 7 for x near 0. They are indistinguishable where x is close to 0. However, as we look at values of x further away from 0 in either direction, the two graphs diverge. To check the accuracy of this approximation numerically, we see how well it approximates $\sin(\pi/3) = \sqrt{3}/2 = 0.8660254\ldots$.

Figure 9.4: Graph of $\sin x$ and its seventh degree Taylor polynomial, $P_7(x)$, for x near 0

---

[1] We define 3! (read *three factorial*) to be equal to $3 \cdot 2 \cdot 1 = 6$. Similarly, $4! = 4 \cdot 3 \cdot 2 \cdot 1 = 24$ and $k! = k(k-1) \cdots 2 \cdot 1$. In addition, $0! = 1$.

When we substitute $\pi/3 = 1.0471976\ldots$ into the polynomial approximation, we obtain $P_7(\pi/3) = 0.8660213\ldots$, which is extremely accurate—to about four parts in a million.

Example 4   Graph the Taylor polynomial of degree 8 approximating $f(x) = \cos x$ for x near 0.

Solution   We can find the coefficients of the Taylor polynomial by the method of the preceding example. We have

$$\cos x \approx P_8(x) = 1 - \frac{x^2}{2!} + \frac{x^4}{4!} - \frac{x^6}{6!} + \frac{x^8}{8!}.$$

Figure 9.5 shows that $P_8(x)$ is close to the cosine function for a larger interval of x-values than the quadratic approximation $P_2(x) = 1 - x^2/2$ that we found in Example 2 on page 427.

Figure 9.5: $P_8(x)$ approximates $\cos x$ better than $P_2(x)$ for x near 0

Example 5   Construct the Taylor polynomial of degree 10 about $x = 0$ for the function $f(x) = e^x$.

Solution   We have $f(0) = 1$. Since the derivative of e^x is equal to e^x, all the higher-order derivatives will be equal to e^x. Consequently, for any $k = 1, 2, \ldots, 10$, $f^{(k)}(x) = e^x$ and $f^{(k)}(0) = e^0 = 1$. Therefore, the Taylor polynomial approximation of degree 10 is given by

$$e^x \approx P_{10}(x) = 1 + x + \frac{x^2}{2!} + \frac{x^3}{3!} + \frac{x^4}{4!} + \cdots + \frac{x^{10}}{10!}, \quad \text{for } x \text{ near 0}.$$

To check the accuracy of this approximation, suppose we use it to approximate $e = e^1$ which is 2.718281828 correct to nine decimal places. If we substitute $x = 1$ into the approximating polynomial, we obtain $P_{10}(1) = 2.718281801$. Thus, this tenth-degree polynomial yields the first seven decimal places for e. For large values of x, however, the accuracy must diminish because e^x grows much faster than any polynomial as $x \to \infty$. Figure 9.6 shows graphs of $f(x) = e^x$ and the Taylor polynomials of degree $n = 0, 1, 2, 3, 4$. Notice that each successive approximation remains close to the exponential curve for a larger interval of x-values.

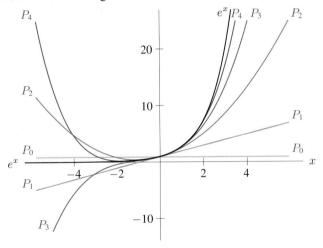

Figure 9.6: For x near 0, e^x is more closely approximated by higher-degree Taylor polynomials

Example 6   Construct the Taylor polynomial of degree n approximating $f(x) = \dfrac{1}{1-x}$ for x near 0.

Solution   You can check, by differentiating, that $f(0) = 1$, $f'(0) = 1$, $f''(0) = 2$, $f'''(0) = 3!$, $f^{(4)}(0) = 4!$, and so on. This means

$$\frac{1}{1-x} \approx P_n(x) = 1 + x + x^2 + x^3 + x^4 + x^5 + \cdots + x^n, \quad \text{for } x \text{ near } 0.$$

This is an example of a geometric series which we will study in Section 9.4.

Taylor Polynomials Around $x = a$

Suppose we want to approximate $f(x) = \ln x$ by a Taylor polynomial. This function has no Taylor polynomial about $x = 0$ because the function is not defined for $x = 0$ or for $x < 0$. Can we find another Taylor polynomial for $f(x) = \ln x$?

Rather than constructing a polynomial approximation for $f(x)$ about $x = 0$, let's construct a polynomial centered about some other point, $x = a$. First, let's look at the equation of the tangent line at $x = a$. Recall that since the tangent line goes through the point $(a, f(a))$ and has slope $f'(a)$, its equation is

$$y = f(a) + f'(a)(x - a).$$

This gives the approximation

$$f(x) \approx f(a) + f'(a)(x - a) \quad \text{for } x \text{ near } a.$$

The $f'(a)(x - a)$ term is a correction term which approximates how much $f(x)$ moves away from $f(a)$ as x moves away from a.

The approximating polynomial $P_n(x)$ centered at $x = a$ will be set up as $f(a)$ plus correction terms which depend on the derivatives of $f(x)$ and which are zero for $x = a$. This is achieved by writing the polynomial in powers of $(x - a)$ instead of powers of x:

$$f(x) \approx P_n(x) = C_0 + C_1(x - a) + C_2(x - a)^2 + \cdots + C_n(x - a)^n.$$

If we require the derivatives of the approximating polynomial $P_n(x)$ and the original function $f(x)$ to agree at $x = a$, we get the following result.

Taylor Polynomial of Degree n Approximating $f(x)$ for x near a

$$f(x) \approx P_n(x)$$
$$= f(a) + f'(a)(x - a) + \frac{f''(a)}{2!}(x - a)^2 + \cdots + \frac{f^{(n)}(a)}{n!}(x - a)^n$$

We call this a Taylor polynomial centered at $x = a$, or a Taylor polynomial about $x = a$.

You can derive the formula for these coefficients in the same way that we did for $a = 0$. (See Problem 33, page 435.)

Example 7   Construct the Taylor polynomial of degree 4 approximating the function $f(x) = \ln x$ for x near 1.

Solution   We have

$$
\begin{aligned}
f(x) &= \ln x & \text{so} \quad f(1) &= \ln(1) = 0 \\
f'(x) &= 1/x & f'(1) &= 1 \\
f''(x) &= -1/x^2 & f''(1) &= -1 \\
f'''(x) &= 2/x^3 & f'''(1) &= 2 \\
f^{(4)}(x) &= -6/x^4, & f^{(4)}(1) &= -6.
\end{aligned}
$$

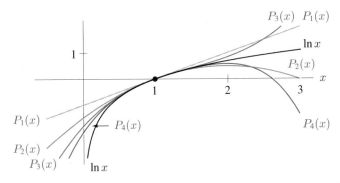

Figure 9.7: Taylor polynomials approximate $\ln x$ closely for x near 1, but not necessarily farther away

The Taylor polynomial is then

$$\ln x \approx P_4(x) = 0 + (x - 1) - \frac{(x-1)^2}{2!} + 2\frac{(x-1)^3}{3!} - 6\frac{(x-1)^4}{4!}$$

$$= (x - 1) - \frac{(x-1)^2}{2} + \frac{(x-1)^3}{3} - \frac{(x-1)^4}{4}, \quad \text{for } x \text{ near 1.}$$

We show $\ln x$ and several Taylor polynomials in Figure 9.7. Notice that $P_4(x)$ stays reasonably close to $\ln x$ for x near 1, but bends away as x gets farther from 1. Also, note that the Taylor polynomials are defined for $x \leq 0$, but $\ln x$ is not.

The examples in this section suggest that the following results are usually true:
- Taylor polynomials are a good approximation for x near a. Farther away, they may or may not be good.
- The higher the degree of the Taylor polynomial, the larger the interval over which it fits the function closely.

Taylor Series Approximations

We have just seen how to approximate a function near a point by Taylor polynomials. Now we define a Taylor series, which can be thought of as a Taylor polynomial that goes on forever. In the next section we will see at which points such a series serves as a good approximation to the function.

We have the following Taylor polynomials centered at $x = 0$ for $\cos x$:

$$\cos x \approx P_0(x) = 1$$

$$\cos x \approx P_2(x) = 1 - \frac{x^2}{2!}$$

$$\cos x \approx P_4(x) = 1 - \frac{x^2}{2!} + \frac{x^4}{4!}$$

$$\cos x \approx P_6(x) = 1 - \frac{x^2}{2!} + \frac{x^4}{4!} - \frac{x^6}{6!}$$

$$\cos x \approx P_8(x) = 1 - \frac{x^2}{2!} + \frac{x^4}{4!} - \frac{x^6}{6!} + \frac{x^8}{8!}.$$

Here we have a sequence of polynomials, $P_0(x)$, $P_2(x)$, $P_4(x)$, $P_6(x)$, $P_8(x)$, ..., each of which is a better approximation to $\cos x$ than the last, at least for x near 0. Notice that when we go to a higher-degree polynomial (say from P_6 to P_8), we add more terms ($x^8/8!$, for example), but the terms of lower degree don't change. Thus each polynomial includes the information from all the previous ones. We will write the *Taylor series* for $\cos x$:

$$T(x) = 1 - \frac{x^2}{2!} + \frac{x^4}{4!} - \frac{x^6}{6!} + \frac{x^8}{8!} - \cdots$$

to represent the whole sequence of Taylor polynomials.

What Does $\cdots$ **Mean?** The three dots indicate that the terms in the series continue further, possibly forever. This is the same idea as in $\pi = 3.1415\ldots$ where the three dots mean that the digits go on forever after the 5. For example,

$$1 - \frac{x^2}{2!} + \frac{x^4}{4!} \qquad \text{is a finite polynomial of degree 4,}$$

whereas

$$1 - \frac{x^2}{2!} + \frac{x^4}{4!} - \cdots \qquad \text{is an infinite series, and continues forever.}$$

Taylor Series for $\sin x$, $\cos x$, e^x

We can define the Taylor series for $\sin x$ and e^x in a similar way to the Taylor series for $\cos x$. In the next section we discuss what it means for a series to converge. It turns out that, for all x, we can write the following:

$$\sin x = x - \frac{x^3}{3!} + \frac{x^5}{5!} - \frac{x^7}{7!} + \frac{x^9}{9!} - \cdots$$

$$\cos x = 1 - \frac{x^2}{2!} + \frac{x^4}{4!} - \frac{x^6}{6!} + \frac{x^8}{8!} - \cdots$$

$$e^x = 1 + x + \frac{x^2}{2!} + \frac{x^3}{3!} + \frac{x^4}{4!} + \cdots$$

These series are also called *Taylor expansions* of the functions $\sin x$, $\cos x$, and e^x about $x = 0$.

Taylor Series in General

Any function f, all of whose derivatives exist at 0, has a Taylor series. However, for many functions, f, the series does not converge to $f(x)$ for all values of x. Assuming that the series does converge to $f(x)$, we have the following general formula.

Taylor Series for $f(x)$ about $x = 0$

$$f(x) = f(0) + f'(0)x + \frac{f''(0)}{2!}x^2 + \frac{f'''(0)}{3!}x^3 + \cdots + \frac{f^{(n)}(0)}{n!}x^n + \cdots$$

In addition, just as we have Taylor polynomials centered at points other than 0, we can also have a Taylor series centered at $x = a$ (provided all the derivatives exist at $x = a$). Assuming the series converges to $f(x)$, we have the following result.

Taylor Series for $f(x)$ about $x = a$

$$f(x) = f(a) + f'(a)(x - a) + \frac{f''(a)}{2!}(x - a)^2 + \frac{f'''(a)}{3!}(x - a)^3 + \cdots + \frac{f^{(n)}(a)}{n!}(x - a)^n + \cdots$$

For many functions f, the Taylor series converges to $f(x)$ only for x near a.

Problems for Section 9.1

For Problems 1–10, find the Taylor polynomials of degree n approximating the given functions for x near 0.

1. $\dfrac{1}{1+x}$, $n = 4, 6, 8$

2. $\dfrac{1}{1-x}$, $n = 3, 5, 7$

3. $\sqrt{1+x}$, $n = 2, 3, 4$

4. $\cos x$, $n = 2, 4, 6$

5. $\arctan x$, $n = 3, 4$

6. $\tan x$, $n = 3, 4$

7. $\sqrt[3]{1-x}$, $n = 2, 3, 4$

8. $\ln(1+x)$, $n = 5, 7, 9$

9. $\dfrac{1}{\sqrt{1+x}}$, $n = 2, 3, 4$

10. $(1+x)^p$, $n = 2, 3, 4$ (p is a constant.)

11. Suppose the function $f(x)$ is approximated near $x = 0$ by a sixth degree Taylor polynomial

$$P_6(x) = 3x - 4x^3 + 5x^6.$$

Give the value of

(a) $f(0)$ (b) $f'(0)$ (c) $f'''(0)$ (d) $f^{(5)}(0)$ (e) $f^{(6)}(0)$

12. Suppose g is a function which has continuous derivatives, and that $g(5) = 3$, $g'(5) = -2$, $g''(5) = 1$, $g'''(5) = -3$.

(a) What is the Taylor polynomial of degree 2 for g near 5? What is the Taylor polynomial of degree 3 for g near 5?

(b) Use the two polynomials that you found in part (a) to approximate $g(4.9)$.

For Problems 13–16, find the Taylor polynomial of degree n for x near the given point a.

13. $\sin x$, $a = \pi/2$, $n = 4$

14. $\cos x$, $a = \pi/4$, $n = 3$

15. e^x, $a = 1$, $n = 4$

16. $\sqrt{1+x}$, $a = 1$, $n = 3$

For Problems 17–20, suppose $P_2(x) = a + bx + cx^2$ is the second degree Taylor polynomial for the function f about $x = 0$. What can you say about the signs of a, b, c if f has the graph given below?

17. $a>0$ $b>0$ $x<0$ 18.   19. $a<0, b>0, c>0$ 20.

For Problems 21–24, find the first four terms of the Taylor series for the function about the point a.

21. $\sin x$, $a = \pi/4$ 22. $\cos \theta$, $a = \pi/4$ 23. $\sin \theta$, $a = -\pi/4$

24. $\tan x$, $a = \pi/4$

25. Suppose that you are told that the Taylor series of $f(x) = x^2 e^{x^2}$ about $x = 0$ is

$$x^2 + x^4 + \frac{x^6}{2!} + \frac{x^8}{3!} + \frac{x^{10}}{4!} + \cdots.$$

Find $\dfrac{d}{dx}\left(x^2 e^{x^2}\right)\Big|_{x=0}$ and $\dfrac{d^6}{dx^6}\left(x^2 e^{x^2}\right)\Big|_{x=0}$.

26. Suppose you know that all the derivatives of some function f exist at 0, and that Taylor series for f about $x = 0$ is

$$x + \frac{x^2}{2} + \frac{x^3}{3} + \frac{x^4}{4} + \cdots + \frac{x^n}{n} + \cdots.$$

Find $f'(0)$, $f''(0)$, $f'''(0)$, and $f^{(10)}(0)$.

27. Find the second-degree Taylor polynomial for $f(x) = 4x^2 - 7x + 2$ about $x = 0$. What do you notice?

28. Find the third-degree Taylor polynomial for $f(x) = x^3 + 7x^2 - 5x + 1$ about $x = 0$. What do you notice?

29. (a) Based on your observations in Problems 27–28, make a conjecture about Taylor approximations in the case when f is itself a polynomial.
    (b) Show that your conjecture is true.

30. Show how you can use the Taylor approximation $\sin x \approx x - \dfrac{x^3}{3!}$, for x near 0, to explain why $\displaystyle\lim_{x \to 0} \dfrac{\sin x}{x} = 1$.

31. Use the fourth-degree Taylor approximation $\cos x \approx 1 - \dfrac{x^2}{2!} + \dfrac{x^4}{4!}$ for x near 0 to explain why $\displaystyle\lim_{x \to 0} \dfrac{1 - \cos x}{x^2} = \dfrac{1}{2}$.

32. Use a fourth degree Taylor approximation for e^h, for h near 0, to evaluate the following limits. Would your answer be different if you used a Taylor polynomial of higher degree?
    (a) $\displaystyle\lim_{h \to 0} \dfrac{e^h - 1 - h}{h^2}$
    (b) $\displaystyle\lim_{h \to 0} \dfrac{e^h - 1 - h - \frac{h^2}{2}}{h^3}$

33. Derive the formulas given in the box on page 431 for the coefficients of the Taylor polynomial approximating a function f for x near a.

34. (a) Find the Taylor polynomial approximation of degree 4 about $x = 0$ for the function $f(x) = e^{x^2}$.
    (b) Compare this result to the Taylor polynomial approximation of degree 2 for the function $f(x) = e^x$ about $x = 0$. What do you notice?
    (c) Use your observation in part (b) to write out the Taylor polynomial approximation of degree 20 to the function in part (a).
    (d) What is the Taylor polynomial approximation of degree 5 for the function $f(x) = e^{-2x}$?

35. (a) Find the terms up to degree 6 of the Taylor series for $f(x) = \sin(x^2)$ about $x = 0$ by taking derivatives.
    (b) Compare your result in part (a) to the series for $\sin x$. How could you have obtained your answer to part (a) from the series for $\sin x$?

36. The integral $\int_0^1 (\sin t/t)\, dt$ is difficult to approximate using, for example, left Riemann sums or the trapezoid rule because the integrand $(\sin t)/t$ is not defined at $t = 0$. However, this integral converges; its value is $0.94608\ldots$. Estimate the integral using Taylor polynomials for $\sin t$ about $t = 0$ of
    (a) Degree 3   (b) Degree 5

9.2 CONVERGENCE OF SERIES

Power Series

In this section we investigate the convergence of Taylor series. We will see the Taylor expansions for more functions, including $\ln(1 + x)$ and $(1 + x)^p$, and determine for which x-values we can safely use these series to make approximations.

We first introduce *power series*. Taylor series are a particular type of power series, just as Taylor polynomials are a particular type of polynomial.

A **power series** is a sum of constants times powers of x:

$$P(x) = C_0 + C_1 x + C_2 x^2 + \cdots + C_n x^n + \cdots = \sum_{n=0}^{\infty} C_n x^n.$$

We introduce the polynomials $P_n(x) = C_0 + C_1 x + C_2 x^2 + \ldots + C_n x^n$. For any particular values of x and n, the polynomial $P_n(x)$ is the sum of $n + 1$ numbers; it is a number and it always exists. Let us fix x and consider the sequence

$$P_0(x), \quad P_1(x), \quad P_2(x), \quad P_3(x), \quad \ldots \quad, P_n(x), \ldots$$

> If this sequence converges to a limit L, that is, if $\lim_{n \to \infty} P_n(x) = L$, then we say that the power series, $P(x)$, **converges** to L for this value of x.

In this section we see that some power series converge for all x, some converge only for $x = 0$, and some converge only on an interval. For a given function f and point x, the Taylor polynomials may or may not converge to a limit as $n \to \infty$, and if they do converge, it might not be to the original function $f(x)$, though it usually is. Fortunately, the Taylor polynomials for e^x, $\cos x$, and $\sin x$ converge to the original function for all x.

Convergence of Taylor Series for $\cos x$

We have already seen that the Taylor polynomials centered at $x = 0$ for $\cos x$ are good approximations for x near 0. (See Figure 9.8.) Amazingly enough, for any value of x, if we take a Taylor polynomial centered at $x = 0$ of high enough degree, its graph is nearly indistinguishable from the graph of the cosine near that point.

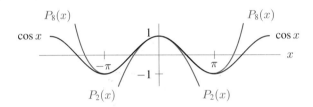

Figure 9.8: Graph of $\cos x$ and two Taylor polynomials for x near 0

Let's see what happens numerically. Suppose we take $x = \pi/2$. The successive Taylor polynomial approximations to $\cos(\pi/2) = 0$ are

$$
\begin{aligned}
P_2(\pi/2) &= \quad 1 - (\pi/2)^2/2! \quad\quad\quad\quad = -0.23370\ldots \\
P_4(\pi/2) &= 1 - (\pi/2)^2/2! + (\pi/2)^4/4! = \quad 0.01997\ldots \\
P_6(\pi/2) &= \quad\quad\quad \cdots \quad\quad\quad\quad = -0.00089\ldots \\
P_8(\pi/2) &= \quad\quad\quad \cdots \quad\quad\quad\quad = \quad 0.00002\ldots,
\end{aligned}
$$

and it appears that the approximations converge to the true value very rapidly. If we take a value of x somewhat further away from 0, say $x = \pi$, then $\cos \pi = -1$ and

$$
\begin{aligned}
P_2(\pi) &= 1 - (\pi)^2/2! = \quad -3.93480\ldots \\
P_4(\pi) &= \quad \cdots \quad = \quad\quad 0.12391\ldots \\
P_6(\pi) &= \quad \cdots \quad = \quad -1.21135\ldots \\
P_8(\pi) &= \quad \cdots \quad = \quad -0.97602\ldots \\
P_{10}(\pi) &= \quad \cdots \quad = \quad -1.00183\ldots \\
P_{12}(\pi) &= \quad \cdots \quad = \quad -0.99990\ldots \\
P_{14}(\pi) &= \quad \cdots \quad = \quad -1.000004\ldots
\end{aligned}
$$

We see that the rate of convergence is somewhat slower; it takes a 14th degree polynomial to approximate $\cos(\pi)$ as well as an 8th degree polynomial approximates $\cos(\pi/2)$. If x were taken still further away from 0, then we would need still more terms to obtain as accurate an approximation of $\cos x$. However it turns out that no matter how large x is, either positive or negative, the approximations $P_n(x)$ will converge to $\cos x$ as $n \to \infty$.

In this case, we are justified in writing an equality:

$$\cos x = 1 - \frac{x^2}{2!} + \frac{x^4}{4!} - \frac{x^6}{6!} + \frac{x^8}{8!} - \cdots \qquad \text{for all } x.$$

We say that the Taylor series converges to $\cos x$ for all x, because for all x, the numbers

$$P_1(x), \ P_2(x), \ P_3(x), \ldots, \ P_n(x), \ldots$$

have a limit of $\cos x$ as $n \to \infty$.

Intervals of Convergence

Let us look again at the Taylor polynomials for $\ln x$ about $x = 1$ that we derived on page 431.

Graphical Viewpoint

What happens if we attempt to improve the approximation to $\ln x$ by using higher-degree Taylor polynomials? Figure 9.9 suggests that for $0 < x < 2$, the polynomials fit the curve well, though outside that interval they are not close at all. In fact for $0 < x < 2$, the higher the degree of the polynomial, the better it fits the curve. It can be shown that, for $0 < x < 2$, the polynomials

$$P_5(x), P_6(x), P_7(x), \ldots, P_n(x), \ldots$$

converge to $\ln x$ as $n \to \infty$. For such x-values, a higher-degree polynomial will, in general, give a better approximation. (See Figure 9.9.)

However, when $x > 2$, the polynomials move away from the curve and the approximations get worse as the degree of the polynomial increases. For such x-values, we say the polynomials *diverge* from the original function.

The Taylor polynomial approximations about $x = 1$ are effective only as approximations to $\ln x$ for values of x between 0 and 2; beyond that, they definitely should not be used. We say the *interval of convergence* (without endpoints) of the Taylor series is $0 < x < 2$. As x gets near 0 or 2, the polynomials converge very slowly. This means we might have to take a polynomial of very high degree to get an accurate value for $\ln x$. At the endpoints of the interval of convergence, in this case $x = 0$ and $x = 2$, the series may or may not converge; we consider this question at the end of this section.

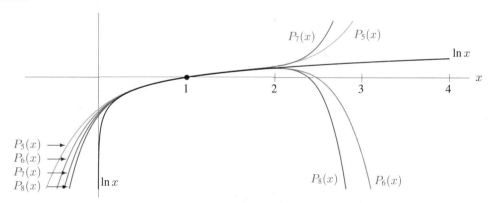

Figure 9.9: Taylor polynomials $P_5(x), P_6(x), P_7(x), P_8(x), \ldots$ converge to $\ln x$ for $0 < x < 2$ and diverge outside that interval

Numerical Viewpoint

To illustrate the interval of convergence numerically, consider the successive Taylor polynomial approximations:

$$\ln x \approx (x - 1) - \frac{(x-1)^2}{2} + \frac{(x-1)^3}{3} - \frac{(x-1)^4}{4} + \cdots + (-1)^{n-1} \frac{(x-1)^n}{n} \qquad \text{for } x \text{ near } 1.$$

If $x = 1.4$, we have the function value $\ln 1.4 = 0.33647\ldots$ and the approximations

$$P_2(1.4) = 0.32 \qquad\qquad P_6(1.4) = 0.33630\ldots$$
$$P_4(1.4) = 0.33493\ldots \qquad P_8(1.4) = 0.33645\ldots$$

So we see that the convergence is quite rapid.

In Table 9.1 we show the results of using $x = 1.9$ and $x = 2.3$ in the Taylor series for $\ln x$. Notice that for $x = 1.9$, which is inside the interval of convergence but close to an endpoint, the approximations converge, though rather slowly. For $x = 2.3$, which is outside the interval of convergence of the Taylor series, the "approximations" diverge: the larger the value for n, the less accurate the "approximations" become. In fact, the contribution of the twenty-fifth term is about 28; the contribution of the hundredth term is about $-2,500,000,000$. Thus, outside the interval of convergence, the "approximations" will oscillate ever more wildly away from the desired value. They are not approximations at all!

TABLE 9.1 *Approximations for*
$\ln 1.9 = 0.64185$ *and* $\ln 2.3 = 0.83291$

n	$P_n(1.9)$	n	$P_n(2.3)$
2	0.495	2	0.455
5	0.69021	5	1.21589
8	0.61802	8	0.28817
11	0.65473	11	1.71710
14	0.63440	14	-0.70701

Since the Talyor series about $x = 1$ for $\ln x$ converges to $\ln x$ for $0 < x < 2$, we write

$$\ln x = (x - 1) - \frac{(x-1)^2}{2} + \frac{(x-1)^3}{3} - \frac{(x-1)^4}{4} + \cdots \qquad \text{for } 0 < x < 2.$$

For $x < 0$ or $x > 2$, the series does not converge. Notice that the interval of convergence is centered at $x = 1$. For most functions f that we will meet, a Taylor series about $x = a$ either converges for all x, or has an interval of convergence centered at $x = a$.

Example 1   Find the Taylor series for $\ln(1 + x)$ about $x = 0$, and estimate its interval of convergence.

Solution   Taking derivatives of $\ln(1 + x)$ and substituting $x = 0$ leads to the Taylor series

$$\ln(1 + x) = x - \frac{x^2}{2} + \frac{x^3}{3} - \frac{x^4}{4} + \cdots.$$

Notice that this is the same series that we get by substituting $(1 + x)$ for x in the series for $\ln x$:

$$\ln x = (x - 1) - \frac{(x-1)^2}{2} + \frac{(x-1)^3}{3} - \frac{(x-1)^4}{4} + \cdots \qquad \text{for } 0 < x < 2.$$

Since the series for $\ln x$ about $x = 1$ converges for $0 < x < 2$, we shouldn't be surprised to see in Figure 9.10 that the interval of convergence for the Taylor series for $\ln(1 + x)$ about $x = 0$ is $-1 < x < 1$. Thus we write

$$\ln(1 + x) = x - \frac{x^2}{2} + \frac{x^3}{3} - \frac{x^4}{4} + \cdots \qquad \text{for } -1 < x < 1.$$

By the way, it should be clear that the series cannot possibly converge to $\ln(1 + x)$ for $x \leq -1$ since $\ln(1 + x)$ is not defined there.

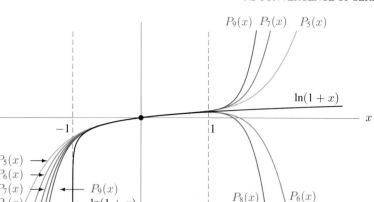

Figure 9.10: Interval of convergence for the Taylor series for $\ln(1+x)$ is $-1 < x < 1$

The Binomial Series Expansion

We now find the Taylor series about $x = 0$ for the function $f(x) = (1 + x)^p$, with p a constant, not necessarily an integer. Taking derivatives:

$$f(x) = (1 + x)^p \qquad\qquad \text{so} \qquad f(0) = 1$$
$$f'(x) = p(1 + x)^{p-1} \qquad\qquad\qquad f'(0) = p$$
$$f''(x) = p(p - 1)(1 + x)^{p-2} \qquad\qquad f''(0) = p(p - 1)$$
$$f'''(x) = p(p - 1)(p - 2)(1 + x)^{p-3}, \qquad f'''(0) = p(p - 1)(p - 2).$$

Thus, the third-degree Taylor polynomial for x near 0 is

$$(1 + x)^p \approx P_3(x) = 1 + px + \frac{p(p - 1)}{2!}x^2 + \frac{p(p - 1)(p - 2)}{3!}x^3.$$

Graphing $P_3(x), P_4(x), \ldots$ for various specific values of p suggests that the Taylor polynomials converge to $f(x)$ for $-1 < x < 1$. (See Problems 13–15, page 443.) The Taylor series for $f(x) = (1 + x)^p$ is as follows.

The Binomial Series

$$(1 + x)^p = 1 + px + \frac{p(p - 1)}{2!}x^2 + \frac{p(p - 1)(p - 2)}{3!}x^3 + \cdots \qquad \text{for } -1 < x < 1.$$

As we see in the next example, the binomial series gives the same result as multiplying $(1 + x)^p$ out when p is a positive integer. For more details, see the section on the Binomial Theorem in the Focus on Theory for Chapter 1. (Newton discovered that the binomial series can be used for noninteger exponents.)

Example 2   Use the binomial series with $p = 3$ to expand $(1 + x)^3$.

Solution   The series is

$$(1 + x)^3 = 1 + 3x + \frac{3 \cdot 2}{2!}x^2 + \frac{3 \cdot 2 \cdot 1}{3!}x^3 + \frac{3 \cdot 2 \cdot 1 \cdot 0}{4!}x^4 + \cdots.$$

The x^4 term and all terms beyond it turn out to be zero, because each coefficient contains a factor of 0. Simplifying gives

$$(1 + x)^3 = 1 + 3x + 3x^2 + x^3,$$

which is the usual expansion obtained by multiplying $(1 + x)(1 + x)(1 + x)$ out.

Example 3   Find the Taylor series about $x = 0$ for $\dfrac{1}{1+x}$.

Solution   Since $\dfrac{1}{1+x} = (1+x)^{-1}$, let $p = -1$. Then

$$\frac{1}{1+x} = (1+x)^{-1} = 1 + (-1)x + \frac{(-1)(-2)}{2!}x^2 + \frac{(-1)(-2)(-3)}{3!}x^3 + \cdots$$
$$= 1 - x + x^2 - x^3 + \cdots \quad \text{for } -1 < x < 1.$$

This series is a special case of the binomial series and so converges for $-1 < x < 1$. It is also an example of a *geometric series,* which we will study in more detail in Section 9.4.

How Do We Compute Intervals of Convergence Analytically?

A graph can suggest an interval of convergence; the following result shows how the interval can be computed exactly.

> **Theorem: The Ratio Test**
>
> For the power series
>
> $$P(x) = C_0 + C_1 x + C_2 x^2 + \cdots + C_n x^n + \cdots,$$
>
> suppose that
>
> $$\lim_{n \to \infty} \frac{|C_n|}{|C_{n+1}|} = R.$$
>
> Then
> - If[2] $R = \infty$, then the series converges for all x.
> - If $0 < R < \infty$, then the series converges for $|x| < R$.
> - If $R = 0$, then the series converges only for $x = 0$.
>
> We call R the **radius of convergence**.

Note that the ratio test does not tell us anything if $\lim_{n \to \infty} |C_n|/|C_{n+1}|$ fails to exist.

Example 4   Show that the Taylor series for e^x converges for all x.

Solution   The Taylor series for e^x is

$$e^x = 1 + x + \frac{x^2}{2!} + \frac{x^3}{3!} + \cdots + \frac{x^n}{n!} + \cdots.$$

We have

$$\lim_{n \to \infty} \frac{|C_n|}{|C_{n+1}|} = \lim_{n \to \infty} \frac{1/n!}{1/(n+1)!} = \lim_{n \to \infty} \frac{(n+1)!}{n!} = \lim_{n \to \infty} (n+1) = \infty.$$

Since $(n+1)$ increases without bound, the series for e^x converges for all x.

Showing that the series for $\sin x$ and $\cos x$ converge for all x requires an extension of the ratio test because each series contains only odd or only even terms. See Example 4 on page 479.

---

[2]That is, if the sequence $|C_n|/|C_{n+1}|$ grows without bound.

Example 5   Determine the exact radius of convergence of the Taylor series for $\ln(1 + x)$ about $x = 0$. What does this tell us about the convergence of this series?

Solution   The general term of the series for $\ln(1 + x)$ is x^n/n if n is odd and $-x^n/n$ if n is even, so we can write

$$\ln(1 + x) = x - \frac{x^2}{2} + \frac{x^3}{3} - \frac{x^4}{4} + \cdots + (-1)^{n-1}\frac{x^n}{n} + \cdots.$$

We have

$$\lim_{n \to \infty} \frac{|C_n|}{|C_{n+1}|} = \lim_{n \to \infty} \frac{|(-1)^n/n|}{|(-1)^{n-1}/(n+1)|} = \lim_{n \to \infty} \frac{n+1}{n} = 1.$$

Therefore the radius of convergence is $R = 1$. This tells us that the Taylor series for $\ln(1 + x)$ converges for $|x| < 1$ and does not converge for $|x| > 1$. Notice that the radius of convergence does not tell us what happens at the endpoints, $x = \pm 1$.

For a Taylor series centered at $x = a$, we compute the radius of convergence using the ratio test. The interval of convergence with this radius is then centered at $x = a$.

Example 6   What does the ratio test tell us about the convergence of the Taylor series for $\ln x$ about $x = 1$?

Solution   The series is

$$\ln x = (x - 1) - \frac{(x-1)^2}{2} + \frac{(x-1)^3}{3} - \frac{(x-1)^4}{4} + \cdots + (-1)^{n-1}\frac{(x-1)^n}{n} + \cdots.$$

The coefficients, C_n, are the same as the coefficients in the series for $\ln(1 + x)$ in the previous example, so by the same computation, $R = 1$. This tells us that the series for $\ln x$

Converges for $|x - 1| < 1$;   that is $0 < x < 2$,

Does not converge for $|x - 1| > 1$;   that is $x < 0, x > 2$.

Note that the ratio test does not tell us what happens at the endpoints $x = 0$ and $x = 2$.

What Happens at the Endpoints of the Interval of Convergence?

The ratio test does not tell us whether the series converges at the endpoints of the interval of convergence, $x = \pm R$. There is no single theorem that will answer this question. Since substituting $x = \pm R$ converts the power series to a series of numbers, the following results are often useful.

Divergence of the Harmonic Series

The *harmonic series* is defined to be the infinite series of constants

$$1 + \frac{1}{2} + \frac{1}{3} + \frac{1}{4} + \frac{1}{5} + \cdots + \frac{1}{n} + \cdots.$$

Convergence of this sum would mean that the sequence of *partial sums*

$$S_1 = 1, \quad S_2 = 1 + \frac{1}{2}, \quad S_3 = 1 + \frac{1}{2} + \frac{1}{3}, \quad \cdots, \quad S_n = 1 + \frac{1}{2} + \frac{1}{3} + \cdots + \frac{1}{n}, \cdots$$

tends to a limit as $n \to \infty$. Let's look at some values:

$$S_1 = 1, \quad S_{10} \approx 2.93, \quad S_{100} \approx 5.19, \quad S_{1000} \approx 7.49, \quad S_{10000} \approx 9.79.$$

The growth of these partial sums is slow, but they do in fact grow without bound, and so the harmonic series does not converge. See the next example and Problem 21.

Example 7   Show that the series $1 + 1/2 + 1/3 + 1/4 + \dots$ does not converge.

Solution   The idea is to think of the terms $1, 1/2, 1/3, \dots$ as the heights of upper rectangles of base 1 being used to approximate $\int_1^\infty (1/x)\, dx$. (See Figure 9.11.)

Figure 9.11: Comparing the harmonic series to $\int_1^\infty 1/x\, dx$

Since the sum of the areas of the rectangles is larger than the area under the curve, we have

$$S_n = 1 + \frac{1}{2} + \frac{1}{3} + \dots + \frac{1}{n} > \int_1^{n+1} \frac{1}{x}\, dx = \ln(n+1)$$

for the nth partial sum. Since $\ln(n+1)$ gets arbitrarily large as $n \to \infty$, so must the partial sums. Thus, the partial sums have no limit, so the series does not converge.

Convergence of an Alternating Series

When we substitute for x in a power series, we often obtain an *alternating series* such as

$$1 - \frac{1}{2} + \frac{1}{3} - \frac{1}{4} + \dots + \frac{(-1)^n}{n} + \dots$$

$$2 - \frac{2}{3} + \frac{2}{9} - \frac{2}{27} + \dots + (-1)^n \frac{2}{3^n} + \dots$$

It is easy to check for the convergence of alternating series using the following test:

Alternating Series Test

A series of the form

$$a_0 - a_1 + a_2 - a_3 + \dots + (-1)^n a_n + \dots \qquad \text{with } a_n \geq 0 \text{ for all } n,$$

converges if $a_{n+1} < a_n$ for all n and

$$\lim_{n \to \infty} a_n = 0.$$

Although we do not prove this result, we can see why it is reasonable. The first partial sum, $S_0 = a_0$, is positive. The second $S_1 = a_0 - a_1$ is still positive, since $a_1 < a_0$. The next sum, $S_2 = a_0 - a_1 + a_2$, is to the right of S_1 but smaller than S_0. (See Figure 9.12.) The partial sums continue to oscillate back and forth, and since the distance between them tends to 0, they eventually converge.

Figure 9.12: Partial sums, S_0, S_1, S_2, S_3 of an alternating series

Example 8   Investigate convergence at the end points of the interval of convergence of the Taylor series for $\ln(1+x)$ about $x = 0$.

Solution   In Example 5, we showed that the radius of convergence of the series

$$x - \frac{x^2}{2} + \frac{x^3}{3} - \frac{x^4}{4} + \cdots + (-1)^{n-1}\frac{x^n}{n} + \cdots$$

is $R = 1$. The endpoints of the interval are $x = \pm 1$. At $x = 1$, we have the series

$$1 - \frac{1}{2} + \frac{1}{3} - \frac{1}{4} + \cdots + \frac{(-1)^{n-1}}{n} + \cdots$$

This is an alternating series with $a_n = 1/n$, so by the alternating series test, it converges. At $x = -1$, we have the series

$$-1 - \frac{1}{2} - \frac{1}{3} - \frac{1}{4} - \cdots - \frac{1}{n} - \cdots$$

This is the negative of the harmonic series, so it does not converge. Therefore the right endpoint is included, and the left endpoint is not, in the interval of convergence, which is $-1 < x \le 1$.

Problems for Section 9.2

Which of the series in Problems 1–4 are power series?

1. $x - x^3 + x^6 - x^{10} + x^{15} - \cdots$

2. $\dfrac{1}{x} + \dfrac{1}{x^2} + \dfrac{1}{x^3} + \dfrac{1}{x^4} + \cdots$

3. $1 + x + (x-1)^2 + (x-2)^3 + (x-3)^4 + \cdots$

4. $x^7 + x + 2$

For Problems 5–8, find the first four terms of the Taylor series for the given function about 0.

5. $\dfrac{1}{1-x}$

6. $\sqrt{1+x}$

7. $\dfrac{1}{\sqrt{1+x}}$

8. $\sqrt[3]{1-y}$

For Problems 9–11, find the first four terms of the Taylor series for the function about the point a.

9. $1/x$, $a = 1$

10. $1/x$, $a = 2$

11. $1/x$, $a = -1$

12. (a) Find the Taylor series for $f(x) = \ln(1 + 2x)$ about $x = 0$ by taking derivatives.
    (b) Compare your result in part (a) to the series for $\ln(1 + x)$. How could you have obtained your answer to part (a) from the series for $\ln(1 + x)$?
    (c) What do you expect the interval of convergence for the series for $\ln(1 + 2x)$ to be?

13. By graphing the function $f(x) = \dfrac{1}{1-x}$ and several of its Taylor polynomials, estimate the interval of convergence of the series you found in Problem 5.

14. By graphing the function $f(x) = \sqrt{1+x}$ and several of its Taylor polynomials, estimate the interval of convergence of the series you found in Problem 6.

15. By graphing the function $f(x) = \dfrac{1}{\sqrt{1+x}}$ and several of its Taylor polynomials, estimate the interval of convergence of the series you found in Problem 7.

Use the ratio test to find the radius of convergence of the power series in Problems 16–20.

16. $x + 4x^2 + 9x^3 + 16x^4 + 25x^5 + \cdots$

17. $x - \dfrac{x^2}{4} + \dfrac{x^3}{9} - \dfrac{x^4}{16} + \dfrac{x^5}{25} - \cdots$

18. $1 + 2x + \dfrac{4x^2}{2!} + \dfrac{8x^3}{3!} + \dfrac{16x^4}{4!} + \dfrac{32x^5}{5!} + \cdots$

19. $\dfrac{x}{3} + \dfrac{2x^2}{5} + \dfrac{3x^3}{7} + \dfrac{4x^4}{9} + \dfrac{5x^5}{11} + \cdots$

20. $1 + 2x + \dfrac{4!x^2}{(2!)^2} + \dfrac{6!x^3}{(3!)^2} + \dfrac{8!x^4}{(4!)^2} + \dfrac{10!x^5}{(5!)^2} + \cdots$

21. Consider the following grouping of terms in the harmonic series:

$$1 + \left(\frac{1}{2}\right) + \left(\frac{1}{3} + \frac{1}{4}\right) + \left(\frac{1}{5} + \frac{1}{6} + \frac{1}{7} + \frac{1}{8}\right) + \left(\frac{1}{9} + \frac{1}{10} + \cdots + \frac{1}{16}\right) + \cdots$$

   (a) Show that the sum of each group of fractions is more than $1/2$.
   (b) Explain why this shows that the harmonic series does not converge.

22. Estimate the sum of the first 100,000 terms of the harmonic series,

$$\sum_{k=1}^{100{,}000} \frac{1}{k},$$

   to the closest integer. [Hint: Use left- and right-hand sums of the function $f(x) = 1/x$ on the interval from 1 to 100,000 with $\Delta x = 1$.]

23. Although the harmonic series does not converge, the partial sums grow very, very slowly. Take a right-hand sum of $f(x) = 1/x$ with $\Delta x = 1$ on the interval $[1, n]$ to show that

$$\frac{1}{2} + \frac{1}{3} + \frac{1}{4} + \cdots + \frac{1}{n} < \ln n.$$

   If a computer could add a million terms of the harmonic series each second, estimate what the sum would be after one year.

Use a computer or calculator to investigate the behavior of the partial sums of the alternating series in Problems 24–26. Which ones converge? If a series converges, estimate its sum.

24. $1 - 0.1 + 0.01 - 0.001 + \cdots + (-1)^n 10^{-n} + \cdots$

25. $1 - 2 + 3 - 4 + 5 + \cdots + (-1)^n(n + 1) + \cdots$

26. $1 - \dfrac{1}{1!} + \dfrac{1}{2!} - \dfrac{1}{3!} + \cdots + (-1)^n \dfrac{1}{n!} + \cdots$

By recognizing each series in Problems 27–31 as a Taylor series evaluated at a particular value of x, find the sum of each of the following convergent series.

27. $1 + \dfrac{2}{1!} + \dfrac{4}{2!} + \dfrac{8}{3!} + \cdots + \dfrac{2^n}{n!} + \cdots$

28. $1 - \dfrac{1}{3!} + \dfrac{1}{5!} - \dfrac{1}{7!} + \cdots + \dfrac{(-1)^n}{(2n + 1)!} + \cdots$

29. $1 + \dfrac{1}{4} + \left(\dfrac{1}{4}\right)^2 + \left(\dfrac{1}{4}\right)^3 + \cdots + \left(\dfrac{1}{4}\right)^n + \cdots$

30. $1 - \dfrac{100}{2!} + \dfrac{10000}{4!} + \cdots + \dfrac{(-1)^n \cdot 10^{2n}}{(2n)!} + \cdots$

31. $\dfrac{1}{2} - \dfrac{\left(\frac{1}{2}\right)^2}{2} + \dfrac{\left(\frac{1}{2}\right)^3}{3} - \dfrac{\left(\frac{1}{2}\right)^4}{4} + \cdots + \dfrac{(-1)^n \cdot \left(\frac{1}{2}\right)^{n+1}}{(n + 1)} + \cdots$

9.3 FINDING AND USING TAYLOR SERIES

Finding a Taylor series for a function means finding the coefficients. Assuming the function has all its derivatives defined, finding the coefficients can always be done, in theory at least, by differentiation. That is how we derived the four most important Taylor series, those for the functions e^x, $\sin x$, $\cos x$, and $(1 + x)^p$.

For many functions, however, computing Taylor series coefficients by differentiation can be a very laborious business. The purpose of this section is to look for easier ways of finding Taylor series, at least if the series we want is closely related to a series that we already know.

New Series by Substitution

Suppose we want to find the Taylor series for e^{-x^2} about $x = 0$. We could find the coefficients by differentiation. Differentiating e^{-x^2} by the chain rule gives $-2xe^{-x^2}$, and differentiating again gives $-2e^{-x^2} + 4x^2e^{-x^2}$. Now, every time we differentiate we will need the product rule, and the number of terms will grow. Finding the tenth or twentieth derivative of e^{-x^2}, and thus the series for e^{-x^2} up to the x^{10} or x^{20} terms, by this method will be tiresome (at least without a computer or calculator that can differentiate).

Fortunately, there's a quicker way. Recall that

$$e^y = 1 + y + \frac{y^2}{2!} + \frac{y^3}{3!} + \frac{y^4}{4!} + \cdots, \quad \text{for all } y.$$

Substituting $y = -x^2$ tells us that

$$e^{-x^2} = 1 + (-x^2) + \frac{(-x^2)^2}{2!} + \frac{(-x^2)^3}{3!} + \frac{(-x^2)^4}{4!} + \cdots, \quad \text{for all } x.$$

Simplifying shows that the Taylor series for e^{-x^2} is

$$e^{-x^2} = 1 - x^2 + \frac{x^4}{2!} - \frac{x^6}{3!} + \frac{x^8}{4!} + \cdots, \quad \text{for all } x.$$

Using this method, it is easy to find the series up to the x^{10} or x^{20} terms.

Example 1   Find the Taylor series about $x = 0$ for $f(x) = \dfrac{1}{1 + x^2}$.

Solution   The binomial series tells us that

$$\frac{1}{1 + y} = (1 + y)^{-1} = 1 - y + y^2 - y^3 + y^4 + \cdots, \quad \text{for } -1 < y < 1.$$

Substituting $y = x^2$ gives

$$\frac{1}{1 + x^2} = 1 - x^2 + x^4 - x^6 + x^8 + \cdots \quad \text{for } -1 < x < 1.$$

which is the Taylor series for $\dfrac{1}{1 + x^2}$.

These examples show that we can get new series from old ones by substitution. Proving that the new series we get is the same one we would have obtained by direct calculation of the derivatives is more difficult, and is done in more advanced texts.

In Example 1, we made the substitution $y = x^2$. We can also substitute an entire series into another one, as in the next example.

Example 2   Find the Taylor series about $\theta = 0$ for $g(\theta) = e^{\sin \theta}$.

Solution   For all y and θ, we know that

$$e^y = 1 + y + \frac{y^2}{2!} + \frac{y^3}{3!} + \frac{y^4}{4!} + \cdots$$

and

$$\sin \theta = \theta - \frac{\theta^3}{3!} + \frac{\theta^5}{5!} - \cdots.$$

Let's substitute the series for $\sin\theta$ for y:

$$e^{\sin\theta} = 1 + \left(\theta - \frac{\theta^3}{3!} + \frac{\theta^5}{5!} - \cdots\right) + \frac{1}{2!}\left(\theta - \frac{\theta^3}{3!} + \frac{\theta^5}{5!} - \cdots\right)^2 + \frac{1}{3!}\left(\theta - \frac{\theta^3}{3!} + \frac{\theta^5}{5!} - \cdots\right)^3 + \cdots.$$

To simplify, we multiply out and collect terms. The only constant term is the 1, and there's only one θ term. The only θ^2 term is the first term you get by multiplying out the square, and it is $\theta^2/2!$. There are two contributors to the θ^3 term: the $-\theta^3/3!$ from within the first parentheses, and the first term you get from multiplying out the cube, which is $\theta^3/3!$. Thus the series starts

$$e^{\sin\theta} = 1 + \theta + \frac{\theta^2}{2!} + \left(-\frac{\theta^3}{3!} + \frac{\theta^3}{3!}\right) + \cdots$$

$$= 1 + \theta + \frac{\theta^2}{2!} + 0\cdot\theta^3 + \cdots \quad\text{for all } \theta.$$

New Series by Integration

Just as we can get new series by substitution, we can also get new series by integration. Here again, proof that the new series so obtained is actually the Taylor series we want, and that it has the same interval of convergence as the original series, can be found in more advanced texts.

Example 3   Find the Taylor series about $x = 0$ for $\arctan x$ from the series for $\dfrac{1}{1+x^2}$.

Solution   We know that $\dfrac{d(\arctan x)}{dx} = \dfrac{1}{1+x^2}$, so we start with the series from Example 1:

$$\frac{d(\arctan x)}{dx} = \frac{1}{1+x^2} = 1 - x^2 + x^4 - x^6 + x^8 - \cdots \quad\text{for } -1 < x < 1.$$

Antidifferentiating term by term (which turns out to be legal) gives

$$\arctan x = C + x - \frac{x^3}{3} + \frac{x^5}{5} - \frac{x^7}{7} + \frac{x^9}{9} - \cdots \quad\text{for } -1 < x < 1,$$

where C is the constant of integration. The fact that $\arctan 0 = 0$ tells us that we must have $C = 0$, so

$$\arctan x = x - \frac{x^3}{3} + \frac{x^5}{5} - \frac{x^7}{7} + \frac{x^9}{9} - \cdots \quad\text{for } -1 < x < 1.$$

Applications of Taylor Series

Example 4   Use a series to estimate the numerical value of π.

Solution   Since $\arctan 1 = \pi/4$, we use the series for $\arctan x$ that we worked out in Example 3. It can be shown that the series does converge to $\pi/4$ at $x = 1$; we will assume this fact. We therefore substitute $x = 1$ into the series for $\arctan x$, getting

$$\pi = 4\arctan 1 = 4\left(1 - \frac{1}{3} + \frac{1}{5} - \frac{1}{7} + \frac{1}{9} - \cdots\right).$$

Table 9.2 shows the value of the sum, S_n, obtained by summing the terms from 1 through n. The values of S_n do seem to converge to $\pi = 3.141\ldots$. Unfortunately, though, this series converges

TABLE 9.2   *Approximating π using the series for* $\arctan x$

n	7	9	25	100	500	1000	10,000
S_n	2.895	3.340	3.218	3.122	3.138	3.140	3.141

very slowly, meaning that we have to take a large number of terms to get an accurate estimate for π. So, this way of calculating π is not particularly practical (a better one is given in Problem 30, page 451). However, the expression for π given by this series is surprising and elegant.

One of the most basic questions we can ask about two functions is which one is larger. The first few terms of the Taylor series for the two functions can often be used to answer this question over a small interval. If the constant terms of the two series are the same, compare the linear terms; if the linear terms are the same, compare the quadratic terms, and so on.

Example 5 By looking at their Taylor series, decide which of the following functions is largest, and which is smallest, for a small positive θ.

(a) $1 + \sin \theta$ (b) e^θ (c) $\dfrac{1}{\sqrt{1 - 2\theta}}$

Solution The Taylor expansion about $\theta = 0$ for $\sin \theta$ is

$$\sin \theta = \theta - \frac{\theta^3}{3!} + \frac{\theta^5}{5!} - \frac{\theta^7}{7!} + \cdots .$$

So

$$1 + \sin \theta = 1 + \theta - \frac{\theta^3}{3!} + \frac{\theta^5}{5!} - \frac{\theta^7}{7!} + \cdots .$$

The Taylor expansion about $\theta = 0$ for e^θ is

$$e^\theta = 1 + \theta + \frac{\theta^2}{2!} + \frac{\theta^3}{3!} + \frac{\theta^4}{4!} + \cdots .$$

The Taylor expansion about $\theta = 0$ for $1/\sqrt{1 + \theta}$ is

$$\frac{1}{\sqrt{1 + \theta}} = (1 + \theta)^{-1/2} = 1 - \frac{1}{2}\theta + \frac{(-\frac{1}{2})(-\frac{3}{2})}{2!}\theta^2 + \frac{(-\frac{1}{2})(-\frac{3}{2})(-\frac{5}{2})}{3!}\theta^3 + \cdots$$

$$= 1 - \frac{1}{2}\theta + \frac{3}{8}\theta^2 - \frac{5}{16}\theta^3 + \cdots .$$

So, substituting -2θ for θ:

$$\frac{1}{\sqrt{1 - 2\theta}} = 1 - \frac{1}{2}(-2\theta) + \frac{3}{8}(-2\theta)^2 - \frac{5}{16}(-2\theta)^3 + \cdots$$

$$= 1 + \theta + \frac{3}{2}\theta^2 + \frac{5}{2}\theta^3 + \cdots .$$

For θ near 0, we can neglect the higher-order terms in these expansions. Keeping the constant, linear, and second-degree terms, we are left with three approximations, valid for θ near 0:

$$1 + \sin \theta \approx 1 + \theta$$

$$e^\theta \approx 1 + \theta + \frac{\theta^2}{2}$$

$$\frac{1}{\sqrt{1 - 2\theta}} \approx 1 + \theta + \frac{3}{2}\theta^2 .$$

Since

$$1 + \theta < 1 + \theta + \frac{1}{2}\theta^2 < 1 + \theta + \frac{3}{2}\theta^2 ,$$

we conclude that, for small positive θ,

$$1 + \sin \theta < e^\theta < \frac{1}{\sqrt{1 - 2\theta}} .$$

Example 6   Two electrical charges of equal magnitude and opposite signs located near one another are called an electrical dipole. Suppose the charges Q and $-Q$ are a distance r apart. (See Figure 9.13.) The electric field, E, at the point P is given by

$$E = \frac{Q}{R^2} - \frac{Q}{(R+r)^2}.$$

Use series to investigate the behavior of the electric field far away from the dipole. Show that when R is large in comparison to r, the electric field is approximately proportional to $1/R^3$.

Figure 9.13: A dipole

Solution   In order to use a series approximation, we need to choose a variable whose value we know to be small. Although we know that r is much smaller than R, we do not know that r is itself small. The quantity r/R is, however, very small—much smaller than 1. Hence we expand $1/(R+r)^2$ in terms of r/R so that we can safely use only the first few terms:

$$\frac{1}{(R+r)^2} = \frac{1}{R^2(1+r/R)^2} = \frac{1}{R^2}\left(1+\frac{r}{R}\right)^{-2}$$

$$= \frac{1}{R^2}\left(1 + (-2)\left(\frac{r}{R}\right) + \frac{(-2)(-3)}{2!}\left(\frac{r}{R}\right)^2 + \frac{(-2)(-3)(-4)}{3!}\left(\frac{r}{R}\right)^3 + \cdots\right)$$

$$= \frac{1}{R^2}\left(1 - 2\frac{r}{R} + 3\frac{r^2}{R^2} - 4\frac{r^3}{R^3} + \cdots\right).$$

So

$$E = \frac{Q}{R^2} - \frac{Q}{(R+r)^2} = Q\left[\frac{1}{R^2} - \frac{1}{R^2}\left(1 - 2\frac{r}{R} + 3\frac{r^2}{R^2} - 4\frac{r^3}{R^3} + \cdots\right)\right]$$

$$= \frac{Q}{R^2}\left(2\frac{r}{R} - 3\frac{r^2}{R^2} + 4\frac{r^3}{R^3} - \cdots\right).$$

Since r/R is smaller than 1, the binomial expansion for $(1+r/R)^{-2}$ will converge. We are interested in the electric field far away from the dipole. The quantity r/R is small there, and $(r/R)^2$ and higher powers are much smaller still. Thus, we approximate by disregarding all terms except the first, giving

$$E \approx \frac{Q}{R^2}\left(\frac{2r}{R}\right), \quad \text{so} \quad E \approx \frac{2Qr}{R^3}.$$

Since Q and r are constants, this means that E is nearly proportional to $1/R^3$.

In the previous example, we say that E is *expanded in terms of* r/R, meaning that the independent variable in the expansion is r/R.

Problems for Section 9.3

Find the first four nonzero terms of the Taylor series about 0 for the functions in Problems 1–12.

1. $\sqrt{1-2x}$ 2. $\cos(\theta^2)$ 3. e^{-x} 4. $\dfrac{t}{1+t}$

5. $\ln(1-2y)$ 6. $\arcsin x$ 7. $\dfrac{1}{\sqrt{1-z^2}}$ 8. $\phi^3\cos(\phi^2)$

9. $\dfrac{z}{e^{z^2}}$ 10. $\sqrt{(1+t)}\sin t$ 11. $e^t\cos t$ 12. $\sqrt{1+\sin\theta}$

13. By looking at the Taylor series, decide which of the following functions is largest, and which is smallest, for small positive θ.
    (a) $1 + \sin\theta$ (b) $\cos\theta$ (c) $\dfrac{1}{1-\theta^2}$

14. For values of y near 0, put the following functions in increasing order, using their Taylor expansions.
    (a) $\ln(1+y^2)$ (b) $\sin(y^2)$ (c) $1 - \cos y$

15. Figure 9.14 shows the graphs of the four functions below for values of x near 0. Use Taylor series to match graphs and formulas.
    (a) $\dfrac{1}{1-x^2}$ (b) $(1+x)^{1/4}$ (c) $\sqrt{1+\dfrac{x}{2}}$ (d) $\dfrac{1}{\sqrt{1-x}}$

Figure 9.14

16. Consider the two functions $y = e^{-x^2}$ and $y = 1/(1+x^2)$.

    (a) Write the Taylor expansions for the two functions about $x = 0$. What is similar about the two series? What is different?

    (b) Looking at the series, which function do you predict will be greater over the interval $(-1, 1)$? Graph both and see.

    (c) Are these functions even or odd? How might you see this by looking at the series expansions?

    (d) By looking at the coefficients, explain why it is reasonable that the series for $y = e^{-x^2}$ converges for all values of x, but the series for $y = 1/(1+x^2)$ converges only on $(-1, 1)$.

For Problems 17–18, expand the quantity about 0 in terms of the variable given. Give four nonzero terms.

17. $\dfrac{1}{2+x}$ in terms of $\dfrac{x}{2}$

18. $\dfrac{a}{\sqrt{a^2+x^2}}$ in terms of $\dfrac{x}{a}$, where $a > 0$

19. Padé approximants are rational functions used to approximate more complicated functions. In this problem, you will derive the Padé approximant to the exponential function.

    (a) Let $f(x) = (1+ax)/(1+bx)$, where a and b are constants. Write down the first three terms of the Taylor series for $f(x)$ about $x = 0$.

    (b) By equating the first three terms of the Taylor series about $x = 0$ for $f(x)$ and for e^x, find a and b so that $f(x)$ approximates e^x as closely as possible near $x = 0$.

20. An electric dipole on the x-axis consists of a charge Q at $x = 1$ and a charge $-Q$ at $x = -1$. The electric field, E, at the point $x = R$ on the x-axis is given (for $R > 1$) by

$$E = \frac{kQ}{(R-1)^2} - \frac{kQ}{(R+1)^2}$$

where k is a positive constant whose value depends on the units. Expand E as a series in $1/R$, giving the first two nonzero terms.

21. Assume a is a positive constant. Suppose z is given by the expression

$$z = \sqrt{a^2+x^2} - \sqrt{a^2-x^2}.$$

Expand z as a series in x as far as the second nonzero term.

22. The electric potential, V, at a distance R along the axis perpendicular to the center of a charged disc with radius a and constant charge density σ, is given by

$$V = 2\pi\sigma(\sqrt{R^2 + a^2} - R).$$

Show that, for large R,

$$V \approx \frac{\pi a^2 \sigma}{R}.$$

23. One of Einstein's most amazing predictions was that light traveling from distant stars would bend around the sun on the way to earth. His calculations involved solving for ϕ in the equation

$$\sin\phi + b(1 + \cos^2\phi + \cos\phi) = 0$$

where b is a very small positive constant.

(a) Explain why the equation could have a solution for ϕ which is near 0.
(b) Expand the left-hand side of the equation in Taylor series about $\phi = 0$, disregarding terms of order ϕ^2 and higher. Solve for ϕ. (Your answer will involve b.)

24. The Michelson-Morley experiment, which contributed to the formulation of the Theory of Relativity, involved the difference between the two times t_1 and t_2 that light took to travel between two points. If v is the velocity of light; l_1, l_2, and c are constants; and $v < c$, then the times t_1 and t_2 are given by

$$t_1 = \frac{2l_2}{c(1 - v^2/c^2)} - \frac{2l_1}{c\sqrt{1 - v^2/c^2}} \qquad t_2 = \frac{2l_2}{c\sqrt{1 - v^2/c^2}} - \frac{2l_1}{c(1 - v^2/c^2)}.$$

(a) Find an expression for $\Delta t = t_1 - t_2$, and give its Taylor expansion in terms of v^2/c^2 up to the second nonzero term.
(b) For small v, to what power of v is Δt proportional? What is the constant of proportionality?

25. A hydrogen atom consists of an electron, of mass m, orbiting a proton, of mass M, where m is much smaller than M. The *reduced mass*, μ, of the hydrogen atom is defined by

$$\mu = \frac{mM}{m + M}.$$

(a) Show that $\mu \approx m$.
(b) To get a more accurate approximation for μ, express μ as m times a series in m/M.
(c) The approximation $\mu \approx m$ is obtained by disregarding all but the constant term in the series. The first-order correction is obtained by including the linear term but no higher terms. If $m \approx M/1836$, by what percentage does including the first-order correction change the estimate $\mu \approx m$?

26. A thin disk of radius a and mass M lies horizontally; a particle of mass m is at a height h directly above the center of the disk. The gravitational force, F, exerted by the disk on the mass m is given by

$$F = \frac{2GMmh}{a^2}\left(\frac{1}{h} - \frac{1}{(a^2 + h^2)^{1/2}}\right).$$

Assume $a < h$ and think of F as a function of a, with the other quantities constant.

(a) Expand F as a series in a/h. Give the first two nonzero terms.
(b) Show that the approximation for F obtained by using only the first nonzero term in the series is independent of the radius, a.
(c) If $a = 0.02h$, by what percentage does the approximation in part (a) differ from the approximation in part (b)?

27. When a body is near the surface of the earth, we usually assume that the force due to gravity on it is a constant mg, where m is the mass of the body and g is the acceleration due to gravity at sea level. For a body at a distance h above the surface of the earth, a more accurate expression for the force F is

$$F = \frac{mgR^2}{(R + h)^2}$$

where R is the radius of the earth. We will consider the situation in which the body is close to the surface of the earth so that h is much smaller than R.

(a) Show that $F \approx mg$.

(b)   Express F as mg multiplied by a series in h/R.

(c)   The first-order correction to the approximation $F \approx mg$ is obtained by taking the linear term in the series but no higher terms. How far above the surface of the earth can you go before the first-order correction changes the estimate $F \approx mg$ by more than 10%? (Assume $R = 6400$ km.)

28.   (a)   Estimate the value of $\int_0^1 e^{-x^2}\,dx$ using Riemann sums for both left-hand and right-hand sums with $n = 5$ subdivisions.

(b)   Approximate the function $f(x) = e^{-x^2}$ with a Taylor polynomial of degree 6.

(c)   Estimate the integral in part (a) by integrating the Taylor polynomial from part (b).

(d)   Indicate briefly how you could improve the results in each case.

29.   Use Taylor series to explain how the following patterns arise:

(a) $\dfrac{1}{0.98} = 1.020408163264\ldots$ (b) $\left(\dfrac{1}{0.99}\right)^2 = 1.020304050607\ldots$

30.   Machin's formula says $\pi/4 = 4\arctan(1/5) - \arctan(1/239)$.

(a)   Verify that $\pi/4$ and the quantity $4\arctan(1/5) - \arctan(1/239)$ agree to as many decimal places as your calculator shows.

(b)   Use the Taylor polynomial approximation of degree 5 to the arctangent function to approximate the value of π. (Note: In 1873 William Shanks used this approach to calculate π to 707 decimal places. Unfortunately, in 1946 it was found that he made an error in the 528th place.)

(c)   Can you explain why the two series for arctangent converge so rapidly here while the series used in Example 4 on page 446 converges so slowly?

9.4   GEOMETRIC SERIES

In the last two sections we expressed a given function by a Taylor series. We started with the function and then derived the series. In this section we will work in the opposite direction, starting with a series and finding its sum, the function. This can't be done for most series, but for some particular types of series it can.

Repeated Drug Dosage

A person with an ear infection is told to take antibiotic tablets regularly for several days. Since the drug is being excreted by the body in between doses, how can we calculate the quantity of the drug remaining in the body at any particular time?

To be specific, let's suppose the drug is ampicillin (a common antibiotic) taken in 250 mg doses four times a day (that is, every six hours). It is known that at the end of six hours, about 4% of the drug is still in the body. What quantity of the drug is in the body right after the tenth tablet? The fortieth?

Let Q_n represent the quantity, in milligrams, of ampicillin in the blood right after the nth tablet. Then

$$Q_1 = 250 \qquad\qquad\qquad\qquad\qquad\qquad\qquad\qquad\qquad = 250 \text{ mg}$$

$$Q_2 = \underbrace{250(0.04)}_{\text{Remnants of first tablet}} + \underbrace{250}_{\text{New tablet}} \qquad\qquad\qquad\qquad = 260 \text{ mg}$$

$$Q_3 = Q_2(0.04) + 250 = \left(250(0.04) + 250\right)(0.04) + 250$$
$$= \underbrace{250(0.04)^2 + 250(0.04)}_{\text{Remnants of first and second tablets}} + \underbrace{250}_{\text{New tablet}} \qquad\qquad = 260.4 \text{ mg}$$

$$Q_4 = Q_3(0.04) + 250 = \left(250(0.04)^2 + 250(0.04) + 250\right)(0.04) + 250$$
$$= \underbrace{250(0.04)^3 + 250(0.04)^2 + 250(0.04)}_{\text{Remnants of first, second, and third tablets}} + \underbrace{250}_{\text{New tablet}} \qquad = 260.416 \text{ mg}$$

Looking at the pattern that is emerging, we can probably guess that

$$Q_5 = 250(0.04)^4 + 250(0.04)^3 + 250(0.04)^2 + 250(0.04) + 250$$
$$Q_{10} = 250(0.04)^9 + 250(0.04)^8 + \cdots + 250(0.04) + 250.$$

Notice that there are 10 terms in this sum—one for every tablet—but that the highest power of 0.04 is the ninth, because no tablet has been in the body for more than 9 six-hour time periods. (Check this: do you see why?) Now suppose we actually want to find the numerical value of Q_{10}. It seems that we have to add 10 terms—and if we want the value of Q_{40}, we would be faced with adding 40 terms:

$$Q_{40} = 250(0.04)^{39} + 250(0.04)^{38} + \cdots + 250(0.04) + 250.$$

Fortunately, there's a better way. Let's start with Q_{10}.

$$Q_{10} = 250(0.04)^9 + 250(0.04)^8 + 250(0.04)^7 + \cdots + 250(0.04)^2 + 250(0.04) + 250.$$

Notice the remarkable fact that if you subtract $(0.04)Q_{10}$ from Q_{10}, a great many terms (all but two, in fact) drop out. First multiplying by 0.04, we get

$$(0.04)Q_{10} = 250(0.04)^{10} + 250(0.04)^9 + 250(0.04)^8 + \cdots + 250(0.04)^3 + 250(0.04)^2 + 250(0.04).$$

Subtracting gives

$$Q_{10} - (0.04)Q_{10} = 250 - 250(0.04)^{10}.$$

Factoring Q_{10} on the left and solving for Q_{10} gives

$$Q_{10}(1 - 0.04) = 250 \left(1 - (0.04)^{10}\right)$$
$$Q_{10} = \frac{250 \left(1 - (0.04)^{10}\right)}{1 - 0.04}.$$

This is called the *closed-form* expression for Q_{10}. It is easy to evaluate on a calculator, giving $Q_{10} = 260.42$ (to two decimal places). Similarly, Q_{40} is given in closed-form by

$$Q_{40} = \frac{250 \left(1 - (0.04)^{40}\right)}{1 - 0.04}.$$

Evaluating this on a calculator shows $Q_{40} = 260.42$, which is the same (to two decimal places) as Q_{10}. Thus after ten tablets, the value of Q_n appears to have stabilized at just over 260 mg.

Looking at the closed-forms for Q_{10} and Q_{40}, we can see that, in general, Q_n must be given by

$$Q_n = \frac{250 \left(1 - (0.04)^n\right)}{1 - 0.04}.$$

What Happens as $n \to \infty$?

What does this closed-form for Q_n predict about the long-run level of ampicillin in the body right after a tablet is taken, assuming that 250 mg continue to be taken every six hours? As $n \to \infty$, the quantity $(0.04)^n \to 0$, so in the long run,

$$Q_n = \frac{250 \left(1 - (0.04)^n\right)}{1 - 0.04} \to \frac{250(1 - 0)}{1 - 0.04} \approx 260.42.$$

The Geometric Series in General

In the previous example we encountered sums of the form $a + ax + ax^2 + \cdots + ax^8 + ax^9$ (with $a = 250$ and $x = 0.04$). Such a sum is called a finite *geometric series*. In general, a geometric series is defined as one in which each term is a constant multiple of the one before. (In our example, the constant multiple was 0.04.) We assume $a \neq 0$ and $x \neq 0$.

> A **finite geometric series** has the form
>
> $$a + ax + ax^2 + \cdots + ax^{n-2} + ax^{n-1}$$
>
> An **infinite geometric series** has the form
>
> $$a + ax + ax^2 + \cdots + ax^{n-2} + ax^{n-1} + ax^n + \cdots$$

The "$\cdots$" at the end of the second series tells us that the series is going on forever—in other words, that it is infinite.

Sum of a Finite Geometric Series

The same remarkable trick that enabled us to find the closed-form for Q_{10} can be used to find the sum of any finite geometric series. Suppose we define S_n to be the sum of the first n terms, which means up to the term in x^{n-1}:

$$S_n = a + ax + ax^2 + \cdots + ax^{n-2} + ax^{n-1}.$$

Multiply S_n by x:

$$xS_n = ax + ax^2 + ax^3 + \cdots + ax^{n-1} + ax^n.$$

Now subtract xS_n from S_n, which cancels out all terms except for two, giving

$$S_n - xS_n = a - ax^n$$
$$(1-x)S_n = a(1 - x^n).$$

Thus, provided $x \neq 1$, we can solve for S_n as follows.

> The **sum of a finite geometric series** is given by
>
> $$S_n = a + ax + ax^2 + \cdots + ax^{n-1} = \frac{a(1-x^n)}{1-x}, \qquad x \neq 1.$$

You may find it helpful to remember that the value of n which appears in the closed-form for S_n is the number of terms in the sum S_n.

Sum of an Infinite Geometric Series

We define convergence for an infinite geometric series in the same way as we defined convergence for a power series. In Example 9.4, we found the sum Q_n and then let $n \to \infty$. We do the same here. Suppose we want to find the sum S of the infinite series $a + ax + ax^2 + \cdots + ax^{n-1} + \cdots$. We consider the *partial sum*, S_n,

$$S_n = a + ax + ax^2 + \cdots + ax^{n-1} = \frac{a(1-x^n)}{1-x}$$

and we look at what happens to S_n as $n \to \infty$. What happens? It depends on the value of x. If $|x| < 1$, then $x^n \to 0$ as $n \to \infty$, and so

$$\lim_{n \to \infty} S_n = \lim_{n \to \infty} \frac{a(1-x^n)}{1-x} = \frac{a(1-0)}{1-x} = \frac{a}{1-x}.$$

Thus if $|x| < 1$, the partial sums approach a limit as $n \to \infty$ and we say the infinite geometric series *converges*.

> For $|x| < 1$, the **sum of the infinite geometric series** is given by
>
> $$S = a + ax + ax^2 + \cdots + ax^n + \cdots = \frac{a}{1-x}.$$

If, on the other hand, $|x| > 1$, then x^n and the partial sums have no limit as $n \to \infty$, and we say that the series does not converge. This corresponds to the fact that when $x > 1$, the terms in the series get larger and larger, so adding up more and more of them couldn't possibly give a finite sum. When $x < -1$, the partial sums oscillate more and more wildly and so do not converge.

What happens when $x = 1$? The series is

$$a + a + a + a \dots,$$

and since $a \neq 0$, the partial sums grow without bound, and the series doesn't converge. When $x = -1$, the series is

$$a - a + a - a + a \dots,$$

and, since $a \neq 0$, the partial sums oscillate between a and 0, and the series does not converge.

Example 1 For each of the following infinite series, find several partial sums and the sum (if the sum exists).

(a) $1 + \dfrac{1}{2} + \dfrac{1}{4} + \dfrac{1}{8} + \cdots$ (b) $1 + 2 + 4 + 8 + \cdots$ (c) $6 - 2 + \dfrac{2}{3} - \dfrac{2}{9} + \dfrac{2}{27} - \cdots$

Solution (a) This series may be written

$$1 + \frac{1}{2} + \left(\frac{1}{2}\right)^2 + \left(\frac{1}{2}\right)^3 + \cdots$$

which we can identify as a geometric series with $a = 1$ and $x = \frac{1}{2}$, so $S = \dfrac{1}{1 - (1/2)} = 2$. Let's check this by finding the partial sums:

$$S_1 = 1$$
$$S_2 = 1 + \frac{1}{2} = \frac{3}{2} = 2 - \frac{1}{2}$$
$$S_3 = 1 + \frac{1}{2} + \frac{1}{4} = \frac{7}{4} = 2 - \frac{1}{4}$$
$$S_4 = 1 + \frac{1}{2} + \frac{1}{4} + \frac{1}{8} = \frac{15}{8} = 2 - \frac{1}{8}$$
$$S_5 = 1 + \frac{1}{2} + \frac{1}{4} + \frac{1}{8} + \frac{1}{16} = \frac{31}{16} = 2 - \frac{1}{16}.$$

Clearly the partial sums are creeping up on the sum $S = 2$, so $S_n \to 2$ as $n \to \infty$.

(b) The partial sums of this geometric series (with $a = 1$ and $x = 2$) grow uncontrollably, so the series has no sum:

$$S_1 = 1$$
$$S_2 = 1 + 2 = 3$$
$$S_3 = 1 + 2 + 4 = 7$$
$$S_4 = 1 + 2 + 4 + 8 = 15$$
$$S_5 = 1 + 2 + 4 + 8 + 16 = 31.$$

(c) This is an infinite geometric series with $a = 6$ and $x = -\frac{1}{3}$. The partial sums,

$$S_1 = 6.00, \quad S_2 = 4.00, \quad S_3 \approx 4.67, \quad S_4 \approx 4.44, \quad S_5 \approx 4.52, \quad S_6 \approx 4.49,$$

appear to be converging to 4.5 from alternatively above and below. The sum of the series is the limit of the partial sums,

$$S = \frac{6}{1 - (-1/3)} = 4.5.$$

Relationship between Geometric and Taylor Series

The geometric series *is* a Taylor series. For example, the fact that

$$1 + x + x^2 + x^3 + \cdots = \frac{1}{1-x}$$

suggests that this geometric series is the Taylor expansion of $f(x) = 1/(1-x)$ for x near 0. We can check that the geometric series really is the Taylor series by taking derivatives. Alternatively, this geometric series can also be obtained from the binomial series:

$$(1+x)^p = 1 + px + \frac{p(p-1)}{2!}x^2 + \frac{p(p-1)(p-2)}{3!}x^3 + \cdots.$$

First we substitute $p = -1$:

$$\frac{1}{1+x} = (1+x)^{-1} = 1 + (-1)x + \frac{(-1)(-2)}{2!}x^2 + \frac{(-1)(-2)(-3)}{3!}x^3 + \cdots$$

$$= 1 - x + x^2 - x^3 + \cdots.$$

Now replace x by $-x$:

$$\frac{1}{1-x} = (1-x)^{-1} = 1 - (-x) + (-x)^2 - (-x)^3 + \cdots$$

$$= 1 + x + x^2 + x^3 + \cdots.$$

Regular Deposits into a Savings Account

People who save money often do so by putting some fixed amount aside regularly. To be specific, suppose \$1000 is deposited every year in a savings account earning 5% a year, compounded annually. What is the balance, B_n, in dollars, in the savings account right after the nth deposit?

As before, let's start by looking at the first few years:

$$B_1 = 1000$$

$$B_2 = B_1(1.05) + 1000 = \underbrace{1000(1.05)}_{\text{Original deposit}} + \underbrace{1000}_{\text{New deposit}}$$

$$B_3 = B_2(1.05) + 1000 = \underbrace{1000(1.05)^2 + 1000(1.05)}_{\text{First two deposits}} + \underbrace{1000}_{\text{New deposit}}$$

$$B_4 = B_3(1.05) + 1000 = \underbrace{1000(1.05)^3 + 1000(1.05)^2 + 1000(1.05)}_{\text{First three deposits}} + \underbrace{1000}_{\text{New deposit}}$$

Observing the pattern, we see

$$B_n = 1000(1.05)^{n-1} + 1000(1.05)^{n-2} + \cdots + 1000(1.05) + 1000.$$

So B_n is a finite geometric series with $a = 1000$ and $x = 1.05$. Thus we have

$$B_n = \frac{1000\left(1 - (1.05)^n\right)}{1 - 1.05}.$$

Rewriting this so both the numerator and denominator of the fraction are positive gives

$$B_n = \frac{1000\left((1.05)^n - 1\right)}{1.05 - 1}.$$

What Happens as $n \to \infty$?

Common sense tells you that if you keep depositing \$1000 in an account and it keeps earning interest, your balance will grow without bound. This is what the formula for B_n shows also: $(1.05)^n \to \infty$ as $n \to \infty$, so B_n has no limit. (Alternatively, observe that the infinite geometric series of which B_n is a partial sum has $x > 1$ and so does not converge.)

Problems for Section 9.4

In Problems 1–10, decide which of the following are geometric series. For those which are, give the first term and the ratio between successive terms. For those which are not, explain why not.

1. $1 - \dfrac{1}{2} + \dfrac{1}{4} - \dfrac{1}{8} + \dfrac{1}{16} + \cdots$

2. $1 + \dfrac{1}{2} + \dfrac{1}{3} + \dfrac{1}{4} + \dfrac{1}{5} + \cdots$

3. $5 - 10 + 20 - 40 + 80 - \cdots$

4. $2 + 1 + \dfrac{1}{2} + \dfrac{1}{4} + \dfrac{1}{8} + \cdots$

5. $1 + x + 2x^2 + 3x^3 + 4x^4 + \cdots$

6. $y^2 + y^3 + y^4 + y^5 + \cdots$

7. $1 - x + x^2 - x^3 + x^4 - \cdots$

8. $1 - y^2 + y^4 - y^6 + \cdots$

9. $3 + 3z + 6z^2 + 9z^3 + 12z^4 + \cdots$

10. $1 + 2z + (2z)^2 + (2z)^3 + \cdots$

11. Find the sum of the series in Problem 6.

12. Find the sum of the series in Problem 7.

13. Find the sum of the series in Problem 8.

14. Find the sum of the series in Problem 10.

15. Find an exact value for each of the following sums.

    (a) $7(1.02)^3 + 7(1.02)^2 + 7(1.02) + 7 + \dfrac{7}{(1.02)} + \dfrac{7}{(1.02)^2} + \cdots + \dfrac{7}{(1.02)^{100}}.$

    (b) $7 + 7(0.1)^2 + \dfrac{7(0.1)^4}{2!} + \dfrac{7(0.1)^6}{3!} + \cdots .$

Find the sum of the series in Problems 16–19.

16. $-2 + 1 - \dfrac{1}{2} + \dfrac{1}{4} - \dfrac{1}{8} + \dfrac{1}{16} - \cdots$

17. $3 + \dfrac{3}{2} + \dfrac{3}{4} + \dfrac{3}{8} + \cdots + \dfrac{3}{2^{10}}$

18. $\displaystyle\sum_{n=4}^{\infty} \left(\dfrac{1}{3}\right)^n$

19. $\displaystyle\sum_{n=0}^{\infty} \dfrac{3^n + 5}{4^n}$

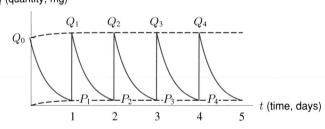

20. Figure 9.15 shows the quantity of the drug atenolol in the blood as a function of time, with the first dose at time $t = 0$. Atenolol is taken in 50 mg doses once a day to lower blood pressure.

Figure 9.15

    (a) If the half-life of atenolol in the blood is 6.3 hours, what percentage of the atenolol present at the start of a 24-hour period is still there at the end?

    (b) Find expressions for the quantities $Q_0, Q_1, Q_2, Q_3, \ldots$, and Q_n shown in Figure 9.15. Write the expression for Q_n in closed-form.

    (c) Find expressions for the quantities $P_1, P_2, P_3, \ldots$, and P_n shown in Figure 9.15. Write the expression for P_n in closed-form.

21. On page 451, you saw how to compute the quantity Q_n mg of ampicillin in the body right after the nth tablet of 250 mg, taken once every six hours.

    (a) Do a similar calculation for P_n, the quantity of ampicillin (in mg) in the body right *before* the nth tablet is taken.

    (b) Express P_n in closed form.

    (c) What is $\displaystyle\lim_{n \to \infty} P_n$? Is this limit the same as $\displaystyle\lim_{n \to \infty} Q_n$? Explain in practical terms why your answer makes sense.

22. Draw a graph like that in Figure 9.15 for 250 mg of ampicillin taken every 6 hours, starting at time $t = 0$. Put on the graph the values of $Q_1, Q_2, Q_3, \ldots$ introduced in the text on page 451 and the values of $P_1, P_2, P_3, \ldots$ calculated in Problem 21.

23. A ball is dropped from a height of 10 feet and bounces. Each bounce is $\frac{3}{4}$ of the height of the bounce before. Thus after the ball hits the floor for the first time, the ball rises to a height of $10(\frac{3}{4}) = 7.5$ feet, and after it hits the floor for the second time, it rises to a height of $7.5(\frac{3}{4}) = 10(\frac{3}{4})^2 = 5.625$ feet.

  (a) Find an expression for the height to which the ball rises after it hits the floor for the nth time.
  (b) Find an expression for the total vertical distance the ball has traveled when it hits the floor for the first, second, third, and fourth times.
  (c) Find an expression for the total vertical distance the ball has traveled when it hits the floor for the nth time. Express your answer in closed-form.

24. You might think that the ball in Problem 23 keeps bouncing forever since it takes infinitely many bounces. This is not true!

  (a) Show that a ball dropped from a height of h feet reaches the ground in $\frac{1}{4}\sqrt{h}$ seconds.
  (b) Show that the ball in Problem 23 stops bouncing after

$$\frac{1}{4}\sqrt{10} + \frac{1}{2}\sqrt{10}\sqrt{\frac{3}{4}}\left(\frac{1}{1 - \sqrt{3/4}}\right) \quad \text{seconds,}$$

  or approximately 11 seconds.

25. This problem illustrates how banks create credit and can thereby lend out more money than has been deposited. Suppose that initially \$100 is deposited in a bank. Experience has shown bankers that on average only 8% of the money deposited is withdrawn by the owner at any time. Consequently, bankers feel free to lend out 92% of their deposits. Thus \$92 of the original \$100 is loaned out to other customers (to start a business, for example). This \$92 will become someone else's income and, sooner or later, will be redeposited in the bank. Then 92% of \$92, or \$92(0.92) = \$84.64, is loaned out again and eventually redeposited. Of the \$84.64, the bank again loans out 92%, and so on.

  (a) Find the total amount of money deposited in the bank as a result of these transactions.
  (b) The total amount of money deposited divided by the original deposit is called the *credit multiplier*. Calculate the credit multiplier for this example and explain what this number tells us.

26. This problem deals with the question of estimating the cumulative effect of a tax cut on a country's economy. Suppose the government proposes a tax cut totaling \$100 million. We assume that all the people who have extra money to spend would spend 80% of it and save 20%. Thus, of the extra income generated by the tax cut, \$100(0.8) million = \$80 million would be spent and so become extra income to someone else. Assume that these people also spend 80% of their additional income, or \$80(0.8) million, and so on. Calculate the total additional spending created by such a tax cut.

27. Suppose the government proposes a tax cut of \$100 million as in Problem 26, but that economists now predict that people will spend 90% of their extra income and save only 10%. How much additional spending would be generated by the tax cut under these assumptions?

28. In an old puzzle, there are two trains, each moving at 10 km/hr toward one another. Initially the trains are 30 kilometers apart. At the same moment a fly, whose velocity is 20 km/hr, starts at one train and flies till it meets the other, then turns around and flies back till it meets the first train, and so on.

  (a) How far has the fly traveled the first time it turns around? The second time? The third time? The fourth time?
  (b) How far has the fly traveled by the nth time it turns around? Write your answer in closed-form.
  (c) Use you answer to part (b) to decide how far the fly has traveled by the time the trains meet in the middle and squash it.
  (d) How long does it take the trains to meet in the middle? Use this to answer part (c) without summing a series.

9.5 FOURIER SERIES

We have seen how to approximate a function by a Taylor polynomial of fixed degree. Such a polynomial is usually very close to the true value of the function near one point (the point at which the Taylor polynomial is centered), but not necessarily at all close anywhere else. In other words, Taylor polynomials are good approximations of a function *locally*, but not necessarily *globally*. In this section, we take another approach: we approximate the function by trigonometric functions, called *Fourier approximations*. The resulting approximation may not be as close to the original function at some points as the Taylor polynomial. However, the Fourier approximation is, in general, close over a larger interval. In other words, a Fourier approximation can be a better approximation globally. In addition, unlike Taylor approximations, Fourier approximations are periodic, so they are useful for approximating periodic functions.

Many processes in nature are periodic or repeating, so it makes sense to approximate them by periodic functions. For example, sound waves are made up of periodic oscillations of air molecules. Heartbeats, the movement of the lungs, and the electrical current that powers our homes are all periodic phenomena. Two of the simplest periodic functions are the square wave in Figure 9.16 and the triangular wave in Figure 9.17. Electrical engineers use the square wave as the model for the flow of electricity when a switch is repeatedly flicked on and off.

Figure 9.16: Square wave   *Figure 9.17:* Triangular wave

Fourier Polynomials

We can express the square wave and the triangular wave by the formulas

$$f(x) = \begin{cases} \vdots & \vdots \\ 0 & -1 \leq x < 0 \\ 1 & 0 \leq x < 1 \\ 0 & 1 \leq x < 2 \\ 1 & 2 \leq x < 3 \\ 0 & 3 \leq x < 4 \\ \vdots & \vdots \end{cases} \qquad g(x) = \begin{cases} \vdots & \vdots \\ -x & -1 \leq x < 0 \\ x & 0 \leq x < 1 \\ 2-x & 1 \leq x < 2 \\ x-2 & 2 \leq x < 3 \\ 4-x & 3 \leq x < 4 \\ \vdots & \vdots \end{cases}$$

However, these formulas are not particularly easy to work with. Worse, the functions are not differentiable at various points. Here we show how to approximate such functions by differentiable, periodic functions.

Since the sine and cosine are the simplest periodic functions, they are the building blocks we use. Because they repeat every 2π, we assume that the function f we want to approximate repeats every 2π. (Later, we deal with the case where f has some other period.) We start by considering the square wave in Figure 9.18. Because of the periodicity of all the functions concerned, we only have to consider what happens in the course of a single period; the same behavior repeats in any other period.

$$f(x) = \begin{cases} 0 & -\pi \leq x < 0 \\ 1 & 0 \leq x < \pi \end{cases}$$

Figure 9.18: Square wave on $[-\pi, \pi]$

We will attempt to approximate f with a sum of trigonometric functions of the form

$$f(x) \approx F_n(x)$$
$$= a_0 + a_1 \cos x + a_2 \cos(2x) + a_3 \cos(3x) + \cdots + a_n \cos(nx)$$
$$+ b_1 \sin x + b_2 \sin(2x) + b_3 \sin(3x) + \cdots + b_n \sin(nx)$$
$$= a_0 + \sum_{k=1}^{n} a_k \cos(kx) + \sum_{k=1}^{n} b_k \sin(kx).$$

$F_n(x)$ is known as a *Fourier polynomial of degree n*, named after the French mathematician Joseph Fourier (1768-1830), who was one of the first to investigate it.[3] The coefficients a_k and b_k are called *Fourier coefficients*. Since each of the component functions $\cos(kx)$ and $\sin(kx)$, $k = 1, 2, \ldots, n$, repeats every 2π, $F_n(x)$ must repeat every 2π and so is a potentially good match for $f(x)$, which also repeats every 2π. The problem is to determine values for the Fourier coefficients that achieve a close match between $f(x)$ and $F_n(x)$. We choose the following values:

The Fourier Coefficients for a Periodic Function f of Period 2π

$$a_0 = \frac{1}{2\pi} \int_{-\pi}^{\pi} f(x)\, dx,$$

$$a_k = \frac{1}{\pi} \int_{-\pi}^{\pi} f(x) \cos(kx)\, dx \quad \text{for } k > 0,$$

$$b_k = \frac{1}{\pi} \int_{-\pi}^{\pi} f(x) \sin(kx)\, dx \quad \text{for } k > 0.$$

Notice that a_0 is just the average value of f over the interval $[-\pi, \pi]$.

For an informal justification for the use of these values, see page 466.

Example 1   Construct successive Fourier polynomials for the square wave function f, with period 2π, given by

$$f(x) = \begin{cases} 0 & -\pi \leq x < 0 \\ 1 & 0 \leq x < \pi. \end{cases}$$

Solution   Since a_0 is the average value of f on $[-\pi, \pi]$, we suspect from the graph of f that $a_0 = \frac{1}{2}$. We can verify this analytically:

$$a_0 = \frac{1}{2\pi} \int_{-\pi}^{\pi} f(x)\, dx = \frac{1}{2\pi} \int_{-\pi}^{0} 0\, dx + \frac{1}{2\pi} \int_{0}^{\pi} 1\, dx = 0 + \frac{1}{2\pi}(\pi) = \frac{1}{2}.$$

Furthermore,

$$a_1 = \frac{1}{\pi} \int_{-\pi}^{\pi} f(x) \cos x\, dx = \frac{1}{\pi} \int_{0}^{\pi} 1 \cos x\, dx = 0$$

and

$$b_1 = \frac{1}{\pi} \int_{-\pi}^{\pi} f(x) \sin x\, dx = \frac{1}{\pi} \int_{0}^{\pi} 1 \sin x\, dx = \frac{2}{\pi}.$$

Therefore, the Fourier polynomial of degree 1 is given by

$$f(x) \approx F_1(x) = \frac{1}{2} + \frac{2}{\pi} \sin x$$

---

[3]The Fourier polynomials are not polynomials in the usual sense of the word.

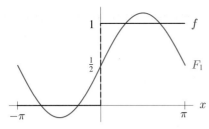

Figure 9.19: First Fourier approximation to the square wave

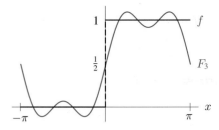

Figure 9.20: Third Fourier approximation to the square wave

and the graphs of the function and the first Fourier approximation are shown in Figure 9.19.

We next construct the Fourier polynomial of degree 2. The coefficients a_0, a_1, b_1 are the same as before. In addition,

$$a_2 = \frac{1}{\pi} \int_{-\pi}^{\pi} f(x) \cos(2x)\, dx = \frac{1}{\pi} \int_0^{\pi} 1 \cos(2x)\, dx = 0$$

and

$$b_2 = \frac{1}{\pi} \int_{-\pi}^{\pi} f(x) \sin(2x)\, dx = \frac{1}{\pi} \int_0^{\pi} 1 \sin(2x)\, dx = 0.$$

Since $a_2 = b_2 = 0$, the Fourier polynomial of degree 2 is identical to the Fourier polynomial of degree 1. Let's look at the Fourier polynomial of degree 3:

$$a_3 = \frac{1}{\pi} \int_{-\pi}^{\pi} f(x) \cos(3x)\, dx = \frac{1}{\pi} \int_0^{\pi} 1 \cos(3x)\, dx = 0$$

and

$$b_3 = \frac{1}{\pi} \int_{-\pi}^{\pi} f(x) \sin(3x)\, dx = \frac{1}{\pi} \int_0^{\pi} 1 \sin(3x)\, dx = \frac{2}{3\pi}.$$

So the approximation is given by

$$f(x) \approx F_3(x) = \frac{1}{2} + \frac{2}{\pi} \sin x + \frac{2}{3\pi} \sin(3x).$$

The graph of F_3 is shown in Figure 9.20. This approximation is better than $F_1(x) = \frac{1}{2} + \frac{2}{\pi} \sin x$, as comparing Figure 9.20 to Figure 9.19 shows.

Without going through the details, we calculate the coefficients for higher-degree Fourier approximations:

$$F_5(x) = \frac{1}{2} + \frac{2}{\pi} \sin x + \frac{2}{3\pi} \sin(3x) + \frac{2}{5\pi} \sin(5x)$$

$$F_7(x) = \frac{1}{2} + \frac{2}{\pi} \sin x + \frac{2}{3\pi} \sin(3x) + \frac{2}{5\pi} \sin(5x) + \frac{2}{7\pi} \sin(7x).$$

Figure 9.21 shows that higher-degree approximations match the step-like nature of the square wave function more and more closely.

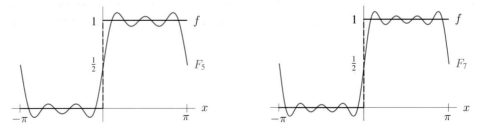

Figure 9.21: Fifth and seventh Fourier approximations to the square wave

We could have used a Taylor series to approximate the square wave, provided we did not center the series at a point of discontinuity. Since the square wave is a constant function on each interval, all its derivatives are zero, and so its Taylor series approximations are the constant functions: 0 or 1, depending on where the Taylor series is centered. They approximate the square wave perfectly on each piece, but they do not do a good job over the whole interval of length 2π. That is what Fourier polynomials succeed in doing: they approximate a curve fairly well everywhere, rather than just near a particular point. The Fourier approximations above look a lot like square waves, so they approximate well *globally*. However, they may not give good values near points of discontinuity. (For example, near $x = 0$, they all give values near $1/2$, which are incorrect.) Thus Fourier polynomials may not be good *local* approximations.

> Taylor polynomials give good *local* approximations to a function;
> Fourier polynomials give good *global* approximations to a function.

Fourier Series

As with Taylor polynomials, the higher the degree of the Fourier approximation, the more accurate it is. Therefore, we carry this procedure on indefinitely by letting $n \to \infty$, and we call the resulting infinite series a *Fourier series*.

> **The Fourier Series for f on $[-\pi, \pi]$**
>
> $$f(x) = a_0 + a_1 \cos x + a_2 \cos 2x + a_3 \cos 3x + \cdots$$
> $$+ b_1 \sin x + b_2 \sin 2x + b_3 \sin 3x + \cdots$$
>
> where a_k and b_k are the Fourier coefficients.

Thus, the Fourier series for the square wave is

$$f(x) = \frac{1}{2} + \frac{2}{\pi} \sin x + \frac{2}{3\pi} \sin 3x + \frac{2}{5\pi} \sin 5x + \frac{2}{7\pi} \sin 7x + \cdots.$$

Harmonics

Let us start with a function $f(x)$ that is periodic with period 2π, expanded in a Fourier series:

$$f(x) = a_0 + a_1 \cos x + a_2 \cos 2x + a_3 \cos 3x + \cdots$$
$$+ b_1 \sin x + b_2 \sin 2x + b_3 \sin 3x + \cdots$$

The function

$$a_k \cos kx + b_k \sin kx$$

is referred to as the kth *harmonic* of f, and it is customary to say that the Fourier series expresses f in terms of its harmonics. The first harmonic, $a_1 \cos x + b_1 \sin x$, is sometimes called the *fundamental harmonic* of f.

Example 2   Find a_0 and the first four harmonics of a *pulse train* function f of period 2π shown in Figure 9.22:

$$f(x) = \begin{cases} 1 & 0 \le x < \pi/2 \\ 0 & \pi/2 \le x < 2\pi \end{cases}$$

Figure 9.22: A train of pulses with period 2π

Solution   First, a_0 is the average value of the function, so

$$a_0 = \frac{1}{2\pi} \int_{-\pi}^{\pi} f(x)\,dx = \frac{1}{2\pi} \int_{0}^{\pi/2} 1\,dx = \frac{1}{4}.$$

Next, we compute a_k and b_k, $k = 1, 2, 3$, and 4. The formulas

$$a_k = \frac{1}{\pi} \int_{-\pi}^{\pi} f(x)\cos(kx)\,dx = \frac{1}{\pi} \int_{0}^{\pi/2} \cos(kx)\,dx$$

$$b_k = \frac{1}{\pi} \int_{-\pi}^{\pi} f(x)\sin(kx)\,dx = \frac{1}{\pi} \int_{0}^{\pi/2} \sin(kx)\,dx$$

lead to the harmonics

$$a_1 \cos x + b_1 \sin x = \frac{1}{\pi}\cos x + \frac{1}{\pi}\sin x$$

$$a_2 \cos(2x) + b_2 \sin(2x) = \frac{1}{\pi}\sin(2x)$$

$$a_3 \cos(3x) + b_3 \sin(3x) = -\frac{1}{3\pi}\cos(3x) + \frac{1}{3\pi}\sin(3x)$$

$$a_4 \cos(4x) + b_4 \sin(4x) = 0.$$

Figure 9.23 shows the graph of the sum of a_0 and these harmonics, which is the fourth Fourier approximation of f.

Figure 9.23: Fourth Fourier approximation to pulse train f equals the sum of a_0 and the first four harmonics

Energy and the Energy Theorem

The quantity $A_k = \sqrt{a_k^2 + b_k^2}$ is called the amplitude of the kth harmonic. The square of the amplitude has a useful interpretation. Adopting terminology from the study of periodic waves, we define the *energy E* of a periodic function f of period 2π to be the number

$$E = \frac{1}{\pi} \int_{-\pi}^{\pi} [f(x)]^2\,dx.$$

Problem 16 on page 469 asks you to verify that for all positive integers k,

$$\frac{1}{\pi} \int_{-\pi}^{\pi} (a_k \cos(kx) + b_k \sin(kx))^2\,dx = a_k^2 + b_k^2 = A_k^2.$$

This shows that the kth harmonic of f has energy A_k^2. The energy of the constant term a_0 of the Fourier series is $\frac{1}{\pi} \int_{-\pi}^{\pi} a_0^2\,dx = 2a_0^2$, so we make the definition

$$A_0 = \sqrt{2}a_0.$$

It turns out that for all reasonable periodic functions f, the energy of f equals the sum of the energy of its harmonics:

The Energy Theorem for a Periodic Function f of Period 2π

$$E = \frac{1}{\pi} \int_{-\pi}^{\pi} [f(x)]^2 \, dx = A_0^2 + A_1^2 + A_2^2 + \cdots$$

where $A_0 = \sqrt{2}a_0$ and $A_k = \sqrt{a_k^2 + b_k^2}$ (for all integers $k \geq 1$).

The graph of A_k^2 against k is called the *energy spectrum* of f. It shows how the energy of f is distributed among its harmonics.

Example 3 (a) Graph the energy spectrum of the square wave of Example 1.

(b) What fraction of the energy of the square wave is contained in the constant term and first three harmonics of its Fourier series?

Solution (a) We know from Example 1 that $a_0 = 1/2$, $a_k = 0$ for $k \geq 1$, $b_k = 0$ for k even, and $b_k = 2/(k\pi)$ for k odd. Thus

$$A_0^2 = 2a_0^2 = \frac{1}{2}$$

$$A_k^2 = 0 \quad \text{if } k \text{ is even}, \quad k \geq 1,$$

$$A_k^2 = \left(\frac{2}{k\pi}\right)^2 = \frac{4}{k^2\pi^2} \quad \text{if } k \text{ is odd}, \quad k \geq 1.$$

The energy spectrum is graphed in Figure 9.24. Notice that it is customary to represent the energy A_k^2 of the kth harmonic by a vertical line of length A_k^2. The graph shows that the constant term and first harmonic carry most of the energy of f.

Figure 9.24: The energy spectrum of a square wave

(b) The energy of the square wave $f(x)$ is

$$E = \frac{1}{\pi} \int_{-\pi}^{\pi} [f(x)]^2 \, dx = \frac{1}{\pi} \int_0^{\pi} 1 \, dx = 1.$$

The energy in the constant term and the first three harmonics of the Fourier series is

$$A_0^2 + A_1^2 + A_2^2 + A_3^2 = \frac{1}{2} + \frac{4}{\pi^2} + 0 + \frac{4}{9\pi^2} = 0.950.$$

The fraction of energy carried by the constant term and the first three harmonics is

$$0.95/1 = 0.95, \text{ or } 95\%.$$

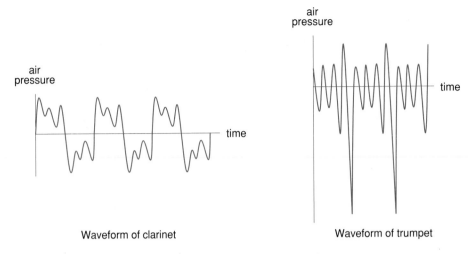

Figure 9.25: Sound waves of a clarinet and trumpet

Musical Instruments

You may have wondered why different musical instruments sound different, even when playing the same note. A first step might be to graph the periodic variations in air pressure that form the sound waves they produce. This has been done for clarinet and trumpet in Figure 9.25.[4] However, it is more revealing to graph the energy spectra of these functions, as in Figure 9.26. The most striking difference is the relative weakness of the second, fourth, and sixth harmonics for the clarinet, with the second harmonic completely absent. The trumpet sounds the second harmonic with as much energy as it does the fundamental.

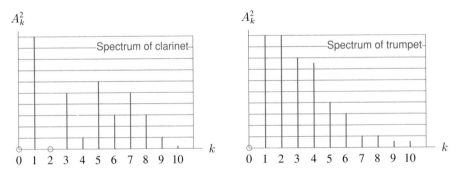

Figure 9.26: Energy spectra of a clarinet and trumpet

What Do We Do if Our Function Does Not Have Period 2π?

We can easily adapt what we have already done by changing variables. Suppose we have a function $f(x)$ that is periodic with period b. If we define $x = bt/2\pi$, then x varies over the interval $[-b/2, b/2]$ when t varies over the interval $[-\pi, \pi]$. Thus, if we substitute $x = bt/2\pi$ into f and define a new function g by

$$g(t) = f\left(\frac{bt}{2\pi}\right) = f(x),$$

then g has period 2π. We can find the Fourier series for g as before, giving

$$g(t) = a_0 + a_1 \cos t + a_2 \cos(2t) + a_3 \cos(3t) + \cdots$$
$$+ b_1 \sin t + b_2 \sin(2t) + b_3 \sin(3t) + \cdots.$$

---

[4]Adapted from C.A. Culver, *Musical Acoustics* (New York: McGraw-Hill, 1956) pp. 204, 220.

Substituting $t = 2\pi x/b$ allows us to convert back:

$$f(x) = g\left(\frac{2\pi x}{b}\right) = a_0 + a_1\cos\left(\frac{2\pi x}{b}\right) + a_2\cos\left(\frac{4\pi x}{b}\right) + a_3\cos\left(\frac{6\pi x}{b}\right) + \cdots$$
$$+ b_1\sin\left(\frac{2\pi x}{b}\right) + b_2\sin\left(\frac{4\pi x}{b}\right) + b_3\sin\left(\frac{6\pi x}{b}\right) + \cdots$$

Note that the terms in the Fourier series of a periodic function of period b are not $\cos(kx)$ and $\sin(kx)$, but instead are $\cos(2\pi kx/b)$ and $\sin(2\pi kx/b)$.

In addition, for a periodic function of period 2π, integrals over $[-\pi, \pi]$ can be replaced by integrals over any interval of length 2π. Thus the integrals for a_k and b_k can be evaluated over any interval of length 2π, not just $[-\pi, \pi]$.

Example 4   Find the fifth-degree Fourier polynomial of the square wave $f(x)$ graphed in Figure 9.27.

Figure 9.27: Square wave f and its fifth Fourier approximation F_5

Solution   Since f has period $b = 2$, we let $x = bt/2\pi = t/\pi$ and consider the function

$$g(t) = f\left(\frac{t}{\pi}\right) = f(x)$$

which has period 2π. Then

$$g(t) = \begin{cases} 0 & -\pi \le t < 0 \\ 1 & 0 \le t < \pi, \end{cases}$$

so $g(t)$ is the square wave of Example 1. We have, therefore,

$$g(t) \approx \frac{1}{2} + \frac{2}{\pi}\sin t + \frac{2}{3\pi}\sin(3t) + \frac{2}{5\pi}\sin(5t).$$

Hence, substituting $t = \pi x$,

$$f(x) = g(t) = g(\pi x) \approx \frac{1}{2} + \frac{2}{\pi}\sin(\pi x) + \frac{2}{3\pi}\sin(3\pi x) + \frac{2}{5\pi}\sin(5\pi x).$$

Seasonal Variation in the Incidence of Measles

Example 5   Fourier approximations have been used to analyze the seasonal variation in the incidence of diseases. One study[5] done in Baltimore, Maryland, for the years 1901–1931, studied $I(t)$, the average number of cases of measles per 10,000 susceptible children in the tth month of the year. The data points in Figure 9.28 show $f(t) = \log I(t)$. The curve in Figure 9.28 shows the second Fourier approximation of $f(t)$. Figure 9.29 contains the graphs of the first and second harmonics of $f(t)$, plotted separately as deviations about a_0, the average logarithmic incidence rate. Describe what these two harmonics tell you about incidence of measles.

[5]From C. I. Bliss and D. L. Blevins, *The Analysis of Seasonal Variation in Measles* (Am. J. Hyg. 70, 1959), reported by Edward Batschelet, *Introduction to Mathematics for the Life Sciences* (Springer-Verlag, Berlin, 1979).

Figure 9.28: Logarithm of incidence of measles per month (dots) and second Fourier approximation (curve)

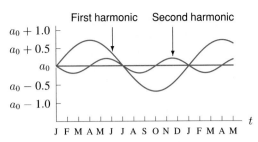

Figure 9.29: First and second harmonics of $f(t)$ plotted as deviations from average log incidence rate, a_0

Solution   Taking the log of $I(t)$ has the effect of reducing the amplitude of the oscillations. However, since the log of a function increases when the function increases, and decreases when it decreases, oscillations in $f(t)$ correspond to oscillations in $I(t)$.

Figure 9.29 shows that the first harmonic in the Fourier series has a period of one year (the same period as the original function); the second harmonic has a period of six months. The graph in Figure 9.29 shows that the first harmonic is approximately a sine with amplitude about 0.7; the second harmonic is approximately the negative of a sine with amplitude about 0.2. Thus, for t in months ($t = 0$ in January),

$$\log I(t) = f(t) \approx a_0 + 0.7 \sin\left(\frac{\pi}{6}t\right) - 0.2 \sin\left(\frac{\pi}{3}t\right),$$

where $\pi/6$ and $\pi/3$ are introduced to make the periods 12 and 6 months, respectively. We can estimate a_0 from the original graph of f: it is the average value of approximately 1.5. Thus

$$f(t) \approx 1.5 + 0.7 \sin\left(\frac{\pi}{6}t\right) - 0.2 \sin\left(\frac{\pi}{3}t\right).$$

Figure 9.28 shows that the second Fourier approximation of $f(t)$ is quite good. The harmonics of $f(t)$ beyond the second must be rather insignificant. This suggests that the variation in incidence in measles comes from two sources, one with a yearly cycle that is reflected in the first harmonic and one with a half-yearly cycle reflected in the second harmonic. At this point the mathematics can tell us no more; we must turn to the epidemiologists for further explanation.

Informal Justification of the Formulas for the Fourier Coefficients

Recall that the coefficients in a Taylor series (which is a good approximation locally) are found by differentiation. In contrast, the coefficients in a Fourier series (which is a good approximation globally) are found by integration.

We will use a $\sum$ sign with ∞ at the top to denote an infinite series:

$$f(x) = a_0 + \sum_{k=1}^{\infty} a_k \cos(kx) + \sum_{k=1}^{\infty} b_k \sin(kx).$$

Consider the integral

$$\int_{-\pi}^{\pi} f(x)\,dx = \int_{-\pi}^{\pi} \left[a_0 + \sum_{k=1}^{\infty} a_k \cos(kx) + \sum_{k=1}^{\infty} b_k \sin(kx) \right] dx.$$

Splitting the integral into separate terms, and assuming we can interchange integration and summation, we get

$$\int_{-\pi}^{\pi} f(x)\,dx = \int_{-\pi}^{\pi} a_0\,dx + \int_{-\pi}^{\pi} \sum_{k=1}^{\infty} a_k \cos(kx)\,dx + \int_{-\pi}^{\pi} \sum_{k=1}^{\infty} b_k \sin(kx)\,dx$$

$$= \int_{-\pi}^{\pi} a_0\,dx + \sum_{k=1}^{\infty} \int_{-\pi}^{\pi} a_k \cos(kx)\,dx + \sum_{k=1}^{\infty} \int_{-\pi}^{\pi} b_k \sin(kx)\,dx.$$

But for $k \geq 1$, thinking of the integral as an area shows that

$$\int_{-\pi}^{\pi} \sin(kx)\,dx = 0 \;\text{ and }\; \int_{-\pi}^{\pi} \cos(kx)\,dx = 0,$$

so all terms drop out except the first, giving

$$\int_{-\pi}^{\pi} f(x)\,dx = \int_{-\pi}^{\pi} a_0\,dx = a_0 x \Big|_{-\pi}^{\pi} = 2\pi a_0$$

and so we get the following result:

$$a_0 = \frac{1}{2\pi} \int_{-\pi}^{\pi} f(x)\,dx.$$

Thus a_0 is the average value of f on the interval $[-\pi, \pi]$.

To determine the values of any of the other a_k (for positive k), we use a rather clever method which depends on the following facts. For all integers k and m,

$$\int_{-\pi}^{\pi} \sin(kx)\cos(mx)\,dx = 0,$$

and, provided $k \neq m$,

$$\int_{-\pi}^{\pi} \cos(kx)\cos(mx)\,dx = 0.$$

(See Problems 23–27 on page 470.) In addition, provided $m \neq 0$, we have

$$\int_{-\pi}^{\pi} \cos^2(mx)\,dx = \pi.$$

To use these facts, we multiply the Fourier series by $\cos(mx)$, where m is any positive integer:

$$f(x)\cos(mx) = a_0 \cos(mx) + \sum_{k=1}^{\infty} a_k \cos(kx)\cos(mx) + \sum_{k=1}^{\infty} b_k \sin(kx)\cos(mx).$$

We integrate this between $-\pi$ and π, term by term:

$$\int_{-\pi}^{\pi} f(x)\cos(mx)\,dx = \int_{-\pi}^{\pi} \left(a_0 \cos(mx) + \sum_{k=1}^{\infty} a_k \cos(kx)\cos(mx) + \sum_{k=1}^{\infty} b_k \sin(kx)\cos(mx) \right) dx$$

$$= a_0 \int_{-\pi}^{\pi} \cos(mx)\,dx + \sum_{k=1}^{\infty} \left(a_k \int_{-\pi}^{\pi} \cos(kx)\cos(mx)\,dx \right)$$

$$+ \sum_{k=1}^{\infty} \left(b_k \int_{-\pi}^{\pi} \sin(kx)\cos(mx)\,dx \right).$$

Provided $m \neq 0$, we have $\int_{-\pi}^{\pi} \cos(mx)\, dx = 0$. Since the integral $\int_{-\pi}^{\pi} \sin(kx)\cos(mx)\, dx = 0$, all the terms in the second sum are zero. Since $\int_{-\pi}^{\pi} \cos kx \cos mx\, dx = 0$ provided $k \neq m$, all the terms in the first sum are zero except where $k = m$. Thus the right-hand side reduces to one term:

$$\int_{-\pi}^{\pi} f(x)\cos(mx)\, dx = a_m \int_{-\pi}^{\pi} \cos(mx)\cos(mx)\, dx = \pi a_m.$$

This leads, for each value of $m = 1, 2, 3 \ldots$, to the following formula:

$$a_m = \frac{1}{\pi} \int_{-\pi}^{\pi} f(x)\cos(mx)\, dx.$$

Using a similar argument, we multiply through by $\sin(mx)$ instead of $\cos(mx)$ and eventually obtain, for each value of $m = 1, 2, 3 \ldots$, the following result:

$$b_m = \frac{1}{\pi} \int_{-\pi}^{\pi} f(x)\sin(mx)dx\, dx.$$

Problems for Section 9.5

Which of the series in Problems 1–4 are Fourier series?

1. $1 + \cos x + \cos^2 x + \cos^3 x + \cos^4 x + \cdots$

2. $\sin x + \sin(x+1) + \sin(x+2) + \cdots$

3. $\cos x + \sin x - \cos(2x) - \dfrac{\sin(2x)}{2} + \cos(3x) + \dfrac{\sin(3x)}{3} - \cdots$

4. $\dfrac{1}{2} - \sin x + \sin(2x) - \sin(3x) + \cdots$

5. Construct the first three Fourier approximations to the square wave function

$$f(x) = \begin{cases} -1 & -\pi \leq x < 0 \\ 1 & 0 \leq x < \pi. \end{cases}$$

   Use a calculator or computer to draw the graph of each approximation.

6. Repeat Problem 5 with the function

$$f(x) = \begin{cases} -x & -\pi \leq x < 0 \\ x & 0 \leq x < \pi. \end{cases}$$

7. What fraction of the energy of the function in Problem 6 is contained in the constant term and first three harmonics of its Fourier series?

For Problems 8–10, find the nth Fourier polynomial for the given functions, assuming them to be periodic with period 2π. Graph the first three approximations with the original function.

8. $f(x) = x^2, \quad -\pi < x \leq \pi.$ 9. $h(x) = \begin{cases} 0 & -\pi < x \leq 0 \\ x & 0 < x \leq \pi. \end{cases}$ 10. $g(x) = x, \quad -\pi < x \leq \pi.$

11. (a) Find and graph the third Fourier approximation of the square wave $g(x)$ of period 2π:

$$g(x) = \begin{cases} 0 & -\pi \leq x < \pi/2 \\ 1 & -\pi/2 \leq x < \pi/2 \\ 0 & \pi/2 \leq x < \pi. \end{cases}$$

   (b) How does the result of part (a) differ from that of the square wave in Example 1?

12. Suppose we have a periodic function f with period 1 defined by $f(x) = x$ for $0 \leq x < 1$. Find the fourth degree Fourier polynomial for f and graph it on the interval $0 \leq x < 1$. [Hint: Remember that since the period is not 2π, you will have to start by doing a substitution. Notice that the terms in the sum are not $\sin(nx)$ and $\cos(nx)$, but instead turn out to be $\sin(2\pi nx)$ and $\cos(2\pi nx)$.]

13. Suppose f has period 2 and $f(x) = x$ for $0 \leq x < 2$. Find the fourth-degree Fourier polynomial and graph it on $0 \leq x < 2$. [Hint: See Problem 12.]

14. Suppose that a spacecraft near Neptune has measured a quantity A and sent it to earth in the form of a periodic signal $A \cos t$ of amplitude A. On its way to earth, the signal picks up periodic noise, containing only second and higher harmonics. Suppose that the signal $h(t)$ actually received on earth is graphed below in Figure 9.30. Determine the signal that the spacecraft originally sent and hence the value A of the measurement.

Figure 9.30

15. Figures 9.31 and 9.32 show the waveforms and energy spectra for notes produced by flute and bassoon.[6] Describe the principal differences between the two spectra.

Figure 9.31

Figure 9.32

16. Show that for positive integers k, the periodic function $f(x) = a_k \cos kx + b_k \sin kx$ of period 2π has energy $a_k^2 + b_k^2$.

---

[6]Adapted from C.A. Culver, *Musical Acoustics* (New York: McGraw-Hill, 1956), pp. 200, 213.

17.   Given the graph of f in Figure 9.33, find the first two Fourier approximations numerically.

Figure 9.33

18.   Justify the formula $b_k = \frac{1}{\pi} \int_{-\pi}^{\pi} f(x) \sin(kx)\, dx$ for the Fourier coefficients, b_k, of a periodic function of period 2π. The argument is similar to that in the text for a_k. In addition to some of the formulas used there, you will need the formulas: $\int_{-\pi}^{\pi} \sin^2(kx)\, dx = \pi$ and $\int_{-\pi}^{\pi} \sin(kx) \sin(mx)\, dx = 0$ for $k \neq m$.

In Problems 19–22, the pulse train of width c is the periodic function f of period 2π given by

$$f(x) = \begin{cases} 0 & -\pi \leq x < -c/2 \\ 1 & -c/2 \leq x < c/2 \\ 0 & c/2 \leq x < \pi. \end{cases}$$

19.   Suppose that f is the pulse train of width 1.

(a)   What fraction of the energy of f is contained in the constant term of its Fourier series? In the constant term and the first harmonic together?

(b)   Find a formula for the energy of the kth harmonic of f. Use it to sketch the energy spectrum of f.

(c)   How many terms of the Fourier series of f are needed to capture 90% of the energy of f?

(d)   Graph f and its fifth Fourier approximation on the interval $[-3\pi, 3\pi]$.

20.   Suppose that f is the pulse train of width 0.4.

(a)   What fraction of the energy of f is contained in the constant term of its Fourier series? In the constant term and the first harmonic together?

(b)   Find a formula for the energy of the kth harmonic of f. Use it to sketch the energy spectrum of f.

(c)   What fraction of the energy of f is contained in the constant term and the first five harmonics of f? (The constant term and the first thirteen harmonics are needed to capture 90% of the energy of f.)

(d)   Graph f and its fifth Fourier approximation on the interval $[-3\pi, 3\pi]$.

21.   Suppose that f is the pulse train of width 2.

(a)   What fraction of the energy of f is contained in the constant term of its Fourier series? In the constant term and the first harmonic together?

(b)   How many terms of the Fourier series of f are needed to capture 90% of the energy of f?

(c)   Graph f and its third Fourier approximation on the interval $[-3\pi, 3\pi]$.

22.   After working Problems 19– 21, write a paragraph about the approximation of pulse trains by Fourier polynomials. Explain how the energy spectrum of a pulse train of width c will change as c gets closer and closer to 0 and how this affects the number of terms required for an accurate approximation.

For Problems 23–27, use the table of integrals inside the back cover to show that the following statements are true for positive integers k and m.

23. $\displaystyle\int_{-\pi}^{\pi} \cos(kx) \cos(mx)\, dx = 0, \quad \text{if } k \neq m.$ 24. $\displaystyle\int_{-\pi}^{\pi} \cos^2(mx)\, dx = \pi.$

25. $\displaystyle\int_{-\pi}^{\pi} \sin^2(mx)\, dx = \pi.$ 26. $\displaystyle\int_{-\pi}^{\pi} \sin(kx) \cos(mx)\, dx = 0.$

27. $\displaystyle\int_{-\pi}^{\pi} \sin(kx) \sin(mx)\, dx = 0, \quad \text{if } k \neq m.$

CHAPTER SUMMARY

- **Taylor series and polynomials**
  General expansion about $x = 0$ or $x = a$; specific series for e^x, $\sin x$, $\cos x$, $(1 + x)^p$; using known Taylor series to find others by substitution, integration, and differentiation; interval of convergence
- **Geometric series**
  Finite sum, infinite sum
- **Power series**
  Ratio test for radius of convergence
- **Series of constants**
  Harmonic series, alternating series
- **Fourier series**

REVIEW PROBLEMS FOR CHAPTER NINE

For Problems 1–3, find the second-degree Taylor polynomial about the given point.

1. e^x, $x = 1$ **2.** $\ln x$, $x = 2$ **3.** $\sin x$, $x = -\pi/4$

4. Find the third-degree Taylor polynomial for $f(x) = x^3 + 7x^2 - 5x + 1$ at $x = 1$.

In Problems 5–8, find the first four nonzero terms of the Taylor series about the origin of the given functions.

5. $\theta^2 \cos \theta^2$ **6.** $\sin t^2$ **7.** $\dfrac{1}{\sqrt{4 - x}}$ **8.** $\dfrac{1}{1 - 4z^2}$

For Problems 9–10, expand the quantity in a Taylor series around the origin in terms of the variable given. Give the first four nonzero terms.

9. $\dfrac{a}{a + b}$ in terms of $\dfrac{b}{a}$ **10.** $\sqrt{R - r}$ in terms of $\dfrac{r}{R}$

Find the exact value of the sums of the series in Problems 11–14.

11. $1 - \dfrac{1}{3} + \dfrac{1}{9} - \dfrac{1}{27} + \dfrac{1}{81} - \cdots$ **12.** $8 + 4 + 2 + 1 + \dfrac{1}{2} + \dfrac{1}{4} + \cdots + \dfrac{1}{2^{10}}$

13. $3 + 3 + \dfrac{3}{2!} + \dfrac{3}{3!} + \dfrac{3}{4!} + \dfrac{3}{5!} + \cdots$ **14.** $(0.1)^2 - \dfrac{(0.1)^4}{3!} + \dfrac{(0.1)^6}{5!} - \dfrac{(0.1)^8}{7!} + \cdots$

15. Suppose x is positive but very small. Arrange the following expressions in increasing order:
$$x, \quad \sin x, \quad \ln(1 + x), \quad 1 - \cos x, \quad e^x - 1, \quad \arctan x, \quad x\sqrt{1 - x}.$$

16. Consider the Taylor expansion
$$f(x) = \frac{1}{1 + x} = 1 - x + x^2 - x^3 + x^4 - \cdots$$
By plotting several Taylor polynomials and the function $f(x) = 1/(1 + x)$, confirm that the interval of convergence of this series is $-1 < x < 1$.

17. Use Taylor series to evaluate $\displaystyle \lim_{x \to 0} \frac{\ln(1 + x + x^2) - x}{\sin^2 x}$.

18. (a) Find $\lim\limits_{\theta \to 0} \dfrac{\sin(2\theta)}{\theta}$. Explain your reasoning.

   (b) Use series to explain why $f(\theta) = \dfrac{\sin(2\theta)}{\theta}$ looks like a parabola near $\theta = 0$. What is the equation of the parabola?

19. (a) Find the Taylor series for $f(t) = te^t$ about $t = 0$.

   (b) Using your answer to part (a), find a Taylor series expansion about $x = 0$ for

$$\int_0^x te^t \, dt.$$

   (c) Using your answer to part (b), show that

$$\frac{1}{2} + \frac{1}{3} + \frac{1}{4(2!)} + \frac{1}{5(3!)} + \frac{1}{6(4!)} + \cdots = 1.$$

20. The theory of relativity predicts that when an object moves at speeds close to the speed of light, the object appears heavier. The apparent, or relativistic, mass, m, of the object when it is moving at speed v is given by the formula

$$m = \frac{m_0}{\sqrt{1 - v^2/c^2}}$$

where c is the speed of light and m_0 is the mass of the object when it is at rest.

   (a) Use the formula for m to decide what values of v are possible.

   (b) Sketch a rough graph of m against v, labeling intercepts and asymptotes.

   (c) Write the first three nonzero terms of the Taylor series for m in terms of v.

   (d) For what values of v do you expect the series to converge?

21. The potential energy, V, of two gas molecules separated by a distance r is given by

$$V = -V_0 \left(2 \left(\frac{r_0}{r} \right)^6 - \left(\frac{r_0}{r} \right)^{12} \right),$$

where V_0 and r_0 are positive constants.

   (a) Show that if $r = r_0$, then V takes on its minimum value, $-V_0$.

   (b) Write V as a series in $(r - r_0)$ up through the quadratic term.

   (c) For r near r_0, show that the difference between V and its minimum value is approximately proportional to $(r - r_0)^2$. In other words, show that $V - (-V_0) = V + V_0$ is approximately proportional to $(r - r_0)^2$.

   (d) The force, F, between the molecules is given by $F = -dV/dr$. What is F when $r = r_0$? For r near r_0, show that F is approximately proportional to $(r - r_0)$.

22. The *gravitational field* at a point in space is the gravitational force that would be exerted on a unit mass placed there. We will assume that the gravitational field strength at a distance d away from a mass M is

$$\frac{GM}{d^2}$$

where G is constant. In this problem you will investigate the gravitational field strength, F, exerted by a system consisting of a large mass M and a small mass m, with a distance r between them. (See Figure 9.34.)

Figure 9.34

   (a) Write an expression for the gravitational field strength, F, at the point P.

   (b) Assuming r is small in comparison to R, expand F in a series in r/R.

   (c) By discarding terms in $(r/R)^2$ and higher powers, explain why you can view the field as resulting from a single particle of mass $M + m$, plus a correction term. What is the position of the particle of mass $M + m$? Explain the sign of the correction term.

23. A repeating decimal can always be expressed as a fraction. This problem shows how writing a repeating decimal as a geometric series enables you to find the fraction. Consider the decimal $0.232323\dots$.

    (a) Use the fact that $0.232323\dots = 0.23 + 0.0023 + 0.000023 + \cdots$ to write $0.232323\dots$ as a geometric series.
    (b) Use the formula for the sum of a geometric series to show that $0.232323\dots = 23/99$.

24. Cephalexin is an antibiotic with a half-life in the body of 0.9 hours, taken in tablets of 250 mg every six hours.

    (a) What percentage of the cephalexin in the body at the start of a six-hour period is still there at the end (assuming no tablets are taken during that time)?
    (b) Write an expression for Q_1, Q_2, Q_3, Q_4, where Q_n mg, is the amount of cephalexin in the body right after the nth tablet is taken.
    (c) Express Q_3, Q_4 in closed-form and evaluate them.
    (d) Write an expression for Q_n and put it in closed-form.
    (e) If the patient keeps taking the tablets, use your answer to part (d) to find the quantity of cephalexin in the body in the long run, right after taking a pill.

25. Expand $f(x + h)$ and $g(x + h)$ in Taylor series and take a limit to confirm the product rule:

$$\frac{d}{dx}(f(x)g(x)) = f'(x)g(x) + f(x)g'(x).$$

26. Use Taylor expansions for $f(y + k)$ and $g(x + h)$ to confirm the chain rule:

$$\frac{d}{dx}(f(g(x))) = f'(g(x)) \cdot g'(x).$$

27. Suppose all the derivatives of g exist at $x = 0$ and that g has a critical point at $x = 0$.
    (a) Write the nth Taylor polynomial for g at $x = 0$.
    (b) What does the Second Derivative test for local maxima and minima say?
    (c) Use the Taylor polynomial to explain why the Second Derivative test works.

28. (Continuation of Problem 27) You may remember that the Second Derivative test tells us nothing when the second derivative is zero at the critical point. In this problem you will investigate that special case.

    Assume g has the same properties as in Problem 27, and that, in addition, $g''(0) = 0$. What does the Taylor polynomial tell you about whether g has a local maximum or minimum at $x = 0$?

29. Use the Fourier polynomials for the square wave

$$f(x) = \begin{cases} -1 & -\pi < x \le 0 \\ 1 & 0 < x \le \pi \end{cases}$$

to explain why

$$1 - \frac{1}{3} + \frac{1}{5} - \frac{1}{7} + \cdots + (-1)^{2n+1}\frac{1}{2n + 1}$$

must approach $\frac{\pi}{4}$ as $n \to \infty$.

30. Find a Fourier polynomial of degree three for $f(x) = e^{2\pi x}$, for $0 \le x < 1$.

31. Suppose that $f(x)$ is a differentiable periodic function of period 2π. Assume the Fourier series of f is differentiable term by term.

    (a) If the Fourier coefficients of f are a_k and b_k, show that the Fourier coefficients of its derivative f' are kb_k and $-ka_k$.
    (b) How are the amplitudes of the harmonics of f and f' related?
    (c) How are the energy spectra of f and f' related?

32. If the Fourier coefficients of f are a_k and b_k, and the Fourier coefficients of g are c_k and d_k, and if A and B are real, show that the Fourier coefficients of $Af + Bg$ are $Aa_k + Bc_k$ and $Ab_k + Bd_k$.

33. Suppose that f is a periodic function of period 2π and that g is a horizontal shift of f, say $g(x) = f(x+c)$. Show that f and g have the same energy.

PROJECTS

1. **Approximating the Derivative**[7]

   In applications, the values of a function $f(x)$ are frequently known only at discrete values x_0, $x_0 \pm h$, $x_0 \pm 2h, \ldots$. Suppose we are interested in approximating the derivative $f'(x_0)$. The definition

   $$f'(x_0) = \lim_{h \to 0} \frac{f(x_0 + h) - f(x_0)}{h}$$

   suggests that for small h we can approximate $f'(x)$ as follows:

   $$f'(x_0) \approx \frac{f(x_0 + h) - f(x_0)}{h}.$$

   Such *finite-difference approximations* are used frequently in programming a computer to solve differential equations.

   Taylor series can be used to analyze the error in this approximation. Substituting

   $$f(x_0 + h) = f(x_0) + f'(x_0)h + \frac{f''(x_0)}{2}h^2 + \cdots$$

   into the approximation for $f'(x_0)$, we find

   $$\frac{f(x_0 + h) - f(x_0)}{h} = f'(x_0) + \frac{f''(x_0)}{2}h + \cdots$$

   This suggests (and it can be proved) that the error in the approximation is bounded as follows:

   $$\left| \frac{f(x_0 + h) - f(x_0)}{h} - f'(x_0) \right| \leq \frac{Mh}{2},$$

   where

   $$|f''(x)| \leq M \qquad \text{for} \qquad |x - x_0| \leq |h|.$$

   Notice that as $h \to 0$, the error also goes to zero, provided M is bounded.

   As an example, we take $f(x) = e^x$ and $x_0 = 0$, so $f'(x_0) = 1$. The error for various values of h are given in Table 9.3. We see that decreasing h by a factor of 10 decreases the error by a factor of about 10, as predicted by the error bound $Mh/2$.

   TABLE 9.3

h	$(f(x_0 + h) - f(x_0))/h$	Error
10^{-1}	1.05171	5.171×10^{-2}
10^{-2}	1.00502	5.02×10^{-3}
10^{-3}	1.00050	5.0×10^{-4}
10^{-4}	1.00005	5.0×10^{-5}

   (a) Using Taylor series, suggest an error bound for each of the following finite-difference approximations.

   (i) $f'(x_0) \approx \dfrac{f(x_0) - f(x_0 - h)}{h}$

   (ii) $f'(x_0) \approx \dfrac{f(x_0 + h) - f(x_0 - h)}{2h}$

   (iii) $f'(x_0) \approx \dfrac{-f(x_0 + 2h) + 8f(x_0 + h) - 8f(x_0 - h) + f(x_0 - 2h)}{12h}$

---

[7]From Mark Kunka

(b)   Use each of the formulas in part (a) to approximate the first derivative of e^x at $x = 0$ for $h = 10^{-1}, 10^{-2}, 10^{-3}, 10^{-4}$. As h is decreased by a factor of 10, how does the error decrease? Does this agree with the error bounds found in part (a)? Which is the most accurate formula?

(c)   Repeat part (b) using $f(x) = 1/x$ and $x_0 = 10^{-5}$. Why are these formulas not good approximations anymore? Continue to decrease h by factors of 10. How small does h have to be before formula (iii) is the best approximation? At these smaller values of h, what changed to make the formulas accurate again?

2.   **Probability of Winning in Sports**

In certain sports, winning a game requires a lead of two points. That is, if the score is tied you have to score two points in a row to win.

(a)   For some sports (e.g. tennis), a point is scored every play. Suppose your probability of scoring the next point is always p. Then, your opponent's probability of scoring the next point is always $1 - p$.

   (i)   What is the probability that you win the next two points?

   (ii)   What is the probability that you and your opponent split the next two points, that is, that neither of you wins both points?

   (iii) What is the probability that you split the next two points but you win the two after that?

   (iv)   What is the probability that you either win the next two points or split the next two and then win the next two after that?

   (v)   Give a formula for your probability w of winning a tied game.

   (vi)   Compute your probability of winning a tied game when $p = 0.5$; when $p = 0.6$; when $p = 0.7$; when $p = 0.4$. Comment on your answers.

(b)   In other sports (e.g. volleyball), you can score a point only if it is your turn, with turns alternating until a point is scored. Suppose your probability of scoring a point when it is your turn is p, and your opponent's probability of scoring a point when it is her turn is q.

   (i)   Find a formula for the probability S that you are the first to score the next point, assuming it is currently your turn.

   (ii)   Suppose that if you score a point, the next turn is yours. Using your answers to part (a) and your formula for S, compute the probability of winning a tied game (if you need two points in a row to win).

      • Assume $p = 0.5$ and $q = 0.5$ and it is your turn.

      • Assume $p = 0.6$ and $q = 0.5$ and it is your turn.

FOCUS ON THEORY

CONVERGENCE THEOREMS

In this section[8] we study some methods for analyzing the convergence of the power series

$$\sum_{n=0}^{\infty} C_n x^n = C_0 + C_1 x + C_2 x^2 + \cdots + C_n x^n + \cdots.$$

For any particular value of x, the power series is a series of constants of the form

$$\sum_{n=1}^{\infty} a_n = a_1 + a_2 + a_3 + \cdots + a_n + \cdots.$$

So we first consider the convergence of series of constants. We can plot the terms of a series of constants as in Figure 9.35. If $a_n \geq 0$ for all n, then each rectangle has area a_n.

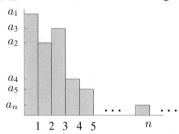

Figure 9.35: Height and area of the nth rectangle is a_n

Convergence of Series of Constants

We define the nth partial sum, S_n, as

$$S_n = a_1 + a_2 + \cdots + a_n,$$

and we say that

> If $S = \lim_{n \to \infty} S_n$ exists, then the series $\sum a_n$ **converges**, and its sum is S. If a series does not converge, we say that it **diverges**.

If $a_n \geq 0$ for all n, the series converges when the total area of the rectangles in Figure 9.35 is finite. In that case, the sum of the series is the total area of all the rectangles. This is similar to an improper integral $\int_0^\infty f(x)\,dx$, where we saw that it was possible for the area under the graph of f to be finite, even on an infinite interval.

Here are some properties that are useful in determining whether or not a series converges.

> ### Theorem: Convergence Properties of Series
> 1. If $\sum a_n$ and $\sum b_n$ converge and if k is a constant, then
>   - $\sum (a_n + b_n)$ converges to $\sum a_n + \sum b_n$.
>   - $\sum k a_n$ converges to $k \sum a_n$.
> 2. Changing a finite number of terms in a series does not change whether or not it converges, although it may change its sum if it does converge.
> 3. If $\sum a_n$ converges, then $\lim_{n \to \infty} a_n = 0$.

---

[8]Based on notes by Lynne Small.

For proofs of these properties, see Problems 1–3. Notice that Property 3 can often be used to tell us that a series does *not* converge. It says that if the terms do not go to 0, the series has no chance of converging. However, knowing that $\lim\limits_{n\to\infty} a_n = 0$ is *not* enough to ensure convergence. For example, in Example 7 on page 442 we showed that the harmonic series

$$1 + \frac{1}{2} + \frac{1}{3} + \frac{1}{4} + \cdots + \frac{1}{n} + \cdots$$

does not converge, even though

$$\lim_{n\to\infty} a_n = \lim_{n\to\infty} \frac{1}{n} = 0.$$

Example 1   Show that the geometric series

$$1 + x + x^2 + \cdots$$

does not converge if $x = \pm 1$.

Solution   We can use Property 3 to see that the series does not converge when $x = \pm 1$. When $x = 1$, the nth term is $a_n = 1$, so $\lim_{n\to\infty} a_n = 1 \neq 0$. When $x = -1$, then $a_n = (-1)^{n-1}$, so $\lim_{n\to\infty} a_n$ does not exist. So, in either case, $\sum a_n$ does not converge.

Calculating the partial sums confirms this. If $x = 1$, the series is $1 + 1 + \cdots$. Thus the nth partial sum is $S_n = n$, so the partial sums have no limit. Therefore the series diverges. If $x = -1$, the series is $1 - 1 + 1 - \cdots$. The nth partial sum is 1 if n is odd and 0 if n is even. So, again the partial sums have no limit and the series diverges.

Comparing Series

When we studied improper integrals in Chapter 7, we saw that it was sometimes useful to compare one integral with another in deciding whether the integral converges. In the same way, we can decide whether a series converges by comparing it with one that we already know converges or diverges.

Theorem: Comparison Test

Suppose $0 \leq a_n \leq b_n$ for all n.
- If $\sum b_n$ converges, then $\sum a_n$ converges.
- If $\sum a_n$ does not converge, then $\sum b_n$ does not converge.

Since $a_n \leq b_n$, the plot of the a_n lies under the plot of the b_n. (See Figure 9.36.) The comparison test says that if the total area for $\sum b_n$ is finite, then the total area for $\sum a_n$ is finite also; and that if the total area for $\sum a_n$ is not finite, then neither is the total area for $\sum b_n$.

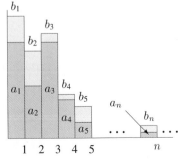

Figure 9.36: Each a_n is represented by the area of a dark rectangle, and each b_n by a dark plus a light rectangle

Example 2   Show that if $0 < p < 1$, then $\sum_{n=1}^{\infty} 1/n^p$ diverges.

Solution   Since $p < 1$, we have $n^p \leq n$, so $1/n^p \geq 1/n$. Since the harmonic series $\sum 1/n$ diverges, $\sum 1/n^p$ diverges also.

Infinite Series and Improper Integrals

Just as we can compare a series with a series, so we can compare a series with an improper integral as we did in Example 7 on page 442 to show that the harmonic series diverges. Here is an example where this type of comparison is used to prove convergence.

Example 3   Show that the series

$$\sum_{n=1}^{\infty} \frac{1}{n^2} = 1 + \frac{1}{4} + \frac{1}{9} + \cdots$$

converges by comparing it to the improper integral $\int_{1}^{\infty} (1/x^2)\, dx$.

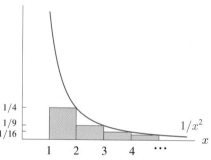

Figure 9.37: $\int_{1}^{\infty} (1/x^2)\, dx$ is greater than $\sum_{n=2}^{\infty} 1/n^2$

Solution   Figure 9.37 shows that the area under the graph of $1/x^2$ for $x \geq 1$ is greater than $1/4 + 1/9 + \cdots$. Since

$$\int_{1}^{\infty} \frac{1}{x^2}\, dx = \lim_{b \to \infty} \int_{1}^{b} \frac{1}{x^2}\, dx = \lim_{b \to \infty} \left(\frac{-1}{b} + 1 \right) = 1,$$

the area under the graph is finite, hence $1/4 + 1/9 + \cdots$ converges. So $1 + 1/4 + 1/9 + \cdots$ also converges.

Series of Both Positive and Negative Terms

If $\sum_n a_n$ has both positive and negative terms, then its plot has rectangles lying both above and below the x-axis. The total area of the rectangles is no longer equal to $\sum a_n$. However, it is still true that if the total area is finite, then the series converges. The area of the nth rectangle is $|a_n|$, so we have:

> If $\sum |a_n|$ converges, then so does $\sum a_n$.

Problem 10 shows how to prove this result.

Example 4   Explain why the series for $\cos x$ converges.

Solution   The series for $\cos x$ can be written

$$1 + 0x - \frac{x^2}{2!} + 0x^3 + \frac{x^4}{4!} + 0x^5 - \frac{x^6}{6!} + \cdots,$$

where the zero terms have been included to enable us to compare with the series for e^x. We consider the series

$$1 + 0|x| + \frac{|x^2|}{2!} + 0|x^3| + \frac{|x^4|}{4!} + 0|x^5| + \frac{|x^6|}{6!} + \cdots.$$

This series converges by comparison with the series for $e^{|x|}$. Hence the series for $\cos x$ converges.

The Ratio Test

In a geometric series, the ratio between the $(n+1)^{\text{st}}$ term and the nth term is a constant, r. If $|r| < 1$, the series converges, and if $|r| \geq 1$, the series diverges. We now state a similar test for a general series of constants, involving the limit of the ratio between the $(n+1)^{\text{st}}$ and n^{th} terms.

The Ratio Test for Series of Constants

For a series $\sum a_n$, suppose

$$\lim_{n \to \infty} \frac{|a_{n+1}|}{|a_n|} = L.$$

- If $L < 1$, then $\sum a_n$ converges.
- If $L > 1$, then $\sum a_n$ does not converge.
- If $L = 1$, the test does not tell us anything.

Proof   The basic idea is to compare the series with a geometric series.

Suppose $L < 1$. Let r be a number between L and 1, so that $L < r < 1$. Since

$$\lim_{n \to \infty} \frac{|a_{n+1}|}{|a_n|} = L,$$

for sufficiently large n, we have

$$\frac{|a_{n+1}|}{|a_n|} < r.$$

Then,

$$|a_{n+1}| < |a_n|r,$$
$$|a_{n+2}| < |a_{n+1}|r < |a_n|r^2,$$
$$|a_{n+3}| < |a_{n+2}|r < |a_n|r^3,$$

and so on. Thus, for a sufficiently large n, we have

$$|a_n| + |a_{n+1}| + |a_{n+2}| + \cdots < |a_n| + |a_n|r + |a_n|r^2 + |a_n|r^3 + \cdots$$
$$= |a_n|(1 + r + r^2 + r^3 + \cdots).$$

Thus, $|a_n| + |a_{n+1}| + |a_{n+2}| + \cdots$ converges by comparison with the convergent geometric series $|a_n|(1 + r + r^2 + r^3 + \cdots)$. Hence, $\sum |a_n|$ converges and so $\sum a_n$ converges.

If $L > 1$, then we choose r so that $1 < r < L$. Then, by a similar argument, the terms in $\sum a_n$ eventually have greater magnitude than the terms in a geometric series with common ratio r. Since $r > 1$, the geometric series does not converge, so $\sum a_n$ does not converge.

Power Series

We are now ready to prove the results about convergence of power series that we used in Section 9.2.

Interval of Convergence

For a power series $\sum C_n x^n$:
- If the series converges for $x = b$, then it converges for all x with $|x| < |b|$.
- If the series does not converge for $x = b$, then it does not converge for all x with $|x| > |b|$.

Proof   We assume[9] $b \neq 0$. If $\sum C_n b^n$ converges, then

$$\lim_{n \to \infty} C_n b^n = 0.$$

In particular, for sufficiently large n, we have

$$|C_n b^n| < 1.$$

For values of x with $|x| < |b|$, we have $|x|/|b| < 1$. Let $r = |x|/|b|$. Then, for sufficiently large n,

$$|C_n x^n| = |C_n b^n| \left| \frac{x^n}{b^n} \right| < \left| \frac{x}{b} \right|^n = r^n.$$

Therefore for x with $|x| < |b|$, the series $\sum |C_n x^n|$ converges by comparison with the convergent geometric series $\sum r^n$. Hence $\sum C_n x^n$ converges too.

For the second part of the theorem, we are told that $\sum C_n b^n$ does not converge. Let us assume that there is a value of x, with $|x| > |b|$, for which the series $\sum C_n x^n$ converges. Then, the first part of the theorem, with the roles of x and b reversed, says that the series $\sum C_n b^n$ would have to converge. But this contradicts the assumption and so completes the proof.

Ratio Test for Power Series

For the power series $\sum C_n x^n$, suppose that

$$\lim_{n \to \infty} \frac{|C_n|}{|C_{n+1}|} = R \quad \text{and } 0 < R < \infty.$$

Then the power series converges for x with $|x| < R$ and does not converge for x with $|x| > R$.

Proof   For $x = 0$, the power series has only one nonzero term and so is convergent. For any fixed x with $|x| \neq 0$, the series $\sum C_n x^n$ is a series of constants. We can apply the ratio test:

$$\lim_{n \to \infty} \frac{|a_{n+1}|}{|a_n|} = \lim_{n \to \infty} \frac{|C_{n+1} x^{n+1}|}{|C_n x^n|} = \lim_{n \to \infty} \frac{|C_{n+1}||x|}{|C_n|} = |x| \lim_{n \to \infty} \frac{|C_{n+1}|}{|C_n|} = \frac{|x|}{R}.$$

By the ratio test,

If $|x|/R < 1$, that is $|x| < R$, then the series converges.

If $|x|/R > 1$, that is, $|x| > R$, then the series does not converge.

---

[9]If $b = 0$, there are no x's where $|x| < |b|$, and $\sum C_n b^n$ must converge.

Problems on Convergence Theorems

1. Show that if $\sum a_n$ and $\sum b_n$ converge and if k is a constant, then $\sum(a_n + b_n)$, $\sum(a_n - b_n)$, and $\sum ka_n$ converge.

2. Let N be a positive integer. Show that if $a_n = b_n$ for $n \geq N$, then $\sum a_n$ and $\sum b_n$ either both converge, or both diverge. In other words, changing a finite number of terms in a series does not change whether or not it converges.

3. Show that if $\sum a_n$ converges, then $\lim_{n \to \infty} a_n = 0$. [Hint: Consider $\lim_{n \to \infty}(S_n - S_{n-1})$, where S_n is the nth partial sum.]

Use the Comparison Test to determine whether the series in Problems 4–5 converge.

4. $\displaystyle\sum_{n=1}^{\infty} \frac{1}{n^3}$

5. $\displaystyle\sum_{n=2}^{\infty} \frac{1}{\ln n}$

By comparing to an improper integral, decide whether the series in Problems 6–9 converge.

6. $1 + \dfrac{1}{5} + \dfrac{1}{9} + \dfrac{1}{13} + \dfrac{1}{17} + \cdots + \dfrac{1}{4n-3} + \cdots$

7. $\dfrac{1}{2} + \dfrac{2}{5} + \dfrac{3}{10} + \dfrac{4}{17} + \dfrac{5}{26} + \cdots + \dfrac{n}{n^2+1} + \cdots$

8. $1 + \dfrac{1}{2^{3/2}} + \dfrac{1}{3^{3/2}} + \dfrac{1}{4^{3/2}} + \dfrac{1}{5^{3/2}} + \cdots + \dfrac{1}{n^{3/2}} + \cdots$

9. $1 + \dfrac{1}{2^p} + \dfrac{1}{3^p} + \cdots + \dfrac{1}{n^p} + \cdots$, where $p > 1$.

10. For a series $\sum a_n$:
    (a) Show that $0 \leq a_n + |a_n| \leq 2|a_n|$.
    (b) Show that if $\sum |a_n|$ converges, then $\sum a_n$ converges.

11. Show that if $C_0 + C_1 x + C_2 x^2 + C_3 x^3 + \cdots$ converges by the ratio test with radius of convergence R, then so does $C_1 + 2C_2 x + 3C_3 x^2 + \cdots$.

THE ERROR IN TAYLOR APPROXIMATIONS

In order to use an approximation intelligently, we need to be able to estimate the size of the error, which is the difference between the exact answer (which we usually do not know) and the approximate value. In general, we cannot find the exact value of the error; if we could, we would not need the approximation. However, we can often find a maximum possible value for the error. This gives us a feel for how good the approximation is.

We now consider how to assess the size of the error involved in using $P_n(x)$, the nth degree Taylor polynomial, to approximate $f(x)$ for values of x near a. The error is the difference

$$E_n(x) = f(x) - P_n(x).$$

If E_n is positive, the approximation is smaller than the true value. If E_n is negative, the approximation is too large. Often we are only interested in the magnitude of the error, $|E_n|$.

The Error in $P_1(x)$

Let's try to bound the error in the first-degree Taylor polynomial approximation to a function $f(x)$. For simplicity, we will assume the Taylor polynomial is centered at $x = 0$. We have

$$f(x) \approx P_1(x) = f(0) + f'(0)x.$$

The error in using $P_1(x)$ depends on how straight or curved the graph of f is for values of x near 0. The more curved f is, the worse the approximation. Thus we expect that the size of the second derivative $f''(x)$ will be important. If we know how big $f''(x)$ is, we should be able to tell how good or bad the approximation can be.

Suppose we know that $f''(x)$ is bounded above by M and bounded below by L for values of x near 0, say for $0 \leq x \leq d$. Since $f''(x) < M$, we have

$$\int_0^x f''(x)\,dx \leq \int_0^x M\,dx, \quad \text{for} \quad 0 \leq x \leq d.$$

Since integrating the second derivative gives the first derivative, the Fundamental Theorem of Calculus tells us that $\int_0^x f''(x)\,dx = f'(x) - f'(0)$. The previous inequality gives us

$$f'(x) - f'(0) \leq M(x - 0),$$

so

$$f'(x) \leq f'(0) + Mx.$$

Integrating again gives

$$\int_0^x f'(x)\,dx \leq \int_0^x (f'(0) + Mx)\,dx.$$

On the left side, use the Fundamental Theorem. On the right side, observe that $f'(0)$ is just a constant, so $\int_0^x f'(0)\,dx = f'(0)(x - 0)$. We now have

$$f(x) - f(0) \leq f'(0)x + \frac{M}{2}x^2.$$

Thus we have shown that if $f''(x) \leq M$ for $0 \leq x \leq d$, then

$$f(x) \leq f(0) + f'(0)x + \frac{M}{2}x^2.$$

Now we consider the lower bound. By reversing the inequalities, it can be shown that if $L \leq f''(x)$ for $0 \leq x \leq d$, then

$$f(0) + f'(0)x + \frac{L}{2}x^2 \leq f(x).$$

If we subtract $P_1(x) = f(0) + f'(0)x$ from both sides of these last two inequalities, we get a bound on the error $E_1 = f(x) - P_1(x)$:

Bounding the Error in $P_1(x)$

If $L \leq f''(x) \leq M$ for $0 \leq x \leq d$,   then

$$\frac{L}{2}x^2 \leq f(x) - P_1(x) \leq \frac{M}{2}x^2 \quad \text{for} \quad 0 \leq x \leq d.$$

Error in $P_2(x)$

Suppose we now want to bound the error in the second-degree Taylor polynomial approximation. By analogy with the error estimate for $P_1(x)$, we expect that the size of $f'''(x)$ will control the size of the error in $P_2(x)$. Suppose we know that

$$f'''(x) \leq M \quad \text{for} \quad 0 \leq x \leq d$$

and try to find a bound on $E = f(x) - P_2(x)$. Integrating as before, we get

$$\int_0^x f'''(x)\,dx \leq \int_0^x M\,dx$$
$$f''(x) - f''(0) \leq M(x-0)$$
$$f''(x) \leq f''(0) + Mx.$$

Integrate again:

$$\int_0^x f''(x)\,dx \leq \int_0^x (f''(0) + Mx)\,dx$$
$$f'(x) - f'(0) \leq f''(0)x + \frac{M}{2}x^2$$
$$f'(x) \leq f'(0) + f''(0)x + \frac{M}{2}x^2,$$

and again:

$$\int_0^x f'(x)\,dx \leq \int_0^x \left(f'(0) + f''(0)x + \frac{M}{2}x^2\right)\,dx$$
$$f(x) - f(0) \leq f'(0)x + \frac{f''(0)}{2}x^2 + \frac{M}{3 \cdot 2}x^3.$$

We conclude that if $f'''(x) \leq M$ for $0 \leq x \leq d$, then

$$f(x) \leq f(0) + f'(0)x + \frac{f''(0)}{2}x^2 + \frac{M}{3!}x^3.$$

Similarly, if $L \leq f'''(x)$ for $0 \leq x \leq d$, then

$$f(0) + f'(0)x + \frac{f''(0)}{2!}x^2 + \frac{L}{3!}x^3 \leq f(x).$$

Thus, the error in the approximation by $P_2(x)$ is bounded as follows:

Bounding the Error in $P_2(x)$

If $\quad L \leq f'''(x) \leq M \quad$ for $\quad 0 \leq x \leq d, \quad$ then

$$\frac{L}{3!}x^3 \leq f(x) - P_2(x) \leq \frac{M}{3!}x^3 \quad \text{for} \quad 0 \leq x \leq d.$$

Error in $P_n(x)$

The error in higher-degree Taylor polynomials follows the same pattern. If $L \leq f^{(n+1)}(x) \leq M$ for $0 \leq x \leq d$, then

$$\frac{L}{(n+1)!}x^{n+1} \leq f(x) - P_n(x) \leq \frac{M}{(n+1)!}x^{n+1}.$$

General Error Formula

When x is to the left of 0, so $-d \leq x \leq 0$, and when the Taylor series is centered at $a \neq 0$, similar calculations give us the following statement:

Error Bound for Taylor Polynomial Approximations

If $|f^{(n+1)}(x)| \leq M$ for $|x - a| \leq d$,   then

$$|E_n(x)| = |f(x) - P_n(x)| \leq \frac{M}{(n+1)!}|x - a|^{n+1} \quad \text{for} \quad |x - a| \leq d.$$

Example 5   Give a bound on the error, E_4, when e^x is approximated by its fourth-degree Taylor polynomial for $-0.5 \leq x \leq 0.5$.

Solution   Let $f(x) = e^x$. Then the fifth derivative is $f^{(5)}(x) = e^x$. Since e^x is increasing,

$$|f^{(5)}(x)| \leq e^{0.5} = \sqrt{e} < 2 \quad \text{for} \quad -0.5 \leq x \leq 0.5.$$

Thus

$$|E_4| = |f(x) - P_4(x)| \leq \frac{2}{5!}|x|^5.$$

This means, for example, that on $-0.5 \leq x \leq 0.5$, the approximation

$$e^x \approx 1 + x + \frac{x^2}{2!} + \frac{x^3}{3!} + \frac{x^4}{4!}$$

has an error of at most $\frac{2}{120}(0.5)^5 < 0.0006$.

Importance of the Error Bound for Taylor Polynomials

The error formula for Taylor polynomials can be used to bound the error in a particular numerical approximation, or to see how the accuracy of the approximation depends on the value of x or the value of n. Observe that the error for a Taylor polynomial of degree n depends on the $(n + 1)^{\text{st}}$ power of x. That means, for example, with a Taylor polynomial of degree n centered at 0, if we decrease x by a factor of 2, the error bound decreases by a factor of 2^{n+1}.

Example 6   Compare the errors in the approximations

$$e^{0.1} \approx 1 + 0.1 + \frac{1}{2!}(0.1)^2 \quad \text{and} \quad e^{0.05} \approx 1 + (0.05) + \frac{1}{2!}(0.05)^2$$

Solution We are approximating e^x by its second-degree Taylor polynomial, first at $x = 0.1$, and then at $x = 0.05$. Since we have decreased x by a factor of 2, we expect that the error has decreased by a factor of $2^3 = 8$. Let's see what actually happens. Here are the approximate values of the two errors:

$$e^{0.1} - \left(1 + 0.1 + \frac{1}{2!}(0.1)^2\right) = 1.105171 - 1.105000 = 0.000171$$

$$e^{0.05} - \left(1 + 0.05 + \frac{1}{2!}(0.05)^2\right) = 1.051271 - 1.051250 = 0.000021$$

Thus the error has decreased by a factor of $(0.000171)/(0.000021) = 8.1$, which is about what we expected.

Convergence of Taylor Series

The error bound allows us to show that in most cases the Taylor series for a function converges to *that* function. Since the error represents the difference between the Taylor polynomial and the function, showing that the series converges to the function for a particular value of x means showing that the error in the nth degree Taylor approximation goes to 0 as $n \to \infty$ for that value of x. We illustrate this idea by showing that the Taylor series for e^x converges to e^x for all x.

Showing the Taylor Series for e^x Converges to e^x

When we write the equation

$$e^x = 1 + x + \frac{x^2}{2!} + \frac{x^3}{3!} + \cdots,$$

the equal sign and the "$\cdots$" mean that as we look at more and more terms of the series on the right, we eventually get arbitrarily close to e^x. Since

$$E_n(x) = e^x - P_n(x) = e^x - \left(1 + x + \frac{x^2}{2!} + \cdots + \frac{x^n}{n!}\right),$$

we can write

$$e^x = 1 + x + \frac{x^2}{2!} + \cdots + \frac{x^n}{n!} + E_n(x).$$

Thus, for the Taylor series to converge to e^x, we must have $E_n(x) \to 0$ as $n \to \infty$. We assumed this to be true when we defined Taylor series in Section 9.2; we will now prove it for any fixed x.

Showing $E_n(x) \to 0$ as $n \to \infty$

Proof Since $f(x) = e^x$, the $(n+1)^{\text{st}}$ derivative $f^{(n+1)}(x)$ is also e^x, no matter what n is. So if we let $M = e^x$, then $|f^{(n+1)}| \leq M$ on the interval between 0 and x. (This works for $x \geq 0$; if $x < 0$ then we actually need to take $M = 1$.) The important observation is that for any x the *same* number M bounds all the higher derivatives $f^{(n+1)}(x)$.

By the error bound formula, we now have

$$|E_n(x)| = |e^x - P_n(x)| \leq \frac{M|x|^{n+1}}{(n+1)!} \quad \text{for every } n.$$

To show that the errors go to zero, we must show that for a fixed x and a fixed number M,

$$\frac{M}{(n+1)!}|x|^{n+1} \to 0 \quad \text{as} \quad n \to \infty.$$

Since M is fixed, we need only show that

$$\frac{1}{(n+1)!}|x|^{n+1} \to 0 \quad \text{as} \quad n \to \infty.$$

To see why this is true, think of what happens when n is much larger than x. Suppose, for example, that $x = 17.3$. Let's look at the value of

$$\frac{1}{(n+1)!}(17.3)^{n+1}$$

for n at least twice as big as 17.3, that is $n = 35$, or $n = 36$, or $n = 37, \ldots$. We get the values

$$\frac{1}{35!}(17.3)^{35} \ ,$$

$$\frac{1}{36!}(17.3)^{36} = \frac{17.3}{36} \cdot \frac{1}{35!}(17.3)^{35},$$

$$\frac{1}{37!}(17.3)^{37} = \frac{17.3}{37} \cdot \frac{17.3}{36} \cdot \frac{1}{35!}(17.3)^{35}, \quad \cdots$$

The $n = 38$ term will be $17.3/38$ times the $n = 37$ term and so on. Each time we increase n by 1, the value of $\frac{1}{(n+1)!}(17.3)^{n+1}$ is multiplied by a number less than $\frac{1}{2}$. No matter what the value of $\frac{1}{35!}(17.3)^{35}$ is, if you keep on halving it, the result will get closer and closer to zero. Thus $\frac{1}{(n+1)!}(17.3)^{n+1}$ goes to 0 as n goes to infinity.

We can generalize this by replacing 17.3 with an arbitrary $|x|$. For $n > 2|x|$, as we increase n, we will keep multiplying $\frac{1}{(n+1)!}|x|^{n+1}$ by factors of $\frac{|x|}{n+1} < \frac{1}{2}$. So

$$\frac{M}{(n+1)!}|x|^{n+1} \to 0 \quad \text{as} \quad n \to \infty.$$

Therefore, the Taylor series $1 + x + x^2/2! + \cdots$ does converge to e^x.

Most functions have Taylor series which converge to the original function for x within the interval of convergence. Functions for which this is true are called *analytic functions*. The Taylor series for such a function can be interpreted when x is replaced by a complex number. This extends the domain of the function. See Problem 16.

Problems on the Error in Taylor Approximations

1. Suppose you approximate $f(t) = e^t$ by a Taylor polynomial of degree 0 about $t = 0$ on the interval $[0, 0.5]$.

    (a) Is the approximation an overestimate or an underestimate?
    (b) Estimate the magnitude of the largest possible error. Check your answer graphically on a computer or calculator.

2. Repeat Problem 1 using the second-degree Taylor approximation, $P_2(t)$, to e^t.

3. Consider the error in using the approximation $\sin \theta \approx \theta$ on the interval $[-1, 1]$.

    (a) Where is the approximation an overestimate, and where is it an underestimate?
    (b) Estimate the magnitude of the largest possible error. Check your answer graphically on a computer or calculator.

4.   Repeat Problem 3 for the approximation $\sin\theta \approx \theta - \theta^3/3!$.

Use the methods of this section to show how you can estimate the magnitude of the error in approximating the quantities in Problems 5–8 using a third-degree Taylor polynomial about $x = 0$.

5. $\tan 1$ 6. $0.5^{1/3}$ 7. $\ln(1.5)$ 8. $1/\sqrt{3}$

9.   Give a bound for the maximum possible error for the nth degree Taylor polynomial about $x = 0$ approximating $\cos x$ on the interval $[0, 1]$. What is the bound for $\sin x$?

10. What degree Taylor polynomial about $x = 0$ do you need to calculate $\cos 1$ to four decimal places? To six decimal places? Justify your answer using the results of Problem 9.

11. Show that the Taylor series about 0 for $\sin x$ converges to $\sin x$ for every x.

12. Show that the Taylor series about 0 for $\cos x$ converges to $\cos x$ for every x.

13. (a) Using a calculator, make a table of the values to four decimal places of $\sin x$ for

$$x = -0.5, -0.4, \ldots, -0.1, 0, 0.1, \ldots, 0.4, 0.5.$$

     (b) Add to your table the values of the error $E_1 = \sin x - x$ for these x-values.
     (c) Using a calculator or computer, draw a graph of the quantity $E_1 = \sin x - x$ showing that

$$|E_1| < 0.03 \quad \text{for} \quad -0.5 \le x \le 0.5.$$

14. In this problem, you will investigate the error in the nth degree Taylor approximation to e^x for various values of n.

     (a) Let $E_1 = e^x - P_1(x) = e^x - (1+x)$. Using a calculator or computer, graph E_1 for $-0.1 \le x \le 0.1$. What shape is the graph of E_1? Use the graph to confirm that

$$|E_1| \le x^2 \quad \text{for} \quad -0.1 \le x \le 0.1.$$

     (b) Let $E_2 = e^x - P_2(x) = e^x - (1 + x + x^2/2)$. Choose a suitable range and graph E_2 for $-0.1 \le x \le 0.1$. What shape is the graph of E_2? Use the graph to confirm that

$$|E_2| \le x^3 \quad \text{for} \quad -0.1 \le x \le 0.1.$$

     (c) Explain why the graphs of E_1 and E_2 have the shapes they do.

15. Graph the error

$$E_0 = \cos x - P_0(x) = \cos x - 1$$

for $|x| \le 0.1$. Explain the shape of the graph, using information given by the Taylor expansion of $\cos x$, and find a bound for $|E_0|$ for $|x| \le 0.1$.

16. Let $i = \sqrt{-1}$. We define $e^{i\theta}$ by substituting $i\theta$ in the Taylor series for e^x. Use this definition[10] to explain Euler's formula

$$e^{i\theta} = \cos\theta + i\sin\theta.$$

---

[10]Complex numbers are discussed in Appendix D.

CHAPTER TEN

DIFFERENTIAL EQUATIONS

The derivative was introduced as the rate of change of a function in Chapter 2. In Chapter 3, we saw how to use the definite integral to compute changes in the original function from the derivative. Chapter 6 introduced antiderivatives, differential equations, and going backward from the derivative to the original function.

Every time we antidifferentiate to find $\int f(x)\,dx$, we are actually solving the differential equation

$$\frac{dy}{dx} = f(x).$$

The new feature of this chapter is that the right-hand side of the differential equation may now be a function of y, or of both x and y, instead of just x.

We will see how to solve differential equations graphically first, then numerically, and, finally, in some of the special cases when it can be done, analytically. In the course of this chapter, we will treat first-order differential equations (involving first derivatives only) and some second-order differential equations (involving first and second derivatives).

10.1 WHAT IS A DIFFERENTIAL EQUATION?

How Fast Does a Person Learn?

Suppose we are interested in how fast an employee learns a new task. One theory claims that the more the employee already knows of the task, the slower he or she learns. In other words, if $y\%$ is the percentage of the task that has already been mastered, and dy/dt the rate at which the employee is learning, then dy/dt decreases as y increases.

What can we say about y as a function of time, t? Figure 10.1 shows three graphs whose slope, dy/dt, decreases as y increases. Figure 10.1(a) represents an employee who starts learning at $t = 0$ and who eventually masters 100% of the task. Figure 10.1(b) represents an employee who starts later but eventually masters 100% of the task. Figure 10.1(c), on the other hand, represents an employee who starts learning at $t = 0$, but who does not master the whole task (since y levels off at some percentage below 100%).

Notice the fact that y levels off does not follow from the fact that dy/dt decreases as y increases. However, a person's knowledge of the task cannot exceed 100%, and all the graphs in Figure 10.1 have been drawn to reflect this fact.

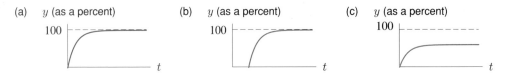

Figure 10.1: Possible graphs showing percentage of task learned, y, as a function of time, t

Setting up a Differential Equation to Model How a Person Learns

To describe more precisely how a person learns, we need more exact information about how dy/dt depends on y. Suppose we are told that if time, t, is measured in weeks, then

$$\frac{\text{Rate a person}}{\text{learns}} = \frac{\text{Percentage of task}}{\text{not yet learned.}}$$

Since y is the percentage learned by time t (in weeks), the percentage not yet learned by that time is $100 - y$. Thus we have

$$\frac{dy}{dt} = 100 - y.$$

Such an equation, which gives information about the rate of change of an unknown function, is called a *differential equation*.

Solving the Differential Equation Numerically

Suppose that the person starts learning at time zero, so $y = 0$ when $t = 0$. Then initially the person is learning at a rate

$$\frac{dy}{dt} = 100 - 0 = 100\% \text{ per week.}$$

In other words, if the person were to continue learning at this rate, the task would be mastered in a week. In fact, however, the rate at which the person learns decreases, so it takes more than a week to get close to mastering the task. Let's assume a five-day work week and that the 100% per week learning rate holds for the whole first day. (It doesn't, but we will assume this for now.) One day is 1/5 of a week, so during the first day the person learns $100(1/5) = 20\%$ of the task. By the end of the first day the rate at which the person learns has therefore been reduced to

$$\frac{dy}{dt} = 100 - 20 = 80\% \text{ per week.}$$

Thus, during the second day the person learns $80(1/5) = 16\%$, so by the end of the second day the person knows $20 + 16 = 36\%$ of the task. Continuing in this fashion, we compute the approximate values[1] of y in Table 10.1. These values for y represent an approximate numerical solution to the differential equation.

TABLE 10.1 *Approximate percentage of task learned as a function of time*

Time (working days)	0	1	2	3	4	5	10	20
Percentage learned	0	20	36	48.8	59.0	67.2	89.3	98.8

A Formula for the Solution to the Differential Equation

A function $y = f(t)$ which satisfies the differential equation

$$\frac{dy}{dt} = 100 - y$$

is called a *solution*. Figure 10.1 contains graphs of possible solutions, and Table 10.1 shows approximate numerical values of a solution. It is also sometimes (but not always) possible to find a formula for the solution. In this particular case there is a formula, which we will derive in Section 10.4. For now, we will simply check that the formula is correct, and look at its general form. The formula for the solution is

$$y = 100 + Ce^{-t},$$

where C is an *arbitrary constant*. Substituting this into the differential equation $dy/dt = 100 - y$ gives:

$$\text{Left side} = \frac{dy}{dt} = -Ce^{-t}$$
$$\text{Right side} = 100 - y = 100 - (100 + Ce^{-t}) = -Ce^{-t}.$$

Since we get the same thing on both sides, we say $y = 100 + Ce^{-t}$ is a solution to this differential equation.

Finding the Arbitrary Constant: Initial Conditions

To find a value for the arbitrary constant C, we need an additional piece of information—usually the initial value of y. If, for example, we are told that $y == 0$ when $t = 0$, then substituting into

$$y = 100 + Ce^{-t}$$

shows us that

$$0 = 100 + Ce^0, \quad \text{so} \quad C = -100.$$

So the function $y = 100 - 100e^{-t}$ satisfies the differential equation *and* the condition that $y = 0$ when $t = 0$.

The Family of Solutions

Later in this chapter we will show that any solution to this differential equation is of the form $y = 100 + Ce^{-t}$ for some constant C. Since any value of C gives a solution, we say that the *general solution* to the differential equation $dy/dt = 100 - y$ is the family of functions $y = 100 + Ce^{-t}$. The solution $y = 100 - 100e^{-t}$ that satisfies the differential equation together with the *initial condition* that $y = 0$ when $t = 0$ is called a *particular solution*. The differential equation and the initial condition together are called an *initial-value problem*. Several members of the family of solutions are graphed in Figure 10.2. Notice that, like a family of antiderivatives, this family contains one arbitrary constant, C.

---

[1]The values of y after 6, 7, $\ldots$, 9, 11, $\ldots$, 19 days were computed by the same method, but omitted from the table.

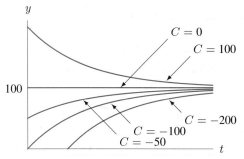

Figure 10.2: Solution curves for $dy/dt = 100 - y$:
Members of the family $y = 100 + Ce^{-t}$

First- and Second-Order Differential Equations

First, some more definitions. The differential equation

$$\frac{dy}{dt} = 100 - y$$

is called *first-order* because it involves the first derivative, but no higher derivatives. By contrast, if s is the height (in feet) of a body moving under the force of gravity and t is time (in seconds), then

$$\frac{d^2s}{dt^2} = -32.$$

This is a *second-order* differential equation because it involves the second derivative of the unknown function, $s = f(t)$, but no higher derivatives.

Example 1   Show that $y = e^{2t}$ is not a solution to the second-order differential equation

$$\frac{d^2y}{dt^2} + 4y = 0.$$

Solution   To decide whether the function $y = e^{2t}$ is a solution, substitute it into the differential equation:

$$\frac{d^2y}{dt^2} + 4y = 2(2e^{2t}) + 4e^{2t} = 8e^{2t}.$$

Since $8e^{2t}$ is not identically zero, $y = e^{2t}$ is not a solution.

How Many Arbitrary Constants Should We Expect in the Family of Solutions?

Since a differential equation involves the derivative of an unknown function, solving it usually involves antidifferentiation, which introduces arbitrary constants. The solution to a first-order differential equation usually involves one antidifferentiation and one arbitrary constant (for example, the C in $y = 100 + Ce^{-t}$). Picking out one particular solution involves knowing an additional piece of information, such as an initial condition. Solving a second-order differential equation generally involves two antidifferentiations and so two arbitrary constants. Consequently, finding a particular solution usually involves two initial conditions.

For example, if s is the height (in feet) of a body above the surface of the earth at time t (in seconds), then

$$\frac{d^2s}{dt^2} = -32.$$

Integrating gives

$$\frac{ds}{dt} = -32t + C_1,$$

and integrating again gives

$$s = -16t^2 + C_1 t + C_2.$$

Thus the general solution for s involves the two arbitrary constants C_1 and C_2. We can find C_1 and C_2 if we are told, for example, that the initial velocity is 100 feet per second upward and that the initial position is 15 feet above the ground. In this case, $C_1 = 100$ and $C_2 = 15$, so

$$s = -16t^2 + 100t + 15.$$

Problems for Section 10.1

1. Match the graphs in Figure 10.3 with the following descriptions.

    (a) The population of a new species introduced onto a tropical island
    (b) The temperature of a metal ingot placed in a furnace and then removed
    (c) The speed of a car traveling at uniform speed and then braking uniformly
    (d) The mass of carbon-14 in a historical specimen
    (e) The concentration of tree pollen in the air over the course of a year.

(I)   (II)   (III)   (IV)   (V)

Figure 10.3

2. Suppose you know that $Q = Ce^{kt}$ satisfies the differential equation

    $$\frac{dQ}{dt} = -0.03Q.$$

    What (if anything) does this tell you about the values of C and k?

3. Show that, for any constant P_0, the function $P = P_0e^t$ satisfies the differential equation

    $$\frac{dP}{dt} = P.$$

4. Find the value(s) of ω for which $y = \cos \omega t$ satisfies

    $$\frac{d^2y}{dt^2} + 9y = 0.$$

5. Show that $y = \sin 2t$ satisfies

    $$\frac{d^2y}{dt^2} + 4y = 0.$$

6. (a) Show that $P = \dfrac{1}{1 + e^{-t}}$ satisfies the *logistic equation*

    $$\frac{dP}{dt} = P(1 - P).$$

    (b) What is the limiting value of P as time, t, tends to infinity?

7. For what values of λ is $y = Ax^\lambda$ a solution of the equation

    $$x^2y'' + 2xy' - 6y = 0?$$

8. Pick out which functions are solutions to which differential equations. (Note: Functions may be solutions to more than one equation or to none; an equation may have more than one solution.)

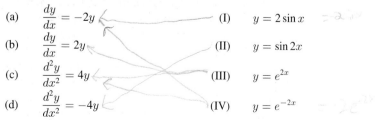

(a) $\dfrac{dy}{dx} = -2y$ (I) $y = 2\sin x$

(b) $\dfrac{dy}{dx} = 2y$ (II) $y = \sin 2x$

(c) $\dfrac{d^2y}{dx^2} = 4y$ (III) $y = e^{2x}$

(d) $\dfrac{d^2y}{dx^2} = -4y$ (IV) $y = e^{-2x}$

9. Which functions are solutions to which differential equations? (Note: a differential equation may have zero, one, or multiple solutions.)

(A) $\dfrac{dy}{dx} = \dfrac{y}{x}$

(B) $\dfrac{dy}{dx} = 3\dfrac{y}{x}$

(C) $\dfrac{dy}{dx} = 3x$

(D) $\dfrac{dy}{dx} = y$

(E) $\dfrac{dy}{dx} = 3y$

(I) $y = x^3$

(II) $y = 3x$

(III) $y = e^{3x}$

(IV) $y = 3e^x$

(V) $y = x$

10. Match the following differential equations and possible solutions. (Note the given functions may satisfy more than one equation or none, and some equations may have more than one solution.)

(a) $y'' = y$
(b) $y' = -y$
(c) $y' = 1/y$
(d) $y'' = -y$
(e) $x^2 y'' - 2y = 0$

(I) $y = \cos x$
(II) $y = \cos(-x)$
(III) $y = x^2$
(IV) $y = e^x + e^{-x}$
(V) $y = \sqrt{2x}$

11. Suppose y is the height of a hanging cable above a fixed horizontal line, as in Figure 10.4. It can be shown that

$$\frac{d^2 y}{dx^2} = k\sqrt{1 + \left(\frac{dy}{dx}\right)^2}.$$

Figure 10.4

(a) Show that $y = \dfrac{e^x + e^{-x}}{2}$ satisfies this differential equation if $k = 1$.

(b) For general k, one solution to this differential equation is of the form

$$y = \frac{e^{Ax} + e^{-Ax}}{2A}.$$

Substitute this expression for y into the differential equation to find A in terms of k.

10.2 SLOPE FIELDS

In this section, we will visualize a first-order differential equation. Let's start with the equation

$$\frac{dy}{dx} = y.$$

We can check that each curve in the family of exponentials, $y = Ce^x$, is a solution to this differential equation.

Any solution to this differential equation has the property that the slope at any point is equal to the y coordinate at that point. (That's what the equation $dy/dx = y$ is telling us!) This means that if the solution goes through the point $(0, 0.5)$, its slope there is 0.5; where it goes through a point with $y = 1$ its slope is 1. A solution which goes through $(0, 1.5)$ has slope 1.5 there. (See Figure 10.5.)

In Figure 10.5 a small line segment is drawn at each of the marked points showing the slope of the curve there. Imagine drawing many of these line segments, but leaving out the curves; this gives

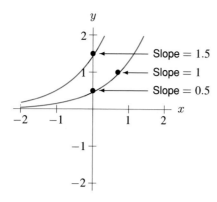

Figure 10.5: Solutions to $\frac{dy}{dx} = y$

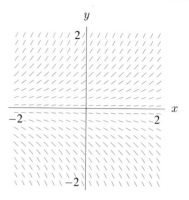

Figure 10.6: Slope field for $\frac{dy}{dx} = y$

the *slope field* for the equation $dy/dx = y$ shown in Figure 10.6. From this picture, we can see that above the x-axis, the slopes are all positive (because y is positive there), and they increase as we move upward (as y increases). Below the x-axis, the slopes are all negative, and get more so as we move downward. Notice that on any horizontal line (where y is constant) the slopes are the same.

In the slope field we can see the ghost of the solution curve lurking. Start anywhere on the plane and move so that the slope lines are tangent to your path; you will trace out one of the solution curves. It may be helpful to think of the slope field as a set of signposts pointing in the direction you should go at each point.

Example 1   Figure 10.7 shows the slope field of the differential equation $\dfrac{dy}{dx} = 2x$.
  (a)   How does the slope field vary as you move around the xy-plane?
  (b)   Compare the solution curves sketched on the slope field with the formula for the solutions to this differential equation.

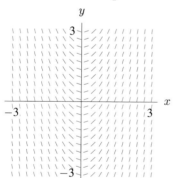

Figure 10.7: Slope field for $\frac{dy}{dx} = 2x$

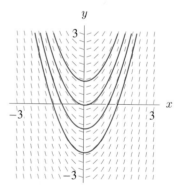

Figure 10.8: Some solutions to $\frac{dy}{dx} = 2x$

Solution   (a)   In Figure 10.7 you will notice that on a vertical line (where x is constant) the slopes are all the same. This is because in this differential equation dy/dx depends on x only. (In the previous example, $dy/dx = y$, the slopes depended on y only.)
  (b)   The solution curves in Figure 10.8 look like parabolas. Since the solution to the differential equation

$$\frac{dy}{dx} = 2x$$

is, by antidifferentiation,

$$y = \int 2x\,dx = x^2 + C,$$

we see that the solution curves really are parabolas.

Example 2   Using the slope field, guess the form of the solution curves of the differential equation

$$\frac{dy}{dx} = -\frac{x}{y}.$$

Solution   The slope field is shown in Figure 10.9. Notice that on the y-axis, where x is 0, the slope is 0. On the x-axis, where y is 0, the line segments are vertical and the slope is infinite. At the origin the slope is undefined, and there is no line segment.

What do the solution curves of this differential equation look like? The slope field suggests that they look like circles centered at the origin. Later we'll see how to obtain the solution analytically, but even without this, we can still check that the circle is a solution. Let's take the circle of radius r,

$$x^2 + y^2 = r^2,$$

and differentiate implicitly, thinking of y as a function of x. Using the chain rule, we get

$$2x + 2y \cdot \frac{dy}{dx} = 0.$$

Solving for dy/dx gives our differential equation,

$$\frac{dy}{dx} = -\frac{x}{y}.$$

This tells us that $x^2 + y^2 = r^2$ is a solution to the differential equation.

Figure 10.9: Slope field for $\frac{dy}{dx} = -\frac{x}{y}$

The previous example shows that the solution to a differential equation may often be an implicit function. (Recall that implicit functions are ones which have not been "solved" for y; in other words, the dependent variable is not expressed as an explicit function of x.)

Example 3   The slope fields for $\frac{dy}{dt} = 2 - y$ and $\frac{dy}{dt} = \frac{t}{y}$ are shown in Figures 10.10 and 10.11. For each slope field:
(a)   Sketch solution curves with initial conditions
      (i) $y = 1$ when $t = 0$ (ii) $y = 0$ when $t = 1$ (iii) $y = 3$ when $t = 0$
(b)   For each solution curve, can you say anything about the long-run behavior of y? For example, does $\lim_{t \to \infty} y$ exist? If so, what is its value?

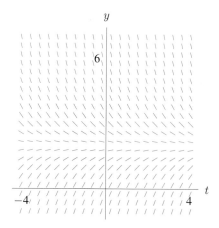

Figure 10.10: Slope field for $\frac{dy}{dt} = 2 - y$

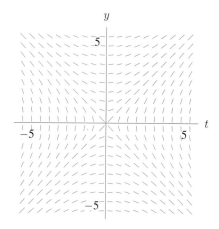

Figure 10.11: Slope field for $\frac{dy}{dt} = \frac{t}{y}$

Solution (a) See Figures 10.12 and 10.13.

(b) For $dy/dt = 2 - y$, all solution curves have $y = 2$ as a horizontal asymptote, so $\lim_{t \to \infty} y = 2$. For $dy/dt = t/y$, as $t \to \infty$, it appears that either $y \to t$ or $y \to -t$.

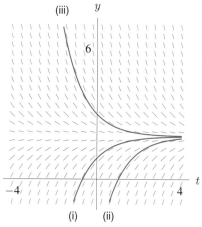

Figure 10.12: Solution curves for $\frac{dy}{dt} = 2 - y$

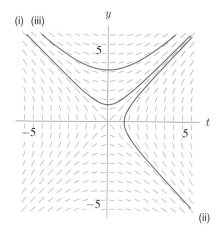

Figure 10.13: Solution curves for $\frac{dy}{dt} = \frac{t}{y}$

Existence and Uniqueness of Solutions

Since differential equations are used to model many real situations, the question of whether a solution is unique can have great practical importance. If we know how the velocity of a satellite is changing, can we know its velocity at any future time? If we know the initial population of a city, and we know how the population is changing, can we predict the population in the future? Common sense says yes: if we know the initial value of some quantity and we know exactly how it is changing, we should be able to figure out the future value of the quantity.

In the language of differential equations, an initial-value problem (that is, a differential equation and an initial condition) almost always has a unique solution. One way to see this is by looking at the slope field. Imagine starting at the point representing the initial condition. Through that point there will usually be a line segment pointing in the direction the solution curve must go. By following

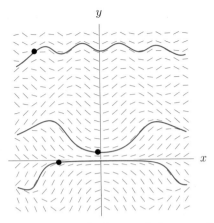

Figure 10.14: There's one and only one solution curve through each
point in the plane for this slope field (Dots represent initial conditions)

these line segments, we trace out the solution curve. See Figure 10.14. In general, at each point there
is one line segment and therefore only one direction for the solution curve to go. The solution curve
exists and is *unique* provided we are given an initial point. Notice that even though we can draw the
solution curves, we may have no simple formula for them.

It can be shown that if the slope field is continuous as we move from point to point in the
plane, we can be sure that a solution curve exists everywhere. Ensuring that each point has only one
solution curve through it requires a slightly stronger condition.

Problems for Section 10.2

1. The slope field for the equation $y' = x + y$ is shown in Figure 10.15.

    (a) Carefully sketch the solutions that pass through the points
        (i) $(0, 0)$ (ii) $(-3, 1)$ (iii) $(-1, 0)$
    (b) From your sketch, write the equation of the solution passing through $(-1, 0)$.
    (c) Verify your solution to part (b) by substituting it into the differential equation.

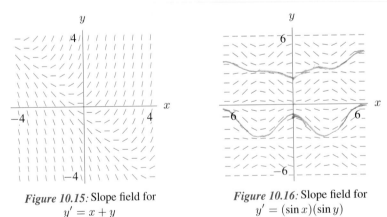

Figure 10.15: Slope field for
$y' = x + y$

Figure 10.16: Slope field for
$y' = (\sin x)(\sin y)$

2. Figure 10.16 shows the slope field for the equation $y' = (\sin x)(\sin y)$.

    (a) Carefully sketch the solutions that pass through the points: (i) $(0, -2)$ (ii) $(0, \pi)$.
    (b) What is the equation of the solution that passes through $(0, n\pi)$, where n is any integer?

3. Match the slope fields in Figure 10.17 with their differential equations:

(a) $y' = -y$ (b) $y' = y$ (c) $y' = x$

(d) $y' = 1/y$ (e) $y' = y^2$

(I)

(II)

(III)

(IV)

(V)

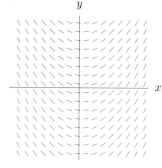

Figure 10.17

4. Match the slope fields in Figure 10.18 with their differential equations:

(a) $y' = 1 + y^2$ (b) $y' = x$ (c) $y' = \sin x$

(d) $y' = y$ (e) $y' = x - y$ (f) $y' = 4 - y$

(I)

(II)

(III)

(IV)

(V)

(VI)

Figure 10.18: Each slope field is graphed for $-5 \le x \le 5$, $-5 \le y \le 5$

5.   One of the slope fields in Figure 10.19 has the equation $y' = (x + y)/(x - y)$. Which one?

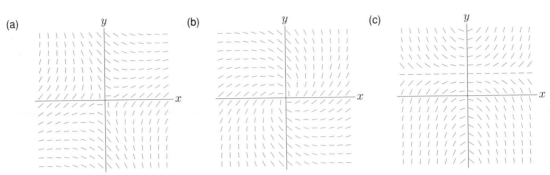

Figure 10.19

6.   Match the slope fields in Figure 10.20 to the corresponding differential equations:

(a) $y' = xe^{-x}$ (b) $y' = \sin x$ (c) $y' = \cos x$

(d) $y' = x^2 e^{-x}$ (e) $y' = e^{-x^2}$ (f) $y' = e^{-x}$

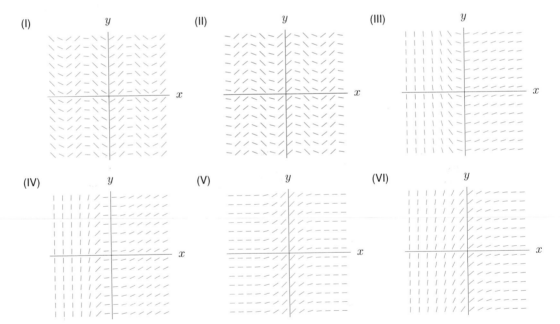

Figure 10.20

7.   The slope field for $y' = 0.5(1 + y)(2 - y)$ is shown in Figure 10.21.

(a)   Plot the following points on the slope field:

(i)   the origin   (ii) $(0, 1)$ (iii) $(1, 0)$
(iv) $(0, -1)$ (v) $(0, -5/2)$ (vi) $(0, 5/2)$

(b)   Plot solution curves through the points in part (a).

(c)   For which regions are all solution curves increasing? For which regions are all solution curves decreasing? When can the solution curves have horizontal tangents? Explain why, using both the slope field and the differential equation.

Figure 10.21: Note: x and y scales are equal

Figure 10.22

8. The slope field for the differential equation $y' = y - x^2$ together with two special solutions are shown in Figure 10.22.

   (a) The two curves shown in the figure are $y = x^2 + 2x + 2$ and $y = x^2 + 2x + 2 - 2e^x$. Verify that these formulas give solutions, and indicate which formula goes with which curve.
   (b) Sketch a few solutions to the differential equation in each of the three regions.
   (c) Describe the solution curves you have sketched in region I.
   (d) Describe the solution curves in region II.
   (e) Describe the solution curves in region III.

10.3 EULER'S METHOD

In the preceding section we saw how to sketch a solution curve to a differential equation using its slope field. In this section we compute points on a solution curve numerically using *Euler's method*. (Leonhard Euler was an eighteenth-century Swiss mathematician.) In some cases—but not many—we can find an equation for the solution curve. That will be the subject of Section 10.4.

Here's the concept behind Euler's method. Think of the slope field as a set of signposts directing you across the plane. Pick a starting point (corresponding to the initial value), and calculate the slope at that point using the differential equation. This slope is a signpost telling you the direction to take. Head off a small distance in that direction. Stop and look at the new signpost. Recalculate the slope from the differential equation, using the coordinates of the new point. Change direction to correspond to the new slope, and move another small distance, and so on.

Example 1 Use Euler's method to construct a solution for $\dfrac{dy}{dx} = y$. Start at the point $P_0 = (0, 1)$ and take $\Delta x = 0.1$.

Solution The slope at the point $P_0 = (0, 1)$ is $dy/dx = 1$. (See Figure 10.23.) As we move from P_0 to P_1, y will increase by Δy, where

$$\Delta y = (\text{slope at } P_0)\Delta x = 1(0.1) = 0.1.$$

So we have

$$y\text{-value at } P_1 = (y \text{ value at } P_0) + \Delta y = 1 + 0.1 = 1.1.$$

TABLE 10.2   *Euler's method for $dy/dx = y$, starting at $(0, 1)$*

	x	y	$\Delta y = (\text{Slope})\Delta x$
P_0	0	1	$0.1 = (1)(0.1)$
P_1	0.1	1.1	$0.11 = (1.1)(0.1)$
P_2	0.2	1.21	$0.121 = (1.21)(0.1)$
P_3	0.3	1.331	$0.1331 = (1.331)(0.1)$
P_4	0.4	1.4641	$0.14641 = (1.4641)(0.1)$
P_5	0.5	1.61051	$0.161051 = (1.61051)(0.1)$

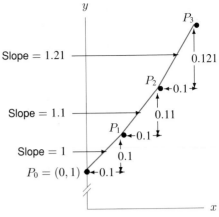

Figure 10.23: Euler's approximate solution to $\frac{dy}{dx} = y$

Thus the point P_1 is $(0.1, 1.1)$. Now, using the differential equation again, we see that

$$\text{slope at } P_1 = 1.1,$$

so if we move to P_2, then y will change by

$$\Delta y = (\text{slope at } P_1)\Delta x = (1.1)(0.1) = 0.11.$$

This means

$$y\text{-value at } P_2 = (y \text{ value at } P_1) + \Delta y = 1.1 + 0.11 = 1.21.$$

Thus P_2 is $(0.2, 1.21)$. Continuing gives the results in Table 10.2.

Since the solution curves of $dy/dx = y$ are exponentials, they are concave up, and they curve upward away from the line segments of the slope field. Therefore, in this case, Euler's method produces y-values which are too small.

Notice that Euler's method calculates approximate values of y for points on a solution curve; it does *not* give a general formula for y in terms of x. The graph in Figure 10.23 approximates the solution curve.

Example 2   Show that Euler's method for $\dfrac{dy}{dx} = y$ starting at $(0, 1)$ and using two steps with $\Delta x = 0.05$ gives $y \approx 1.1025$ when $x = 0.1$.

Solution   At $(0, 1)$, the slope is 1 and $\Delta y = (1)(0.05) = 0.05$, so new $y = 1 + 0.05 = 1.05$. At $(0.05, 1.05)$, the slope is 1.05 and $\Delta y = (1.05)(0.05) = 0.0525$, so new $y = 1.05 + 0.0525 = 1.1025$ at $x = 0.1$.

In general, dy/dx may be a function of both x and y. Euler's method still works then, as the next example shows.

Example 3   Approximate four points on the solution curve to $\dfrac{dy}{dx} = -\dfrac{x}{y}$, starting at $(0, 1)$ and using $\Delta x = 0.1$.

Solution   The results from Euler's method are in Table 10.3, along with the exact y-values (to two decimals) calculated from the equation of the circle $x^2 + y^2 = 1$, which is the solution curve through $(0, 1)$. Since the curve is concave down, the approximate y-values are above the exact ones. (See Figure 10.24.)

TABLE 10.3 *Euler's method for $dy/dx = -x/y$, starting at $(0, 1)$*

x	Approximate y-value	$\Delta y = (\text{Slope})\Delta x$	True y-value
0	1	$0 = (0)(0.1)$	1
0.1	1	$-0.01 = (-0.1/1)(0.1)$	0.99
0.2	0.99	$-0.02 = (-0.2/0.99)(0.1)$	0.98
0.3	0.97		0.95

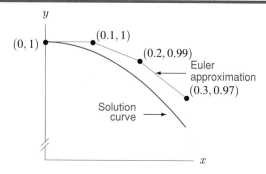

Figure 10.24: Euler's approximate solution to $\frac{dy}{dx} = -\frac{x}{y}$

The Accuracy of Euler's Method

To improve the accuracy of Euler's method, we choose Δx smaller. Let's go back to the differential equation $dy/dx = y$ and compare the exact and approximate values for different Δx's. The exact solution going through the point $(0, 1)$ is $y = e^x$, so the exact values are calculated using this function. (See Figure 10.25.) Where $x = 0.1$,

$$\text{Exact } y\text{-value } = e^{0.1} \approx 1.1051709.$$

In Example 1 we had $\Delta x = 0.1$, and where $x = 0.1$,

$$\text{Approximate } y\text{-value} = 1.1, \quad \text{so the error} \approx 0.005.$$

In Example 2 we decreased Δx to 0.05. After two steps, $x = 0.1$, and we had

$$\text{Approximate } y\text{-value} = 1.1025, \quad \text{so error} \approx 0.00267.$$

Thus, it appears that halving the step size has approximately halved the error.

The *error* in using Euler's method is the difference between the approximate value and the exact value. If the number of steps used is n, the error is approximately proportional to $1/n$.

Just as there are more accurate numerical integration methods than left and right Riemann sums, there are more accurate methods than Euler's for approximating solution curves. However, Euler's method will be all we need.

Figure 10.25: Euler's approximate solution to $\frac{dy}{dx} = y$

Problems for Section 10.3

1. Consider the solution of the differential equation $y' = y$ passing through $y(0) = 1$.

    (a) Sketch the slope field for this differential equation, and sketch the solution passing through the point $(0,1)$.
    (b) Use Euler's method with step size $\Delta x = 0.1$ to estimate the solution at $x = 0.1, 0.2, \ldots, 1$.
    (c) Plot the estimated solution on the slope field; compare the solution and the slope field.
    (d) Check that $y = e^x$ is the solution of $y' = y$ with $y(0) = 1$.

2. Consider the differential equation $y' = (\sin x)(\sin y)$.

    (a) Calculate approximate y-values using Euler's method with three steps and $\Delta x = 0.1$, starting at each of the following points: (i) $(0, 2)$ (ii) $(0, \pi)$.
    (b) Use the slope field in Figure 10.16 on page 498 to explain your solution to part (a)(ii).

3. Consider the differential equation $y' = x + y$ whose slope field is in Figure 10.15 on page 498. Use Euler's method with $\Delta x = 0.1$ to estimate y when $x = 0.4$ for the solution curves satisfying

    (a) $y(0) = 1$ (b) $y(-1) = 0$.

4. (a) Use ten steps of Euler's method to determine an approximate solution for the differential equation $y' = x^3$, $y(0) = 0$, using a step size $\Delta x = 0.1$.
    (b) Guess the exact solution and compare it to the computed approximation.
    (c) Use a sketch of the slope field for this equation to explain the results of part (b).

5. (a) Use ten steps of Euler's method to approximate y-values for $dy/dt = 1/t$, starting at $(1, 0)$ and using $\Delta t = 0.1$.
    (b) Using integration, solve the differential equation to find the exact value of y at the end of these ten steps.
    (c) Is your approximate value of y at the end of ten steps bigger or smaller than the exact value? Use a slope field to explain your answer.

6. (a) Use Euler's method to approximate the value of y at $x = 1$ on the solution curve to the differential equation

    $$\frac{dy}{dx} = x^3 - y^3$$

    that passes through $(0, 0)$. Use $\Delta x = 1/5$ (i.e., 5 steps).
    (b) Figure 10.26 shows the slope field for the differential equation. Sketch the solution that passes through $(0, 0)$. Show on the graph the approximation you made in part (a).
    (c) From the graph of the slope field, can you say whether your answer to part (a) is an overestimate or an underestimate?

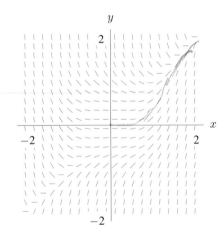

Figure 10.26: Slope field for $\frac{dy}{dx} = x^3 - y^3$

7. Consider the differential equation

    $$\frac{dy}{dx} = 2x, \quad \text{with initial condition } y(0) = 1.$$

    (a) Use Euler's method with two steps to estimate y when $x = 1$.
    (b) Use Euler's method with four steps to estimate y when $x = 1$.
    (c) What is the formula for the exact value of y?
    (d) Does the error in Euler's approximation behave as predicted in the box on page 503?

8. Consider the differential equation

$$\frac{dy}{dx} = \sin(xy), \qquad \text{with initial condition } y(1) = 1.$$

Estimate $y(2)$, using Euler's method with step sizes $\Delta x = 0.2, 0.1, 0.05$.

Plot the computed approximations for $y(2)$ against Δx. What do you conclude? Use your observations to estimate the exact value of $y(2)$.

9. Use Euler's method to solve

$$\frac{dB}{dt} = 0.05B$$

with initial value $B = 1000$ when $t = 0$. Take:

(a) $\Delta t = 1$ and 1 step.   (b) $\Delta t = 0.5$ and 2 steps.   (c) $\Delta t = 0.25$ and 4 steps.

(d) Suppose B is the balance in a bank account earning interest. Explain why the result of your calculation in part (a) is equivalent to compounding the interest once a year instead of continuously.

(e) Interpret the result of your calculations in parts (b) and (c) in terms of compound interest.

10. Consider the differential equation $dy/dx = f(x)$ with initial value $y(0) = 0$. Explain why using Euler's method to approximate the solution curve gives the same results as using left Riemann sums to approximate $\int_0^x f(t)\, dt$.

10.4 SEPARATION OF VARIABLES

We have seen how to sketch solution curves of a differential equation using a slope field and how to calculate approximate numerical solutions. Now we see how to find the equation of a solution curve for certain differential equations.

First, we'll look at a familiar example, the differential equation

$$\frac{dy}{dx} = -\frac{x}{y},$$

whose solution curves are the circles

$$x^2 + y^2 = C.$$

We can check that these circles are solutions by differentiation; the question now is how they were obtained. The method of *separation of variables* works by putting all the x's on one side of the equation and all the y's on the other, giving

$$y\, dy = -x\, dx.$$

We then integrate each side separately:

$$\int y\, dy = -\int x\, dx,$$

$$\frac{y^2}{2} = \frac{-x^2}{2} + k.$$

This gives the circles we were expecting:

$$x^2 + y^2 = C \qquad \text{where } C = 2k.$$

You might worry about whether it is legitimate to separate the dx and the dy. The reason it can be done is explained at the end of this section.

The Exponential Growth and Decay Equations

Let's use separation of variables on an equation that occurs frequently in practice:

$$\frac{dy}{dx} = ky.$$

Separating variables,

$$\frac{1}{y} \, dy = k \, dx,$$

and integrating,

$$\int \frac{1}{y} \, dy = \int k \, dx,$$

gives

$$\ln|y| = kx + C \quad \text{for some constant } C.$$

Solving for $|y|$ leads to

$$|y| = e^{kx+C} = e^{kx}e^C = Ae^{kx}$$

where $A = e^C$, so A is positive. Thus

$$y = (\pm A)e^{kx} = Be^{kx}$$

where $B = \pm A$, so B is any nonzero constant. Even though there's no C leading to $B = 0$, B can be 0 because $y = 0$ is a solution to the differential equation. We lost this solution when we divided through by y at the first step.

The general solution to $\dfrac{dy}{dx} = ky$ is $y = Be^{kx}$ for any B.

The differential equation $dy/dx = ky$ always leads to exponential growth (if $k > 0$) or exponential decay (if $k < 0$). Graphs of solution curves for some fixed $k > 0$ are in Figure 10.27. For $k < 0$, the graphs are reflected in the y-axis.

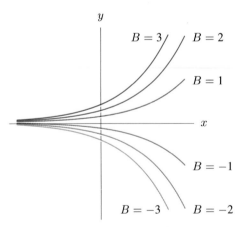

Figure 10.27: Graphs of $y = Be^{kx}$, which are solutions to $\frac{dy}{dx} = ky$, for some fixed $k > 0$

Example 1   For $k > 0$, find and graph solutions of

$$\frac{dH}{dt} = -k(H - 20).$$

Solution   The slope field in Figure 10.28 shows the qualitative behavior of the solutions. To find the equation of the solution curves, we separate variables and integrate:

$$\int \frac{1}{H - 20} \, dH = - \int k \, dt.$$

This gives

$$\ln |H - 20| = -kt + C.$$

Solving for H leads to:

$$|H - 20| = e^{-kt+C} = e^{-kt}e^{C} = Ae^{-kt}$$

or

$$H - 20 = (\pm A)e^{-kt} = Be^{-kt}$$

$$H = 20 + Be^{-kt}.$$

Graphs for $k = 1$ and $B = -10, 0, 10$, with $t \geq 0$, are in Figure 10.28.

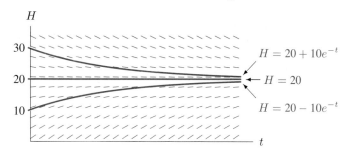

Figure 10.28: Slope field and some solution curves for $\frac{dH}{dt} = -k(H - 20)$, with $k = 1$

Example 2   Find and sketch the solution to

$$\frac{dP}{dt} = 2P - 2Pt$$

satisfying $P = 5$ when $t = 0$.

Solution   Factoring the right-hand side gives

$$\frac{dP}{dt} = P(2 - 2t).$$

Separating variables, we get

$$\int \frac{dP}{P} = \int (2 - 2t) \, dt,$$

so

$$\ln |P| = 2t - t^2 + C.$$

Solving for P leads to

$$|P| = e^{2t-t^2+C} = e^{C} e^{2t-t^2} = Ae^{2t-t^2}$$

with $A = e^{C}$, so $A > 0$. Thus the general solution to the differential equation is

$$P = Be^{2t-t^2} \quad \text{for any } B.$$

To find the value of B for this example, substitute $P = 5$ and $t = 0$ into the general solution, giving

$$5 = Be^{2(0)-0^2} = B$$

so

$$P = 5e^{2t-t^2}.$$

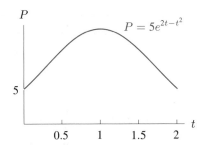

Figure 10.29: Bell-shaped solution curve

The graph of this function is in Figure 10.29.

Since the solution can be rewritten as

$$P = 5e^{1-1+2t-t^2} = 5e^1 e^{-1+2t-t^2} = (5e)e^{(t-1)^2},$$

the graph has the same shape as the graph of $y = e^{-t^2}$, the bell-shaped curve of statistics. Here the maximum, normally at $t = 0$, is shifted one unit to the right and occurs at $t = 1$.

Justification For Separation of Variables

Suppose a differential equation can be written in the form

$$\frac{dy}{dx} = g(x)f(y).$$

Then, provided $f(y) \neq 0$, we write $f(y) = 1/h(y)$ so that the right-hand side can be thought of as a fraction,

$$\frac{dy}{dx} = \frac{g(x)}{h(y)}.$$

If we multiply through by $h(y)$ we get

$$h(y)\frac{dy}{dx} = g(x).$$

Thinking of y as a function of x, so $y = y(x)$, and $dy/dx = y'(x)$, we can write the line above as

$$h(y(x)) \cdot y'(x) = g(x).$$

Now integrate both sides with respect to x:

$$\int h(y(x)) \cdot y'(x)\, dx = \int g(x)\, dx.$$

The form of the integral on the left suggests that we use the substitution $y = y(x)$. Since $dy = y'(x)\, dx$, we get

$$\int h(y)\, dy = \int g(x)\, dx.$$

If antiderivatives of h and g can be found, then this will give the equation of the solution curve.

Note that transforming the original differential equation,

$$\frac{dy}{dx} = \frac{g(x)}{h(y)},$$

into

$$\int h(y)\, dy = \int g(x)\, dx$$

looks as though we have treated dy/dx as a fraction, cross-multiplied and then integrated. Although that's not exactly what we've done, you may find this a helpful way of remembering the method. In fact, the dy/dx notation was introduced by Leibniz to allow shortcuts like this (more specifically, to make the chain rule look like cancelation).

Problems for Section 10.4

Find the solutions to the differential equations in Problems 1–22, subject to the given initial condition.

1. $\dfrac{dP}{dt} = 0.02P, \quad P(0) = 20$

2. $\dfrac{dQ}{dt} = \dfrac{Q}{5}, \quad Q = 50$ when $t = 0$

3. $\dfrac{dm}{dt} = 3m, \quad m = 5$ when $t = 1$

4. $\dfrac{dI}{dx} = 0.2I, \quad I = 6$ where $x = -1$

5. $\dfrac{dy}{dx} + \dfrac{y}{3} = 0, \quad y(0) = 10$

6. $\dfrac{1}{z}\dfrac{dz}{dt} = 5, \quad z(1) = 5$

7. $\dfrac{dP}{dt} = P + 4, \quad P = 100$ when $t = 0$

8. $\dfrac{dy}{dx} = 2y - 4, \quad$ through $(2, 5)$

9. $\dfrac{dm}{dt} = 0.1m + 200, \quad m(0) = 1000$

10. $\dfrac{dB}{dt} + 2B = 50, \quad B(1) = 100$

11. $\dfrac{dy}{dt} = 0.5(y - 200), \quad y = 50$ when $t = 0$

12. $\dfrac{dQ}{dt} = 0.3Q - 120, \quad Q = 50$ when $t = 0$

13. $\dfrac{dR}{dr} + R = 1, \quad R(1) = 0.1$

14. $\dfrac{dy}{dt} = \dfrac{y}{3 + t}, \quad y(0) = 1$

15. $\dfrac{dz}{dt} = te^z, \quad$ through the origin

16. $\dfrac{dy}{dx} = \dfrac{5y}{x}, \quad y = 3$ where $x = 1$

17. $\dfrac{dy}{dt} = y^2(1 + t), \quad y = 2$ when $t = 1$

18. $\dfrac{dz}{dt} = z + zt^2, \quad z = 5$ when $t = 0$

19. $\dfrac{dw}{d\theta} = \theta w^2 \sin \theta^2, \quad w(0) = 1$

20. $x(x + 1)\dfrac{du}{dx} = u^2, \quad u(1) = 1$

21. $\dfrac{dz}{dr} = z + zr^2, \quad z(0) = 1$

22. $\dfrac{dw}{d\psi} = \psi w^2 \cos(\psi^2), \quad w(0) = 1$

Solve the differential equations in Problems 23–28. Assume a, b, and k are constants.

23. $\dfrac{dR}{dt} = kR$

24. $\dfrac{dQ}{dt} - \dfrac{Q}{k} = 0$

25. $\dfrac{dP}{dt} = P - a$

26. $\dfrac{dQ}{dt} = b - Q$

27. $\dfrac{dP}{dt} = k(P - a)$

28. $\dfrac{dR}{dt} = aR + b$

Solve the differential equations in Problems 29–32. Assume $x \geq 0$, $y \geq 0$.

29. $\dfrac{dy}{dt} = y(2 - y), \quad y(0) = 1.$

30. $t\dfrac{dx}{dt} = (1 + 2\ln t)\tan x.$

31. $\dfrac{dx}{dt} = \dfrac{x \ln x}{t}.$

32. $\dfrac{dy}{dt} = -y \ln\left(\dfrac{y}{2}\right), \quad y(0) = 1.$

33. (a) Sketch the slope field for $y' = x/y$.
    (b) Sketch several solution curves.
    (c) Solve the differential equation analytically.

34. (a) Sketch the slope field for $y' = -y/x$.
    (b) Sketch several solution curves.
    (c) Solve the differential equation analytically.

35. Compare the slope field for $y' = x/y$, Problem 33, with that for $y' = -y/x$, Problem 34. Show that the solution curves of Problem 33 intersect the solution curves of Problem 34 at right angles.

36. The differential equation $\dfrac{dy}{dx} = \dfrac{x + y}{x - y}$ is not separable. Show that if we put $y(x) = xv(x)$, then the resulting differential equation is separable.

37. The differential equation $\dfrac{dy}{dx} = \dfrac{x+y}{x-y}$ can be rewritten as $\dfrac{dy}{dx} = \dfrac{1+(y/x)}{1-(y/x)} = f\left(\dfrac{y}{x}\right)$. In general, any differential equation that can be rewritten as $\dfrac{dy}{dx} = f\left(\dfrac{y}{x}\right)$ is called *homogeneous*. Show that every homogeneous differential equation can be reduced to a separable equation (in terms of x and v) by putting $y(x) = xv(x)$.

38. The differential equation $\dfrac{dt}{ds} = \dfrac{s+t+5}{s-t-1}$ is neither separable nor homogeneous (as defined in Problem 37). Show that making a change of variable of the form $s = x+a$, $t = y+b$, and choosing a and b appropriately, yields a homogeneous equation. The homogeneous equation can then be solved with the method of Problem 37.

10.5 GROWTH AND DECAY

In this section we look at exponential growth and decay equations. Consider the population of a region. If there is no immigration or emigration, the rate at which the population is changing is often proportional to the population. In other words, the larger the population, the faster it is growing, because there are more people to have babies. If the population at time t is P, and its continuous growth rate is 2% per unit time, then we know

$$\text{Rate of growth of population} = 2\%(\text{Current population})$$

and we can write this as

$$\frac{dP}{dt} = 0.02P.$$

The 2% growth rate is called the *relative growth rate* to distinguish it from the absolute growth rate, or rate of change of the population, dP/dt. Notice they are in different units. Since

$$\text{Relative growth rate} = 2\% = \frac{1}{P}\frac{dP}{dt}$$

the relative growth rate is in percent per unit time, while

$$\text{Absolute growth rate} = \text{Rate of change of population} = \frac{dP}{dt}$$

is in people per unit time.

The equation $dP/dt = 0.02P$ is of the form $dP/dt = kP$ for $k = 0.02$ and therefore has the solution

$$P = P_0 e^{0.02t}.$$

Other processes are described by differential equations similar to that for population growth, but with negative values for k. In summary, we have the following result from the preceding section:

Every solution to the equation

$$\frac{dP}{dt} = kP$$

can be written in the form

$$P = P_0 e^{kt},$$

where P_0 is the initial value of P, and $k > 0$ represents growth, while $k < 0$ represents decay.

Recall that the *doubling time* of an exponentially growing quantity is the time required for it to double. The *half-life* of an exponentially decaying quantity is the time for half of it to decay.

Continuously Compounded Interest

Continuous compounding means that, at any time, interest is being accrued at a rate that is a fixed percentage of the balance at that moment[2]. Thus, the larger the balance the faster interest is being earned and the faster the balance grows.

Example 1   A bank account earns interest continuously at a rate of 5% of the current balance per year. Assume that the initial deposit is $1000, and that no other deposits or withdrawals are made.

(a) Write the differential equation satisfied by the balance in the account.

(b) Solve the differential equation and graph the solution.

Solution   (a) In this problem we are looking for B, the balance in the account in dollars, as a function of t, time in years. The fact that interest is being earned continuously means that interest is continuously being added to the account at a rate of 5% of the balance at that moment. Thus, at any instant,

$$\text{Rate at which interest is earned} = 5\%(\text{Current balance}).$$

If B is the balance at time t, and if no deposits or withdrawals are made, then the rate of change of the balance is exactly the rate the interest is earned, so

$$\text{Rate at which balance is increasing} = 5\%(\text{Current balance}),$$

which we write as

$$\frac{dB}{dt} = 0.05B.$$

This is the differential equation that describes the process. Notice that it does not involve the $1000, because the initial deposit does not affect the process by which interest is earned. The $1000 is the initial condition.

(b) Solving the differential equation by separation of variables gives

$$B = B_0 e^{0.05t},$$

where B_0 is the initial value of B, so $B_0 = 1000$. Thus

$$B = 1000 e^{0.05t}$$

and this function is graphed in Figure 10.30.

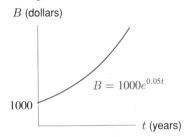

Figure 10.30: Bank balance against time

You may wonder how we can represent an amount of money by a differential equation, since money can only take on discrete values (we can't have fractions of a cent). In fact, the differential equation is only an approximation, but for large amounts of money, it is a pretty good approximation.

The Difference Between Continuous and Annual Percentage Growth Rates

If $P = P_0(1+r)^t$ with t in years, we say that r is the *annual* growth rate, while if $P = P_0 e^{kt}$, we say that k is the *continuous* or *instantaneous* growth rate.

---

[2]Appendix B introduces continuous compounding as the limiting case in which interest is added more and more often.

It is important to note that the constant k in the differential equation $dP/dt = kP$ is not the annual growth rate, but rather the continuous growth rate. In Example 1, with a continuous interest rate of 5%, we obtain a balance at time t of $B = B_0 e^{0.05t}$, where t is measured in years. At the end of one year the balance is $B_0 e^{0.05}$. In that one year, our balance has changed from B_0 to $B_0 e^{0.05}$; that is, it has changed by the factor $e^{0.05} = 1.0513$. Thus the annual growth rate is 5.13%. This is what the bank means when it says "5% compounded continuously for an effective annual yield of 5.13%." Since $P_0 e^{0.05t} = P_0 (1.0513)^t$, we are talking about two different ways to represent the same function of t.

Since most growth is measured over discrete time intervals, a continuous growth rate is an idealized concept. A demographer who says the population of some country is growing at the rate of 2% per year usually means just what you think: After one year the population will have increased by a factor of 1.02; and after t years the population will be given by $P = P_0 (1.02)^t$. To find the continuous growth rate, k, for the population we proceed as follows. Suppose the population is expressed as $P = P_0 e^{kt}$. At the end of one year $P = P_0 e^k$. To get 2% growth in that year we need $e^k = 1.02$. Thus $k = \ln 1.02 \approx 0.0198$. The continuous growth rate $k = 1.98\%$ is close to the annual growth rate of 2%, but it is not the same. Again, we have two different representations of the same function, since $P_0 (1.02)^t = P_0 e^{0.0198t}$.

Pollution in the Great Lakes

In the 1960s pollution in the Great Lakes became an issue of public concern. We set up a model for how long it would take for the lakes to flush themselves clean, assuming no further pollutants are being dumped in the lake.

Suppose Q is the total quantity of pollutant in a lake of volume V at time t. Suppose that clean water is flowing into the lake at a constant rate r and that water flows out at the same rate. Assume that the pollutant is evenly spread throughout the lake, and that the clean water coming into the lake immediately mixes with the rest of the water.

We investigate how Q varies with time. Since pollutants are being taken out of the lake but not added, Q decreases, and the water leaving the lake becomes less polluted, so the rate at which the pollutants leave decreases. This tells us that Q is decreasing and concave up. In addition, the pollutants are never completely removed from the lake though the quantity remaining becomes arbitrarily small: in other words, Q is asymptotic to the t-axis. (See Figure 10.31.)

Setting Up a Differential Equation for the Pollution

To model exactly how Q changes with time, we write an equation for the rate at which Q changes. We know that

$$\text{Rate } Q \text{ changes} = -\left(\text{Rate pollutants leave in outflow} \right).$$

where the negative sign represents the fact that Q is decreasing. At time t, the concentration of pollutants is Q/V and water containing this concentration is leaving at rate r. Thus

$$\text{Rate pollutants leave in outflow} = \text{Rate of outflow} \times \text{Concentration} = r \cdot \frac{Q}{V}.$$

Figure 10.31: Pollutant in lake versus time

So the differential equation is

$$\frac{dQ}{dt} = -\frac{r}{V}Q$$

and its solution is

$$Q = Q_0 e^{-rt/V}.$$

Table 10.4 contains values of r and V for four of the Great Lakes.[3] We use this data to calculate how long it would take for certain fractions of the pollution to be removed.

TABLE 10.4   *Volume and outflow in Great Lakes*

	V (thousands of km^3)	r (km^3/year)
Superior	12.2	65.2
Michigan	4.9	158
Erie	0.46	175
Ontario	1.6	209

Example 2   According to this model, how long will it take for 90% of the pollution to be removed from Lake Erie? For 99% to be removed?

Solution   Substituting r and V for Lake Erie into the differential equation for Q gives

$$\frac{dQ}{dt} = -\frac{r}{V}Q = \frac{-175}{0.46 \times 10^3}Q = -0.38Q$$

where t is measured in years. Thus Q is given by

$$Q = Q_0 e^{-0.38t}.$$

When 90% of the pollution has been removed, 10% remains, so $Q = 0.1Q_0$. Substituting gives

$$0.1Q_0 = Q_0 e^{-0.38t}.$$

Canceling Q_0 and solving for t, we get

$$t = \frac{-\ln(0.1)}{0.38} \approx 6 \text{ years.}$$

When 99% of the pollution has been removed, $Q = 0.01Q_0$, so t satisfies

$$0.01Q_0 = Q_0 e^{-0.38t}.$$

Solving for t gives

$$t = \frac{-\ln(0.01)}{0.38} \approx 12 \text{ years.}$$

Newton's Law of Heating and Cooling

Newton proposed that the temperature of a hot object decreases at a rate proportional to the difference between its temperature and that of its surroundings. Similarly, a cold object heats up at a rate proportional to the temperature difference between the object and its surroundings.

For example, a hot cup of coffee standing on the kitchen table cools at a rate proportional to the temperature difference between the coffee and the surrounding air. As the coffee cools down, the rate at which it cools decreases, because the temperature difference between the coffee and the air decreases. In the long run, the rate of cooling tends to zero, and the temperature of the coffee

[3]Data from William E. Boyce and Richard C. DiPrima, *Elementary Differential Equations* (New York: Wiley, 1977).

approaches room temperature—as we would expect. Figure 10.32 shows the temperature of two cups of coffee against time, one starting at a higher temperature than the other, but both tending toward room temperature in the long run.

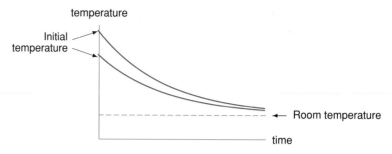

Figure 10.32: Temperature of coffee versus time

Example 3   When a murder is committed, the body, originally at 37°C, cools according to Newton's Law of Cooling. Suppose that after two hours the temperature is 35°C, and that the temperature of the surrounding air is a constant 20°C.

(a) Find the temperature, H, of the body as a function of t, the time in hours since the murder was committed.

(b) Sketch a graph of temperature against time.

(c) What happens to the temperature in the long run? Show this on the graph and algebraically.

(d) If the body is found at 4 pm at a temperature of 30°C, when was the murder committed?

Solution   (a) We first find a differential equation for the temperature of the body as a function of time. Newton's Law of Cooling says that for some constant α,

$$\text{Rate of change of temperature} = \alpha(\text{Temperature difference}).$$

If H is the temperature of the body, then

$$\text{Temperature difference} = H - 20$$

so

$$\frac{dH}{dt} = \alpha(H - 20).$$

What about the sign of α? If the temperature difference is positive (i.e., $H > 20$), then H is falling, so the rate of change must be negative. Thus α should be negative, so we write:

$$\frac{dH}{dt} = -k(H - 20), \qquad \text{for some } k > 0.$$

Separating variables and solving, as in Example 1 on page 506, gives:

$$H - 20 = Be^{-kt}.$$

To find B, substitute the initial condition that $H = 37$ when $t = 0$:

$$37 - 20 = Be^{-k(0)} = B,$$

so $B = 17$. Thus,

$$H - 20 = 17e^{-kt}.$$

To find k, we use the fact that after 2 hours, the temperature is 35°C, so

$$35 - 20 = 17e^{-k(2)}.$$

Dividing by 17 and taking natural logs, we get:

$$\ln\left(\frac{15}{17}\right) = \ln(e^{-2k})$$
$$-0.125 = -2k$$
$$k \approx 0.063.$$

Therefore, the temperature is given by

$$H - 20 = 17e^{-0.063t}$$

or

$$H = 20 + 17e^{-0.063t}.$$

(b) The graph of $H = 20 + 17e^{-0.063t}$ has a vertical intercept of $H = 37$, because the temperature of the body starts at 37°C. The temperature decays exponentially with $H = 20$ as the horizontal asymptote. (See Figure 10.33.)

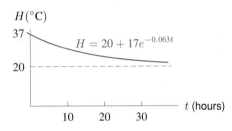

Figure 10.33: Temperature of dead body

(c) "In the long run" means as $t \to \infty$. The graph shows that as $t \to \infty$, $H \to 20$. Algebraically, since $e^{-0.063t} \to 0$ as $t \to \infty$,

$$H = 20 + \underbrace{17e^{-0.063t}}_{\text{goes to 0 as } t \to \infty} \longrightarrow 20 \quad \text{as } t \to \infty.$$

(d) We want to know when the temperature reaches 30°C. Substitute $H = 30$ and solve for t:

$$30 = 20 + 17e^{-0.063t}$$
$$\frac{10}{17} = e^{-0.063t}.$$

Taking natural logs:

$$-0.531 = -0.063t,$$

which gives

$$t \approx 8.4 \, \text{hours}.$$

Thus the murder must have been committed about 8.4 hours before 4 pm. Since 8.4 hours = 8 hours 24 minutes, the murder was committed at about 7:30 am.

Equilibrium Solutions

Figure 10.34 shows the temperature of several objects in a 20°C room. One is initially hotter than 20°C and cools down toward 20°C; another is initially cooler and warms up toward 20°C. All these curves are solutions to the differential equation

$$\frac{dH}{dt} = -k(H - 20)$$

for some fixed $k > 0$, and all the solutions have the form

$$H = 20 + Ae^{-kt}$$

for some A. Notice that $H \to 20$ as $t \to \infty$ because $e^{-kt} \to 0$ as $t \to \infty$. In other words, in the long run, the temperature of the object always tends toward 20°C, the temperature of the room, no matter what the initial temperature. This means that what happens in the long run is independent of the initial condition.

In the special case when $A = 0$, we have the *equilibrium solution*

$$H = 20$$

for all t. This means that if the object starts at 20°C, it remains at 20°C for all time. Notice that such a solution can be found directly from the differential equation by solving $dH/dt = 0$:

$$\frac{dH}{dt} = -k(H - 20) = 0$$

giving

$$H = 20.$$

Regardless of the initial temperature, H always gets closer and closer to 20 as $t \to \infty$. As a result, $H = 20$ is called a *stable* equilibrium[4] for H.

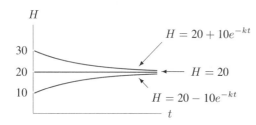

Figure 10.34: $H = 20$ is stable equilibrium $(k > 0)$

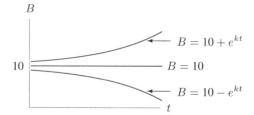

Figure 10.35: $B = 10$ is unstable equilibrium $(k > 0)$

A different situation is displayed in Figure 10.35, which shows solutions to the differential equation

$$\frac{dB}{dt} = k(B - 10)$$

for some fixed $k > 0$. Solving $dB/dt = 0$ gives the equilibrium $B = 10$, which is *unstable* because if B starts near 10, it moves away as $t \to \infty$.

In general, we have the following definitions.

- An **equilibrium solution** is constant for all values of the independent variable. The graph is a horizontal line. Equilibrium solutions can be identified by setting the derivative of the function to zero.

- An equilibrium is **stable** if a small change in the initial conditions gives a solution which tends toward the equilibrium as the independent variable tends to positive infinity.

- An equilibrium is **unstable** if a small change in the initial conditions gives a solution curve which veers away from the equilibrium as the independent variable tends to positive infinity.

---

[4]In more advanced work, this behavior is described as asymptotic stability.

Problems for Section 10.5

1. The amount of arable land (land that can be used for growing crops) in use increases as the world's population increases. Suppose $A(t)$ represents the total number of hectares of arable land in use in year t. (A hectare is about $2\frac{1}{2}$ acres.)

   (a) Explain why it is plausible that $A(t)$ satisfies the equation $A'(t) = kA(t)$. What assumptions are you making about the world's population and its relation to the amount of arable land used?

   (b) In 1950 about $1 \cdot 10^9$ hectares of arable land were in use; in 1980 the figure was $2 \cdot 10^9$. If the total amount of arable land available is thought to be $3.2 \cdot 10^9$ hectares, when does this model predict it is exhausted? (Let $t = 0$ in 1950.)

2. A yam is put in a $200°$C oven and heats up according to the differential equation

$$\frac{dH}{dt} = -k(H - 200),$$

   where k is positive.

   (a) If the yam is at $20°$C when it is put in the oven, solve the differential equation.

   (b) Find k using the fact that after 30 minutes the temperature of the yam is $120°$C.

3. Each of the curves in Figure 10.36 represents the balance in a bank account into which a single deposit was made at time zero. Assuming interest is compounded continuously, find

   (a) The curve representing the largest initial deposit.

   (b) The curve representing the largest interest rate.

   (c) Two curves representing the same initial deposit.

   (d) Two curves representing the same interest rate.

Figure 10.36

4. The graphs in Figure 10.37 represent the temperature, $H°$C, of four eggs as a function of time, t, measured in minutes. Match three of the graphs with the descriptions below. Write a similar description for the fourth graph, including an interpretation of any intercepts and asymptotes.

Figure 10.37

   (a) An egg is taken out of the refrigerator (just above $0°$C) and put into boiling water.

   (b) Twenty minutes after the egg in part (a) is taken out of the fridge and put into boiling water, the same thing is done with another egg.

   (c) An egg is taken out of the refrigerator at the same time as the egg in part (a) and left to sit on the kitchen table.

5. (a) What are the equilibrium solutions for the differential equation

$$\frac{dy}{dt} = 0.2(y - 3)(y + 2)?$$

   (b) Use a slope field to determine whether each equilibrium solution is stable or unstable.

6.   The slope field for a differential equation is given in Figure 10.38. Estimate all equilibrium solutions for this differential equation, and indicate whether each is stable or unstable.

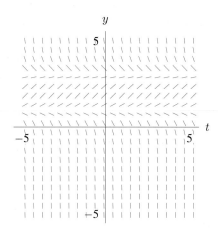

Figure 10.38

7.   (a)   Find the equilibrium solution to the differential equation

$$\frac{dy}{dt} = 0.5y - 250.$$

(b)   Find the general solution to this differential equation.

(c)   Sketch the graphs of several solutions to this differential equation, using different initial values for y.

(d)   Is the equilibrium solution stable or unstable?

8.   Money in a bank account grows continuously at an annual rate of r (when the interest rate is 5%, $r = 0.05$, and so on). Suppose $1000 is put into the account in 1970.

(a)   Write a differential equation satisfied by M, the amount of money in the account at time t, measured in years since 1970.

(b)   Solve the equation.

(c)   Sketch the solution until the year 2000 for interest rates of 5% and 10%.

9.   (a)   If $B = f(t)$ is the balance of a bank account at time t that earns interest at a rate of r%, compounded continuously, what is the differential equation that describes the rate at which the balance changes? What is the constant of proportionality, in terms of r?

(b)   What is the solution to this differential equation?

(c)   Sketch the graph of $B = f(t)$ for an account that starts with $1000 and earns interest at the following rates:

       (i)   4%   (ii)   10%   (iii)   15%

10.   In Example 2 on page 513 we saw that it would take about 6 years for 90% of the pollution in Lake Erie to be removed, and about 12 years for 99% to be removed. Explain why one time is double the other.

11.   Using the model in the text and data in Table 10.4 on page 513, find how long it would take for 90% of the pollution to be removed from Lake Michigan and from Lake Ontario, assuming no new pollutants are added. Explain how you can tell which lake will take longer to be purified just by looking at the data.

12.   Use the model in the text and the data in Table 10.4 on page 513 to determine which of the Great Lakes would require the longest time and which would require the shortest time for 80% of the pollution to be removed, assuming no new pollutants are being added. Find the ratio of these two times.

13.   Warfarin is a drug used as an anticoagulant. After stopping administration of the drug, the quantity remaining in a patient's body decreases at a rate proportional to the quantity remaining. The half-life of Warfarin in the body is 37 hours.

    (a)   Sketch a rough graph of the quantity, Q, of warfarin in a patient's body as function of the time, t, since stopping administration of the drug. Mark the 37 hours on your graph.

    (b)   Write a differential equation satisfied by Q.

    (c)   How many days does it take for the drug level in the body to be reduced to 25% of the original level?

14.   Morphine is often used as a pain-relieving drug. The half-life of morphine in the body is 2 hours. Suppose morphine is administered to a patient intravenously at a rate of 2.5 mg per hour, and the rate at which the morphine is eliminated is proportional to the amount present.

    (a)   Show that the constant of proportionality for the rate at which morphine leaves the body (in mg/hour) is $k = -0.347$.

    (b)   Write a differential equation for the quantity, Q, of morphine in the blood after t hours.

    (c)   Use the differential equation to find the equilibrium solution. (This is the long-term amount of morphine in the body, once the system has stabilized.)

15.   When the electromotive force (emf) is removed from a circuit containing inductance and resistance but no capacitors, the rate of decrease of current is proportional to the current. If the initial current is 30 amps but decays to 11 amps after 0.01 seconds, find an expression for the current as a function of time.

16.   In some chemical reactions, the rate at which the amount of a substance changes with time is proportional to the amount present. For example, this is the case as δ-glucono-lactone changes into gluconic acid.

    (a)   Write a differential equation satisfied by y, the quantity of δ-glucono-lactone present at time t.

    (b)   If 100 grams of δ-glucono-lactone is reduced to 54.9 grams in one hour, how many grams will remain after 10 hours?

17.   The rate at which barometric pressure decreases with altitude is proportional to the barometric pressure at that altitude. If the barometric pressure is measured in inches of mercury, and the altitude in feet, then the constant of proportionality is 3.7×10^{-5}. Suppose the barometric pressure at sea level is 29.92 inches of mercury.

    (a)   Calculate the barometric pressure at the top of Mount Whitney, 14,500 feet (the highest mountain in the US outside Alaska), and at the top of Mount Everest, 29,000 feet (the highest mountain in the world).

    (b)   People cannot easily survive at a pressure below 15 inches of mercury. What is the highest altitude to which people can safely go?

18.   The rate (per foot) at which light is absorbed as it passes through water is proportional to the intensity, or brightness, at that point.

    (a)   Find the intensity as a function of the distance the light has traveled through the water.

    (b)   If 50% of the light is absorbed in 10 feet, how much is absorbed in 20 feet? 25 feet?

19.   A detective finds a murder victim at 9 am. The temperature of the body is measured at $90.3°$F. One hour later, the temperature of the body is $89.0°$F. The temperature of the room has been maintained at a constant $68°$F.

    (a)   Assuming the temperature, T, of the body obeys Newton's Law of Cooling, write a differential equation for T.

    (b)   Solve the differential equation to estimate the time the murder occurred.

20.   Suppose that at 1:00 pm one winter afternoon, there is a power failure at your house in Wisconsin, and your heat does not work without electricity. When the power goes out, it is $68°$F in your house. At 10:00 pm, it is $57°$F in the house, and you notice that it is $10°$F outside.

    (a)   Assuming that the temperature, T, in your home obeys Newton's Law of Cooling, write the differential equation satisfied by T.

    (b)   Solve the differential equation to estimate the temperature in the house when you get up at 7:00 am the next morning. Should you worry about your water pipes freezing?

    (c)   What assumption did you have to make in part (a) about the temperature outside? Given this (probably incorrect) assumption, would you revise your estimate up or down? Why?

21. The amount of radioactive carbon-14 in a sample is measured using a Geiger counter, which records each disintegration of an atom. Living tissue disintegrates at a rate of about 13.5 atoms per minute per gram of carbon. In 1977 a charcoal fragment found at Stonehenge, England, recorded 8.2 disintegrations per minute per gram of carbon. Assuming that the half-life of carbon-14 is 5730 years and that the charcoal was formed during the building of the site, estimate the date at which Stonehenge was built.

22. The radioactive isotope carbon-14 is present in small quantities in all life forms, and it is constantly replenished until the organism dies, after which it decays to stable carbon-12 at a rate proportional to the amount of carbon-14 present, with a half-life of 5730 years. Suppose $C(t)$ is the amount of carbon-14 present at time t.

    (a) Find the value of the constant k in the differential equation $C' = -kC$.
    (b) In 1988 three teams of scientists found that the Shroud of Turin, which was reputed to be the burial cloth of Jesus, contained 91% of the amount of carbon-14 contained in freshly made cloth of the same material.[5] How old is the Shroud of Turin, according to these data?

23. Before Galileo discovered that the speed of a falling body with no air resistance is proportional to the time since it was dropped, he mistakenly conjectured that the speed was proportional to the distance it had fallen.

    (a) Assume the mistaken conjecture to be true and write an equation relating the distance fallen, $D(t)$, at time t, and its derivative.
    (b) Using your answer to part (a) and the correct initial conditions, show that D would have to be equal to 0 for all t, and therefore the conjecture must be wrong.

10.6 APPLICATIONS AND MODELING

Much of this book involves functions that represent real processes, such as how the temperature of a yam or the population of Mexico is changing with time. You may wonder where such functions come from. In some cases, we fit functions to experimental data by trial and error. In other cases, we take a more theoretical approach, leading to a differential equation whose solution is the function we want. In this section we give examples of the more theoretical approach.

How a Layer of Ice Forms

When ice forms on a lake, the water on the surface freezes first. As heat from the water travels up through the ice and is lost to the air, more ice is formed. The question we will consider is: How thick is the layer of ice as a function of time? Since the thickness of the ice increases with time, the thickness function is increasing. In addition, as the ice gets thicker, it insulates better, therefore we expect the layer of ice to form more slowly as time goes on. Hence the thickness function is increasing at a decreasing rate, so its graph is concave down.

A Differential Equation for the Thickness of the Ice

To get more detailed information about the thickness function, we have to make some assumptions. Suppose y represents the thickness of the ice as a function of time, t. Since the thicker the ice, the longer it takes the heat to get through it, we will assume that the rate at which ice is formed is inversely proportional to the thickness. In other words, we assume that for some constant k,

$$\text{Rate thickness is increasing} = \frac{k}{\text{Thickness}},$$

so

$$\frac{dy}{dt} = \frac{k}{y} \quad \text{where} \quad k > 0.$$

_____

[5]*The New York Times*, October 18, 1988

Figure 10.39: Thickness of ice as a function of time

This differential equation enables us to find a formula for the function y. Using separation of variables:

$$\int y \, dy = \int k \, dt$$

$$\frac{y^2}{2} = kt + C.$$

If we measure time so that $y = 0$ when $t = 0$, then $C = 0$. Since y must be non-negative, we have

$$y = \sqrt{2kt}.$$

Graphs of y against t are in Figure 10.39.

Notice that the larger y is, the more slowly y increases. In addition, this model suggests that y increases indefinitely as time passes. (Of course, the value of y cannot increase beyond the depth of the lake.)

The Net Worth of a Company

In the preceding section, we saw an example in which money in a bank account was earning interest (Example 1, page 511). Consider a company whose revenues are proportional to its net worth (like interest on a bank account) but which must also make payroll payments. The question is: under what circumstances does the company make money and under what circumstances does it go bankrupt?

Common sense says that if the payroll exceeds the rate at which revenue is earned, the company will eventually be in trouble, whereas if revenue exceeds payroll, the company should do well. In order to make this more precise, we make several assumptions. We assume that revenue is earned continuously and that payments are made continuously. (In practice, revenue and payments are not made continuously, but for a large company this is a good approximation.) We also assume that the only factors affecting net worth are revenue and payroll.

Example 1   Suppose a company's revenue is earned at a continuous annual rate of 5% of its net worth. At the same time, the company's payroll obligations amount to \$200 million a year, paid out continuously.
(a)   Write a differential equation that governs the net worth of the company, W million dollars.
(b)   Solve the differential equation, assuming an initial net worth of W_0 million dollars.
(c)   Sketch the solution for $W_0 = 3000$, 4000, and 5000.

Solution   First, let's see what we can learn without writing a differential equation. For example, we can ask if there is any initial net worth W_0 which will exactly keep the net worth constant. If there's such an equilibrium, the rate at which revenue is earned must exactly balance the payments made, so

$$\begin{array}{ccc} \text{Rate at which} & = & \text{Rate at which payroll} \\ \text{revenue is earned} & & \text{payments are made.} \end{array}$$

If net worth is a constant W_0, revenue must be earned at a constant rate of $0.05W_0$ per year, so we must have

$$0.05W_0 = 200 \quad \text{giving} \quad W_0 = 4000.$$

Therefore, if the net worth starts at \$4000 million, the revenue and payments are equal, and the net worth will remain constant. Therefore \$4000 million is an equilibrium solution.

Suppose, however, the initial net worth is above \$4000 million. Then, the revenue earned will be more than the payroll expenses, and the net worth of the company will increase, thereby increasing the revenue still further. Thus the net worth will increase faster and faster. On the other hand, if the initial net worth is below \$4000 million, the revenue will not be enough to meet the payments, and the net worth of the company will decline. This decreases the revenue; making the net worth decrease still faster. Hence the net worth will eventually go to zero, and the company will go bankrupt.

(a)   Now we set up a differential equation for the net worth, using the fact that

$$\begin{array}{ccc} \text{Rate at which} & = & \text{Rate revenue} \quad - \quad \text{Rate payroll payments} \\ \text{net worth is increasing} & & \text{is earned} \qquad\qquad \text{are made} \end{array}$$

In millions of dollars per year, revenue is earned at a rate of $0.05W$, and payments are made at a rate of 200 per year, so we have

$$\frac{dW}{dt} = 0.05W - 200$$

where t is in years. Notice that the initial net worth does not enter into the differential equation. The equilibrium solution, $W = 4000$, is obtained by setting $dW/dt = 0$.

(b)   We solve this equation by separation of variables. It is helpful to factor out 0.05 before separating, so that the W moves over to the left-hand side without a coefficient. This makes the manipulations less messy. So we write

$$\frac{dW}{dt} = 0.05(W - 4000).$$

Separating and integrating gives

$$\int \frac{dW}{W - 4000} = \int 0.05 \, dt,$$

so

$$\ln|W - 4000| = 0.05t + C,$$

or

$$|W - 4000| = e^{0.05t + C} = e^C e^{0.05t}.$$

This means

$$W - 4000 = Ae^{0.05t} \quad \text{where } A = \pm e^C.$$

To find A we use the initial condition that $W = W_0$ when $t = 0$.

$$W_0 - 4000 = Ae^0 = A.$$

Substituting this value for A into $W = 4000 + Ae^{0.05t}$ gives

$$W = 4000 + (W_0 - 4000)e^{0.05t}.$$

(c)   If $W_0 = 4000$, then $W = 4000$, the equilibrium solution.
If $W_0 = 5000$, then $W = 4000 + 1000e^{0.05t}$.
If $W_0 = 3000$, then $W = 4000 - 1000e^{0.05t}$. Substituting $t \approx 27.7$ gives $W = 0$, so this solution means that the company goes bankrupt in its twenty-eighth year.

The graphs of these functions are in Figure 10.40. Notice that if the net worth starts with W_0 near, but not equal to, \$4000 million, then W moves further away. Thus, $W = 4000$ is an unstable equilibrium.

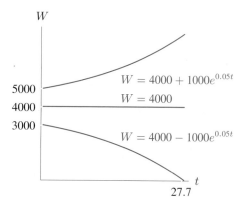

Figure 10.40: Solutions to $\frac{dW}{dt} = 0.05W - 200$

The Velocity of a Falling Body

Imagine a sky-diver jumping out of a plane. Think about how the person's velocity varies during the time before his parachute opens. When the sky-diver first jumps, his velocity is zero. Then the pull of gravity comes into play, making the velocity increase. As the sky-diver speeds up, the air resistance also increases. Since the air resistance partly balances the pull of gravity, as the sky-diver speeds up, the force causing him to accelerate decreases. Thus the velocity must be an increasing function of time, and it must be increasing at a decreasing rate. If the air resistance increases until it balances gravity, then the sky-diver's velocity will stop increasing—the velocity levels off. Thus we expect the velocity to be increasing, and to have a graph that is concave down with a horizontal asymptote.

A Differential Equation: Air Resistance Proportional to Velocity

In order to compute the velocity function, we must make some assumptions about the forces acting on the falling body and the accelerations they produce. It is reasonable to assume that the faster the body is falling, the larger the air resistance. However, to decide whether air resistance is proportional to the velocity, say, or is some other function of velocity, requires either lab experiments or a theoretical idea of how the air resistance is created. In the case of a very small object, such as a dust particle settling on a computer component during manufacturing,[5] we assume that air resistance is proportional to velocity.

Newton's Second Law of Motion, which states that

$$\text{Force} = \text{Mass} \times \text{Acceleration},$$

relates force to velocity, since acceleration is dv/dt. The force is the net force on the object, which has two components: the gravitational force, mg, which acts downward, and the air resistance, kv, which acts upward, in the opposite direction to the motion (we assume $k > 0$). (See Figure 10.41.) Thus

$$\text{Gravity} - \text{Air resistance} = \text{Mass} \times \text{Acceleration},$$

so

$$mg - kv = m\frac{dv}{dt}.$$

This differential equation can be solved by separation of variables. It is easier if we factor out $-k/m$ before separating, giving

$$\frac{dv}{dt} = -\frac{k}{m}\left(v - \frac{mg}{k}\right).$$

Separating and integrating gives

$$\int \frac{dv}{v - mg/k} = -\frac{k}{m}\int dt$$

$$\ln\left|v - \frac{mg}{k}\right| = -\frac{k}{m}t + C.$$

[5]Example suggested by Howard Stone.

Solving for v:

$$\left| v - \frac{mg}{k} \right| = e^{-kt/m+C} = e^C e^{-kt/m}$$

$$v - \frac{mg}{k} = Ae^{-kt/m},$$

where A is an arbitrary constant. We find A from the initial condition that the object starts from rest, so $v = 0$ when $t = 0$. Substituting

$$0 - \frac{mg}{k} = Ae^0$$

gives

$$A = -\frac{mg}{k}.$$

Thus

$$v = \frac{mg}{k} - \frac{mg}{k}e^{-kt/m} = \frac{mg}{k}(1 - e^{-kt/m}).$$

The graph of this function is in Figure 10.42. As expected, at the start the object is moving slowly, which means the air resistance is small, so it speeds up. As the object starts to move faster, the air resistance builds up until it balances the gravitational force. At this point the object has reached its *terminal velocity*, mg/k. In this model, the object never actually reaches terminal velocity (because $e^{-kt/m}$ is never actually equal to zero).

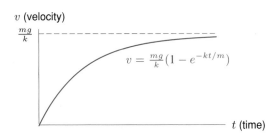

Figure 10.41: Forces acting on a falling object

Figure 10.42: Velocity under the assumption that air resistance $= kv$

Alternative Method of Finding the Terminal Velocity

Notice that the terminal velocity can also be obtained from the differential equation by setting $dv/dt = 0$ and solving for v:

$$m\frac{dv}{dt} = mg - kv = 0$$

so

$$v = \frac{mg}{k}.$$

Compartmental Analysis: A Reservoir

Many processes can be modeled as a container with various solutions flowing in and out—for example, drugs given intravenously or the discharge of pollutants into a lake. We will consider a city's water reservoir, fed partly by clean water from a spring and partly by run-off from the surrounding land. In New England, or many other areas with much snow in the winter, the run-off contains salt which has been put on the roads to make them safe for driving. We will consider the concentration of salt in the reservoir. If there is no salt in the reservoir initially, we expect the concentration to build up until the rate at which the salt is being brought into the reservoir by the run-off is balanced by the rate at which salt flows out. If, on the other hand, the reservoir starts with a great deal of salt in it, then initially the rate at which the salt is flowing in is less than the rate at which it is flowing out, and the quantity of salt in the lake should decrease. In either case, the salt concentration levels off at an equilibrium value.

A Differential Equation for Salt Concentration

Suppose the water reservoir holds 100 million gallons of water and supplies a city with 1 million gallons a day. The reservoir is partly refilled by a spring which provides 0.9 million gallons a day, and the rest of the water, 0.1 million gallons a day, comes from run-off from the surrounding land. The spring is clean, but the run-off contains salt with a concentration of 0.0001 pound per gallon. Assume that there was no salt in the reservoir initially and that the reservoir is well mixed (that is, that the water taken by the city residents contains the concentration of salt in the tank at that instant). We find the concentration of salt in the reservoir as a function of time, assuming the reservoir remains full.

In this type of problem, it is very important to distinguish between the total quantity and the concentration of salt, where

$$\text{Concentration} = \frac{\text{Quantity of salt}}{\text{Volume of water}}.$$

If C is the concentration of salt (in pounds/gallon) and Q is the quantity (in pounds), then since the volume of the reservoir is 100 million gallons:

$$C = \frac{Q}{100 \text{ million}} \left(\frac{\text{lb}}{\text{gal}} \right).$$

The differential equation we use is based on the relation

$$\begin{array}{c} \text{Rate of change of} \\ \text{quantity of salt in reservoir} \end{array} = \text{Rate salt entering} - \text{Rate salt leaving,}$$

which describes how the quantity is changing with time. We will find Q first, and then C.

To find the rate salt is entering, notice that all the salt is entering through the run-off. The run-off is 0.1 million gallons per day, with each gallon containing 0.0001 pound of salt. Therefore

$$\begin{aligned} \text{Rate salt entering} &= \text{Concentration} \times \text{Volume per day} \\ &= 0.0001 \left(\frac{\text{lb}}{\text{gal}} \right) \times 0.1 \left(\frac{\text{million gal}}{\text{day}} \right) \\ &= 0.00001 \left(\frac{\text{million lb}}{\text{day}} \right) = 10 \left(\frac{\text{lb}}{\text{day}} \right). \end{aligned}$$

Thus

$$\text{Rate salt entering } = 10 \text{ lb/day.}$$

Salt is leaving in the water used by the city. The city uses a million gallons of water each day with a concentration of $\dfrac{Q}{100 \text{ million}} \dfrac{\text{lb}}{\text{gal}}$. Thus

$$\begin{aligned} \text{Rate salt leaving} &= \text{Concentration} \times \text{Volume per day} \\ &= \frac{Q}{100 \text{ million}} \left(\frac{\text{lb}}{\text{gal}} \right) \times 1 \left(\frac{\text{million gal}}{\text{day}} \right) \\ &= \frac{Q}{100} \left(\frac{\text{lb}}{\text{day}} \right). \end{aligned}$$

So

$$\text{Rate salt leaving } = \frac{Q}{100} \text{ lb/day.}$$

Therefore Q must satisfy

$$\frac{dQ}{dt} = 10 - \frac{Q}{100}.$$

We factor out $-1/100$:

$$\frac{dQ}{dt} = -\frac{1}{100}(Q - 1000) = -0.01(Q - 1000),$$

and separate variables:

$$\int \frac{dQ}{Q - 1000} = -\int 0.01 \, dt$$
$$\ln|Q - 1000| = -0.01t + k,$$

so

$$Q - 1000 = Ae^{-0.01t}.$$

There is no salt initially, so we substitute $Q = 0$ when $t = 0$:

$$0 - 1000 = Ae^0 \quad \text{giving} \quad A = -1000.$$

Thus

$$Q - 1000 = -1000e^{-0.01t}$$

so

$$Q = 1000(1 - e^{-0.01t}) \quad \text{pounds.}$$

Therefore

$$\text{Concentration} = C = \frac{Q}{100 \text{ million}} = \frac{1000}{10^8}(1 - e^{-0.01t}) = 10^{-5}(1 - e^{-0.01t}) \text{ lbs/gal.}$$

A sketch of concentration against time is in Figure 10.43.

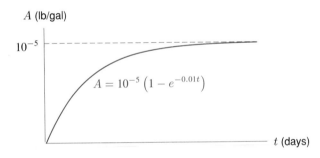

Figure 10.43: Concentration of salt in reservoir

Qualitative Information from a Differential Equation

In the last part of this section we see how information can sometimes be obtained from a differential equation without solving it. The example we take is a second-order differential equation. We first integrate to get a first-order equation, and then investigate qualitative behavior of the solution.

The Expansion of the Universe

Observations show that the universe is currently expanding, so astronomers are interested in exactly how fast it is expanding. One model[6] assumes that the universe is a giant sphere of radius $R(t)$, where t is time, and that

$$R'' = -\frac{GM_0}{R^2}$$

---

[6]From M. Rowan-Robinson, *Cosmology,* Oxford University Press (1977).

where G is the universal gravitational constant and M_0 is the mass of the universe. This is a second-order differential equation for R. Even without solving the equation, we can get information about the expansion of the universe.

The first step is to get a first-order equation from this second-order equation. In this case we multiply both sides of the equation for R'' by R', giving

$$R'' R' = -\frac{GM_0}{R^2} R'.$$

Since $\frac{d}{dt}\left[(R')^2\right] = 2R'' R'$ and $\frac{d}{dt}\left(R^{-1}\right) = -R^{-2} R'$, we can write the preceding equation as

$$\frac{1}{2}\frac{d}{dt}\left[(R')^2\right] = GM_0 \frac{d}{dt}\left(\frac{1}{R}\right).$$

Multiplying through by 2 and integrating with respect to t gives

$$(R')^2 = \frac{2GM_0}{R} + C.$$

We use this relation between R' and R to determine the behavior of R over time. We look only at the case in which $C > 0$. The cases where $C < 0$ and $C = 0$ are investigated in Problems 19 and 20 on page 530. Since

$$R' = \pm\sqrt{\frac{2GM_0}{R} + C}$$

and since the universe is currently expanding, we know that R' must be positive, so R' must be given by the positive root:

$$R' = \sqrt{\frac{2GM_0}{R} + C}.$$

As R gets larger, R' remains positive, so the universe continues to expand. However, R', the rate of expansion, decreases to $\sqrt{C}$ as $R \to \infty$. This model predicts that the universe will expand forever, but that the rate of the expansion is slowing down to $\sqrt{C}$.

Problems for Section 10.6

1. Dead leaves accumulate on the ground in a forest at a rate of 3 grams per square centimeter per year. At the same time, these leaves decompose at a continuous rate of 75% a year. Write a differential equation for the total quantity of dead leaves (per square centimeter) at time t. Sketch a solution showing that the quantity of dead leaves tends towards an equilibrium level. What is that equilibrium level?

2. According to a simple physiological model, an athletic adult needs 20 calories per day per pound of body weight to maintain his weight. If he consumes more or fewer calories than those required to maintain his weight, his weight will change at a rate proportional to the difference between the number of calories consumed and the number needed to maintain his current weight; the constant of proportionality is $1/3500$ pounds per calorie. Suppose that a particular person has a constant caloric intake of I calories per day. Let $W(t)$ be the person's weight in pounds at time t (measured in days).

   (a) What differential equation has solution $W(t)$?
   (b) Solve this differential equation.
   (c) Draw a graph of $W(t)$ if the person starts out weighing 160 pounds and consumes 3000 calories a day. Label your axes and any intercepts and asymptotes clearly.

3. When a gas expands without gain or loss of heat, the rate of change of pressure with respect to volume is proportional to pressure divided by volume. Find a law connecting pressure and volume in this case.

4. In 1692, Johann Bernoulli was teaching the Marquis de l'Hopital calculus in Paris. Solve the following problem, which is similar to the one that they did. What is the equation of the curve which has subtangent (distance BC in Figure 10.44) equal to twice its abscissa (distance OC)?

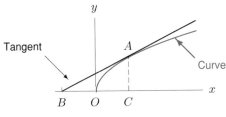

Figure 10.44

5. Water leaks out of a barrel at a rate proportional to the square root of the depth of the water at that time. If the water level starts at 36 inches and drops to 35 inches in 1 hour, how long will it take for all of the water to leak out of the barrel?

6. In Brazil, interest rates increased steadily in the second half of the 1970s. As 1975 began, the interest rate was 50%, and it increased 25% a year after that (to 75% in 1976, to 100% in 1977 and so on). Assuming both that the interest rate climbed continuously and that interest accrued continuously:

   (a) Write a differential equation for the amount of money, M, in a bank account as a function of time, t, measured from 1975. Assume the only deposit was 100,000 cruzieros on January 1, 1975.
   (b) Solve this differential equation.
   (c) How much money was in the account on January 1, 1980?

7. (Continuation of Problem 6)

   (a) Suppose now Brazilian banks had offered a constant interest rate between 1975 and 1980. What would that interest rate have had to be to produce the same growth in the bank balance between 1975 and 1980?
   (b) Sketch a graph from 1975 into the early 1980's showing the balance in the account under both schemes, the rising and the constant interest rates.

8. Consider the sky-diver whose velocity is described by the differential equation

$$m\frac{dv}{dt} = mg - kv, \quad v(0) = 0.$$

Find and solve a differential equation for a, the acceleration of the sky-diver.

9. Suppose a chemical reaction involves one molecule of a substance A combining with one molecule of substance B to form one molecule of substance C, written $A + B \rightarrow C$. The Law of Mass Action states that the rate at which C is formed is proportional to the product of the quantities of A and B present. Assume a and b are the initial quantities of A and B, and x is the quantity of C present at time t.

   (a) Write a differential equation for x.
   (b) Solve the equation with $x(0) = 0$.

10. If the initial quantities, a and b, in Problem 9 are the same, write and solve a differential equation for x, with $x(0) = 0$.

11. A patient is given the drug theophylline intravenously at a rate of 43.2 mg/hour to relieve acute asthma. You can imagine the drug as entering a compartment of volume 35,000 ml. (This is the volume of the part of the body through which the drug circulates.) The rate at which the drug leaves the patient is proportional to the quantity there, with proportionality constant 0.082. Assume the patient's body contains none of the drug initially.

    (a) Describe in words how you would expect the concentration of theophylline in the patient to vary with time and sketch an approximate graph.
    (b) Write a differential equation satisfied by the concentration of the drug, $c(t)$.
    (c) Solve the differential equation and graph the solution. What happens to the concentration in the long run?

12. A drug is administered intravenously at a constant rate of r mg/hour and is excreted at a rate proportional to the quantity present, with constant of proportionality $\alpha > 0$.

    (a) Solve a differential equation for the quantity, Q, in milligrams, of the drug in the body at time t hours. Your answer will contain r and α. Sketch a graph of Q against t. What is Q_∞, the limiting long-run value of Q?
    (b) What effect does doubling r have on Q_∞? What effect does doubling r have on the time to reach half the limiting value, $\frac{1}{2}Q_\infty$?
    (c) What effect does doubling α have on Q_∞? On the time to reach $\frac{1}{2}Q_\infty$?

13. As you know, when a course ends, students start to forget the material they have learned. One model (called the Ebbinghaus model) assumes that the rate at which a student forgets material is proportional to the difference between the material he or she currently remembers and some positive constant, a.

    (a) Let $y = f(t)$ be the fraction of the original material remembered t weeks after the course has ended. Set up a differential equation for y. Your equation will contain two constants; the constant a is less than y for all t.
    (b) Solve the differential equation.
    (c) Describe the practical meaning (in terms of the amount remembered) of the constants in the solution $y = f(t)$.

14. A certain commodity is currently being sold at a price of $\$p$ per unit. Over a period of time market forces will make this price tend toward the equilibrium price, which we call $\$p_0$, at which supply exactly balances demand. The rate at which the price changes is described by the Evans Price Adjustment model, which says that the rate of change in the actual market price ($\$p$) is proportional to the difference between the actual market price and the equilibrium price.

    (a) Write a differential equation for p as a function of t.
    (b) Solve for p.
    (c) Sketch solutions for various different initial prices, both above and below the equilibrium price.
    (d) What happens to p as $t \to \infty$?

15. When people smoke, carbon monoxide is released into the air. Suppose, in a room of volume 60 m^3, air containing 5% carbon monoxide is introduced at a rate of 0.002 m^3/min. (This means that 5% of the volume of the incoming air is carbon monoxide.) Assume that the carbon monoxide mixes immediately with the rest of the air, and that the mixture leaves the room at the same rate as it enters.

    (a) Write a differential equation for $c(t)$, the concentration of carbon monoxide at time t, in minutes.
    (b) Solve the differential equation, assuming there was no carbon monoxide in the room initially.
    (c) What happens to the value of $c(t)$ in the long run?

16. (Continuation of Problem 15.) Medical texts[7] warn that exposure to air containing 0.1% carbon monoxide for some time can lead to a coma. How long does it take for the concentration of carbon monoxide in the room in Problem 15 to reach this level?

17. An aquarium pool has volume 2×10^6 liters. The pool initially contains pure fresh water. At $t = 0$ minutes, water containing 10 grams/liter of salt is poured into the pool at a rate of 60 liters/minute. The salt water instantly and totally mixes with the fresh water, and the excess mixture is drained out of the bottom of the pool at the same rate (60 liters/minute). Let $S(t)$ = the mass of salt in the pool at time t.

    (a) Write a differential equation for the amount of salt in the pool.
    (b) Solve the differential equation to find $S(t)$.
    (c) What happens to $S(t)$ as $t \to \infty$?

18. Imagine an object of mass m thrown vertically upward from the surface of the earth with initial velocity v_0. In this problem, we calculate the value of v_0, called the *escape velocity*, with which the object can escape the pull of the gravity and never return to earth. Suppose v is the velocity of the object (measured upward) at time t.

    Since the object is moving far from the surface of the earth, we must take into account the variation of gravity with altitude. If the acceleration due to gravity at sea level is g, the gravitational force, F, on

---

[7] See, for example, R.G. Petersdorf et al., *Harrison's Principles of Internal Medicine* (New York: McGraw Hill, 1983).

the object of mass m at an altitude h above the surface of the earth is given by

$$F = \frac{mgR^2}{(R+h)^2},$$

where R is the radius of the earth.

(a) Use Newton's Law of Motion to show that

$$\frac{dv}{dt} = -\frac{gR^2}{(R+h)^2}.$$

(b) Rewrite this equation with h instead of t as the independent variable using the chain rule $\frac{dv}{dt} = \frac{dv}{dh} \cdot \frac{dh}{dt}$. Hence show that

$$v\frac{dv}{dh} = -\frac{gR^2}{(R+h)^2}.$$

(c) Solve the differential equation in part (b).

(d) Find the escape velocity, the value of v_0 such that v is never zero.

Problems 19–21 refer to the model of the expansion of the universe given on page 526 by the equations

$$R'' = -\frac{GM_0}{R^2} \quad \text{and} \quad (R')^2 = \frac{2GM_0}{R} + C,$$

where $R(t)$ is the radius of the universe (assumed spherical), t is time, G is the universal gravitational constant, and M_0 is the mass of the universe. (In the text we assumed that $C > 0$.)

19. In this problem we look at the case where $C < 0$. Writing $C = -K$, where $K > 0$, we have

$$R' = \pm\sqrt{\frac{2GM_0}{R} - K}.$$

Since the universe is currently expanding, at this time $R' > 0$, so R' is currently given by the positive root. Show that R increases to some value $R_{\max}$, and then decreases again. In addition, show that when this happens and R again approaches zero, then R' has a large negative value; in other words, a "big crunch" happens.

20. In this problem we look at the case where $C = 0$. Then we know that

$$R' = \sqrt{\frac{2GM_0}{R}}.$$

Solve this differential equation, assuming $R = 0$ when $t = 0$. Give R as a function of t. The resulting formula for R is called the "flat universe model." What does this model predict about the expansion of the universe?

21. (a) Einstein, who formulated the model of the universe described before Problem 19, wanted the universe to be stable—neither expanding nor shrinking. Why don't these differential equations allow for a stable universe?

(b) One estimate for the age of the universe is $R(t_0)/R'(t_0)$ where t_0 is the current time. (This number is called the *Hubble constant*.) Why is this a reasonable estimate for the age of the universe? Is it an overestimate or an underestimate?

10.7 MODELS OF POPULATION GROWTH

Population projections have been important to political philosophers since at least the late eighteenth century. As concern for scarce resources has grown, so has interest in accurate population projections. In this section we will look at two differential equations which are used to model both human and animal population growth. These differential equations have applications in economics and medicine, such as modeling the spread of an innovation or the growth of a tumor.

Relative versus Absolute Growth Rates

When describing population growth, we often use percentages rather than absolute numbers. For example, we say the population of the world is now about 5.8 billion people and growing at a continuous rate of about 1.7% a year, meaning that the relative growth rate is 1.7%:

$$\frac{dP/dt}{P} = \frac{1}{P}\frac{dP}{dt} = 0.017.$$

The quantity dP/dt is called the *absolute growth rate* and measures the growth rate in, say, people per year. The quantity $(dP/dt)/P$ is called the *relative growth rate* and represents the absolute growth rate as a fraction of the whole population. Its units are, say, % per year. When talking about populations we often use the relative growth rate.

The US Population: 1790--1860

Every ten years the population of the United States is recorded by a census. The first such census was in 1790. Table 10.5 contains the census data from 1790 to 1940.

TABLE 10.5   *US Population in millions, 1790–1940*

Year	Population	Year	Population	Year	Population
1790	3.9	1850	23.1	1910	92.0
1800	5.3	1860	31.4	1920	105.7
1810	7.2	1870	38.6	1930	122.8
1820	9.6	1880	50.2	1940	131.7
1830	12.9	1890	62.9		
1840	17.1	1900	76.0		

You might note that the population is given only to the nearest 0.1 million (or 100,000), although the population is reported by the US Census Bureau down to the last digit (for example, 131,669,275 in 1940). We have rounded off because census figures are notoriously inaccurate. For example, in the census of 1990, New York City claimed the census had missed a million people in that city alone. Thus, giving more digits in the population does not necessarily give more accuracy.

Let us concentrate first on the relative growth rate,$(dP/dt)/P$, of the US population from 1790 to 1860. If we want to estimate $(dP/dt)/P$ in 1830, we take the rate of change in the population and divide it by the population itself:[8]

$$\frac{1}{P}\frac{dP}{dt} \approx \frac{1}{\text{Population in 1830}} \cdot \frac{\text{Population in 1840} - \text{Population in 1830}}{10 \text{ years}}$$

$$= \frac{1}{12.9} \cdot \frac{17.1 - 12.9}{10} = 0.0326 = 3.26\%.$$

Similar calculations for 1790, 1800, ..., 1850 give the percentages in Table 10.6:

TABLE 10.6   *Rough estimates of yearly growth rate of US population*

Year	1790	1800	1810	1820	1830	1840	1850
Relative growth rate	3.59%	3.58%	3.33%	3.44%	3.26%	3.57%	3.53%

These percentages are pretty close. The relative growth rate, while not precisely constant, is nearly so. In fact, political and economic events such as war or recession affect the population, so we don't expect the growth rate to be exactly constant.

The simplest model for population growth is to assume that the relative growth rate *is* constant, in other words

$$\frac{1}{P}\frac{dP}{dt} = k,$$

---

[8]In Problem 12 at the end of this section we look at an alternative way of estimating the relative growth rate in 1830, using data from 1820 as well as from 1840.

where k is the continuous growth rate. This is equivalent to assuming that the population grows exponentially:

$$P = P_0 e^{kt}.$$

What should we take for the value of k? One possibility would be the average of the percentages we just calculated, namely 3.47%. However, there is a serious objection to using this percentage as an estimate for k. Remember that k is a continuous growth rate, but the populations are given at 10-year intervals. A 10-year population growth of 34.7% doesn't come from a continuous yearly rate of 3.47%. (This would be ignoring the effects of compounding.) If the population increases by a factor of 34.7% in 10 years, then $P(10)/P_0 = 1.347$, where $P(10)$ is the population when $t = 10$. Assuming that $P = P_0 e^{kt}$, we need k to satisfy

$$\frac{P(10)}{P_0} = e^{k \cdot 10} = 1.347.$$

Now we can find the continuous growth rate, k:

$$k = \frac{\ln(1.347)}{10} = 0.0298.$$

Let's compare predicted and actual values if we model the US population by the differential equation

$$\frac{dP}{dt} = 0.0298P.$$

We start with initial population $P_0 = 3.9$ in 1790. Notice that this says we will consider 1790 as time $t = 0$, so 1800 is $t = 10$, and 1810 is $t = 20$, etc. The solution to the differential equation is

$$P = 3.9 e^{0.0298t}.$$

If we put $t = 0, 10, 20, \ldots, 70$ into this function we get the populations predicted by our model for the years 1790, 1800, $\ldots$, 1860. Table 10.7 contains the comparison to the actual populations.

TABLE 10.7 *Predicted versus actual US population 1790–1860 (exponential model)*

Year	Actual	Predicted	Year	Actual	Predicted
1790	3.9	3.9	1830	12.9	12.8
1800	5.3	5.3	1840	17.1	17.3
1810	7.2	7.1	1850	23.1	23.3
1820	9.6	9.5	1860	31.4	31.4

The agreement is remarkable. Of course, since we used the data from the entire 70 year period to estimate k, we should expect good agreement throughout that period. What is surprising is that if we had used only the populations in 1790 and 1800 to estimate k, the predictions are still quite good. Let's find a new value of k using only the 1790 and 1800 data and compare predictions. The 10 year growth from 1790 to 1800 is 35.9% so $k = \ln(1.359)/10 = 0.0307$. We would then predict the population in 1860 to be

$$3.9 e^{0.0307(70)} = 33.4,$$

which is within about 6% of the actual population of 31.4. It is remarkable that a person in 1800 could accurately predict what the population of the US would be 60 years later, especially considering all the wars, recessions, epidemics, additions of new territory, and immigration that took place from 1800 to 1860.

Predictions From The Exponential Model

The grim predictions of the exponential model are reflected in the ideas of Thomas Malthus, an early nineteenth-century clergyman and political philosopher, who believed that, if unchecked, a population would grow exponentially, whereas the food supply would grow linearly, and therefore the population would eventually outstrip the food supply.

Interestingly enough, an exponential model has fit the growth of world population and the population of many regions remarkably well for decades, even centuries. However, the model must break down at some point because it predicts that the population will continue to grow without bound as time goes on—and this cannot be true forever. Eventually the effects of crowding, emigration, disease, war, and lack of food will have to curb growth. In searching for an improvement, then, we should look for a model whose solution is approximately an exponential function for small values of the population, but which levels off later.

How to Estimate dP/dt from Data

If, as is often the case, all we know about a population, P, is its values at certain points in time, we have to approximate dP/dt by $\Delta P/\Delta t$. However there are several different ways this approximation can be made. As in the example above, we can say

$$\frac{dP}{dt} \text{ at } 1830 \approx \frac{\text{Population in 1840} \ - \ \text{Population in 1830}}{10}.$$

However we could equally well have said

$$\frac{dP}{dt} \text{ at } 1830 \approx \frac{\text{Population in 1830} \ - \ \text{Population in 1820}}{10}.$$

Both of these are called *one-sided estimates* because they involve using the population to one side of 1830 but not the other. The first one involving 1840 is the forward, or right-hand, estimate; the second one involving 1820 is the backward, or left-hand, estimate. In general, both are equally good or bad estimates for dP/dt. A more accurate approximation can be obtained by averaging the two one-sided estimates, giving

$$\frac{dP}{dt} \text{ at } 1830 \approx \frac{1}{2}\left(\frac{\text{Population in 1840} \ - \ \text{Population in 1830}}{10} + \frac{\text{Population in 1830} \ - \ \text{Population in 1820}}{10} \right).$$

When we add the two fractions the population in 1830 cancels and we get the two-sided estimate, so called because it involves data on both sides of 1830:

$$\frac{dP}{dt} \text{ at } 1830 \approx \frac{1}{2}\left(\frac{\text{Population 1840} \ - \ \text{Population 1820}}{10} \right) = \left(\frac{\text{Population 1840} \ - \ \text{Population 1820}}{20} \right).$$

In the exponential model above we used a one-sided estimate. This turned out to give good results after adjusting from 10-year to continuous growth rates. If we had been unable to get accurate predictions using one-sided estimates, we might have tried two-sided estimates instead. In the logistic model below, we use two-sided estimates, as they turn out to give noticeably better predictions.

The US Population: 1790--1940

The exponential model for the US population works well for reasonable periods of time, but exponential growth cannot go on forever. For example, if we tried to predict the population of the US in 1990 using the exponential model we developed above with $k = 0.0298$, we would get

$$\text{US Population in 1990} = 3.9e^{(0.0298)(200)} = 1{,}512 \text{ million,}$$

which is far from the actual figure of around 250 million.

The problem is that growth rate in the US population between 1790 and 1860 did not stay constant in later decades. The 10-year percentage growths[9] from 1860 to 1930 are listed in Table 10.8:

TABLE 10.8   *Estimated 10-year growth rate of US population, 1860–1930*

Year	1860	1870	1880	1890	1900	1910	1920	1930
Relative growth rate	22.9%	30.1%	25.3%	20.8%	21.1%	14.9%	16.2%	7.2%

[9]Calculated using one-sided forward estimates, so that $\frac{1}{P}\frac{dP}{dt}$ at 1860 $\approx \frac{\text{Population in 1870} - \text{Population in 1860}}{10(\text{Population in 1860})}$.

These figures are nothing like those during the period 1790 to 1860, where they hovered around 34%. The dramatic drop to 22.9% for the decade 1860–1870 is explainable by the Civil War (but don't ascribe the entire drop to deaths of the war—see Problem 15, page 541). The growth rate goes back up during the decade 1870–1880, but by 1890–1900 it has dropped below the rate during the Civil War. There is a slight increase in the rate during the immigrations of 1900 to 1910, another drop during World War I, a small bounce back, and finally the rate plummets during the recession of the 1930s. Notice that the effect of the recession is more dramatic than the effect of wars, which suggests that it is a decrease in birth rate rather than an increase in death rate that is a major factor in declines in the relative population growth rate.

Our exponential model gives accurate predictions up to 1860. But for the years following 1860, the exponential model is inadequate. We look for a new model that will take into account the effects of overcrowding. Because of the effects of crowding, we expect the relative growth rate to decrease as the population increases. Thus we look at how $(dP/dt)/P$ changes as P changes. This time we use a two-sided estimate for dP/dt. For example:

$$\frac{dP}{dt} \text{ at } 1830 \approx \frac{\text{Population in } 1840 - \text{Population } 1820}{20}.$$

Table 10.9 contains estimates for $(dP/dt)/P$ computed this way for some years between 1790 and 1940. Comparing the last two columns suggests that the values in the last column may be an approximately linear function of P. To get a better picture of how $(dP/dt)/P$ varies with P, we plot $(dP/dt)/P$ versus P and see whether the points lie on a line. Figure 10.45 shows the scatterplot of the points together with the line that best fits the points. The equation for the line, which fits quite well, is

$$\frac{1}{P}\frac{dP}{dt} = 0.0318 - 0.000170P.$$

TABLE 10.9 *Some estimates for $\dfrac{1}{P}\dfrac{dP}{dt}$.*

Year	P	$\left(P(t+10) - P(t-10)\right)/(20P)$
1860	31.4	0.0245
1890	62.9	0.0205
1910	92.0	0.0161
1930	122.8	0.0106

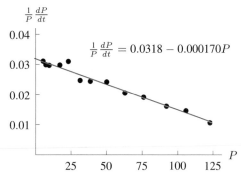

Figure 10.45: Scatterplot for US census data of $\frac{1}{P}\frac{dP}{dt}$ versus P.

Therefore, in our new model P satisfies the differential equation

$$\frac{dP}{dt} = 0.0318P - 0.000170P^2.$$

This is known as a *logistic equation.* Its slope field is shown in Figure 10.46, together with the solution with $P(0) = 3.9$ superimposed. (Notice that $t = 0$ in 1790.)

The most striking difference in this model compared with the exponential model is that it predicts that the US population will level off somewhere below 200 million. The population will continue to grow until $dP/dt = 0$, which occurs when

$$0 = 0.0318P - 0.000170P^2$$

so

$$P = 0 \quad \text{or} \quad P \approx 187 \text{ million.}$$

Figure 10.46: Solution to $\frac{dP}{dt} = 0.0318P - 0.000170P^2$ with $P(0) = 3.9$

Looking at the shape of the solution curve in Figure 10.46, we see that initially the population grows faster and faster and then slows down as the limiting value of 187 is approached; the fastest growth rate appears to be about half-way to the limiting value.

Later on in this section we derive the formula for the solution to the logistic equation. For now, you can check by substitution that the function

$$P = \frac{187}{1 + 47e^{-0.0318t}}$$

is a solution to our logistic equation modeling the US population. (The numbers 187 and the 47 are not exact values, but have been rounded.) The values predicted by this equation for P agree very well with the actual populations up to 1940. See Table 10.10. During the period from 1700 to 1940, the largest deviation is about 3% in 1840 and 1870 (the Civil War accounts for the second one). All other errors are less than 2%.

Of course, the final test is how well our model, based on data from 1790 to 1940, predicts the population in the "future," 1950 to 1990. Table 10.10 contains the predicted and actual data.

TABLE 10.10 *Predicted versus actual US population in millions, 1790–1980 (logistic model)*

Year	Actual	Predicted	Year	Actual	Predicted	Year	Actual	Predicted
1790	3.9	3.9	1860	31.4	30.8	1930	122.8	120.8
1800	5.3	5.3	1870	38.6	39.9	1940	131.7	133.7
1810	7.2	7.2	1880	50.2	50.7	1950	150.7	145.0
1820	9.6	9.8	1890	62.9	63.3	1960	179.3	154.4
1830	12.9	13.2	1900	76.0	77.2	1970	203.3	162.1
1840	17.1	17.7	1910	92.0	91.9	1980	226.5	168.2
1850	23.1	23.5	1920	105.7	106.7	1990	248.7	172.9

The fit between predicted and actual population values is clearly not good from 1950 on. Despite World War II, which undoubtedly depressed population growth between 1942 and 1945, in the last half of the 1940s the US population surged, wiping out in five years a deficit caused by 15 years of depression and war. The 1950s saw a population growth of 28 million, leaving our logistic model in the dust. This surge in population is referred to as the baby boom. All one can say is that based on 150 years of data, what happened in the US in the 20 years after World War II was completely without precedent. The baby boom could well end up being one of the most important sociological events of the twentieth century in the United States, and its consequences will be felt for many years to come.

Once again we have reached a point where our model is no longer useful. This should not lead you to believe that a reasonable mathematical model cannot be found; rather it should serve to point out that no model is perfect and that when one model fails, we seek a better one. Just as we abandoned the exponential model in favor of the logistic model for the US population, we could look further. (See Problems 9 and 10 on page 540.)

The Logistic Model

The logistic model we used to model the US population from 1790 to 1940 assumed that the relative growth rate of the population was a linearly decreasing function of P:

$$\frac{1}{P}\frac{dP}{dt} = k - aP.$$

(We took $k = 0.0318$ and $a = 0.000170$). For small P, we have approximately

$$\frac{1}{P}\frac{dP}{dt} \approx k.$$

The solution to this equation is an exponential function. This is why an exponential model fit the US population well during the years 1790-1860 when the population was relatively small. In the logistic model, as P increases, the relative growth rate decreases to zero; it reaches zero when P is given by

$$k - aP = 0.$$

Solving for P, we get

$$P = \frac{k}{a}.$$

This is the limiting value of the population, which we call L:

$$L = \frac{k}{a}.$$

The value L is called the *carrying capacity* of the environment, and represents the largest population the environment can support. Writing $a = k/L$, the logistic equation becomes

$$\frac{1}{P}\frac{dP}{dt} = k - \frac{k}{L}P$$
$$\frac{dP}{dt} = kP - \frac{k}{L}P^2$$

or

$$\frac{dP}{dt} = kP\left(1 - \frac{P}{L}\right).$$

This is the general logistic differential equation, first proposed as a model for population growth by the Belgian mathematician P. F. Verhulst in the 1830s.

Qualitative Solution to the Logistic Equation

Figure 10.47 shows the slope field and characteristic *sigmoid*, or *S*-shaped, solution curve for the logistic model. Notice that for each fixed value of P, that is, along each horizontal line, the slopes are all the same because dP/dt depends only on P and not on t. The slopes are small near $P = 0$ and near $P = L$; they are steepest around $P = L/2$. For $P > L$, the slopes are negative, meaning that if the population is above the carrying capacity, the population will decrease.

Figure 10.47: Slope field for $\frac{dP}{dt} = kP(1 - \frac{P}{L})$

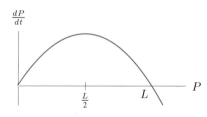

Figure 10.48: $\frac{dP}{dt} = kP(1 - \frac{P}{L})$

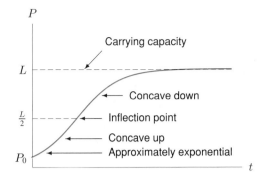

Figure 10.49: Logistic growth with inflection point

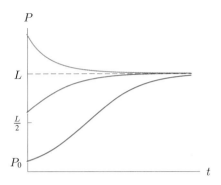

Figure 10.50: Solutions to the logistic equation

We can locate precisely the inflection point where the slopes are greatest using the graph of dP/dt against P in Figure 10.48. The graph is a parabola because dP/dt is a quadratic function of P. The horizontal intercepts are at $P = 0$ and $P = L$, so the maximum, where the slope is greatest, is at $P = L/2$. The graph in Figure 10.48 also tells us that for $0 < P < L/2$, the slope dP/dt is positive and increasing, so the graph of P against t is concave up. (See Figure 10.49.) For $L/2 < P < L$, the slope dP/dt is positive and decreasing, so the graph of P against t is concave down. For $P > L$, the slope dP/dt is negative, so the graph of P against t is decreasing.

If $P = 0$ or $P = L$, there is an equilibrium solution (not a very interesting one if $P = 0$). Figure 10.50 shows that $P = 0$ is an unstable equilibrium because solutions which start near 0 move away. However, $P = L$ is a stable equilibrium.

The Analytic Solution to the Logistic Equation

We have already obtained a lot of information about logistic growth without finding a formula for the solution. However, the equation can be solved analytically by separating variables:

$$\frac{dP}{dt} = kP\left(1 - \frac{P}{L}\right) = kP\left(\frac{L - P}{L}\right)$$

giving

$$\int \frac{dP}{P(L - P)} = \int \frac{k}{L}\,dt.$$

We can integrate the left side using the integral tables (Formula 26), or by rewriting

$$\frac{1}{P(L - P)} = \frac{1}{L}\left(\frac{1}{P} + \frac{1}{L - P}\right).$$

Thus, we have

$$\int \frac{1}{L}\left(\frac{1}{P} + \frac{1}{L - P}\right) dP = \int \frac{k}{L}\,dt.$$

Canceling the constant L, we get

$$\int \left(\frac{1}{P} + \frac{1}{L - P}\right) dP = \int k\,dt$$

which can be integrated to give

$$\ln|P| - \ln|L - P| = kt + C.$$

Multiplying through by (-1) and using the fact that $\ln M - \ln N = \ln(M/N)$, we have

$$\ln\left|\frac{L - P}{P}\right| = -kt - C.$$

Exponentiating both sides gives

$$\left| \frac{L-P}{P} \right| = e^{-kt-C} = e^{-C}e^{-kt},$$

so

$$\frac{L-P}{P} = Ae^{-kt} \quad \text{where} \quad A = \pm e^{-C}.$$

We find A by substituting $P = P_0$ when $t = 0$, which gives

$$\frac{L-P_0}{P_0} = Ae^0 = A.$$

Thus

$$\frac{L-P}{P} = Ae^{-kt} \quad \text{where} \quad A = \frac{L-P_0}{P_0}.$$

Since $(L-P)/P = (L/P) - 1$, we have

$$\frac{L}{P} = 1 + Ae^{-kt}$$

giving the formula for the logistic curve:

$$P = \frac{L}{1 + Ae^{-kt}} \quad \text{where} \quad A = \frac{L-P_0}{P_0}.$$

Problems for Section 10.7

1. Assuming that Switzerland's population is growing exponentially at a continuous rate of 0.2% a year and that its 1988 population was 6.6 million, write an expression for the population as a function of time in years. (Let $t = 0$ in 1988.)

2. Consider the logistic model

$$\frac{dP}{dt} = 3P - 3P^2.$$

(a) On the slope field in Figure 10.51, sketch three solution curves showing different types of behavior.

(b) Is there a stable value of the population? If so, what is it?

(c) Describe the meaning of the shape of the solution curves for the population: Where is P increasing? Decreasing? What happens in the long run? Are there any inflection points? Where? What do they mean for the population?

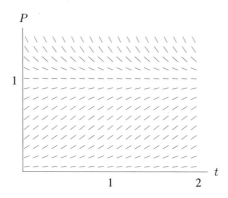

Figure 10.51: Slope field for $dP/dt = 3P - 3P^2$.

(d) Sketch a graph of dP/dt against P. Where is dP/dt positive? Negative? Zero? Maximum? How do your observations about dP/dt explain the shapes of your solution curves?

3. The total number of people infected with a virus often grows like a logistic curve. Suppose that 10 people originally have the virus, and that in the early stages of the virus (with time, t, measured in weeks), the number of people infected is increasing exponentially with $k = 1.78$. It is estimated that, in the long run, approximately 5000 people become infected.

    (a) Use this information to find a logistic function to model this situation.
    (b) Sketch a graph of your answer to part (a).
    (c) Use your graph to estimate the length of time until the rate at which people are becoming infected starts to decrease. What is the vertical coordinate at this point?

4. Table 10.11 gives the percentage, P, of households with a VCR, as a function of year.

    (a) Explain why a logistic model is a reasonable one to use for this data.
    (b) Use the data to estimate the point of inflection of P. What limiting value L does this point of inflection predict? Does this limiting value appear to be accurate given the percentages for 1990 and 1991?
    (c) The best logistic equation for this data turns out to be the following. What limiting value does this model predict?

$$P = \frac{75}{1 + 316.75e^{-0.699t}}.$$

    TABLE 10.11   *Percentage of households with a VCR*

Year	1978	1979	1980	1981	1982	1983	1984
$P(\%)$	0.3	0.5	1.1	1.8	3.1	5.5	10.6
Year	1985	1986	1987	1988	1989	1990	1991
$P(\%)$	20.8	36.0	48.7	58.0	64.6	71.9	71.9

5. The growth of a certain animal population is governed by the equation

$$\frac{1000}{P}\frac{dP}{dt} = 100 - P,$$

    where $P(t)$ is the number of individuals in the colony at time t. The initial population is known to be 200 individuals. Sketch a graph of $P(t)$. Will there ever be more than 200 individuals in the colony? Will there ever be fewer than 100 individuals? Explain.

6. It is of considerable interest to policy makers to model the spread of information through a population. For example, various agricultural ministries use models to help them understand the spread of technical innovations or new seed types through their countries. Two models, based on how the information is spread, are given below. Assume the population is of a constant size M.

    (a) If the information is spread by mass media (TV, radio, newspapers), the rate at which information is spread is believed to be proportional to the number of people not having the information at that time. Write a differential equation for the number of people having the information by time t. Sketch a solution assuming that no one (except the mass media) has the information initially.
    (b) If the information is spread by word of mouth, the rate of spread of information is believed to be proportional to the product of the number of people who know and the number who don't. Write a differential equation for the number of people having the information by time t. Sketch the solution for the cases in which
        (i)   no one
        (ii) 5% of the population
        (iii) 75% of the population
        knows initially. In each case, when is the information spreading fastest?

7. The population of a species of elk on Reading Island in Canada has been monitored for some years. When the population was 600, the relative birth rate was found to be 35% and the relative death rate was 15%. As the population grew to 800, the corresponding figures were 30% and 20%. The island is isolated so there is no hunting or migration.

    (a) Write a differential equation to model the population as a function of time. Assume that relative growth rate is a linear function of population.

    (b)   Find the equilibrium size of the population. Today there are 900 elk on Reading Island. How do you expect the population to change in the future?

    (c)   Oil has been discovered on a neighboring island and the oil companies want to move 450 elk of the same species to Reading Island. What effect would this move have on the elk population on Reading Island in the future?

    (d)   Assuming the elk are moved to Reading Island, sketch the population on Reading Island as a function of time. Start before the elk are transferred and continue for some time into the future. Comment on the significance of your results.

8.   Many organ pipes in old European churches are made of tin. In cold climates such pipes can be affected with *tin pest*, when the tin becomes brittle and crumbles into a grey powder. This transformation can appear to take place very suddenly because the presence of the grey powder encourages the reaction to proceed. At the start, when there is little grey powder, the reaction proceeds slowly. Similarly, toward the end, when there is little metallic tin left, the reaction is also slow. In between, however, when there is plenty of both metallic tin and powder, the reaction can be alarmingly fast.

    Suppose that the rate of the reaction is proportional to the product of the amount of tin left and the quantity of grey powder, p, present at time t. Assume also that when metallic tin is converted to grey powder, its mass does not change.

    (a)   Write a differential equation for p. Let the total quantity of metallic tin present originally be B.

    (b)   Sketch a graph of the solution $p = f(t)$ if there is a small quantity of powder initially. How much metallic tin has crumbled when it is crumbling fastest?

    (c)   Suppose there is no grey powder initially. (For example, suppose the tin is completely new.) What does this model predict will happen? How do you reconcile this with the fact that many organ pipes do get tin pest?

9.   (a)   In the text we fitted a logistic model to the US population from 1790–1940. In this problem, we try to fit a logistic equation to the US population all the way from 1790 to 1990. No logistic equation fits the data exactly over this entire period, but we can use the method of page 534 to find an equation that does reasonably well throughout.

    To fit a logistic equation to the data in Table 10.12, we estimate the relative growth rate, $(dP/dt)/P$, and plot it against P. To approximate $(dP/dt)/P$, calculate $(\Delta P/\Delta t)/P$ from the data for seven fairly spread-out points. Draw a reasonable line (by eye) through your points, and thus estimate k and a for the equation $(dP/dt)/P = k - aP$.

TABLE 10.12   *US population in millions, 1790–1990*

Year	Population	Year	Population	Year	Population
1790	3.9	1860	31.4	1930	122.8
1800	5.3	1870	38.6	1940	131.7
1810	7.2	1880	50.2	1950	150.7
1820	9.6	1890	62.9	1960	179.0
1830	12.9	1900	76.0	1970	205.0
1840	17.1	1910	92.0	1980	226.5
1850	23.1	1920	105.7	1990	248.7

    (b)   What does this model predict about the US population in the long run?

10.   (a)   In Problem 9 we saw that a logistic model cannot be made to fit the US population very closely, because the points on the graph of $(dP/dt)/P$ against P are not exactly on a line. In this problem we try another model. This time we assume that $(dP/dt)/P$ is a linear function of t, so you plot $(dP/dt)/P$ against t (using the same approximate values you calculated in Problem 9). Put a line through the points by eye, and estimate a and b to fit the equation

$$\frac{1}{P}\frac{dP}{dt} = a - bt.$$

    (b)   When, if ever, does this model predict that the US population will be at its maximum?

    (c)   Solve the differential equation and sketch its solution.

11. An alternative method of finding the analytic solution to the logistic equation

$$\frac{dP}{dt} = kP\left(1 - \frac{P}{L}\right)$$

uses the substitution $P = 1/u$.

   (a) Show that

$$\frac{dP}{dt} = -\frac{1}{u^2}\frac{du}{dt}$$

   (b) Rewrite the logistic equation in terms of u and t, and solve for u in terms of t.
   (c) Using your answer to part (b), find P as a function of t.

12. On page 531, we used one-sided estimates for dP/dt to fit an exponential model to the US population from 1790–1860. In this problem we use two-sided estimates for the years 1800–1850. Suppose P is the US population in millions.

   (a) Estimate dP/dt by the symmetric difference quotient $\left(P(t+10) - P(t-10)\right)/20$. Then compute $(dP/dt)/P$ for each of these years and average them to estimate k for the exponential model $dP/dt = kP$. Compare your value of k with the estimate $k \approx 3.47\%$ obtained on page 532 by using $\left(P(t+10) - P(t)\right)/10$ for dP/dt.
   (b) Using your answer to part (a), compute k as the continuous rate of change that leads to the observed 10-year percentage change. Then compare your k with the value $k \approx 2.98\%$ obtained on page 532.

13. This section suggested two ways of estimating dP/dt: the one-sided $\left(P(t+h) - P(t)\right)/h$, and the two-sided $\left(P(t+h) - P(t-h)\right)/(2h)$. Let $f(x) = x^3$. Consider the approximations $f'(2) \approx \left(f(2+h) - f(2)\right)/h$ and $f'(2) \approx \left(f(2+h) - f(2-h)\right)/2h$ for $h = 0.1, 0.01, 0.001$. Which is the better approximation? Is there a pattern to the errors in the approximation as you decrease h? If so, describe it.

14. Show that if $f(x) = x^2$, then $(f(x+h) - f(x-h))/(2h) = f'(x)$ for any value of h. This shows that if $f(x) = x^2$, the two-sided estimate for the derivative of $f(x)$ is exactly equal to the derivative.

15. Estimate the US population in 1870 from the 1860 population of 31.4 million, assuming that the population increased at the same percentage rate during the 1860s as it did in the decades previous to 1860 (about 34.7% each decade). Compare your estimate with the actual 1870 population of 38.6 million. Find some estimate of the number of people who died in the Civil War. Does the number of deaths in the Civil War explain the shortfall in the actual population in 1870? What else might influence the shortfall?

16. Another way to estimate the limiting population L for the logistic differential equation is to note that the derivative dP/dt is largest when $P = L/2$. Compute dP/dt for the US census data for 1790–1940 using the two-sided difference quotient $\left(P(t+10) - P(t-10)\right)/20$. Estimate L by doubling the population when dP/dt is largest. Compare this estimate with the estimate $L = 187$ given in the text.

Any population, P, for which we can ignore immigration, satisfies

$$\frac{dP}{dt} = \text{Birth rate} - \text{Death rate}.$$

For organisms which need a partner for reproduction but rely on a chance encounter for meeting a mate, the birth rate is proportional to the square of the population. Thus, the population of such a type of organism satisfies a differential equation of the form

$$\frac{dP}{dt} = aP^2 - bP \quad \text{with } a, b > 0.$$

Problems 17–19 investigate the solutions to such an equation.

17. Consider the equation

$$\frac{dP}{dt} = 0.02P^2 - 0.08P.$$

   (a) Sketch the slope field for this differential equation for $0 \le t \le 50$, $0 \le P \le 8$.
   (b) Use your slope field to sketch the general shape of the solutions to the differential equation satisfying the following initial conditions:

   (i) $P(0) = 1$ (ii) $P(0) = 3$ (iii) $P(0) = 4$ (iv) $P(0) = 5$

   (c) Are there any equilibrium values of the population? If so, are they stable?

18. Consider the equation

$$\frac{dP}{dt} = P^2 - 6P.$$

   (a) Sketch a graph of dP/dt against P for positive P.
   (b) Use the graph you drew in part (a) to sketch the approximate shape of the solution curve with $P(0) = 5$. To do this, consider the following question. For $0 < P < 6$, is dP/dt positive or negative? What does this tell you about the graph of P against t? As you move along the solution curve with $P(0) = 5$, how does the value of dP/dt change? What does this tell you about the concavity of the graph of P against t?
   (c) Use the graph you drew in part (a) to sketch the solution curve with $P(0) = 8$.
   (d) Describe the qualitative differences in the behavior of populations with initial value less than 6 and initial value more than 6. Why do you think $P = 6$ is called the *threshold population*?

19. Consider a population satisfying

$$\frac{dP}{dt} = aP^2 - bP \quad \text{with constants } a, b > 0.$$

   (a) Sketch a graph of dP/dt against P.
   (b) Use this graph to sketch the shape of solution curves with various initial values. Use your graph from part (a) to decide where dP/dt is positive or negative, and where it is increasing or decreasing. What does this tell you about the graph of P against t?
   (c) Why is $P = b/a$ called the threshold population? What happens if $P(0) = b/a$? What happens in the long-run if $P(0) > b/a$? What if $P(0) < b/a$?

10.8 SECOND-ORDER DIFFERENTIAL EQUATIONS: OSCILLATIONS

A Second-Order Differential Equation

When a body moves freely under gravity, we know that

$$\frac{d^2 s}{dt^2} = -g,$$

where s is the height of the body above ground at time t and g is the acceleration due to gravity. To solve this equation, we first integrate to get the velocity, $v = ds/dt$:

$$\frac{ds}{dt} = -gt + v_0,$$

where v_0 is the initial velocity. Then we integrate again, giving

$$s = -\frac{1}{2}gt^2 + v_0 t + s_0,$$

where s_0 is the initial height.

The differential equation $d^2 s/dt^2 = -g$ is called *second order* because the equation contains a second derivative but no higher derivatives. The general solution to a second-order differential equation will be a family of functions with two parameters, here v_0 and s_0. Finding values for the two constants corresponds to picking a particular function out of this family.

A Mass on a Spring

Not every second-order differential equation can be solved simply by integrating twice. Consider a mass m attached to the end of a spring hanging from the ceiling. We assume that the mass of the spring itself is negligible in comparison with the mass m. (See Figure 10.52.)

Figure 10.52: Spring and mass in equilibrium position and after upward displacement

When the system is left undisturbed, no net force acts on the mass. The force of gravity is balanced by the force the spring exerts on the mass, and the spring is in the *equilibrium position*. If we pull down on the mass, we feel a force pulling upwards. If instead, we push upward on the mass, the opposite happens: a force pushes the mass down.[10]

What happens if we push the mass upward and then release it? We'd expect the mass to oscillate up and down around the equilibrium position.

Springs and Hooke's Law

In order to figure out how the mass moves, we need to know the exact relationship between its displacement, s, from the equilibrium position and the net force, F, exerted on the mass. (See Figure 10.52.) We would expect that the further the mass is from the equilibrium position, and the more the spring is stretched or compressed, the larger the force. In fact, provided that the displacement is not large enough to deform the spring permanently, experiments show that the net force, F, is approximately proportional to the displacement, s:

$$F = -ks,$$

where k is the *spring constant* ($k > 0$) and the negative sign means that the net force is in the opposite direction to the displacement. The value of k depends on the physical properties of the particular spring. This relationship is known as *Hooke's Law*. Suppose we push the mass upward some distance and then release it. After we let go the net force causes the mass to accelerate toward the equilibrium position. By Newton's Second Law of Motion we have

Force = Mass × Acceleration.

Since acceleration is d^2s/dt^2 and force is $F = -ks$ by Hooke's law, we have

$$-ks = m\frac{d^2s}{dt^2},$$

which is equivalent to the following differential equation.

Equation for Oscillations of a Mass on a Spring

$$\frac{d^2s}{dt^2} = -\frac{k}{m}s$$

---

[10]Pulling down on the mass stretches the spring, increasing the tension, so the combination of gravity and spring force is upward. Pushing up the spring decreases tension in the spring, so the combination of gravity and spring force is downward.

Thus, the motion of the mass is described by a second-order differential equation. Since we expect the mass to oscillate, we guess that the solution to this equation involves trigonometric functions.

Solving the Differential Equation by Guess-and-Check

Guessing is the tried-and-true method for solving differential equations. It may surprise you to learn that there is no systematic method for solving most differential equations analytically, so guesswork is often extremely important. Before looking for a general solution to the equation

$$\frac{d^2s}{dt^2} = -\frac{k}{m}s,$$

we consider the special case $k/m = 1$.

Example 1   Find the general solution to the equation

$$\frac{d^2s}{dt^2} = -s.$$

Solution   We want to find functions whose second derivative is the negative of the original function. We are already familiar with two functions that have this property: $s(t) = \cos t$ and $s(t) = \sin t$. We check that they are solutions by substituting:

$$\frac{d^2}{dt^2}(\cos t) = \frac{d}{dt}(-\sin t) = -\cos t,$$

and

$$\frac{d^2}{dt^2}(\sin t) = \frac{d}{dt}(\cos t) = -\sin t.$$

The remarkable thing is that starting from these two particular solutions, we can build up *all* the solutions to our equation. Here's how: If C is a constant, then $C \sin t$ and $C \cos t$ are also solutions to the differential equation. (Why?) Also, $\sin t + \cos t$ is a solution to the differential equation. (Why?) In fact, given two constants C_1 and C_2, the function

$$s(t) = C_1 \cos t + C_2 \sin t$$

satisfies the differential equation, since

$$\frac{d^2}{dt^2}(C_1 \cos t + C_2 \sin t) = \frac{d}{dt}(-C_1 \sin t + C_2 \cos t)$$
$$= -C_1 \cos t - C_2 \sin t$$
$$= -(C_1 \cos t + C_2 \sin t).$$

It can be shown (though we will not do it) that $s(t) = C_1 \cos t + C_2 \sin t$ is the most general form of the solution. As expected, it contains two constants, C_1 and C_2.

If the differential equation represents a physical problem, then C_1 and C_2 are often determined by certain conditions of that physical problem. For example if the mass is pushed up to an initial displacement of s_0 and then released, then C_1 and C_2 can be computed, as shown in the next example.

Example 2   Find the solution to

$$\frac{d^2s}{dt^2} = -s$$

if the mass is displaced by a distance of s_0 and then released.

Solution   The position of the mass is given by the equation

$$s(t) = C_1 \cos t + C_2 \sin t.$$

We also know that the initial position is s_0; thus,

$$s(0) = C_1 \cos 0 + C_2 \sin 0 = C_1 \cdot 1 + C_2 \cdot 0 = s_0,$$

so $C_1 = s_0$, the initial displacement. What is C_2? To find it, we use the fact that at $t = 0$, when the mass has just been released, its velocity is 0. Velocity is the derivative of the displacement, so

$$\left.\frac{ds}{dt}\right|_{t=0} = \left.(-C_1 \sin t + C_2 \cos t)\right|_{t=0} = -C_1 \cdot 0 + C_2 \cdot 1 = 0,$$

so $C_2 = 0$. Therefore the solution is $s = s_0 \cos t$.

Solution to the General Spring Equation

Having found the general solution to the equation $\dfrac{d^2 s}{dt^2} = -s$, let us return to the equation

$$\frac{d^2 s}{dt^2} = -\frac{k}{m} s.$$

To solve this we let $\omega = \sqrt{\dfrac{k}{m}}$ (why will be clear in a moment); then $\dfrac{d^2 s}{dt^2} = -\omega^2 s$ or $\dfrac{d^2 s}{dt^2} + \omega^2 s = 0$.
Unfortunately, this equation no longer has $\sin t$ and $\cos t$ as solutions. For example,

$$\frac{d^2}{dt^2}(\sin t) = -\sin t \neq -\omega^2 \sin t.$$

However, we can adapt our guess as follows: we know that the functions $\sin t$ and $\cos t$ are *almost* solutions; we just have to deal with the extra ω^2 factor somehow. Ideally, we would like to be able to obtain a factor of ω^2 in the process of differentiating the sine or cosine function twice. This would be accomplished if, each time we differentiated, we obtained a factor of ω. The chain rule suggests an answer: try the function $\sin \omega t$. Checking this:

$$\frac{d^2}{dt^2}(\sin \omega t) = \frac{d}{dt}(\omega \cos \omega t) = -\omega^2 \sin \omega t$$

we see that $\sin \omega t$ is a solution, and you can check that $\cos \omega t$ is a solution, too.

The general solution to the equation

$$\frac{d^2 s}{dt^2} + \omega^2 s = 0$$

is of the form

$$s(t) = C_1 \cos \omega t + C_2 \sin \omega t,$$

where C_1 and C_2 are arbitrary constants. (We assume $\omega > 0$.) The period of this oscillation is

$$T = \frac{2\pi}{\omega}.$$

Such oscillations are called **simple harmonic motion**.

The solution to our original equation, $\dfrac{d^2 s}{dt^2} + \dfrac{k}{m} s = 0$, is thus $s = C_1 \cos \sqrt{\dfrac{k}{m}} t + C_2 \sin \sqrt{\dfrac{k}{m}} t$.

Initial-Value and Boundary-Value Problems

A problem in which the initial position and the initial velocity are used to determine the particular solution is called an *initial-value problem*. (See Example 2.) Alternatively, we may be given the position at two known times. Such a problem is known as a *boundary-value problem*.

Example 3   Find a solution to the differential equation satisfying each set of conditions below

$$\frac{d^2 s}{dt^2} + 4s = 0.$$

(a)   The boundary conditions $s(0) = 0$, $s(\pi/4) = 20$.
(b)   The initial conditions $s(0) = 1$, $s'(0) = -6$.

Solution   Since $\omega^2 = 4$, $\omega = 2$, the general solution to the differential equation is

$$s(t) = C_1 \cos 2t + C_2 \sin 2t.$$

(a)   Substituting the boundary condition $s(0) = 0$ into the general solution gives

$$s(0) = C_1 \cos(2 \cdot 0) + C_2 \sin(2 \cdot 0) = C_1 \cdot 1 + C_2 \cdot 0 = C_1 = 0.$$

Thus $s(t)$ must have the form $s(t) = C_2 \sin 2t$. The second condition yields the value of C_2:

$$s\left(\frac{\pi}{4}\right) = C_2 \sin\left(2 \cdot \frac{\pi}{4}\right) = C_2 = 20.$$

Therefore, the solution satisfying the boundary conditions is

$$s(t) = 20 \sin 2t.$$

(b)   For the initial-value problem, we start from the same general solution: $s(t) = C_1 \cos 2t + C_2 \sin 2t$. Substituting 0 for t once again, we find

$$s(0) = C_1 \cos(2 \cdot 0) + C_2 \sin(2 \cdot 0) = C_1 = 1.$$

Differentiating $s(t) = \cos 2t + C_2 \sin 2t$ gives

$$s'(t) = -2 \sin 2t + 2C_2 \cos 2t,$$

and applying the second initial condition gives us C_2:

$$s'(0) = -2 \sin(2 \cdot 0) + 2C_2 \cos(2 \cdot 0) = 2C_2 = -6,$$

so $C_2 = -3$ and our solution is

$$s(t) = \cos 2t - 3 \sin 2t.$$

What do the Graphs of Our Solutions Look Like?

Since the general solution of the equation $d^2 s/dt^2 + \omega^2 s = 0$ is of the form

$$s(t) = C_1 \cos \omega t + C_2 \sin \omega t$$

it would be useful to know what the graph of such a sum of sines and cosines looks like. We start with the example $s(t) = \cos t + \sin t$, which is graphed in Figure 10.53.

Interestingly, the graph in Figure 10.53 looks precisely like another sine function, and in fact it is one. If we measure the graph carefully, we find that its period is 2π, the same as $\sin t$ and $\cos t$, but the amplitude is approximately 1.414 (In fact, it's $\sqrt{2}$.), and the graph is shifted along the t-axis (in fact by $\pi/4$). If we plot $C_1 \cos t + C_2 \sin t$ for any C_1 and C_2, the resulting graph is always a sine function with period 2π, though the amplitude and shift can vary. For example, the graph of $s = 6\cos t - 8\sin t$ is in Figure 10.54; the graph has period 2π.

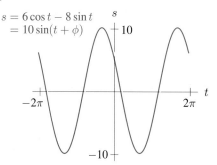

Figure 10.53: Graph of the sum:
$s(t) = \cos t + \sin t = \sqrt{2}\sin(t + \pi/4)$

Figure 10.54: Graph of a sum of sine and cosine. Amplitude $A = 10 = \sqrt{6^2 + 8^2}$

These graphs suggest that we can write the sum of a sine and a cosine of the same argument as one single sine function.[11] It turns out that this can always be done using the following relations. (See Problem 17.)

If $C_1 \cos\omega t + C_2 \sin\omega t = A\sin(\omega t + \phi)$, the *amplitude*, A, is given by

$$A = \sqrt{C_1^2 + C_2^2}.$$

The angle ϕ is called the *phase shift* and satisfies

$$\tan\phi = \frac{C_1}{C_2}.$$

We choose ϕ in $(-\pi, \pi]$ such that if $C_1 > 0$, ϕ is positive, and if $C_1 < 0$, ϕ is negative.

The reason that it is often useful to write the solution to the differential equation as a single sine function $A\sin(\omega t + \phi)$, as opposed to the sum of a sine and a cosine, is that the amplitude A and the phase shift ϕ are easier to recognize in this form.

Warning: Phase Shift Is Not the Same as Horizontal Translation

Let's look at $s = \sin(3t + \pi/2)$. If we rewrite this as $s = \sin(3(t + \pi/6))$, we can see from Figure 10.55 that the graph of this function is the graph of $s = \sin 3t$ shifted to the left a distance of $\pi/6$. (Remember that replacing x by $x - 2$ shifts a graph by 2 to the right.) But $\pi/6$ is *not* the phase shift; the phase shift[12] is $\pi/2$. From the point of view of a scientist, the important question is often not the distance the curve has shifted, but the relation between the distance shifted and the period.

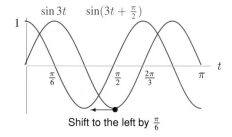

Figure 10.55: Phase shift is $\frac{\pi}{2}$; horizontal translation is $\frac{\pi}{6}$

---

[11]The sum of a sine and a cosine can also be written as a single cosine function.

[12]This definition of phase shift is the one usually used in the sciences.

Problems for Section 10.8

1. Check by differentiation that $y = 2\cos t + 3\sin t$ is a solution to $y'' + y = 0$.

2. Check that $y = A\cos t + B\sin t$ is a solution to $y'' + y = 0$ for any constants A and B.

3. What values of α and A make $y = A\cos \alpha t$ a solution to $y'' + 5y = 0$ such that $y'(1) = 3$?

The functions in Problems 4–6 describe the motion of a mass on a spring satisfying the differential equation $y'' = -9y$, where y is the displacement of the mass from the equilibrium position at time t, with upwards as positive. In each case, describe in words how the motion starts when $t = 0$. For example, is the mass at the highest point, the lowest point, or in the middle? Is it moving up or down or is it at rest?

4. $y = 2\cos 3t$ 5. $y = -0.5\sin 3t$ 6. $y = -\cos 3t$

7. The following differential equations represent oscillating springs.

    $$
    \begin{array}{llll}
    \text{(i)} & s'' + 4s = 0 & s(0) = 5, & s'(0) = 0 \\
    \text{(ii)} & 4s'' + s = 0 & s(0) = 10, & s'(0) = 0 \\
    \text{(iii)} & s'' + 6s = 0 & s(0) = 4, & s'(0) = 0 \\
    \text{(iv)} & 6s'' + s = 0 & s(0) = 20, & s'(0) = 0
    \end{array}
    $$

    Which differential equation represents:

    (a) The spring oscillating most quickly (with the shortest period)?
    (b) The spring oscillating with the largest amplitude?
    (c) The spring oscillating most slowly (with the longest period)?
    (d) The spring with largest maximum velocity?

8. (a) Find the general solution of the differential equation

    $$y'' + 9y = 0$$

    (b) For each of the following initial conditions find a particular solution.
       (i) $y(0) = 0, y'(0) = 1$ (ii) $y(0) = 1, y'(0) = 0$
       (iii) $y(0) = 1, y(1) = 0$ (iv) $y(0) = 0, y(1) = 1$
    (c) Sketch a graph of the solutions found in (b).

9. Each graph in Figure 10.56 represents a solution to one of the differential equations:
    (a) $x'' + x = 0$,   (b) $x'' + 4x = 0$,   (c) $x'' + 16x = 0$.

    Assuming the t-scales on the four graphs are the same, which graph represents a solution to which equation? Find an equation for each graph.

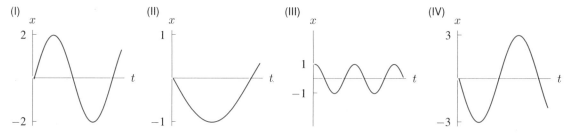

Figure 10.56: Graphs of solutions to $x'' + kx = 0$, for various values of k

10. A pendulum of length l makes an angle of x (radians) with the vertical (see Figure 10.57). When x is small, it can be shown that, approximately:

$$\frac{d^2x}{dt^2} = -\frac{g}{l}x,$$

where g is the acceleration due to gravity.

(a) Solve this equation assuming that $x(0) = 0$ and $x'(0) = v_0$.

(b) Solve this equation assuming that the pendulum is let go from the position where $x = x_0$. ("Let go" means that the velocity of the pendulum is zero when $x = x_0$. Measure t from the moment when the pendulum is let go.)

Figure 10.57

11. Look at the pendulum motion in Problem 10. What effect does it have on x as a function of time if:

(a) x_0 is increased?   (b) l is increased?

12. (a) Using a calculator or computer, graph $s = 4\cos t + 3\sin t$ for $-2\pi \le t \le 2\pi$.

(b) If $4\cos t + 3\sin t = A\sin(t + \phi)$, use your graph to estimate the values of A and ϕ.

(c) Calculate A and ϕ analytically.

13. Find the amplitude of $3\sin 2t + 7\cos 2t$.

14. Find the amplitude of the oscillation $y = 3\sin(2t) + 4\cos(2t)$.

15. Write $y(t) = 5\sin(2t) + 12\cos(2t)$ in the form $y(t) = A\sin(\omega t + \psi)$.

16. Write $7\sin \omega t + 24\cos \omega t$ in the form $A\sin(\omega t + \phi)$.

17. (a) Expand $A\sin(\omega t + \phi)$ using the trigonometric identity $\sin(x + y) = \sin x \cos y + \cos x \sin y$.

(b) Assume $A > 0$. If $A\sin(\omega t + \phi) = C_1\cos \omega t + C_2\sin \omega t$, show that we must have

$$A = \sqrt{C_1^2 + C_2^2} \quad \text{and} \quad \tan \phi = C_1/C_2.$$

The potential energy, U, of a molecule in various configurations can be expressed by

$$U = \frac{V_0}{2}\left(1 + \cos(n\phi - \phi_0)\right)$$

where ϕ is the (variable) angle which describes the configuration and V_0, n, ϕ_0 are constants, with $n = 2$ or 3. For each of the graphs[13] of U against ϕ in Problems 18-19:

(a) Give reasonable values for V_0, n, and ϕ_0.

(b) Find a differential equation (satisfied by U) of the form

$$\frac{d^2U}{d\phi^2} + \omega^2(U - K) = 0 \qquad U(0) = A \text{ and } U'(0) = B,$$

where ω^2, K, A, and B are constants to be determined.

18.

Figure 10.58

19.

Figure 10.59

[13] Adapted from Ignacio Tinoco Jr., Kenneth Sauer, James C. Wang, *Physical Chemistry*. (New Jersey: Prentice Hall, 1978).

Problems 20–22 show how a second order differential equation can be used to describe an electric circuit. A charged capacitor connected to an inductor (as shown in Figure 10.60) will cause a current to flow through the inductor until the capacitor is fully discharged. The current in the inductor will, in turn, charge up the capacitor until the capacitor is fully charged again. We will let $Q(t)$ be the amount of charge on the capacitor at time t. We can then find the current, the rate at which charge is moving in the circuit, by looking at the rate of change of Q. Thus we have the relation between the current I and the charge Q:

$$I = \frac{dQ}{dt}.$$

The current can be positive or negative; its sign tells us which way the current is flowing. Q can also be positive or negative; its sign tells us whether the charge on the capacitor is positive or negative.

If we assume the circuit resistance is zero, then it can be shown that the charge Q and the current I in the circuit satisfy the differential equation

$$L\frac{dI}{dt} + \frac{Q}{C} = 0.$$

So

$$L\frac{d^2Q}{dt^2} + \frac{Q}{C} = 0$$

where C, the capacitance, and L, the inductance, are constants depending on the particular capacitor and inductor. The unit of charge is the coulomb, the unit of capacitance the farad, the unit of inductance the henry, the unit of current is the ampere, and time is measured in seconds. This equation is similar to the differential equations we have seen for springs based on Hooke's law.

Capacitor

Inductor

Figure 10.60

20. If $L = 36$ henry and $C = 9$ farad, find a formula for $Q(t)$ if

   (a) $Q(0) = 0 \quad I(0) = 2$ (b) $Q(0) = 6 \quad I(0) = 0$

21. Suppose we wanted to set up our circuit so that $Q(0) = 0, Q'(0) = I(0) = 4$, and the maximum possible charge is $2\sqrt{2}$ coulombs. What should the capacitance of our capacitor be if we have an inductor with inductance 10 henry?

22. What happens to the charge and current as t goes to infinity? What does it mean that the charge and current are sometimes positive and sometimes negative?

10.9 LINEAR SECOND-ORDER DIFFERENTIAL EQUATIONS

A Spring with Friction: Damped Oscillations

The differential equation $d^2s/dt^2 = -(k/m)s$, which we used to describe the motion of a spring, disregards friction. But there is friction in every real system. For a mass on a spring, the frictional force from air resistance increases with the velocity of the mass. The frictional force is often approximately proportional to velocity, and so we introduce a *damping term* of the form $a(ds/dt)$, where a is a constant called the *damping coefficient* and ds/dt is the velocity of the mass.

Remember that without damping, the differential equation was obtained from

$$\text{Force} = \text{Mass} \times \text{Acceleration},$$

or

$$F = m\frac{d^2s}{dt^2}.$$

With damping, the force $-ks$ is replaced by $-ks - a(ds/dt)$, where a is positive and the $a(ds/dt)$ term is subtracted because the frictional force is in the direction opposite to the motion. The new differential equation is therefore

$$-ks - a\frac{ds}{dt} = m\frac{d^2s}{dt^2}$$

which is equivalent to the following differential equation:

Equation for Damped Oscillations of a Spring

$$\frac{d^2s}{dt^2} + \frac{a}{m}\frac{ds}{dt} + \frac{k}{m}s = 0$$

We expect the solution to this equation to die away with time, as friction brings the motion to a stop.

The General Solution to a Linear Differential Equation

The equation for damped oscillations is an example of a *linear second-order differential equation with constant coefficients*. This section explains an analytic method of solving any such equation,

$$\frac{d^2y}{dt^2} + b\frac{dy}{dt} + cy = 0,$$

where b and c are constants. As we have seen for the spring equation, if $f_1(t)$ and $f_2(t)$ satisfy the differential equation, then the *principle of superposition* says that, for any constants C_1 and C_2, the function

$$y(t) = C_1 f_1(t) + C_2 f_2(t)$$

is also a solution. It can be shown that the general solution is of this form, provided $f_1(t)$ is not a multiple of $f_2(t)$.

Finding Solutions: The Characteristic Equation

We will now see how to use complex numbers to solve the differential equation

$$\frac{d^2y}{dt^2} + b\frac{dy}{dt} + cy = 0.$$

The method is a form of guess-and-check. This time we ask what kind of function might satisfy a differential equation which says that the second derivative d^2y/dt^2 is a sum of multiples of dy/dt and y. One possibility is the exponential function, so we try to find a solution of the form:

$$y = Ce^{rt},$$

where r may be a complex number.[14] In order to find r, we substitute into the differential equation:

$$\frac{d^2y}{dt^2} + b\frac{dy}{dt} + cy = r^2 Ce^{rt} + b \cdot rCe^{rt} + c \cdot Ce^{rt} = Ce^{rt}(r^2 + br + c) = 0.$$

We can divide this equation through by $y = Ce^{rt}$ so long as $C \neq 0$, because the exponential function is never zero. If $C = 0$, then $y = 0$, which is not a very interesting solution (though it is a solution). So we assume $C \neq 0$. Then $y = Ce^{rt}$ will be a solution to the differential equation if

$$r^2 + br + c = 0.$$

---

[14]See Appendix D on complex numbers.

This quadratic equation is called the *characteristic equation* of the differential equation. Its solutions are

$$r = -\frac{1}{2}b \pm \frac{1}{2}\sqrt{b^2 - 4c}.$$

There are three different types of solutions to the differential equation, depending on whether the solutions to the characteristic equation are real and distinct, complex, or repeated. The sign of $b^2 - 4c$ determines the type of solutions.

The Case with $b^2 - 4c > 0$

There are two real solutions r_1 and r_2 to the characteristic equation, and so the following two functions satisfy the differential equation:

$$C_1 e^{r_1 t} \quad \text{and} \quad C_2 e^{r_2 t}.$$

The sum of these two solutions is the general solution to the differential equation:

If $b^2 - 4c > 0$, the general solution to

$$\frac{d^2y}{dt^2} + b\frac{dy}{dt} + cy = 0$$

is

$$y(t) = C_1 e^{r_1 t} + C_2 e^{r_2 t}$$

where r_1 and r_2 are the solutions to the characteristic equation.
If $r_1, r_2 < 0$, the motion is called **overdamped**.

A physical system satisfying a differential equation of this type is said to be overdamped because it occurs when there is a lot of friction in the system. For example, a spring moving in a thick fluid such as oil or molasses is overdamped: it will not oscillate.

Example 1 Suppose a spring is placed in oil, where it satisfies the differential equation

$$\frac{d^2s}{dt^2} + 3\frac{ds}{dt} + 2s = 0.$$

Solve this equation with the initial conditions $s = -0.5$ and $\dfrac{ds}{dt} = 3$ when $t = 0$.

Solution The characteristic equation is

$$r^2 + 3r + 2 = 0,$$

with solutions $r = -1$ and $r = -2$, and so the general form of the solution to the differential equation is

$$s(t) = C_1 e^{-t} + C_2 e^{-2t}.$$

We use the initial conditions to find C_1 and C_2. At $t = 0$, we have

$$s = C_1 e^{-0} + C_2 e^{-2(0)} = C_1 + C_2 = -0.5.$$

Furthermore, since $ds/dt = -C_1 e^{-t} - 2C_2 e^{-2t}$, we have

$$\left.\frac{ds}{dt}\right|_{t=0} = -C_1 e^{-0} - 2C_2 e^{-2(0)} = -C_1 - 2C_2 = 3.$$

Solving these equations simultaneously, we find $C_1 = 2$ and $C_2 = -2.5$, so that the solution is

$$s(t) = 2e^{-t} - 2.5e^{-2t}.$$

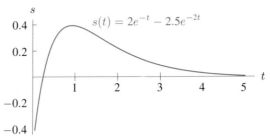

Figure 10.61: Solution to overdamped equation
$$\frac{d^2 s}{dt^2} + 3\frac{ds}{dt} + 2s = 0$$

The graph of this function is in Figure 10.61. The mass is so slowed by the oil that it passes through the equilibrium point only once (when $t \approx 1/4$) and for all practical purposes, it comes to rest after a short time. The motion has been "damped out" by the oil.

The Case with $b^2 - 4c = 0$

In this case, the characteristic equation has only one solution, $r = -b/2$. By substitution, we can check that both $y = e^{-bt/2}$ and $y = te^{-bt/2}$ are solutions.

If $b^2 - 4c = 0$,
$$\frac{d^2 y}{dt^2} + b\frac{dy}{dt} + cy = 0$$

has general solution
$$y(t) = (C_1 t + C_2)e^{-bt/2}.$$

If $b > 0$, the system is said to be **critically damped**.

The Case with $b^2 - 4c < 0$

Example 2   An object of mass $m = 10$ kg is attached to a spring with spring constant $k = 20$ kg/sec^2, and the object experiences a frictional force proportional to the velocity, with constant of proportionality $a = 20$ kg/sec. At time $t = 0$, the object is released from rest 2 meters above the equilibrium position. Write the differential equation that describes the motion.

Solution   The differential equation that describes the motion can be obtained from the following general expression for the damped motion of a spring:

$$\underbrace{m\frac{d^2 s}{dt^2}}_{\text{Mass}\times\text{Acceleration}} = \underbrace{F}_{\text{Net force}} = -\underbrace{ks}_{\text{Spring force}} - \underbrace{a\frac{ds}{dt}}_{\text{Frictional force}}.$$

Substituting $m = 10$ kg, $k = 20$ kg/sec^2, and $a = 20$ kg/sec, we obtain the differential equation for the motion:

$$\frac{d^2 s}{dt^2} + 2\frac{ds}{dt} + 2s = 0.$$

At $t = 0$, the object is at rest 2 meters above equilibrium, so the initial conditions are $s(0) = 2$ and $s'(0) = 0$, where s is in meters and t in seconds.

Notice that this is the same differential equation as in Example 1 except that the coefficient of ds/dt has decreased from 3 to 2, which means that the frictional force has been reduced. This time, the roots of the characteristic equation have imaginary parts which lead to oscillations.

Example 3   Solve the differential equation

$$\frac{d^2s}{dt^2} + 2\frac{ds}{dt} + 2s = 0,$$

subject to $s(0) = 2$, $s'(0) = 0$.

Solution   The characteristic equation is

$$r^2 + 2r + 2 = 0 \quad \text{giving} \quad r = -1 \pm i.$$

The solution to the differential equation is

$$s(t) = A_1 e^{(-1+i)t} + A_2 e^{(-1-i)t},$$

where A_1 and A_2 are arbitrary complex numbers. The initial condition $s(0) = 2$ gives

$$2 = A_1 e^{(-1+i)\cdot 0} + A_2 e^{(-1-i)\cdot 0} = A_1 + A_2.$$

Also,

$$s'(t) = A_1(-1+i)e^{(-1+i)t} + A_2(-1-i)e^{(-1-i)t},$$

so $s'(0) = 0$ gives

$$0 = A_1(-1+i) + A_2(-1-i).$$

Solving the simultaneous equations for A_1 and A_2 gives (after some algebra)

$$A_1 = 1 - i \quad \text{and} \quad A_2 = 1 + i.$$

The solution is therefore

$$s(t) = (1-i)e^{(-1+i)t} + (1+i)e^{(-1-i)t} = (1-i)e^{-t}e^{it} + (1+i)e^{-t}e^{-it}.$$

Using Euler's formula[15], $e^{it} = \cos t + i \sin t$ and $e^{-it} = \cos t - i \sin t$, we get:

$$s(t) = (1-i)e^{-t}(\cos t + i\sin t) + (1+i)e^{-t}(\cos t - i\sin t).$$

Multiplying out and simplifying, all the complex terms drop out, giving

$$s(t) = e^{-t}\cos t + ie^{-t}\sin t - ie^{-t}\cos t + e^{-t}\sin t$$
$$+ e^{-t}\cos t - ie^{-t}\sin t + ie^{-t}\cos t + e^{-t}\sin t$$
$$= 2e^{-t}\cos t + 2e^{-t}\sin t.$$

The $\cos t$ and $\sin t$ terms tell us that the solution oscillates. However the oscillations are damped extremely quickly by the factor e^{-t}. (See Figure 10.62.) Alternatively, you can see the behavior of the solution by rewriting it in the form

$$s(t) = 2e^{-t}(\cos t + \sin t) = 2\sqrt{2}e^{-t}\sin\left(t + \frac{\pi}{4}\right).$$

Physically the most interesting feature of the solution is that the period of the oscillations does not change as the amplitude decreases. This is why a spring-driven clock can keep accurate time even as it is running down.

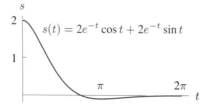

Figure 10.62: Solution to underdamped equation $\frac{d^2s}{dt^2} + 2\frac{ds}{dt} + 2s = 0$

In Example 3, notice that although the coefficients A_1 and A_2 are complex, the solution, $s(t)$, is real. (We expect this, since $s(t)$ represents a real displacement.) In general, provided the coefficients

---

[15]See Appendix D on complex numbers

b and c in the original differential equation and the initial values are real, the solution will always be real too. The complex coefficients A_1 and A_2 will always be complex conjugates.

If $b^2 - 4c < 0$, to solve

$$\frac{d^2y}{dt^2} + b\frac{dy}{dt} + cy = 0,$$

- Find the solutions $r = \alpha \pm i\beta$ to the characteristic equation $r^2 + br + c = 0$.
- The general solution to the differential equation is, for some real C_1 and C_2,

$$y = C_1 e^{\alpha t} \cos \beta t + C_2 e^{\alpha t} \sin \beta t.$$

If $\alpha < 0$, such oscillations are called **underdamped**.

Example 4   Find the general solution of the equations   (a) $y'' = 9y$ (b) $y'' = -9y$.

Solution   (a)   The characteristic equation is $r^2 - 9 = 0$, so $r = \pm 3$. Thus the general solution is
$$y = C_1 e^{3t} + C_2 e^{-3t}.$$

(b)   The characteristic equation is $r^2 + 9 = 0$, so $r = 0 \pm 3i$. The general solution is
$$y = C_1 e^{0t} \cos 3t + C_2 e^{0t} \sin 3t = C_1 \cos 3t + C_2 \sin 3t.$$

Notice that we have seen the solution to this equation in Section 10.8; it's the equation of undamped simple harmonic motion.

Example 5   Solve the initial value problem
$$y'' + 4y' + 13y = 0, \quad y(0) = 0, \ y'(0) = 30.$$

Solution   We solve the characteristic equation
$$r^2 + 4r + 13 = 0, \quad \text{getting} \quad r = -2 \pm 3i.$$

The general solution to the differential equation is
$$y = C_1 e^{-2t} \cos 3t + C_2 e^{-2t} \sin 3t.$$

Substituting $t = 0$ gives
$$y(0) = C_1 \cdot 1 \cdot 1 + C_2 \cdot 1 \cdot 0 = 0, \quad \text{so} \quad C_1 = 0.$$

Differentiating $y(t) = C_2 e^{-2t} \sin 3t$ gives
$$y'(t) = C_2(-2e^{-2t} \sin 3t + 3e^{-2t} \cos 3t).$$

Substituting $t = 0$ gives
$$y'(0) = C_2(-2 \cdot 1 \cdot 0 + 3 \cdot 1 \cdot 1) = 3C_2 = 30 \quad \text{so} \quad C_2 = 10.$$

The solution is therefore
$$y(t) = 10e^{-2t} \sin 3t.$$

Summary of Solutions to $y'' + by' + cy = 0$

If $b^2 - 4c > 0$, then $y = C_1 e^{r_1 t} + C_2 e^{r_2 t}$
If $b^2 - 4c = 0$, then $y = (C_1 t + C_2)e^{-bt/2}$
If $b^2 - 4c < 0$, then $y = C_1 e^{\alpha t} \cos \beta t + C_2 e^{\alpha t} \sin \beta t$

Problems for Section 10.9

For Problems 1–12, find the general solution to the given differential equation.

1. $y'' + 4y' + 3y = 0$
2. $y'' + 4y' + 4y = 0$
3. $y'' + 4y' + 5y = 0$
4. $s'' - 7s = 0$
5. $s'' + 7s = 0$
6. $y'' - 3y' + 2y = 0$
7. $4z'' + 8z' + 3z = 0$
8. $\dfrac{d^2x}{dt^2} + 4\dfrac{dx}{dt} + 8x = 0$
9. $\dfrac{d^2p}{dt^2} + \dfrac{dp}{dt} + p = 0$
10. $z'' + 2z = 0$
11. $z'' + 2z' = 0$
12. $P'' + 2P' + P = 0$

For Problems 13–16, solve the initial value problem.

13. $y'' + 6y' + 5y = 0,$ $y(0) = 1,$ $y'(0) = 0$
14. $y'' + 6y' + 5y = 0,$ $y(0) = 5,$ $y'(0) = 5$
15. $y'' + 6y' + 10y = 0,$ $y(0) = 0,$ $y'(0) = 2$
16. $y'' + 6y' + 10y = 0,$ $y(0) = 0,$ $y'(0) = 0$

For Problems 17–18, solve the boundary value problem.

17. $p'' + 2p' + 2p = 0,$ $p(0) = 0,$ $p(\pi/2) = 20$
18. $p'' + 4p' + 5p = 0,$ $p(0) = 1,$ $p(\pi/2) = 5$
19. Match the graphs of solutions in Figure 10.63 with the differential equations below.
    (a) $x'' + 4x = 0$
    (b) $x'' - 4x = 0$
    (c) $x'' - 0.2x' + 1.01x = 0$
    (d) $x'' + 0.2x' + 1.01x = 0$

(I)   (II)   (III)   (IV)

Figure 10.63

20. Match the differential equations to the solution graphs (I)-(IV). Use each graph only once.
    (a) $y'' + 5y' + 6y = 0$
    (b) $y'' + y' - 6y = 0$
    (c) $y'' + 4y' + 9y = 0$
    (d) $y'' = -9y$

(I)   (II)   (III)   (IV)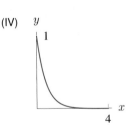

21. If $y = e^{2t}$ is a solution to the differential equation

$$\frac{d^2y}{dt^2} - 5\frac{dy}{dt} + ky = 0,$$

find the value of the constant k and the general solution to this equation.

22. Assuming $b, c > 0$, explain how you know that the solutions of an underdamped differential equation must go to 0 as $t \to \infty$.

Each of the differential equations below represents the position of a 1 gram mass oscillating on the end of a damped spring. For Problems 23–27 below, pick the differential equation representing the system which answers the question.

(i) $s'' + s' + 4s = 0$

(iii) $s'' + 3s' + 3s = 0$

(ii) $s'' + 2s' + 5s = 0$

(iv) $s'' + 0.5s' + 2s = 0$

23. Which spring has the largest coefficient of damping?

24. Which spring exerts the smallest restoring force for a given displacement?

25. In which system does the mass experience the frictional force of smallest magnitude for a given velocity?

26. Which oscillation has the longest period?

27. Which spring is the stiffest? [Hint: You need to determine what it means for a spring to be stiff. Think of an industrial strength spring and a slinky.]

For each of the differential equations in Problems 28–30, find the values of c that make the general solution:
(a)   overdamped,   (b)   underdamped,   (c)   critically damped.

28. $s'' + 4s' + cs = 0$ 29. $s'' + 2\sqrt{2}s' + cs = 0$ 30. $s'' + 6s' + cs = 0$

31. Find a solution to the following equation which satisfies $z(0) = 3$ and does not tend to infinity as $t \to \infty$:

$$\frac{d^2z}{dt^2} + \frac{dz}{dt} - 2z = 0.$$

32. Consider an overdamped differential equation with $b, c > 0$.

    (a)   Show that both roots of the characteristic equation are negative.
    (b)   Show that any solution to the differential equation goes to 0 as $t \to \infty$.

Recall the discussion of electric circuits in Section 10.8 on page 550. Just as a spring can have a damping force which affects its motion, so can a circuit. Problems 33–36 illustrate this situation. The damping force is caused by a resistor shown in Figure 10.64.

Figure 10.64

Just as the damping force for a mass on a spring is proportional to the velocity of the mass, the damping force in the circuit is proportional to the current. We get an extra term $RI = R(dQ/dt)$ in the differential equation. The resistance, R, is a constant that depends on the resistor, and is measured in ohms. The equation for the charge Q on a capacitor in a circuit with inductance L, capacitance C, and resistance R satisfies the differential equation

$$L\frac{d^2Q}{dt^2} + R\frac{dQ}{dt} + \frac{1}{C}Q = 0.$$

33. If $L = 1$ henry, $R = 2$ ohms, and $C = 4$ farads, find a formula for the charge when
    (a) $Q(0) = 0, Q'(0) = 2$.   (b) $Q(0) = 2, Q'(0) = 0$.

34. If $L = 1$ henry, $R = 1$ ohm, and $C = 4$ farads, find a formula for the charge when
    (a) $Q(0) = 0, Q'(0) = 2$.   (b) $Q(0) = 2, Q'(0) = 0$.
    (c)   How did reducing the resistance affect the charge? Compare with your solution to Problem 33.

35. If $L = 8$ henry, $R = 2$ ohm, and $C = 4$ farads, find a formula for the charge when
    (a) $Q(0) = 0, Q'(0) = 2$.   (b) $Q(0) = 2, Q'(0) = 0$.
    (c)   How did increasing the inductance affect the charge? Compare with your solution to Problem 33.

36. Given any positive values for R, L and C, what happens to the charge as t goes to infinity?

37. Could the graph in Figure 10.65 show the position of a mass oscillating at the end of an overdamped spring? Why or why not?

s (position)

t (time)

Figure 10.65

38. Consider the system of differential equations

$$\frac{dx}{dt} = -y \qquad \frac{dy}{dt} = -x.$$

(a) Convert this system to a second order differential equation in y by differentiating the second equation with respect to t and substituting for x from the first equation.

(b) Solve the equation you obtained for y as a function of t; hence find x as a function of t.

CHAPTER SUMMARY

- **Differential equations terminology**
  Order, initial conditions, families of solutions, stable/unstable equilibrium solutions
- **Solving first-order differential equations**
  Slope fields (graphical), Euler's method (numerical), separation of variables (analytical)
- **Modeling with differential equations**
  Growth and decay, Newton's law of heating and cooling, compartment models
- **Second-order differential equations**
  Initial and boundary conditions, oscillations (with and without damping), characteristic equation, solutions using complex numbers

REVIEW PROBLEMS FOR CHAPTER TEN

For the differential equations in Problems 1–16, find a solution which passes through the given point.

1. $\dfrac{dy}{dt} = y(10 - y)$, $\quad y(0) = 1$

2. $\dfrac{dP}{dt} = 0.03P + 400$, $\quad P(0) = 0$

3. $\dfrac{dy}{dx} = \dfrac{y(3 - x)}{x(0.5y - 4)}$, $\quad y(1) = 5$

4. $\dfrac{dy}{dx} = e^{x-y}$, $\quad y(0) = 1$

5. $\dfrac{df}{dx} = \sqrt{xf(x)}$, $\quad f(1) = 1$

6. $1 + y^2 - \dfrac{dy}{dx} = 0$, $\quad y(0) = 0$

7. $2\sin x - y^2\dfrac{dy}{dx} = 0$, $\quad y(0) = 3$

8. $\dfrac{dk}{dt} = (1 + \ln t)k$, $\quad k(1) = 1$

9. $\dfrac{dy}{dx} + xy^2 = 0$, $\quad y(1) = 1$

10. $\dfrac{dy}{dx} = e^{x+y}$, $\quad y = 0$ where $x = 1$

11. $\dfrac{dz}{dt} = z(z-1), \quad z(0) = 10$

12. $\dfrac{dy}{dx} = \dfrac{y(100-x)}{x(20-y)}, \quad (1, 20)$

13. $\dfrac{dy}{dt} = 2^y \sin^3 t, \quad y(0) = 0$

14. $e^{-\cos\theta}\dfrac{dz}{d\theta} = \sqrt{1-z^2}\sin\theta, \quad z(0) = \tfrac{1}{2}$

15. $(1+t^2)y\dfrac{dy}{dt} = 1 - y, \quad y(1) = 0$

16. $\dfrac{dy}{dx} = \dfrac{0.2y(18+0.1x)}{x(100+0.5y)}, \quad (10, 10)$

Solve the differential equations in Problems 17–20. Assume $x \geq 0$, $y \geq 0$.

17. $\dfrac{dQ}{dt} + t^2 Q^2 + Q^2 - 4t^2 - 4 = 0$

18. $\dfrac{dy}{dt} = \dfrac{y \ln y}{t^2}$

19. $\dfrac{x}{y}\dfrac{dx}{dy} = e^{(x/a)^2}\ln y \quad (a \text{ is constant})$

20. $(y\sqrt{x^3+1})\dfrac{dy}{dx} + x^2 y^2 + x^2 = 0$

21. Consider the initial value problem
$$y' = 5 - y, \quad y(0) = 1.$$

   (a) Use Euler's method with five steps to estimate $y(1)$.

   (b) Sketch the slope field for this differential equation in the first quadrant, and use it to decide if your estimate is an over- or underestimate.

   (c) Find the exact solution to the differential equation and hence find $y(1)$ exactly.

   (d) Without doing the calculation, roughly what would you expect the approximation for $y(1)$ to be with ten steps?

22. Solve the differential equation
$$\frac{dy}{dx} = \frac{1}{(\cos x)(\cos y)}$$
by Euler's method, for the solution passing through $(0,0)$. Find y when $x = \tfrac{1}{2}$.

   (a) Find the solution from Euler's method when $\Delta x = \tfrac{1}{2}$ (1 step),

   (b) $\Delta x = \tfrac{1}{4}$ (2 steps),

   (c) $\Delta x = \tfrac{1}{8}$ (4 steps).

   (d) Solve the differential equation by separation of variables for y passing through $(0,0)$ and find $y(\tfrac{1}{2})$. Compare with your results in parts (a)–(c).

Find a general solution to the differential equations in Problems 23–26.

23. $y'' + 6y' + 8y = 0$

24. $9z'' + z = 0$

25. $9z'' - z = 0$

26. $x'' + 2x' + 10x = 0$

For each of the differential equations in Problems 27–28, find the values of b that make the general solution:
(a) overdamped,   (b) underdamped,   (c) critically damped.

27. $s'' + bs' + 5s = 0$

28. $s'' + bs' - 16s = 0$

29. Suppose that the rate at which a drug leaves the bloodstream and passes into the urine is proportional to the quantity of the drug in the blood at that time. If an initial dose of Q_0 is injected directly into the blood, only 20% is left in the blood after 3 hours.

   (a) Write and solve a differential equation for the quantity, Q, of the drug in the blood at time, t, measured in hours.

   (b) How much of this drug is in a patient's body after 6 hours if the patient is given 100 mg initially?

30. Water leaks from a vertical cylindrical tank through a small hole in its base at a rate proportional to the square root of the volume of water remaining. If the tank initially contains 200 liters and 20 liters leak out during the first day, when will the tank be half empty? How much water will there be after 4 days?

31. (a) A cup of coffee is made with boiling water and stands in a room where the temperature is $20°$ C. If $T(t)$ is the temperature of the coffee at time t, explain what the differential equation

$$\frac{dT}{dt} = -k(T - 20)$$

says in everyday terms. What is the sign of k?

(b) Solve this differential equation. If the coffee cools to $90°$ C in 2 minutes, how long will it take to cool to $60°$ C degrees?

32. A roast is taken from the refrigerator (where the temperature is $40°$F) and put in a $350°$F oven. One hour later, the meat thermometer shows a temperature of $90°$F. If the roast is done when its temperature reaches $140°$F, what is the total time the roast should be in the oven? Newton's Law of Cooling says that the rate of change of temperature is proportional to the difference between the temperature of the object and the temperature of the surrounding air.

33. A bank account earns 10% annual interest, compounded continuously. Money is deposited in a continuous cash flow at a rate of $1200 per year into the account.

(a) Write a differential equation that describes the rate at which the balance $B = f(t)$ is changing.
(b) Solve the differential equation given an initial balance $B_0 = 0$.
(c) Find the balance after 5 years.

34. (Continuation of Problem 33.) Now suppose the money is deposited once a month (instead of continuously) but still at a rate of $1200 per year.

(a) Write down the sum that gives the balance after 5 years, assuming the first deposit is made one month from today, and today is $t = 0$.
(b) The sum you wrote in part (a) is a Riemann sum approximation to the integral

$$\int_0^5 1200e^{0.1t}\,dt.$$

Determine whether it is a left sum or right sum, and determine what Δt and N are. Then use your calculator to evaluate the sum.

(c) Compare your answer in part (b) to your answer to Problem 33(c).

35. The spread of a non-fatal disease through a population of fixed size M can be modeled as follows. The rate that healthy people are infected, in people per day, is proportional to the product of the numbers of healthy and infected people. The constant of proportionality is $0.01/M$. The rate of recovery, in people per day, is 0.009 times the number of people infected. Construct a differential equation which models the spread of the disease. Assuming that initially only a small number of people are infected, plot a graph of the number of infected people against time. What fraction of the population is infected in the long run?

36. A model for the population, $P(t)$, of carp in a landlocked lake is given by the differential equation

$$\frac{dP}{dt} = 0.25P(1 - 0.0004P).$$

(a) What is the long-term equilibrium population of carp in the lake?
(b) Ten years ago a census was taken and there were found to be 1000 carp in the lake. Estimate the current size of the population.
(c) It is planned to join the lake to a nearby river so that the fish will be able to leave the lake. It is estimated that there will be a net loss of 10% of the carp each year but that the patterns of birth and death are not expected to change. Revise the differential equation to take this into account. Use the revised differential equation to predict the future development of the carp population.

37. Juliet is in love with Romeo, who happens (in our version of this story) to be a fickle lover. The more Juliet loves him, the more he begins to dislike her. When she hates him, his feelings for her warm up. On the other hand, her love for him grows when he loves her and withers when he hates her. A model for their ill-fated romance is

$$\frac{dj}{dt} = Ar, \qquad \frac{dr}{dt} = -Bj,$$

where A and B are positive constants, $r(t)$ represents Romeo's love for Juliet at time t, and $j(t)$ represents Juliet's love for Romeo at time t. (Negative love is hate.)

(a) The constant on the right-hand side of Juliet's equation (the one including dj/dt) has a positive sign, whereas the constant in Romeo's equation is negative. Explain why these signs follow from the story.

(b) Derive a second-order differential equation for $r(t)$ and solve it. (Your equation should involve r and its derivatives, but not j and its derivatives.)

(c) Express $r(t)$ and $j(t)$ as functions of t, given $r(0) = 1$ and $j(0) = 0$. Your answer will contain A and B.

(d) As you may have discovered, the outcome of the relationship is a never-ending cycle of love and hate. Find what fraction of the time they both love one another.

38. The behavior of an electron is described by Schrödinger's equation. Under certain conditions, this equation becomes

$$\frac{-h^2}{8\pi^2 m} \frac{d^2\Psi}{dx^2} = E\Psi,$$

where Ψ, called a *wave function*, describes the electron's behavior as a function of x, the distance from a fixed point; the constants h, m, E are positive. In addition, Ψ satisfies the boundary conditions

$$\Psi(0) = \Psi(l) = 0$$

where l is another positive constant.

(a) Solutions to Schrödinger's equation are of the form $\Psi = C_1 \cos(\omega x) + C_2 \sin(\omega x)$ for some ω. Show that if Ψ satisfies the boundary conditions, then Ψ must be of the form

$$\Psi = C_2 \sin\left(\frac{n\pi x}{l}\right) \quad \text{for some positive integer } n.$$

(b) Using part (a), find E in terms of other constants.

(c) The constant E represents the energy level of the electron. Assuming that only positive integral values of n are possible, explain why not all values of E are possible. (This is described by saying that energy is quantized.) In addition, show that the possible energy levels are one of the following multiples of the lowest energy level: $1, 4, 9, 16 \ldots$.

PROJECTS

1. Pareto's Law

In analyzing a society, sociologists are often interested in how incomes are distributed through the society. Pareto's law asserts that each society has a constant k such that the average income of all people wealthier than you is k times your income ($k > 1$). If $p(x)$ is the number of people in the society with an income of x or above, we define $\Delta p = p(x + \Delta x) - p(x)$, for small Δx.

(a) Explain why the number of people with incomes between x and $x + \Delta x$ is represented by $-\Delta p$. Then show that the total amount of money earned by people with incomes between x and $x + \Delta x$ is approximated by $-x\Delta p$.

(b) Use Pareto's law to show that the total amount of money earned by people with incomes of x or above is $kxp(x)$. Then show that the total amount of money earned by people with incomes between x and $x + \Delta x$ is approximated by $-kp\Delta x - kx\Delta p$.

(c) Using your answers to parts (a) and (b), show that p satisfies the differential equation

$$(1 - k)xp' = kp.$$

(d) Solve this differential equation for $p(x)$.

(e) Sketch a graph of $p(x)$ for various values of k, some large and some near 1. How does the value of k alter the shape of the graph?

2. **Vibrations in a Molecule**

Suppose the force, F, between two atoms a distance r apart in a molecule is given by

$$F(r) = b\left(\frac{a^2}{r^3} - \frac{a}{r^2}\right)$$

where a and b are positive constants.

(a) Find the equilibrium distance, r, at which $F = 0$.

(b) Expand F in a series about the equilibrium position. Give the first two nonzero terms.

(c) Suppose one atom is fixed. Let x be the displacement of the other atom from the equilibrium position. Give the first two nonzero terms of the series for F in terms of x.

(d) Using Newton's Second Law, show that if x is small, the second atom oscillates about the equilibrium position. Find the period of the oscillation.

3. **Harvesting and Logistic Growth**

In this project, we look at the effects of *harvesting* a population which is growing logistically. Harvesting could be, for example, fishing or logging. An important question is what level of harvesting leads to a *sustainable yield*. In other words, how much can be harvested without having the population depleted in the long run?

(a) When there is no fishing, suppose a population of fish is governed by the differential equation

$$\frac{dN}{dt} = 2N - 0.01N^2,$$

where N is the number of fish at time t in years. Sketch a graph of dN/dt against N. Mark on your graph the equilibrium values of N. Now sketch N against t for various different initial values.

(b) Suppose now that fish are removed by fishermen at a continuous rate of 75 fish/year. Let P be the number of fish at time t with harvesting. Explain why P satisfies the differential equation

$$\frac{dP}{dt} = 2P - 0.01P^2 - 75.$$

(c) Sketch dP/dt against P. Find and label the intercepts.

(d) Sketch the slope field for the differential equation for P.

(e) Use the graph you drew in part (c) to sketch graphs of P against t, with the initial values:

(i) $P(0) = 40$ (ii) $P(0) = 60$ (iii) $P(0) = 160$ (iv) $P(0) = 140$

(f) Use the slope field in part (d) to sketch graphs of P against t, with the initial values:

(i) $P(0) = 40$ (ii) $P(0) = 50$ (iii) $P(0) = 60$
(iv) $P(0) = 150$ (v) $P(0) = 170$

(g) Using the graphs you drew, decide what the equilibrium values of the populations are and whether or not they are stable.

(h) We now look at the effect of different levels of fishing on a fish population. If fishing takes place at a continuous rate of H fish/year, the fish population P satisfies the differential equation

$$\frac{dP}{dt} = 2P - 0.01P^2 - H.$$

(i) For each of the values $H = 75, 100, 200$, plot a graph of dP/dt against P.

(ii) For which of the three values of H that you considered in part (i) is there an initial condition such that the fish population does not die out eventually?

(iii) Looking at your answer to part (ii), decide for what values of H there is an initial value for P such that the population does not die out eventually.

(iv) Recommend a policy to ensure long-term survival of the fish population.

FOCUS ON MODELING

SYSTEMS OF DIFFERENTIAL EQUATIONS

In Section 10.7 we modeled the growth of a single population over time. We now consider the growth of two populations which interact, such as a population of sick people infecting the healthy people around them. This will involve not just one differential equation, but a system of two.

Diseases and Epidemics

Differential equations can be used to predict when an outbreak of a disease will become so severe that it is called an *epidemic*,[16] and to decide what level of vaccination will be necessary to prevent an epidemic. Let's consider a specific example.

Flu in a British Boarding School

In January 1978, 763 students returned to a boys' boarding school after their winter vacation. A week later, one boy developed the flu, followed immediately by two more. By the end of the month, nearly half the boys were sick. Most of the school had been affected by the time the epidemic was over in mid-February.[17]

Being able to predict how many people will get sick, and when, is an important step toward controlling an epidemic. This is one of the responsibilities of Britain's Communicable Disease Surveillance Centre and the US's Center for Disease Control and Prevention.

The *S-I-R* model

We will apply one of the most commonly used models for an epidemic, called the *S-I-R* model, to the boarding school flu example. Imagine the population of the school divided into three groups:

S = the number of *susceptibles*, the people who are not yet sick but who could become sick

I = the number of *infecteds*, the people who are currently sick

R = the number of *recovered*, or *removed*, the people who have been sick and can no longer infect others or be reinfected.

In this model, the number of susceptibles decreases with time, as people become infected. We will assume that the rate people become infected is proportional to the number of contacts between susceptible and infected people. We expect the number of contacts between the two groups to be proportional to both S and I. (If S doubles, we expect the number of contacts to double; similarly, if I doubles, we expect the number of contacts to double.) Thus we assume that the number of contacts is proportional to the product, SI. In other words, we assume that for some constant $a > 0$,

$$\frac{dS}{dt} = -\left(\begin{array}{c} \text{Rate susceptibles} \\ \text{get sick} \end{array} \right) = -aSI.$$

(The negative sign is used because S is decreasing.)

---

[16]Exactly when a disease should be called an epidemic is not always clear. The medical profession generally classifies a disease an epidemic when the frequency is higher than usually expected—leaving open the question of what is usually expected. See, for example, *Epidemiology in Medicine* by C. H. Hennekens and J. Buring (Boston: Little, Brown, 1987)

[17]Data from the Communicable Disease Surveillance Centre (UK); reported in "Influenza in a Boarding School," *British Medical Journal* March 4, 1978, and by J. D. Murray in *Mathematical Biology* (New York: Springer Verlag, 1990)

The number of infecteds is changing in two ways: newly sick people are added to the infected group, and others are removed. The newly sick people are exactly those people leaving the susceptible group and so accrue at a rate of aSI (with a positive sign this time). People leave the infected group either because they recover (or die), or because they are physically removed from the rest of the group and can no longer infect others. We assume that people are removed at a rate proportional to the number sick, or bI, where b is a positive constant. Thus,

$$\frac{dI}{dt} = \begin{array}{c}\text{Rate susceptibles}\\ \text{get sick}\end{array} - \begin{array}{c}\text{Rate infecteds}\\ \text{get removed}\end{array} = aSI - bI.$$

Assuming that those who have recovered from the disease are no longer susceptible, the recovered group increases at the rate of bI, so

$$\frac{dR}{dt} = bI.$$

We are assuming that having the flu confers immunity on a person, that is, that the person cannot get the flu again. (This is true for a given strain of flu, at least in the short run.)

In analyzing the flu, we can use the fact that the total population $S + I + R$ is not changing. (The total population, the total number of boys in the school, did not change during the epidemic.) Thus, once we know S and I, we can calculate R. So we will restrict our attention to the two equations

$$\frac{dS}{dt} = -aSI$$
$$\frac{dI}{dt} = aSI - bI.$$

The Constants a and b

The constant a measures how infectious the disease is—that is, how quickly it is transmitted from the infecteds to the susceptibles. In the case of the flu, we know from medical accounts that the epidemic started with one sick boy, with two more becoming sick roughly a day later. Thus, when $I = 1$ and $S = 762$, we have $dS/dt \approx -2$, enabling us to roughly[18] approximate a:

$$a = -\frac{dS/dt}{SI} = \frac{2}{(762)(1)} = 0.0026.$$

The constant b represents the rate at which infected people are removed from the infected population. In this case of the flu, boys were generally taken to the infirmary within one or two days of becoming sick. About half the infected population was removed each day, so we take $b \approx 0.5$. Thus, our equations are:

$$\frac{dS}{dt} = -0.0026SI$$
$$\frac{dI}{dt} = 0.0026SI - 0.5I.$$

The Phase Plane

We can get a good idea of the progress of the disease from graphs. You might expect that we would look for graphs of S and I against t, and eventually we will. However, we first look at a graph of I against S. If we plot a point (S, I) representing the number of susceptibles and the number of infecteds at any moment in time, then, as the numbers of susceptibles and infecteds change, the point moves. The SI-plane on which the point moves is called the *phase plane*. The path along which the point moves is called the *phase trajectory*, or *orbit*, of the point.

---

[18]The values of a and b are close to those obtained by J. D. Murray in *Mathematical Biology* (New York: Springer Verlag, 1990).

To find the phase trajectory, we need a differential equation relating S and I directly. Thinking of I as a function of S, and S as a function of t, we use the chain rule to get

$$\frac{dI}{dt} = \frac{dI}{dS} \cdot \frac{dS}{dt},$$

giving

$$\frac{dI}{dS} = \frac{dI/dt}{dS/dt}.$$

Substituting for dI/dt and dS/dt, we get

$$\frac{dI}{dS} = \frac{0.0026SI - 0.5I}{-0.0026SI}.$$

Assuming I is not zero, this equation simplifies to approximately

$$\frac{dI}{dS} = -1 + \frac{192}{S}.$$

The slope field of this differential equation is shown in Figure 10.66. The trajectory with initial condition $S_0 = 762$, $I_0 = 1$ is shown in Figure 10.67. Time is represented by the arrow showing the direction that a point moves on the trajectory. The disease starts at the point $S_0 = 762$, $I_0 = 1$. At first, more people become infected and fewer are susceptible. In other words, S decreases and I increases. Later, I decreases as S continues to decrease.

Figure 10.66: Slope field for $dI/dS = -1 + 192/S$

Figure 10.67: Trajectory for $S_0 = 762$, $I_0 = 1$

What does the *SI*-Phase Plane Tell Us?

To learn how the disease progresses, look at the shape of the curve in Figure 10.67. The value of I increases to about 300 (the maximum number infected and infectious at any one time); then I decreases to zero. This peak value of I occurs when $S \approx 200$. We can determine exactly when the peak value occurs by solving

$$\frac{dI}{dS} = -1 + \frac{192}{S} = 0,$$

which gives

$$S = 192.$$

Notice that the peak value for I always occurs at the same value of S, namely $S = 192$. The graph shows that if a trajectory starts with $S_0 > 192$, then I first increases and then decreases to zero. On the other hand, if $S_0 < 192$, there is no peak as I decreases right away.

For this example, the value $S_0 = 192$ is called a *threshold value*. If S_0 is around or below 192, there is no epidemic. If S_0 is significantly greater than 192, an epidemic occurs.[19]

[19]Here we are using J. D. Murray's definition of an epidemic as an outbreak in which the number of infecteds increases from the initial value, I_0. See *Mathematical Biology* (New York: Springer Verlag, 1990)

The phase diagram makes clear that the maximum value of I is about 300. Another question answered by the phase plane diagram is the total number of students who are expected to get sick during the epidemic. (This is not the maximum value reached by I, which gives the maximum number infected at any one time.) The point at which the trajectory crosses the S-axis represents the time when the epidemic has passed (since $I = 0$). Thus, the S-intercept shows how many boys never get the flu and, hence, how many do get it.

How Many People Should Be Vaccinated?

Faced with an outbreak of the flu or, as happened on several US campuses in the 1980s, of the measles, many institutions consider a vaccination program. How many students must be vaccinated in order to control an outbreak? To answer this, we can think of vaccination as removing people from the S category (without increasing I), which amounts to moving the initial point on the trajectory to the left, parallel to the S-axis. To avoid an epidemic, the initial value of S_0 should be around or below the threshold value. Therefore, for the boarding school, all but about 192 students need to be vaccinated to avoid an epidemic.

Graphs of S and I against t

To find out exactly when I reaches its maximum, we need numerical methods. A modification of Euler's method was used to generate the solution curves of S and I against t in Figure 10.68. Notice that, as we would expect, the number of susceptibles drops throughout the disease as healthy people get sick. The number of infecteds peaks after about 6 days and then drops. The epidemic has run its course in 20 days.

Analytical Solution for the *SI*-Phase Trajectory

The differential equation

$$\frac{dI}{dS} = -1 + \frac{192}{S}$$

can be integrated directly giving

$$I = -S + 192 \ln S + C.$$

Using $S_0 = 762$, $I_0 = 1$, gives $1 = -762 + 192 \ln 762 + C$, so we get $C = 763 - 192 \ln(762)$. Putting this value for C into the equation for I, we get:

$$I = -S + 192 \ln S - 192 \ln(762) + 763$$

$$I = -S + 192 \ln\left(\frac{S}{762}\right) + 763.$$

This is the equation of the solution curve in Figure 10.67.

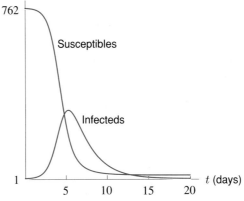

Figure 10.68: Progress of the flu over time.

Two Interacting Populations: Predator-Prey

We now consider two populations which interact. For example, they may compete for food, one may prey on the other, or they may enjoy a symbiotic relationship in which each helps the other. We model a predator-prey system using what are called the Lotka-Volterra equations.

Robins and Worms

Let's look at a simplified and idealized case[20] in which robins are the predators and worms the prey. Suppose there are r thousand robins and w million worms. If there were no robins, the worms would increase exponentially according to the equation

$$\frac{dw}{dt} = aw \quad \text{where } a \text{ is a constant and } a > 0.$$

If there were no worms, the robins would have no food and so their population would decrease according to the equation[21]

$$\frac{dr}{dt} = -br \quad \text{where } b \text{ is a constant and } b > 0.$$

Now we account for the effect of the two populations on one another. Clearly, the presence of the robins is bad for the worms, so

$$\frac{dw}{dt} = aw - \text{Effect of robins on worms.}$$

On the other hand, the robins do better with the worms around, so

$$\frac{dr}{dt} = -br + \text{Effect of worms on robins.}$$

How exactly do the two populations interact? Let's assume the effect of one population on the other is proportional to the number of "encounters." (An encounter is when a robin eats a worm.) The number of encounters is likely to be proportional to the product of the populations because the more there are of either population, the more encounters there will be. So we assume

$$\frac{dw}{dt} = aw - cwr \quad \text{and} \quad \frac{dr}{dt} = -br + kwr,$$

where c and k are positive constants.

To analyze this system of equations, let's look at the specific example with $a = b = c = k = 1$:

$$\frac{dw}{dt} = w - wr \quad \text{and} \quad \frac{dr}{dt} = -r + wr.$$

To visualize the solutions to these equations, we look for trajectories in the phase plane. First we use the chain rule,

$$\frac{dr}{dw} = \frac{dr/dt}{dw/dt},$$

to obtain

$$\frac{dr}{dw} = \frac{-r + wr}{w - wr}.$$

---

[20]Based on ideas from Thomas A. McMahon.

[21]You might criticize this assumption because it predicts that the number of robins will decay exponentially, rather than die out in finite time.

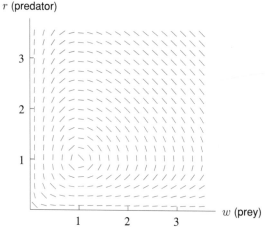

Figure 10.69: Slope field for $\dfrac{dr}{dw} = \dfrac{-r + wr}{w - wr}$

The Slope Field and Equilibrium Points

We can get an idea of what solutions of this equation look like from the slope field in Figure 10.69. At the point $(1, 1)$ there is no slope drawn because at this point the rate of change of the worm population with respect to time is zero:

$$\frac{dw}{dt} = 1 - (1)(1) = 0.$$

The rate of change of the robin population with respect to time is also zero:

$$\frac{dr}{dt} = -1 + (1)(1) = 0.$$

Thus dr/dw is undefined. In terms of worms and robins, this means that if at some moment $w = 1$ and $r = 1$ (that is, there are 1 million worms and 1 thousand robins), then w and r remain constant forever. The point $w = 1$, $r = 1$ is therefore an equilibrium solution. The slope field suggests that there are no other equilibrium points except the origin.

> At an **equilibrium point**, both w and r are constant, so
> $$\frac{dw}{dt} = 0 \qquad \text{and} \qquad \frac{dr}{dt} = 0.$$

Therefore, we look for equilibrium points by solving

$$\frac{dw}{dt} = w - wr = 0 \quad \text{and} \quad \frac{dr}{dt} = -r + rw = 0,$$

which has $w = 0$, $r = 0$ and $w = 1$, $r = 1$ as the only solutions.

Trajectories in the wr-Phase Plane

Let's look at the trajectories in the phase plane. Remember that a point on a curve represents a pair of populations (w, r) existing at the same time t (though t is not shown on the graph). At a short time later, the pair of populations will be represented by a nearby point. As time passes, the point

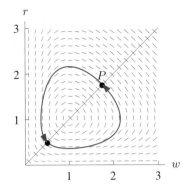

Figure 10.70: Solution curve is closed

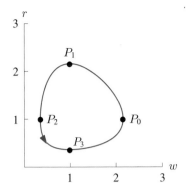

Figure 10.71: A trajectory

traces out a trajectory. The direction is marked on the curve by an arrow. (See Figure 10.70.)

How do we figure out which way to move on the trajectory? Approximating the solution numerically shows that the trajectory is traversed counterclockwise. Alternatively, look at the original pair of differential equations. They involve time and so tell us how w and r change with time. Imagine, for example, that we are at the point P_0 in Figure 10.71, where $w > 1$ and $r = 1$; then

$$\frac{dr}{dt} = -r + wr = -1 + w > 0.$$

Therefore, r is increasing, so the point is moving counterclockwise around the closed curve.

Now let's think about why the solution curves are closed curves (that is, why they come back and meet themselves). Notice that the slope field is symmetric about the line $w = r$. We can confirm this by observing that interchanging w and r does not alter the differential equation for dr/dw. This means that if we start at point P on the line $w = r$ and travel once around the point $(1, 1)$, we arrive back at the same point P. The reason is that the second half of the path is the reflection of the first half in the line $w = r$. (See Figure 10.70.) If we did not end up at P again, the second half of our path would have a different shape from the first half.

The Populations as Functions of Time

The shape of trajectories tells us how the populations vary with time. Suppose we start at $t = 0$ at the point P_0 in Figure 10.71. Then suppose we move to P_1 at time t_1, to P_2 at time t_2, to P_3 at time t_3, and so on. At time t_4 we will be back at P_0, and the whole cycle will repeat. The fact that the trajectory is a closed curve means that both populations oscillate periodically. Both populations have the same period, and the worms (the prey) are at their maximum a quarter of a cycle before the robins. (See Figure 10.72.)

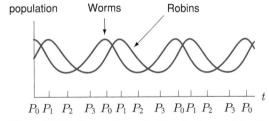

Figure 10.72: Populations of robins (in thousands) and worms (in millions) over time

Lynxes and Hares

A predator-prey system for which there are long-term data is the Canadian lynx and the hare. Since both animals were of interest to fur trappers, records of the Hudson Bay Company shed some light on their populations through much of the last century. These records show that both populations oscillated up and down, quite regularly, with a period of about ten years. In addition, the maximum values of the two populations did not occur at the same time; the hare was a quarter of a cycle ahead of the lynx. This is exactly the behavior predicted by Lotka-Volterra equations.

Problems on Systems of Differential Equations

1. Show that if S, I, and R satisfy the differential equations on page 563, the total population, $S + I + R$, is a constant.

For Problems 2–5, suppose x and y are the populations of two different species. Describe in words how each population changes with time.

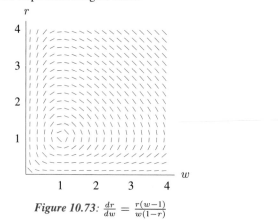

2. y 3. y 4. y 5. y

For Problems 6–13 let w be the number of worms (in millions) and r the number of robins (in thousands) living on an island. Suppose w and r satisfy the differential equations

$$\frac{dw}{dt} = w - wr$$

$$\frac{dr}{dt} = -r + wr.$$

6. Explain why this might be a reasonable way to model the interaction between the two populations. Why have the signs been chosen this way?

7. Solve these equations in the two special cases when there are no robins and when there are no worms living on the island.

For Problems 8–13, you may use the slope field in Figure 10.73.

Figure 10.73: $\dfrac{dr}{dw} = \dfrac{r(w-1)}{w(1-r)}$

8. Describe and explain the symmetry you observe in the slope field. What consequences does this symmetry have for the solution curves?

9. Assume $w = 2$ and $r = 2$ when $t = 0$. Do the numbers of robins and worms increase or decrease at first? What happens in the long run?

10. For the case discussed in Problem 9, estimate the maximum and the minimum values of the robin population. How many worms are there at the time when the robin population reaches its maximum?

11. On the same axes, graph w and r (the worm and the robin populations) against time. Use initial values of 1.5 for w and 1 for r. You may do this without units for t.

12. People on the island like robins so much that they decide to import 200 robins all the way from England, to increase the initial population to $r = 2.2$ when $t = 0$. Does this make sense? Why or why not?

13. Assume that $w = 3$ and $r = 1$ when $t = 0$. Do the numbers of robins and worms increase or decrease initially? What happens in the long run?

14. Two companies share the market for a new technology. They have no competition except each other. Let $A(t)$ be the net worth of one company and $B(t)$ the net worth of the other at time t. Assume that net worth cannot be negative, because a company goes out of business if its net worth is zero. Suppose A and B satisfy the differential equations

$$A' = 2A - AB$$
$$B' = B - AB.$$

(a) What do these equations predict about the net worth of each company if the other were not present? What effect do the companies have on each other?

(b) Are there any equilibrium points? If so, what are they?

(c) Sketch a slope field for these equations (using a computer or calculator), and hence describe the different possible long-run behaviors.

For Problems 15–17, consider a conflict between two armies of x and y soldiers, respectively. During World War I, F. W. Lanchester assumed that if both armies are fighting a conventional battle within sight of one another, the rate at which soldiers in one army are put out of action (killed or wounded) is proportional to the amount of fire the other army can concentrate on them, which is in turn proportional to the number of soldiers in the opposing army. Thus Lanchester assumed that if there are no reinforcements and t represents time since the start of the battle, then x and y obey the differential equations

$$\frac{dx}{dt} = -ay$$
$$\frac{dy}{dt} = -bx \quad a, b > 0.$$

15. Near the end of World War II a fierce battle took place between US and Japanese troops over the island of Iwo Jima, off the coast of Japan. Applying Lanchester's analysis to this battle, with x representing the number of US troops and y the number of Japanese troops, it has been estimated[22] that $a = 0.05$ and $b = 0.01$.

(a) Using these values for a and b, and ignoring reinforcements, write a differential equation involving dy/dx and sketch its slope field.

(b) Assuming that the initial strength of the US forces was 54,000 and that of the Japanese was 21,500, draw the trajectory which describes the battle. What outcome is predicted? (That is, which side do the differential equations predict will win?)

(c) Would knowing that the US in fact had 19,000 reinforcements, while the Japanese had none, alter the outcome predicted?

16. (a) For two armies of strengths x and y fighting a conventional battle governed by Lanchester's differential equations, write a differential equation involving dy/dx and the constants of attrition a and b.

(b) Solve the differential equation and hence show that the equation of the phase trajectory is

$$ay^2 - bx^2 = C.$$

for some constant C. This equation is called *Lanchester's square law*. The value of C depends on the initial sizes of the two armies.

17. Consider the battle of Iwo Jima, described in Problem 15. Take $a = 0.05$, $b = 0.01$ and assume the initial strength of the US troops to be 54,000 and that of the Japanese troops to be 21,500. (Again, ignore reinforcements.)

(a) Using Lanchester's square law derived in Problem 16, find the equation of the trajectory describing the battle.

(b) Assuming that the Japanese fought without surrendering until they had all been killed, as was the case, how many US troops does this model predict would be left when the battle ended?

[22] See Martin Braun, *Differential Equations and their Applications*, 2nd ed. (New York: Springer and Verlag, 1975).

18. In this problem we adapt Lanchester's model for a conventional battle (described before Problem 15), to the case in which one or both of the armies is a guerrilla force. We assume that the rate at which a guerrilla force is put out of action is proportional to the product of the strengths of the two armies.

    (a) Give a justification for the assumption that the rate at which a guerrilla force is put out of action is proportional to the product of the strengths of the two armies.
    (b) Write the differential equations which describe a conflict between a guerrilla army of strength x and a conventional army of strength y, assuming all the constants of proportionality are 1.
    (c) Find a differential equation involving dy/dx and solve it to find equations of phase trajectories.
    (d) If $C > 0$, which side wins? If $C < 0$, which side wins? What if $C = 0$?
    (e) Use your solution to part (d) to divide the phase plane into regions according to which side wins.

19. To model a conflict between two guerrilla armies, we assume that the rate that each one is put out of action is proportional to the product of the strengths of the two armies.

    (a) Write the differential equations which describe a conflict between two guerrilla armies, of strengths x and y, respectively.
    (b) Find a differential equation involving dy/dx and solve to find equations of phase trajectories.
    (c) If $C > 0$, which side wins? If $C < 0$, which side wins? What if $C = 0$?
    (d) Use your solution to part (b) to divide the phase plane into regions according to which side wins.

20. The concentrations of two chemicals A and B as functions of time are denoted by x and y respectively. Each alone decays at a rate proportional to its concentration. Put together, they also interact to form a third substance, at a rate proportional to the product of their concentrations. All this is expressed in the equations:

$$\frac{dx}{dt} = -2x - xy, \qquad \frac{dy}{dt} = -3y - xy.$$

    (a) Find a differential equation describing the relationship between x and y, and solve it.
    (b) Show that the only equilibrium state is $x = y = 0$. (Note that the concentrations are nonnegative.)
    (c) Show that when x and y are positive and very small, y^2/x^3 is roughly constant.
       [Hint: When x is small, x is negligible compared to $\ln x$.]

    If now the initial concentrations are $x(0) = 4$, $y(0) = 8$:

    (d) Find the equation of the phase trajectory.
    (e) What would be the concentrations of each substance if they become equal?
    (f) If $x = e^{-10}$, find an approximate value for y.

The systems of differential equations in Problems 21–23 model the interaction of two populations x and y. In each case, answer the following two questions:

(a) What kinds of interaction (symbiosis[23], competition, predator-prey) do the equations describe?
(b) What happens in the long run? (For one of the systems, your answer will depend on the initial populations.) Use a calculator or computer to draw slope fields.

21. $\dfrac{1}{x}\dfrac{dx}{dt} = y - 1$

    $\dfrac{1}{y}\dfrac{dy}{dt} = x - 1$

22. $\dfrac{1}{x}\dfrac{dx}{dt} = 1 - \dfrac{x}{2} - \dfrac{y}{2}$

    $\dfrac{1}{y}\dfrac{dy}{dt} = 1 - x - y$

23. $\dfrac{1}{x}\dfrac{dx}{dt} = y - 1 - 0.05x$

    $\dfrac{1}{y}\dfrac{dy}{dt} = 1 - x - 0.05y$

ANALYZING THE PHASE PLANE

In the previous section we analyzed a system of differential equations using a slope field. In this section we will again analyze a system of differential equations, using curves called *nullclines*. The problem we will consider is of two similar species in competition for food and space. In practice, one of the two species usually becomes extinct, and the other takes over completely. Biologists say that if two species occupy similar *niches*, or ways of living, then one will drive the other to extinction. This phenomenon is called the *Principle of Competitive Exclusion*. We will see how differential equations predict this in a particular case.

---

[23]Symbiosis takes place when the interaction of two species benefits both. An example is the pollination of plants by insects.

Competitive Exclusion: Citrus Tree Parasites

The citrus farmers of Southern California are interested in controlling the insects that live on their trees. Some of these insects can be controlled by parasites which live on the trees too. Scientists are therefore interested in understanding under what circumstances these parasites flourish or die out. One such parasite was introduced accidentally from the Mediterranean; later, other parasites were introduced from China and India; in each case the previous parasite became extinct over part of its habitat. In 1963 a lab experiment was carried out to determine which one of a pair of species became extinct when they were in competition with each other. The data on one pair of species, called *A. fisheri* and *A. melinus*, with populations P_1 and P_2 respectively, is given in Table 10.13 and shows that *A. melinus* (P_2) became extinct after 8 generations.[24]

TABLE 10.13   *Population (in thousands) of two species of parasite as a function of time*

Generation number	1	2	3	4	5	6	7	8
Population P_1 (thousands)	0.193	1.093	1.834	5.819	13.705	16.965	18.381	16.234
Population P_2 (thousands)	0.083	0.229	0.282	0.378	0.737	0.507	0.13	0

Data from the same experimenters indicates that, when alone, each population grows logistically. In fact, their data suggests that, when alone, the population of P_1 might grow according to the equation

$$\frac{dP_1}{dt} = 0.05P_1 \left(1 - \frac{P_1}{20}\right),$$

and when alone, the population of P_2 might grow according to the equation

$$\frac{dP_2}{dt} = 0.09P_2 \left(1 - \frac{P_2}{15}\right).$$

Now suppose both parasites are present. Each tends to reduce the growth rate of the other, so each differential equation is modified by subtracting a term on the right. The experimental data shows that together P_1 and P_2 can be well described by the equations

$$\frac{dP_1}{dt} = 0.05P_1 \left(1 - \frac{P_1}{20}\right) - 0.002P_1P_2$$

$$\frac{dP_2}{dt} = 0.09P_2 \left(1 - \frac{P_2}{15}\right) - 0.15P_1P_2.$$

The fact that P_2 dies out with time is reflected in these equations: the coefficient of P_1P_2 is so much larger in the equation for P_2 than in the equation for P_1. This indicates that the interaction has a much more devastating effect upon the growth of P_2 than on the growth of P_1.

The Phase Plane and Nullclines

We consider the phase plane with the P_1 axis horizontal and the P_2 axis vertical. To find the trajectories in the P_1P_2 phase plane, we could draw a slope field as in the previous section. Instead, we use a method which gives a good qualitative picture of the behavior of the trajectories even without a calculator or computer. We find the *nullclines* or curves along which $dP_1/dt = 0$ or $dP_2/dt = 0$. At points where $dP_2/dt = 0$, the population P_2 is momentarily constant, so only population P_1 is changing with time. Therefore, at this point the trajectory is horizontal. (See Figure 10.74.) Similarly, at points where $dP_1/dt = 0$, the population P_1 is momentarily constant and population P_2 is the only

[24]Data adapted from Paul DeBach and Ragnhild Sundby, "Competitive Displacement between Ecological Homologues," *Hilgardia 34*:17 (1963).

one changing, so the trajectory is vertical there. A point where both $dP_1/dt = 0$ and $dP_2/dt = 0$ is called an *equilibrium point* because P_1 and P_2 both remain constant if they reach these values.

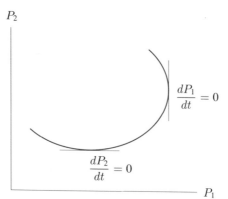

Figure 10.74: Points on a trajectory where
$dP_1/dt = 0$ or $dP_2/dt = 0$

On the $P_1 P_2$ phase plane:

- If $\dfrac{dP_1}{dt} = 0$, the trajectory is vertical.

- If $\dfrac{dP_2}{dt} = 0$, the trajectory is horizontal.

- If $\dfrac{dP_1}{dt} = \dfrac{dP_2}{dt} = 0$, there is an equilibrium point.

Using Nullclines to Analyze the Parasite Populations

In order to see where $dP_1/dt = 0$ or $dP_2/dt = 0$, we factor the right side of our differential equations:

$$\frac{dP_1}{dt} = 0.05P_1 \left(1 - \frac{P_1}{20}\right) - 0.002P_1 P_2 = 0.001P_1(50 - 2.5P_1 - 2P_2)$$

$$\frac{dP_2}{dt} = 0.09P_2 \left(1 - \frac{P_2}{15}\right) - 0.15P_1 P_2 = 0.001P_2(90 - 150P_1 - 6P_2).$$

Thus $dP_1/dt = 0$ where $P_1 = 0$ or where $50 - 2.5P_1 - 2P_2 = 0$. Graphing these equations in the phase plane gives two lines, which are nullclines. Since the trajectory is vertical where $dP_1/dt = 0$, in Figure 10.75 we draw small vertical line segments on these nullclines to represent the direction of the trajectories as they cross the nullcline. Similarly $dP_2/dt = 0$ where $P_2 = 0$ or where $90 - 150P_1 - 6P_2 = 0$. These equations are graphed in Figure 10.75 with small horizontal line segments on them.

The equilibrium points are where both $dP_1/dt = 0$ and $dP_2/dt = 0$, namely the points $P_1 = 0, P_2 = 0$ (meaning that both species die out); $P_1 = 0, P_2 = 15$ (where P_1 is extinct); and $P_1 = 20, P_2 = 0$ (where P_2 is extinct).

What Happens in the Regions Between the Nullclines?

Nullclines are useful because they divide the plane into regions in which the signs of dP_1/dt and dP_2/dt remain the same. Thus, in each region, the direction of every trajectory remains roughly the same.

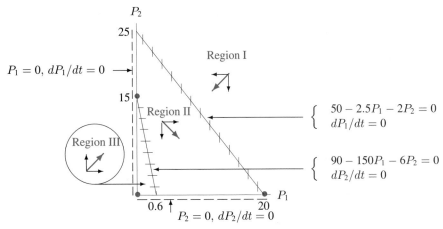

Figure 10.75: Analyzing three regions in the phase plane using nullclines (axes distorted) with equilibrium points represented by dots

In region I, for example, we might try the point $P_1 = 20, P_2 = 25$. Then

$$\frac{dP_1}{dt} = 0.001(20)(50 - 2.5(20) - 2(25)) < 0$$

$$\frac{dP_2}{dt} = 0.001(25)(90 - 150(20) - 6(25)) < 0$$

Now $dP_1/dt < 0$, so P_1 is decreasing, which can be represented by an arrow in the direction $\leftarrow$. Also, $dP_2/dt < 0$, so P_2 is decreasing, as represented by the arrow $\downarrow$. Combining these directions, we know that the trajectories in this region go approximately in the diagonal direction $\swarrow$ (See region I in Figure 10.75.)

In region II, try, for example, $P_1 = 1, P_2 = 1$. Then we have

$$\frac{dP_1}{dt} = 0.001(1)(50 - 2.5 - 2) > 0$$

$$\frac{dP_2}{dt} = 0.001(1)(90 - 150 - 6) < 0$$

So here, P_1 is increasing while P_2 is decreasing. (See region II in Figure 10.75.)

In region III, try $P_1 = 0.1, P_2 = 0.1$:

$$\frac{dP_1}{dt} = 0.001(0.1)(50 - 2.5(0.1) - 2(0.1)) > 0$$

$$\frac{dP_2}{dt} = 0.001(0.1)(90 - 150(0.1) - 6(0.1)) > 0$$

So here, both P_1 and P_2 are increasing. (See region III in Figure 10.75.)

Notice that the behavior of the populations in each region makes biological sense. In region I both populations are so large that overpopulation is a problem, so both populations decrease. In region III both populations are so small that they are effectively not in competition, so both grow. In region II competition between the species does come into play. The fact that P_1 increases while P_2 decreases in region II means that P_1 wins.

Solution Trajectories

Suppose the system starts with some of each population. This means that the initial point of the trajectory is not on one of the axes, and so it is in region I, II, or III. Then the point moves on a trajectory like one of those computed numerically and shown in Figure 10.76. Notice that *all* these

Figure 10.76: Trajectories showing exclusion of population P_2 (not to scale)

trajectories tend toward the point $P_1 = 20, P_2 = 0$, corresponding to a population of 20,000 for P_1 and extinction for P_2. Consequently, this model predicts that no matter what the initial populations are, provided $P_1 \neq 0$, the population of P_2 will be excluded by P_1, and P_1 will tend to a constant value. This makes biological sense: in the absence of P_2, we would expect P_1 to settle down to the carrying capacity of the niche, which is 20,000.

Problems on Analyzing the Phase Plane

1. The equations describing the flu epidemic in a boarding school are

    $$\frac{dS}{dt} = -0.0026SI$$
    $$\frac{dI}{dt} = 0.0026SI - 0.5I.$$

    (a) Find the nullclines and equilibrium points in the SI phase plane.
    (b) Find the direction of the trajectories in each region.
    (c) Sketch some typical trajectories and describe their behavior in words.

2. Use the idea of nullclines dividing the plane into sectors to analyze the equations describing the interactions of robins and worms.

    $$\frac{dw}{dt} = w - wr$$
    $$\frac{dr}{dt} = -r + rw.$$

For Problems 3–8, analyze the phase plane of the differential equations for $x, y \geq 0$. Show the nullclines and equilibrium points, and sketch the direction of the trajectories in each region.

3. $\dfrac{dx}{dt} = x(2 - x - y)$
    $\dfrac{dy}{dt} = y(1 - x - y)$

4. $\dfrac{dx}{dt} = x(2 - x - 3y)$
    $\dfrac{dy}{dt} = y(1 - 2x)$

5. $\dfrac{dx}{dt} = x(2 - x - 2y)$
    $\dfrac{dy}{dt} = y(1 - 2x - y)$

6. $\dfrac{dx}{dt} = x(1 - y - \dfrac{x}{3})$
    $\dfrac{dy}{dt} = y(1 - \dfrac{y}{2} - x)$

7. $\dfrac{dx}{dt} = x\left(1 - x - \dfrac{y}{3}\right)$
    $\dfrac{dy}{dt} = y\left(1 - y - \dfrac{x}{2}\right)$

8. $\dfrac{dx}{dt} = x\left(1 - \dfrac{x}{2} - y\right)$
    $\dfrac{dy}{dt} = y\left(1 - \dfrac{y}{3} - x\right)$

9. In the 1930s, the Soviet ecologist G. F. Gause performed a series of experiments on competition among two yeasts with populations P_1 and P_2, respectively. By performing population studies at low density in

large volumes, he determined what he called the *coefficients of geometric increase* (and we would call continuous exponential growth rates). These coefficients described the growth of each yeast alone:

$$\frac{1}{P_1}\frac{dP_1}{dt} = 0.2$$

$$\frac{1}{P_2}\frac{dP_2}{dt} = 0.06$$

where P_1 and P_2 are measured in units that Gause established.

He also determined that, in his units, the carrying capacity of P_1 was 13 and the carrying capacity of P_2 was 6. He then observed that one P_2 occupies the niche space of 3 P_1 and that one P_1 occupied the niche space of 0.4 P_2. This led him to the following differential equations to describe the interaction of P_1 and P_2:

$$\frac{dP_1}{dt} = 0.2P_1\left(\frac{13 - (P_1 + 3P_2)}{13}\right),$$

$$\frac{dP_2}{dt} = 0.06P_2\left(\frac{6 - (P_2 + 0.4P_1)}{6}\right).$$

When both yeasts were growing together, Gause recorded the data in Table 10.14.

TABLE 10.14 *Gause's yeast populations*

Time (hours)	6	16	24	29	48	53
P_1	0.375	3.99	4.69	6.15	7.27	8.30
P_2	0.29	0.98	1.47	1.46	1.71	1.84

(a) Carry out a phase plane analysis of Gause's equations.

(b) Mark the data points on the phase plane and describe what would have happened had Gause continued the experiment.

10. When two countries are in an arms race, the rate at which each one spends money on arms is determined by its own current level of spending and by its opponent's level of spending. We expect that the more a country is already spending on armaments, the less willing it will be to increase its military expenditures. On the other hand, the more a country's opponent spends on armaments, the more rapidly the country will arm. If x billion is the country's yearly expenditure on arms, and y billion is its opponent's, then the *Richardson arms race model* proposes that x and y are determined by differential equations. Suppose that for some particular arms race the equations are

$$\frac{dx}{dt} = -0.2x + 0.15y + 20$$

$$\frac{dy}{dt} = 0.1x - 0.2y + 40.$$

(a) Explain the signs of the three terms on the right side of the equations for dx/dt.

(b) Find the equilibrium points for this system.

(c) Analyze the direction of the trajectories in each region.

(d) Are the equilibrium points stable or unstable?

(e) What does this model predict will happen if both countries disarm?

(f) What does this model predict will happen in the case of unilateral disarmament (one country disarms, and the other country does not)?

(g) What does the model predict will happen in the long run?

11.   In the 1930's L. F. Richardson proposed that an arms race between two countries could be modeled by a system of differential equations. One arms race which can be reasonably well described by differential equations is the US-Soviet Union arms race between 1945 and 1960. If x represents the annual Soviet expenditures on armaments (in billions of dollars) and y represents the corresponding US expenditures, it has been suggested[25] that x and y obey the following differential equations

$$\frac{dx}{dt} = -0.45x + 10.5,$$

$$\frac{dy}{dt} = 8.2x - 0.8y - 142.$$

(a)   Find the nullclines and equilibrium points for these differential equations. Which direction do the trajectories go in each region?

(b)   Sketch some typical trajectories in the phase plane.

(c)   What do these differential equations predict will be the long term outcome of the US-Soviet arms race?

(d)   Discuss these predictions in the light of the actual expenditures in Table 10.15.

TABLE 10.15   *Arms budgets of the United States and the Soviet Union for the years 1945–1960 (billions of dollars)*

	USSR	USA		USSR	USA
1945	14	97	1953	25.7	71.4
1946	14	80	1954	23.9	61.6
1947	15	29	1955	25.5	58.3
1948	20	20	1956	23.2	59.4
1949	20	22	1957	23.0	61.4
1950	21	23	1958	22.3	61.4
1951	22.7	49.6	1959	22.3	61.7
1952	26.0	69.6	1960	22.1	59.6

[25]R. Taagepera, G. M. Schiffler, R. T. Perkins and D. L. Wagner, *Soviet-American and Israeli-Arab Arms Races and the Richardson Model* (General Systems, XX, 1975).

APPENDICES

A ROOTS, ACCURACY, AND BOUNDS

It is often necessary to find the zeros of a polynomial or the points of intersection of two curves. So far, you have probably used algebraic methods, such as the quadratic formula, to solve such problems. Unfortunately, however, mathematicians' search for similar solutions to more complicated equations, has not been all that successful. The formulas for the solutions to third- and fourth-degree equations are so complicated that you'd never want to use them. Early in the nineteenth century, it was proved that there is no algebraic formula for the solutions to equations of degree 5 and higher. Most nonpolynomial equations cannot be solved using a formula either.

However, we can still find roots of equations, provided we use approximation methods, not formulas. In this section we will discuss three ways to find roots: algebraic, graphical, and numerical. Of these, only the algebraic method gives exact solutions.

First, let's get some terminology straight. Given the equation $x^2 = 4$, we call $x = -2$ and $x = 2$ the *roots*, or *solutions of the equation*. If we are given the function $f(x) = x^2 - 4$, then -2 and 2 are called the *zeros of the function*; that is, the zeros of the function f are the roots of the equation $f(x) = 0$.

The Algebraic Viewpoint: Roots by Factoring

If the product of two numbers is zero, then one or the other or both must be zero, that is, if $AB = 0$, then $A = 0$ or $B = 0$. This observation lies behind finding roots by factoring. You may have spent a lot of time factoring polynomials. Here you will also factor expressions involving trigonometric and exponential functions.

Example 1 Find the roots of $x^2 - 7x = 8$.

Solution Rewrite the equation as $x^2 - 7x - 8 = 0$. Then factor the left side: $(x + 1)(x - 8) = 0$. By our observation about products, either $x + 1 = 0$ or $x - 8 = 0$, so the roots are $x = -1$ and $x = 8$.

Example 2 Find the roots of $\dfrac{1}{x} - \dfrac{x}{(x + 2)} = 0$.

Solution Rewrite the left side with a common denominator:

$$\frac{x + 2 - x^2}{x(x + 2)} = 0.$$

Whenever a fraction is zero, the numerator must be zero. Therefore we must have

$$x + 2 - x^2 = (-1)(x^2 - x - 2) = (-1)(x - 2)(x + 1) = 0.$$

We conclude that $x - 2 = 0$ or $x + 1 = 0$, so 2 and -1 are the roots. They can be checked by substitution.

Example 3 Find the roots of $e^{-x} \sin x - e^{-x} \cos x = 0$.

Solution Factor the left side: $e^{-x}(\sin x - \cos x) = 0$. The factor e^{-x} is never zero; it is impossible to raise e to a power and get zero. Therefore, the only possibility is that $\sin x - \cos x = 0$. This equation is equivalent to $\sin x = \cos x$. If we divide both sides by $\cos x$, we get

$$\frac{\sin x}{\cos x} = \frac{\cos x}{\cos x} \quad \text{so} \quad \tan x = 1.$$

The roots of this equation are

$$\ldots, \frac{-7\pi}{4}, \frac{-3\pi}{4}, \frac{\pi}{4}, \frac{5\pi}{4}, \frac{9\pi}{4}, \frac{13\pi}{4}, \ldots.$$

Warning: Using factoring to solve an equation only works when one side of the equation is 0. It is not true that if, say, $AB = 7$ then $A = 7$ or $B = 7$. For example, you *cannot* solve $x^2 - 4x = 2$ by factoring $x(x - 4) = 2$ and then assuming that either x or $x - 4$ equals 2.

The problem with factoring is that factors are not easy to find. For example, the left side of the quadratic equation $x^2 - 4x - 2 = 0$ does not factor, at least not into "nice" factors with integer coefficients. For the general quadratic equation

$$ax^2 + bx + c = 0,$$

there is the quadratic formula for the roots:

$$x = \frac{-b \pm \sqrt{b^2 - 4ac}}{2a}.$$

Thus the roots of $x^2 - 4x - 2 = 0$ are $(4 \pm \sqrt{24})/2$, or $2 + \sqrt{6}$ and $2 - \sqrt{6}$.

Notice that in each of these examples, we have found the roots exactly.

The Graphical Viewpoint: Roots by Zooming

To find the roots of an equation $f(x) = 0$, it helps to draw the graph of f. The roots of the equation, that is the zeros of f, are *the values of x where the graph of f crosses the x-axis*. Even a very rough sketch of the graph can be useful in determining how many zeros there are and their approximate values. If you have a computer or graphing calculator, then finding solutions by graphing is the easiest method, especially if you use the zoom feature. However, a graph can never tell you the exact value of a root, only an approximate one.

Example 4   Find the roots of $x^3 - 4x - 2 = 0$.

Solution   Attempting to factor the left side with integer coefficients will convince you it cannot be done, so we cannot easily find the roots by algebra. We know the graph of $f(x) = x^3 - 4x - 2$ will have the usual cubic shape; see Figure A.1.

There are clearly three roots: one between $x = -2$ and $x = -1$, another between $x = -1$ and $x = 0$, and a third between $x = 2$ and $x = 3$. Zooming in on the largest root with a graphing calculator or computer shows that it lies in the following interval:

$$2.213 < x < 2.215.$$

Thus, the root is $x = 2.21$, accurate to two decimal places. Zooming in on the other two roots shows them to be $x = -1.68$ and $x = -0.54$, accurate to two decimal places.

Useful trick: Suppose you want to solve the equation $\sin x - \cos x = 0$ graphically. Instead of graphing $f(x) = \sin x - \cos x$ and looking for zeros, you may find it easier to rewrite the equation as $\sin x = \cos x$ and graph $g(x) = \sin x$ and $h(x) = \cos x$. (After all, you already know what these two graphs look like. See Figure A.2.) The roots of the original equation are then precisely the x coordinates of the points of intersection of the graphs of $g(x)$ and $h(x)$.

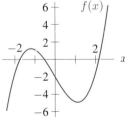

Figure A.1: The cubic $f(x) = x^3 - 4x - 2$

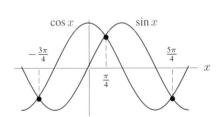

Figure A.2: Finding roots of $\sin x - \cos x = 0$

Example 5   Find the roots of $2 \sin x - x = 0$.

Solution   Rewrite the equation as $2 \sin x = x$, and graph both sides. Since $g(x) = 2 \sin x$ is always between -2 and 2, there are no roots of $2 \sin x = x$ for $x > 2$ or for $x < -2$. We need only consider the graphs between -2 and 2 (or between $-\pi$ and π, which makes graphing the sine function easier). Figure A.3 shows the graphs. There are three points of intersection: one appears to be at $x = 0$, one between $x = \pi/2$ and $x = \pi$, and one between $x = -\pi/2$ and $x = -\pi$. You can tell that $x = 0$ is the exact value of one root because it satisfies the original equation exactly. Zooming in shows that there is a second root $x \approx 1.9$, and the third root is $x \approx -1.9$ by symmetry.

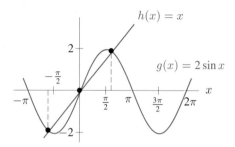

Figure A.3: Finding roots of $2 \sin x - x = 0$

The Numerical Viewpoint: Roots by Bisection

We now look at a numerical method of approximating the solutions to an equation. This method depends on the idea that if the value of a function $f(x)$ changes sign in an interval, and if we believe there is no break in the graph of the function there, then there is a root of the equation $f(x) = 0$ in that interval.

Let's go back to the problem of finding the root of $f(x) = x^3 - 4x - 2 = 0$ between 2 and 3. To locate the root, we close in on it by evaluating the function at the midpoint of the interval, $x = 2.5$. Since $f(2) = -2$, $f(2.5) = 3.625$, and $f(3) = 13$, the function changes sign between $x = 2$ and $x = 2.5$, so the root is between these points. Now we look at $x = 2.25$.

Since $f(2.25) = 0.39$, the function is negative at $x = 2$ and positive at $x = 2.25$, so there is a root between 2 and 2.25. Now we look at 2.125. We find $f(2.125) = -0.90$, so there is a root between 2.125 and 2.25, ... and so on. (You may want to round the decimals as you work.) See Figure A.4. The intervals containing the root are listed in Table A.1 and show that the root is $x = 2.21$ to two decimal places.

Figure A.4: Locating a root of $x^3 - 4x - 2 = 0$

TABLE A.1   *Intervals containing root of $x^3 - 4x - 2 = 0$ (Note [2, 3] means $2 \leq x \leq 3$)*

[2, 3]

[2, 2.5]

[2, 2.25]

[2.125, 2.25]

[2.1875, 2.25]   So $x = 2.2$ rounded to one decimal place

[2.1875, 2.21875]

[2.203125, 2.21875]

[2.2109375, 2.21875]

[2.2109375, 2.2148438]   So $x = 2.21$ rounded to two decimal places

> This method of estimating roots is called the **Bisection Method**:
> - To solve an equation $f(x) = 0$ using the bisection method, we need two starting values for x, say, $x = a$ and $x = b$, such that $f(a)$ and $f(b)$ have opposite signs and f is continuous on $[a, b]$.
> - Evaluate f at the midpoint of the interval $[a, b]$, and decide in which half-interval the root lies.
> - Repeat, using the new half-interval instead of $[a, b]$.

There are some problems with the bisection method:
- The function may not change signs near the root. For example, $f(x) = x^2 - 2x + 1 = 0$ has a root at $x = 1$, but $f(x)$ is never negative because $f(x) = (x - 1)^2$, and a square cannot be negative. (See Figure A.5.)
- The function f must be continuous between the starting values $x = a$ and $x = b$.
- If there is more than one root between the starting values $x = a$ and $x = b$, the method will find only one of the roots. For example, if we had tried to solve $x^3 - 4x - 2 = 0$ starting at $x = -12$ and $x = 10$, the bisection method would zero in on the root between $x = -2$ and $x = -1$, not the root between $x = 2$ and $x = 3$ that we found earlier. (Try it! Then see what happens if you use $x = -10$ instead of $x = -12$.)
- The bisection method is slow and not very efficient. Applying bisection three times in a row only traps the root in an interval $(\frac{1}{2})^3 = \frac{1}{8}$ as large as the starting interval. Thus, if we initially know that a root is between, say, 2 and 3, then we would need to apply the bisection method at least four times to know the first digit after the decimal point.

There are much more powerful methods available for finding roots, such as Newton's method, which are more complicated but which avoid some of these difficulties.

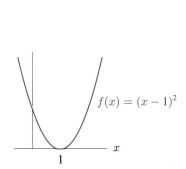

Figure A.5: f doesn't change sign at the root

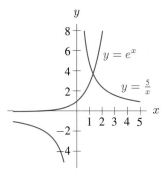

Figure A.6: Intersection of $y = e^x$ and $y = 5/x$

TABLE A.2   *Bisection method for* $f(x) = xe^x - 5 = 0$ *(Note that* $[1, 2]$ *means the interval* $1 \le x \le 2$*)*

Interval Containing Root
$[1, 2]$
$[1, 1.5]$
$[1.25, 1.5]$
$[1.25, 1.375]$
$[1.3125, 1.375]$
$[1.3125, 1.34375]$

Example 6   Find all the roots of $xe^x = 5$ to at least one decimal place.

Solution   If we rewrite the equation as $e^x = 5/x$ and graph both sides, as in Figure A.6, it is clear that there is exactly one root, and it is somewhere between 1 and 2. Table A.2 shows the intervals obtained by the bisection method. After five iterations, we have the root trapped between 1.3125 and 1.34375, so we can say the root is $x = 1.3$ to one decimal place.

Iteration

Both zooming in and bisection as discussed here are examples of *iterative* methods, in which a sequence of steps is repeated over and over again, using the results of one step as the input for the next. We can use such methods to locate a root to any degree of accuracy. In bisection, each iteration traps the root in an interval that is half the length of the previous one. Each time you zoom in on a calculator, you trap the root in a smaller interval; how much smaller depends on the settings on the calculator.

Accuracy and Error

In the previous discussion, we used the phrase "accurate to 2 decimal places." For an iterative process where we get closer and closer estimates of some quantity, we take a common-sense approach to accuracy: we watch the numbers carefully, and when a digit stays the same for several iterations, we assume it has stabilized and is correct, especially if the digits to the right of that digit also stay the same. For example, suppose 2.21429 and 2.21431 are two successive estimates for a zero of $f(x) = x^3 - 4x - 2$. Since these two estimates agree to the third digit after the decimal point, we probably have at least 3 decimal places correct.

There is a problem with this, however. Suppose we are estimating a root whose true value is 1, and the estimates are converging to the value from below—say, 0.985, 0.991, 0.997 and so on. In this case, not even the first decimal place is "correct," even though the difference between the estimates and the actual answer is very small—much less than 0.1. To avoid this difficulty, we say that an estimate a for some quantity r is *accurate to p decimal places* if the error, which is the absolute value of the difference between a and r, or $|r - a|$, is as follows:

Accuracy to p decimal places	means	Error less than
$p = 1$		0.05
2		0.005
3		0.0005
$\vdots$		$\vdots$
n		$0.\underbrace{000\ldots0}_{n}5$

This is the same as saying that r must lie in an interval of length twice the maximum error, centered on a. For example, if a is accurate to 1 decimal place, r must lie in the following interval:

$$a - 0.05 \qquad\qquad a \qquad\qquad a + 0.05$$

Since both the graphing calculator and the bisection method give us an interval in which the root is trapped, this definition of decimal accuracy is a natural one for these processes.

Example 7   Suppose the numbers $\sqrt{10}$, $22/7$, and 3.14 are given as approximations to $\pi = 3.1415\ldots$. To how many decimal places is each approximation accurate?

Solution   Using $\sqrt{10} = 3.1622\ldots$,

$$|\sqrt{10} - \pi| = |3.1622\ldots - 3.1415\ldots| = 0.0206\ldots < 0.05,$$

so $\sqrt{10}$ is accurate to one decimal place. Similarly, using $22/7 = 3.1428\ldots$,

$$\left|\frac{22}{7} - \pi\right| = |3.1428\ldots - 3.1415\ldots| = 0.0013\ldots < 0.005,$$

so 22/7 is accurate to two decimal places. Finally,

$$|3.14 - 3.1415\ldots| = 0.0015\ldots < 0.005,$$

so 3.14 is accurate to two decimal places.

Warning:

- Saying that an approximation is accurate to, say, 2 decimal places does *not* guarantee that its first two decimal places are "correct," that is, that the two digits of the approximation are the same as the corresponding two digits in the true value. For example, an approximate value of 5.997 is accurate to 2 decimal places if the true value is 6.001, but neither of the 9s in the approximation agrees with the 0s in the true value (nor does the digit 5 agree with the digit 6).

- When finding a root r of an equation, the number of decimal places of accuracy refers to the number of digits that have stabilized in the root. It does *not* refer to the number of digits of $f(r)$ that are zero. For example, Table A.1 on page 582 shows that $x = 2.2$ is a root of $f(x) = x^3 - 4x - 2 = 0$, accurate to one decimal place. Yet, $f(2.2) = -0.152$, so $f(2.2)$ does not have one zero after the decimal point. Similarly, $x = 2.21$ is the root accurate to two decimal places, but $f(2.21) = -0.046$ does not have two zeros after the decimal point.

Example 8   Is $x = 2.2143$ a zero of $f(x) = x^3 - 4x - 2$ accurate to four decimal places?

Solution   We want to know whether r, the exact value of the zero, lies in the interval

$$2.2143 - 0.00005 < r < 2.2143 + 0.00005$$

which is the same as

$$2.21425 < r < 2.21435.$$

Since $f(2.21425) < 0$ and $f(2.21435) > 0$, the zero does lie in this interval, and so $r = 2.2143$ is accurate to four decimal places.

How to Write a Decimal Answer

The graphing calculator and bisection method naturally give an interval for a root or a zero. However, other numerical techniques do not give a pair of numbers bounding the true value, but rather a single number near the true value. What should you do if you want a single number, rather than an interval, for an answer? In general, averaging the endpoint of the interval is the best solution.

When giving a single number as an answer and interpreting it, be careful about giving rounded answers. For example, suppose you know a root lies in the interval between 0.81 and 0.87. Averaging gives 0.84 as a single number estimating the root. But it would be wrong to round 0.84 to 0.8 and say that the answer is 0.8 accurate to one decimal place; the true value could be 0.86, which is not within 0.05 of 0.8. The right thing to say is that the answer is 0.84 accurate to one decimal place. Similarly, to give an answer accurate to, say, 2 decimal places, you may have to show 3 decimal places in your answer.

Bounds of a Function

Knowing how big or how small a function gets can sometimes be useful, especially when you can't easily find exact values of the function. You can say, for example, that $\sin x$ always stays between -1 and 1 and that $2 \sin x + 10$ always stays between 8 and 12. But 2^x is not confined between any two numbers, because 2^x will exceed any number you can name if x is large enough. We say that $\sin x$ and $2 \sin x + 10$ are *bounded* functions, and that 2^x is an *unbounded* function.

A function f is **bounded** on an interval if there are numbers L and U such that

$$L \leq f(x) \leq U$$

for all x in the interval. Otherwise, f is **unbounded** on the interval.

We say that L is a **lower bound** for f on the interval, and that U is an **upper bound** for f on the interval.

Example 9 Use Figures A.7 and A.8 to decide which of the following functions are bounded.

(a) x^3 on $-\infty < x < \infty$; on $0 \leq x \leq 100$.
(b) $2/x$ on $0 < x < \infty$; on $1 \leq x < \infty$.

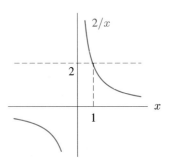

Figure A.7: Is x^3 bounded?   *Figure A.8:* Is $2/x$ bounded?

Solution (a) The graph of x^3 in Figure A.7 shows that x^3 will exceed any number, no matter how large, if x is big enough, so x^3 does not have an upper bound on $-\infty < x < \infty$. Therefore, x^3 is unbounded on $-\infty < x < \infty$. But on the interval $0 \leq x \leq 100$, x^3 stays between 0 (a lower bound) and $100^3 = 1{,}000{,}000$ (an upper bound). Therefore, x^3 is bounded on the interval $0 \leq x \leq 100$. Notice that upper and lower bounds, when they exist, are not unique. For example, -100 is another lower bound and $2{,}000{,}000$ another upper bound for x^3 on $0 \leq x \leq 100$.

(b) $2/x$ is unbounded on $0 < x < \infty$, since it has no upper bound on that interval. But $0 \leq 2/x \leq 2$ for $1 \leq x < \infty$, so $2/x$ is bounded, with lower bound 0 and upper bound 2, on $1 \leq x < \infty$. (See Figure A.8.)

Best Possible Bounds

Consider a group of people whose height in feet, h, ranges from 5 feet to 6 feet. Then 5 feet is a lower bound for the people in the group and 6 feet is an upper bound:

$$5 \leq h \leq 6.$$

But the people in this group are also all between 4 feet and 7 feet, so it is also true that

$$4 \leq h \leq 7.$$

So, there are many lower bounds and many upper bounds. However, the 5 and the 6 are considered the best bounds because they are the closest together of all the possible pairs of bounds.

The **best possible bounds** for a function, f, over an interval are numbers A and B such that, for all x in the interval,

$$A \leq f(x) \leq B$$

and where A and B are as close together as possible. A is called the **greatest lower bound** and B is the **least upper bound**.

What Do Bounds Mean Graphically?

Upper and lower bounds can be represented on a graph by horizontal lines. See Figure A.9.

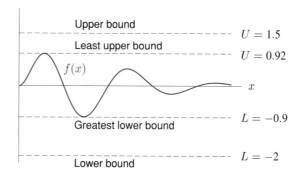

Figure A.9: Upper and lower bounds for the function f

Problems for Section A

1. Use a calculator or computer graph of $f(x) = 13 - 20x - x^2 - 3x^4$ to determine:
   (a) The range of this function;
   (b) The number of zeros of this function.

For Problems 2–12, determine the roots or points of intersection to an accuracy of one decimal place.

2. (a) The root of $x^3 - 3x + 1 = 0$ between 0 and 1
   (b) The root of $x^3 - 3x + 1 = 0$ between 1 and 2
   (c) The smallest root of $x^3 - 3x + 1 = 0$

3. The root of $x^4 - 5x^3 + 2x - 5 = 0$ between -2 and -1

4. The root of $x^5 + x^2 - 9x - 3 = 0$ between -2 and -1

5. The largest real root of $2x^3 - 4x^2 - 3x + 1 = 0$

6. All real roots of $x^4 - x - 2 = 0$

7. All real roots of $x^5 - 2x^2 + 4 = 0$

8. The smallest positive root of $x \sin x - \cos x = 0$

9. The left-most point of intersection between $y = 2x$ and $y = \cos x$

10. The left-most point of intersection between $y = 1/2^x$ and $y = \sin x$

11. The point of intersection between $y = e^{-x}$ and $y = \ln x$

12. All roots of $\cos t = t^2$

13. Estimate all real zeros of the following polynomials, accurate to 2 decimal places:
    (a) $f(x) = x^3 - 2x^2 - x + 3$
    (b) $f(x) = x^3 - x^2 - 2x + 2$

14. Find the largest zero of
$$f(x) = 10xe^{-x} - 1$$
    to two decimal places, using the bisection method. Make sure to demonstrate that your approximation is as good as you claim.

15. (a) Find the smallest positive value of x where the graphs of $f(x) = \sin x$ and $g(x) = 2^{-x}$ intersect.
    (b) Repeat with $f(x) = \sin 2x$ and $g(x) = 2^{-x}$.

16. Use a graphing calculator to sketch $y = 2 \cos x$ and $y = x^3 + x^2 + 1$ on the same set of axes. Find the positive zero of $f(x) = 2 \cos x - x^3 - x^2 - 1$. A friend claims there is one more real zero. Is your friend correct? Explain.

17. Use the table below to investigate the zeros of the function

$$f(\theta) = (\sin 3\theta)(\cos 4\theta) + 0.8$$

in the interval $0 \le \theta \le 1.8$.

θ	0	0.2	0.4	0.6	0.8	1.0	1.2	1.4	1.6	1.8
$f(\theta)$	0.80	1.19	0.77	0.08	0.13	0.71	0.76	0.12	−0.19	0.33

(a) Decide how many zeros the function has in the interval $0 \le \theta \le 1.8$.
(b) Locate each zero, or a small interval containing each zero.
(c) Are you sure you have found all the zeros in the interval $0 \le \theta \le 1.8$? Graph the function on a calculator or computer to decide.

18. (a) Use Table A.3 to locate approximate solution(s) to

$$(\sin 3x)(\cos 4x) = \frac{x^3}{\pi^3}$$

in the interval $1.07 \le x \le 1.15$. Give an interval of length 0.01 in which each solution lies.
(b) Make an estimate for each solution accurate to two decimal places.

TABLE A.3

x	x^3/π^3	$(\sin 3x)(\cos 4x)$
1.07	0.0395	0.0286
1.08	0.0406	0.0376
1.09	0.0418	0.0442
1.10	0.0429	0.0485
1.11	0.0441	0.0504
1.12	0.0453	0.0499
1.13	0.0465	0.0470
1.14	0.0478	0.0417
1.15	0.0491	0.0340

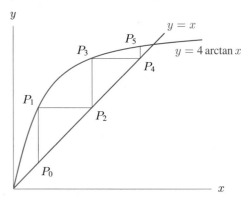

Figure A.10

19. (a) With your calculator in radian mode, take the arctangent of 1 and multiply that number by 4. Now, take the arctangent of the result and multiply it by 4. Continue this process 10 times or so and record each result as in the accompanying table. At each step, you get 4 times the arctangent of the result of the previous step.

1
3.14159...
5.05050...
5.50129...
$\vdots$

(b) Your table allows you to find a solution of the equation

$$4 \arctan x = x.$$

Why? What is that solution?

(c) What does your table in part (a) have to do with Figure A.10?
[Hint: The coordinates of P_0 are $(1, 1)$. Find the coordinates of $P_1, P_2, P_3, \ldots$]

(d) In part (a), what happens if you start with an initial guess of 10? Of −10? What types of behavior do you observe? (That is, for which initial guesses is the sequence increasing, and for which is it decreasing; does the sequence approach a limit?) Explain your answers graphically, as in part (c).

20.   Using radians, apply the iteration method of Problem 19 to the equation

$$\cos x = x.$$

Represent your results graphically, as in Figure A.10.

For Problems 21–23, draw a graph to decide if the function is bounded on the interval given. Give the best possible upper and lower bounds for any function which is bounded.

21. $f(x) = 4x - x^2$ on $[-1, 4]$ 22. $h(\theta) = 5 + 3\sin\theta$ on $[-2\pi, 2\pi]$

23. $f(t) = \dfrac{\sin t}{t^2}$ on $[-10, 10]$

B COMPOUND INTEREST

If you have some money, you may decide to invest it to earn interest. The interest can be paid in many different ways—for example, once a year or many times a year. If the interest is paid more frequently than once per year and the interest is not withdrawn, there is a benefit to the investor since the interest earns interest. This effect is called *compounding*. You may have noticed banks offering accounts that differ both in interest rates and in compounding methods. Some offer interest compounded annually, some quarterly, and others daily. Some even offer continuous compounding.

What is the difference between a bank account advertising 8% compounded annually (once per year) and one offering 8% compounded quarterly (four times per year)? In both cases 8% is an annual rate of interest. The expression 8% *compounded annually* means that at the end of each year, 8% of the current balance is added. This is equivalent to multiplying the current balance by 1.08. Thus, if $100 is deposited, the balance, B, in dollars, will be

$$B = 100(1.08) \quad \text{after one year,}$$
$$B = 100(1.08)^2 \quad \text{after two years,}$$
$$B = 100(1.08)^t \quad \text{after } t \text{ years.}$$

The expression 8% *compounded quarterly* means that interest is added four times per year (every three months) and that $\frac{8}{4} = 2\%$ of the current balance is added each time. Thus, if $100 is deposited, at the end of one year, four compoundings will have taken place and the account will contain $100(1.02)^4$. Thus, the balance will be

$$B = 100(1.02)^4 \quad \text{after one year,}$$
$$B = 100(1.02)^8 \quad \text{after two years,}$$
$$B = 100(1.02)^{4t} \quad \text{after } t \text{ years.}$$

Note that 8% is *not* the rate used for each three month period; the annual rate is divided into four 2% payments. Calculating the total balance after one year under each method shows that

$$\text{Annual compounding:} \quad B = 100(1.08) = 108.00,$$
$$\text{Quarterly compounding:} \quad B = 100(1.02)^4 = 108.24.$$

Thus, more money is earned from quarterly compounding, because the interest earns interest as the year goes by. In general, the more often interest is compounded, the more money will be earned (although the increase may not be very large).

We can measure the effect of compounding by introducing the notion of *effective annual yield*. Since $100 invested at 8% compounded quarterly grows to $108.24 by the end of one year, we say that the *effective annual yield* in this case is 8.24%. We now have two interest rates which describe the same investment: the 8% compounded quarterly and the 8.24% effective annual yield.

Banks call the 8% the *annual percentage rate*, or APR. We may also call the 8% the *nominal rate* (nominal means "in name only"). However, it is the effective yield which tells you exactly how much interest the investment really pays. Thus, to compare two bank accounts, simply compare the effective annual yields. The next time that you walk by a bank, look at the advertisements, which should (by law) include both the APR, or nominal rate, and the effective annual yield. We will often abbreviate *annual percentage rate* to *annual rate*.

Using the Effective Annual Yield

Example 1   Which is better: Bank X paying a 7% annual rate compounded monthly or Bank Y offering a 6.9% annual rate compounded daily?

Solution   We find the effective annual yield for each bank.

Bank X: There are 12 interest payments in a year, each payment being $0.07/12 = 0.005833$ times the current balance. If the initial deposit were \$100, then the balance B will be

$$B = 100(1.005833) \quad \text{after one month,}$$
$$B = 100(1.005833)^2 \quad \text{after two months,}$$
$$B = 100(1.005833)^t \quad \text{after } t \text{ months.}$$

To find the effective annual yield, we look at one year, or 12 months, which gives $B = 100(1.005833)^{12}$ $= 100(1.072286)$, so the effective annual yield $\approx 7.23\%$.

Bank Y: There are 365 interest payments in a year (assuming it is not a leap year), each being $0.069/365 = 0.000189$ times the current balance. Then the balance is

$$B = 100(1.000189) \quad \text{after one day,}$$
$$B = 100(1.000189)^2 \quad \text{after two days,}$$
$$B = 100(1.000189)^t \quad \text{after } t \text{ days.}$$

so at the end of one year we have multiplied the initial deposit by

$$(1.000189)^{365} = 1.071413$$

so the effective annual yield for Bank Y $\approx 7.14\%$.

Comparing effective annual yields for the banks, we see that Bank X is offering a better investment, by a small margin.

Example 2   If \$1000 is invested in each bank in Example 1, write an expression for the balance in each bank after t years.

Solution   For Bank X, the effective annual yield $\approx 7.23\%$, so after t years the balance, in dollars, will be
$$B = 1000(1.0723)^t.$$

For Bank Y, the effective annual yield $\approx 7.14\%$, so after t years the balance, in dollars, will be
$$B = 1000(1.0714)^t.$$

(Again, we are ignoring leap years.)

> If interest at an annual rate of r is compounded n times a year, then r/n times the current balance is added n times a year. Therefore, with an initial deposit of \$$P$, the balance t years later is
> $$B = P\left(1 + \frac{r}{n}\right)^{nt}.$$
> Note that r is the nominal rate; for example, $r = 0.05$ when the annual rate is 5%.

Increasing the Frequency of Compounding: Continuous Compounding

Example 3   Find the effective annual yield for a 7% annual rate compounded
(a)   1000 times a year.   (b)   10,000 times a year.

Solution   (a)

$$\left(1 + \frac{0.07}{1000}\right)^{1000} \approx 1.0725056$$

giving an effective annual yield of about 7.25056%.

(b)

$$\left(1 + \frac{0.07}{10,000}\right)^{10,000} \approx 1.0725079$$

giving an effective annual yield of about 7.25079%.

You can see that there's not a great deal of difference between compounding 1000 times each year (about three times per day) and 10,000 times each year (about 30 times per day). What happens if we compound more often still? Every minute? Every second? You may be surprised to know that the effective annual yield does not increase indefinitely, but tends to a finite value. The benefit of increasing the frequency of compounding becomes negligible beyond a certain point.

For example, if you were to compute the effective annual yield on a 7% investment compounded n times per year for values of n larger than 100,000, you would find that

$$\left(1 + \frac{0.07}{n}\right)^{n} \approx 1.0725082.$$

So the effective annual yield is about 7.25082%. Even if you take $n = 1,000,000$ or $n = 10^{10}$, the effective annual yield will not change appreciably. The value 7.25082% is an upper bound which is approached as the frequency of compounding increases.

When the effective annual yield is at this upper bound, we say that the interest is being *compounded continuously*. (The word *continuously* is used because the upper bound is approached by compounding more and more frequently.) Thus, when a 7% nominal annual rate is compounded so frequently that the effective annual yield is 7.25082%, we say that the 7% is compounded *continuously*. This represents the most one can get from a 7% nominal rate.

Where Does the Number e Fit In?

It turns out that e is intimately connected to continuous compounding. To see this, use your calculator to check that $e^{0.07} \approx 1.0725082$, which is the same number we obtained when we compounded 7% a large number of times. So you have discovered that for very large n

$$\left(1 + \frac{0.07}{n}\right)^{n} \approx e^{0.07}.$$

As n gets larger, the approximation gets better and better, and we write

$$\left(1 + \frac{0.07}{n}\right)^{n} \longrightarrow e^{0.07},$$

meaning that as n increases, the value of $\left(1 + 0.07/n\right)^{n}$ approaches $e^{0.07}$.

If P is deposited at an annual rate of 7% compounded continuously, the balance, B, will be given by

$$B = P(1.0725082) = Pe^{0.07} \qquad \text{after one year,}$$

$$B = P(1.0725082)^2 = P\left(e^{0.07}\right)^2 = Pe^{(0.07)2} \qquad \text{after two years,}$$

$$B = P(1.0725082)^t = P\left(e^{0.07}\right)^t = Pe^{0.07t} \qquad \text{after } t \text{ years.}$$

If interest on an initial deposit of P is *compounded continuously* at an annual rate r, the balance t years later can be calculated using the formula

$$B = Pe^{rt}.$$

Again, r is the nominal rate, and, for example, $r = 0.05$ when the annual rate is 5%.

In solving a problem involving compound interest, it is important to be clear whether interest rates are nominal rates or effective yields, as well as whether compounding is continuous or not.

Example 4   Find the effective annual yield of a 6% annual rate, compounded continuously.

Solution   In one year, an investment of P becomes $Pe^{0.06}$. Using a calculator, we see that

$$Pe^{0.06} = P(1.0618365).$$

So the effective annual yield is about 6.18%.

Example 5   Suppose you want to invest money in a certificate of deposit (CD) for your child's education. You need it to be worth $120,000 in 10 years. How much should you invest if the CD pays interest at a 9% annual rate compounded quarterly? Continuously?

Solution   Suppose you invest P initially. A 9% annual rate compounded quarterly has an effective annual yield given by $(1 + 0.09/4)^4 = 1.0930833$, or 9.30833%. So after 10 years you will have

$$P(1.0930833)^{10} = 120000.$$

Therefore, you should invest

$$P = \frac{120000}{(1.0930833)^{10}} = \frac{120000}{2.4351885} = 49,277.50.$$

On the other hand, if the CD pays 9% compounded continuously, after 10 years you will have

$$Pe^{(0.09)10} = 120000.$$

So you would need to invest

$$P = \frac{120000}{e^{(0.09)10}} = \frac{120000}{2.4596031} = 48,788.36.$$

Notice that to achieve the same result, continuous compounding requires a smaller initial investment than quarterly compounding. This is to be expected since the effective annual yield is higher for continuous than for quarterly compounding.

Problems for Section B

1. Use a graph of $y = \left(1 + 0.07/x\right)^x$ to estimate the value of $\left(1 + 0.07/x\right)^x$ as $x \to \infty$. Confirm that the value you get is $e^{0.07}$.

2. If you deposit $10,000 in an account earning interest at an 8% annual rate compounded continuously, how much money is in the account after five years?

3. (a) Find the effective annual yield for a 5% annual interest rate compounded
   (i) 1000 times/year, (ii) 10,000 times/year, (iii) 100,000 times/year.
   (b) Look at the sequence of answers in part (a), and predict the effective annual yield for a 5% annual rate compounded continuously.
   (c) Compute $e^{0.05}$. How does this confirm your answer to part (b)?

4. (a) Find $\left(1 + 0.04/n\right)^n$ for $n = 10,000$, and $100,000$, and $1,000,000$. Use the results to predict the effective annual yield of a 4% annual rate compounded continuously.
   (b) Confirm your answer by computing $e^{0.04}$.

5. Use the number e to find the effective annual yield of a 6% annual rate, compounded continuously.

6. A bank account is earning interest at 6% per year compounded continuously.

   (a) By what percentage has the bank balance in the account increased over one year? (This is the effective annual yield.)
   (b) How long does it take the balance to double?
   (c) Assuming now that the interest rate is r, find a formula giving the doubling time in terms of the interest rate.

7. Suppose $1000 is invested at 6% annual interest compounded continuously.

   (a) How long will it take for the investment to double?
   (b) Use your answer to part (a) to express the value of the investment after t years in terms of a base 2 exponential function.

8. What is the effective annual yield of an investment paying a 12% annual rate, compounded continuously?

9. (a) The Banque Nationale du Zaïre pays 100% nominal interest on deposits, compounded monthly. You invest 1 million zaïre. (The "zaïre" is the unit of currency of the Republic of Zaïre.) How much money do you have after one year?
   (b) How much money do you have after one year if you invest 1 million zaïre with interest compounded daily? Hourly? Each minute?
   (c) Does this amount increase without bound as interest is compounded more and more often, or does it level off? If it levels off, provide a close "upper" estimate for the total after one year.

10. Explain how you can match the interest rates (a)–(e) with the effective annual yields (I)–(V) without doing any calculations.
    (a) 5.5% annual rate, compounded continuously.   (I) 5%
    (b) 5.5% annual rate, compounded quarterly.   (II) 5.06%
    (c) 5.5% annual rate, compounded weekly.   (III) 5.61%
    (d) 5% annual rate, compounded yearly.   (IV) 5.651%
    (e) 5% annual rate, compounded twice a year.   (V) 5.654%

11. When you rent an apartment, you are often required to give the landlord a security deposit which is returned if you leave the apartment undamaged. In Massachusetts the landlord is required to pay the tenant interest on the deposit once a year, at a 5% annual rate, compounded annually. The landlord, however, may invest the money at a higher (or lower) interest rate. Suppose the landlord invests a $1000 deposit at an annual rate of
    (a) 6%, compounded continuously   (b) 4%, compounded continuously.

    In each case, determine the net gain or loss by the landlord at the end of the first year. (Give your answer to the nearest cent.)

C   POLAR COORDINATES

A point P with Cartesian coordinates (x, y) can be referred to by its *polar coordinates, r* and θ. The number r is the distance between P and the origin, and θ is the angle between the positive x-axis and the line joining P to the origin (with the convention that counterclockwise is positive). Figure C.11 shows the connection between Cartesian and polar coordinates:

Relation between Cartesian and Polar Coordinates

$$x = r\cos\theta \qquad r = \sqrt{x^2 + y^2}$$
$$y = r\sin\theta \qquad \tan\theta = \frac{y}{x}$$

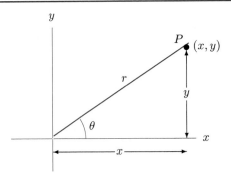

Figure C.11: Cartesian and polar coordinates

Example 1   Convert the point $(x, y) = (-3, 4)$ to polar coordinates.

Solution   The formula for r gives

$$r = \sqrt{x^2 + y^2} = \sqrt{(-3)^2 + 4^2} = 5.$$

The formula involving the angle θ is

$$\tan\theta = \frac{y}{x} = -\frac{4}{3}.$$

This means $\theta = \arctan(-4/3) \approx -0.927$ or $\theta = \arctan(-4/3) + \pi \approx 2.21$. Since the point is in the second quadrant, we choose $\theta \approx 2.21$.

Problems for Section C

For Problems 1–7, give Cartesian coordinates for the points with the following polar coordinates (r, θ). The angles are measured in radians.

1. $(1, 0)$ 2. $(0, 1)$ 3. $(2, \pi)$ 4. $(\sqrt{2}, 5\pi/4)$
5. $(5, -\pi/6)$ 6. $(3, \pi/2)$ 7. $(1, 1)$

For Problems 8–15, give polar coordinates for the points with the following Cartesian coordinates. Choose $0 \le \theta < 2\pi$.

8. $(1, 0)$ 9. $(0, 2)$ 10. $(1, 1)$ 11. $(-1, 1)$
12. $(-3, -3)$ 13. $(0.2, -0.2)$ 14. $(3, 4)$ 15. $(-3, 1)$

In Problems 16–21, graph the functions $r = g(\theta)$ in the xy-plane using a calculator. Explain how the equations are related to the shapes of your graphs.

16. $r = 1$

17. $\theta = \dfrac{\pi}{3}$

18. $r = \dfrac{\theta}{10}$

19. $r = \dfrac{2}{\cos \theta}$

20. $r = 2 \cos \theta$

21. $r = \dfrac{\sin \theta}{\cos^2 \theta}$

22. Every point in the plane can be represented by some pair of polar coordinates, but are the polar coordinates (r, θ) uniquely determined by the Cartesian coordinates (x, y)? In other words, for each pair of Cartesian coordinates, is there one and only one pair of polar coordinates for that point? Why or why not?

D COMPLEX NUMBERS

The quadratic equation

$$x^2 - 2x + 2 = 0$$

is not satisfied by any real number x. If you try applying the quadratic formula, you get

$$x = \frac{2 \pm \sqrt{4 - 8}}{2} = 1 \pm \frac{\sqrt{-4}}{2}.$$

Apparently, you need to take a square root of -4. But -4 doesn't have a square root, at least, not one which is a real number. Let's give it a square root.

We define the imaginary number i to be a number such that

$$i^2 = -1.$$

Using this i, we see that $(2i)^2 = -4$, so

$$x = 1 \pm \frac{\sqrt{-4}}{2} = 1 \pm \frac{2i}{2} = 1 \pm i.$$

This solves our quadratic equation. The numbers $1 + i$ and $1 - i$ are examples of complex numbers.

A **complex number** is defined as any number that can be written in the form

$$z = a + bi,$$

where a and b are real numbers and $i = \sqrt{-1}$.
The *real part* of z is the number a; the *imaginary part* is the number b.

Calling the number i imaginary makes it sound as if i doesn't exist in the same way that real numbers exist. In some cases, it is useful to make such a distinction between real and imaginary numbers. For example, if we measure mass or position, we want our answers to be real numbers. But the imaginary numbers are just as legitimate mathematically as the real numbers are.

As an analogy, consider the distinction between positive and negative numbers. Originally, people thought of numbers only as tools to count with; their concept of "five" or "ten" was not far removed from "five arrows" or "ten stones." They were unaware that negative numbers existed at all. When negative numbers were introduced, they were viewed only as a device for solving equations like $x + 2 = 1$. They were considered "non-numbers," or, in Latin, "negative numbers." Thus, even though people started to use negative numbers, they did not view them as existing in the same way that positive numbers did. An early mathematician might have reasoned: "The number 5 exists because I can have 5 coins in my hand. But how can I have -5 coins in my hand?" Today we have an answer: "I have -5 coins" means I owe somebody 5 coins. We have realized that negative numbers are just as real as positive ones, and that in some cases, negative numbers can have physical meaning, even though they are useless for measuring length or keeping baseball scores. As we will see, complex numbers can have physical meaning as well. For example, complex numbers are used in studying wave motion in electric circuits.

Algebra of Complex Numbers

Numbers such as 0, 1, $\frac{1}{2}$, π, and $\sqrt{2}$ are called *purely real* because they contain no imaginary components. Numbers such as i, $2i$, and $\sqrt{2}i$ are called *purely imaginary* because they contain only the number i multiplied by a nonzero real coefficient.

Two complex numbers are called *conjugates* if their real parts are equal and if their imaginary parts are opposites. The complex conjugate of the complex number $z = a + bi$ is denoted $\bar{z}$ (pronounced "z bar"), so we have

$$\bar{z} = a - bi.$$

(Note that z is real if and only if $z = \bar{z}$.) Complex conjugates have the following remarkable property: if $f(x)$ is any polynomial with real coefficients ($x^3 + 1$, say) and $f(z) = 0$, then $f(\bar{z}) = 0$. This means that if z is the solution to a polynomial equation with real coefficients, then so is $\bar{z}$.

- Adding two complex numbers is done by adding real and imaginary parts separately:

$$(a + bi) + (c + di) = (a + c) + (b + d)i.$$

- Subtracting is similar:

$$(a + bi) - (c + di) = (a - c) + (b - d)i.$$

- Multiplication works just like for polynomials, using $i^2 = -1$:

$$\begin{aligned}(a + bi)(c + di) &= a(c + di) + bi(c + di) \\ &= ac + adi + bci + bdi^2 \\ &= ac + adi + bci - bd = (ac - bd) + (ad + bc)i.\end{aligned}$$

- Powers of i: We know that $i^2 = -1$; then $i^3 = i \cdot i^2 = -i$, and $i^4 = (i^2)^2 = (-1)^2 = 1$. Then $i^5 = i \cdot i^4 = i$, and so on. Thus we have

$$(bi)^n = b^n i^n = \begin{cases} b^n i & \text{for } n = 1, 5, 9, 13, \ldots \\ -b^n & \text{for } n = 2, 6, 10, 14, \ldots \\ -b^n i & \text{for } n = 3, 7, 11, 15, \ldots \\ b^n & \text{for } n = 4, 8, 12, 16, \ldots \end{cases}$$

- The product of a number and its conjugate is always real and nonnegative:

$$z \cdot \bar{z} = (a + bi)(a - bi) = a^2 - abi + abi - b^2 i^2 = a^2 + b^2.$$

- Dividing is done by multiplying the denominator by its conjugate, thereby making the denominator real:

$$\frac{a + bi}{c + di} = \frac{a + bi}{c + di} \cdot \frac{c - di}{c - di} = \frac{ac - adi + bci - bdi^2}{c^2 + d^2} = \frac{ac + bd}{c^2 + d^2} + \frac{bc - ad}{c^2 + d^2}i.$$

Example 1   Compute $(2 + 7i)(4 - 6i) - i$.

Solution $(2 + 7i)(4 - 6i) - i = 8 + 28i - 12i - 42i^2 - i = 8 + 15i + 42 = 50 + 15i.$

Example 2   Compute $\dfrac{2 + 7i}{4 - 6i}$.

Solution

$$\frac{2 + 7i}{4 - 6i} = \frac{2 + 7i}{4 - 6i} \cdot \frac{4 + 6i}{4 + 6i} = \frac{8 + 12i + 28i + 42i^2}{4^2 + 6^2} = \frac{-34 + 40i}{52} = \frac{-17}{26} + \frac{10}{13}i.$$

The Complex Plane and Polar Coordinates

It is often useful to picture a complex number $z = x + iy$ in the plane, with x along the horizontal axis and y along the vertical. The xy-plane is then called the *complex plane*. Figure D.12 shows the complex numbers $-2i$, $1 + i$, and $-2 + 3i$.

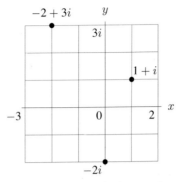

Figure D.12: Points in the complex plane

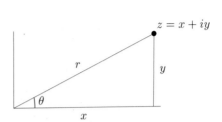

Figure D.13: The point $z = x + iy$ in the complex plane, showing polar coordinates

The triangle in Figure D.13 shows that a complex number can be written using polar coordinates as follows:

$$z = x + iy = r \cos \theta + ir \sin \theta.$$

Example 3   Express $z = -2i$ and $z = -2 + 3i$ using polar coordinates. (See Figure D.12.)

Solution   For $z = -2i$, the distance of z from the origin is 2, so $r = 2$. Also, one value for θ is $\theta = 3\pi/2$. Using polar coordinates, $-2i = 2 \cos(3\pi/2) + i\, 2(\sin 3\pi/2)$.
For $z = -2 + 3i$, we have $x = -2$, $y = 3$. So $r = \sqrt{(-2)^2 + 3^2} \approx 3.61$, and one solution of $\tan \theta = 3/(-2)$ is $\theta \approx 2.16$. So $-2 + 3i \approx 3.61 \cos(2.16) + i\, 3.61 \sin(2.16)$.

Example 4   Consider the point with polar coordinates $r = 5$ and $\theta = 3\pi/4$. What complex number does this point represent?

Solution   Since $x = r \cos \theta$ and $y = r \sin \theta$ we see that $x = 5 \cos 3\pi/4 = -5/\sqrt{2}$, and $y = 5 \sin 3\pi/4 = 5/\sqrt{2}$, so $z = -5/\sqrt{2} + i\, 5/\sqrt{2}$.

Euler's Formula

Consider the complex number z lying on the unit circle in Figure D.14. Writing z in polar coordinates, and using the fact that $r = 1$, we have

$$z = f(\theta) = \cos \theta + i \sin \theta.$$

It turns out that there is a particularly beautiful and compact way of rewriting $f(\theta)$ using complex exponentials. Let's take the derivative of f, treating i like any other constant but using the fact that $i^2 = -1$:

$$f'(\theta) = -\sin \theta + i \cos \theta = i \cos \theta + i^2 \sin \theta.$$

Factoring out an i gives

$$f'(\theta) = i(\cos \theta + i \sin \theta) = i \cdot f(\theta).$$

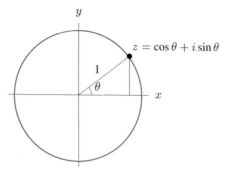

Figure D.14: Complex number represented by
a point on the unit circle

As you know from Chapter 10, page 510, the only real-valued function whose derivative is proportional to the function itself is the exponential function. In other words, we know that if

$$g'(x) = k \cdot g(x), \quad \text{then} \quad g(x) = Ce^{kx}$$

for some constant C. If we assume that a similar result holds for complex-valued functions, then we have

$$f'(\theta) = i \cdot f(\theta), \quad \text{so} \quad f(\theta) = Ce^{i\theta}$$

for some constant C. To find C we substitute $\theta = 0$. Now $f(0) = Ce^{i \cdot 0} = C$, and since $f(0) = \cos 0 + i \sin 0 = 1$, we must have $C = 1$. Therefore $f(\theta) = e^{i\theta}$. Thus we have *Euler's formula:*

$$e^{i\theta} = \cos \theta + i \sin \theta$$

This elegant and surprising relationship was discovered by the Swiss mathematician Leonhard Euler in the eighteenth century, and it is particularly useful in solving second-order differential equations. Another derivation of Euler's formula (using Taylor series) is given in Problem 16 on page 487. It allows us to write the complex number represented by the point with polar coordinates (r, θ) in the following form:

$$z = r(\cos \theta + i \sin \theta) = re^{i\theta}.$$

Similarly, since $\cos(-\theta) = \cos \theta$ and $\sin(-\theta) = -\sin \theta$, we have

$$re^{-i\theta} = r\left(\cos(-\theta) + i \sin(-\theta)\right) = r(\cos \theta - i \sin \theta).$$

Example 5   Evaluate $e^{i\pi}$.

Solution   Using Euler's formula, $e^{i\pi} = \cos \pi + i \sin \pi = -1$.

Example 6   Express the complex number represented by the point $r = 8$, $\theta = 3\pi/4$ in Cartesian form and polar form, $z = re^{i\theta}$.

Solution   Using Cartesian coordinates, the complex number is

$$z = 8\left(\cos\left(\frac{3\pi}{4}\right) + i \sin\left(\frac{3\pi}{4}\right)\right) = \frac{-8}{\sqrt{2}} + i\frac{8}{\sqrt{2}}.$$

Using polar coordinates, we have

$$z = 8e^{i\,3\pi/4}.$$

Among its many benefits, the polar form of complex numbers makes finding powers and roots of complex numbers much easier. Using the polar form, $z = re^{i\theta}$, for a complex number, we may find any power of z as follows:

$$z^p = (re^{i\theta})^p = r^p e^{ip\theta}.$$

To find roots, we let p be a fraction, as in the following example.

Example 7   Find a cube root of the complex number represented by the point with polar coordinates $(8, 3\pi/4)$.

Solution   In Example 6, we saw that this complex number could be written as $z = 8e^{i3\pi/4}$. So,

$$\sqrt[3]{z} = \left(8e^{i3\pi/4}\right)^{1/3} = 8^{1/3}e^{i(3\pi/4)\cdot(1/3)} = 2e^{\pi i/4} = 2\left(\cos(\pi/4) + i\sin(\pi/4)\right)$$

$$= 2\left(1/\sqrt{2} + i/\sqrt{2}\right) = \sqrt{2}(1 + i).$$

Problems for Section D

For Problems 1–8, express the given complex number in polar form, $z = re^{i\theta}$.

1. $2i$ 2. -5 3. $1 + i$ 4. $-3 - 4i$

5. 0 6. $-i$ 7. $-1 + 3i$ 8. $5 - 12i$

For Problems 9–18, perform the indicated calculations. Give your answer in Cartesian form, $z = x + iy$.

9. $(2 + 3i) + (-5 - 7i)$ 10. $(2 + 3i)(5 + 7i)$

11. $(2 + 3i)^2$ 12. $(1 + i)^2 + (1 + i)$

13. $(0.5 - i)(1 - i/4)$ 14. $(2i)^3 - (2i)^2 + 2i - 1$

15. $(e^{i\pi/3})^2$ 16. $\sqrt{e^{i\pi/3}}$

17. $(5e^{i7\pi/6})^3$ 18. $\sqrt[4]{10e^{i\pi/2}}$

By writing the complex numbers in polar form, $z = re^{i\theta}$, find a value for the quantities in Problems 19–28. Give your answer in Cartesian form, $z = x + iy$.

19. $\sqrt{i}$ 20. $\sqrt{-i}$ 21. $\sqrt[3]{i}$ 22. $\sqrt{7i}$

23. $(1 + i)^{100}$ 24. $(1 + i)^{2/3}$ 25. $(-4 + 4i)^{2/3}$ 26. $(\sqrt{3} + i)^{1/2}$

27. $(\sqrt{3} + i)^{-1/2}$ 28. $(\sqrt{5} + 2i)^{\sqrt{2}}$

Solve the simultaneous equations in Problems 29–30 for A_1 and A_2.

29. $A_1 + A_2 = 2$ 30. $A_1 + A_2 = 2$
     $(1 - i)A_1 + (1 + i)A_2 = 3$ $(i - 1)A_1 + (1 + i)A_2 = 0$

31. Let $z_1 = -3 - i\sqrt{3}$ and $z_2 = -1 + i\sqrt{3}$.

   (a) Find $z_1 z_2$ and z_1/z_2. Give your answer in Cartesian form, $z = x + iy$.
   (b) Put z_1 and z_2 into polar form, $z = re^{i\theta}$. Find $z_1 z_2$ and z_1/z_2 using the polar form, and verify that you get the same answer as in part (a).

32. If the roots of the equation $x^2 + 2bx + c = 0$ are the complex numbers $p \pm iq$, find expressions for p and q in terms of b and c.

Are the statements in Problems 33–38 true or false? Explain your answer.

33. Every nonnegative real number has a real square root.

34. For any complex number z, the product $z \cdot \bar{z}$ is a real number.

35. The square of any complex number is a real number.

36. If f is a polynomial, and $f(z) = i$, then $f(\bar{z}) = i$.

37. Every nonzero complex number z can be written in the form $z = e^w$, where w is another complex number.

38. If $z = x + iy$, where x and y are positive, then $z^2 = a + ib$ has a and b positive.

For Problems 39–43, use Euler's formula to derive the following relationships. (Note that if a, b, c, d are real numbers, $a + bi = c + di$ means that $a = c$ and $b = d$.)

39. $\sin^2 \theta + \cos^2 \theta = 1$

40. $\sin 2\theta = 2 \sin \theta \cos \theta$

41. $\cos 2\theta = \cos^2 \theta - \sin^2 \theta$

42. $\dfrac{d}{d\theta} \sin \theta = \cos \theta$

43. $\dfrac{d^2}{d\theta^2} \cos \theta = -\cos \theta$

E NEWTON'S METHOD

Many problems in mathematics involve finding the root of an equation. For example, we might have to locate the zeros of a polynomial, or determine the point of intersection of two curves. Here we will see a numerical method for approximating solutions which cannot be calculated exactly.

One such method, bisection, is described in Appendix A. Although it is very simple, the bisection method has two major drawbacks. First, it cannot locate a root where the curve is tangent to, but does not cross, the x-axis. Second, it is relatively slow in the sense that it requires a considerable number of iterations to achieve a desired level of accuracy. Although speed may not be important in solving a single equation, a practical problem may involve solving thousands of equations as a parameter changes. In such a case, any reduction in the number of steps can be important.

Using Newton's Method

We now consider a powerful root-finding method developed by Newton. Suppose we have a function $y = f(x)$. The equation $f(x) = 0$ has a root at $x = r$, as shown in Figure E.15. We begin with an initial estimate, x_0, for this root. (This can be a guess.) We will now obtain a better estimate x_1. To do this, construct the tangent line to the graph of f at the point $x = x_0$, and extend it until it crosses the x-axis, as shown in Figure E.15. The point where it crosses the axis is usually much closer to r, and we use that point as the next estimate, x_1. Having found x_1, we now repeat the process starting with x_1 instead of x_0. We construct a tangent line to the curve at $x = x_1$, extend it until it crosses the x-axis, use that x-intercept as the next approximation, x_2, and so on. The resulting sequence of x-intercepts usually converges rapidly to the root r.

Let's see how this looks algebraically. We know that the slope of the tangent line at the initial estimate x_0 is $f'(x_0)$, and so the equation of the tangent line is

$$y - f(x_0) = f'(x_0)(x - x_0).$$

At the point where this tangent line crosses the x-axis, we have $y = 0$ and $x = x_1$, so that

$$0 - f(x_0) = f'(x_0)(x_1 - x_0).$$

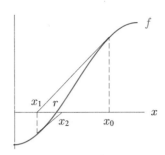

Figure E.15: Newton's method: successive approximations $x_0, x_1, x_2, \ldots$ to the root, r

Solving for x_1, we obtain

$$x_1 = x_0 - \frac{f(x_0)}{f'(x_0)}$$

provided that $f'(x_0)$ is not zero. We now repeat this argument and find that the next approximation is

$$x_2 = x_1 - \frac{f(x_1)}{f'(x_1)}.$$

Summarizing, for any $n = 0, 1, 2, \ldots$, we obtain the following result.

Newton's Method to Solve the Equation $f(x) = 0$

Choose x_0 near a solution and compute the sequence $x_1, x_2, x_3 \ldots$ using the rule

$$x_{n+1} = x_n - \frac{f(x_n)}{f'(x_n)}$$

provided that $f'(x_n)$ is not zero. For large n, the solution is well-approximated by x_n.

Example 1   Use Newton's method to find the fifth root of 23. (By calculator, this is 1.872171231, correct to nine decimal places.)

Solution   To use Newton's method, we need an equation of the form $f(x) = 0$ having $23^{1/5}$ as a root. Since $23^{1/5}$ is a root of $x^5 = 23$ or $x^5 - 23 = 0$, we take $f(x) = x^5 - 23$. The root of this equation is between 1 and 2 (since $1^5 = 1$ and $2^5 = 32$), so we will choose $x_0 = 2$ as our initial estimate. Now $f'(x) = 5x^4$, so we can set up Newton's method as

$$x_{n+1} = x_n - \frac{x_n^5 - 23}{5x_n^4}.$$

In this case, we can simplify using a common denominator, to obtain

$$x_{n+1} = \frac{4x_n^5 + 23}{5x_n^4}.$$

Therefore, starting with $x_0 = 2$, we find that $x_1 = 1.8875$. This leads to $x_2 = 1.872418193$ and $x_3 = 1.872171296$. These values are in Table E.4. Since $f(1.872171231) > 0$ and $f(1.872171230) < 0$, the root lies between 1.872171230 and 1.872171231. Therefore, in just four iterations of Newton's method, we have achieved eight-decimal accuracy.

TABLE E.4 *Newton's method:* $x_0 = 2$

n	x_n	$f(x_n)$
0	2	9
1	1.8875	0.957130661
2	1.872418193	0.015173919
3	1.872171296	0.000004020
4	1.872171231	0.000000027

TABLE E.5 *Newton's method:* $x_0 = 10$

n	x_n	n	x_n
0	10	6	2.679422313
1	8.000460000	7	2.232784753
2	6.401419079	8	1.971312452
3	5.123931891	9	1.881654220
4	4.105818871	10	1.872266333
5	3.300841811	11	1.872171240

As a general guideline for Newton's method, once the first correct decimal place is found, each successive iteration approximately doubles the number of correct digits.

What happens if we select a very poor initial estimate? In the preceding example, suppose x_0 were 10 instead of 2. The results are in Table E.5. Notice that even with $x_0 = 10$, the sequence of values moves reasonably quickly toward the solution: we achieve six-decimal place accuracy by the eleventh iteration.

Example 2 Find the first point of intersection of the curves given by $f(x) = \sin x$ and $g(x) = e^{-x}$.

Solution The graphs in Figure E.16 make it clear that there are an infinite number of points of intersection, all with $x > 0$. In order to find the first one numerically, we consider the function

$$F(x) = f(x) - g(x) = \sin x - e^{-x}$$

whose derivative is $F'(x) = \cos x + e^{-x}$. From the graph, we see that the point we want is fairly close to $x = 0$, so we start with $x_0 = 0$. The values in Table E.6 are approximations to the root. Since $F(0.588532744) > 0$ and $F(0.588532743) < 0$, the root lies between 0.588532743 and 0.588532744. (Remember, your calculator must be set in radians.)

TABLE E.6 *Successive approximations to root of* $\sin x = e^{-x}$

n	x_n
0	0
1	0.5
2	0.585643817
3	0.588529413
4	0.588532744
5	0.588532744

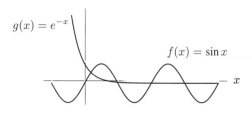

Figure E.16: Root of $\sin x = e^{-x}$

When Does Newton's Method Fail?

In most practical situations, Newton's method works well. Occasionally, however, the sequence x_0, x_1, x_2, ... fails to converge or fails to converge to the root you want. Sometimes, for example, the sequence can jump from one root to another. This is particularly likely to happen if the magnitude of the derivative $f'(x_n)$ is small for some x_n. In this case, the tangent line is nearly horizontal and so x_{n+1} will be far from x_n. (See Figure E.17.)

If the equation $f(x) = 0$ has *no* root, then the sequence will not converge. In fact, the sequence obtained by applying Newton's method to $f(x) = 1 + x^2$ is one of the best known examples of *chaotic behavior* and has attracted considerable research interest recently. (See Figure E.18.)

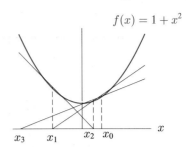

$f(x) = 1 + x^2$

Figure E.17: Problems with Newton's
method: Converges to wrong root

Figure E.18: Problems with Newton's
method: Chaotic behavior

Problems for Section E

1. Suppose you want to find a solution of the equation

   $$x^3 + 3x^2 + 3x - 6 = 0.$$

   Consider $f(x) = x^3 + 3x^2 + 3x - 6$.

   (a) Find $f'(x)$, and use it to show that $f(x)$ increases everywhere.
   (b) How many roots does the original equation have?
   (c) For each root, find an interval which contains it.
   (d) Find each root to two decimal places, using Newton's method.

   For Problems 2–4, use Newton's method to find the given quantities to two decimal places:

2. $\sqrt[3]{50}$ 3. $\sqrt[4]{100}$ 4. $10^{-1/3}$

   For Problems 5–8, solve each equation and give each answer to two decimal places:

5. $\sin x = 1 - x$ 6. $\cos x = x$

7. $e^{-x} = \ln x$ 8. $e^x \cos x = 1$, for $0 < x < \pi$

9. Find, to two decimal places, all solutions of $\ln x = 1/x$.

10. How many zeros do the following functions have? For each zero, find an upper and a lower bound which differ by no more than 0.1.
    (a) $f(x) = x^3 + x - 1$ (b) $f(x) = \sin x - \frac{2}{3}x$ (c) $f(x) = 10xe^{-x} - 1$

11. Find the largest zero of

    $$f(x) = x^3 + x - 1$$

    to six decimal places, using Newton's method. How do you know your approximation is as good as you claim?

12. For any positive number, a, the problem of calculating the square root, $\sqrt{a}$, is often done by applying Newton's method to the function $f(x) = x^2 - a$. Apply the method to obtain an expression for x_{n+1} in terms of x_n. Use this to approximate $\sqrt{a}$ for $a = 2, 10, 1000$, and π, correct to four decimal places, starting at $x_0 = a/2$ in each case.

F PARAMETRIC EQUATIONS

How Do We Represent Motion?

To represent the motion of a particle in the xy-plane we use two equations, one for the x-coordinate of the particle, $x = f(t)$, and another for the y-coordinate, $y = g(t)$. Thus at time t the particle is at the point $(f(t), g(t))$. The equation for x describes the right-left motion; the equation for y describes the up-down motion. The two equations for x and y are called *parametric equations* with *parameter t*.

Example 1   Describe the motion of the particle whose coordinates at time t are $x = \cos t$ and $y = \sin t$.

Solution   Since $(\cos t)^2 + (\sin t)^2 = 1$, we have $x^2 + y^2 = 1$. That is, at any time t the particle is at a point (x, y) somewhere on the unit circle $x^2 + y^2 = 1$. We plot points at different times to see how the particle moves on the circle. (See Figure F.19 and Table F.7.) The particle moves at a uniform speed, completing one full trip counterclockwise around the circle every 2π units of time. Notice how the x-coordinate goes repeatedly back and forth from -1 to 1 while the y-coordinate goes repeatedly up and down from -1 to 1. The two motions combine to trace out a circle.

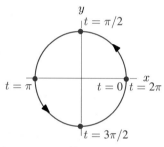

Figure F.19: The circle parameterized by $x = \cos t, y = \sin t$

TABLE F.7   *Points on the circle with* $x = \cos t, y = \sin t$

t	x	y
0	1	0
$\pi/2$	0	1
π	-1	0
$3\pi/2$	0	-1
2π	1	0

Example 2   Figure F.20 shows the graphs of two functions, $f(t)$ and $g(t)$. Describe the motion of the particle whose coordinates at time t are $x = f(t)$ and $y = g(t)$.

Figure F.20: Graphs of $x = f(t)$ and $y = g(t)$ used to trace out the path $(f(t), g(t))$ in Figure F.21

Solution   Between times $t = 0$ and $t = 1$, the x-coordinate goes from 0 to 1, while the y-coordinate stays fixed at 0. So the particle moves along the x-axis from $(0, 0)$ to $(1, 0)$. Then, between times $t = 1$ and $t = 2$, the x-coordinate stays fixed at $x = 1$, while the y-coordinate goes from 0 to 1. Thus, the particle moves along the vertical line from $(1, 0)$ to $(1, 1)$. Similarly, between times $t = 2$ and $t = 3$, it moves horizontally back to $(0, 1)$, and between times $t = 3$ and $t = 4$ it moves down the y-axis to $(0, 0)$. Thus, it traces out the square in Figure F.21.

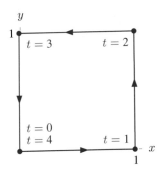

Figure F.21: The square parameterized by $(f(t), g(t))$

Different Motions Along the Same Path

Example 3   Describe the motion of the particle whose x and y coordinates at time t are given by the equations

$$x = \cos(3t), \quad y = \sin(3t).$$

Solution   Since $(\cos(3t))^2 + (\sin(3t))^2 = 1$, we have $x^2 + y^2 = 1$, giving motion around the unit circle. But if we plot points at different times, we see that in this case the particle is moving three times as fast as in Example 1 on page 604. (See Figure F.22 and Table F.8.)

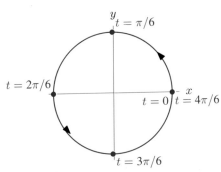

Figure F.22: The circle parameterized by $x = \cos(3t), y = \sin(3t)$

TABLE F.8   *Points on circle with* $x = \cos(3t)$, $y = \sin(3t)$

t	x	y
0	1	0
$\pi/6$	0	1
$2\pi/6$	-1	0
$3\pi/6$	0	-1
$4\pi/6$	1	0

Example 3 is obtained from Example 1 by replacing t by $3t$; this is called a *change in parameter*. If we make a change in parameter, the particle traces out the same curve (or a part of it) but at a different speed or in a different direction.

Example 4   Describe the motion of the particle whose x and y coordinates at time t are

$$x = \cos(e^{-t^2}), \quad y = \sin(e^{-t^2}).$$

Solution   As in Examples 1 and 3, we have $x^2 + y^2 = 1$ so the motion lies on the unit circle. As time t goes from $-\infty$ (way back in the past) to 0 (the present) to ∞ (way off in the future), e^{-t^2} goes from near 0 to 1 back to near 0. So $(x, y) = (\cos(e^{-t^2}), \sin(e^{-t^2}))$ goes from near $(1, 0)$ to $(\cos 1, \sin 1)$

and back to near $(1, 0)$. The particle does not actually reach the point $(1, 0)$. (See Figure F.23 and Table F.9.)

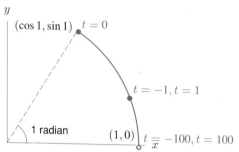

Figure F.23: The circle parameterized by $x = \cos\left(e^{-t^2}\right)$, $y = \sin\left(e^{-t^2}\right)$

TABLE F.9   *Points on circle with* $x = \cos(e^{-t^2})$, $y = \sin(e^{-t^2})$

t	x	y
-100	~ 1	~ 0
-1	0.93	0.36
0	0.54	0.84
1	0.93	0.36
100	~ 1	~ 0

Parametric Representations of Curves in the Plane

Sometimes we are more interested in the curve traced out by the particle than we are in the motion itself. In that case we will call the parametric equations a *parameterization* of the curve. As we can see by comparing Examples 1 and 3, two different parameterizations can describe the same curve in 2-space. Though the parameter, which we usually denote by t, may not have physical meaning, it is still helpful to think of it as time.

Example 5   Give a parameterization of the semicircle of radius 1 shown in Figure F.24.

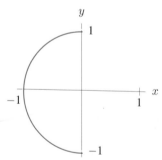

Figure F.24: Find a parameterization of this semicircle

Figure F.25: Find a parameterization of the ellipse $4x^2 + y^2 = 1$

Solution   We can use the equations $x = \cos t$ and $y = \sin t$ for counterclockwise motion in a circle, from Example 1 on page 604. The particle passes $(0, 1)$ at $t = \pi/2$, moves counterclockwise around the circle, and reaches $(0, -1)$ at $t = 3\pi/2$. So a parameterization is

$$x = \cos t, \; y = \sin t, \quad \frac{\pi}{2} \leq t \leq \frac{3\pi}{2}.$$

Example 6   Give a parameterization of the ellipse $4x^2 + y^2 = 1$ shown in Figure F.25.

Solution   Since $(2x)^2 + y^2 = 1$, we adapt the parameterization of the circle in Example 1. Replacing x by $2x$ gives the equations $2x = \cos t, y = \sin t$. A parameterization of the ellipse is thus

$$x = \tfrac{1}{2}\cos t, \qquad y = \sin t, \qquad 0 \leq t \leq 2\pi.$$

We usually require that the parameterization of a curve go from one end of the curve to the other without retracing any portion of the curve. This is different from parameterizing the motion of a particle, where, for example, a particle may move around the same circle many times.

Parameterizing the Graph of a Function

The graph of any function $y = f(x)$ can be parameterized by letting the parameter t be x:

$$x = t, \quad y = f(t).$$

Example 7   Give parametric equations for the curve $y = x^3 - x$. In which direction does this parameterization trace out the curve?

Solution   Let $x = t$, $y = t^3 - t$. Thus, $y = t^3 - t = x^3 - x$. Since $x = t$, as time increases the x-coordinate moves from left to right, so the particle traces out the curve $y = x^3 - x$ from left to right.

Curves Given Parametrically

Some complicated curves can be graphed more easily using parametric equations; the next example shows such a curve.

Example 8   Assume t is time in seconds. Sketch the curve traced out by the particle whose motion is given by

$$x = \cos(3t), \quad y = \sin(5t).$$

Solution   The x-coordinate oscillates back and forth between 1 and -1, completing 3 oscillations every 2π seconds. The y-coordinate oscillates up and down between 1 and -1, completing 5 oscillations every 2π seconds. Since both the x- and y-coordinates return to their original values every 2π seconds, the curve is retraced every 2π seconds. The result is a pattern called a Lissajous figure. (See Figure F.26.) Problems 23–26 concern Lissajous figures $x = \cos(at)$, $y = \sin(bt)$ for other values of a and b.

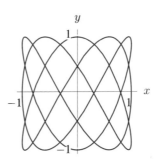

Figure F.26: A Lissajous figure: $x = \cos(3t), y = \sin(5t)$

Problems for Section F

For Problems 1–4, describe the motion of a particle whose position at time t is $x = f(t)$, $y = g(t)$, where the graphs of f and g are as shown.

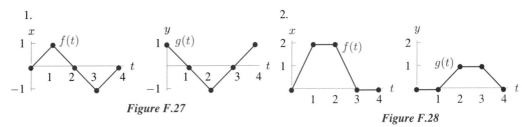

1.

Figure F.27

2.

Figure F.28

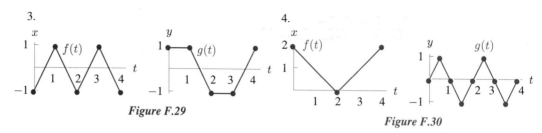

Figure F.29

Figure F.30

Problems 5–10 give parameterizations of the unit circle or a part of it. In each case, describe in words how the circle is traced out, including when and where the particle is moving clockwise and when and where the particle is moving counterclockwise.

5. $x = \cos t, \quad y = -\sin t$

6. $x = \sin t, \quad y = \cos t$

7. $x = \cos(t^2), \quad y = \sin(t^2)$

8. $x = \cos(t^3 - t), \quad y = \sin(t^3 - t)$

9. $x = \cos(\ln t), \quad y = \sin(\ln t)$

10. $x = \cos(\cos t), \quad y = \sin(\cos t)$

11. Describe the similarities and differences among the motions in the plane given by the following three pairs of parametric equations:
   (a) $x = t, \quad y = t^2$ (b) $x = t^2, \quad y = t^4$ (c) $x = t^3, \quad y = t^6$.

Write a parameterization for each of the curves in the xy-plane in Problems 12–18.

12. A circle of radius 3 centered at the origin and traced out clockwise.

13. A vertical line through the point $(-2, -3)$.

14. A circle of radius 5 centered at the point $(2, 1)$ and traced out counterclockwise.

15. A circle of radius 2 centered at the origin traced clockwise starting from $(-2, 0)$ when $t = 0$.

16. The line through the points $(2, -1)$ and $(1, 3)$.

17. An ellipse centered at the origin and crossing the x-axis at ± 5 and the y-axis at ± 7.

18. An ellipse centered at the origin, crossing the x-axis at ± 3 and the y-axis at ± 7. Start at the point $(-3, 0)$ and trace out the ellipse counterclockwise.

19. As t varies, the following parametric equations trace out a line in the plane

$$x = 2 + 3t, \quad y = 4 + 7t.$$

   (a) What part of the line is obtained by restricting t to nonnegative numbers?
   (b) What part of the line is obtained if t is restricted to $-1 \leq t \leq 0$?
   (c) How should t be restricted to give the part of the line to the left of the y-axis?

20. Suppose $a, b, c, d, m, n, p, q > 0$. Match each pair of parametric equations with one of the lines l_1, l_2, l_3, l_4 in Figure F.31.

   I. $\begin{cases} x = a + ct, \\ y = -b + dt. \end{cases}$ II. $\begin{cases} x = m + pt, \\ y = n - qt. \end{cases}$

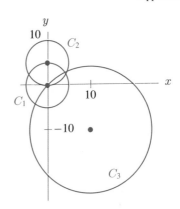

Figure F.31   Figure F.32

21. What can you say about the values of a, b and k if the equations

$$x = a + k \cos t, \quad y = b + k \sin t, \qquad 0 \le t \le 2\pi,$$

trace out the following circles in Figure F.32? (a) C_1 (b) C_2 (c) C_3

22. Describe in words the curve represented by the parametric equations

$$x = 3 + t^3, \quad y = 5 - t^3.$$

Graph the Lissajous figures in Problems 23–26 using a calculator or computer.

24. $x = \cos 3t, \quad y = \sin 7t$

23. $x = \cos 2t, \quad y = \sin 5t$

25. $x = \cos 2t, \quad y = \sin 4t$

26. $x = \cos 2t, \quad y = \sin \sqrt{3}t$

27. Motion along a straight line is given by a single equation, say, $x = t^3 - t$ where x is distance along the line. It is difficult to see the motion from a plot; it just traces out the x-line, as in Figure F.33. To visualize the motion, we introduce a y-coordinate and let it increase. If we let $y = t$, we get Figure F.34. What does Figure F.34 tell you about the particle's motion given by $x = t^3 - t, y = t$?

Figure F.33   Figure F.34

For Problems 28–30, plot the motion along the x-line by the method of Problem 27. What does the plot tell you about the particle's motion?

28. $x = \cos t, \quad -10 \le t \le 10$

29. $x = t^4 - 2t^2 + 3t - 7, \quad -3 \le t \le 2$

30. $x = t \ln t, \quad 0.01 \le t \le 10$

G PARAMETRIC EQUATIONS AND CALCULUS

Motion in a Straight Line

Consider an object moving with constant speed along a straight line through the point (x_0, y_0). Both the x- and y-coordinates have a constant rate of change. Let $a = dx/dt$ and $b = dy/dt$. Then at time t the object has coordinates $x = x_0 + at$, $y = y_0 + bt$. (See Figure G.35.) Notice that a represents the horizontal run in one unit of time, and b represents the vertical rise. Thus the line has slope $m = b/a$.

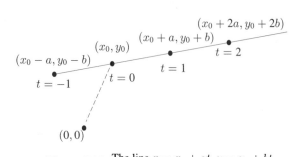

Figure G.35: The line $x = x_0 + at$, $y = y_0 + bt$

This yields the following:

Parametric Equations for a Straight Line

An object moving along a line through the point (x_0, y_0), with $dx/dt = a$ and $dy/dt = b$, has parametric equations

$$x = x_0 + at, \quad y = y_0 + bt.$$

The slope of the line is $m = b/a$.

Example 1   Find the parametric equation for

  (a)   The line passing through the points $(2, -1)$ and $(-1, 5)$.
  (b)   The line segment from $(2, -1)$ to $(-1, 5)$.

Solution   (a)   Imagine an object moves with constant speed along a straight line from $(2, -1)$ to $(-1, 5)$, making the journey from the first point to the second in one unit of time. Then $dx/dt = ((-1) - 2)/1 = -3$ and $dy/dt = (5 - (-1))/1 = 6$. Thus the parametric equation is

$$x = 2 - 3t, \quad y = -1 + 6t.$$

  (b)   In the parameterization in part (a), $t = 0$ corresponds to the point $(2, -1)$ and $t = 1$ corresponds to the point $(-1, 5)$. So the parameterization of the segment is

$$x = 2 - 3t, \quad y = -1 + 6t, \qquad 0 \leq t \leq 1.$$

Speed

Suppose an object moves along a straight line at a constant speed, with $dx/dt = a$ and $dy/dt = b$. In one unit of time, the object moves a units horizontally and b units vertically. Thus, by the Pythagorean Theorem, it travels a distance $\sqrt{a^2 + b^2}$. So its speed is

$$\text{Speed} = \frac{\text{Distance Traveled}}{\text{Time Taken}} = \frac{\sqrt{a^2 + b^2}}{1} = \sqrt{a^2 + b^2}.$$

For general motion along a curve with varying speed, we say

> The *instantaneous speed* of a moving object is defined to be
> $$v = \sqrt{\left(\frac{dx}{dt}\right)^2 + \left(\frac{dy}{dt}\right)^2}.$$

Example 2   A child is sitting on a ferris wheel of diameter 10 meters, making one revolution every 2 minutes. Find the speed of the child

    (a)   Using geometry.

    (b)   Using a parameterization of the motion.

Solution   (a)   The child moves at a constant speed around a circle of radius 5 meters, completing one revolution every 2 minutes. One revolution around a circle of radius 5 is a distance of 10π, so the child's speed is $10\pi/2 = 5\pi \approx 15.7$ m/min.

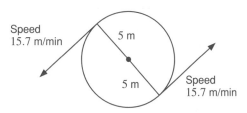

Figure G.36: Speed of a child on a ferris wheel

    (b)   The ferris wheel has radius 5 meters and completes 1 revolution counterclockwise every 2 minutes. The motion is parameterized by an equation of the form

$$x = 5\cos(\omega t), \quad y = 5\sin(\omega t),$$

where ω is chosen to make the period 2 minutes. Since the period of $\cos(\omega t)$ and $\sin(\omega t)$ is $2\pi/\omega$, we must have

$$\frac{2\pi}{\omega} = 2, \quad \text{so} \quad \omega = \pi.$$

Thus, the motion is described by the equation

$$x = 5\cos(\pi t), \quad y = 5\sin(\pi t),$$

where t is in minutes. So the speed is given by

$$v = \sqrt{\left(\frac{dx}{dt}\right)^2 + \left(\frac{dy}{dt}\right)^2}$$
$$= \sqrt{(-5\pi)^2 \sin^2(\pi t) + (5\pi)^2 \cos^2(\pi t)} = 5\pi\sqrt{\sin^2(\pi t) + \cos^2(\pi t)} = 5\pi \approx 15.7,$$

which agrees with the speed we calculated in part (a).

Tangent Lines

To find the tangent line at a point (x_0, y_0) to a curve given parametrically, we find the straight line motion through (x_0, y_0) with the same velocity in the x and y directions.

Example 3   Find the tangent line at the point $(1, 2)$ to the curve defined by the parametric equation

$$x = t^3, \quad y = 2t.$$

Solution   At time $t = 1$ the particle is at the point $(1, 2)$. The velocity in the x-direction at time t is $dx/dt = 3t^2$, and the velocity in the y-direction is $dx/dt = 2$. So at $t = 1$ the velocity in the x-direction is 3 and the velocity in the y-direction is 2. Thus the tangent line has parametric equations

$$x = 1 + 3t, \quad y = 2 + 2t.$$

The Length of a Curve

The speed of a particle is given by:

$$v = \sqrt{\left(\frac{dx}{dt}\right)^2 + \left(\frac{dy}{dt}\right)^2}.$$

As in one dimension, we can find the distance traveled by a particle along a curve by integrating its speed. Thus,

$$\text{Distance traveled} = \int_a^b v(t)\, dt.$$

If the particle never stops or reverses its direction as it moves along the curve, the distance it travels will be the same as the length of the curve. This suggests the following formula, which is justified in Problem 16:

> If the curve C is given parametrically for $a \le t \le b$ by differentiable functions and if the velocity v is not 0 for $a < t < b$, then
>
> $$\text{Length of } C = \int_a^b v\, dt = \int_a^b \sqrt{\left(\frac{dx}{dt}\right)^2 + \left(\frac{dy}{dt}\right)^2}\, dt.$$

Example 4   Find the circumference of the ellipse given by the parametric equations

$$x = 2\cos t, \quad y = \sin t, \quad 0 \le t \le 2\pi.$$

Solution   The circumference of this curve is given by an integral which must be calculated numerically:

$$\text{Circumference} = \int_0^{2\pi} \sqrt{\left(\frac{dx}{dt}\right)^2 + \left(\frac{dy}{dt}\right)^2}\, dt = \int_0^{2\pi} \sqrt{(-2\sin t)^2 + (\cos t)^2}\, dt$$

$$= \int_0^{2\pi} \sqrt{4\sin^2 t + \cos^2 t}\, dt = 9.69.$$

Since the ellipse is inscribed in a circle of radius 2 and circumscribes a circle of radius 1, we would expect the length of the ellipse to be between $2\pi(2) \approx 12.57$ and $2\pi(1) \approx 6.28$, so the value of 9.69 is reasonable.

Problems for Section G

1. (a) Explain how you know that the following two pairs of equations,
        $x = 2 + t, y = 4 + 3t$ and $x = 1 - 2t, y = 1 - 6t$, parameterize the same line.
    (b) What are the slope and y intercept of this line?

2. The equations $x = 10 + t$ and $y = 2t$ parameterize a line.

    (a) Suppose we restrict ourselves to $t < 0$. What part of the line do we get?
    (b) Suppose we restrict ourselves to $0 \le t \le 1$. What part of the line do we get?

3. (a) Sketch the parameterized curve $x = t \cos t$, $y = t \sin t$ for $0 \le t \le 4\pi$.
    (b) By calculating the position at $t = 2$ and $t = 2.01$, estimate the speed at $t = 2$.
    (c) Use derivatives to calculate the speed at $t = 2$ and compare your answer to part (b).

For Problems 4–7, find the speed for the given motion of a particle. Also find any times when the particle comes to a stop.

4. $x = t^2, \quad y = t^3$

5. $x = \cos(t^2), \quad y = \sin(t^2)$

6. $x = \cos 2t, \quad y = \sin t$

7. $x = t^2 - 2t, y = t^3 - 3t$

8. Find parametric equations for the tangent line at $t = 2$ for Problem 4.

Find the length of the curves in Problems 9–11.

9. $x = 3 + 5t, y = 1 + 4t$ for $1 \le t \le 2$. Explain your answer.

10. $x = \cos(e^t), y = \sin(e^t)$ for $0 \le t \le 1$. Explain why your answer is reasonable.

11. $x = \cos(3t), y = \sin(5t)$ for $0 \le t \le 2\pi$.

12. Consider the motion of the particle given by the parametric equations

    $$x = t^3 - 3t, \quad y = t^2 - 2t,$$

    where the y-axis is vertical and the x-axis is horizontal.
    (a) Does the particle ever come to a stop? If so, when and where?
    (b) Is the particle ever moving straight up or down? If so, when and where?
    (c) Is the particle ever moving straight horizontally right or left? If so, when and where?

13. Emily is standing on the outer edge of a merry-go-round, 10 meters from the center. The merry-go-round completes one full revolution every 20 seconds. As Emily passes over a point P on the ground, she drops a ball from 3 meters above the ground.

    (a) How fast is Emily going?
    (b) How far from P does the ball hit the ground? (The acceleration due to gravity is 9.8 m/sec^2.)
    (c) How far from Emily does the ball hit the ground?

14. A hypothetical moon orbits a planet which in turn orbits a star. Suppose that the orbits are circular and that the moon orbits the planet 12 times in the time it takes for the planet to orbit the star once. In this problem we will investigate whether the moon could come to a stop at some instant. (See Figure G.37.)

    (a) Suppose the radius of the moon's orbit around the planet is 1 unit and the radius of the planet's orbit around the star is R units. Explain why the motion of the moon relative to the star can be described by the parametric equations

    $$x = R \cos t + \cos(12t), \quad y = R \sin t + \sin(12t).$$

    (b) Find values for R and t such that the moon stops relative to the star at time t.
    (c) On a graphing calculator, plot the path of the moon for the value of R you obtained in part (b). Experiment with other values for R.

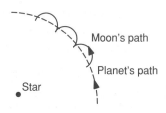

Figure G.37:

15. Suppose $F(x, y) = 1/(x^2 + y^2 + 1)$ gives the temperature at the point (x, y) in the plane. A ladybug moves along a parabola according to the parametric equations

$$x = t, \quad y = t^2.$$

Find the rate of change in the temperature of the ladybug at time t.

16. In this problem we justify the formula for the length of a curve given on page 612. Suppose the curve C is given by smooth parametric equations $x = x(t)$, $y = y(t)$ for $a \le t \le b$. By dividing the parameter interval $a \le t \le b$ at points $t_1, \ldots, t_{n-1}$ into small segments of length $\Delta t = t_{i+1} - t_i$, we get a corresponding division of the curve C into small pieces. See Figure G.38, where the points $P_i = (x(t_i), y(t_i))$ on the curve C correspond to parameter values $t = t_i$. Let C_i be the portion of the curve C between P_i and P_{i+1}.

   (a) Use local linearity to show that

   $$\text{Length of } C_i \approx \sqrt{x'(t_i)^2 + y'(t_i)^2}\, \Delta t.$$

   (b) Use part (a) and a Riemann sum to explain why

   $$\text{Length of } C = \int_a^b \sqrt{x'(t)^2 + y'(t)^2}\, dt.$$

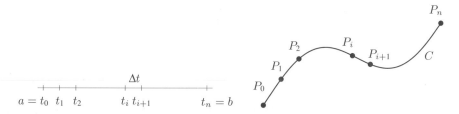

Figure G.38: A subdivision of the parameter interval and the corresponding subdivision of the curve C

H PROJECTS

1. AN ORBITING SATELLITE   Chapter 3

Overview: Consider a NASA satellite which orbits the earth every 90 minutes. During an orbit, the satellite's electric power comes either from solar array wings, when these are illuminated by the sun, or from batteries. The batteries discharge whenever the satellite uses more electricity than the solar array can provide or whenever the satellite is in the shadow of the earth (where the solar array cannot be used). If the batteries are overused, however, they can be damaged.

The Problem:[1] You are to determine whether the batteries could be damaged in either of the following operations. You are told that the battery capacity is 50 ampere-hours. If the total battery discharge does not exceed 40% of battery capacity, the batteries will not be damaged.

(a)   Operation 1 is performed by the satellite while orbiting the earth. At the beginning of a given 90-minute orbit, the satellite performs a 15-minute maneuver which requires more current than the solar array can deliver, causing the batteries to discharge. The maneuver causes a sinusoidally varying battery discharge of period 30 minutes with a maximum discharge of ten amperes at 7.5 minutes. For the next 45 minutes the solar array meets the total satellite current demand, and the batteries do not discharge. During the last 30 minutes, the satellite is in the shadow of the earth and the batteries supply the total current demand of 30 amperes.

(i)   The battery current in amperes is a function of time. Plot the function, showing the current in amperes as a function of time for the 90-minute orbit. Write a formula (or formulas) for the battery current function.

(ii)   Calculate the total battery discharge (in units of ampere-hours) for the 90-minute orbit for Operation 1.

(iii)   What is your recommendation regarding the advisability of Operation 1?

(b)   Operation 2 is simulated at NASA's laboratory in Houston. The following graph was produced by the laboratory simulation of the current demands on the battery during the 90-minute orbit required for Operation 2.

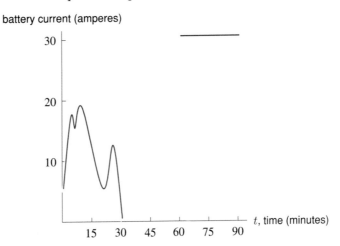

Figure H.39: Battery discharge simulation graph for Operation 2

(i)   Calculate the total battery discharge (in units of ampere-hours) for the 90-minute orbit for Operation 2.

(ii)   What is your recommendation regarding the advisability of Operation 2?

2.   THE EARTH'S PATH   **Chapter 3**

Overview: The orbits of the earth and Halley's comet are both ellipses. In this project we identify some parameters of these orbits.

The Problem:[2] The equation of an ellipse is

$$\frac{x^2}{a^2} + \frac{y^2}{b^2} = 1,$$

---

[1] Adapted from Amy C. R. Gerson, "Electrical Engineering: Space Systems," in *She Does Math! Real Life Problems from Women on the Job*, ed. Marla Parker, p. 61 (Washington, DC: Mathematical Association of America, 1995).

[2] This project is adapted from M.E. Moody and K. Shannon, *Microcomputer Exercises for Calculus*, (St. Paul: West, 1988).

where a and b are positive constants, $a \geq b$. The foci of this ellipse are located at $(c, 0)$ and $(-c, 0)$, where $c = \sqrt{a^2 - b^2}$. The eccentricity[3] of the ellipse is defined as $e = c/a$. The length of the ellipse's perimeter is given by

$$\int_0^{\pi/2} 4a(1 - e^2 \sin^2 \theta)^{1/2} d\theta.$$

The path of the earth is an ellipse, with the sun at one of its foci. The closest the earth comes to the sun is 91.5 million miles and the furthest is 94.5 million miles.

The path of Halley's comet is also an ellipse, with the sun at one of its foci. The semi-major axis, a, of the ellipse is 1674 million miles and its eccentricity is 0.97. It takes about 76 years for Halley's comet to orbit the sun. In 1705 Halley (1656-1742) predicted correctly that the comet would be seen in 1758.

(a)   What is the eccentricity of a circle?
(b)   What does the magnitude of e say about the shape of an ellipse?
(c)   For the earth and sun configuration, what are the values of a, b, c, and e?
(d)   How far does the earth travel in one orbit of the sun?
(e)   What is the average speed of the earth around the sun in miles per second?
(f)   If the earth's path were a circle but it still traveled the same distance as in part (c), what would the radius have to be?
(g)   How close does Halley's comet come to the sun? How far does it travel in one orbit? What is the average speed around the sun? Compare this to part (e).
(h)   In what year did Halley see the comet? Did he know whether his prediction came true?

3.   **GAS TANKS**   **Chapter 3**

Overview: Gasoline is stored at a gas station in an underground tank which is a circular cylinder lying on its side. You may have seen a gas station attendant dropping a long measuring stick into the tank to measure the amount of gasoline remaining. Suppose the remaining fuel comes to a height of h inches and the diameter of the tank is d inches. It might be tempting to guess that the tank is h/d full. The purpose of this project is to see how far off such a guess is.

The Problem: Assume a cylindrical gas tank has diameter d and length l, as in Figure H.40.

Figure H.40

(a)   Suppose another gas tank has the shape of a w by d by l rectangular box. What is w if this tank holds the same amount as the cylindrical tank? Does h/d give an exact measure of how full the rectangular tank is?
(b)   When computing the accuracy of the stick method for determining what fraction of the tank is full, why can we ignore the length of the tank?
(c)   Draw cross-sections of both tanks on one set of axes, with the x-axis measuring height and with both tanks centered at the origin.
(d)   Let $E(h)$ be the error, as a function of h, when using h/d to measure how full the cylindrical tank is. Use the picture you drew in part (c) to write $E(h)$ as an integral.
(e)   For what value of h is the error maximal?

---

[3]Although it is traditionally represented by the letter e, the eccentricity has nothing to do with the number $2.71828\ldots$.

4. **THE GRAPHS OF $y = b^x$ AND $y = x^b$**   **Chapter 4**

   Overview: This project investigates how the number of intersections of the curves $y = b^x$ and $y = x^b$ depends on the value of the parameter b. Throughout the project, we restrict our attention to $x, y > 0$ and $b > 1$.

   The Problem:[4]

   (a) Where do the graphs of $y = 2^x$ and $y = x^2$ cross?
   (b) Where do the graphs of $y = 3^x$ and $y = x^3$ cross?
   (c) Where do the graphs of $y = 4^x$ and $y = x^4$ cross?

   We now consider the graphs of $y = b^x$ and $y = x^b$ for any $b > 1$.

   (d) There is one obvious crossing point for the graphs of $y = b^x$ and $y = x^b$. Find this point.
   (e) Using parts (a)-(c), decide if there is a second crossing point for the graphs of $y = b^x$ and $y = x^b$. If so, is it to the left or right of the point you found in part (d)?
   (f) Is there a value of b for which the graphs of $y = b^x$ and $y = x^b$ touch only once? Guess an answer and confirm it using a calculator or computer.
   (g) For the value of b you chose in part (f), show analytically that the graphs of $y = b^x$ and $y = x^b$ touch.
   (h) (Optional; Uses Chapter 5.) Show that for any $b > 1$, the graphs of $y = b^x$ and $y = x^b$ cross. Do this by considering the fact that $b^x = x^b$ whenever the function $f(x) = x \ln b - b \ln x$ is zero.

5. **THE ROAD AND THE RIVER**   **Chapter 5**

   Overview: The equation of a road, a river, and a dam are given; we will find the location of a spillway and a road.

   The Problem: A particular river follows the quadratic curve $f(x) = x^2$, while a nearby road follows the cubic curve

   $$g(x) = ax^3 + bx^2 + cx + d.$$

   There are three bridges on the road and they cross the river at the points $(-1, 1)$, $(0, 0)$ and $(1, 1)$. (The units are kilometers.) The bridge at $(1, 1)$ is at right angles to the river. A dam is located at $(1, 2)$. The dam is not on this river. There is a straight spillway from the dam to the nearest point on the river.

   (a) What are the values of a, b, c, and d?
   (b) What are the distances between the bridges, as the crow flies?
   (c) At what angles (in radians and degrees) do the other two bridges cross the river?
   (d) Where does the spillway from the dam intersect the river?
   (e) A straight access road is to be built from the existing road to the dam. Where on the original road should it start to have minimum length?
   (f) A farmer owns the land between the river and the road, bounded by the first and last bridges. How many square kilometers does the farmer own?

6. **INTERSTATE TRUCKING**   **Chapter 5**

   Overview: In this project we will investigate the effect of the price of diesel fuel and drivers' wages on a trucker's optimal driving speed. We will analyze the relationship between fuel cost and wages and the 55 mph speed limit.

   The Problem:[5] After an extensive study, the Interstate Commerce Commission has concluded that the primary variable expenses for any over-the-road freight hauler are the wages for the driver and the cost of fuel. The study concluded that maintenance and replacement costs for

---

[4]Based on an idea from Gil Strang.

[5]Adapted from L. Carl Leinbach, *Calculus Laboratories Using Derive*, (Belmont, CA: Wadsworth, 1991). Reprinted by permission of Brooks/Cole Publishing Company, a division of International Thompson Publishing, Inc.

vehicles, although considerable, did not vary significantly from carrier to carrier. On the other hand, the fuel costs and wages did vary. Given this result, your project is to determine how these two variables affect the cost of transporting freight. In particular, you are to determine if there is an optimal wage to fuel cost relationship for which most haulers would encourage their drivers to abide by a 55 mile per hour speed limit.

Reading the study tells you that under ideal conditions, an interstate freight hauling vehicle (18 wheeler) gets 6 miles per gallon of fuel. This mileage is affected by the speed of the vehicle and its weight. In general, the miles per gallon of fuel decreases by 0.2 for each increase of 10,000 pounds in the weight of the truck and freight over 25,000 pounds. Also, the miles per gallon of fuel decreases by 0.1 mile per gallon for each mile per hour the truck averages above 45 miles per hour.

(a)   Using this information, create an expression for the cost per mile of driving, taking into account only the drivers' wages and fuel cost.

(b)   Suppose that the national average for diesel fuel is \$1.25 per gallon, that drivers on the average earn \$15.00 per hour, and that the average weight of a loaded truck is 75,000 pounds. What is the optimal average speed under these conditions?

(c)   Compute the cost per mile for 55 and 60 miles per hour. Suppose you hauled produce from California to Maine. What average speed would you choose?

(d)   In solving parts (a)-(c), you found an equation relating cost, wages, fuel cost, weight of the truck, and the average speed of the truck. This equation is a mathematical model of interstate freight hauling costs. Using this model, determine the fuel cost that makes 55 miles per hour the optimum average speed for trucking companies.

(e)   Each year drivers' wages change as a result of contract renewals and inflation. The cost of fuel also fluctuates, but can be adjusted by surcharges and fuel taxes. Thus, the relationship between wages and the cost of fuel can be altered by various government agencies. Assuming that 75,000 pounds remains the national average for the weight of a truck involved in interstate freight hauling, what should be the relationship between fuel cost and wages to maintain 55 miles per hour as the optimal speed?

(f)   There have been some suggestions of changing the road use tax to lower the average weight of over-the-road freight haulers. Assuming the relationship of wages to fuel cost found in part (e) is maintained, how would lowering the average weight affect the optimal average speed?

7.   APPROXIMATING π WITH INTEGRALS   Chapter 7

Overview: Since π is irrational, it is often useful to approximate it by a rational number. We derive an explicit formula for finding rational approximations to π. The inequality we derive was first discovered in about 1650 by John Wallis, though using a different method than the one we shall follow.

The Problem:

(a)   Use integration by parts to show that if n is an even positive integer, then

$$\int_0^{\pi/2} \sin^n x \, dx = \frac{n-1}{n} \cdot \frac{n-3}{n-2} \cdots \frac{3}{4} \cdot \frac{1}{2} \cdot \frac{\pi}{2}.$$

(b)   Use integration by parts to show that if n is an odd integer and $n > 1$, then

$$\int_0^{\pi/2} \sin^n x \, dx = \frac{n-1}{n} \cdot \frac{n-3}{n-2} \cdots \frac{4}{5} \cdot \frac{2}{3}.$$

(c)   Explain why the following inequality is true for any integer k:

$$\int_0^{\pi/2} \sin^{2k+1} x \, dx < \int_0^{\pi/2} \sin^{2k} x \, dx < \int_0^{\pi/2} \sin^{2k-1} x \, dx.$$

Use this inequality to show, for any integer $k > 1$, that

$$2 \cdot \frac{(2k)(2k)(2k-2)(2k-2)\cdots 4 \cdot 4 \cdot 2 \cdot 2}{(2k+1)(2k-1)(2k-1)\cdots 5 \cdot 5 \cdot 3 \cdot 3 \cdot 1} < \pi < 2 \cdot \frac{(2k)(2k-2)(2k-2)\cdots 4 \cdot 4 \cdot 2 \cdot 2}{(2k-1)(2k-1)\cdots \cdot 5 \cdot 5 \cdot 3 \cdot 3 \cdot 1}.$$

(d)   How large is the gap between the upper and the lower bounds for π given by these inequalities? Try different values of k and make a table showing the size of the gap against the value of k. Does the size of the gap appear proportional to some function of k? What does this tell us about the accuracy of the approximation?

8.   **SURFACE AREA OF AN UNPAINTABLE CAN OF PAINT**   **Chapter 8**

Overview: This project introduces the formula for the surface area of a volume of revolution. It then shows that it is possible to have a solid whose volume is finite but whose surface area is infinite.

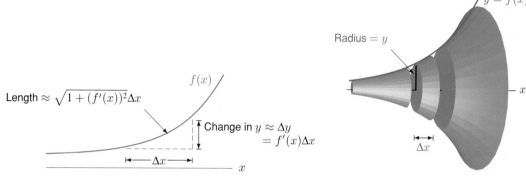

Figure H.41   **Figure H.42**

You know that the arclength of the curve $y = f(x)$ from a to b can be found using the integral

$$\text{Arc length} = \int_a^b \sqrt{1 + (f'(x))^2}\, dx.$$

Figure H.41 shows the corresponding arc length of a small piece of the curve.

Similarly, it can be shown that the surface area from a to b of the solid obtained by revolving $y = f(x)$ around the x-axis is given by

$$\text{Surface area} = 2\pi \int_a^b f(x)\sqrt{1 + (f'(x))^2}\, dx.$$

We can see this why this might be true by looking at Figure H.42.

We approximate the small piece of the surface by a slanted cylinder of radius y and "height" equal to the arclength of the curve, so that

$$\text{Surface area of edge of slice} \approx 2\pi y \sqrt{1 + (f'(x))^2}\, \Delta x.$$

Integrating this expression from $x = a$ to $x = b$ gives the surface area of the solid.

The Problem:

(a)   Calculate the surface area of a sphere of radius r.
(b)   Calculate the surface area of a cone of radius r and height h.
(c)   Rotate the curve $y = 1/x$ for $x \geq 1$ around the x-axis. Find the volume of this solid.
(d)   Show that the surface area of the solid in part (c) is infinite. [Hint: You might not be able to find an antiderivative of the integrand in the surface area formula; can you get a lower bound on the integral?]
(e)   (Optional. Requires Chapter 10.) Find a curve such that when the portion of the curve from $x = a$ to $x = b$ is rotated around the x-axis (for any a and b), the volume of the solid of revolution is equal to its surface area. You may assume $dy/dx \geq 0$.

9.   **QUININE**   **Chapter 9**

Overview: Malaria is a parasitic infection transmitted by mosquito bites, mainly in tropical areas of the world. The disease has existed since ancient times, and currently there are hundreds of millions of cases each year, with millions of deaths. Around 1630, Jesuits in Peru introduced the bark of the cinchona tree to the West as the first treatment for malaria. The drug quinine is the active ingredient in the bark, and it is still used today. However, the concentration in the body of quinine, and of most drugs, must be kept within certain parameters. If the concentration is too low, the drug is ineffective. If the concentration is too high, toxic side effects can result.

The Problem: Suppose you are a doctor who prescribes quinine for a 70 kg malaria patient. At 8 am each day, she is to receive 50 mg of the drug.[6] To be effective, the average concentration in the body must be at least 0.4 mg/kg. However, concentrations above 3.0 mg/kg can be fatal. The half-life of quinine in the body is 11.5 hours.

(a)   What is the continuous rate of decay of quinine (in units of %/min)?

(b)   How much quinine is in the patient's system just before and just after the first day's dose? After the second day's dose?

(c)   How much is in her system after a few more doses? After the nth dose? Explain what happens in the long run.

(d)   Graph the concentration of the drug versus time showing what the steady-state looks like. Find a formula for the repeating function.

(e)   What is the average (over time) of the concentration of quinine in the patient's body?

(f)   Is this treatment both effective and safe?

(g)   Suppose instead that the patient were to receive two doses of 25 mg each day, one at 8 am and one at 8 pm. Is this a safe and effective treatment?

(h)   Determine the average value of an exponentially decaying function between two points (x_0, y_0) and (x_1, y_1).

(i)   Notice that the average concentration of quinine is the same for the 50 mg once per day and 25 mg twice per day treatments. Use part (h) to explain why. Would the average concentration be the same for a 100 mg dose once every two days?

(j)   Suppose the original quinine regimen is stopped. How long after the last dose will the amount of quinine be less than 10^{-10} times the patient's body mass?

10.   **US ELECTRICITY CONSUMPTION**   **Chapter 10**

Overview: In this project we are given US electricity consumption over much of the 20th century and asked to predict consumption in the future. Several different predictions are made by fitting various differential equations to the data.

The Problem: The following data shows US electricity consumption, measured in billions of kilowatt hours per year, for much of this century.

Year	1912	1917	1920	1929	1936	1945
Electricity consumption (E)	12	25	39	92	109	222

Year	1955	1960	1965	1970	1980	1987
Electricity consumption (E)	547	755	1055	1531	2286	2455

(a)   Decide how well each of the following differential equations model the growth of electricity consumption, E, over time, t, measured in years. Do this by using the data to

---

[6]This is a simplified model; actual treatments involve different drugs and more complicated dosage regimens. See, for example, *The Pharmacological Basis of Therapeutics*, 9th Ed., ed. Joel G. Hardman, Alfred Goodman Gilman, and Lee E. Limbird, (New York: McGraw Hill, 1996).

approximate dE/dt. For each differential equation, make a plot to which you can fit a line and use it to estimate the parameters a, b, or c.

(i) $\dfrac{1}{E}\dfrac{dE}{dt} = c$

(ii) $\dfrac{dE}{dt} = a - bE$

(iii) $\dfrac{1}{E}\dfrac{dE}{dt} = a - bE$

(iv) $\dfrac{1}{E}\dfrac{dE}{dt} = a - bt$

(b) Use each model to predict energy consumption for the year 2000. Which prediction do you think is most reliable? Why? Discuss the long-run predictions from each model. Which differential equation do you think fits best?

11. TERMINAL VELOCITY $\qquad$ Chapter 10

Overview: The motion of falling objects is often modeled by the equation $dv/dt = -g$. However, since the earth is surrounded by air, any accurate description of falling objects must take account of air resistance. We frequently use the model:

$$m\frac{dv}{dt} = mg - f(v),$$

where t is time, v is the speed of the object measured downwards and m is its mass. The term mg represents the force of gravity; it is positive and acts to speed up the motion of a falling object. The value of g varies slightly from place to place, but is very close to 9.8 m/sec^2 everywhere on the earth. The function $f(v)$ represents the force exerted by the air. It will oppose the motion of a falling object and so we have written it with a negative sign. Common sense says $f(0) = 0$. Let us assume $f(v)$ strictly increases with v; that is, the faster the object moves, the greater the air resistance. The actual function f may be obtained either experimentally or theoretically. If the falling object is irregular, like a person, the air resistance depends on the orientation of the object as it falls. To avoid this complication, we assume that the object is a ball.

The Problem:

(a) Make up a few possible f's. The only requirements are that f be strictly increasing and that $\lim_{v\to\infty} f = \infty$. Look at the corresponding slope fields (on v vs. t axes) and sketch solutions for each. What do these solutions have in common? Explain, in terms of derivatives, the concavity of your solutions. [Hint: Take $m = 1$. Consider f proportional to v, v^2, v^3.]

(b) Under the assumptions of part (a), the differential equation has exactly one constant solution. If $f(v) = kv$, then what is the constant solution? How about for $f(v) = kv^2$? How about for a general increasing f? Explain why this velocity is called the terminal velocity.

(c) Suppose now that we have a ball which can be opened and filled with various materials to change its mass. We drop this ball from the Washington Monument (height 160 m). Let's assume that this is high enough to ensure the ball will reach terminal velocity, and that f has one of the forms $f(v) = kv$, $f(v) = kv^{3/2}$, or $f(v) = kv^2$. When the ball weighs 50 gm, it hits the ground at 24 m/sec, and when it weighs 100 gm it hits the ground at 33.5 m/sec. What is the correct formula for f?

For the rest of the problem, assume f is given by the formula you found in part (c).

(d) Use the ball with a mass of 50 gm. Suppose the falling ball increases its speed from 12 m/sec to 13 m/sec in a short time interval. How far has the ball fallen in this time interval? (Give lower and upper bounds).

(e) Suppose instead the speed of this ball goes from v to $v + \Delta v$. Approximately how long is this time interval? About how far does the ball fall? Using this result, write integrals for how long it takes the ball to go from a speed of v_1 to a speed of v_2 and for how far the ball travels during this interval.

(f)   Use the integral from part (e) for the time to go from speed v_1 to speed v_2 to show that the balls falling according to this model never exactly reach their terminal velocity. [Hint: It is easier if you retain the symbols for the various constants and use the expression for terminal velocity you got in part (b).]

(g)   How long did it take the 50 gm ball to fall from the top of the Washington Monument? Was it close to its terminal velocity? [Hint: Realize that you don't know the exact velocity with which the ball hit the ground.]

(h)   Assume our ball has the same dimensions as a baseball and that a baseball has a mass of 145 gm. What is its terminal velocity? How fast would it be going when it hits the ground after being dropped from the top of the Washington Monument?

ANSWERS TO ODD NUMBERED PROBLEMS

Section 1.1

1. (I)   None
   (II) (b)
   (III) (c)
   (IV) (a)

5. $F = k/d^2$

7. $V = kr^3$

15. p_1 = maximum price any consumer would pay
    q_1 = quantity given away if item were free

17. Domain: $0 \le x \le 5$
    Range: $0 \le y \le 4$

19. Domain: all x
    Range: $y \ge 2$

21. Domain: all x
    Range: $0 < y \le 1/2$

23. All $x \ne 0, -1$
    $x = \pm 2$

Section 1.2

1. Slope: $-12/7$
   Vertical intercept: $2/7$

3. Slope: 2
   Vertical intercept: $-2/3$

5. $y = (1/2)x + 2$

7. $y = -\frac{1}{5}x + \frac{7}{5}$

9. Parallel: $y = m(x - a) + b$
   Perpendicular:
   $y = (-1/m)(x - a) + b$

11. (a) (V)
    (b) (VI)
    (c) (I)
    (d) (IV)
    (e) (III)
    (f) (II)

13. $W = (5/3)R + 10$

15. (a) \$0.025/cubic foot
    (b) $c = 65 + 0.025w$
        c = cost of water
        w = cubic feet of water
    (c) $w = 2600$ cubic feet

17. (a) Slope = 1.8
    (b) $°F = 1.8(°C) + 32$
    (c) 68° Fahrenheit
    (d) $-40°$

19. $9 \cdot 10^9$ newton/meter^2

21. (a) $d = 5t$
    (b) No, in the last 200 years the average speed is:
        $700/200 = 3.5$ mi/yr.
    (c) Unlikely

23. (a) $R = k(350 - H)$
        $(k \ge 0)$

Section 1.3

1. Concave up

3. Neither

5. Exponential growth

7. Exponential growth
   $g(t) = 2^t$

9. Exponential decay
   $h(t) = 16768(1/2)^t$

15. $P = 5$

17. $Z = 3$

19. (a) 1.05
    (b) 5%

21. $y = 4(2^{-x})$

23. $y = 4(1 - 2^{-x})$

25. $f(s) = 2(1.1)^s$
    $g(s) = 3(1.05)^s$
    $h(s) = (1.03)^s$

27. (a) $Q = Q_0 \left(\frac{1}{2}\right)^{(t/1620)}$
    (b) 80.7%

29. 25.5%

31. (a) 47.3%
    (b) 11054.4%

Section 1.4

1. 16

3. 100,000

5. 1/3

7. 1/4

9. $10 \cdot 2^x$

11. $0.25\sqrt{x}$

13. As $x \to \infty$, $y \to \infty$.
    As $x \to -\infty$, $y \to -\infty$.

17. $f(x) = x^3$

19. (a) 1.3 m^2
    (b) 86.8 kg
    (c) $h = 112.6s^{4/3}$

21. 44.25 ft and 708 ft

23. $h(t) = ab^t$
    $g(t) = kt^3$
    $f(t) = ct^2$

25. (a) $0 \le x \le 5$,
        $0 \le y \le 9$
    (b) $0 \le x \le 5$,
        $0 \le y \le 625$
    (c) $0 \le x \le 10$,
        $0 \le x \le 50,000$

Section 1.5

3. Length of column of mercury when temperature is 75°F

5. Probably not invertible

7. Not invertible

9. Not invertible

11. Not invertible

13. (a) $q = \frac{C-100}{2} = f^{-1}(C)$
    (b) Number of articles produced at given cost

17. (a) -1

Section 1.6

3. $(\log 2)/(\log 17) \approx 0.24$

5. $(\log(2/5))/(\log 1.04) \approx -23.4$

7. $(\log(2/11))/(\log(7/5)) \approx -5.07$

9. $(\log a)/(\log b)$

11. $(\log Q - \log Q_0)/(n \log a)$

13. $(10/3)\log A + \log B$

15. 2

17. $(A/B)^2$

19. The graph is a straight line, with slope 1, to the right of the origin.

21. $\log x$

23. 14.21 years

25. 1990

27. (a) $A = 10.32e^{-0.057762t}$
    (b) 40.41 days

29. $p^{-1}(t) \approx 58.708 \log t$

Section 1.7

3. -0.347

5. 0.515

7. 1.01, 1.17

9. $\ln(P/P_0)/k$

11. 1/2

13. 2AB

15. $-1 + \ln A + \ln B$

17. $3\ln A + \ln B$

19. $P = 2(0.61)^t$; decay

21. $P = 79(0.0821)^t$; decay

23. $P = 7(0.0432)^t$; decay

25. $P = 10e^{0.5306t}$

27. $P = 4e^{-0.6t}$

29. $f^{-1}(t) = 10\ln\left(\frac{t}{50}\right)$

31. (a) Decreasing
    (b) $g^{-1}(x) = \ln(\frac{2}{x} - 5)$

33 (a) $P = 10^6(e^{0.02t})$

35 (a) 47.5%

    (b) 24%

37 2010

39 7.925 hours

41 (a) 5 kg

    (b) 38.2 years

43 (a) $P = 3.6(1.034)^t$

    (b) $P = 3.6e^{0.0334t}$

    (c) Annual = 3.4%

        Continuous = 3.3%

45 It is a fake

Section 1.8

3 $2\ln(x+3) + 1$

5 $2e^{4x+7} + 1$

7 $\ln(e^{4x+7} + 3)$

9 $4x + 7$

11 $e^7(2x+4)^4$

13 Even

15 Neither

17 Neither

19 $f(x) = x + 1$

    $g(x) = x^3$

21 $f(x) = x^3$

    $g(x) = \ln x$

23 $2zh + h^2$

25 $4hz$

27 $-60 < x < -40$

    $-25100 < y < -24000$

31 $g(f(2)) \approx 1.1$

Section 1.9

1 Negative

  0

  Undefined

3 Positive

  Positive

  Positive

5 Positive

  Positive

  Positive

7 Positive

  Negative

  Negative

9 Negative

  Positive

  Negative

11 0.259

13 0.966

15 20.94 to 52.36 rad/sec

17 If $f(x) = \sin x$ and

    $g(x) = x^2$ then

    $\sin x^2 = f(g(x))$

    $\sin^2 x = g(f(x))$

    $\sin(\sin x) = f(f(x))$

19 (a) $h(t)$

    (b) $f(t)$

    (c) $g(t)$

21 $f(x) = 2\sin(x/4)$

23 $f(x) = 2\sin(x/4) + 2$

25 $f(x) = \sin\left(2(\pi/5)x\right)$

27 $f(x) = 3\sin(\pi x/9)$

29 (a) $\frac{1}{60}$ second

    (b) V_0 represents the amplitude of oscillation.

31 $f(t) = 75 - 15\cos\left(\frac{2\pi}{12}t\right)$

33 (b) Domain:

      $-1 \leq x \leq 1$

      Range:

      $-\frac{\pi}{2} \leq x \leq \frac{\pi}{2}$

35 $(0,0)$, $(1.31, 2.29)$,

    $(-1.31, -2.29)$

37 $\theta = \pi/4$; $R = v_0^2/g$

Section 1.10

1 (a) $f(x) \to \infty$

      as $x \to \infty$;

      $f(x) \to -\infty$

      as as $x \to -\infty$

    (b) $f(x) \to -\infty$

      as as $x \to \pm\infty$

    (c) $f(x) \to 0$ as $x \to \pm\infty$

    (d) $f(x) \to 6$ as $x \to \pm\infty$

7 a, c, f and h are even;

  b, d and g are odd;

  e and i are neither

9 $x = 2$: vertical asymptote

  No horizontal asymptote

  $y \to -\infty$ as $x \to +\infty$

  $y \to +\infty$ as $x \to -\infty$

  $y \to -\infty$ as $x \to 2^+$

  $y \to +\infty$ as $x \to 2^-$

11 $x = \pm 2$: vertical asymptotes

  $y = 1$: horizontal asymptote

  $y \to +\infty$ as $x \to 2^+$

  $y \to -\infty$ as $x \to 2^-$

  $y \to -\infty$ as $x \to -2^+$

  $y \to +\infty$ as $x \to -2^-$

13 (a) 0

    (b) $t = 2v_0/g$

    (c) $t = v_0/g$

    (d) $(v_0)^2/(2g)$

15 (a) $1 = a + b + c$

    (b) $b = -2a$ and $c = 1 + a$

    (c) $c = 6$

    (d) $y = 5x^2 - 10x + 6$

17 $y = -\frac{1}{2}(x+2)^2(x-2)$

19 (a) $f(x) = kx(x+3)(x-4)$

      $(k < 0)$

    (b) Increasing:

      $-1.5 < x < 2.5$

      Decreasing:

      $x < -1.5$ and $x > 2.5$

21 (a) $f(x) = k(x+2)(x-2)^2(x-5)$

      $(k < 0)$

    (b) Increasing:

      $x < -1$ and

      $2 < x < 4$

      Decreasing:

      $-1 < x < 2$ and

      $4 < x$

23 (a) III

    (b) IV

    (c) I

    (d) II

25 (a) $a(v) = \frac{1}{m}(F_E - kv^2)$

      $(k > 0)$

27 (a) $S = 2\pi r^2 + 2V/r$

    (b) $S \to \infty$ as $r \to \infty$

29 $y = x$

Section 1.11

1 No

3 Yes

5 No

7 Yes

9 Not continuous

Chapter 1 Review

1 (a) $[0, 7]$

    (b) $[-2, 5]$

    (c) 5

    (d) $(1, 7)$

    (e) Concave up

    (f) 1

    (g) No

3 (b) Concave down

7 $x = 3\left(1 - a^{-t}\right)$, $a > 1$

9 $y = (x+1)^3 + 1$

11 (a) $h(x) = 31 - 3x$

    (b) $g(x) = 36(1.5^x)$

13 $Q(m) = T + L + Pm$

  $T =$ fuel for take-off

  $L =$ fuel for landing

  $P =$ fuel per mile in the air

  $m =$ length of the trip (miles)

15 (a) $h^2 + 6h + 11$

    (b) 11

    (c) $h^2 + 6h$

17 (a) $f(n) + g(n) = 3n^2 + n - 1$

    (b) $f(n)g(n) = 3n^3 + 3n^2 - 2n - 2$

    (c) $n \neq -1$

    (d) $f(g(n)) = 3n^2 + 6n + 1$

    (e) $g(f(n)) = 3n^2 - 1$

19 Not continuous

21 Not continuous

23 $P = (5 \cdot 10^{-3})(0.9812)^t$

25 (a) $Q = 25e^{0.5423t}$
    (b) 1.3 months
    (c) 6.8 months

27 10 hours

29 38.87 hours

31 (a) 44%
    (b) about two years

33 Mercury: 87.8 days
    Earth: 1 year
    Pluto: 253 years

35 $y = e^{0.4621x}$

37 Simplest is $y = 1 - e^{-x}$

39 $y = k(x^3 + 2x^2 - x - 2)$
    $(k > 0)$

41 $y = 5 \sin\left(\pi t/20\right)$

43 $z = 1 - \cos\theta$

45 (b) It becomes a vertical asymptote at $x = -4$.

47 (a) $f(15) \approx 48$
    (b) f is invertible because the graph of f is cut by a horizontal line at most once.
    (c) $f^{-1}(120) \approx 35$.
         At a depth of 120 meters down, the rock is 35 million years old.

49 (a) $R = k - aG$

51 (a) Domain: $(0, 4000)$
         Range: $(0, 10^8)$
    (b) Domain: $(0, 3000)$
         Range: $(0, 10^7)$
    (c) Domain: $(0, 0.2)$
         Range: $(0, 0.04)$

Theory: Binomial Theorem

1 Must be integer

Section 2.1

7 $F < B < E < 0 < D < A < C$

9 27

11 $1.9 \ldots$

Section 2.2

1 negative

9 From smallest to largest:
   $0, f'(3), f(3) - f(2), f'(2), 1$

11 (a) $(f(b) - f(a))/(b - a)$
    (b) Slopes same
    (c) Yes

13 -1

15 12

17 $-1/4$

19 $y = 12x + 16$

21 $y = x$

23 $f'(3) = 6, y = 6x - 8$

25 (a) $f'(0) \approx 0.01745$

27 $f'(0) \approx -1$
    $f'(1) \approx 3.5$

29 (a) $\text{erf}'(0) \approx 1.12$
    (b) $\text{erf}'(0) \approx 1.1283$

31 16.0 million people/year
    16.4 million people/year

Section 2.3

11 $f'(x)$ positive:
    $4 \leq x \leq 8$
    $f'(x)$ negative:
    $0 \leq x \leq 3$
    $f'(x)$ greatest:
    at $x \approx 8$

13 $4x$

15 $-2/x^3$

29 (a) x_3
    (b) x_4
    (c) x_5
    (d) x_3

31 (a) $t = 3$
    (b) $t = 9$
    (c) $t = 14$

Section 2.4

1 feet/mile; negative

3 dollars/year; negative

5 dollars/year

7 (a) Quarts; dollars.
    (b) Quarts; dollars/quart.

11 (a) Gal/minute
    (b) (i) 0
         (ii) Negative
         (iii) 0

13 mpg/mph

15 (a) $f'(a)$ is always positive
    (c) $f'(100) = 2$: more
         $f'(100) = 0.5$: less

Section 2.5

1 $f'(x) > 0$
   $f''(x) > 0$

3 $f'(x) < 0$
   $f''(x) = 0$

5 $f'(x) > 0; f''(x) < 0$

7 $0 \leq t \leq 1$:
   acceleration $= 30$ ft/sec^2
   $1 \leq t \leq 2$:
   acceleration $= 22$ ft/sec^2

9 (b) $\frac{dN}{dt}$ is positive.
        $\frac{d^2N}{dt^2}$ is negative.

17 (a) B and E
    (b) A and D

Chapter 2 Review

3 (b) 0.24
   (c) 0.22

5 $f'(1) = 1$
   $f'(2) = 0.5$
   $f'(5) = 0.2$
   $f'(10) = 0.1$
   $f'(x) = \frac{1}{x}$

13 $10x + 1$

17 (a) Negative
    (b) F$^\circ$/min

19 (b) Student C
    (c) $f'(x) = \frac{f(x+h) - f(x-h)}{2h}$

23 (a) 0.25
    (b) $y = 0.25x + 1$
    (c) $k = \frac{1}{8}$
    (d) $(-2, 0.5)$ and $(4, 2)$

25 (b) One
    (c) It must be between
         $x = 1$ and $x = 5$.
    (d) $\lim_{x \to -\infty} f(x) = -\infty$
    (e) Yes
    (f) No

27 (a) Concave down
    (b) $120° < T < 140°$
    (c) $135° < T < 140°$
    (d) $45 < t < 50$

Theory: Limits

1 0.05

3 2

5 $0.01745 \ldots$

7 The limit appears to be 1.

9 (a) $x = 1/(n\pi)$,
        $n = 1, 2, 3, \ldots$
    (b) $x = 2/(n\pi)$,
        $n = 1, 5, 9, \ldots$
    (c) $x = 2/(n\pi)$,
        $n = 3, 7, 11, \ldots$

19 0.46, 0.21, 0.09

Theory: Differentiability

1 (a) (i) $x = 1$
         (ii) $x = 1, 2, 3$
    (b) (i) None
         (ii) $x = 2, 4$

3 Yes

5 Yes

7 (a) $B = B_0$ at $r = r_0$
        B_0 is max
    (b) Yes
    (c) No

9 (a) Yes
    (b) Yes

11 (a) $y = 2 + (x - 4)/4$
    (b) $2.02485\ldots, 2.025$
    (c) True $= 4$, Approx$= 5$

13 $2x - 1$

Section 3.1

1 (a) Lower est = 122 ft
Upper est = 298 ft

3 (a) Lower est. = 5.25 mi
Upper est. = 5.75 mi
(b) Lower est. = 11.5 mi
Upper est. = 14.5 mi
(c) Every 30 seconds

5 Left sum = 46 m
Right sum = 118 m
Average = 82 m

9 Between 140 and 150 meters

11 (a) 14.73 minutes
15.93 minutes
(b) 0.60 minutes

Section 3.2

1 (a) 224
(b) 96
(c) 200
(d) 136

3 limit $= \frac{1}{4}$
True value is between
0.248004 and 0.252004

5 Since $\sin(t^2)$ is *not* monotonic on
$[2, 3]$, we cannot be sure of the true
value.

7 Since $\sin\left(1/x\right)$ is *not* monotonic on
$[0.2, 3]$, we cannot be sure of the true
value.

9 Left-hand sum = 1.96875
Right-hand sum = 2.71875
The most the estimate could be off is
0.375.

11 396

13 24.7

15 4.39

17 0.0833

19 (b) 3.084
(c) 2.250

21 Positive

23 93.47

25 (a) $\sum_{i=1}^{n} i^4/n^5$
(b) $(6n^4+15n^3+10n^2-1)/30n^4$
(c) 1/5

27 $a = 2$, $b = 6$, $f(x) = x^2$; other
answers possible

Section 3.3

1 (a) Car 1: 1031.25 ft
Car 2: 562.5 ft
(b) 1.6 minutes

3 (b) Twice
At each intersection point,
the distance between the two
vehicles is at a local extremum.

5 Dollars

7 Total amount $= \int_0^{60} f(t)\, dt$.

9 15

11 (a) 8.5
(b) 1.7

13 About $13,800

17 (a) 0.79

19 (a) 22°C
(b) 183°C
(c) Smaller

21 (a) 9.9 hours
(b) 14.4 hours
(c) 12.0 hours

23 12 newton · meters

25 (a) III
(b) I
(c) II and IV

Section 3.4

3 $F(0) = 0$
$F(1) = 1$
$F(2) = 1.5$
$F(3) = 1$
$F(4) = 0$
$F(5) = -1$
$F(6) = -1.5$

5 $f(3) - f(2)$,
$\frac{f(4)-f(2)}{2}$,
$f(4) - f(3)$

7 45.8°C.

9 (a) 0.375 thousand/hour
(b) 1.75 thousands

15 9

17 8c

19 8

21 (a) Positive, since $e^{x^2} > 0$

23 (a) 0
(b) 0

25 30/7

Chapter 3 Review

1 (a) 260 ft
(b) Every 0.5 sec

3 60

5 0.40

7 2.00

9 $\pi - 2 \approx 1.14$

11 (iii) < (ii) < (i) < (iv)

15 (b) Largest to smallest:
$n = 1$, $n = 3$,
$n = 4$, and $n = 2$.

17 (a) ≈ 76.8 million
(b) $= 77.24$ million

19 (a) 18 appliances per month
(b) 17 appliances per month
(c) Close, but not equal
(d) Integral is easier to calculate

21 (a) 300 m³/sec
(b) 250 m³/sec
(c) 1996: 1250 m³/sec
1957: 3500 m³/sec
(d) 1996: 10 days
1957: 4 months
(e) 10^9 meter³
(f) $2 \cdot 10^{10}$ meter³

23 (a) At $t = 17, 23, 27$ seconds
(b) Right: $t = 10$ seconds
Left: $t = 40$ seconds
(c) Right: $t = 17$ seconds
Left: $t = 40$ seconds
(d) $t = 10$ to 17 seconds,
20 to 23 seconds, and
24 to 27 seconds
(e) At $t = 0$ and $t = 35$

25 (a) $\int_0^5 f(x)\, dx -$
$\frac{1}{2} \int_{-2}^2 f(x)\, dx$
(b) $\int_{-2}^5 f(x)\, dx -$
$2 \int_{-2}^0 f(x)\, dx$
(c) $\frac{1}{2}(\int_{-2}^5 f(x)\, dx -$
$\int_2^5 f(x)\, dx)$

27 0; Positive; 0; Negative

Theory: Definite Integral

5 2.852, 2.919, 2.886, $n = 30$

7 0.465, 0.474, 0.470, $n = 10$

9 0.0045, 0.0276, 0.016, $n = 10$

11 0.825, 0.905, 0.865, $n = 20$

Section 4.1

3 (a) -10
(b) $\frac{d}{dx}(f[g(x)]) = f'(x)g'(x)$

5 $12x^{11}$

7 $3.2x^{2.2}$

9 $4x^{1/3}/3$

11 $-3x^{-7/4}/4$

13 $x^{-3/4}/4$

15 $6x^{1/2} - \frac{5}{2}x^{-1/2}$

17 $-12x^3 - 12x^2 - 6$

19 $6t - 6/t^{\frac{3}{2}} + 2/t^3$

21 $1 - x^{-2}$

23 $-2t^{-3} - t^{-2} + 4t^{-5}$

25 Problems 8, 10, 11, 12, 13, 15

27 Do not apply.

29 $-2/3z^3$
(power rule and sum rule)

31 Do not apply.

33 $12/\sqrt[6]{x^5} - 18/(x^{5/3})$

35 $f'(t) = 6t^2 - 8t + 3$
$f''(t) = 12t - 8$

37 For $x < 0$ or $2 < x < 3$

39 $r = 3\sqrt{2}$

41 $y = 2x - 1$

43 $n = 4, a = 3/32$

45 $y = 2x$ and $y = -6x$

47 $-2GMm/r^3$

49 At $x = 1$:
$y = -x + 2$
(tangent line approx)
$f(2) \approx 0$
At $x = 2$:
$y = -0.0001x + 0.02$
(tangent line approx)
$f(2) \approx 0.0198$

51 $V(r) = 4\pi r^3/3$
$\frac{dV}{dr} = 4\pi r^2 = $ surface area of a sphere

Section 4.2

1 $2e^x + 2x$

3 $(\ln 5)5^x$

5 $10x + (\ln 2)2^x$

7 $4(\ln 10)10^x - 3x^2$

9 $((\ln 3)3^x)/3 - (33x^{-3/2})/2$

11 e^{1+x}

13 $e^{\theta - 1}$

15 $(\ln 4)^2 4^x$

17 $(\ln(\ln 3))(\ln 3)^t$

19 $5 \cdot 5^t \ln 5 + 6 \cdot 6^t \ln 6$

21 $\pi^2 x^{(\pi^2 - 1)} + (\pi^2)^x \ln(\pi^2)$

23 $1/(2\sqrt{x}) + \ln 2(1/2)^x$

25 Our rules do not apply here.

27 $5e^{5x}$

29 $(\ln 2)2^z$

31 $\approx -444.3 \frac{\text{people}}{\text{year}}$

33 $22.5(1.35)^t$

35 (a) $f'(0) = -1$
(b) $y = -x$
(c) $y = x$

37 $g(x) = x^2/2 + x + 1$

39 $x = 1$ and $x = 2$

Section 4.3

1 $5x^4 + 10x$

3 $e^x(x + 1)$

5 $2^x/(2\sqrt{x}) + \sqrt{x}(\ln 2)2^x$

7 $4s^3 - 1$

9 $(3t^2 + 5)(t^2 - 7t + 2) + (t^3 + 5t)(2t - 7)$

11 $(1 - x)/e^x$

13 $(3.2w^{2.2} - (\ln 5)w^{3.2})/(5^w)$

15 $1/(5t + 2)^2$

17 $1 - 3/x^2$

19 $\frac{(t^2+1)/2\sqrt{t} - \sqrt{t}(2t)}{(t^2+1)^2}$

21 $17e^x(1 - \ln 2)/2^x$

23 $\frac{(-4x^2 - 8x - 1)}{(2 + 3x + 4x^2)^2}$

25 $x < 2$

27 $f'(t) = -e^{-t}$

29 $f'(x) = 3e^{3x}$

31 (a) $\frac{d}{dx}\left(\frac{e^x}{x}\right) = \frac{e^x}{x} - \frac{e^x}{x^2}$
$\frac{d}{dx}\left(\frac{e^x}{x^2}\right) = \frac{e^x}{x^2} - \frac{2e^x}{x^3}$
$\frac{d}{dx}\left(\frac{e^x}{x^3}\right) = \frac{e^x}{x^3} - \frac{3e^x}{x^4}$
(b) $\frac{d}{dx}\left(\frac{e^x}{x^n}\right) = \frac{e^x}{x^n} - \frac{ne^x}{x^{n+1}}$

35 (a) 19
(b) -11

37 (a) $f(140) = 15,000$:
If the cost \$140 per board then 15,000 skateboards are sold.
$f'(140) = -100$:
Every dollar of increase from \$140 will decrease the total sales by 100 boards.
(b) $\frac{dR}{dp}|_{p=140} = 1000$
(c) Positive
Increase by \$1000.

39 (c) -3776.63

43 $f'(x) = f(x)(\frac{1}{x-r_1} + \frac{1}{x-r_2} + \cdots + \frac{1}{x-r_n})$

Section 4.4

1 $99(x + 1)^{98}$

3 $200t(t^2 + 1)^{99}$

5 $50(\sqrt{t} + 1)^{99}/\sqrt{t}$

7 $(\ln 2)2^{(x+2)}$

9 $4(x^3 + e^x)^3(3x^2 + e^x)$

11 $(1 - 2z\ln 2)/(2^{z+1}\sqrt{z})$

13 $\frac{3}{2}e^{\frac{3}{2}w}$

15 $3s^2/2\sqrt{s^3 + 1}$

17 $e^{-t^2}(1 - 2t^2)$

19 $e^{-z}/2\sqrt{z} - \sqrt{z}e^{-z}$

21 $e^{5-2t}(1 - 2t)$

23 $e^{-\theta}/(1 + e^{-\theta})^2$

25 $2we^{w^2}(5w^2 + 8)$

27 $-(\ln 10)(10^{\frac{5}{2} - \frac{y}{2}})/2$

29 $2ye^{[e^{(y^2)} + y^2]}$

31 $y = 4451.66x - 3560.81$

33 $f'(x) = [(2x + 1)^9(3x - 1)^6] \times (102x + 1)$
$f''(x) = [9(2x + 1)^8(2)(3x - 1)^6 + (2x + 1)^9(6)(3x - 1)^5(3)] \times (102x + 1) + (2x + 1)^9(3x - 1)^6(102)$

35 (a) $\pi\sqrt{2}$
(b) $7e$
(c) πe

37 (a) $g'(1) = 3/4$
(b) $h'(1) = 3/2$

39 $e^{(x^6)}/6$

41 (a) $dQ/dt = -0.000121e^{-0.000121t}$

43 (a) $dB/dt = P(1 + r/100)^t \ln(1 + r/100)$
(b) $dB/dr = Pt(1 + r/100)^{t-1}/100$

45 (a) $\frac{dm}{dv} = \frac{m_0 v}{c^2\sqrt{(1 - v^2/c^2)^3}}$
(b) The rate of change of mass with respect to the speed v

47 $T = RC$

Section 4.5

3 $\cos^2\theta - \sin^2\theta = \cos 2\theta$

5 $-4\sin(4\theta)$

7 $e^t \cos(e^t)$

9 $-\sin x e^{\cos x}$

11 $\sin x/2\sqrt{1 - \cos x}$

13 $\cos x/\cos^2(\sin x)$

15 $(\ln 2)2^{\sin x}\cos x$

17 $e^{\cos\theta} - \theta(\sin\theta)e^{\cos\theta}$

19 $2\cos(2x)\sin(3x) + 3\sin(2x)\cos(3x)$

21 $e^{-2x}[\cos x - 2\sin x]$

23 $5\sin^4\theta\cos\theta$

25 $-3e^{-3\theta}/\cos^2(e^{-3\theta})$

27 $\cos t - t\sin t + 1/\cos^2 t$

29 $\theta^2 \cos\theta$

31 $e^x/\sin x$

33 (a) $dy/dt = -\frac{4.9\pi}{6}\sin\left(\frac{\pi}{6}t\right)$
Represents rate of change of depth of water
(b) Occurs when $\sin(\frac{\pi}{6}t) = 0$, or at $t = 6$ am, 12 noon, 6 pm, and 12 midnight
$dy/dt = 0$ means that the water level is at either high tide or low tide

35 (a) $t = (\pi/2)(m/k)^{\frac{1}{2}}$;
$t = 0$;
$t = (3\pi/2)(m/k)^{\frac{1}{2}}$
(b) $T = 2\pi(m/k)^{\frac{1}{2}}$
(c) $dT/dm = \pi/\sqrt{km}$;
Positive sign means an increase in mass causes the period to increase

37 $2/\cos^2\theta$

Section 4.6

1 $\frac{2t}{(t^2+1)}$

3 2

5   $-1/z(\ln z)^2$

7   $\frac{e^{-x}}{1-e^{-x}}$

9   $\frac{e^x}{e^x+1}$

11   7

13   $-\tan(w-1)$

15   $2y/\sqrt{1-y^4}$

17   1

19   $-\sin(\ln t)/t$

21   $\arcsin w + \frac{w}{\sqrt{1-w^2}}$

23   $-1 < x < 1$

25   $\frac{d}{dx}(\log x) = \frac{1}{(\ln 10)x}$

27   (a)   10.59 million
     (b)   0.02 million/year

29   (a) $f'(x)=0$
     (b) f is a constant function.

31   (a) $y=-x^2/2+2x-3/2$
     (b)   From graph, notice that around $x=1$, the values of $\ln x$ and its approximation are very close.
     (c)   At $x=1.1$, $y\approx 0.095$
         At $x=2$, $y=0.5$

33   (a) $z=\sqrt{0.25+x^2}$
     (b)   0.693 km/min
     (c)   0.4 radians/min

35   (a) $k\approx 0.067$
     (b) $t\approx 10.3$ hours
     (c)   Formula:
         $T(24)\approx 74.1°$ F,
         rule of thumb: 73.6° F

Section 4.7

1   $dy/dx = -x/y$

3   $1/25$

5   $-y/(2x)$

7   $\frac{y^2+x^4y^4-2xy}{x^2-2xy-2x^5y^3}$

9   $dy/dx = \frac{2-y\cos(xy)}{x\cos(xy)}$

11   $\frac{dy}{dx} = \frac{3x^2\arctan y}{-e^{\cos y}\sin y - x^3(1+y^2)^{-1}}$

13   $y=x/2-3/2$

15   $y=2$

17   $dy/dx = (m/n)x^{m/n-1}$

19   (a) $dy/dx = -9x/25y$
     (b)   The slope is not defined anywhere along the line $y=0$.

21   Tangent line is $y=2x-1$; circle is $(x-8)^2+y^2=45$.

Section 4.8

1   $1/x \approx 2-x$

5   (b)   This estimate is about right.
     (c)   Below

7   Positive

9   Negative

11   (a)   Not defined
     (b)   0
     (c)   0
     (d)   Not defined

13   $0.1x^7$

15   $x^{0.2}$

17   -2

19   $y=2/3$

21   (c)   0

Chapter 4 Review

1   2/3

3   (a) $h(4)=1$
     (b) $h'(4)=2$
     (c) $h(4)=3$
     (d) $h'(4)=3$
     (e) $h'(4)=-5/16$
     (f) $h'(4)=13$

5   These functions look like the line $y=0$:
$\sin x - \tan x$, $\frac{x^2}{x^2+1}$
$x-\sin x$, $\frac{1-\cos x}{\cos x}$
These functions look like the line $y=x$:
$\arcsin x$, $\frac{\sin x}{1+\sin x}$
$\arctan x$, e^x-1
$\frac{x}{x+1}$, $\frac{x}{x^2+1}$
These functions are undefined at the origin:
$\frac{\sin x}{x}-1$, $-x\ln x$
Defined at the origin but with a vertical tangent:
$x^{10}+\sqrt[10]{x}$

7   (a) $dg/dr = -2GM/r^3$
     (b) dg/dr is rate of change of acceleration due to pull of gravity.
         The further away from the earth's center, the weaker gravity's pull.
     (c) -3.05×10^{-6}
     (d)   It is reasonable because the magnitude of $\frac{dg}{dr}$ is so small (compared to $g=9.8$) that for r near 6400 km, g is not varying much at all.

9   (a) $v(t)=10e^{t/2}$
     (b) $v(t)=s(t)/2$

11   (a)   Falling, 0.38 m/hr
     (b)   Rising, 3.76 m/hr
     (c)   Rising, 0.75 m/hr
     (d)   Falling, 1.12 m/hr

13   (a) $v = -2\pi\omega y_0\sin(2\pi\omega t)$
         $a = -4\pi^2\omega^2 y_0\cos(2\pi\omega t)$
     (b)   Amplitudes: different $(y_0, 2\pi\omega y_0, 4\pi^2\omega^2 y_0)$
         Periods $=1/\omega$

15   $1/\pi \approx 0.32\mu$m/day

17   233.2 miles/hour

19   (a)   Decreases
     (b) -0.25 cm^3/min

Practice: Differentiation

1   $6t-4$

3   $-(5x^4+2)/2$

5   $20x^3 - 2/x^3$

7   $2e^{2x}(x^2-x+1)/(x^2+1)^2$

9   $4x(x^2+2)/9$

11   $-3\cos(2-3x)$

13   $(z^2-1)/3z^2$

15   $1/(\sqrt{\sin(2z)}\sqrt{\cos^3(2z)})$

17   $ae^{ax}/(e^{ax}+b)$

19   $\cos(\tan\theta)/\cos^2\theta$

21   $\cos(\cos x+\sin x)\cdot(\cos x-\sin x)$

23   $5\sin^4\alpha\cos^4\alpha - 3\sin^6\alpha\cos^2\alpha$

25   $-2/(\sqrt{t}(3+\sqrt{t})^2)$

27   $-(3e^{3x}+2x)/(e^{3x}+x^2)^2$

29   $\frac{(4\theta-2\sin(2\theta)\cos(2\theta))}{\sqrt{4\theta^2-\sin^2(2\theta)}}$

31   $3w^2\ln(10w)+w^2$

33   π

35   $2r^3/\sqrt{r^4+1}$

37   $\frac{\sqrt{x+3}(x^2+6x-9)}{[2\sqrt{x^2+9}(x+3)^2]}$

39   $1/(1+2u+2u^2)$

41   $(2t-ct^2)e^{-ct}$

43   $\ln x/(1+\ln x)^2$

45   $8/\sin t$

47   $3\cos(3\theta)e^{\sin(3\theta)}$

49   $(\ln\pi)\pi^{(x+2)}$

51   $(\cos\theta)e^{\sin\theta}$

53   $e^{2x}\left[2x^2+2x+(\ln 5+2)5^x\right]$

55   $-8/(4+t)^2$

57   $-8b^4z/(a+z^2)^5$

59   $(\ln(\ln 2))(\ln 2)^z$

61   $-3\sin(\arctan 3x)/(1+9x^2)$

63   $6(1+3t)e^{(1+3t)^2}$

65   $2r(r+1)/(2r+1)^2$

67   $2e^t+2te^t+1/(2t^{3/2})$

69   $-\frac{2^w\ln 2+e^w}{(2^w+e^w)^2}$

71   $x^2\ln x$

73   $6(3\theta-\pi)\cos[(3\theta-\pi)^2]$

75   $(-e^{-t}-1)/(e^{-t}-t)$

77   $1/\sin^2\theta - 2\theta\cos\theta/\sin^3\theta$

79   $-1/(1+(2-x)^2)$

81 $e^n \cos(e^n)$

83 $\cos(\sqrt{t}e^t) - t\sin(\sqrt{t}e^t) \cdot$
$(\sqrt{t}e^t + e^t/(2\sqrt{t}))$

85 0

87 $e^{\tan x} + xe^{\tan x}/\cos^2 x$

89 $6x/\left(9x^4 + 6x^2 + 2\right)$

91 a

93 $ke^{k\theta}$

95 $e^{-4kt}(\cos t - 4k\sin t)$

97 $-4a^2 x/(a^2 + x^2)^2$

99 $(-3a^2 s - s^3)/(a^2 + s^2)^{3/2}$

101 $(-cat^2 + 2at - bc)e^{-ct}$

103 $-2/(x^2 + 4)$

105 $ae^{au}/(a^2 + b^2)$

107 $4/(e^x + e^{-x})^2$

109 $(-4 - 6x)(6x^e - 3\pi) +$
$(2 - 4x - 3x^2)(6ex^{e-1})$

111 0

113 $4x - 2 - 4x^{-2} + 8x^{-3}$

115 (a) $f'(2) = 20$
(b) $f'(2) = 11/9$
(c) $f'(2) = -4$
(d) $f'(2) = -24$
(e) $f'(2) = \sin 3 - 8\cos 3$
(f) $f'(2) = 4\ln 3 - 16/3$

117 $(1 - y)/(x - 3)$

119 ax/by

121 $\dfrac{8xy - 3x^2}{3y^2 - 4x^2}$

Section 5.1

3 (a) $x \approx 2.5$
(or any $2 < x < 3$)
$x \approx 6.5$
(or any $6 < x < 7$)
$x \approx 9.5$
(or any $9 < x < 10$)
(b) $x \approx 2.5$: local max;
$x \approx 6.5$: local min;
$x \approx 9.5$: local max

7 $x = -1, 1/2$

9 Increasing for all x, no critical point

11 Alternately incr/decr

13 Oscillating: $x < 0$
Increasing: $x > 0$

15 Local min: $(0.37, -0.37)$

19 $a = -1/3$

21 (a) Local max
(b) Local min
(c) Neither
(d) Local min

33 (a) Decreasing.
(b) Local minimum at x_1.
(c) Concave up at x_2.

Section 5.2

3 Increasing $|a|$ stretches the graph horizontally.

5 (a) $b = 20°\text{C}, a = 180°\text{C}$
(b) $k = \frac{1}{90}\min^{-1}$

7 (a) $x = 0$ and
$x = \pm\sqrt{-a/2}$
(b) For any a or b, $x = 0$ is critical point. Only a critical point if $a \geq 0$
Local minimum.
(c) If a is negative:
$x = 0$ is local max,
$x = \pm\sqrt{-\frac{a}{2}}$ are local minima
(d) No

11 $(1/b, 1/be)$

13 A determines y intercept; stretches or flattens bell-shaped graph.
B affects width.

17 (a) Larger $|A|$, steeper
(b) Shifted horizontally by B. Left for $B > 0$; right for $B < 0$. Vertical asymptote $x = -B$

Section 5.3

3 (a) $f(1)$ local minimum;
$f(0), f(2)$ local maxima
(b) $f(1)$ global minimum
$f(2)$ global maximum

5 (a) $f(\frac{2\pi}{3})$ local maximum
$f(0)$ and $f(\pi)$ local minima
(b) $f(\frac{2\pi}{3})$ global maximum
$f(0)$ global minimum

7 (b) Yes, at $x = 0$
(c) $5 > g(0) > g(2)$

9 44.1 feet

11 $r = \frac{2}{3}R$

13 $\theta = -\mu + \sqrt{\mu^2 + 1}$

15 Minimum: $x = -r_0/\sqrt{2}$
Maximum: $x = r_0/\sqrt{2}$

17 (c) No

19 $0 \leq y$; No upper bound.

21 $0 \leq y \leq 16$

23 (a) 10
(b) 9

25 (b) 160 mph or 320 mph; no; yes
(c) 220 mph

Section 5.4

3 (a) Fixed costs
(b) Decreases slowly, then increases

5 (a) $\$0$
(b) $\$96.56$
(c) Raise the price by $\$5$

7 (b) (i) $N'(x) = 20$
(ii) $\dfrac{N(x)}{x} = \dfrac{100}{x} + 20$

11 (b) $q = \left[Fa/(K(1 - a))\right]^a$

Section 5.5

1 $2000 - 1200/\sqrt{5}$

3 Min $v = \sqrt{2}k$; no max

5 (a) $V = Ax/4 - x^3/2$
(c) $(A/6)^{3/2}$

7 40 feet by 80 feet

9 $h = \sqrt{50}$ meters

11 $(0.59, 0.35)$

13 When the rectangle is a square.

15 15 miles/hour

17 (a) The arithmetic mean unless $a = b$, in which case the two means are equal.
(b) The arithmetic mean unless $a = b = c$, in which case the two means are equal.

19 (a) $E = 500e\left(\dfrac{200 - \cos\theta}{\sin\theta}\right) + 2000e$
$(\arctan\left(\dfrac{500}{2000}\right) \leq \theta \leq \dfrac{\pi}{2})$
(b) $\theta = \pi/3$
(c) Independent of e, but dependent on $\overline{AB}/\overline{AL}$.

21 (a) $T = \sqrt{a^2 + (c - x)^2}/v_1$
$+ \sqrt{b^2 + x^2}/v_2$

Section 5.6

5 $\cosh(2x) = \cosh^2 x + \sinh^2 x$

7 $2\sinh(2x)$

9 $2te^{t^2}\sinh(e^{t^2})$

11 $\tanh(1 + \theta)$

13 (b) $A = 6.325$
(stretch factor)
$c = 0.458$
(horizontal shift)

15 (b) U-shaped
(c) Incr $(A > 0)$
or Decr $(A < 0)$
(d) Max: $A < 0, B < 0$
Min: $A > 0, B > 0$

17 $y \approx 715 - 100\cosh(x/100)$

Chapter 5 Review

3 (a) Increasing: $(0, \infty)$
Decreasing: $(-\infty, 0)$
(b) Local and global min: $f(0)$

5 (a) Increasing: $(0, 4)$
Decreasing: $(-\infty, 0), (4, \infty)$
(b) Local max: $f(4)$
Local min: $f(0)$

7   (a) $f'(x) = 3x(x-2)$
           $f''(x) = 6(x-1)$
  (b) $x = 0$
        $x = 2$
  (c)   Inflection point: $x = 1$
  (d)   Endpoints:
        $f(-1) = -4$
        $f(3) = 0$
        Critical Points:
        $f(0) = 0$
        $f(2) = -4$
        Global max:
        $f(0) = 0$ and $f(3) = 0$
        Global min:
        $f(-1) = -4$ and
        $f(2) = -4$
  (e) f increasing:
        for $x < 0$ and $x > 2$
        f decreasing:
        for $0 < x < 2$
        f concave up:
        for $x > 1$
        f concave down:
        for $x < 1$.

9   (a) $f'(x) =$
        $-e^{-x} \sin x + e^{-x} \cos x$
        $f''(x) = -2e^{-x} \cos x$
  (b)   Critical points:
        $x = \frac{\pi}{4}$ and $\frac{5\pi}{4}$
  (c)   Inflection points:
        $x = \frac{\pi}{2}$ and $\frac{3\pi}{2}$
  (d)   Endpoints:
        $f(0) = 0$
        $f(2\pi) = 0$
        Global max:
        $f(\frac{\pi}{4}) = (e^{-\frac{\pi}{4}})(\frac{\sqrt{2}}{2})$
        Global min:
        $f(\frac{5\pi}{4}) = -e^{\frac{-5\pi}{4}}(\frac{\sqrt{2}}{2})$
  (e) f increasing:
        $0 < x < \frac{\pi}{4}$ and
        $\frac{5\pi}{4} < x < 2\pi$
        f decreasing:
        $\frac{\pi}{4} < x < \frac{5\pi}{4}$
        f concave down:
        for $0 \le x < \frac{\pi}{2}$
        and $\frac{3\pi}{2} < x \le 2\pi$
        f concave up:
        for $\frac{\pi}{2} < x < \frac{3\pi}{2}$

11    $\lim\limits_{x \to \infty} f(x) = \infty$
     $\lim\limits_{x \to -\infty} f(x) = -\infty$
  (a) $f'(x) = 6(x-2)(x-1)$
        $f''(x) = 6(2x-3)$
  (b) $x = 1$ and $x = 2$
  (c) $x = \frac{3}{2}$
  (d)   Critical points:
        $f(1) = 6$, $f(2) = 5$ Local
        max: $f(1) = 6$
        Local min: $f(2) = 5$
        Global max and min: none
  (e) f increasing: $x < 1$ and $x > 2$
        Decreasing: $1 < x < 2$
        f concave up: $x > \frac{3}{2}$
        f concave down: $x < \frac{3}{2}$

13    $\lim\limits_{x \to -\infty} f(x) = -\infty$
     $\lim\limits_{x \to \infty} f(x) = 0$
  (a) $f'(x) = (1-x)e^{-x}$
        $f''(x) = (x-2)e^{-x}$
  (b)   Only critical point is at $x = 1$.
  (c)   Inflection point: $f(2) = \frac{2}{e^2}$
  (d)   Global max: $f(1) = \frac{1}{e}$
        Local and global min: none
  (e) f increasing: $x < 1$
        f decreasing: $x > 1$
        f concave up: $x > 2$
        f concave down: $x < 2$

15   Local max: $f(-3) = 12$
     Local min: $f(1) = -20$
     Inflection pt: $x = -1$
     Global max and min: none

17   Global and local min: $x = 2$
     Global and local max: none
     Inflection pts: none

19   Global and local max:
       $f(0) = 1$
     Local minimum: none
     Inflection pts: $x = \pm\frac{1}{\sqrt{2}}$

25    $x = \sqrt[3]{2V}$
     $y = \sqrt[3]{V/4}$

27    $-4.81 \le f(x) \le 1.82$

29   (a) $g(e)$ is a global maximum.
        There is no minimum.
  (b)   There are exactly two solutions.
  (c) $x = 5$ and $x \approx 1.75$

31   (a) x-intercept: $(a, 0)$
        y-intercept: $(0, 1/(a^2 + 1))$
  (b)   Area $= a / \left(2(a^2 + 1)\right)$
  (c) $a = 1$
  (d) $A = \frac{1}{4}$
  (e) $a = 2$ and $a = \frac{1}{2}$

33   (a)   The concavity changes at y_1 and y_3.
  (b) $f(t)$ grows most quickly where the vase is skinniest and most slowly where the vase is widest.

Section 6.1

5   (a) $x = 1$, $x = 3$
  (b)   Local min at $x = 1$, local max at $x = 3$

7   (a) $x = 1$, $x = 4$
  (b)   Local max at $x = 1$, neither at $x = 4$

11   (a) $f(3) = 1$; $f(7) = 0$
  (b) $x = 0, 5.5, 7$

15   Acceleration is zero at points A and C.

Section 6.2

1    $5x$

3    $x^3/3$

5    $\sin t$

7    $\ln|z|$

9    $-1/2z^2$

11    $-\cos t$

13    $t^4/4 - t^3/6 - t^2/2$

15    $5x^2/2 - 2x^{3/2}/3$

17    $-\cos 2\theta$

19    $(t+1)^3/3$

21    $\sin t + \tan t$

23    $F(x) = x^2$
     (only possibility)

25    $F(x) = x^2/8$
     (only possibility)

27    $F(x) = \frac{2}{3}x^{3/2}$
     (only possibility)

29    $F(x) = -\cos x + 1$
     (only possibility)

31    $2t^2 + 7t + C$

33    $5e^z + C$

35    $-\cos t + C$

37    $2t^{5/2}/5 - 2t^{-1/2} + C$

39    $e^{2r}/2 + C$

41    $y^3/3 - 2y - 1/y + C$

43    $\frac{1}{2}\ln|2x-1| + C$

45    $\frac{1}{2}\sin 2x + 2\cos x + C$

47    $1 - \cos 1 \approx 0.460$

49    $16/3 \approx 5.333$

51    $-8/9 \approx -0.889$

53    $\tan \frac{\pi}{4} = 1$

55    $1 - \frac{\sqrt{3}}{4} \approx 0.567$

57    $\sqrt{2} - 1$

59   c=3

61   (a)   0
  (b) $2/\pi$

63   (b)   7 years
  (c) $69\frac{1}{3}$ cubic yards

Section 6.3

1    $x^4/4 + 5x + C$

3    $8t^{3/2}/3 + C$

5    $2x^3 + 2x^2 - 14$

7    $-16t^2 + 100t + 50$

11    $y = 2kt^{3/2}/3$

13   (c)   200 ft
  (d)   200 ft

15   (a)   6 seconds
  (b)   Left sum: 97.5 ft
        overestimate
        Right sum: 82.5 ft
        underestimate
  (c)   90 ft
  (d) $s(t) = 30t - \frac{5}{2}t^2$; $s(6) = 90$ ft
        Distance by antidiff = Average of left and right hand sums

17 (c) height = 400 feet
    (d) $v(t) = -32t + 160$
         height = 400 feet

19 (a) $v(t) = 1.6t$
    (b) $s(t) = 0.8t^2 + s_0$

21 77 mph; 113.1 ft/sec

Section 6.4

1 500

7 (a) Si(4) ≈ 1.76
        Si(5) ≈ 1.55
    (b) $(\sin x)/x$ is negative on that
         interval.

9 $(1 + x)^{200}$

11 $-\cos(t^3)$

13 $(2 \sin x^2)/x$

15 (a) $F'(x) = 1/(\ln x)$
    (b) $F(x)$ is increasing and con-
         cave down for $x \geq 2$.

17 $\frac{1}{\sqrt{\pi x}} e^{-x}$

19 $3x^2 e^{-x^6} - e^{-x^2}$

Chapter 6 Review

1 $\frac{5}{2}x^2 + 7x + C$

3 $e^x + 5x + C$

5 $\tan x + C$

7 $(x + 1)^3/3 + C$

9 $\frac{1}{10}(x + 1)^{10} + C$

11 $\frac{1}{2}x^2 + x + \ln|x| + C$

13 $3 \sin x + 7 \cos x + C$

15 $2e^x - 8 \sin x + C$

17 $\ln|x| - 1/x - 1/(2x^2) + C$

19 $G(t) = 5t + \sin t + C$

21 $H(r) = 4r^{3/2}/3 + r^{1/2} + C$

23 $H(t) = t - 2 \ln|t| - 1/t + C$

25 $F(z) = e^z + 3z + C$

27 $e^x + e^{1+x} + C$

29 $P(r) = \pi r^2 + C$

31 $(1 + \sin t)^{30}/30 + C$

33 $\sin(t^2)/2 + C$

35 $e^2 y + 2^y / \ln 2 + C$

37 $e^{x^2}/2 + C$

39 $\sqrt{3} - \pi/9$

41 c=3/4

49 (a) 14,000 revs/min^2
    (b) 180 revolutions

51 (b) Highest pt: $t = 2.5$ sec
         Hits ground: $t = 5$ sec
    (c) Left sum:
            136 ft (an overest.)
         Right sum:
            56 ft (an underest.)
    (d) 100 ft

53 (b) $t = 6$ hours
    (c) $t = 11$ hrs

55 Positive, zero, negative,
    positive, zero

Modeling: Motion

1 $t = 5; v = -160$ ft/sec

3 400 feet

7 (a) First second: $-g/2$
         Second: $-3g/2$
         Third: $-5g/2$
         Fourth: $-7g/2$
    (b) Galileo seems to have been
         correct.

Section 7.1

1 (a) (i) $2x \cos(x^2 + 1)$
         (ii) $3x^2 \cos(x^3 + 1)$
    (b) (i) $\frac{1}{2} \sin(x^2 + 1) + C$
         (ii) $\frac{1}{3} \sin(x^3 + 1) + C$
    (c) (i) $-\frac{1}{2} \cos(x^2 + 1) + C$
         (ii) $-\frac{1}{3} \cos(x^3 + 1) + C$

3 $-\frac{1}{3} \cos 3x + C$

5 $4 \sin(t^3) + C$

7 $e^{\sin x} + C$

9 $\frac{1}{2} \ln(x^2 + 1) + C$

11 $\frac{1}{18}(y^2 + 5)^9 + C$

13 $\frac{1}{9}(x^2 - 4)^{9/2} + C$

15 $\frac{1}{6}(x^2 + 3)^3 + C$

17 $\frac{1}{148}(2t - 7)^{74} + C$

19 $-\frac{1}{8}(\cos \theta + 5)^8 + C$

21 $-\frac{1}{2} e^{-x^2} + C$

23 $\frac{1}{35} \sin^7 5\theta + C$

25 $\frac{1}{4} \sin^4 \alpha + C$

27 $\ln|e^t + t| + C$

29 $-\frac{1}{2} \ln|\cos 2x| + C$

31 $2e^{\sqrt{y}} + C$

33 $\ln(2 + e^x) + C$

35 $\frac{1}{5}y^5 + \frac{1}{2}y^4 + \frac{1}{3}y^3 + C$

37 $\frac{1}{6} \ln(1 + 3t^2) + C$

39 $t + 2 \ln|t| - \frac{1}{t} + C$

41 $e^{x^2+x} + C$

43 (a) $x^4 + 2x^2 + C$
    (b) $(x^2 + 1)^2 + C$
    (c) Both correct but they differ by
         a constant.

45 $\frac{mg}{k}t - \frac{m^2 g}{k^2}(1 - e^{-kt/m}) + h_0$

Section 7.2

1 (a) $(\ln 2)/2$
    (b) $(\ln 2)/2$

3 $1/\pi$

5 $e(e^3 - 1)$

7 1

9 2/5

11 40

13 $\ln 3$

15 14/3

17 $49932\frac{1}{6}$

19 $\frac{1}{3}(1 - \frac{\sqrt{2}}{2})$

21 $3/2 + \ln 2$

23 $(\ln 2)/2$

25 $\pi/12$

27 $\ln 5$

29 Substitute $w = \ln x$

31 Substitute $w = x + 1, w = 1 + \sqrt{x}$

33 $\arctan(x + 2) + C$

35 $\frac{1}{2}(e^4 - 1)$

37 (a) 0
    (b) 2/3

39 $\frac{1}{2} \ln 3$

41 (a) 30,500 pennies
    (b) Two 15,000-word novels

43 $-\frac{1}{k} \ln \left(\frac{e^{t\sqrt{gk}} + e^{-t\sqrt{gk}}}{2} \right) + h_0$

Section 7.3

1 $x \cdot \arctan x -$
   $\frac{1}{2} \ln(1 + x^2) + C$

3 $\frac{1}{5}t^2 e^{5t} - \frac{2}{25} t e^{5t}$
   $+ \frac{2}{125} e^{5t} + C$

5 $-t \cos t + \sin t + C$

7 $\frac{x^4}{4} \ln x - \frac{x^4}{16} + C$

9 $-(z + 1)e^{-z} + C$

11 $\frac{1}{3}\theta^2 \sin 3\theta + \frac{2}{9}\theta \cos 3\theta$
    $- \frac{2}{27} \sin 3\theta + C$

13 $-(\theta + 1) \cos(\theta + 1)$
    $+ \sin(\theta + 1) + C$

15 $-x^{-1} \ln x - x^{-1} + C$

17 $\frac{2}{3}y(y + 3)^{3/2}$
    $- \frac{4}{15}(y + 3)^{5/2} + C$

19 $-2y(5 - y)^{1/2}$
    $- \frac{4}{3}(5 - y)^{3/2} + C$

21 $t(\ln t)^2 - 2t \ln t + 2t + C$

23 $w \arcsin w + \sqrt{1 - w^2} + C$

25 $\frac{1}{2}x^2 \arctan x^2$
    $- \frac{1}{4} \ln(1 + x^4) + C$

27 $\frac{1}{3}x^3 \sin x^3 + \frac{1}{3} \cos x^3 + C$

29 $\cos 5 + 5 \sin 5 - \cos 3 - 3 \sin 3 \approx$
    -3.944

31 $\frac{9}{2} \ln 3 - 2 \approx 2.944$

33   $6 \ln 6 - 5 \approx 5.751$

35   $\frac{\pi}{2} - 1 \approx 0.571$

37   $\frac{1}{2}\theta - \frac{1}{4}\sin 2\theta + C$

39   $\frac{1}{2}e^x(\sin x - \cos x) + C$

41   $\frac{1}{2}xe^x(\sin x - \cos x) + \frac{1}{2}e^x \cos x + C$

43   Integrate by parts choosing $u = x^n$, $v' = e^x$.

45   Integrate by parts, choosing $u = x^n$, $v' = \sin ax$.

47   π

49   (a) $-a^2 e^{-a} - 2ae^{-a} - 2e^{-a} + 2$
    (b)   Increasing
    (c)   Concave up

51   (a) $A = a/(a^2 + b^2)$,
        $B = -b/(a^2 + b^2)$
    (b) $e^{ax}(b \sin bx + a \cos bx)/(a^2 + b^2) + C$

53   (a) $C_1 = \sqrt{2}$
    (b) $C_n = \sqrt{2}$

Section 7.4

1   $\frac{1}{10}e^{(-3\theta)}(\sin\theta - 3\cos\theta) + C$

3   $-\frac{1}{5}\cos^5 w + C$

5   $\frac{1}{\sqrt{3}}\arctan\frac{y}{\sqrt{3}} + C$

7   $(\frac{1}{2}x^3 - \frac{3}{4}x^2 + \frac{3}{4}x - \frac{3}{8})e^{2x} + C$

9   $\frac{3}{16}\cos 3\theta \sin 5\theta - \frac{5}{16}\sin 3\theta \cos 5\theta + C$

11   $\frac{1}{3}e^{x^3} + C$

13   $\frac{u^6}{6}\ln 5u - \frac{1}{36}u^6 + C$

15   $-\frac{1}{2}x^2 \cos x^2 + \frac{1}{2}\sin x^2 + C$

17   $-\frac{1}{2}y^2 \cos 2y + \frac{1}{2}y \sin 2y + \frac{1}{4}\cos 2y + C$

19   $-\frac{1}{2\tan 2\theta} + C$

21   $\frac{1}{21}\frac{\tan 7x}{\cos^2 7x} + \frac{2}{21}\tan 7x + C$

23   $\frac{1}{3}\frac{\sin x}{\cos^3 x} - \frac{4}{3}\frac{\sin x}{\cos x} + x + C$

25   $-\frac{1}{4}(\ln|y - 2| - \ln|y + 2|) + C$

27   $\arctan(y + 2) + C$

29   $-\frac{1}{9}(\cos^3 3\theta) + \frac{1}{15}(\cos^5 3\theta) + C$

33   $2 \ln|x| + \ln|x + 3| + C$

35   $\ln|x - 1| - \ln|x| + C$

37   $\frac{1}{3}\ln|P/(1 - P)| + C$

39   (a) 0
    (b) $V_0/\sqrt{2}$
    (c)   156 volts

41   0

Section 7.5

1   (a)   LEFT(2)= 12;
       RIGHT(2)= 44
    (b)   LEFT(2) underestimate;
       RIGHT(2) overestimate

5   (a)   (i)   LEFT(32) = 13.6961
          RIGHT(32) = 14.3437
          TRAP(32) = 14.0199
          Exact value: $(x \ln x - x)|_1^{10}$
             ≈ 14.02585093
      (ii)   LEFT(32) = 50.3180
          RIGHT(32) = 57.0178
          TRAP(32) = 53.6679
          Exact value: $e^x|_0^4$
             ≈ 53.59815003
    (b)   (i)   Smallest to largest:
          LEFT(32)
          TRAP(32)
          Actual value
          RIGHT(32)
      (ii)   Smallest to largest:
          LEFT(32)
          Actual value
          TRAP(32)
          RIGHT(32)

7   (a) $0.664 =$ LEFT
       $0.633 =$ TRAP
       $0.632 =$ MID
       $0.601 =$ RIGHT
    (b)   Between 0.632 and 0.633

9   (a)   TRAP(4); 1027.5
    (b)   Underestimate

11   MID: over; TRAP: under

13   TRAP: over; MID: under

15   (b)   LEFT(5) ≈ 1.32350
       error ≈ -0.03810
       RIGHT(5) ≈ 1.24066
       error ≈ 0.04474
       TRAP(5) ≈ 1.28208
       error ≈ 0.00332
       MID(5) ≈ 1.28705
       error ≈ -0.001656

17   445 lbs. of fertilizer

23   RIGHT(10) = 5.556
    TRAP(10) = 4.356
    LEFT(20) = 3.199
    RIGHT(20) = 4.399
    TRAP(20) = 3.799

Section 7.6

1   (a)   76/3
    (b)   76/3
    (c)   0

3   4.2365, $n = 10$

5   1.0894, $n = 10$

7   0.904524, $n = 10$

9   (a) ≈ 53.598
    (b)   LEFT(2)= 16.778;
       error= 36.820
       RIGHT(2)= 123.974;
       error= −70.376
       TRAP(2)= 70.376;
       error= −16.778
       MID(2)= 45.608;
       error= 7.990
       SIMP(2)= 53.864;
       error= −0.266
    (c)   LEFT(4)= 31.193;
       error=22.405
       RIGHT(4)= 84.791;
       error=−31.193
       TRAP(4)= 57.992;
       error=−4.394
       MID(4)= 51.428;
       error=2.170
       SIMP(4)= 53.616;
       error=−0.018

11   (a)   3.449
    (b)   3.816
    (c)   3.980

13   (a)   9.5 years
    (b)   8.33 hours
    (c)   5 minutes

15   0.272

17   (a) ≈ 4.78
    (b) ≈ 4.7962

19   (a)   Mostly increasing; mostly concave down

Section 7.7

1   $e^{-2}/2$

3   Does not converge

5   $\pi/5$

7   $2^{3/4}$

9   Does not converge

11   Does not converge

13   $\pi/8$

15   Does not converge

17   Does not converge

19   $\frac{1}{\ln 3}$

21   1/3

23   Does not converge

25   1

27   (a)   Does not converge
    (b)   Not differentiable

29   (a) $\Gamma(1) = 1$
       $\Gamma(2) = 1$
    (c) $\Gamma(n) = (n - 1)!$

31   9×10^9 joules

33 Converges for $p > -1$
to $pe^{p+1}/(p+1)^2$

Section 7.8

1 (a) diverges
(b) $f(x)$: converges
$g(x)$: impossible to tell
$h(x)$: diverges
$k(x)$: diverges

3 Converges

5 Converges

7 Does not converge

9 Converges

11 Converges

13 Converges

15 Does not converge

17 0.139

19 $a = 0.399$

21 (a) $\int_3^\infty e^{-x^2}\,dx \le \frac{e^{-9}}{3}$
(b) $\int_n^\infty e^{-x^2}\,dx \le \frac{1}{n}e^{-n^2}$

23 Converges for $p < 1$
Diverges for $p \ge 1$

25 (a) e^t is concave up for all t

Chapter 7 Review

1 $\int_0^b h\,dx = hb$

3 $2\int_{-r}^r \sqrt{r^2 - x^2}\,dx = \pi r^2$

5 (a) (i) 0
(ii) $\frac{2}{\pi}$
(iii) $\frac{1}{2}$
(b) Smallest to largest:
Average value of $f(t)$
Average value of $k(t)$
Average value of $g(t)$

7 (a) $\frac{1}{2}\arcsin 2x + C$
(b) ≈ 0.45167
(c) SIMP[100] ≈ 0.45167

9 Substitute $w = x + 2$, $w = x^2 + 1$

11 Substitute $w = 1 - x^2$, $w = \ln x$

13 Converges
$\int_4^\infty t^{-3/2}\,dt = 1$

15 Converges
$\int_0^\infty we^{-w}\,dw = 1$

17 Converges
$\int_{-\pi/4}^{\pi/4} \tan\theta\,d\theta = 0$

19 Converges
$\int_{10}^\infty \frac{1}{z^2-4}\,dz = (\ln(3/2))/4$

21 Does not converge

23 Does not converge

25 Does not converge

27 11/3

29 Area $= 2\sqrt{2}$

31 False

33 True

35 (a) Trapezoid: overestimate.
Midpoint: underestimate.
(b) Trapezoid: underestimate.
Midpoint: overestimate.
(c) Trapezoid and midpoint give exact value.

37 $\approx 7.4175.$

39 45 years

Practice: Integration

1 $-\cos t + C$

3 $(1/5)e^{5z} + C$

5 $(-1/2)\cos 2\theta + C$

7 $2x^{5/2}/5 + 3x^{5/3}/5 + C$

9 $(r+1)^4/4 + C$

11 $x^2/2 + \ln|x| - x^{-1} + C$

13 $\frac{1}{2}e^{t^2} + C$

15 $-\frac{2}{9}(2 + 3\cos x)^{3/2} + C$

17 $\frac{2}{5}(1-x)^{5/2} - \frac{2}{3}(1-x)^{3/2} + C$

19 $-y\cos y + \sin y + C$

21 $2x\ln x - 2x + C$

23 $(1/3)\sin^3\theta + C$

25 $\frac{1}{2}u^2 + 3u + 3\ln|u| - \frac{1}{u} + C$

27 $\tan z + C$

29 $(1/2)t^{12} - (10/11)t^{11} + C$

31 ≈ 3.8875

33 -133.8724

35 $-5/64$

37 $2e(e - 1)$

39 $(1/3)(\ln x)^3 + C$

41 $(1/3)x^3 + x^2 + \ln|x| + C$

43 $(1/2)e^{t^2+1} + C$

45 $(1/10)\sin^2(5\theta) + C$ (other forms of answer are possible)

47 $\arctan z + C$

49 $-\frac{1}{8}\cos^4 2\theta + C$

51 $(-1/4)\cos^4 z + (1/6)\cos^6 z + C$

53 $(2/3)(1 + \sin\theta)^{3/2} + C$

55 $t^3 e^t - 3t^2 e^t + 6te^t - 6e^t + C$

57 $(3z+5)^4/12 + C$

59 $\arctan(\sin w) + C$

61 $-\cos(\ln x) + C$

63 $-\sqrt{16 - w^2} + C$

65 $2\sqrt{1 - \cos w} + C$

67 $(1/3)\ln|3u + 8| + C$

69 $\sqrt{1 + t^2}(t^2 - 2)/3 + C$

71 $(w+5)^6/6 - (w+5)^5 + C$

73 $r^2[(\ln r)^2 - \ln r + (1/2)]/2 + C$

75 $(u^3 \ln u)/3 - u^3/9 + C$

77 $-\cos(2x)/(4\sin^2(2x)) + \frac{1}{8}\ln\left|\frac{(\cos(2x)-1)}{(\cos(2x)+1)}\right| + C$

79 $-y^2\cos(cy)/c + 2y\sin(cy)/c^2 + 2\cos(cy)/c^3 + C$

81 $(1/34)\left[e^{5x}(5\cos(3x) + 3\sin(3x))\right] + C$

83 $(\sqrt{3}/4)(2x\sqrt{1+4x^2} + \ln|2x + \sqrt{1+4x^2}|) + C$

85 $(\ln|x+1| - \ln|x+4|)/3 + C$

87 $x^2/2 - 3x - \ln|x+1| + 8\ln|x+2| + C$

89 $\frac{1}{b}\left(\ln|x| - \ln|x + b/a|\right) + C$

91 $\ln|z| - \ln|z+1| + C$

93 $(1/\ln 2)\ln|2^t + 1| + C$

95 $(1/7)x^7 + 3x^5 + 25x^3 + 125x + C$

97 $\sin^3(2\theta)/6 - \sin^5(2\theta)/10 + C$

99 $\frac{1}{2}x\sqrt{4 - x^2} + 2\arcsin(x/2) + C$

101 $-(1/2)\ln|1 + \cos^2 w| + C$

103 $x\tan x + \ln|\cos x| + C$

105 $(2/3)(\sqrt{x+1})^3 - 2\sqrt{x+1} + C$

107 $(1/2)\ln|e^{2y} + 1| + C$

109 $-1/(z-5) - 5/(2(z-5)^2) + C$

111 $e^{x^2-x} + C$

113 $2\sin(2x) + x^3\sin(2x) + 3x^2\cos(2x)/2 + C$

Section 8.1

3 $V = (\pi r^2 h)/3$

5 $V = (4/3)\pi ab^2$

7 $V = (8/15)\pi \approx 1.68$

9 $V = (\pi(e^2 - 1)/2) \approx 10.036$

11 $V \approx 65.54$

13 $V = (\pi(e^2 - 1)/16) \approx 1.25$

15 (a) Volume ≈ 152 in^3
(b) About 15 apples

17 3509 cubic inches

19 π

21 (a) $4\int_0^r \sqrt{1 + (-\frac{x}{y})^2}\,dx$
(b) $2\pi r$

23 2.35

Section 8.2

1 (a) $\sum_{i=1}^N (2 + 6x_i)\Delta x$
(b) 16 grams

3 (a) $\sum_{k=0}^{n-1} \frac{5\Delta x}{1+x_k^4}$

   (b) 5.5

5 (a) $\int_0^5 2\pi r(0.115e^{-2r})dr$

   (b) 181 cubic meters

7 Total mass = 12; $\bar{x} = 2.06$

9 (a) Right

   (b) $2/(1 + 6e - e^2) \approx 0.2$

13 (a) $\pi r^2 l/2$

   (b) $2klr^3/3$

15 (a) $\sum_{i=0}^{N-1} 4\pi(r_e + h_i)^2$
       $\times\, 1.28e^{-0.000124h_i}\Delta h$

   (b) 6.48×10^{16}

Section 8.3

1 (a) Approximately 1.9
      seconds

   (b) Approximately 103.5
      newton-seconds

   (c) Newton-seconds

   (d) A

   (e) B

3 11,000 ft-lbs

5 1,058,591.1 ft-lb.

7 518,363 ft-lbs

9 1.489×10^{10} joules

11 $v \approx 2360$ m/sec

13 (a) Bottom: 187,200 lbs.

   (b) 15×10 side: 70,200 lbs.

   (c) 15×20 side: 140,400 lbs.

15 60 joules

17 Potential
     $= 2\pi\sigma(\sqrt{R^2 + a^2} - R)$

19 $\frac{2GMmy}{a^2}\left(y^{-1} - \frac{1}{(a^2+y^2)^{1/2}}\right)$

21 $\frac{GM_1M_2}{l_1 l_2}\ln\left[\frac{(a+l_1)(a+l_2)}{a(a+l_1+l_2)}\right]$

Section 8.4

3 (a) Option 1

   (b) Option 1: \$10.929 million;
      Option 2: \$10.529 million

5 (a) \$5820 per year

   (b) \$36,787.94

7 \$46,800

9 (a) 10.6 years

   (b) 624.9 million dollars

11 \$85,750,000

15 (a) Less

   (b) Can't tell

   (c) Less

Chapter 8 Review

1 (b) $\sum_{i=1}^{N} \pi x_i \Delta x$

   (c) Volume $= \pi/2$

3 (a) $\sum_{i=1}^{N} \pi \frac{9x_i}{4}\Delta x$

   (b) 18π

5 (a) 39 cu.in.

   (b) 94¢

7 $\int_0^\pi \sqrt{1 + \cos^2 x}\, dx$

9 16.34 kg

11 1000 ft-lb

13 1,170,000 lbs

15 (a) $\sum_{i=0}^{n-1} (2000 - 100t_i)$
       $\times\, e^{-0.1t_i}\Delta t$

   (b) $\int_0^T e^{-0.10t} \times$
      $(2000 - 100t)\, dt$

   (c) After 20 years
      \$11,353.35

17 (a) $\sum_{i=1}^{N} \pi\left(\frac{3.5\cdot 10^5}{\sqrt{h}+600}\right)^2 \Delta h$

   (b) $1.05 \cdot 10^{12}$ cubic feet

19 0.43m^3

21 (a) $\pi h^2/(2a)$

   (b) $\pi h/a$

   (c) $dh/dt = -k$

   (d) h_0/k

23 The thin spherical shell.

Theory: Distributions

5 (a) 0.9m - 1.1m

9 (a) Cum. increasing

   (b) Vert. 0.2, Horiz. 2

11 (a) 22.1%

   (b) 33.0%

   (c) 30.1%

   (d) $C(h) = 1 - e^{-0.4h}$

13 (a) 21%

   (b) 2%

   (c) 13%

   (e) 20 - 25 mins

15 (b) About 3/4

17 (a) $f(r) = 0.2\ (0 < r < 5)$
      $f(r) = 0\ (5 \leq r)$

   (b) $F(r) = 0.2r$
      $(0 \leq r \leq 5)$
      $F(r) = 1\ (5 < r)$

   (c) $G(v) = 0.124r^{1/3}$
      $(0 \leq v \leq 523.6)$
      $G(v) = 1\ (523.6 < v)$

   (d) $g(v) = 0.0413v^{-2/3}\ (0 <$
      $v < 523.6)$
      $g(v) = 1\ (523.6 \leq v)$

Theory: Probability

3 (a) $c = 0.0176$

   (b) 9%

5 (a) $a = 0.122$

   (b) $P(x) = 1 - e^{-0.122x}$

   (c) Median = 5.68 seconds
      Mean = 8.20 seconds

9 (c) μ represents the mean of the distribution, while σ is the standard deviation.

11 (a) $-e^{-2} + 1 \approx 0.865$

   (b) $-(\ln 0.05)/2 \approx 1.5$ km

13 (a) $6/25 = 24\%$

   (b) \$12,600

   (c) $\approx$ \$8000

15 (a) $p(r) = 4r^2 e^{-2r}$

   (b) Mean: 1.5 Bohr radii
      Median: 1.33 Bohr radii
      Most likely:
      1 Bohr radius

Section 9.1

1 $P_4(x) = 1 - x + x^2 - x^3$
   $+ x^4$
   $P_6(x) = 1 - x + x^2 - x^3$
   $+ x^4 - x^5 + x^6$
   $P_8(x) = 1 - x + x^2 - x^3$
   $+ x^4 - x^5 + x^6 - x^7 + x^8$

3 $P_2(x) = 1 + \frac{1}{2}x - \frac{1}{8}x^2$
   $P_3(x) = 1 + \frac{1}{2}x - \frac{1}{8}x^2$
   $+ \frac{1}{16}x^3$
   $P_4(x) = 1 + \frac{1}{2}x - \frac{1}{8}x^2$
   $+ \frac{1}{16}x^3 - \frac{5}{128}x^4$

5 $P_3(x) = P_4(x) = x - \frac{1}{3}x^3$

7 $P_2(x) = 1 - \frac{1}{3}x - \frac{1}{9}x^2$
   $P_3(x) = 1 - \frac{1}{3}x - \frac{1}{9}x^2$
   $- \frac{5}{81}x^3$
   $P_4(x) = 1 - \frac{1}{3}x - \frac{1}{9}x^2$
   $- \frac{5}{81}x^3 - \frac{10}{243}x^4$

9 $P_2(x) = 1 - \frac{1}{2}x + \frac{3}{8}x^2$
   $P_3(x) = 1 - \frac{1}{2}x + \frac{3}{8}x^2$
   $- \frac{5}{16}x^3$
   $P_4(x) = 1 - \frac{1}{2}x + \frac{3}{8}x^2$
   $- \frac{5}{16}x^3 + \frac{35}{128}x^4$

11 (a) 0

   (b) 3

   (c) -24

   (d) 0

   (e) 3600

13 $P_4(x) = 1 - \frac{1}{2!}\left(x - \frac{\pi}{2}\right)^2$
   $+ \frac{1}{4!}\left(x - \frac{\pi}{2}\right)^4$

15 $P_4(x) = e[1 + (x - 1)$
   $+ \frac{1}{2}(x-1)^2 + \frac{1}{6}(x-1)^3$
   $+ \frac{1}{24}(x-1)^4]$

17 $c < 0, b > 0, a > 0$

19 $a < 0, b > 0, c > 0$

21 $\sin x = \frac{\sqrt{2}}{2} + \frac{\sqrt{2}}{2}\left(x - \frac{\pi}{4}\right)$
   $- \frac{\sqrt{2}}{4}\left(x - \frac{\pi}{4}\right)^2 - \frac{\sqrt{2}}{12}\left(x - \frac{\pi}{4}\right)^3$
   $- \cdots$

23 $\sin\theta = -\frac{\sqrt{2}}{2} + \frac{\sqrt{2}}{2}\left(\theta + \frac{\pi}{4}\right)$
   $+ \frac{\sqrt{2}}{4}\left(\theta + \frac{\pi}{4}\right)^2$
   $- \frac{\sqrt{2}}{12}\left(\theta + \frac{\pi}{4}\right)^3 + \cdots$

25 $\frac{d}{dx}(x^2 e^{x^2})|_{x=0} = 0$
$\frac{d^6}{dx^6}(x^2 e^{x^2})|_{x=0} = \frac{6!}{2} = 360$

27 $P_2(x) = 4x^2 - 7x + 2$
$f(x) = P_2(x)$

29 (a) If $f(x)$ is a polynomial of degree n, then $P_n(x)$, the nth degree Taylor polynomial for $f(x)$ about $x = 0$, is $f(x)$ itself.

35 (a) $f(x) = x^2 - \frac{1}{3!}x^6 + \cdots$
   (b) Substitute x^2 for x in the Taylor expansion of $\sin x$.

Section 9.2

1 Yes

3 No

5 $f(x) = 1 + x + x^2 + x^3 + \cdots$

7 $f(x) = 1 - \frac{x}{2} + \frac{3x^2}{8} - \frac{5x^3}{16} + \cdots$

9 $\frac{1}{x} = 1 - (x-1) + (x-1)^2 - (x-1)^3 + \cdots$

11 $\frac{1}{x} = -1 - (x+1) - (x+1)^2 - (x+1)^3 - \cdots$

13 $-1 < x < 1$

15 $-1 < x < 1$

17 1

19 1

23 32

25 Does not converge

27 e^2

29 4/3

31 $\ln(3/2)$

Section 9.3

1 $\sqrt{1-2x} = 1 - x - \frac{x^2}{2} - \frac{x^3}{2} - \cdots$

3 $e^{-x} = 1 - x + \frac{x^2}{2!} - \frac{x^3}{3!} + \cdots$

5 $\ln(1-2y) = -2y - 2y^2 - \frac{8}{3}y^3 - 4y^4 - \cdots$

7 $\frac{1}{\sqrt{1-z^2}} = 1 + \frac{1}{2}z^2 + \frac{3}{8}z^4 + \frac{5}{16}z^6 + \cdots$

9 $\frac{z}{e^{z^2}} = z - z^3 + \frac{z^5}{2!} - \frac{z^7}{3!} + \cdots$

11 $e^t \cos t = 1 + t - \frac{t^3}{3} - \frac{t^4}{6} + \cdots$

13 Smallest: $\cos\theta$
   Largest: $1 + \sin\theta$

15 (a) I
   (b) IV
   (c) III
   (d) II

17 $\frac{1}{2+x} = \frac{1}{2}(1 - \frac{x}{2} + (\frac{x}{2})^2 - (\frac{x}{2})^3 + \cdots)$

19 (a) $f(x) = 1 + (a-b)x + (b^2 - ab)x^2 + \cdots$
   (b) $a = \frac{1}{2}, b = -\frac{1}{2}$

21 $(x^2/a) + x^6/(8a^5) + \cdots$

23 (b) $\phi \approx -3b$

25 (a) If $M \gg m$, then $\mu \approx \frac{mM}{M} = m$.
   (b) $\mu = m[1 - \frac{m}{M} + (\frac{m}{M})^2 - (\frac{m}{M})^3 + \cdots]$
   (c) -0.0545%

27 (b) $F = mg(1 - \frac{2h}{R} + \frac{3h^2}{R^2} - \frac{4h^3}{R^3} + \cdots)$
   (c) 300 km

Section 9.4

1 Yes, $a = 1$, ratio $= -1/2$

3 Yes, $a = 5$, ratio $= -2$

5 No

7 Yes, $a = 1$, ratio $= -x$

9 No

11 $y^2/(1-y), |y| < 1$

13 $1/(1+y^2), |y| < 1$

15 (a) $\frac{7(1.02^{104}-1)}{(0.02(1.02)^{100})}$
   (b) $7e^{0.01}$

17 $(3(2^{11}-1))/(2^{10})$

19 32/3

21 (a) $P_n = 250(0.04) + 250(0.04)^2 + 250(0.04)^3 + \cdots + 250(0.04)^{n-1}$
   (b) $P_n = 250 \cdot 0.04 \frac{(1-(0.04)^{n-1})}{(1-0.04)}$
   (c) $\lim_{n\to\infty} P_n \approx 10.42$
   We'd expect the difference between them to be 250 mg.

23 (a) $h_n = 10(3/4)^n$
   (b) $D_1 = 10$ feet
   $D_2 = h_0 + 2h_1 = 25$ feet
   $D_3 = h_0 + 2h_1 + 2h_2 = 36.25$ feet
   $D_4 = h_0 + 2h_1 + 2h_2 + 2h_3 \approx 44.69$ feet
   (c) $D_n = 10 + 60\left(1 - (3/4)^{n-1}\right)$

25 (a) \$1250
   (b) 12.50

27 \$900 million

Section 9.5

1 Not a Fourier series

3 Fourier series

5 $F_1(x) = F_2(x) = \frac{4}{\pi}\sin x$
$F_3(x) = \frac{4}{\pi}\sin x + \frac{4}{3\pi}\sin 3x$

7 99.942% of the total energy

9 $H_n(x) = \frac{\pi}{4} + \sum_{i=1}^{n}\frac{(-1)^{i+1}\sin(ix)}{i} + \sum_{i=1}^{[n/2]}\frac{-2}{(2i-1)^2\pi}\cos((2i-1)x)$,
where $[n/2]$ denotes the biggest integer smaller than or equal to $n/2$.

11 (a) $F_3(x) = \frac{1}{2} + \frac{2}{\pi}\cos x - \frac{2}{3\pi}\cos 3x$.
   (b) There are cosines instead of sines, but the energy spectrum remains the same

13 $F_4(x) = 1 - \frac{4}{\pi}\sin(\pi x) - \frac{2}{\pi}\sin(2\pi x) - \frac{4}{3\pi}\sin(3\pi x) - \frac{1}{\pi}\sin(4\pi x)$

19 (a) $15.9155\%, 0.451808\%$
   (b) $(4\sin^2\frac{k}{2})/(k^2\pi^2)$
   (c) The constant term and the first five harmonics are needed.
   (d) $F_5(x) = \frac{1}{2\pi} + \frac{2\sin(1/2)}{\pi}\cos x + \frac{\sin 1}{\pi}\cos 2x + \frac{2\sin(3/2)}{3\pi}\cos 3x + \frac{\sin 2}{2\pi}\cos 4x + \frac{2\sin(5/2)}{5\pi}\cos 5x$

21 (a) $31.83\%, 76.91\%$
   (b) 90.07%
   (c) $F_3(x) = \frac{1}{\pi} + \frac{2\sin 1}{\pi}\cos x + \frac{\sin 2}{\pi}\cos 2x + \frac{2\sin 3}{3\pi}\cos 3x$

Chapter 9 Review

1 $e^x \approx 1 + e(x-1) + \frac{e}{2}(x-1)^2$

3 $\sin x \approx -\frac{1}{\sqrt{2}} + \frac{1}{\sqrt{2}}(x + \frac{\pi}{4}) + \frac{1}{2\sqrt{2}}(x + \frac{\pi}{4})^2$

5 $\theta^2 \cos\theta^2 = \theta^2 - \frac{\theta^6}{2!} + \frac{\theta^{10}}{4!} - \frac{\theta^{14}}{6!} + \cdots$

7 $\frac{1}{\sqrt{4-x}} = \frac{1}{2} + \frac{1}{8}x + \frac{3}{64}x^2 + \frac{5}{256}x^3 + \cdots$

9 $\frac{a}{a+b} = 1 - \frac{b}{a} + (\frac{b}{a})^2 - (\frac{b}{a})^3 \cdots$

11 3/4

13 $3e$

15 Smallest to largest:
$1 - \cos x, x\sqrt{1-x}, \ln(1+x), \arctan x, \sin x, x, e^x - 1$

17 1/2

19 (a) $f(t) = t + t^2 + \frac{t^3}{2!} + \frac{t^4}{3!} + \cdots$
   (b) $\int_0^x f(t)\,dt = \frac{x^2}{2} + \frac{x^3}{3} + \frac{x^4}{4\cdot 2!} + \frac{x^5}{5\cdot 3!} + \cdots$
   (c) Substitute $x = 1$, and integrate by parts.

21 (a) Set $\frac{dV}{dr} = 0$, solve for r. Check for max or min.

    (b) $V(r) = -V_0 + 72V_0 r_0^{-2} \cdot (r - r_0)^2 \cdot \frac{1}{2} + \cdots$

    (d) $F = 0$ when $r - r_0$

23 (a) $0.232323\ldots = 0.23 + 0.23(0.01) + 0.23(0.01)^2 + \cdots$

    (b) $0.23/(1 - 0.01) = (23)/(99)$

27 (a) $g(x) \approx P_n(x) = g(0) + \frac{g''(0)}{2!}x^2 + \frac{g'''(0)}{3!}x^3 + \cdots + \frac{g^{(n)}(0)}{n!}x^n$

    (b) If $g''(0) > 0$:
      0 is a local minimum
      If $g''(0) < 0$:
      0 is a local maximum.

31 (b) If the amplitude of the kth harmonic of f is A_k, then the amplitude of the kth harmonic of f' is kA_k.

    (c) The energy of the kth harmonic of f' is k^2 times the energy of the kth harmonic of f.

Theory: Convergence

5 Does not converge

7 Not convergent

9 Converges

Theory: Errors

1 (a) Underestimate
    (b) 1

3 (a) Overestimate:
      $0 < \theta \le 1$
      Underestimate:
      $-1 \le \theta < 0$
    (b) $|E_2| \le 0.17$

5 $|E_3| \le 16.5$

7 $|E_4| \le 0.016$

9 For $\sin x$ and $\cos x$,
    $|E_n| \le \frac{1}{(n+1)!}$

15 $|E_0| < 0.01$ for $|x| \le 0.1$

Section 10.1

1 (a) (III)
    (b) (V)
    (c) (I)
    (d) (II)
    (e) (IV)

7 $\lambda = -3$ or 2.

9 (A) (II), (V)
    (B) (I)
    (C) No solutions
    (D) (IV)
    (E) (III)

11 (b) $A = k$

Section 10.2

1 (b) $y = -x - 1$

3 (a) (II)
    (b) (I)
    (c) (V)
    (d) (III)
    (e) (IV)

5 Slope field (b)

7 (c) Increasing: $-1 < y < 2$
    Decreasing: $y > 2$ or $y < -1$
    Horizontal: $y = 2$ or $y = -1$

Section 10.3

3 (a) $y(0.4) \approx 1.5282$
    (b) $y(0.4) = -1.4$

5 (b) $y = \ln|t|$, so at $t = 2$, $y \approx 0.693$
    (c) Bigger.

7 (a) At $x = 1$, $y \approx 1.5$
    (b) At $x = 1$, $y \approx 1.75$
    (c) $y = x^2 + 1$, so $y(1) = 2$
    (d) Yes

9 (a) $B \approx 1050$
    (b) $B \approx 1050.63$
    (c) $B \approx 1050.94$

Section 10.4

1 $P = 20e^{0.02t}$

3 $m = 5e^{3t-3}$

5 $y = 10e^{-x/3}$

7 $P = 104e^t - 4$

9 $m = 3000e^{0.1t} - 2000$

11 $y = 200 - 150e^{t/2}$

13 $R(r) = 1 - 0.9e^{1-r}$

15 $z = -\ln(1 - t^2/2)$

17 $y = -2/(t^2 + 2t - 4)$

19 $w = 2/(\cos\theta^2 + 1)$

21 $z(r) = e^{r+(1/3)r^3}$

23 $R = Ae^{kt}$

25 $P = a + Ae^t$

27 $P = a + Ae^{kt}$

29 $y = 2/(1 + e^{-2t})$

31 $x = e^{At}$

33 (c) $y^2 - x^2 = 2C$

Section 10.5

1 (b) 2001

3 (a) (I)
    (b) (IV)
    (c) (II) and (IV)
    (d) (II) and (III)

5 (a) $y = 3$ and $y = -2$
    (b) $y = 3$ unstable;
      $y = -2$ stable

7 (a) $y = 500$
    (b) $y = Ae^{0.5t} + 500$
    (d) Unstable

9 (a) $\frac{dB}{dt} = rB/100$
      Constant $= r/100$
    (b) $B = Ae^{rt/100}$

11 Michigan: 72 years
    Ontario: 18 years

13 (b) $\frac{dQ}{dt} = -0.0187Q$
    (c) 3 days

15 $C(t) = 30e^{-100.33t}$

17 (a) Mt. Whitney:
      17.50 inches
      Mt. Everest:
      10.23 inches
    (b) 18,661.5 feet

19 (a) $\frac{dT}{dt} = -k(T - A)$
      $(A = 68^\circ F)$
    (b) $T = 68 + 22.3e^{-kt}$; 3:45 am

21 about 2150 B.C.

23 (a) $\frac{dD}{dt} = kD$

Section 10.6

1 $\frac{dD}{dt} = -0.75(D - 4)$
    Equilibrium $= 4$ g/cm^2

3 $P = AV^k$

5 about 3 days

7 (a) 112.5%

9 (a) $dx/dt = k(a - x)(b - x)$
    (b) $x = ab(e^{bkt} - e^{akt})/(be^{bkt} - ae^{akt})$

11 (b) $dc/dt = (43.2/35,000) - 0.082c$
    (c) $c = 0.015(1 - e^{-0.082t})$
      As $t \to \infty$,
      $c \to 0.015$ mg/ml

13 (a) $\frac{dy}{dt} = -k(y - a)$
    (b) $y = (1 - a)e^{-kt} + a$
    (c) a: fraction remembered in long run; k: rate material forgotten

15 (a) $\frac{dc}{dt} = \frac{0.0001}{60} - \frac{0.002}{60}c$
    (b) $c = 0.05 - 0.05e^{-3 \times 10^{-5}t}$
    (c) $c \to 0.05$

17 (a) $\frac{dS}{dt} = 600 - \frac{3S}{100,000}$
      (S in grams)
    (b) $S(t) = 2 \times 10^7(1 - e^{-\frac{3}{100,000}t})$
    (c) As $t \to \infty$
      $S(t) \to 2 \times 10^7$

21 (b) Overestimate

Section 10.7

1 $P = (6.6 \times 10^6)e^{0.002t}$

3 (a) $P = 5000/(1 + 499e^{-1.78t})$
    (c) $t \approx 3.5$; $P \approx 2500$

5 $100 < P < 200$

9　(a)　$k \approx 0.0286$
　　　　$a \approx 0.0001$
　　(b)　Increases to 286 million

11　(a)　Use chain rule.
　　(b)　$\frac{du}{dt} - k(u - \frac{1}{L})$;
　　　　$u = \frac{1}{L} + Ae^{-kt}$
　　(c)　$P = L/(1 + LAe^{-kt})$

13　Two-sided better

15　42.3 million people; off by 3.7

17　(c)　$P = 0$ (stable)
　　　　$P = 4$ (unstable)

19　(b)　$dP/dt < 0$ for $P < b/a$
　　　　$dP/dt > 0$ for $P > b/a$
　　(c)　$P > b/a$: increase
　　　　$P < b/a$: extinction

Section 10.8

3　$A \approx -1.705$
　$\alpha = \pm\sqrt{5}$

5　Middle point, moving down

7　(a)　Spring (iii)
　　(b)　Spring (iv)
　　(c)　Spring (iv)
　　(d)　Spring (i)

9　(a) goes with (II)
　(b) goes with (I) and (IV)
　(c) goes with (III)
　(I) $x = 2\sin 2t$
　(II) $x = -\sin t$
　(III) $x = \cos 4t$
　(IV) $x = -3\sin 2t$

11　(a)　Starts higher; period same.
　　(b)　Period increases.

13　$\sqrt{58}$

15　$\omega = 2$, $A = 13$,
　$\psi = \tan^{-1} 12/5$

17　(a)　$A\sin(\omega t + \phi) =$
　　　　$(A\sin\phi)\cos\omega t +$
　　　　$(A\cos\phi)\sin\omega t$

19　(a)　$V_0 \approx 25$, $n = 2$,
　　　　$\phi_0 = -\pi$
　　(b)　$K = 25$, $\omega^2 = 4$,
　　　　$A = B = 0$

21　$C = \frac{1}{20}$ farads

Section 10.9

1　$y(t) = C_1 e^{-t} + C_2 e^{-3t}$

3　$y(t) =$
　$C_1 e^{-2t}\cos t + C_2 e^{-2t}\sin t$

5　$s(t) =$
　$C_1 \cos\sqrt{7}t + C_2\sin\sqrt{7}t$

7　$z(t) = C_1 e^{-t/2} + C_2 e^{-3t/2}$

9　$p(t) = C_1 e^{-t/2}\cos\frac{\sqrt{3}}{2}t +$
　$C_2 e^{-t/2}\sin\frac{\sqrt{3}}{2}t$

11　$y(t) = A + Be^{-2t}$

13　$y(t) = \frac{5}{4}e^{-t} - \frac{1}{4}e^{-5t}$

15　$y(t) = 2e^{-3t}\sin t$

17　$p(t) = 20e^{(\pi/2)-t}\sin t$

19　(a)　(IV)
　　(b)　(II)
　　(c)　(I)
　　(d)　(III)

21　$k = 6$
　$y(t) = C_1 e^{2t} + C_2 e^{3t}$

23　(iii)

25　(iv)

27　(ii)

29　overdamped: $c < 2$
　critically damped: $c = 2$
　underdamped: $c > 2$

31　$z(t) = 3e^{-2t}$

33　(a)　$Q(t) = \frac{2}{\sqrt{3}}(e^{(-1+\frac{\sqrt{3}}{2})t}$
　　　　$- e^{(-1-\frac{\sqrt{3}}{2})t})$
　　(b)　$Q(t) =$
　　　　$\frac{1}{\sqrt{3}}((2+\sqrt{3})e^{(-1+\frac{\sqrt{3}}{2})t} -$
　　　　$(2-\sqrt{3})e^{(-1-\frac{\sqrt{3}}{2})t})$

35　(a)　$Q(t) = 16e^{-\frac{1}{8}t}\sin\frac{t}{8}$
　　(b)　$Q(t) =$
　　　　$2e^{-\frac{1}{8}t}\left(\sin\frac{t}{8} + \cos\frac{t}{8}\right)$
　　(c)　Goes from overdamped to underdamped

37　No

Chapter 10 Review

1　$y = 10/(1 + 9e^{-10t})$

3　$\frac{1}{2}y - 4\ln|y| = 3\ln|x| - x$
　$+ \frac{7}{2} - 4\ln 5$.

5　$f(x) = (\frac{1}{3}x^{3/2} + \frac{2}{3})^2$
　(for $x \geq 0$)

7　$y = \sqrt[3]{33 - 6\cos x}$

9　$y = 2/(x^2 + 1)$

11　$z(t) = 1/(1 - 0.9e^t)$

13　$y = -\ln(\frac{\ln 2}{3}\sin^2 t\cos t$
　$2\ln 23\cos t - \frac{2\ln 2}{3} + 1)/\ln 2$

15　$-y - \ln|1 - y| = \arctan t - \frac{\pi}{4}$

17　$Q = 4/(1 - Ae^{-(4t^3/3)-4t}) - 2$
　or $Q = -2$

19　$-\frac{a^2}{2}e^{-\frac{x^2}{a^2}} =$
　$\frac{y^2}{2}\ln y - \frac{y^2}{4} + C$

21　(a)　$y(1) \approx 3.689$
　　(b)　Overestimate
　　(c)　$y = 5 - 4e^{-x}$
　　　　$y(1) = 5 - 4e^{-1} \approx$
　　　　3.528
　　(d)　≈ 3.61

23　$y(t) = C_1 e^{-2t} + C_2 e^{-4t}$

25　$y(t) = C_1 e^{\frac{1}{3}t} + C_2 e^{-\frac{1}{3}t}$

27　overdamped if:
　　$b > 2\sqrt{5}$ or $b < -2\sqrt{5}$
　critically damped if:
　　$b = \pm 2\sqrt{5}$
　underdamped if:
　　$-2\sqrt{5} < b < 2\sqrt{5}$

29　(a)　$\frac{dQ}{dt} = -0.5365Q$
　　　　$Q = Q_0 e^{-0.5365t}$
　　(b)　4 mg

31　(a)　Positive
　　(b)　Approximately 10 minutes

33　(a)　$dB/dt = 0.1B + 1200$
　　(b)　$B = 12000(e^{0.1t} - 1)$
　　(c)　7784.66

35　$dI/dt = 0.001I(1 - 10(I/M))$;
　Limiting value $= M/10$

37　(b)　$d^2r/dt^2 = -ABr$
　　　　$r(t) = C_1\cos\sqrt{AB}t$
　　　　$+ C_2\sin\sqrt{AB}t$
　　(c)　$r(t) = \cos\sqrt{AB}t$
　　　　$j(t) = \sqrt{\frac{A}{B}}\sin\sqrt{AB}t$
　　(d)　One quarter of the time.

Modeling: Systems of D.E.

3　Initially $x = 0$; y decreases, x increases. Then x increases, y increases. Finally y increases, x decreases.

5　Populations oscillate

7　$r = r_0 e^{-t}$, $w = w_0 e^t$

9　Worms decrease, robins increase. Long run: populations oscillate.

13　Robins increase, worms decrease. Oscillate

15　(a)　$dy/dx = x/5y$
　　(b)　US victory
　　(c)　No

17　(a)　$x^2 - 5y^2 = 604.75$
　　(b)　about 25,000 troops

19　(a)　$dx/dt = -axy$
　　　　$dy/dt = -bxy$
　　(b)　$dy/dx = b/a$
　　　　soln:$y = (b/a)x + C$
　　(c)　if $C > 0$, y wins
　　　　if $C < 0$, x wins
　　　　if $C = 0$,
　　　　　mutual annihilation

21　(a)　Symbiosis
　　(b)　Both $\to \infty$ or both $\to 0$

23　(a)　predator-prey
　　(b)　x, y tend to ≈ 1

Modeling: Phase Planes

1. (a) $\frac{dS}{dt} = 0$ where
   $S = 0$ or $I = 0$.
   $\frac{dI}{dt} = 0$ where
   $I = 0$ or $S = 192$.

   (b) Where $S > 192$,
   $\frac{dS}{dt} < 0$ and $\frac{dI}{dt} > 0$.
   Where $S < 192$,
   $\frac{dS}{dt} < 0$ and $\frac{dI}{dt} < 0$.

3. Vertical nullclines:
   $x = 0$, $x + y = 2$.
   Horizontal nullclines:
   $y = 0$, $x + y = 1$.
   Equilibrium points:
   $(0,0)$, $(0,1)$, $(2,0)$.

5. Horizontal nullclines;
   $y = 0$, $y = 1 - 2x$
   Vertical nullclines;
   $x = 0$, $y = \frac{1}{2}(2 - x)$
   Equilibrium points;
   $(0,0)$, $(0,1)$, $(2,0)$

7. $\frac{dx}{dt} = 0$ when
   $x = 0$ or $x + y/3 = 1$.
   $\frac{dy}{dt} = 0$ when
   $y = 0$ or $y + x/2 = 1$.
   Equilibrium points:
   $(0,0)$, $(0,1)$, $(1,0)$, $(\frac{4}{5}, \frac{3}{5})$

9. (a) $\frac{dP_1}{dt} = 0$ where $P = 0$ or
   $P_1 + 3P_2 = 13$.
   $\frac{dP_2}{dt} = 0$ where $P = 0$ or
   $P_2 + 0.4P_1 = 6$.

11. (a) $dx/dt = 0$ where $x = 23.3$
    $dy/dt = 0$ where
    $8.2x - 0.8y - 142 = 0$
    Equilibrium point:
    $(23.3, 61.7)$

Appendix A

1. (a) $y \le 30$
   (b) two zeros

3. -1.05

5. 2.5

7. $x = 1.05$

9. 0.45

11. 1.3

13. (a) $x = -1.15$
    (b) $x = 1$, $x = 1.41$,
    and $x = -1.41$

15. (a) $x \approx 0.7$
    (b) $x \approx 0.4$

17. (a) 4 zeros
    (b) $[0.65, 0.66]$, $[0.72, 0.73]$,
    $[1.43, 1.44]$, $[1.7, 1.71]$.

19. (b) $x \approx 5.573$

21. Bounded $-5 \le f(x) \le 4$

23. Not bounded.

Appendix B

3. (a) (i) $5.126978\ldots\%$
   (ii) $5.127096\ldots\%$
   (iii) $5.127108\ldots\%$

   (b) 5.127%

   (c) $e^{0.05} = 1.05127109\ldots$

5. $\approx 6.18365\%$

7. (a) 11.55 years
   (b) $P = P_0(2)^{t/11.55}$

9. (a) 2,613,035 zaïre
   (b) (i) $2,714,567$ zaïre
   (ii) $2,718,127$ zaïre
   (iii) $2,718,280$ zaïre

   (c) Levels off; over 2,718,000 zaïre.

11. (a) Makes $11.83
    (b) Loses $9.19

Appendix C

1. $(1,0)$

3. $(-2,0)$

5. $(\frac{5\sqrt{3}}{2}, -\frac{5}{2})$

7. $(\cos 1, \sin 1)$

9. $(2, \pi/2)$

11. $(\sqrt{2}, 3\pi/4)$

13. $(0.28, 7\pi/4)$

15. $(3.16, 2.82)$

Appendix D

1. $2e^{\frac{i\pi}{2}}$

3. $\sqrt{2}e^{\frac{i\pi}{4}}$

5. $0e^{i\theta}$, for any θ.

7. $\sqrt{10}e^{i\arctan(-3)}$

9. $-3 - 4i$

11. $-5 + 12i$

13. $\frac{1}{4} - \frac{9i}{8}$

15. $-\frac{1}{2} + i\frac{\sqrt{3}}{2}$

17. $-125i$

19. $\frac{\sqrt{2}}{2} + i\frac{\sqrt{2}}{2}$

21. $\frac{\sqrt{3}}{2} + \frac{i}{2}$

23. -2^{50}

25. $8i\sqrt[3]{2}$

27. $\frac{1}{\sqrt{2}}\cos(\frac{-\pi}{12}) + i\frac{1}{\sqrt{2}}\sin(\frac{-\pi}{12})$

29. $A_1 = 1 - i$, $A_2 = 1 + i$

31. (a) $z_1 z_2 = 6 - i2\sqrt{3}$
    $\frac{z_1}{z_2} = i\sqrt{3}$
    (b) same as (a)

33. True

35. False

37. True

Appendix E

1. (a) $f'(x) = 3x^2 + 6x + 3$
   (b) At most one
   (c) $[0, 1]$
   (d) $x \approx 0.913$

3. $\sqrt[4]{100} \approx 3.162$

5. $x \approx 0.511$

7. $x \approx 1.310$

9. $x \approx 1.763$

11. $x \approx 0.682328$

Appendix F

1. (a) Both parameterize line
   $y = 3x - 2$
   (b) Slope $= 3$
   y-intercept $= -2$

3. (b) $v \approx 2.2$
   (c) $v = 2.2363$

5. Speed $= 2|t|$, stops at $t = 0$

7. Speed $= ((2t-2)^2 + (3t^2-3)^2)^{1/2}$,
   stops at $t = 1$

9. $\sqrt{42}$

11. ≈ 24.6

13. (a) π m/sec
    (b) 2.45 m
    (c) 3.01 m

15. $-(2t + 4t^3)/(1 + t^2 + t^4)^2$

INDEX